Integrated Mathematics

COURSE I Second Edition

Isidore Dressler
Former Chairman
Department of Mathematics
Bayside High School, New York City

Edward P. Keenan
Curriculum Associate, Mathematics
East Williston Union Free School District
East Williston, New York

Ann Xavier Gantert
Department of Mathematics
Nazareth Academy
Rochester, New York

Marilyn Occhiogrosso
Former Assistant Principal
Mathematics
Erasmus Hall High School, New York City

Integrated Mathematics

COURSE I

Second Edition

AUTHORS

Isidore Dressler

Edward P. Keenan

REVISERS

Ann Xavier Gantert

Marilyn Occhiogrosso

AMSCO SCHOOL PUBLICATIONS, INC.
315 Hudson Street New York, N.Y. 10013

When ordering this book, please specify:
R 481 P *or* INTEGRATED MATHEMATICS: COURSE I, 2ND ED., PAPERBACK
or
R 481 H *or* INTEGRATED MATHEMATICS: COURSE I, 2ND ED., HARDBOUND

ISBN 0-87720-265-6 (Softbound edition)
ISBN 0-87720-266-4 (Hardbound edition)

PRINTED IN THE UNITED STATES OF AMERICA

Preface

INTEGRATED MATHEMATICS: COURSE I, *Second Edition*, is a thorough revision of a textbook that has been a leader in presenting high school mathematics in a contemporary, integrated manner. Over the last decade, this integrated approach has undergone further changes and refinements. Amsco's Second Edition reflects these developments.

The Amsco book parallels the integrated approach to the teaching of high school mathematics that is being promoted by the National Council of Teachers of Mathematics (NCTM) in its STANDARDS FOR SCHOOL MATHEMATICS. Moreover, the Amsco book implements many of the suggestions set forth in the NCTM Standards, which are the acknowledged guidelines for achieving a higher level of excellence in the study of mathematics.

In this new edition:

- **Problem solving** has been expanded by (1) adding a new chapter that focuses on a four-step general technique and its application to selected strategies, (2) adding many nonroutine problems throughout the book, and (3) further emphasizing a problem-solving approach in the text and model problems.

- **Integration** of Algebra, Geometry, and other branches of mathematics, for which the First Edition was well known, has been broadened by the inclusion of new topics and the earlier introduction of selected concepts.

- **New topics,** such as the introduction of Transformation Geometry and the expanded presentation of Coordinate Geometry, ensure that the Second Edition fully satisfies the revised requirements of the New York State syllabus and also provides the framework for a richer course of study.

- **Enrichment** has been extended by (1) increasing the number of challenging exercises, (2) adding a variety of optional topics, and (3) adding to the Teacher's Manual more thought-provoking aspects of topics in the text, and supplementary material that reflects current thinking in mathematics education.

- **Dimensional analysis** has been promoted by (1) including units of measure in applicable model problems and exercises and (2) providing suggestions in the Teacher's Manual for discussions to heighten awareness of unit measure.

- **Review exercises** are included at the end of each of the twenty chapters.

The First Edition of the text had been written to provide effective teaching materials for a unified program appropriate for 9th-grade mathematics students, including topics not previously contained in a traditional Elementary Algebra course. These topics—Logic, Probability, Statistics, and Numerical Geometry—are retained and expanded in the Second Edition.

An intent of the authors was to make the original book of greatest service to average students. Since its publication, however, the text has been used successfully with students of varying ability levels. To maintain this broad spectrum of use, the basic elements of the original work have been preserved in the Second Edition. Once again:

- Concepts are carefully developed using appropriate language and mathematical symbolism.

- General principles and procedures are stated clearly and concisely.

- The numerous model problems are solved through detailed step-by-step explanations.

- Varied and carefully graded exercises, in abundance, test student understanding of manipulative and arithmetic skills.

This new edition is offered so that teachers may effectively continue to help students comprehend, master, and enjoy mathematics from an integrated point of view.

INTEGRATED MATHEMATICS: COURSE I is dedicated to Hilda Dressler and Joan Keenan.

The Authors

Contents

CHAPTER 4 Simple Equations and Problems

CHAPTER 5 Introducing Logic

CHAPTER 6 Using Logic

CHAPTER 7 Signed Numbers

CHAPTER 8 Operations With Monomials

CHAPTER 9 Operations With Polynomials

CHAPTER 10 First-Degree Equations and Inequalities in One Variable

CHAPTER 11 Geometry

CHAPTER 12 Ratio and Proportion

CHAPTER 13 Special Products and Factoring

CHAPTER 14 Fractions, and First-Degree Equations and Inequalities Involving Fractions

CHAPTER 15 Probability

CHAPTER 16 Statistics

CHAPTER 17 The Coordinate Plane

CHAPTER 20 Quadratic Equations

Chapter 1

Numbers, Sets, and Operations

1-1 THE BASIC OPERATIONS

In arithmetic, you have studied numbers and performed computations involving operations on numbers. It is important to review certain ideas from arithmetic because they will aid you in the study of many other branches of mathematics this year. Arithmetic helps you to understand concepts in algebra. The numbers in arithmetic are used in measuring geometric objects. Also, arithmetic plays a key role in the study of probability and statistics. Although arithmetic is only one small part of mathematics, it is an important thread that binds together many different branches of mathematics.

Symbols for Numbers

A ***number*** is really an idea; it is something that we can talk about and think about. We represent numbers in writing by using the symbols 1, 2, 3, 4, and so on. These symbols, called ***numerals*** or ***numerical expressions***, are not numbers but are used to represent numbers.

Counting Numbers or Natural Numbers

The ***counting numbers***, which are also called ***natural numbers***, are represented by the symbols:

1, 2, 3, 4, 5, 6, 7, 8, 9, 10, 11, 12, . . .

The three dots after the 12 indicate that the numbers continue *in the same pattern* without end. The smallest counting number is 1. Every counting number has a ***successor*** that is one more than the number. The successor of 1 is 2, the successor of 2 is 3, and so on. Since this process of counting is endless, there is no last counting number.

Whole Numbers

Zero is not a counting number. By combining zero with all the counting numbers, we form the set of ***whole numbers***. The whole numbers are represented by the symbols:

0, 1, 2, 3, 4, 5, 6, 7, 8, 9, 10, 11, 12, . . .

The smallest whole number is 0. There is no largest whole number.

Subsets of the Whole Numbers

A ***set*** is a collection of distinct objects or elements, such as the set of whole numbers.

When all the elements of a set are whole numbers only, no matter how many or how few, we call this set a ***subset*** of the set of whole numbers.

Some subsets of the whole numbers are:

1. The odd whole numbers: 1, 3, 5, 7, . . .
2. The even whole numbers: 0, 2, 4, 6, . . .
3. One-digit whole numbers: 0, 1, 2, 3, . . . , 9
4. Whole-number multiples of 5: 0, 5, 10, 15, 20, . . .
5. Whole numbers less than 6: 0, 1, 2, 3, 4, 5

Symbols for Operations

The basic operations in arithmetic are addition, subtraction, multiplication, and division. An ***operation*** is a procedure or a rule that relates elements in a set.

We commonly use an operation to work with two elements at a time. Some operations performed on two elements are ***binary operations***. In a binary operation, two elements from a set are replaced by *exactly one* element from the set.

For example, in the addition of the whole numbers 8 and 7, we replace 8 and 7 by exactly one whole number, 15. Here, the set is the set of whole numbers, and the operation is addition. In a binary operation, every problem must have *an* answer, and there must be *only one* answer. We use the word *unique* to mean "one and only one."

Using the + symbol for addition, we write $8 + 7 = 15$. Since adding two whole numbers always gives a unique whole-number result, addition is a binary operation for whole numbers.

Note that even when we are finding the sum of three or more numbers, we still add just two numbers at a time:

$$2 + 3 + 7 = (2 + 3) + 7 = 5 + 7 = 12$$

In the subtraction of the whole numbers 8 and 7, we replace 8 and 7 by a unique whole number, 1. Using the − symbol for subtraction, we write $8 - 7 = 1$. If the order of the elements 8 and 7 were reversed, the replacement would *not* be found in the set of whole numbers. There is no whole number associated with $7 - 8$. Because there are pairs of whole numbers whose difference is not a whole number, subtraction is *not* a binary operation for whole numbers. In the future, you will study a set of numbers for which subtraction is a binary operation.

In a binary operation, the ***order*** of the elements is important. Two elements in a particular order form an ***ordered pair***. Later in this book, we will use the notation (a, b) to show that a is the first element and b is the second element of an ordered pair. For now, we will show a binary operation in symbols as:

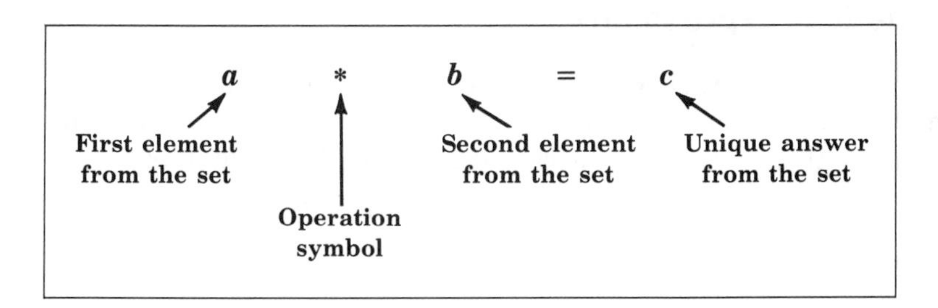

● **Definition.** A ***binary operation*** in a set assigns to every ordered pair of elements from the set a unique answer from the set.

In the study of mathematics, we will identify many different binary operations. Notice now how we have used letters to illustrate a general statement we wish to make. In this way, we are not restricted to examples using specific numbers. Also, we have used $*$ as the symbol for a general operation. The symbol for any particular binary operation can take its place.

Symbols for Arithmetic Operations

Since the symbols for arithmetic operations are used in the study of other branches of mathematics such as algebra, geometry, and probability, it is important to know the words used to describe these operations and their results.

Operation	Symbol	Result
addition	$+$	sum
subtraction	$-$	difference
multiplication	$\times$	product
division	$\div$	quotient

A particular operation can sometimes be shown by the use of different symbols. For example, to indicate multiplication, a centered dot, $\cdot$, can be used or the second number or both numbers can be placed in parentheses with no multiplication symbol.

$$6 \times 2 = 12 \quad \text{or} \quad 6 \cdot 2 = 12 \quad \text{or} \quad 6(2) = 12 \quad \text{or} \quad (6)(2) = 12$$

Division can also be represented symbolically by using a fraction where the numerator, or first element of the ordered pair, is divided by the denominator, or second element of the pair. We can write:

$$6 \div 2 = 3 \quad \text{or} \quad \frac{6}{2} = 3$$

Numerical Expressions

A ***numerical expression*** is a way of writing a number in symbols. The expression can be a single numeral or it can be a collection of numerals with one or more operation symbols. For example:

$$6 + 2 \qquad 18 - 10 \qquad 4 \times 2 \qquad 640 \div 80$$

$$2 \times 2 \times 2 \qquad 2 + 2 + 2 + 2 \qquad 1 \times 7 + 1 \qquad 8$$

Each of these numerical expressions is a way to indicate the number 8. In general, every numerical expression is a way of naming a number.

MODEL PROBLEMS

In 1 and 2, each row contains four numerical expressions. Three of the four represent the same whole number. Which expression does *not* represent that number?

1. $8 + 4$ $\qquad$ 3×4 $\qquad$ $12 \div 1$ $\qquad$ $5 + 6$

Solution: In order, the expressions represent 12, 12, 12, and 11. Therefore, $5 + 6$ does not represent the same number that the other expressions represent. *Answer:* $5 + 6$

2. $0 + 8$ $\quad$ $8 - 0$ $\quad$ $8 \div 0$ $\quad$ 8×1

Solution: The first, second, and fourth expressions represent 8. The expression $8 \div 0$ does not represent any number, since division by zero is not possible. Remember, however, that $0 \div 8 = 0$.
Answer: $8 \div 0$

3. From the quotient of 10 and 2, subtract 1.

Solution:

First, divide 10 by 2. $\quad 10 \div 2 = 5$

Then, from the result 5, subtract 1. $\quad 5 - 1 = 4$ $\quad$ *Answer:* 4

4. Add 6 to the product of 3 and 9.

Solution:

First, multiply 3 by 9. $\quad 3 \times 9 = 27$

Then, add 6 to the product 27. $\quad 27 + 6 = 33$ $\quad$ *Answer:* 33

EXERCISES

In 1–16, use the set of whole numbers, symbolized by $\{0, 1, 2, 3, \ldots\}$, and the indicated binary operations to find whole-number answers. If no whole-number answer is possible, write the word *None.*

1. $4 + 8$	**2.** $4 - 8$	**3.** 4×8	**4.** $4 \div 8$
5. $20 + 20$	**6.** $20 - 20$	**7.** 20×20	**8.** $20 \div 20$
9. $6 + 0$	**10.** $6 - 0$	**11.** 6×0	**12.** $6 \div 0$
13. $78 + 97$	**14.** $93 - 19$	**15.** 27×38	**16.** $594 \div 11$

In 17–26, use the set of whole numbers to answer the question or to perform the operations.

17. What is the product of 2 and 8?
18. What is the difference of 12 and 3?
19. What is the sum of 18 and 6?
20. What is the quotient of 10 and 2?
21. From the quotient of 20 and 4, subtract 2.
22. From the difference of 20 and 4, subtract 4.
23. To the product of 20 and 4, add 8.

24. To the sum of 20 and 4, add 8.
25. Subtract 3 from the product of 4 and 8.
26. Add 7 to the quotient of 6 and 3.

In 27–31, each row contains four numerical expressions. Three of the four expressions result in the same numerical value. Which expression has a numerical value different from the other three?

27.	$3 + 6$	$12 - 3$	$3 \cdot 3$	$3 - 12$
28.	$10 + 0$	$10 - 0$	10×0	$0 + 10$
29.	$14 - 14$	14×0	$0 \div 14$	$14 \div 14$
30.	75×80	25×200	40×125	8×625
31.	$522 \div 29$	$697 \div 41$	$\frac{476}{28}$	$\frac{612}{36}$

In 32–41, write the numeral that answers each of the following. If no answer is possible, explain why.

32. Name the first counting number.
33. Name the first whole number.
34. Name the successor of the whole number 75.
35. Name the successor of the counting number 999.
36. Name the successor of the natural number 1,909.
37. Name a whole number that is *not* a counting number.
38. Name a counting number that is *not* a whole number.
39. Name the largest whole number.
40. Name the product of the smallest whole number and the largest one-digit counting number.
41. Name the product of the smallest counting number and the largest one-digit whole number.

42. In Column I, sets of numbers are described in words. In Column II, the sets are listed using patterns and dots. Match the lists from Column II with their correct sets in Column I.

Column I	*Column II*
1. Counting numbers	a. 0, 1, 2, . . ., 9
2. Whole numbers	b. 0, 1, 2, . . .
3. Even whole numbers	c. 0, 2, 4, 6, . . .
4. Odd whole numbers	d. 0, 2, 4, 6, 8
5. Even counting numbers	e. 1, 2, 3, 4, . . .
6. One-digit whole numbers	f. 1, 2, 3, . . ., 9
7. One-digit counting numbers	g. 1, 3, 5, 7, . . .
8. Odd whole numbers less than 10	h. 1, 3, 5, 7, 9
9. Even whole numbers less than 10	i. 2, 4, 6, 8, . . .

1-2 THE NUMBERS OF ARITHMETIC

In the set of whole numbers, division is not a binary operation because some quotients, such as $3 \div 2$, do not represent whole numbers. If we consider the set of all quotients of whole numbers for which the divisor is not zero, we can define another set of numbers, called *the numbers of arithmetic*, for which division *is* a binary operation.

● **If a is a whole number and b is a counting number, $a \div b$ is a unique member of the set of *numbers of arithmetic*.**

The whole numbers and the counting numbers are subsets of the numbers of arithmetic.

The numbers of arithmetic can be symbolized in many different ways.

Fractions: The quotient of any whole number divided by any counting number can be written as a fraction.

$3 \div 4$ can be written as $\frac{3}{4}$.

$12 \div 5$ can be written as $\frac{12}{5}$.

$21 \div 7$ can be written as $\frac{21}{7}$ or $\frac{3}{1}$.

Mixed Numbers: If the quotient of any whole number divided by a smaller counting number cannot be written as a whole number, then it can be written as a mixed number, that is, the sum of a whole number and a fraction.

$4 \div 3$ can be written as $1\frac{1}{3}$.

$12 \div 5$ can be written as $2\frac{2}{5}$.

Decimal Fractions: If the quotient is the result of dividing a whole number by 10 or the product of 10's, the quotient can be written as a decimal fraction.

$4 \div 10$ can be written as .4 or 0.4.

$203 \div 100$ can be written as 2.03.

Percents: If the quotient is the result of dividing some number by 100, the quotient can be written as a percent.

$27 \div 100$ can be written as 27%.

$35 \div 10 = 350 \div 100$ can be written as 350%.

Many of the numbers of arithmetic can be written in all of these ways.

$$29 \div 4 = \frac{29}{4} = 7\frac{1}{4} = 7.25 = 725\%$$

The following are examples of some fractions represented as equivalent decimals and percents:

$$\frac{1}{4} = .25 = 25\% \qquad \frac{1}{10} = .10 = 10\%$$

$$\frac{1}{2} = .50 = 50\% \qquad \frac{4}{10}\left(\text{or } \frac{2}{5}\right) = .4 \text{ (or } .40) = 40\%$$

$$\frac{3}{4} = .75 = 75\% \qquad \frac{8}{10}\left(\text{or } \frac{4}{5}\right) = .8 \text{ (or } .80) = 80\%$$

Other Operations in Arithmetic

There are many operations in arithmetic other than addition, subtraction, multiplication, and division. Remember that a binary operation is simply a way of assigning, to some ordered pair of numbers in a set, a unique answer that is a number from that set.

Students are often concerned about their average in a subject. To find an ***average*** of two numbers, add the numbers and divide the sum by 2.

For example, the average of 75 and 87 $= \dfrac{75 + 87}{2} = \dfrac{162}{2} = 81.$

This is a binary operation. We could rewrite this problem in the form of the binary operation, $a * b = c$. Replacing $*$, the symbol for the operation, with *avg*, we obtain the result 75 avg 87 = 81.

● **Definition.** If a and b are numbers of arithmetic:

$$\boldsymbol{a} \textbf{ avg } \boldsymbol{b} = \frac{\boldsymbol{a} + \boldsymbol{b}}{\mathbf{2}}$$

MODEL PROBLEMS

1. Compute 98 avg 84.

Solution: (1) Add the numbers given. $98 + 84 = 182$

(2) Divide the sum by 2. $182 \div 2 = 91$ *Ans.*

2. The high temperature of the day was 33.5° Celsius and the low temperature was 19°C. What was the recorded average temperature for the day?

Solution: $33.5 \text{ avg } 19 = \dfrac{33.5 + 19.0}{2} = \dfrac{52.5}{2} = \dfrac{52.50}{2} = 26.25$

Answer: The records should show 26.25° Celsius.

Sometimes we have to discover which of two numbers is the larger. Finding the larger of two numbers can be thought of as a binary operation called ***maximum***. For example, the maximum of 13 and 17 is 17. To rewrite this problem in the general form of a binary operation, $a * b = c$, replace $*$ with *max*. The result is 13 max 17 = 17.

● **Definition.** If a and b are numbers of arithmetic:

a max b = the larger of the numbers a and b

MODEL PROBLEMS

1. Is there more gasoline in the tank of a car when the tank is $\frac{2}{3}$ full or when the tank is $\frac{3}{5}$ full?

Solution: Answer this problem by evaluating $\frac{2}{3} \text{ max } \frac{3}{5}$.

Step 1. Change the fractions to equivalent forms with a common denominator.

$$\frac{2}{3} \text{ max } \frac{3}{5}$$
$$= \frac{2}{3} \times \frac{5}{5} \text{ max } \frac{3}{5} \times \frac{3}{3}$$
$$= \frac{10}{15} \text{ max } \frac{9}{15}$$

Step 2. Of the two fractions that have the same denominator, select the fraction that has the larger numerator.

$$= \frac{10}{15}$$

Step 3. Rewrite this fraction in its original form.

$$= \frac{2}{3} \quad \textit{Ans.}$$

2. Reliable Company's school buses each hold 33 passengers. To transport students to a ball game, how many of these buses are needed if there are:

a. 50 students? **b.** 66 students? **c.** 24 students? **d.** s students?

Solution: To begin, divide the total number of students by the number a bus can hold. Then, where necessary, round up; that is, find the smallest whole number greater than or equal to the quotient.

	Answers
a. $\frac{50}{33} = 1\frac{17}{33}$	2 buses
b. $\frac{66}{33} = 2$	2 buses
c. $\frac{24}{33}$	1 bus
d. $\frac{s}{33}$	the smallest whole number that is greater than or equal to $\frac{s}{33}$

EXERCISES

In 1–5, write three fractions that are different names for the given number.

1. 4 **2.** 0 **3.** 0.5 **4.** $\frac{1}{3}$ **5.** $1\frac{1}{4}$

6. Answer the questions using the symbols 0, 1, $\frac{0}{1}$, $\frac{1}{0}$.

a. Which represent counting numbers?
b. Which represent whole numbers?
c. Which represent numbers that can be written as fractions?
d. Which are meaningless?

In 7–22, write the number that each numerical expression represents.

7. $\frac{3}{5} + \frac{7}{5}$ **8.** $.75 + 1.25$ **9.** $\frac{3}{8} + \frac{5}{8}$ **10.** $\frac{3}{2} + \frac{7}{4}$

11. $\frac{12}{7} - \frac{5}{7}$ **12.** $4.65 - 2.25$ **13.** $\frac{7}{8} - \frac{2}{8}$ **14.** $\frac{15}{3} - \frac{1}{2}$

15. $12 \times \frac{1}{3}$ **16.** $\frac{21}{2} \times \frac{1}{7}$ **17.** $\frac{1}{4} \times 4$ **18.** $2.5 \times .64$

19. $6.5 - .35$ **20.** 1.25×4 **21.** $6.28 \div 4$ **22.** $60 \div 1.25$

In 23–54, perform the indicated operations within the set of all numbers of arithmetic.

23. $\frac{3}{5} + \frac{1}{2}$	**24.** $1\frac{2}{5} + \frac{3}{10}$	**25.** $\frac{5}{9} + 3\frac{1}{2}$	**26.** $\frac{7}{8} + \frac{3}{4}$
27. $3\frac{1}{3} - \frac{1}{5}$	**28.** $7\frac{1}{3} - \frac{1}{2}$	**29.** $5\frac{4}{5} - 2\frac{1}{2}$	**30.** $6\frac{1}{4} - 3\frac{7}{8}$
31. $2\frac{1}{4} \times \frac{2}{3}$	**32.** $3\frac{1}{5} \times \frac{1}{4}$	**33.** $2\frac{2}{5} \times 2\frac{1}{2}$	**34.** $6\frac{2}{3} \times 2\frac{2}{5}$
35. $2\frac{1}{2} \div 3\frac{3}{4}$	**36.** $3\frac{1}{8} \div 1\frac{1}{4}$	**37.** $2\frac{1}{3} \div 7$	**38.** $7 \div 2\frac{1}{3}$
39. $3.7 + .37$	**40.** $1.9 + .09$	**41.** $1.9 - .09$	**42.** $3.2 - .31$
43. $.8 \times .5$	**44.** $.375 \times .8$	**45.** $.125 \div 5$	**46.** $.008 \div .4$
47. $3 - 1\frac{1}{5}$	**48.** $3 - 1.2$	**49.** $12\frac{3}{4} - 8$	**50.** $12.75 - 8$
51. $1\frac{1}{2} \times .6$	**52.** $\frac{2}{3} \div 1.5$	**53.** $\frac{4}{5} - .45$	**54.** $1.8 + \frac{1}{4}$

In 55–66, find the unique answer using the numbers of arithmetic.

55. $1\frac{1}{2}$ avg $4\frac{1}{2}$	**56.** 3.4 avg 6.6	**57.** $1\frac{1}{4}$ avg $\frac{3}{4}$
58. 8 avg 3	**59.** 80 avg 85	**60.** 2.9 avg 4.1
61. $\frac{3}{8}$ avg $\frac{7}{8}$	**62.** $\frac{5}{6}$ avg $\frac{3}{6}$	**63.** $\frac{6}{8}$ avg $\frac{1}{2}$
64. .5 avg .2	**65.** 1.8 avg .9	**66.** .24 avg 2.4

In 67–72, each row contains five numerals. Four of the five are equivalent. Which numeral is *not* equivalent to the other four?

67. .5	50%	.50	.05	$\frac{5}{10}$
68. $\frac{3}{4}$	.75	$\frac{3}{4}$%	75%	$\frac{6}{8}$
69. 1	100%	.1	1.0	$\frac{1}{1}$
70. .2	2%	$\frac{1}{5}$	20%	.20
71. 90%	.90	$\frac{9}{10}$	.9	.09
72. $1\frac{1}{4}$%	125%	$\frac{5}{4}$	1.25	$1\frac{1}{4}$

In 73–84, find the unique answer using the numbers of arithmetic. (We will define a max $a = a$, as in 3 max 3 = 3.)

73. 8 max 3	**74.** 8 max 11	**75.** 8 max 8
76. 1.2 max .12	**77.** .2 max .15	**78.** .8 max .30

79. .4 max .40 **80.** $\frac{1}{5}$ max $\frac{2}{5}$ **81.** $\frac{1}{5}$ max $\frac{1}{10}$

82. $\frac{8}{5}$ max 1.3 **83.** $\frac{8}{5}$ max 1.6 **84.** $\frac{3}{5}$ max $\frac{5}{8}$

85. Elaine is offering her bicycle for sale and agrees to sell it to the person who makes the best offer. Mary Rose offers her $35 and Barbara offers her $27.50. To whom will Elaine sell the bicycle? What binary operation did you use to answer the question?

86. On Monday, Pam walked 3 miles in the morning and 1.5 miles in the afternoon. Anne walked the same total distance as Pam, but she walked equal distances in the morning and the afternoon. How far did Anne walk in the morning? What binary operation did you use to answer the question?

87. A stock clerk is packaging books in cartons for shipment. A carton can hold 24 books. How many cartons are needed to package:
a. 40 books? **b.** 75 books? **c.** 18 books? **d.** b books?

88. In a school store, pens cost 19 cents each. How many pens can be bought for:
a. 50 cents? **b.** 38 cents? **c.** 12 cents? **d.** c cents?

1-3 BASES, EXPONENTS, AND POWERS

Factors

When two or more numbers are multiplied to give a certain product, each number is called a ***factor*** of the product. For example:

Since $1 \times 16 = 16$, then 1 and 16 are factors of 16.
Since $2 \times 8 = 16$, then 2 and 8 are factors of 16.
Since $4 \times 4 = 16$, then 4 is a factor of 16.

The numbers 1, 2, 4, 8, and 16 are all factors of 16.

Bases, Exponents, Powers

In the statement $4 \times 4 = 16$, the number 4 is used as a factor two times. We can rewrite $4 \times 4 = 16$ as $4^2 = 16$. We read 4^2 as "four squared," "four raised to the second power," or "the second power of four." In 4^2, the small ^{2} above and to the right of 4 tells us that 4 is to be used as a factor two times. In $4^2 = 16$, 4 is called the ***base***, 2 is called the ***exponent***, and 4^2 or 16 is called the ***power***. The exponent is always written in smaller size, to the upper right of the base.

A ***base*** is a number that is used as a factor in the product.

An ***exponent*** is a number that tells how many times the base is to be used as a factor.

A ***power*** is a number that can be expressed as a product of equal factors.

In the same way, the product $4 \times 4 \times 4$ may be written 4^3, which is read as "four cubed," "four raised to the third power," or "the third power of four." Since $4 \times 4 \times 4 = 64$, then $4^3 = 64$. Here, 4 is the base, 3 is the exponent, and 4^3 or 64 is the power. The exponent 3 tells us that the base 4 is to be used as a factor 3 times.

The product $3 \times 3 \times 3 \times 3$ may be written 3^4, which is read "three to the fourth power." Since $3 \times 3 \times 3 \times 3 = 81$, 3^4 names the same number as 81. We say that the value of 3^4 is 81, and also that 81 is the fourth power of 3.

The Fourth Power of 3

$3^4 = 3 \times 3 \times 3 \times 3 = 81$

exponent

base → $3^4 = 81$ ← power

A number raised to the first power is equal to the number itself. For example, $5^1 = 5$ and $4^1 = 4$.

Raising a number to a power is an arithmetic operation, just as addition and multiplication are operations. Although raising to a power is defined as repeated multiplication of a base, it can be thought of as a binary operation. The base is the first element of the ordered pair; the exponent is the second element of the ordered pair; the power is the result.

In the general form of a binary operation, $a * b = c$, the base would be a, the exponent would be b, and the power would be c. Some hand calculators and computers use a symbol to show raising to a power: $3 \uparrow 4 = 81$ or $3 ** 4 = 81$. However, it is more convenient to write the expression without an operation symbol simply by raising the exponent to the upper right of the base: $3^4 = 81$. Examples like $3^4 = 81$ and $4^3 = 64$, in which interchanging the 3 and the 4 gives different results, point out that order is important in this binary operation.

MODEL PROBLEMS

Compute the value of the given expression.

1. 5^4 **2.** $(.4)^3$ **3.** $\left(\frac{2}{3}\right)^2$ **4.** $\left(1\frac{1}{2}\right)^2$

Solution:

1. $5^4 = 5 \times 5 \times 5 \times 5 = 625$ *Ans.*

2. $(.4)^3 = .4 \times .4 \times .4 = .064$ *Ans.*

3. $\left(\frac{2}{3}\right)^2 = \frac{2}{3} \times \frac{2}{3} = \frac{4}{9}$ *Ans.*

4. $\left(1\frac{1}{2}\right)^2 = 1\frac{1}{2} \times 1\frac{1}{2} = \frac{3}{2} \times \frac{3}{2} = \frac{9}{4} = 2\frac{1}{4}$ *Ans.*

EXERCISES

In 1–24, find the value of the given expression.

1. 9^2	**2.** 10^3	**3.** 5^3	**4.** 15^2
5. 10^5	**6.** 4^5	**7.** $\left(\frac{1}{3}\right)^2$	**8.** $\left(\frac{1}{2}\right)^2$
9. $\left(\frac{1}{10}\right)^3$	**10.** $\left(\frac{3}{4}\right)^2$	**11.** $\left(\frac{1}{10}\right)^4$	**12.** $\left(\frac{8}{9}\right)^1$
13. $(.8)^2$	**14.** $(.5)^3$	**15.** $(.3)^2$	**16.** $(.2)^4$
17. $(.1)^5$	**18.** $(.25)^2$	**19.** $(1.1)^3$	**20.** $(3.1)^2$
21. $(2.5)^2$	**22.** $\left(2\frac{1}{2}\right)^2$	**23.** $\left(3\frac{1}{10}\right)^2$	**24.** $\left(1\frac{1}{3}\right)^3$

In 25–32: **a.** Find the value of each of the three given expressions. **b.** Name the expression that has the greatest value.

25. $(5)^2$	$(5)^3$	$(5)^4$	**26.** $(.5)^2$	$(.6)^2$	$(.7)^2$
27. $(1.1)^2$	$(1.1)^3$	$(1.1)^4$	**28.** $(1.0)^2$	$(1.2)^2$	$(1.4)^2$
29. $\left(\frac{1}{5}\right)^2$	$\left(\frac{1}{6}\right)^2$	$\left(\frac{1}{7}\right)^2$	**30.** $(.1)^3$	$(.2)^2$	$(.3)^1$
31. $(.5)^3$	$(.4)^2$	$(.2)^4$	**32.** $(1.5)^3$	$(1.4)^2$	$(1.2)^4$

1-4 ORDER OF OPERATIONS

When a numerical expression involves two or more operations, we need to agree upon the order in which these operations will be performed. To evaluate

$$5 + 3 \times 2$$

One person may wish to multiply first. Then:

$$5 + 3 \times 2 = 5 + 6$$
$$= 11$$

Another person may wish to add first. Then:

$$5 + 3 \times 2 = 8 \times 2$$
$$= 16$$

Who is right?

In order to give a single meaning to the expression $5 + 3 \times 2$ and others like it, mathematicians have agreed on the following order of operations:

1. Simplify powers.
2. Do multiplications and divisions.
3. Do additions and subtractions.

We use the above order to evaluate numerical expressions that contain two or more operations. Thus:

$$\begin{aligned} 5 + 3 \times 2 &= 5 + 6 \\ &= 11 \end{aligned}$$

$$\begin{aligned} 3 \cdot 2^2 - 8 &= 3 \cdot 4 - 8 \\ &= 12 - 8 \\ &= 4 \end{aligned}$$

Expressions With Grouping Symbols

In mathematics, *parentheses* ( ) are used to indicate the meaning of an expression. For example, $(4 \times 6) + 7$ and $4 \times (6 + 7)$ have two different meanings. If 7 is to be added to the product of 4 and 6, we would write $(4 \times 6) + 7$. But if the sum of 6 and 7 is to be multiplied by 4, we would write $4 \times (6 + 7)$ or $4(6 + 7)$. When an operation is enclosed in parentheses, that operation is performed first. Therefore, parentheses act as a grouping symbol and can change the order in which operations are performed.

In simplifying any numerical expression, we first perform the operations indicated on the numbers within parentheses. For example:

$$\begin{aligned} 30 - (6 + 5) &= 30 - 11 \\ &= 19 \end{aligned}$$

$$\begin{aligned} 6(4 + 1)^2 &= 6(5)^2 \\ &= 6(25) \\ &= 150 \end{aligned}$$

Besides parentheses, there are other symbols to indicate grouping, such as *brackets*, [ ]. The expressions $2(5 + 9)$ and $2[5 + 9]$ have the same meaning, that is, 2 is multiplied by the sum of 5 and 9. A *bar*, or *fraction line*, also acts as a symbol of grouping, telling us to perform the operations in the numerator and denominator first.

$$\frac{20 - 8}{3} = \frac{12}{3} = 4 \qquad \frac{6}{3 + 1} = \frac{6}{4} = \frac{3}{2} = 1\frac{1}{2}$$

When there are *two or more* grouping symbols in an expression, we perform the operations on the numbers in the *innermost symbol first.* For example:

$$\begin{aligned} &5 + 2[6 + (3 - 1)^3] \\ &= 5 + 2[6 + 2^3] \\ &= 5 + 2[6 + 8] \\ &= 5 + 2[14] \\ &= 5 + 28 \\ &= 33 \end{aligned}$$

PROCEDURE. To simplify a numerical expression:

1. Simplify any numerical expressions within parentheses or within other symbols of grouping, starting with the innermost.
2. Simplify any powers.
3. Do all multiplications and divisions in order from left to right.
4. Do all additions and subtractions in order from left to right.

MODEL PROBLEMS

1. Simplify the numerical expression $80 - 4(6 - 4)$.

How to Proceed	*Solution*
(1) Write the expression.	$80 - 4(6 - 4)$
(2) Simplify the expression within the parentheses.	$= 80 - 4(2)$
(3) Do the multiplication.	$= 80 - 8$
(4) Do the subtraction.	$= 72$ *Ans.*

2. Evaluate: $5(6 - 4)^3 - 5$

How to Proceed	*Solution*
(1) Write the expression.	$5(6 - 4)^3 - 5$
(2) Simplify the expression within parentheses.	$= 5(2)^3 - 5$
(3) Evaluate the power.	$= 5(8) - 5$
(4) Do the multiplication.	$= 40 - 5$
(5) Do the subtraction.	$= 35$ *Ans.*

3. Evaluate: $28 \div 4 - 2(8 - 7)^2 + 5 \times 3$

How to Proceed	*Solution*
(1) Write the expression.	$28 \div 4 - 2(8 - 7)^2 + 5 \times 3$
(2) Simplify the expression within the parentheses.	$= 28 \div 4 - 2(1)^2 + 5 \times 3$
(3) Evaluate the power.	$= 28 \div 4 - 2(1) + 5 \times 3$
(4) Do multiplication and division from left to right.	$= 7 - 2 + 15$
(5) Do addition and subtraction from left to right.	$= 5 + 15$
	$= 20$ *Ans.*

EXERCISES

In 1–8, state the meaning of the expression in part **a** and the meaning of the expression in part **b** and simplify each expression.

1.	**a.** $20 + (6 + 1)$	**b.** $20 + 6 + 1$
2.	**a.** $18 - (4 + 3)$	**b.** $18 - 4 + 3$
3.	**a.** $12 - \left(3 - \frac{1}{2}\right)$	**b.** $12 - 3 - \frac{1}{2}$
4.	**a.** $15 \times (2 + 1)$	**b.** $15 \times 2 + 1$
5.	**a.** $(12 + 8) \div 4$	**b.** $12 + 8 \div 4$
6.	**a.** $48 \div (8 - 4)$	**b.** $48 \div 8 - 4$
7.	**a.** $7 + 5^2$	**b.** $(7 + 5)^2$
8.	**a.** 4×3^2	**b.** $(4 \times 3)^2$

In 9–14, use parentheses to express the sentence in symbols.

9. The sum of 10 and 8 is to be found, and then 5 is to be subtracted from this sum.
10. 15 is to be subtracted from 25, and 7 is to be added to the difference.
11. 8 is to be multiplied by the difference of 6 and 2.
12. 12 is to be subtracted from the product of 10 and 5.
13. The difference of 12 and 2 is to be multiplied by the sum of 3 and 4.
14. The quotient of 20 and 5 is to be subtracted from the product of 16 and 3.

Basic Operations

In 15–20, simplify the numerical expression.

15. $6 \times 5 - 8 \times 2$
16. $20 + 20 \div 5 + 5$
17. $36 - 12 \div 4 - 1$
18. $36 + \frac{1}{2} \times 10$
19. $24 - 4 \div \frac{1}{2}$
20. $28 + 0 \div 4 - 10 \times .2$

Basic Operations and Powers

In 21–35, simplify the numerical expression.

21. 2×3^2
22. 4×5^2
23. $81 \times \left(\frac{1}{3}\right)^3$
24. $64 \times (.5)^2$
25. $2^3 \times 1^2$
26. $10^2 \times 3^3$
27. $1^4 \times 9^2$
28. $(2^5)\left(\frac{1}{2}\right)^3$
29. $5^2 + 12^2$
30. $16^2 + 9^2$
31. $13^2 - 5^2$
32. $20^2 - 12$
33. $6 + 4(5)^2$
34. $3(2)^2 + 6$
35. $120 - 6(2)^4$

Grouping Symbols

In 36–59, simplify the numerical expression.

36. $10 + (1 + 4)$
37. $13 - [9 + 1]$
38. $36 - (10 - 8)$
39. $7(5 + 2)$
40. $(6 - 1)10$
41. $20 \div [7 + 3]$
42. $\frac{48}{15 - 3}$
43. $\frac{24 - 8}{2}$
44. $\frac{17 + 13}{25 - 10}$
45. $15 - (15 \div 5)$
46. $3(6 + 3) - 4$
47. $25 + 3(10 - 4)$
48. $26 - 4[7 - 5]$
49. $25 \div (6 - 1) + 3$
50. $3(6 + 4)(6 - 4)$
51. $100(6)^2 - 75$
52. $(4 + 6)^2$
53. $2[14 - (1 + 7)]$
54. $(20 - 15)^2$
55. $[7 - (2 + 3)]^3$
56. $2(4 + 6)^2 - 10$
57. $200 - 3(5 - 1)^3$
58. $12[5^2 - 4^2]$
59. $(7^2 - 6^2)(1^2 + 2^2)$

In 60–63: **a.** Write a numerical expression for each of the following. **b.** Evaluate the numerical expression written in answer to part **a**.

60. The cost of 2 chocolate chip cookies and 3 peanut butter cookies if each cookie costs 8 cents.

61. The number of miles traveled by Ms. McCarthy if she drove 30 miles per hour for $\frac{3}{4}$ hour and 55 miles per hour for $1\frac{1}{2}$ hours.

62. The cost of 2 pens at $.38 each and 3 notebooks at $.69 each.

63. The cost of 5 pens at $.29 each and 3 notebooks at $.75 each if ordered from a mail order company that adds $1.75 in postage and handling charges.

64. Insert operational symbols (+, −, ×, ÷) and parentheses to make the following statements true.

a. 3 __ 2 __ 1 = 4
b. 1 __ 3 __ 1 = 4
c. 1 __ 2 __ 3 __ 4 = 5
d. 4 __ 3 __ 2 __ 1 = 5
e. 6 __ 6 __ 6 __ 6 = 5
f. 6 __ 6 __ 6 __ 6 = 6

1-5 PROPERTIES OF OPERATIONS

When numbers behave in a certain way for an operation, we describe this behavior as a ***property***.

Commutative Property of Addition

When we add numbers of arithmetic, we assume that we may change the order in which two numbers are added without changing the sum.

For example, $4 + 5 = 5 + 4$, and $\frac{1}{2} + \frac{1}{4} = \frac{1}{4} + \frac{1}{2}$. These examples illustrate the ***commutative property of addition.***

In general, we assume that for every number a and every number b:

$$\boldsymbol{a + b = b + a}$$

Commutative Property of Multiplication

In the same way, when we multiply numbers of arithmetic, we assume that we may change the order of the factors without changing the product. For example, $5 \times 4 = 4 \times 5$, and $\frac{1}{2} \times \frac{1}{4} = \frac{1}{4} \times \frac{1}{2}$. These examples illustrate the ***commutative property of multiplication.***

In general, we assume that for every number a and every number b:

$$\boldsymbol{a \cdot b = b \cdot a}$$

The General Commutative Property

Recall that $\boldsymbol{a * b = c}$ was the general form used for a binary operation where a, b, and c were numbers in a set and where $*$ was the symbol used for the operation. If we can change the order of every pair of numbers, a and b, in the set without changing the answer obtained, c, then the *commutative property holds for the operation* $*$.

In general, when for every number a and every number b:

$$\boldsymbol{a * b = b * a}$$

then $*$ is a commutative operation. However, if we find even one case where $a * b$ and $b * a$ produce different answers, then the operation $*$ is not commutative. For example:

The operation *subtraction* is not commutative because $5 - 4 \neq 4 - 5$. (The symbol $\neq$ means *is not equal to.*)

The operation *division* is not commutative because $8 \div 4 \neq 4 \div 8$.

Although we are not ready at this time to show that the commutative property for a particular operation always holds, we can say *it appears* that an operation is commutative. For example:

3 avg 11 = 7 and 11 avg 3 = 7, so 3 avg 11 = 11 avg 3.

2 avg 9 = $5\frac{1}{2}$ and 9 avg 2 = $5\frac{1}{2}$, so 2 avg 9 = 9 avg 2.

It appears that the commutative property holds for the operation of averaging two numbers. Or, for every a and b:

$$\boldsymbol{a} \textbf{ avg } \boldsymbol{b} = \boldsymbol{b} \textbf{ avg } \boldsymbol{a}$$

Associative Property of Addition

Addition is a binary operation; that is, we add two numbers at a time. If we wish to add three numbers, we find the sum of two and add that sum to the third. For example:

$$\begin{aligned} 2 + 5 + 8 &= (2 + 5) + 8 \\ &= \quad 7 \quad + 8 \\ &= \quad 15 \end{aligned} \quad \text{or} \quad \begin{aligned} 2 + 5 + 8 &= 2 + (5 + 8) \\ &= 2 + \quad 13 \\ &= \quad 15 \end{aligned}$$

Therefore, we see that $(2 + 5) + 8 = 2 + (5 + 8)$. This example illustrates the ***associative property of addition***.

In general, we assume that for every number a, every number b, and every number c:

$$(\boldsymbol{a} + \boldsymbol{b}) + \boldsymbol{c} = \boldsymbol{a} + (\boldsymbol{b} + \boldsymbol{c})$$

Associative Property of Multiplication

In a similar way, to find a product that involves three factors, we first multiply any two factors and then multiply this result by the third factor. We assume that we do not change the product when we change the grouping. For example:

$$\begin{aligned} 5 \times 4 \times 2 &= (5 \times 4) \times 2 \\ &= \quad 20 \quad \times 2 \\ &= \quad 40 \end{aligned} \quad \text{or} \quad \begin{aligned} 5 \times 4 \times 2 &= 5 \times (4 \times 2) \\ &= 5 \times \quad 8 \\ &= \quad 40 \end{aligned}$$

Therefore, $(5 \times 4) \times 2 = 5 \times (4 \times 2)$.
This example illustrates the ***associative property of multiplication***.

In general, we assume that for every number a, every number b, and every number c:

$$(\boldsymbol{a} \cdot \boldsymbol{b}) \cdot \boldsymbol{c} = \boldsymbol{a} \cdot (\boldsymbol{b} \cdot \boldsymbol{c})$$

The General Associative Property

In a binary operation, symbolized by $*$, we work with two numbers at a time. According to the order of operations, we must simplify operations within parentheses first.

In general, when for every number a, for every number b, and for every number c:

$$(\boldsymbol{a} * \boldsymbol{b}) * \boldsymbol{c} = \boldsymbol{a} * (\boldsymbol{b} * \boldsymbol{c})$$

then $*$ is an associative operation.

Remember that we need to find only one case where $(a * b) * c$ and $a * (b * c)$ produce different answers to say that the operation $*$ is not associative. For example, the operation *subtraction* is not associative because $(10 - 8) - 2 \neq 10 - (8 - 2)$. Also, the operation *division* is not associative because $(8 \div 4) \div 2 \neq 8 \div (4 \div 2)$.

Although we are not ready at this time to show that the associative property is true for a particular operation, we can say *it appears* that the operation is associative.

The Distributive Property

We know $4(3 + 2) = 4(5) = 20$ and also $4(3) + 4(2) = 12 + 8 = 20$. Therefore, we see that $4(3 + 2) = 4(3) + 4(2)$.

This result can be illustrated geometrically. Remember that the area of a rectangle is equal to the product of its length and its width.

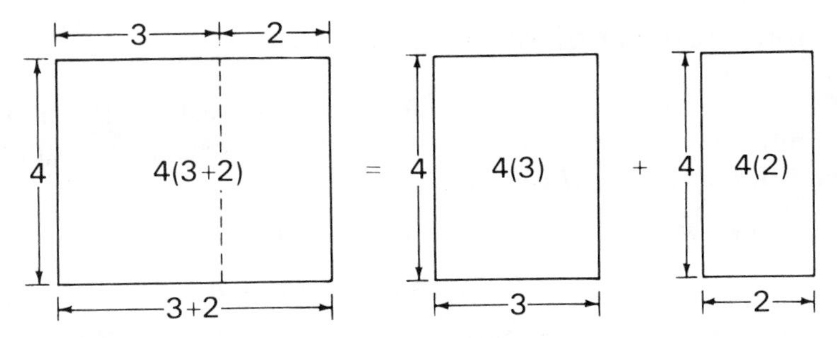

This example illustrates the ***distributive property of multiplication over addition***, also called the ***distributive property***. This means that the product of one number times the sum of a second and a third number equals the product of the first and second numbers plus the product of the first and third numbers.

Thus, $4(3 + 2) = 4(3) + 4(2)$.

In general, we assume that for every number a, every number b, and every number c:

$$\boldsymbol{a(b + c) = ab + ac} \quad \text{and} \quad \boldsymbol{ab + ac = a(b + c)}$$

The distributive property is also assumed to be true for subtraction:

$$a(b - c) = ab - ac \quad \text{and} \quad ab - ac = a(b - c)$$

Observe how we can use the distributive property to find the following products:

1. $6 \times 23 = 6(20 + 3) = 6 \times 20 + 6 \times 3 = 120 + 18 = 138$

2. $9 \times 3\frac{1}{3} = 9\left(3 + \frac{1}{3}\right) = 9 \times 3 + 9 \times \frac{1}{3} = 27 + 3 = 30$

3. $6.5 \times 8 = (6 + .5)8 = 6 \times 8 + .5 \times 8 = 48 + 4 = 52$

Notice how we can use the distributive property to change the form of an expression from a product to a sum or a difference:

1. $5(a + b) = 5a + 5b$ **2.** $9(a - b) = 9a - 9b$

Working backward, we can also use the distributive property to change the form of an expression from a sum or difference to a product:

1. $3c + 3d = 3(c + d)$ **2.** $9y - 4y = (9 - 4)y = 5y$

Note that the ***substitution principle*** allows us to replace one expression by another expression that represents the same number, as in $(9 - 4)y = 5y$ where $(9 - 4)$ was replaced by 5.

Addition Property of Zero

The true sentences $5 + 0 = 5$ and $0 + 8 = 8$ illustrate that the sum of any number and zero is that number itself. These examples illustrate the ***addition property of zero***.

In general, we assume that for every number a:

$$\boldsymbol{a + 0 = a} \quad \text{and} \quad \boldsymbol{0 + a = a}$$

We call 0 the ***identity element of addition***, or the ***additive identity element***, for the numbers of arithmetic. We will also agree that if $a + x = a$ for any number a, then $x = 0$.

Multiplication Property of Zero

The true sentences $4 \times 0 = 0$ and $0 \times 3 = 0$ illustrate that the product of any number and zero is zero. These examples illustrate the ***multiplication property of zero***.

In general, we assume that for every number a:

$$\boldsymbol{a \cdot 0 = 0} \quad \text{and} \quad \boldsymbol{0 \cdot a = 0}$$

Multiplication Property of One

The true sentences $5 \times 1 = 5$ and $1 \times 9 = 9$ illustrate that the product of any number and one is that number itself. These examples illustrate the ***multiplication property of one.***

In general, we assume that for every number a:

$$\boldsymbol{a \cdot 1 = a} \quad \text{and} \quad \boldsymbol{1 \cdot a = a}$$

We call 1 the ***identity element of multiplication***, or the ***multiplicative identity element***, for the numbers of arithmetic.

MODEL PROBLEMS

1. Evaluate the number expression $3 \times 1 + (8 + 0)$ and give the reason for each step of the procedure.

Solution:

Step	*Reason*
(1) $3 \times 1 + (8 + 0) = 3 \times 1 + 8$	(1) Addition property of 0.
(2) $= 3 + 8$	(2) Multiplication property of 1.
(3) $= 11$	(3) Substitution principle.

Answer: 11

2. Express $6t + t$ as a product and give the reason for each step of the procedure.

Solution:

Step	*Reason*
(1) $6t + t = 6t + 1t$	(1) Multiplication property of 1.
(2) $= (6 + 1)t$	(2) Distributive property.
(3) $= 7t$	(3) Substitution principle.

Answer: $7t$

EXERCISES

1. Name the number that is the additive identity element for the numbers of arithmetic.

2. Name the identity element of multiplication for the numbers of arithmetic.

3. Give the value of each number expression.

a. $9 + 0$ **b.** 9×0 **c.** 9×1 **d.** $\frac{2}{3} \times 0$ **e.** $0 + \frac{2}{3}$ **f.** $1 \times \frac{2}{3}$

In 4–20: **a.** Give a replacement for the question mark that makes the sentence true. **b.** Name the property illustrated in the sentence that is formed when the replacement is made.

4. $8 + 6 = 6 + ?$
5. $17 \times 5 = ? \times 17$
6. $(3 \times 9) \times 15 = 3 \times (9 \times ?)$
7. $6(5 + 8) = 6(5) + ?(8)$
8. $\left(\frac{1}{3} + \frac{1}{6}\right) + \frac{1}{2} = ? + \left(\frac{1}{6} + \frac{1}{2}\right)$
9. $4 + 0 = ?$
10. $3(x + 5) = 3x + 3(?)$
11. $(3 \times 7) + 5 = (? \times 3) + 5$
12. $(19 \times 2) \times 50 = ? \times (19 \times 2)$
13. $(19 \times ?) + 50 = 19 + 50$
14. $3(8 + 2) = (? + 2)3$
15. $(5 + ?) + 2 = 5 + (6 + 2)$
16. $\frac{1}{2}$ avg $6 = ?$ avg $\frac{1}{2}$
17. $\frac{1}{2} \times 10 = 10 \times ?$
18. $3(y + ?) = 3y$
19. $?x = x$
20. $(7 \max 9) \max 4 = 7 \max (? \max 4)$

In 21–30, state whether the sentence is a correct application of the distributive property. If you believe that it is not, state your reason.

21. $6(5 + 8) = 6 \times 5 + 6 \times 8$
22. $10\left(\frac{1}{2} + \frac{1}{5}\right) = 10 \times \frac{1}{2} + \frac{1}{5}$
23. $(7 + 9)5 = 7 + 9 \times 5$
24. $3(x + 5) = 3x + 3 \times 5$
25. $2(y + 6) = 2y + 6$
26. $(b + 2)a = ba + 2a$
27. $4a(b + c) = 4ab + 4ac$
28. $4b(c - 2) = 4bc - 2$
29. $8m + 6m = (8 + 6)m$
30. $14x - 4x = (14 - 4)x$

In 31–36, **a.** Tell whether the sentence is true or false. **b.** Tell whether the commutative property holds for the given operation.

31. $357 + 19 = 19 + 357$
32. $2 \div 1 = 1 \div 2$
33. $25 - 7 = 7 - 25$
34. $(18)\left(2\frac{1}{2}\right) = \left(2\frac{1}{2}\right)(18)$
35. $\frac{2}{5} + \frac{3}{10} = \frac{3}{10} + \frac{2}{5}$
36. $5 - 0 = 0 - 5$

In 37–42, **a.** Tell whether the sentence is true or false. **b.** Tell whether the associative property holds for the given operation.

37. $(73 \times 68) \times 92 = 73 \times (68 \times 92)$
38. $(24 \div 6) \div 2 = 24 \div (6 \div 2)$

39. $(19 - 8) - 5 = 19 - (8 - 5)$
40. $(9 + .3) + .7 = 9 + (.3 + .7)$
41. $(8 \div 4) \div 2 = 8 \div (4 \div 2)$
42. $(40 - 20) - 10 = 40 - (20 - 10)$

43. If $r + s = r$: **a.** What is the numerical value of s? **b.** What is the numerical value of rs?
44. If $xy = x$, what is the numerical value of y?
45. If $xy = 0$ and $x \neq 0$, what is the numerical value of y?

In 46–51, complete the sentence so that it is an application of the distributive property.

46. $9(7 + 3) =$ ___ **47.** ___ $= 12 \times \frac{1}{2} + 12 \times \frac{1}{3}$

48. $4(p + q) =$ ___ **49.** ___ $= 2x - 2y$

50. $8t + 13t =$ ___ **51.** ___ $= (15 - 7)m$

52. Which is an illustration of the commutative property of addition?
(1) $a + 0 = a$ (2) $a(b + c) = ab + ac$
(3) $a + b = b + a$ (4) $(a + b) + c = a + (b + c)$

53. Which statement illustrates the associative property of multiplication?
(1) $2\left(\frac{1}{2}\right) = 1$ (2) $2(3 + 4) = 2(3) + 2(4)$
(3) $2(1) = 2$ (4) $2(3 \cdot 4) = (2 \cdot 3)4$

54. **a.** Find the unique solution of (10 avg 14) avg 2.
b. Find the unique solution of 10 avg (14 avg 2).
c. Are the solutions in parts **a** and **b** the same or different?
d. Does it appear that the operation of averaging two numbers is associative?

55. Place parentheses to make each statement true.
a. $3 \times 2 + 1 \div 3 = 3$ **b.** $4 \times 3 \div 2 + 2 = 3$
c. $8 + 8 \div 8 - 8 \times 8 = 8$ **d.** $3 \div 3 + 3 \times 3 - 3 = 1$
e. $3 \div 3 + 3 \times 3 - 3 = 0$ **f.** $0 \times 12 \times 3 - 16 \div 8 = 0$

1-6 COMPARING NUMBERS

In our daily lives, we are often asked to compare quantities. Which is cheaper? Which weighs more? Who is taller? Which will last longer? Are they the same size? The answers to these questions are given by comparing quantities that are stated in numerical terms.

Symbols of Inequality

If two numbers are not equal, the relationship between them can be expressed in several different ways.

Symbol	Example	Read
$>$	$8 > 3$	8 is greater than 3
$<$	$3 < 8$	3 is less than 8
$\ngtr$	$3 \ngtr 8$	3 is not greater than 8
$\nless$	$8 \nless 3$	8 is not less than 3
$\geq$	$8 \geq 3$	8 is greater than or equal to 3
$\leq$	$3 \leq 8$	3 is less than or equal to 8
$\neq$	$8 \neq 3$	8 is not equal to 3

Notice that in an ***inequality*** the symbols $>$ and $<$ point to the smaller number.

When we say 8 max 7 = 8, we are comparing the numbers 8 and 7, and saying that the first element, 8, is greater than the second element, 7. Thus:

$$8 \text{ max } 7 = 8 \quad \text{and} \quad 8 > 7$$

are different ways of stating the relationship between 8 and 7.

In the case of 3 max 8 = 8, the first element 3 is less than the second element 8. Thus:

$$3 \text{ max } 8 = 8 \quad \text{and} \quad 3 < 8$$

are different ways of stating the relationship between 3 and 8.

In general, if a max $b = a$ and $a \neq b$, then $a > b$ and $b < a$.

MODEL PROBLEMS

In 1–6, tell whether the statement is true or false.

1. $6 + 7 \neq 15$ True *Ans.*

2. $0 \neq 8 - 8$ False *Ans.*

3. $8 + 6 \geq 10$ True *Ans.*

4. $3 \times 6 \leq 10$ False *Ans.*

5. $20 \div 5 < 8$ True *Ans.*

6. $15 - 13 \nless 7$ False *Ans.*

In 7 and 8, write three true statements to compare the numbers in the order given.

7. 8 and 2 *Answer:* $8 > 2$; $8 \neq 2$; $8 \nless 2$

8. 12 and 12 *Answer:* $12 = 12$; $12 \ngtr 12$; $12 \nless 12$

EXERCISES

In 1–5, state whether the inequality is true or false.

1. $8 + 5 \neq 6 + 4$ **2.** $9 + 2 \neq 2 + 9$ **3.** $6 \times 0 \neq 4 \times 0$
4. The sum of 8 and 12 is not equal to the product of 24 and 4.
5. The product of 5 and 4 is not equal to 5 divided by 4.

In 6–11, write the inequality using the symbol $>$ or the symbol $<$.

6. 25 is greater than 20.
7. $12 + 3$ is less than 20.
8. $6 - 3$ is less than $5 + 4$.
9. $80 \div 4$ is greater than $6 + 3$.
10. The sum of 9 and 4 is less than the product of 10 and 5.
11. The sum of 8 and 7 is greater than the quotient 20 divided by 5.

In 12–14, express each inequality in words.

12. $9 + 8 > 16$ **13.** $12 - 2 \leq 4 \times 7$ **14.** $5 + 24 \not< 90 \div 3$

In 15–23, state whether the inequality is true or false.

15. $20 - 4 < 5 + 8$ **16.** $6 \times 0 \geq 3 + 5$ **17.** $18 + 0 > 4 + 0$

18. $2.05 < 20.5$ **19.** $\frac{1}{2} + \frac{1}{8} < .8$ **20.** $3 - .25 > 2\frac{1}{2}$

21. $4.6 - 2.1 > 1.5 + .9$ **22.** $8 \times .5 \leq 6 \div \frac{1}{2}$

23. $\frac{1}{2} + \frac{1}{3} \not> 1 - \frac{1}{4}$

In 24–32, replace the question mark with a numeral that will make the resulting statement true.

24. $5 + ? \neq 11$ **25.** $? - 7 > 3$ **26.** $15 \div 5 < ? \div 2$

27. $8 \times \frac{1}{2} \neq 6 - ?$ **28.** $10 \div ? \leq 10$ **29.** $4 - .7 \not< 1 \times ?$

30. 3 max ? = 7 max 5 **31.** 3 avg ? = 7 avg 5

32. $7 - ? = 7 \times 1$

In 33–41, write three true statements to show the comparison of the numerals, using the order in which they are given.

33. 8 and 14 **34.** 9 and 3 **35.** 15 and 15

36. .11 and .6 **37.** .3 and .21 **38.** .8 and .80

39. .8 and .08 **40.** $\frac{2}{3}$ and $\frac{3}{4}$ **41.** $\frac{1}{5}$ and $\frac{1}{8}$

42. Paperbacks can be purchased from a local bookstore for $1.35 each. A mail order catalog has the same selection of books for $1.10 each plus $2.00 per order for postage and handling.

a. Which is the less expensive place to purchase books if 5 books are to be purchased?

b. Which is the less expensive place to purchase books if 12 books are to be purchased?

c. Which is the less expensive place to purchase books if 8 books are to be purchased?

43. A local photo shop will develop pictures for $.20 for each print. A roll of film can be mailed to a company that charges $.12 for each print plus $1.50 for postage.

a. Which offer is better if a roll of 15 pictures is to be developed?

b. Which offer is better if a roll of 24 pictures is to be developed?

44. A typist is offered a wage of $1.50 per page or $7.50 per hour.

a. What is the better offer if he can type 5 pages in an hour?

b. What is the better offer if he can type 6 pages in an hour?

1-7 NUMBER LINES

If we think of a straight line as a set of points, we can assign all the numbers of arithmetic to points on the line. This line, called a ***number line***, is a useful tool in comparing numbers and in understanding some of the operations of arithmetic.

To build a number line, we begin with two points. We label the first point 0 and the second point 1. Then, using the length of the segment from 0 to 1 as a ***unit measure***, and going in the direction of 1, we continue along a straight line, marking equally spaced points. Since every segment between adjacent points is equal in measure to the unit segment, we assign the numbers 2, 3, 4, and so on, to these points. An arrowhead indicates that the number line extends without end, just as the set of whole numbers is endless.

A number line can go in any direction, but ***horizontal*** and ***vertical*** lines are most commonly used. On a horizontal number line, the numbers increase as we move to the right. On a vertical number line, the numbers increase as we move up the line.

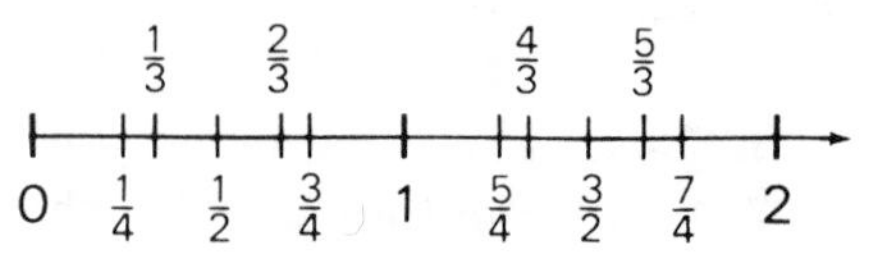

Consider a number line on which points have been associated with whole numbers. If we divide the intervals between whole numbers into halves, thirds, quarters, etc., we can label additional points as shown.

In this manner, we can assign fractions, decimals, mixed numbers, and, in fact, all of the numbers of arithmetic to specific points on the number line. The number that is associated with a point on the number line is called the ***coordinate*** of that point. The point on the number line that is associated with a number is called the ***graph*** of that number.

No matter how close two points may be on a number line, there is always an endless number of points between them. Since there are infinitely many points between any two points, there are infinitely many numbers that can be named between any two given numbers.

While it is true that every number of arithmetic can be associated with a point on a number line, there are some points that are not associated with the numbers of arithmetic. You will study these points and the numbers with which they are associated later on in this course.

Ordering Numbers on a Number Line

Consider the binary operation 4 max 2. When the two numbers involved are graphed on a standard horizontal number line, the larger of the two numbers appears to the right of the smaller; the smaller of the two numbers appears to the left of the larger.

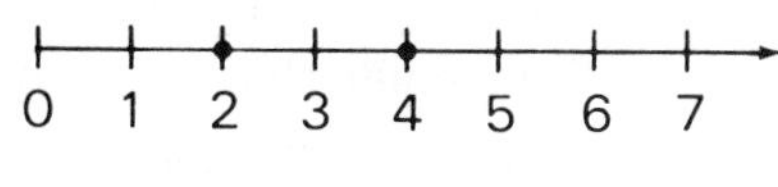

Thus, since $2 < 4$, the graph of 2 is to the left of the graph of 4. Also, since $4 > 2$, the graph of 4 is to the right of the graph of 2. We say that the phrases *less than* and *greater than* express an ***order relation*** illustrated by the order in which the numbers appear on a number line.

The statement "4 is *between* 2 and 6" means $4 > 2$ and $4 < 6$. These two inequalities can be combined into a single expression, $2 < 4 < 6$, which is read as "2 is less than 4 and 4 is less than 6" or as "4 is between 2 and 6." We may also express this relationship as $6 > 4 > 2$.

MODEL PROBLEMS

1. Use the number line in the figure to answer parts **a–d**.

N U M B E R S

0 $\frac{1}{2}$ 1 $\frac{3}{2}$ 2 $\frac{5}{2}$ 3

	Answers
a. Name the number assigned to point *S*.	3
b. Name the coordinate of point *M*.	1
c. Name the point that is the graph of $\frac{1}{2}$.	*U*
d. Name the point that is the graph of $1\frac{1}{2}$.	*B*

2. On the number line shown at the right, locate the points that can be associated with the following numbers:

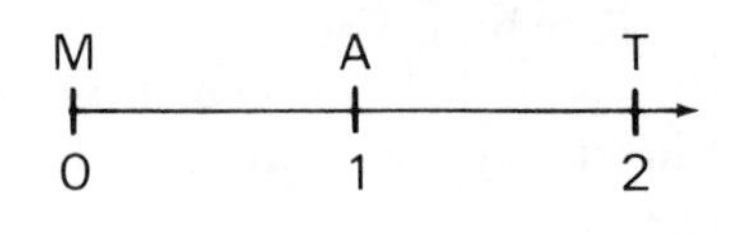

a. $\frac{1}{3}$ **b.** .5 **c.** $\frac{20}{20}$ **d.** $1\frac{2}{3}$ **e.** 1.8

Answer:

a. Point G, which is $\frac{1}{3}$ of the way from M to A.

b. Point R, which is $\frac{1}{2}$ of the way from M to A. $\left(\text{Remember that .5 names the same number as } \frac{1}{2}.\right)$

c. Point A. (Remember that $\frac{20}{20}$ names the same number as 1.)

d. Point P, which is $\frac{2}{3}$ of the way from A to T.

e. Point H, which is $\frac{4}{5}$ of the way from A to T. $\left(\text{Remember that 1.8 names the same number as } 1\frac{4}{5}.\right)$

	Answers
3. a. How many whole numbers are there between 3 and 6?	Two
b. List these numbers.	4, 5
4. a. How many numbers are there between 3 and 6?	Infinitely many
b. List four numbers between 3 and 6.	3.5, $\frac{13}{3}$, 5, $5\frac{1}{2}$ (There are many other possible answers.)

EXERCISES

In 1–3, name the number that can be associated with each of the labeled points on the number line.

1.

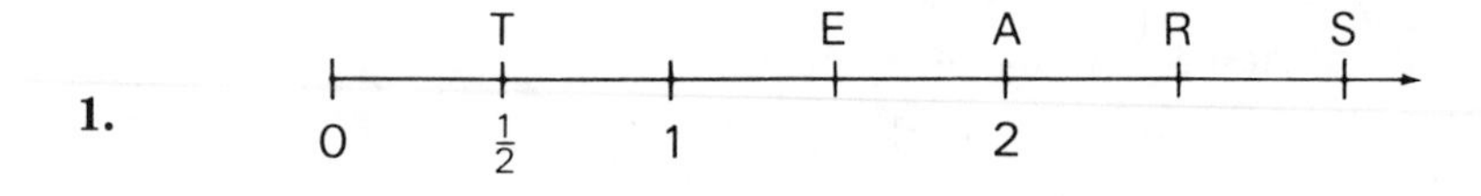

2.

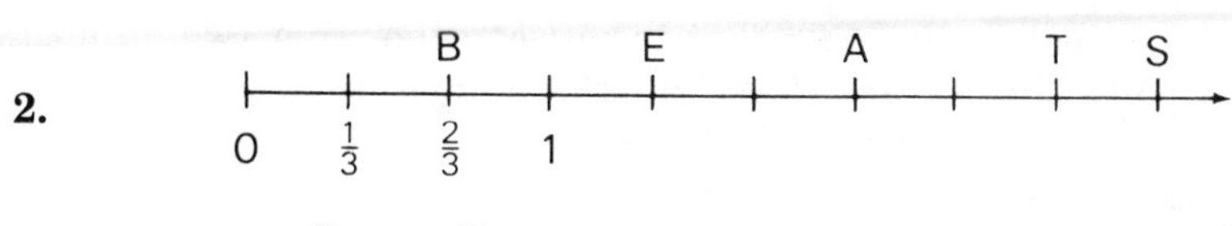

3.

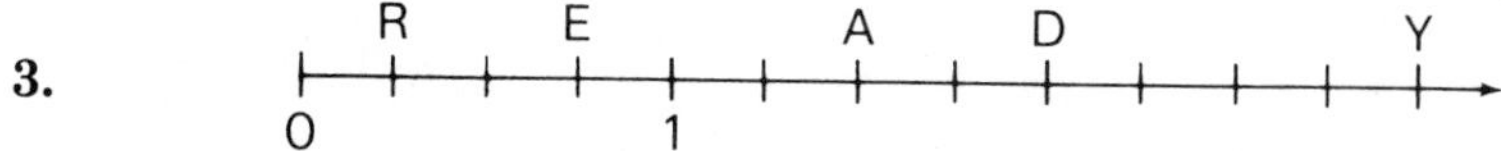

4. Draw a number line and on it locate the points whose coordinates are:

a. $\frac{1}{2}, \frac{3}{2}, \frac{6}{2}, \frac{9}{2}, \frac{11}{2}$ **b.** $\frac{1}{4}, \frac{3}{4}, \frac{6}{4}, \frac{8}{4}, \frac{13}{4}, \frac{16}{4}$

c. $\frac{1}{5}, \frac{3}{5}, \frac{9}{5}, \frac{10}{5}, 2\frac{1}{5}, 3\frac{2}{5}$ **d.** .1, .3, .7, 1.0, 2.7, 3.4

e. $\frac{1}{4}, \frac{1}{2}, \frac{3}{8}, 1\frac{1}{4}$, 2.25, $\frac{25}{8}$ **f.** $\frac{1}{4}, \frac{1}{3}, \frac{5}{12}, \frac{9}{6}, 2\frac{1}{2}$, 3.75

In 5, use the following number line:

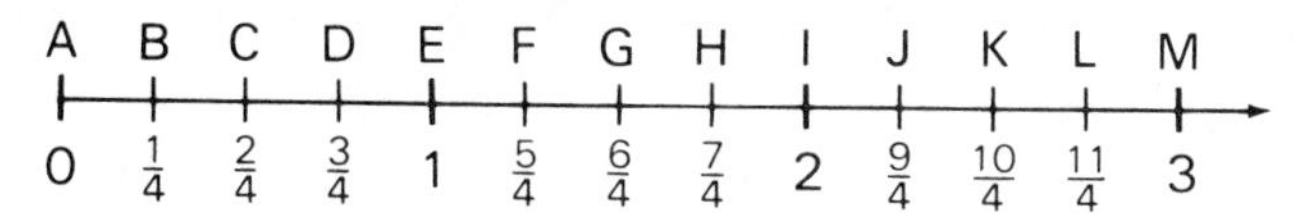

5. Name the point that is the graph of the number:

a. 1 **b.** $\frac{8}{4}$ **c.** $1\frac{3}{4}$ **d.** $2\frac{1}{2}$ **e.** .5 **f.** 1.25 **g.** 2.75

In 6-17: **a.** State whether on a standard number line the graph of the first number lies to the left or to the right of the graph of the second number. **b.** State whether the first number is smaller or greater than the second by writing the two numbers with the symbol $<$ or $>$ between them.

6. 12, 18 **7.** 29, 23 **8.** $\frac{9}{2}, \frac{4}{2}$ **9.** $\frac{1}{4}, \frac{1}{8}$

10. $3\frac{2}{3}, 5\frac{1}{3}$ **11.** 3.9, 1.3 **12.** 3.1, 9.3 **13.** .5, .05

14. 11, 110 **15.** .47, 4.7 **16.** 6.4, 6.45 **17.** .95, .905

In 18–26, arrange the numbers in proper order so that they will appear from left to right on a number line.

18. 16, 4 **19.** $\frac{1}{2}, \frac{1}{4}$ **20.** 2.5, 3.2

21. $2\frac{3}{5}$, 2.8 **22.** 2, 3, 9 **23.** 9, 6, 11

24. $\frac{1}{4}, \frac{7}{8}, \frac{2}{3}$ **25.** 3.2, 2.6, 4.3 **26.** $3\frac{1}{3}$, 3.75, $3\frac{1}{4}$

In 27–35, select the number that is between the other two numbers. Then, rearrange the numbers to show the proper ordering:
(a) using the symbol < (as in $2 < 3 < 5$);
(b) using the symbol > (as in $5 > 3 > 2$).

27. 13, 17, 9

28. $5\frac{1}{3}$, $4\frac{1}{2}$, $6\frac{1}{4}$

29. 4.7, 6.6, 5.3

30. $4\frac{7}{8}$, 4.5, $5\frac{1}{4}$

31. 0, .1, .12

32. .5, .05, .55

33. $\frac{1}{4}$, $\frac{1}{5}$, $\frac{1}{3}$

34. $\frac{1}{3}$, .3, .13

35. .909, .9, .91

1-8 OPERATIONS IN GEOMETRY

Just as there are operations in arithmetic, there are operations in other branches of mathematics. Many of the binary operations in geometry depend on arithmetic. In this section, we will discuss only a few examples of operations in geometry. Many more will be found in later work.

Sets of Points

A ***line*** is a set of points. Unless it is otherwise stated, the word *line* will mean a *straight line.* We usually think of a straight line as the set of points that is suggested by a stretched string or the edge of a ruler. But a stretched string and a ruler are limited in length, whereas a line extends endlessly in both directions.

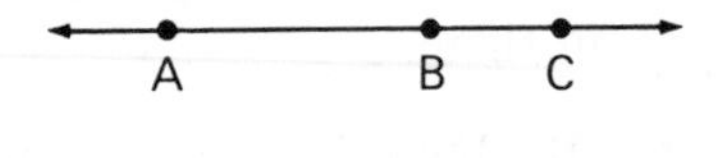

To name a line, we use two capital letters that name any two points on the line. Thus, the line shown is the line AB, the line AC, or the line BC. Line AB can be written $\overleftrightarrow{AB}$. The arrowheads on the diagram of the line and on the symbol for line tell us that the line continues without end.

Line Segments

A line segment can be thought of as a part of a line.

● **Definition.** A ***line segment*** is the set of points consisting of two points on a line, called ***endpoints,*** and all points on the line between the endpoints.

A line segment whose endpoints are A and B is called segment AB or segment BA. Segment AB can be written as $\overline{AB}$ and segment BA can be written as $\overline{BA}$.

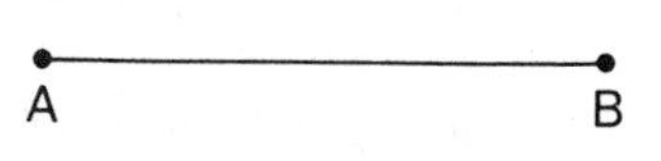

Measure of a Line Segment

To find the distance between two points, A and B, we place a ruler so that it touches both points. We read the numbers that are associated with points A and B. Or we think of them as points on a number line to which coordinates have been assigned.

In the figure, when 0 is associated with A, 3 is associated with B. The distance between A and B is 3 units. This can be written $AB = 3$. We can find the distance by subtracting:

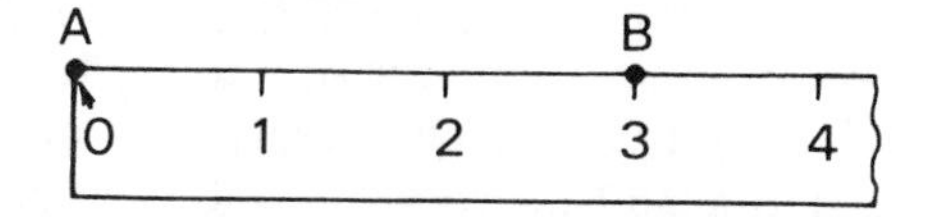

$$AB = 3 - 0 = 3$$

If we use the same ruler but place A at 5, then B is at 8. We can find the distance 3 by subtracting 5 from 8:

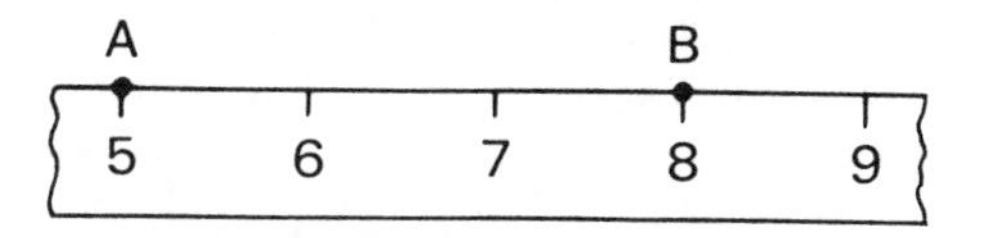

$$AB = 8 - 5 = 3$$

In each case, we subtracted the smaller number from the larger one. It does not matter which coordinates are associated with A and B. The distance will be 3 if we use a ruler or number line having this same scale. Note that the distance from A to B is exactly the same as the distance from B to A.

● **Definition.** The ***distance*** between any two points on a number line is the difference between the coordinates of the two points, always subtracting the smaller coordinate from the larger one.

● **Definition.** The ***measure of a line segment*** is the distance between its endpoints.

Finding the distance between two points is a binary operation that makes use of subtraction. However, we may have to change the order of the numbers to place the larger coordinate first and the smaller coordinate second.

Midpoint of a Line Segment

A line segment is a set of points. In this set, there is a unique point (one and only one) called the midpoint.

● **Definition.** The ***midpoint*** of a line segment is a point of that segment that divides the segment into two segments that have the same length.

We say that the midpoint ***bisects*** the segment.

In the figure below, if M is the midpoint of $\overline{AB}$, then $AM = MB$.

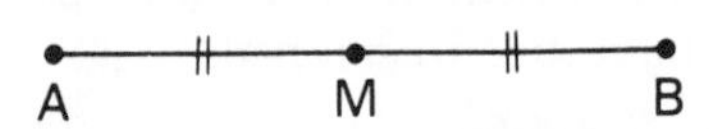

We can use a number line to see that this operation in geometry relates directly to a binary operation in arithmetic. For example, place a number line so that coordinate 0 is at A, coordinate 6 is at B, and $AB = 6$.

If we wish to find the midpoint M of $\overline{AB}$, we must find the point M such that $AM = MB$. Since the length of segment AB is $6 - 0$, or 6, point M must be 3 units from A and 3 units from B. Hence, the coordinate of M must be 3.

Notice that the binary operation *average* takes the ordered pair 0 and 6 and assigns the number 3 as the result: 0 avg 6 = 3.

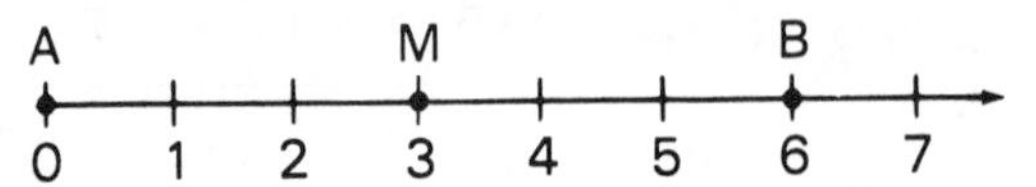

Using the same number line, let us place $\overline{AB}$ so that the coordinate of A is 5 and the coordinate of B is 11. AB is still 6 because $11 - 5 = 6$. Once again, point M must be 3 units from A and 3 units from B. Hence, the coordinate of M must be 8.

Here, too, the binary operation *average* takes the ordered pair 5 and 11 and assigns the number 8 as the result: 5 avg 11 = 8.

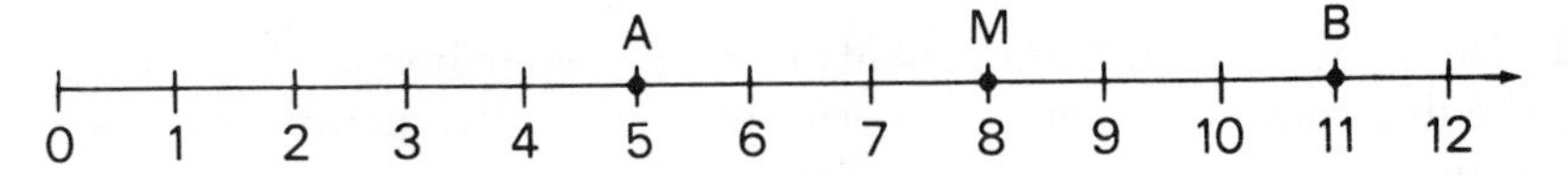

Half-Lines and Rays

Each point on a line separates the line into two opposite sets of points called ***half-lines.*** Since opposite half-lines have no points in common, the point of division does not belong to either half-line. In the diagram, point C divides $\overleftrightarrow{AB}$ into the half-line that contains A and the half-line that contains B.

A C B

● **Definition.** A ***ray*** is a part of a line that consists of a point on the line, called the endpoint, and all of the points on one side of the endpoint.

A ray is named by the endpoint and any point in the half-line. The diagram shows ray AB (written $\overrightarrow{AB}$). Since C is also a point of the same ray, $\overrightarrow{AB}$ may also be called $\overrightarrow{AC}$. Note that the endpoint of the ray is always written first.

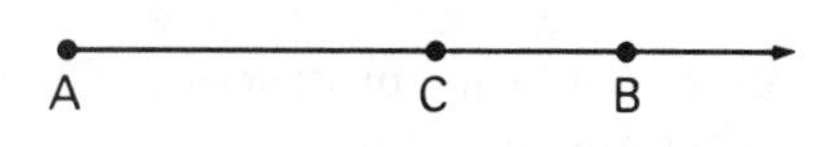

Angles

● **Definition.** An ***angle*** is a set of points formed by two rays having the same endpoint. The common endpoint of the two rays is called the ***vertex*** of the angle. Each ray is called a ***side*** of the angle.

In the diagram, $\overrightarrow{AB}$ and $\overrightarrow{AD}$ have a common endpoint, A. We call the angle formed by these rays angle BAD or angle DAB. Notice that in each case, the letter that names the vertex is between two letters, one from each of the rays that form the angle.

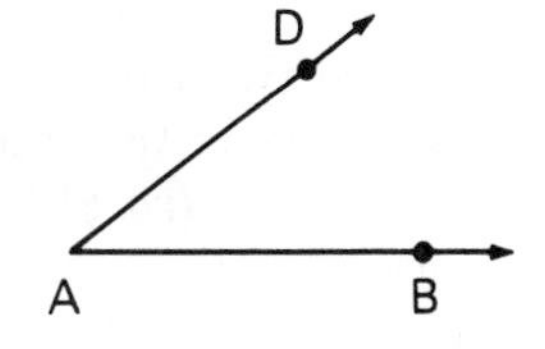

When only one angle is shown at a given vertex, we may also name the angle by using just the letter of the vertex. Using the symbol $\angle$ for angle, we may also call the angle shown above $\angle A$.

Measuring Angles

We use a ***protractor*** to determine the measure of an angle. Most protractors use a ***degree*** as the unit of measure. Although there are other units used to measure angles, in this book the measure of an angle will always be given in degrees.

Place the protractor so that the vertex of the angle is at the center of the protractor and one side of the angle is aligned with 0 on either the inner or the outer scale of the protractor. On the same scale, read the number at the point where the other side of the angle meets the protractor.

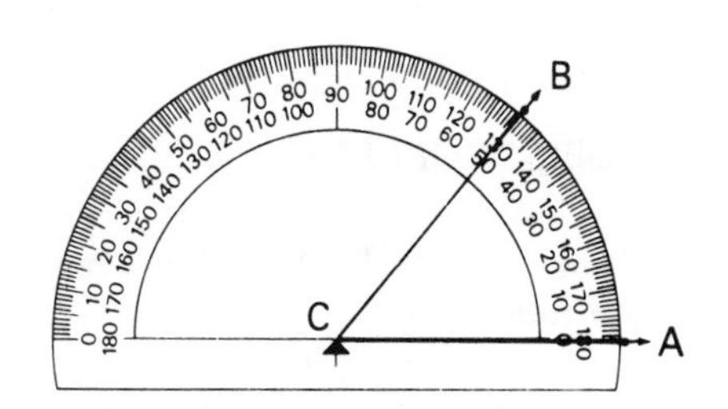

The measure of the angle shown in the figure is 50°, which is written m∠BCA = 50.

Notice that even if one side of the angle is not aligned with 0, we can find the angle measure with the protractor. Place the protractor so that the vertex of the angle is at the center of the protractor. Choose either the inner or the outer scale. Read the numbers at the points where the rays of the angle meet the protractor. The measure of the angle is the difference between these numbers, always subtracting the smaller number from the larger. Thus, the measure of ∠BCA is (110 − 60)° or (120 − 70)° or 50°.

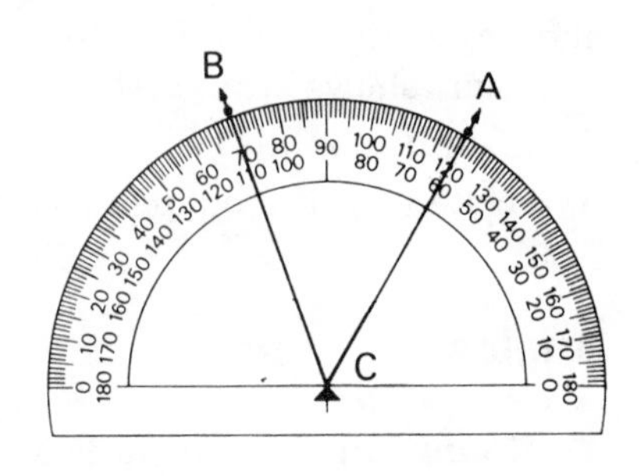

We can think of finding the measure of an angle as a binary operation that makes use of subtraction. However, we may have to change the order of the numbers to place the larger number first.

If the two rays of an angle are different rays of the same line, the angle is called a ***straight angle.*** When we use a protractor to measure this angle, the rays will meet the protractor at 180 and 0. Therefore, the measure of a straight angle is (180 − 0)° or 180°.

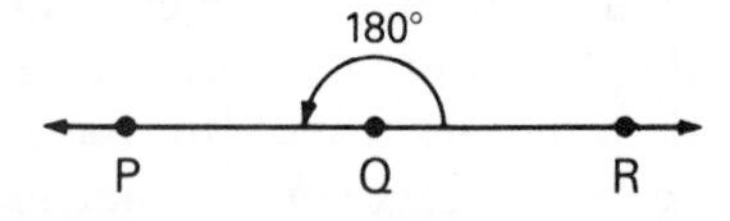

m∠PQR = 180 Angle PQR is a straight angle.

With Q as the endpoint, draw ray $\overrightarrow{QS}$ that divides ∠PQR into two angles of the same measure. Since the measure of ∠PQR is 180°, the measure of each of the angles into which it is divided is 90°. An angle whose measure is 90° is called a ***right angle.***

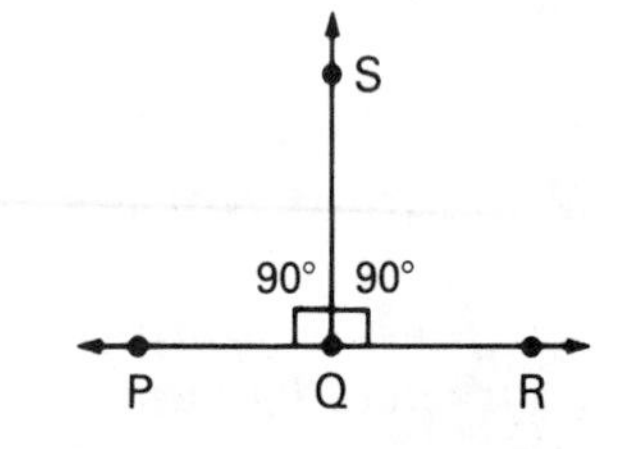

m∠PQS = 90 Angle PQS is a right angle.

m∠RQS = 90 Angle RQS is a right angle.

MODEL PROBLEMS

In 1–3, use the number line and the associated points to find the distances named.

Solutions

1. CD Since 2 is at point C and 10 is at point D, the distance $CD = 10 - 2 = 8$.

2. DC Again, 2 is at C and 10 is at D. Subtract the smaller number from the larger: $DC = 10 - 2 = 8$

3. CR $CR = 7 - 2 = 5$

4. The coordinate of P is 12 and the coordinate of T is 15. Find the coordinate of M, the midpoint of $\overline{PT}$.

   Solution: The coordinate of M is the average of 12 and 15.

$$12 \text{ avg } 15 = \frac{12 + 15}{2} = \frac{27}{2} = 13\frac{1}{2} \quad \textit{Ans.}$$

5. The midpoint of $\overline{RS}$ is K. The coordinates of R and K are 5 and 12, respectively. Find the coordinate of S.

   Solution: Since K is the midpoint, $RK = KS$. $RK = 12 - 5 = 7$. Then, KS must equal 7. Add this distance 7 to the coordinate 12 to obtain the coordinate 19 at point S.

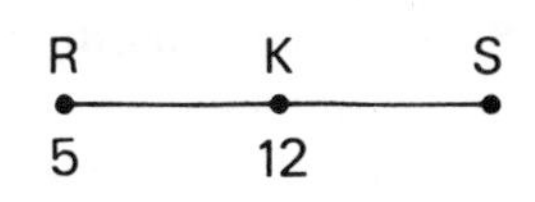

   Answer: 19

6. In the diagram, E is a point on $\overleftrightarrow{DF}$.
   a. Name three angles.
   b. Name the straight angle.

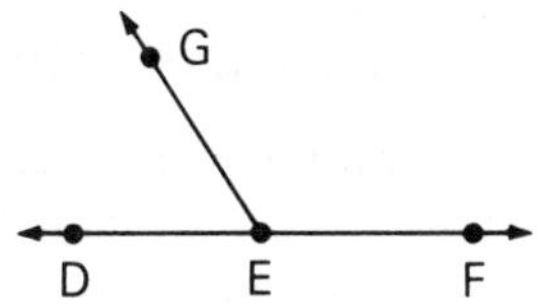

   Solution:

   a. $\angle DEG$ or $\angle GED$, $\angle GEF$ or $\angle FEG$, $\angle DEF$ or $\angle FED$
   b. $\angle DEF$ is the straight angle.

EXERCISES

In 1–6, use the number line shown.

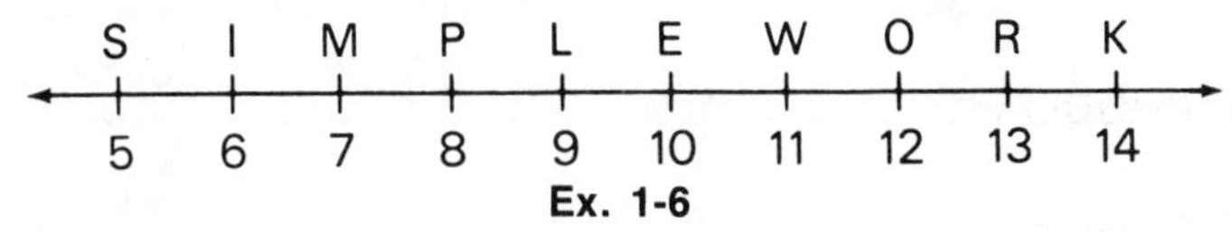

Ex. 1-6

1. Find the distances named:
   a. SL **b.** MR **c.** MS **d.** OK **e.** KS **f.** PI
2. Name the midpoint of the segments given:
   a. $\overline{SL}$ **b.** $\overline{MR}$ **c.** $\overline{MS}$ **d.** $\overline{OP}$ **e.** $\overline{RS}$ **f.** $\overline{PI}$

3. Name four segments on the line whose midpoint is L.
4. If X is a point such that E is the midpoint of $\overline{PX}$, what coordinate is assigned to point X?
5. If Y is a point such that R is the midpoint of $\overline{PY}$, what coordinate is assigned to point Y?
6. If Z is a point such that K is the midpoint of $\overline{SZ}$, what coordinate is assigned to point Z?

In 7 and 8, the points named are spaced at equal intervals so that $TO = OU = UG$, and so on. The coordinate 15 is assigned to point G in every problem.

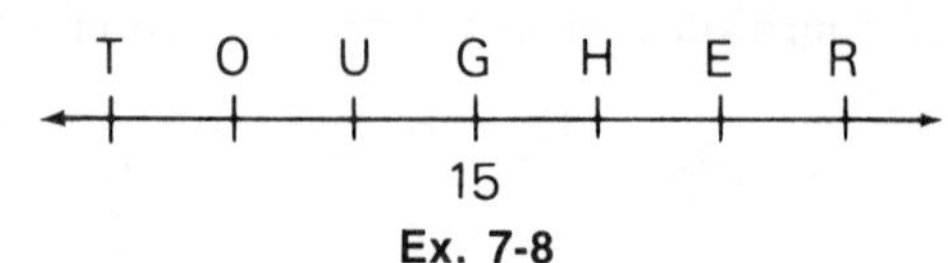

Ex. 7-8

7. If the coordinate 12 is assigned to point U, find the distances named.
   a. UG **b.** GH **c.** UH **d.** TE **e.** RO **f.** GO
8. If the coordinate 13.5 is assigned to point U, find the distances named.
   a. UG **b.** GH **c.** UH **d.** TE **e.** RO **f.** GO

In 9–16, the coordinates of A and B, the endpoints of $\overline{AB}$, are given. Find the coordinate assigned to X, the midpoint of $\overline{AB}$.

9. 3 and 5 **10.** 8 and 2 **11.** 7 and 26 **12.** 40 and 17

13. 8 and 12.8 **14.** 8 and 2.8 **15.** $\frac{2}{3}$ and $\frac{1}{3}$ **16.** $1\frac{1}{4}$ and 5

In 17–20, $\overline{AB}$ has a midpoint X. The coordinate of X is 15. Find the coordinate of endpoint B when the coordinate of endpoint A is the given number.

17. 14 **18.** 10 **19.** 20 **20.** $10\frac{1}{3}$

In 21–28, $\overline{AB}$ has the coordinate 12 assigned to endpoint A. Find the possible coordinate(s) assigned to endpoint B for the given length of $\overline{AB}$.

21. $AB = 1$ **22.** $AB = 3$ **23.** $AB = 10$ **24.** $AB = 12$

25. $AB = \frac{1}{2}$ **26.** $AB = 1.5$ **27.** $AB = .8$ **28.** $AB = \frac{2}{3}$

29. Using the figure at the right:
   a. Name three rays.
   b. Name three angles.
   c. Name a straight angle.
   d. What is the degree measure of $\angle ABC$?

30. A protractor is placed as shown in the diagram. Find the measure of:

a. $\angle ABP$ b. $\angle ABQ$
c. $\angle PBQ$ d. $\angle RBP$
e. $\angle ABC$ f. $\angle PBC$

31. Name two right angles using the rays given in the figure.

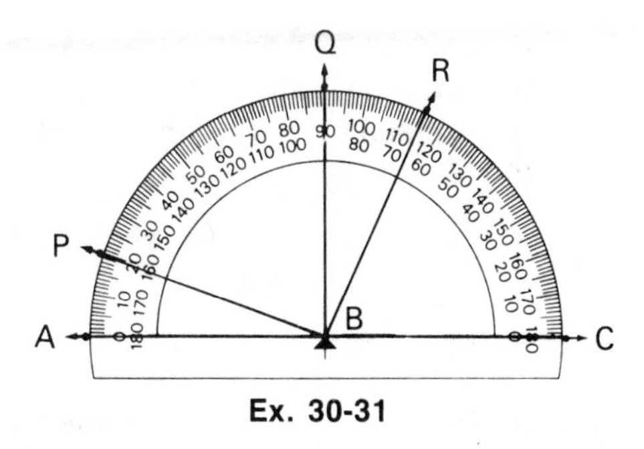

Ex. 30-31

1-9 SETS

We speak of sets of numbers in arithmetic and sets of points in geometry. Sets will play an important role in the algebra, geometry, and probability you will be studying this year.

The Elements of a Set

When a collection of distinct objects is so clearly described that we can always tell whether or not an object belongs to it, we call the collection a *well-defined collection*, or more simply a ***set.***

A set is often represented by a capital letter. The set can be described in words, or its members or elements can be listed within braces, { }. For example, if A is the set of odd counting numbers less than 10, then we can write:

By description: A = the set of all odd counting numbers less than ten
By listing: $A = \{1, 3, 5, 7, 9\}$

To show that 3 is an element of A, write: $3 \in A$
To show that 2 is not an element of A, write: $2 \notin A$

If the elements of a set form a pattern, we can make use of the pattern and three dots to represent the set. For example, $\{1, 2, 3, 4, \ldots\}$ names the set of counting numbers while $\{1, 2, 3, 4, \ldots, 200\}$ names the set of the first two hundred counting numbers.

There is still another way of describing the set of the first two hundred counting numbers. We can use ***set-builder notation:***

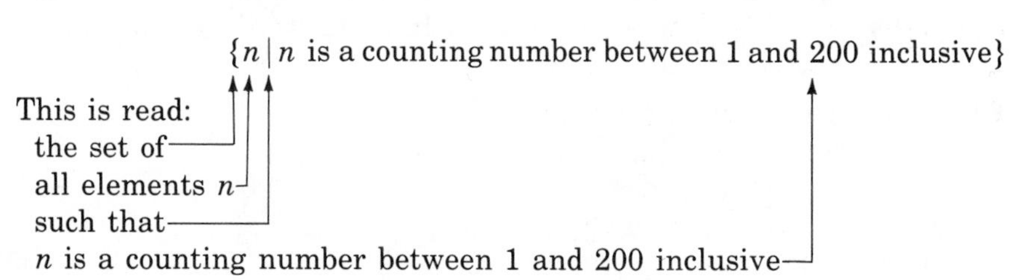

Notice that the word "inclusive" is used to emphasize that the set includes the end numbers 1 and 200 as well as the numbers between.

Other examples of sets described by the set-builder notation are:

1. $\{x \mid x \text{ is a natural number}\} = \{1, 2, 3, 4, 5, \ldots\}$
2. $\{n \mid n \text{ is an odd whole number}\} = \{1, 3, 5, 7, 9, \ldots\}$

Kinds of Sets

A ***finite set*** is a set whose elements can be counted, and in which the counting process comes to an end. Some examples of finite sets are:

1. the set of all students in your mathematics class
2. $\{2, 4, 6, 8, \ldots, 9{,}200\}$
3. $\{x \mid x \text{ is a whole number less than } 20\}$

An ***infinite set*** is a set whose elements cannot be counted, and in which the counting process does not come to an end. Some examples of infinite sets are:

1. the set of counting numbers
2. the set of points that are on a straight line
3. $\{2, 4, 6, 8, \ldots\}$

The ***empty set,*** or ***null set,*** is the set that has no elements. Some descriptions of the empty set are:

1. the set of months that have names beginning with the letter Q
2. $\{x \mid x \text{ is an odd number exactly divisible by } 2\}$

The empty set may be represented by the symbol $\varnothing$ or by a pair of empty braces, $\{\ \}$. By drawing a slash mark through the symbol normally used for zero, we might think of $\varnothing$ as indicating that *not even zero* is a member of this set.

Note that the set $\{0\}$ is *not* the empty set because it does contain an element, the number 0.

Relationships Between Sets

Set A is ***equal*** to set B if every element of A is an element of B and every element of B is an element of A. In other words, set A and set B contain exactly the same elements.

Example:
$A = \{1, 3, 5, 7, 9\}$
$B = \{\text{odd counting numbers less than } 10\}$
$C = \{2, 4, 6, 8, 10\}$
$A = B$, but $A \neq C$ and $B \neq C$

Set A and set B are ***matching sets,*** or ***equivalent sets,*** denoted by $A \sim B$, if each element of set A can be matched with exactly one element of set B, and each element of set B can be matched with exactly one element of set A. When two sets can be matched in this way, we say that there is a ***one-to-one correspondence*** between the two sets.

Example: $A = \{1, 3, 5, 7, 9\}$
$B = \{$odd counting numbers less than 10$\}$
$C = \{2, 4, 6, 8, 10\}$
$A \sim C, B \sim C,$ and $A \sim B$

Note that equivalent sets have the same number of elements but that the elements may be different.

If every student in a class were assigned to a seat and there were no empty seats left over, there would be a match of students to seats and a match of seats to students. The set of students would be equivalent to the set of seats.

If there were unassigned seats, however, or if there were students left standing, then there would not be a match of students to seats, and the two sets would not be equivalent.

Notice that if two sets are equal sets, they are also equivalent sets. However, if two sets are equivalent sets, they are not necessarily equal sets.

The ***universal set,*** or the ***universe,*** is the entire set of elements under consideration in a given situation and is usually denoted by the letter U. For example:

1. In arithmetic, $U = \{$numbers of arithmetic$\}$
2. In given situations, like scores on a classroom test, the universal set may be whole numbers from 0 to 100. $U = \{0, 1, 2, 3, \ldots, 100\}$

Subsets

Set A is a ***subset*** of set B, denoted by $A \subset B$, if every element of set A is an element of set B. For example:

1. The set $A = \{$Harry, Paul$\}$ is a subset of the set $B = \{$Sue, Harry, Mary, Paul$\}$.
2. The set of odd whole numbers $\{1, 3, 5, 7, \ldots\}$ is a subset of the set of whole numbers, $\{0, 1, 2, 3, \ldots\}$.
3. $\overline{AB}$ is a segment containing a set of points. If M is the midpoint of $\overline{AB}$, then set M contains one point. Set M is a subset of the set of points found on the segment $\overline{AB}$.

Since a subset of a set may contain all the elements of the set itself, a subset can have the very same elements as the set itself. When this is true, the two sets are equal. We see, therefore, that every set is a subset of itself.

Mathematicians can show that the empty set $\varnothing$ is a subset of every set.

Consider the set $A = \{1, 2, 3\}$. To list all the subsets of set A, we can follow a simple pattern:

(1) List all subsets of three elements. $\{1, 2, 3\}$

(2) List all subsets of two elements. $\{1, 2\}\ \{1, 3\}\ \{2, 3\}$

(3) List all subsets of one element. $\{1\}\ \{2\}\ \{3\}$

(4) List all subsets of no elements. $\{\ \}$

Eight subsets can be formed from the set $\{1, 2, 3\}$. Notice that $\{1, 2\}$ and $\{2, 1\}$ are not both listed, because they are simply two ways to name the same set.

EXERCISES

In 1–3, list the elements of the set that is described.

1. {days of the week that begin with the letter T}
2. {even numbers greater than 3 and less than 12}
3. {natural numbers less than 100 that are the squares of natural numbers}

In 4 and 5, tabulate the elements of the set that is described. Use three dots when convenient or necessary.

4. the set of all even counting numbers
5. the set of all whole numbers greater than 10 and less than 1,000

In 6–16, state whether the set is a finite nonempty set, an infinite set, or the empty set.

6. the set of all the people who live in the United States today
7. the set of all women who are 120 cm tall
8. the set of all women who are 120 feet tall
9. the set of points on a line segment, $\overline{AB}$
10. the set of endpoints on a line segment, $\overline{AB}$
11. the set of midpoints on a line segment, $\overline{AB}$
12. the set of segments that contain the same midpoint M
13. the set of natural numbers less than 1 billion

14. the set of natural numbers greater than 1 billion
15. the set of natural numbers between 0 and 1
16. the set of rectangles

In 17–19, give a description of the given set.

17. {January, June, July}
18. $\{2, 4, 6, 8, 10, \ldots\}$
19. $\{3, 6, 9, 12, \ldots, 999\}$

In 20–22, use the symbol = or ≠ to write a true sentence about the two sets.

20. $A = \{5, 10, 15, 20\}$ and $B = \{20, 15, 10, 5\}$
21. $C = \{1, 3, 5\}$ and $D = \{1, 3, 5, 7\}$
22. $K = \{0\}$ and $L = \varnothing$

In 23–25: **a.** Tell whether or not there is a one-to-one correspondence between the two sets. **b.** State whether the two sets are equivalent.

23. $\{6, 7, 8, 9\}$ and $\{1, 2, 3, 4\}$
24. $\{a, b, c, d\}$ and $\{x, y, z\}$
25. {Tom, Dick, Harry} and {Sally, Mary, Sue}

26. **a.** If set A has three elements and set B has four elements, can set A and set B be put into one-to-one correspondence?
b. Explain the answer given in part **a.**

In 27–37, tell whether the sentence is true or false. Justify the answer.

27. {Sam, Bill} is a subset of {Joan, Bill, Sam}
28. $\{6, 7, 8, 9\}$ is a subset of $\{6, 7, 8\}$
29. $\varnothing$ is a subset of $\{10, 11, 12\}$
30. $\{2, 4, 6\} \subset \{1, 2, 3, 4, 5, 6\}$
31. {odd natural numbers} ⊂ {odd natural numbers}
32. {squares} ⊂ {rectangles}
33. $\{Y, E, A, S\} \subset \{E, A, S, Y\}$
34. $4 \in \{n \mid n \text{ is a counting number}\}$
35. $4 \in \{x \mid x \text{ is an odd number}\}$
36. $5 \notin \{y \mid y \text{ is an even number}\}$
37. $5 \in \{y \mid y \text{ is an odd number greater than } 10\}$

In 38–41, $A = \{13, 14, 15\}$. Write all the subsets of A that meet the indicated condition.

38. contain one element
39. contain two elements
40. contain three elements
41. contain no elements

In 42–47, list all the subsets of B.

42. $B = \{5\}$ **43.** $B = \{3, 7\}$ **44.** $B = \{m, l, g\}$
45. $B = \{0\}$ **46.** $B = \{\text{true, false}\}$ **47.** $B = \{1, 2, 3, 4\}$

In 48–50, represent the set by listing its members.

48. $\{x \mid x \text{ is a whole number less than } 6\}$
49. $\{n \mid n \text{ is an even whole number between 1 and } 9\}$
50. $\{n \mid n \text{ is a natural number greater than 4 and less than } 12\}$

1-10 OPERATIONS WITH SETS

Just as we have seen operations in arithmetic and in geometry, there are operations with sets.

Intersection of Sets

The ***intersection of two sets,*** A and B, denoted by $A \cap B$, is the set of all elements that belong to both sets, A and B. For example:

1. When $A = \{1, 2, 3, 4, 5\}$ and $B = \{2, 4, 6, 8, 10\}$, then $A \cap B$ is $\{2, 4\}$.
2. In the figure, two lines called $\overleftrightarrow{AB}$ and $\overleftrightarrow{CD}$ intersect. The intersection is a set that has one element, point E. We write the intersection of the lines in the example shown as $\overleftrightarrow{AB} \cap \overleftrightarrow{CD} = E$.

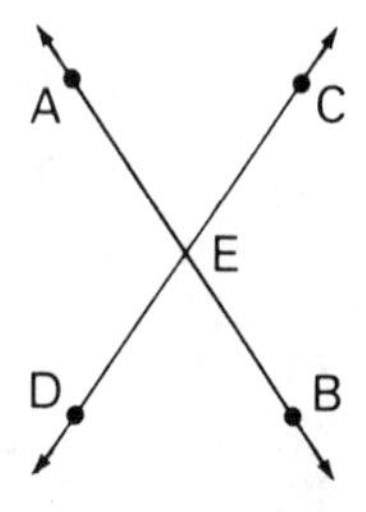

Two sets are ***disjoint sets*** if they do not intersect; that is, if they do not have a common element. For example: when $C = \{1, 3, 5, 7\}$ and $D = \{2, 4, 6, 8\}$, then $C \cap D = \varnothing$.

Remember that earlier a *binary operation* was shown symbolically as $a * b = c$, where two elements a and b, under some operation $*$, were replaced by a unique element c. When we write, for example, $\{1, 2, 3\} \cap \{1, 3, 5\} = \{1, 3\}$, the elements a, b, and c are sets themselves and the operation is intersection, $\cap$. We can think of $\{1, 2, 3\}$, $\{1, 3, 5\}$, and $\{1, 3\}$ as elements of a larger set. In this way, we are taking two subsets from a universal set and performing an operation that results in assigning a unique subset of that universal set. Thus, intersection is a binary operation for a universal set.

Union of Sets

The ***union of two sets,*** A and B, denoted by $A \cup B$, is the set of all elements that belong to set A or to set B, or to both set A and set B. For example:

1. If $A = \{1, 2, 3, 4\}$ and $B = \{2, 4, 6\}$, then $A \cup B = \{1, 2, 3, 4, 6\}$.

Note that an element is not repeated in the union of two sets even if it is an element of each set.

2. In the figure, both region R (vertical shading) and region S (horizontal shading) represent sets of points. The shaded parts of both regions represent $R \cup S$, and the cross-hatched part where the regions overlap represents $R \cap S$.

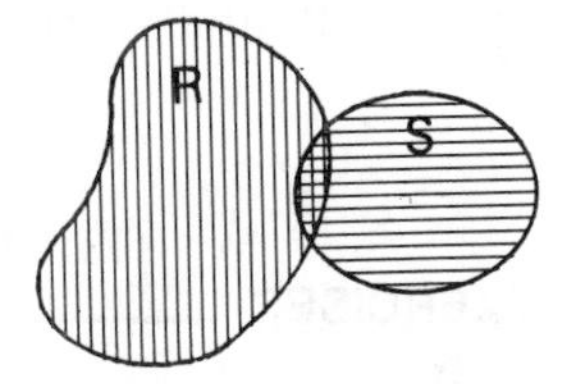

3. If $A = \{1, 2\}$ and $B = \{1, 2, 3, 4, 5\}$, then the union of A and B is $\{1, 2, 3, 4, 5\}$. We can write $A \cup B = \{1, 2, 3, 4, 5\}$, or $A \cup B = B$. Once again we have an example of a binary operation, where the elements are taken from a universal set and where the operation here is union.

Complement of a Set

The ***complement*** of a set A, denoted by $\overline{A}$, is the set of all elements that belong to the universe U but do not belong to set A. Therefore, before we can determine the complement of A, we must know U.

For example:

1. If $A = \{3, 4, 5\}$ and $U = \{1, 2, 3, 4, 5\}$, then $\overline{A} = \{1, 2\}$ because 1 and 2 belong to the universal set U but do not belong to set A.
2. If the universe is {whole numbers} and A = {even whole numbers}, then $\overline{A}$ = {odd whole numbers} because the odd whole numbers belong to the universal set but do not belong to set A.
3. In the figure, a rectangular plate, before it is stamped out by a machine, represents the universe. If the keyhole punched out represents set A, then the shaded region represents the complement of A. $\overline{A}$ corresponds to the metal plate found on the door after the keyhole has been stamped out.

Although it seems at first that only one set is being considered in writing the complement of A as $\overline{A}$, there are two sets. This suggests a binary operation, where the universe U and set A are the pair of elements, where *complement* is the operation, and where the unique result is $\overline{A}$. The complement of any universe is the empty set.

MODEL PROBLEM

If $U = \{1, 2, 3, 4, 5, 6, 7\}$, $A = \{6, 7\}$, and $B = \{3, 5, 7\}$, determine $\overline{A} \cap \overline{B}$.

Solution

$U = \{1, 2, 3, 4, 5, 6, 7\}$
Since $A = \{6, 7\}$, then $\overline{A} = \{1, 2, 3, 4, 5\}$.
Since $B = \{3, 5, 7\}$, then $\overline{B} = \{1, 2, 4, 6\}$.
Since 1, 2, and 4 are elements in both $\overline{A}$ and $\overline{B}$, we can write:

$$\overline{A} \cap \overline{B} = \{1, 2, 4\} \quad \textit{Ans.}$$

EXERCISES

In 1–8, $A = \{1, 2, 3\}$, $B = \{3, 4, 5, 6\}$, and $C = \{1, 3, 4, 6\}$. List the elements of the given sets.

1. $A \cap B$ **2.** $A \cap C$ **3.** $A \cup B$ **4.** $A \cup C$
5. $B \cap C$ **6.** $B \cap \varnothing$ **7.** $B \cup C$ **8.** $B \cup \varnothing$

9. Using the sets A, B, and C given for Exercises 1–8, list the elements of the smallest possible universal set of which A, B, and C are all subsets.

In 10–17, the universe $U = \{1, 2, 3, 4, 5\}$ and subsets $A = \{1, 5\}$, $B = \{2, 5\}$, and $C = \{2\}$. List the elements of the given sets.

10. $\overline{A}$ **11.** $\overline{B}$ **12.** $\overline{C}$ **13.** $A \cup B$
14. $A \cap B$ **15.** $A \cup \overline{B}$ **16.** $\overline{A} \cap \overline{B}$ **17.** $\overline{A} \cup \overline{B}$

In 18–33, the universe $U = \{2, 4, 6, 8\}$. Subsets include $A = \{2, 4\}$, $B = \{2, 4, 8\}$, $C = \{8\}$, and $D = \{6, 8\}$. Indicate the answer to each set operation by writing the capital letter that names the resulting set. (Example: $A \cap B = A$.) If the result is the null set, write $\varnothing$.

18. $A \cup C$ **19.** $B \cup C$ **20.** $C \cup D$ **21.** $A \cup A$
22. $A \cap C$ **23.** $B \cap C$ **24.** $C \cap D$ **25.** $A \cap A$
26. $B \cap D$ **27.** $\overline{A}$ **28.** $\overline{A} \cap D$ **29.** $\overline{D} \cup B$
30. $A \cup D$ **31.** $B \cup D$ **32.** $B \cap U$ **33.** $B \cup U$

34. Suppose that set A has two elements and set B has three elements.
a. What is the greatest number of elements that $A \cup B$ can have?
b. What is the least number of elements that $A \cup B$ can have?
c. What is the greatest number of elements that $A \cap B$ can have?
d. What is the least number of elements that $A \cap B$ can have?

1-11 REVIEW EXERCISES

In 1–14, perform the indicated operation within the set of the numbers of arithmetic.

1. $8.5 + .64$ **2.** $8.5 - .64$ **3.** $8.5 \times .64$ **4.** $6 \div 0.2$

5. $\frac{2}{3} + \frac{1}{6}$ **6.** $3\frac{1}{2} - \frac{4}{9}$ **7.** $5\frac{1}{3} \times 4\frac{1}{2}$ **8.** $1\frac{1}{2} \div \frac{3}{8}$

9. 7 avg 31 **10.** 1.2 avg 2 **11.** 3 max 18 **12.** .3 max .18

13. $\frac{3}{8}$ max $\frac{2}{5}$ **14.** $2\frac{1}{3}$ avg $\frac{1}{3}$

In 15–20, simplify the numerical expression.

15. $20 - 3 \times 4$ **16.** $(20 - 3)4$ **17.** $8 + 16 \div 4 - 2$

18. $(.16)^2$ **19.** $6^2 + 8^2$ **20.** $7(9 - 7)^3$

In 21–23, state whether the inequality is true or false.

21. $18 \times 0 > 6 \div 12$ **22.** $\left(\frac{1}{3}\right)^2 \ngtr \frac{1}{3}$ **23.** $3\frac{1}{4} - .5 < 2.75$

In 24–26, arrange the numbers in order, using the symbol <.

24. 1.4, 14, .14 **25.** $\frac{1}{7}, \frac{1}{3}, \frac{1}{5}$ **26.** .102, .012, .2

In 27–32: **a.** Replace the question mark with a number that makes the sentence true. **b.** Name the property illustrated in the sentence that is formed when the replacement is made.

27. $8 + (2 + 9) = 8 + (9 + ?)$ **28.** $8 + (2 + 9) = (8 + ?) + 9$

29. $3(?) = 3$ **30.** $3(?) = 0$

31. $5(7 + 4) = 5(7) + ?(4)$ **32.** $5(7 + 4) = (7 + 4)?$

In 33 and 34, points on the number line are spaced at equal intervals such that $ME = ET = TR$, and so on.

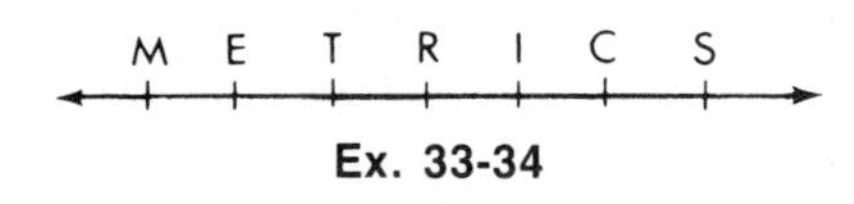

Ex. 33-34

33. If the coordinate 6 is assigned to T and 9 is assigned to R, find the distances named:
a. IC **b.** MR **c.** MS **d.** TM

34. Name the midpoint of the segment: **a.** $\overline{IM}$ **b.** $\overline{IT}$

35. In the diagram, C is a point on $\overleftrightarrow{AB}$ and $\angle ACD$ is a right angle.

a. What is the measure of $\angle ACD$?

b. What is the measure of a straight angle?

c. Name a straight angle in the diagram.

d. What is the measure of $\angle DCB$?

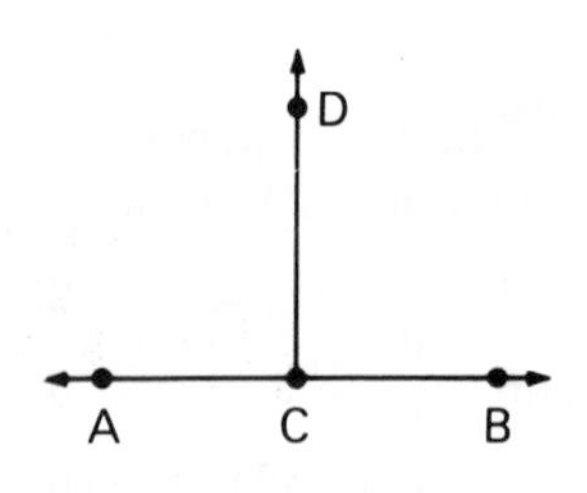

36. If $\overline{AB}$ is a segment whose length is 5 and if the coordinate of B is 17, find the possible coordinate(s) for point A.

In 37–41, the universe $U = \{1, 2, 3, 4, 5, 6\}$, set $A = \{1, 2, 4, 5\}$, and set $B = \{2, 4, 6\}$. List the elements of the given sets.

37. $A \cap B$ **38.** $A \cup B$ **39.** $\overline{A}$ **40.** $\overline{A} \cap B$ **41.** $A \cap \overline{A}$

42. List the members: $\{x \mid x \text{ is an odd whole number less than } 13\}$.

In 43–50, for each property named in Column I, match the correct application of the property found in Column II.

Column I	Column II
43. Associative Property of Multiplication	**a.** $3 + 4 = 4 + 3$
44. Associative Property of Addition	**b.** $3 \cdot 1 = 3$
45. Commutative Property of Addition	**c.** $0 \cdot 4 = 0$
46. Commutative Property of Multiplication	**d.** $3 + 0 = 3$
47. Identity Element of Multiplication	**e.** $3 \cdot 4 = 4 \cdot 3$
48. Identity Element of Addition	**f.** $3(4 + 5) = 3 \cdot 4 + 3 \cdot 5$
49. Distributive Property	**g.** $3(4 \cdot 5) = (3 \cdot 4)5$
50. Multiplication Property of Zero	**h.** $(3 + 4) + 5 = 3 + (4 + 5)$

Chapter 2

Problem Solving

We are living in exciting times. Our world is changing at a faster pace than ever before. Since you will probably see advances in science and technology that today are only ideas in the minds of creative persons, you cannot foresee some of the kinds of problems you will have to solve on your job or in your life.

This chapter is about learning to solve problems. The emphasis is not on any particular type of problem. Rather, the focus is on the technique that you can use to solve any problem. As you will discover, the strategies developed here can help you to become a better problem solver.

General Technique for Solving Problems

In order to solve any problem, it is useful to work through these four basic steps:

1. *Understand the problem.*

   Read and reread the problem. Be sure that you understand what information is given and what the problem is asking you to do.

2. *Make a plan.*

   Gather and organize the information, discarding whatever is not necessary for the solution of the problem.

   Decide on an approach that may lead to a solution. If several steps are involved, list them in order.

   If possible and appropriate, make an estimate to determine what a reasonable answer might be.

3. *Solve the problem.*

   Carry out the steps of your plan.

4. *Check the solution.*

   Test your result. Is it reasonable? Does it satisfy the given conditions? If your check shows that you have correctly solved the problem, you may write your solution as the answer.

Not every problem is solved in the first attempt. If carrying out your plan does not result in a solution that fulfills all of the conditions or if it is impossible to carry out your plan, start again. Reread the problem for information or ideas that you may have overlooked. Try another strategy—a different plan or approach that may lead to a solution.

It is impossible to list all the ways in which a problem may be solved. The following strategies are some of the more commonly used ones. They will help you to formulate your plan for solving a problem.

1. Guessing and Checking
2. Using a Simpler Related Problem
3. Working Backward
4. Discovering Patterns
5. Drawing Pictures and Diagrams
6. Making Lists and Charts

There is no one "right" strategy for a given problem. Often, a problem can be solved by any one of several strategies or by combining strategies. Problems sometimes require that you use familiar ideas in new ways.

As you work through the remaining chapters of this text and develop algebraic skills, you will be able to add to these six strategies a seventh one: using an algebraic equation. Because algebra is important in the development of advanced mathematics, this algebraic strategy is the one that you will most often be asked to use in this course.

2-1 GUESSING AND CHECKING

This strategy is often called trial and error. It is particularly useful when the answer must be a whole number between limits.

MODEL PROBLEMS

1. Find two whole numbers whose sum is 45 and whose difference is 23.

Understand: Read the problem to be sure that you understand what is given and what you are to find. Recall that sum means that you add and difference means that you subtract. If the sum of two whole numbers is 45, the numbers cannot be larger than 45.

Plan: Pick a whole number less than 45. Use this number as a first trial, and increase or decrease the number as needed.

Solution: Start with 20, a convenient number that is about half the required sum. In order that the sum be 45, the other number must be 25. Their difference is 5. Since you need a much larger difference, 23, you need to make the smaller number much smaller.

Try 10. To have a sum of 45, the other number must be 35. The difference is 25. This is closer. The difference is now too big, but only by a little. Make the smaller number a little bigger.

Try 11. To have a sum of 45, the other number must be 34. Since the difference is 23, these are the numbers that satisfy the conditions of the problem.

Check: $34 + 11 = 45$
$34 - 11 = 23$

Answer: The two numbers are 34 and 11.

2. Ann has 8 bills that are five-dollar bills and ten-dollar bills worth $50. What bills does she have?

Plan: Try possible combinations of five-dollar bills and ten-dollar bills until you find the combination whose value is $50.

If all 8 bills were five-dollar bills, the value would be $40. Since this total is just a little too small, you may estimate that you need a small number of ten-dollar bills.

Solution: Try 7 five-dollar bills = $35
1 ten-dollar bill = $10
total value = $45

Since you need to increase the total value, you need another ten-dollar bill.

Try 6 five-dollar bills = $30
2 ten-dollar bills = $20
total value = $50

Check: You checked as part of the solution. Note that the small number of ten-dollar bills agrees with the estimate.

Answer: Ann has 6 five-dollar bills and 2 ten-dollar bills.

EXERCISES

1. Find two whole numbers whose sum is 60 and whose difference is 12.

2. Cynthia is 5 years older than her sister Sylvia. The sum of their ages is 13. How old is each girl?

3. Anthony has 24 coins, all nickels and dimes, worth \$2.20. How many of each coin has he?
4. Mr. Strapp makes stools with 3 legs and with 4 legs. One week, he used 70 legs to make 18 stools. How many stools of each kind did he make?
5. The freshmen made 50 corsages to present to their mothers at Open House. Each corsage used either 2 gardenias or 3 roses. How many of each kind of corsage were made if 130 flowers were used?
6. Two different numbers use the same two digits. The sum of the digits is 7 and the difference of the numbers is 27. Find the numbers.
7. Place 40 checkers in two stacks so that, if 5 checkers are moved from the taller stack to the shorter one, there will be the same number of checkers in each stack. How many checkers are in the taller stack?
8. If Shelly gives John \$5, they will have the same amount of money. If John gives Shelly \$5, she will have twice as much money as he will have. How much money does each have?

2-2 USING A SIMPLER RELATED PROBLEM

If a problem uses large numbers or consists of many cases, it is often possible to find the solution by first finding the solution to a similar problem with smaller numbers or with fewer cases.

MODEL PROBLEMS

1. Find the sum of the whole numbers from 30 through 39.

Plan: First, find the sum of the whole numbers from 0 through 9. Since each number in the sum required is 30 larger, add 10(30) to the sum of the numbers from 0 through 9.

To estimate an answer, use the fact that, since all but one of the 10 numbers are larger than 30, the sum should be larger than 10(30), or 300. Since all 10 numbers are less than 40, the sum should be smaller than 10(40), or 400.

Solution: $0 + 1 + 2 + 3 + 4 + 5 + 6 + 7 + 8 + 9 = 45$
$45 + 10(30) = 45 + 300 = 345$

Check: You can use a calculator to check the sum. Notice that the sum is within the range that was estimated.

Answer: The sum of the whole numbers from 30 through 39 is 345.

2. A man has 100 pennies. He tries to place them all in three stacks so that the second stack has twice as many pennies as the first, and the third stack has twice as many pennies as the second. What is the largest number of pennies that can be placed in each stack and how many pennies are left over?

Plan: Make the smallest possible stacks and determine how many such stacks could be made with 100 pennies.

Solution: To make the smallest possible stacks (1, 2, and 4 pennies), you need 7 pennies.

To find how many such stacks can be made from the 100 pennies, divide:

$$100 \div 7 = 14, \text{ with a remainder of } 2$$

Therefore, you can make each of these small stacks 14 times as large and use all but two of the pennies:

$$\begin{aligned} 14(1) &= 14 \quad \text{first stack} \\ 14(2) &= 28 \quad \text{second stack} \\ 14(4) &= \underline{56} \quad \text{third stack} \\ & \ 98 \quad \text{pennies} \end{aligned}$$

There would be 2 pennies left over.

Check: The second stack has twice as many as the first stack $(2 \cdot 14 = 28)$, and the third stack has twice as many as the second stack $(2 \cdot 28 = 56)$.

Answer: The three stacks contain 14, 28, and 56 pennies. There are 2 pennies left over.

EXERCISES

1. Dora has a bank that contains 4 times as many quarters as dimes. The bank contains more than 10 dollars and less than 12 dollars. How many dimes and quarters are in the bank?
2. Agnes won $10,000. She wants to spend part for a trip and three times as much as that to make repairs on her house. She also plans to put twice as much as she spends on her house into a savings account. How much money does she plan to save?
3. A man wishes to divide his herd of horses among his children so that each child gets half as many horses as the next older child. Any extra horses will be sold. The man has 4 children and 48 horses. How many horses will each child get and how many horses will be sold?

4. Find the sum of the whole numbers from 51 through 59.
5. Cookies cost 15 cents each and brownies cost 25 cents each. What can you buy for exactly $1.50? (*Hint:* Divide each number by 5 in order to have smaller numbers.)
6. Donuts cost 30 cents each and pastries cost 24 cents each. What can you buy for exactly $3.00?
7. In the town of Tranquility, each of the 5,285 families has 0, 1, or 2 children in the local school. The majority of families have 1 child. Half of the remaining families have 2 children. How many children from the town attend the local school? (*Hint:* Try small numbers of families and assign a number at random to the number of families who have 1 child.)

2-3 WORKING BACKWARD

This strategy is often useful when you know the result of a series of events and want to find a value present at the beginning of the series.

MODEL PROBLEMS

1. Bob runs the elevator in an apartment building. He took Mr. Sloan up 6 floors from the middle floor, on which he lives. Then, Bob went down 5 floors, where he picked up Mrs. Rice. He took her down 10 floors to the first-floor lobby. What is the number of the floor on which Mr. Sloan lives?

   Plan: Work backward, reversing the course of the elevator. Then, when you know the number of the middle floor, you will know where Mr. Sloan lives.

   Solution: From the lobby, go up 10 floors to where Mrs. Rice got on. This is the 11th floor.

   Go up 5 more floors to where Mr. Sloan got off. This is the 16th floor.

   From the 16th floor, go down 6 floors to where Mr. Sloan got on. Thus, the middle floor is the 10th floor.

   Check: Let Mr. Sloan get on at the 10th floor and follow the original course of the elevator. The elevator went from the 10th floor up 6 floors to the 16th. From there, it went down 5 floors, to the 11th, and then down the remaining 10 floors to the 1st floor, or lobby.

   Answer: Mr. Sloan lives on the 10th floor.

2. Ms. McCarthy has misplaced her bank statement for May. Since there was little activity in her account that month, she is sure that she can figure out her May 1 balance even before she receives the duplicate statement that she has requested. She knows that she withdrew one-third of her funds early in May, and later deposited a total of $150 on three separate days. She also remembers that her June 1 balance was $672. What was her May 1 balance?

Plan: To solve the problem, you will need to organize the facts:

1. Ms. McCarthy had withdrawn (subtracted) one-third of the May 1 balance.
2. She deposited (added) a total of $150. It is not important that she made the deposits at different times.
3. Her June 1 balance was $672.

Estimate an answer, using rounded numbers and working backward.

1. She had about $650 − $150, or $500, before the deposits.
2. This $500 represents two-thirds of the May 1 balance. Therefore, one-third of the May 1 balance that she withdrew was about $250. The May 1 balance must have been about $500 + $250, or $750.

Solution:

	$672	June 1 balance
−	150	total deposits
	$522	two-thirds of the May 1 balance
+	261	one-third of the May 1 balance (half of $522)
	$783	May 1 balance

Check:

	$783	May 1 balance
−	261	withdrawal (one-third of $783)
	$522	balance after the withdrawal
+	150	total deposits
	$672	June 1 balance

Note that $783 is reasonably close to the estimated $750.

Answer: Ms. McCarthy's bank balance on May 1 was $783.

EXERCISES

1. Yolanda was standing on the middle step of a staircase, painting a wall. She went up 4 steps to touch up a spot. Then, she went down 5 steps to continue painting. When she finished painting everything within reach, she went down 6 steps to the 1st step. Which step is the middle step?

2. After 4 pickup stops, every passenger seat in a school bus was taken. Half as many students got on at the second stop as at the first stop, and half as many got on at the third stop as at the second stop. At the fourth stop, 5 students got on, the same number as at the third stop. How many passenger seats were there on the bus?
3. Dolores bought a box of pastries at the bakery. She gave half of them to friends that she met on her way home. At home, she gave 1 to her brother, ate 1 herself, and had 2 left. How many pastries did she buy?
4. Kim won some money. She spent $25 to have her hair cut, and loaned one-fifth of the remaining money to a friend. After she deposited two-thirds of what was left in the bank, she still had $40. How much money did she win?
5. Marsha had some math exercises to do for homework. She did one-half during study period, two-thirds of those remaining while waiting for her friend after school, and had 3 to finish at home that evening. How many exercises did she have to do for homework?
6. Dan scored the same number of points in each of the first two quarters of a basketball game, 5 points during the third quarter, and 2 points during the fourth quarter. If he scored 19 points in the game, how many points did he score during each of the first two quarters?
7. Of 100 students surveyed, 28 had neither a dog nor a cat, 18 had both a dog and a cat, and 32 had a dog but not a cat. How many students had cats?
8. Of the 200 students in the 9th grade at West High School, 70 study French and biology, 10 study French but not biology, and 12 study neither French nor biology. How many students in the 9th grade of West High School study biology?

2-4 DISCOVERING PATTERNS

Many problems about sets of numbers that follow a pattern can be solved by making use of the patterns involved.

MODEL PROBLEMS

1. What is the next number in the sequence?

$$1, 2, 4, 7, 11, \ldots$$

Plan: Since the numbers are increasing whole numbers, look for patterns that add whole numbers or multiply by whole numbers.

Solution: Look for a multiplication pattern. The second number is twice the first, and the third number is twice the second. However, the fourth number is not twice the third. Thus, the sequence is not the result of multiplication by the same number.

Look for an addition pattern.

$$\begin{array}{ccccccccc} 1 & & 2 & & 4 & & 7 & & 11 \\ & +1 & & +2 & & +3 & & +4 & \end{array}$$

Each number is obtained by adding a number larger by 1 than the number previously added. If this pattern is continued, the next number is obtained by adding 5 to 11. Using this pattern, the next number is 16.

Check:

$$\begin{array}{ccccccccccc} 1 & & 2 & & 4 & & 7 & & 11 & & 16 \\ & +1 & & +2 & & +3 & & +4 & & +5 & \end{array}$$

Answer: The next number is 16.

2. What is the sum of the whole numbers 1 through 80?

Plan: One possible solution is simply to add the eighty numbers.

To estimate an answer, consider that of these 80 numbers, the middle is about 40. Since the numbers below 40 are less than 40 by the same amount as the numbers above 40 are greater than 40, we can think of adding 40 eighty times. Thus, 80(40), or 3,200, is a good estimate of the sum of the first 80 whole numbers.

Now, look for a pattern. If you write the sum of the numbers in increasing order, each number is 1 more than the one before.

$$1 + 2 + 3 + 4 + 5 + \ldots + 76 + 77 + 78 + 79 + 80$$

If you write the sum of the numbers in decreasing order, the sum is not changed but each number is 1 less than the one before.

$$80 + 79 + 78 + 77 + 76 + \ldots + 5 + 4 + 3 + 2 + 1$$

If you put these two ways of writing the numbers together, the effect of increasing in one set and decreasing in the other set will result in no change in the sums of the pairs of numbers from the two sets together.

$$\begin{array}{l} 1 + 2 + 3 + 4 + 5 + \ldots + 76 + 77 + 78 + 79 + 80 \\ \underline{80} + \underline{79} + \underline{78} + \underline{77} + \underline{76} + \ldots + \underline{5} + \underline{4} + \underline{3} + \underline{2} + \underline{1} \\ 81 + 81 + 81 + 81 + 81 + \ldots + 81 + 81 + 81 + 81 + 81 \end{array}$$

Notice that by adding the pairs of numbers as shown, each pair has the same sum.

Solution: The sum of each pair is 81.
There are 80 such pairs in the double sum.
The double sum is 80(81), or 6,480.
The sum is 6,480 ÷ 2, or 3,240.

Check: You could use a calculator to check this sum. Notice that the result is close to the estimate.

Answer: The sum of the whole numbers 1 through 80 is 3,240.

EXERCISES

1. Find the sum of the counting numbers 1 through 100.
2. Find the sum of the first 50 even counting numbers.
3. Find the sum of the first 10 odd counting numbers.
4. If a clock strikes once on the half hour and strikes the hour on the hour (that is, strikes once at one o'clock, twice at two o'clock, and so on), how many times does the clock strike from 12:15 A.M. to 12:15 P.M.?
5. Find the next number in the sequence 1, 3, 7, 13, 21, 31, 43, ...
6. Find the next number in the sequence 2, 6, 18, 54, 162, 486, ...
7. Find the next number in the sequence 1, 3, 7, 15, 31, 63, 127, ...
8. If $4 * 3 = 24$, $8 * 2 = 32$, and $1 * 5 = 10$, then what is the value of $6 * 7$?
9. If $2 * 3 = 12$, $3 * 5 = 45$, and $4 * 5 = 80$, then what is the value of $3 * 4$?

2-5 DRAWING PICTURES AND DIAGRAMS

A picture can often help you visualize a problem and may suggest a way to solve it.

MODEL PROBLEMS

1. Mr. Vroman had a rectangular vegetable garden. He decided to increase the size by making the length twice that of the original garden and the width three times that of the original garden. How many times as large as the original garden is the new garden?

Plan: Draw a diagram of the original garden and of the new garden and compare them.

Solution: Sketch the original garden.

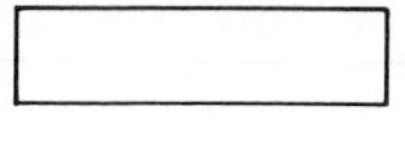

Double the length.

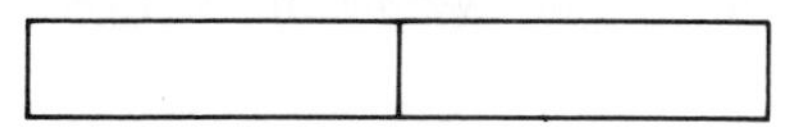

Triple the width.

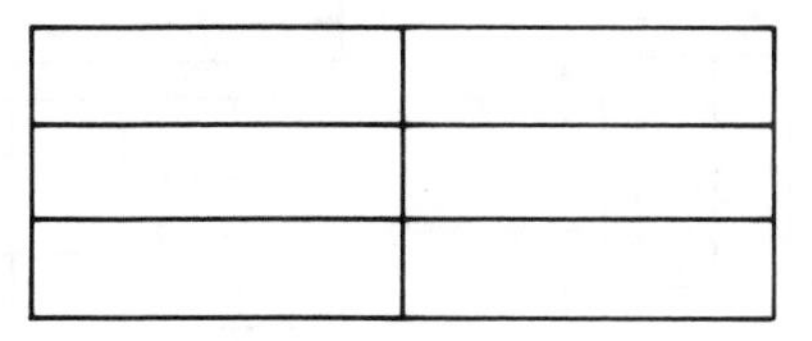

The new garden consists of six gardens of the same size as the original one.

Check: Assign numbers to the length and width to show that the relationship does hold.

Original:	1 by 5	Area 5	3 by 4	Area 12
New:	3 by 10	Area 30	9 by 8	Area 72
		30 = 6(5)		72 = 6(12)

Answer: The new garden is six times as large as the original one.

2. Saleem planted some trees alongside his family's driveway. If the distance from the first tree to the last tree was 200 feet and he planted the trees 50 feet apart, how many trees did he plant?

Solution:

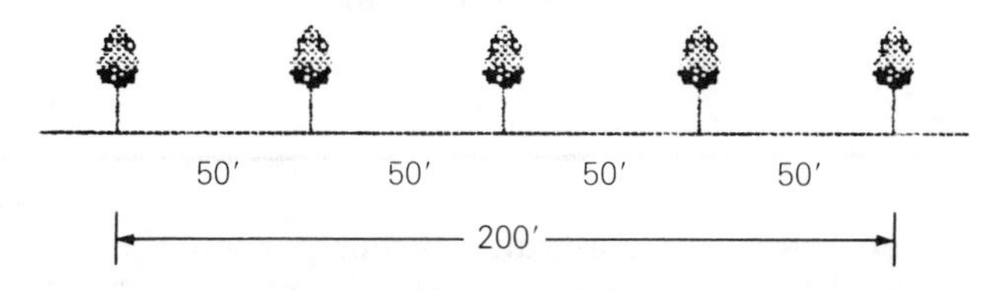

Answer: Saleem planted 5 trees.

3. Boxes measuring 3″ by 2″ by 5″ are to be packed in a carton whose dimensions are 9″ by 15″ by 5″.

a. If all the boxes must be aligned the same way, how many boxes will fit into the carton and still allow the carton to be closed?

b. If the boxes can be placed in *any* way, how many can fit and still allow the carton to be closed?

Plan: Since both the carton and the boxes to be put into the carton are 5″ high, draw diagrams to show how the 3″-by-2″ boxes will fit in the 9″-by-15″ space.

Solution:

a.

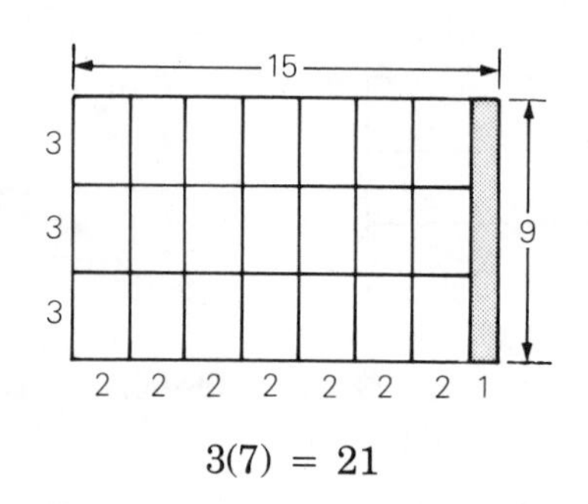

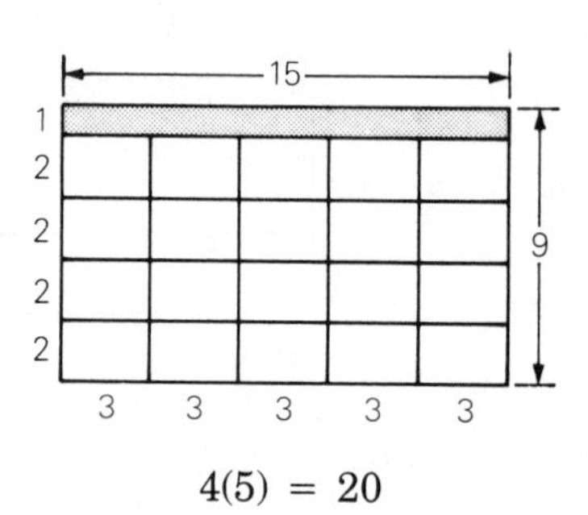

$3(7) = 21$ $4(5) = 20$

The diagrams show that, at most, 21 boxes will fit.

Answer: 21 boxes can fit into the carton.

b.

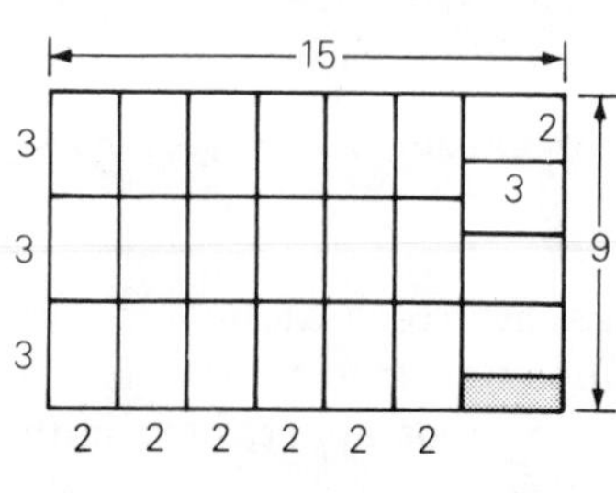

$3(6) + 4 = 18 + 4 = 22$

Answer: 22 boxes can fit into the carton.

EXERCISES

1. Mrs. Rodriguez has a rectangular flower garden that she plans to enlarge by making the garden four times as long and twice as wide. How many times as large as the original garden will the new garden be?

2. One side of a 30-foot walkway is to be fenced using fence posts placed 6 feet apart. How many posts are needed?
3. The McMahons decided to build a fence along the back of their property. They planned to use a post every 8 feet. Since the posts were more expensive than they had expected, however, they bought 3 fewer posts than planned and placed the posts 10 feet apart. How many posts did they purchase?
4. A parking lot is 100 feet wide. Traffic markers are placed in rows so that they are 3 feet apart and so that no marker is closer than 20 feet to the sides of the lot. How many markers are there in a row?
5. A man built some pens to house his dogs. If he puts 1 dog in each pen, he will have 2 dogs left over. If he puts 2 dogs in each pen, he will have 1 empty pen. How many dogs and pens has he?
6. Cora wants to arrange her doll collection on the bookshelves in her room. If she puts 2 dolls on each shelf, she will have 1 doll left over. If she puts 3 dolls on each shelf, she will have 2 extra shelves. How many dolls and how many shelves has she?
7. A milk crate holds bottles in 4 rows of 5. How can you place 14 bottles of milk in the crate so that each row and each column has an even number of bottles?
8. A milk crate holds bottles in 5 rows of 5. How can you place 11 bottles of milk so that each row and each column has an odd number of bottles?

2-6 MAKING LISTS AND CHARTS

Problems in which different possible solutions are to be investigated can often be solved by making organized lists or charts.

MODEL PROBLEMS

1. Three friends, Jane, Rose, and Phyllis, study different languages and have different career goals. One wants to be an artist, one a doctor, and the third a lawyer.
   (1) The girl who studies Italian does not plan to be a lawyer.
   (2) Jane studies French and does not plan to be an artist.
   (3) The girl who studies Spanish plans to be a doctor.
   (4) Phyllis does not study Italian.

   Can you find the language and career goal of each girl?

Plan: Make a chart and fill in the information, starting with the most definite clues first.

Solution: Make a chart.

	Jane	Rose	Phyllis
Language			
Career			

Clue (2) gives the most definite information about language. After using that, use Clue (4) to determine the other languages.

	Jane	Rose	Phyllis
Language	French	Italian	Spanish
Career			

Now that you know which girl studies which language, Clue (3) gives definite information about a career choice. After using that, Clues (1) and (2) determine the other career choices.

	Jane	Rose	Phyllis
Language	French	Italian	Spanish
Career	lawyer	artist	doctor

Check: Compare each clue with the chart.

Rose studies Italian and does not plan to be a lawyer.
Jane studies French and does not plan to be an artist.
Phyllis studies Spanish and plans to be a doctor.
Phyllis does not study Italian.

Answer: The chart displays the answer.

2. Magda needs 50 cents for a coin machine that takes nickels, dimes, and quarters. In how many different ways can she have the correct change?

Plan: List the coins that can be used and all possible combinations of these coins that make 50 cents.

Solution:

Nickels	10	8	6	5	4	3	2	1	0	0
Dimes	0	1	2	0	3	1	4	2	5	0
Quarters	0	0	0	1	0	1	0	1	0	2

Check: Determine the value of each set of coins to verify that it is 50 cents.

Answer: Magda could have the correct change in 10 different ways.

EXERCISES

1. Beta has 3 boxes of paper. The first is labeled "white," the second is labeled "blue," and the third is labeled "assorted." She knows that these 3 labels correctly describe the contents of the boxes, but the covers of the boxes were replaced carelessly so that each box has an incorrect label. She opens the box marked "assorted" and finds that the top sheet of paper is white. Which incorrect label is on which box?
2. Mark, Jay, and Fred have each invited a sister of one of the other two boys to the senior dance. From the clues given below, determine the name of the girl that each boy has invited to the dance.
   (1) No girl is going to the dance with her brother.
   (2) Sally is not Mark's sister and is not going to the dance with Joan's brother.
   (3) Marion is Jay's sister.
3. A florist charges $3 for a rose and $2 for three carnations. Carnations cannot be purchased separately. One customer paid $20 for flowers. What assortments could have been purchased?
4. In how many ways can change for a $20 bill be made in $1, $5, and/or $10 bills?
5. The 8 bills in Marilyn's purse are worth less than $30. What bills could she have?
6. The 8 bills in Marilyn's purse are worth less than $30. She makes a $12 purchase and gives the exact amount in payment. What bills could she have had in her purse? (Consider the combinations you found in answer to Exercise 5.)
7. Find the smallest number that will leave a remainder of 1 when divided by 3, a remainder of 3 when divided by 4, and a remainder of 4 when divided by 5.
8. Find the smallest number that will leave a remainder of 1 when divided by 2, a remainder of 2 when divided by 3, a remainder of 3 when divided by 4, and a remainder of 4 when divided by 5.

2-7 CHOOSING AND COMBINING STRATEGIES

Many problems can be solved by several different strategies or by a combination of strategies.

MODEL PROBLEMS

1. Brian and Linda work at a restaurant that is open 7 days a week. Brian works for 5 days and then has 2 days off. Linda works for 3 days and then has 1 day off. Neither Brian nor Linda worked on Sunday, April 1, and both worked on April 2. What other days in April do both Linda and Brian have off?

Plan 1: Make a chart in the form of a calendar on which you can mark the days on which each person works.

Solution: Start the chart with a blank for April 1, on which neither Brian nor Linda worked.

Mark Brian's schedule, beginning with April 2.
Mark Linda's schedule, beginning with April 2.

S	M	T	W	Th	F	S
1	2 B L	3 B L	4 B L	5 B	6 B L	7 L
8 L	9 B	10 B L	11 B L	12 B L	13 B	14 L
15 L	16 B L	17 B	18 B L	19 B L	20 B L	21
22 L	23 B L	24 B L	25 B	26 B L	27 B L	28 L
29	30 B L					

Answer: Brian and Linda will both be off on April 21 and 29.

There are other strategies that you can use to solve this problem.

Plan 2: Make a list of the days each has off and find the dates that occur in both lists.

Solution:

Brian is off every 6th and 7th day after April 1. Add 6 and 7 to the last date that Brian had off.

Brian: 7 8 14 15 21 22 28 29

Linda is off every 4th day after April 1. Add 4 to the last date that Linda had off.

Linda: 5 9 13 17 21 25 29

April 21 and April 29 appear in both lists.

Answer: Brian and Linda will both be off on April 21 and 29.

2. Mr. Breiner has a machine that can harvest his corn in 40 hours. His neighbor has a larger machine that can harvest the same number of acres of corn in 30 hours. If they work together, how long will it take to harvest Mr. Breiner's corn using both machines?

Plan: Use trial and error. To determine a starting guess, estimate from the facts. The smaller machine needs 20 hours to do half of the job, and the larger machine needs 15 hours to do half of the job. Working together, the larger machine will do more than half, and the smaller machine will do less than half. A good estimate would be between 15 and 20 hours.

Solution: Try 17 hours and test the result. If the smaller machine takes 40 hours to do the job, it can do $\frac{1}{40}$ of the job every hour and $\frac{17}{40}$ in 17 hours. If the larger machine takes 30 hours to do the job, it can do $\frac{1}{30}$ of the job every hour and $\frac{17}{30}$ in 17 hours. Together, the two machines must do 1 whole job.

$$\frac{17}{40} + \frac{17}{30} = \frac{51}{120} + \frac{68}{120} = \frac{119}{120}$$

The result is very close but not exact. The answer is probably a fractional part of an hour and, therefore, the number of possibilities to try is endless. This is not a good strategy. However, what you have done is not lost. It suggests that you need to consider what part of the work can be completed in 1 hour. Use the strategy of a simpler related problem. Use the same problem but change the question: What part of the work can be done by both machines in 1 hour?

You have already seen that, in 1 hour, the smaller machine can do $\frac{1}{40}$ and the larger machine can do $\frac{1}{30}$ of the job. Together, in 1 hour, the two machines can do $\frac{1}{40} + \frac{1}{30}$ of the job.

$$\frac{1}{40} + \frac{1}{30} = \frac{3}{120} + \frac{4}{120} = \frac{7}{120}$$

Now, use the strategy of working backward.

$$\frac{7}{120} \text{ of what number is 1?}$$

$$1 \div \frac{7}{120} = 1 \cdot \frac{120}{7} = \frac{120}{7} = 17\frac{1}{7}$$

Check: Smaller machine: $\frac{1}{40} \cdot 17\frac{1}{7} = \frac{1}{40} \cdot \frac{120}{7} = \frac{3}{7}$

Larger machine: $\frac{1}{30} \cdot 17\frac{1}{7} = \frac{1}{30} \cdot \frac{120}{7} = \frac{4}{7}$

Together: $\frac{3}{7} + \frac{4}{7} = \frac{7}{7} = 1$

Answer: Working together, the two machines will do the job in $17\frac{1}{7}$ hours.

EXERCISES

1. A nine-inch square is divided into 81 one-inch squares that are colored alternately red and black. How many red and how many black squares are there if the corner squares are red?
2. A kitten in a bucket weighs 3 kilograms. A rabbit in the same bucket weighs 5 kilograms. The kitten and the rabbit together weigh 6 kilograms. What is the weight of the bucket?
3. A group of people enter a room in which there are some benches. If 1 person sits on each bench, 4 persons remain standing. If 2 people sit on each bench, 1 bench is empty. How many people and how many benches are there?
4. At a movie, adults pay $4 and children pay $2.50. The cashier collected $300 for 99 tickets. How many adults and how many children attended the movie?
5. Greg agreed to work for 1 year for $10,000 and a car. After 6 months, he quit and was paid $1,000 and the car. What was the value of the car?
6. At the end of an hour, a cook in a diner noted that $\frac{1}{3}$ of the customers had ordered sandwiches, $\frac{1}{2}$ had ordered hamburgers, $\frac{1}{8}$ had ordered steak, and the remaining 2 had ordered just a salad. If no customer ordered more than one of these items, how many customers were served?
7. Draw two squares that intersect in
   a. exactly 1 point **b.** exactly 2 points **c.** exactly 3 points
   d. exactly 4 points **e.** exactly 5 points **f.** exactly 6 points
   g. exactly 7 points **h.** exactly 8 points.

8. Both Rosa and Tony work Monday through Friday of each week. Rosa is paid \$32 per day, including holidays. Tony is paid \$680 per month.
   a. Who is paid more in a month? **b.** Who is paid more in a year?
9. Gumdrops cost 4 for a penny and chocolate drops cost 4 cents each. Glen bought 20 pieces of candy for 20 cents. What did he buy?
10. How many triangles are there in the figure?

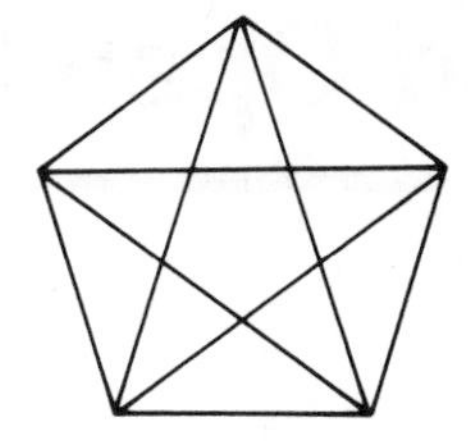

2-8 REVIEW EXERCISES

1. There were 5 students in the cafeteria line today. Bernie was first in line. Joel was 2 places behind Edith. Lester was ahead of Pierre, who was fifth. Who was second in line?
2. Carl is 4 years older than Chris. Five years ago, Carl was twice as old as Chris. How old is Carl now?
3. A student wants to give 25 cents to each of several charities but finds that she is 5 cents short. If she gives 20 cents to each, she will have 15 cents left. How much money does she have to start with?
4. Ernestine is 3 times as old as her sister Lucy. In 5 years, Ernestine will be twice as old as Lucy. How old are Ernestine and Lucy now?
5. The 400 voters in the town of Euclid voted on two issues. There were 225 in favor of the first issue and 355 in favor of the second. If there were 40 persons who voted against both issues, how many voted in favor of both issues?
6. If $2 * 4 = 12$, $3 * 5 = 16$, and $1 * 2 = 6$, what is $2 * 3$?
7. Each letter in the words at the right stands for a number. The same letter always stands for the same number and different letters stand for different numbers. Find numbers that will make a correct addition. There may be more than one solution.

$$\begin{array}{r} \text{COME} \\ +\text{HOME} \\ \hline \text{QUICK} \end{array}$$

8. How many boxes 3″ by 4″ by 5″ will fit in a carton whose dimensions are 9″ by 15″ by 10″?
9. A group of Americans and Canadians were on a bus tour of Niagara Falls. There were 8 boys, 5 American children, 9 men, 6 Canadian boys, 10 Americans, 3 American males, and 15 Canadian females. How many persons were on the bus tour?

Chapter 3

Algebraic Expressions and Open Sentences

3-1 TRANSLATING VERBAL PHRASES INTO ALGEBRAIC LANGUAGE

A machine can produce 25 articles in one day. The number of articles the machine can produce would be:

25×1	in one day	25×3	in three days
25×2	in two days	25×4	in four days

Each of these expressions is of the form $25 \times n$, where n may represent a member of $\{1, 2, 3, 4\}$. We call n a ***variable*** or a ***placeholder*** because it represents any member of a given set. The set whose members the variable may represent is called the ***domain*** or the ***replacement set*** of the variable.

An expression or mathematical phrase that contains one or more variables is called an ***open expression***, an ***open phrase***, or an ***algebraic expression***.

In Chapter 1, you saw how number relationships could be expressed with mathematical symbols. Now you will see how verbal phrases can be translated into the language of algebra by using letters as variables and by using symbols to represent operations.

Verbal Phrases Involving Addition

The algebraic expression $a + b$ may be used to represent several different verbal phrases, such as:

a plus b	a and b are added	a is increased by b
the sum of a and b	b is added to a	b more than a

The word *exceeds* means *is more than*. Thus:

7 exceeds 5 by 2 means 7 is 2 more than 5
or $7 = 5 + 2$.

The number that exceeds 5 by 2 is $5 + 2$ or 7.

The number that exceeds a by b is $a + b$.

Verbal Phrases Involving Subtraction

The algebraic expression $a - b$ may be used to represent several different verbal phrases, such as:

a minus b
the difference between a and b
b subtracted from a
a decreased by b
a diminished by b
b less than a
a reduced by b

Verbal Phrases Involving Multiplication

The algebraic expressions $a \times b$, $a \cdot b$, and ab may be used to represent several different verbal phrases, such as:

a times b the product of a and b b multiplied by a

Since the multiplication symbol $\times$ can be confused with the letter x, which is often used as a variable, and since the raised dot can be confused with a decimal point, we prefer not to use these symbols. In algebra, a product such as ab is commonly indicated by using no symbol between the factors.

If the factors are numbers, all but the first must be in parentheses: 3(5)(2). If the factors are variables, parentheses may be used but are not needed: $3(b)(h) = 3bh$.

Verbal Phrases Involving Division

The algebraic expressions $a \div b$ and $\frac{a}{b}$ may be used to represent several different verbal phrases, such as:

a divided by b the quotient of a and b

The symbols $a \div 4$ and $\frac{a}{4}$ mean one-fourth of a as well as a divided by 4.

In some verbal phrases, using a comma can prevent misreading. For example, in "the product of x and y, decreased by 2," the comma after y makes it clear that the phrase means $(xy) - 2$ and not $x(y - 2)$.

MODEL PROBLEMS

1. Use mathematical symbols to translate each of the following verbal phrases into algebraic language:

	Answers
a. w more than 3	$3 + w$
b. w less than 3	$3 - w$
c. r decreased by 2	$r - 2$
d. the product of $5r$ and s	$5rs$
e. twice x, decreased by 10	$2x - 10$
f. 25, diminished by 4 times n	$25 - 4n$
g. the sum of t and u, divided by 6	$\frac{t + u}{6}$
h. 100 decreased by twice $(x + 5)$	$100 - 2(x + 5)$

2. Represent in algebraic language:

a. a number that exceeds 5 by m	$5 + m$
b. a number that x exceeds by 5	$x - 5$
c. twice the sum of x and y	$2(x + y)$

EXERCISES

In 1–28, use mathematical symbols to translate the verbal phrase into algebraic language.

1. y plus 8
2. 8 plus y
3. r minus 4
4. 4 minus r
5. 7 times x
6. x times 7
7. x divided by 10
8. 10 divided by x
9. the product of 6 and d
10. c decreased by 6
11. 15 added to b
12. one-tenth of w
13. the sum of b and 8
14. x diminished by y
15. the product of x and y
16. the quotient of s and t
17. 12 increased by a
18. 5 less than d
19. 8 divided by y
20. y multiplied by 10
21. the product of $2c$ and $3d$
22. t more than w
23. one-third of z
24. twice the difference of p and q

25. a number that exceeds m by 4
26. one-half of the sum of L and W
27. 5 times x, increased by 2
28. 10 decreased by twice a

In 29–39, using the letter n to represent a number, write the verbal phrase in algebraic language.

29. a number increased by 2
30. 20 more than a number
31. 8 increased by a number
32. a number decreased by 6
33. 2 less than a number
34. 3 times a number
35. three-fourths of a number
36. 4 times a number, increased by 3
37. 10 times a number, decreased by 2
38. 3 less than twice a number
39. the product of 5 more than a number, and 4

In 40–45, translate the verbal phrase into algebraic language, representing the two numbers by L and W, with L being the larger.

40. the sum of the two numbers
41. the product of the two numbers
42. the larger number decreased by the smaller number
43. the smaller number divided by the larger number
44. the sum of twice the larger number and twice the smaller number
45. 10 times the smaller number, decreased by 6 times the larger number

3-2 USING LETTERS TO REPRESENT VARIABLES

A knowledge of arithmetic is important in algebra. Since the variables represent numbers that are familiar to you, it will be helpful to apply the strategy of using a simpler related problem; that is, relate similar arithmetic problems to the given algebraic ones.

PROCEDURE. To write an algebraic expression involving variables:

1. Write a similar problem in arithmetic.
2. Write an expression for the arithmetic problem.
3. Write a similar expression involving the letters.

MODEL PROBLEMS

Represent by an algebraic expression:

1. a distance that is 20 meters shorter than x meters

How to Proceed	*Solution*
(1) Write a similar problem in arithmetic.	a distance that is 20 meters shorter than 50 meters
(2) Write an expression for the arithmetic problem.	$50 - 20$
(3) Write a similar expression involving the letter.	$(x - 20)$ meters *Ans.*

2. the value of n hats, each worth d dollars

How to Proceed	*Solution*
(1) Write a similar problem in arithmetic.	the value of 5 hats, each worth 10 dollars
(2) Write an expression for the arithmetic problem.	$5(10)$
(3) Write a similar expression involving the letters.	nd dollars *Ans.*

After some practice, you will be able to do steps (1) and (2) mentally.

	Answers
3. a weight that is 40 lb. heavier than p lb.	$(p + 40)$ lb.
4. an amount of money that is twice d dollars	$2d$ dollars

EXERCISES

In 1–18, represent the answer in algebraic language, using the variable mentioned in the problem.

1. The number of kilometers traveled by a bus is represented by x. If a train traveled 200 kilometers farther than the bus, represent the number of kilometers traveled by the train.

2. Mr. Gold invested $1,000 in stocks. If he lost d dollars when he sold them, represent the amount he received for them.

3. The cost of a fur coat is 5 times the cost of a cloth coat. If the cloth coat costs x dollars, represent the cost of the fur coat.

4. The length of a rectangle is represented by L. If the width of the rectangle is one-half of its length, represent its width.

5. After 12 centimeters had been cut from a piece of lumber, there were c centimeters left. Represent the length of the original piece of lumber.
6. Paul and Martha saved 100 dollars. If the amount saved by Paul is represented by x, represent the amount saved by Martha.
7. The sum of two numbers is s. If one number is represented by x, represent the other number in terms of s and x.
8. A suit costs $150. Represent the cost of n suits.
9. A ballpoint pen sells for 39 cents. Represent the cost of x pens.
10. Represent the cost of t feet of lumber that sells for g cents a foot.
11. If Hilda weighed 45 kilograms, represent her weight after she had lost x kilograms.
12. Ronald, who weighs c pounds, is d pounds overweight. Represent the number of pounds Ronald should weigh.
13. A man spent $250 for a suit and a coat. If he spent y dollars for the coat, represent the amount he spent for the suit.
14. A man bought an article for c dollars and sold it at a profit of $25. Represent the amount for which he sold it.
15. The width of a rectangle is represented by W meters. Represent the length of the rectangle if it exceeds the width by 8 meters.
16. The width of a rectangle is x centimeters. Represent the length of the rectangle if it exceeds twice the width by 3 centimeters.
17. If a plane travels 550 kilometers per hour, represent the distance it will travel in h hours.
18. If an auto traveled for 5 hours at an average rate of r kilometers per hour, represent the distance it has traveled.

19. Represent algebraically the number of:
    a. centimeters in m meters
    b. meters in i centimeters
    c. days in w weeks
    d. weeks in d days
    e. hours in d days
    f. days in h hours
    g. feet in c inches
    h. grams in k kilograms

20. Represent the total number of days in w weeks and d days.
21. Represent the number of baseballs you can buy with c dollars if each baseball costs m dollars.
22. Represent the total number of calories in x peanuts and y potato chips if each peanut contains 15 calories and each potato chip contains 18 calories.
23. The charges for a long-distance telephone call are $.45 for the first 3 minutes and $.09 for each additional minute. Represent the cost of a telephone call that lasts m minutes when m is greater than 3.
24. A printing shop charges a 50-cent minimum for the first 8 photocopies of a flyer. Additional copies cost 6 cents each. Represent the cost of c copies if c is greater than 8.

25. A utilities company measures gas consumption by the hundred cubic feet, CCF. The company has a 3-step rate schedule for gas customers. First, there is a minimum charge of $5.00 per month for up to 3 CCF of gas used. Then, for the next 6 CCF, the charge is $.75 per CCF. Finally, after 9 CCF, the charge is $.55 per CCF. Represent the cost of g CCF of gas, if g is greater than 9.

3-3 UNDERSTANDING THE MEANING OF SOME VOCABULARY USED IN ALGEBRA

Term

A ***term*** is a numeral, a variable, or the indicated product or quotient of numerals and variables. For example, 5, x, $4y$, $8ab$, and $\frac{x}{y}$ are terms.

An algebraic expression such as $4a + 2b - 5c$ has three terms. The terms $4a$, $2b$, and $5c$ are separated by + and − signs.

Factors of a Product

If an indicated product involves two or more numbers, each of the numbers, and the product of any of them, are ***factors*** of the product. Also, any product has 1 as a factor. For example, the factors of $3xy$ are 1, 3, x, y, $3x$, $3y$, xy, and $3xy$. Note that when we factor, we usually concern ourselves only with factors that are whole numbers.

Coefficient

Any factor of a product is the ***coefficient*** of the remaining factor or product of factors.

For example, in the product $4ab$:

4 is the coefficient of ab $4a$ is the coefficient of b
$4b$ is the coefficient of a ab is the coefficient of 4

When a numeral and variables are factors of a product, the numeral is called the ***numerical coefficient*** of the product. For example, in $8y$ the numerical coefficient is 8; in $4ab$ the numerical coefficient is 4.

When the word *coefficient* is used alone, it usually means the numerical coefficient. For example, in the term $7rs$ the coefficient is 7.

Since x names the same number as $1x$, the coefficient of x is understood to be 1. Likewise, the coefficient of ab is understood to be 1.

Base, Exponent, Power

A *power* is a product of equal factors:

$$4 \cdot 4 = 4^2 \qquad c \cdot c \cdot c \cdot c = c^4$$

4^2, or 16, and c^4 are powers.

An *exponent* refers only to the number or variable that is directly to the left of it:

$$5d^2 \text{ means } 5 \cdot d \cdot d$$

To use the product $5d$ as a factor 2 times, enclose the product in parentheses:

$$(5d)^2 \text{ means } (5d)(5d)$$

To use a sum, difference, product, or quotient as a *base*, enclose the base in parentheses:

$$(a + 2)(a + 2) = (a + 2)^2 \qquad \left(\frac{a}{4}\right)\left(\frac{a}{4}\right)\left(\frac{a}{4}\right) = \left(\frac{a}{4}\right)^3$$

MODEL PROBLEM

Name the numerical coefficient, base, and exponent in the term $4x^5$.

Answer: The numerical coefficient is 4, the base is x, and the exponent is 5.

EXERCISES

In 1–6, name the factors (other than 1) of each product.

1. xy **2.** $3a$ **3.** $5n$ **4.** $7mn$ **5.** $13xy$ **6.** $11st$

In 7–12, name the numerical coefficient of x.

7. $8x$ **8.** $(5 + 2)x$ **9.** $\frac{1}{2}x$ **10.** x **11.** $1.4x$ **12.** $2 + 7x$

In 13–18, name the base and exponent of the power.

13. m^2 **14.** s^3 **15.** t **16.** 10^6 **17.** $(5y)^4$ **18.** $(x + y)^5$

In 19–30, write each expression, using exponents.

19. $m \cdot m \cdot m$

20. $b \cdot b \cdot b \cdot b \cdot b$

21. $4 \cdot x \cdot x \cdot x \cdot x \cdot x$

22. $\pi \cdot r \cdot r$

23. $a \cdot a \cdot a \cdot a \cdot b \cdot b$

24. $7 \cdot r \cdot r \cdot r \cdot s \cdot s$

25. $9 \cdot c \cdot c \cdot c \cdot d$

26. $(6a)(6a)(6a)$

27. $(x + y)(x + y)$

28. $(a - b)(a - b)(a - b)$

29. the square of $(b - 5)$

30. the fourth power of $(m + 2n)$

In 31–36, write the term as a product without using exponents.

31. r^6 **32.** $5x^4$ **33.** x^3y^5 **34.** $4a^4b^2$ **35.** $3c^2d^3e$ **36.** $(3y)^5$

3-4 EVALUATING ALGEBRAIC EXPRESSIONS

The algebraic expression $3n + 1$ represents an unspecified number. It is only when we replace the variable n by a specific number that $3n + 1$ represents a specific number. For example, suppose that the domain of n is $\{1, 2, 3\}$. The specific numbers that $3n + 1$ represents (the values of $3n + 1$) can be found as follows:

If $n = 1$, $3n + 1 = 3(1) + 1 = 3 + 1 = 4$.
If $n = 2$, $3n + 1 = 3(2) + 1 = 6 + 1 = 7$.
If $n = 3$, $3n + 1 = 3(3) + 1 = 9 + 1 = 10$.

When we determine the number that an algebraic expression represents for specified values of its variables, we are ***evaluating the algebraic expression***; that is, we are finding its value or values.

PROCEDURE. To evaluate an algebraic expression, use the following steps that follow the rules for the order of operations.

1. Replace the variables by their specific values.
2. Evaluate the expressions included within symbols of grouping such as parentheses, simplifying the expression in the innermost symbols of grouping first.
3. Simplify any powers and roots. (Roots will be studied later.)
4. Do all multiplications and divisions in order from left to right.
5. Do all additions and subtractions in order from left to right.

MODEL PROBLEMS

1. Evaluate $50 - 3x$ when $x = 7$.

How to Proceed	*Solution*
(1) Write the expression.	$50 - 3x$
(2) Replace the variable by its given value.	$= 50 - 3(7)$
(3) Do the multiplication.	$= 50 - 21$
(4) Do the subtraction.	$= 29$ *Ans.*

2. Evaluate $2x^2 - 5x + 4$ when $x = 7$.

How to Proceed	*Solution*
(1) Write the expression.	$2x^2 - 5x + 4$
(2) Replace the variable by its given value.	$= 2(7)^2 - 5(7) + 4$
(3) Evaluate the power.	$= 2(49) - 5(7) + 4$
(4) Do the multiplication.	$= 98 - 35 + 4$
(5) Do the addition and subtraction.	$= 67$ *Ans.*

3. Evaluate $\frac{2a}{5} + (n - 1)d$ when $a = 40$, $n = 10$, and $d = 3$.

How to Proceed	*Solution*
(1) Write the expression.	$\frac{2a}{5} + (n - 1)d$
(2) Replace the variables by their given values.	$= \frac{2(40)}{5} + (10 - 1)(3)$
(3) Simplify the expression grouped by parentheses or a fraction bar.	$= \frac{80}{5} + (9)(3)$
(4) Do the multiplication and division.	$= 16 + 27$
(5) Do the addition.	$= 43$ *Ans.*

4. Evaluate $(2x)^2 - 2x^2$ when $x = 4$.

How to Proceed	*Solution*
(1) Write the expression.	$(2x)^2 - 2x^2$
(2) Replace the variable by its given value.	$= (2 \cdot 4)^2 - 2(4)^2$
(3) Simplify the expression within parentheses.	$= (8)^2 - 2(4)^2$
(4) Evaluate the powers.	$= 64 - 2(16)$
(5) Do the multiplication.	$= 64 - 32$
(6) Do the subtraction.	$= 32$ *Ans.*

EXERCISES

Basic Operations

In 1–21, find the numerical value of the expression. Use $a = 8$, $b = 6$, $d = 3$, $x = 4$, $y = 5$, and $z = 1$.

1. $5a$ **2.** $9b$ **3.** $\frac{1}{2}x$ **4.** $.3y$ **5.** $a + 3$ **6.** $7 + y$

7. $a - 2$ **8.** $5 - y$ **9.** $a - b$ **10.** ax **11.** $3xy$ **12.** $\frac{2b}{3}$

13. $\frac{3bd}{9}$ **14.** $2x + 9$ **15.** $3y - b$

16. $20 - 4z$ **17.** $5x + 2y$ **18.** $ab - dx$

19. $a + 5d + 3x$ **20.** $9y + 6b - d$ **21.** $ab - d - xy$

Basic Operations and Powers

In 22–51, find the numerical value of the expression. Use $a = 8$, $b = 6$, $d = 3$, $x = 4$, $y = 5$, and $z = 1$.

22. a^2 **23.** x^2 **24.** b^3 **25.** y^3 **26.** d^4 **27.** z^5

28. $2x^2$ **29.** $3b^2$ **30.** $4d^3$ **31.** $6z^5$ **32.** $\frac{b^2}{9}$ **33.** $\frac{1}{2}x^3$

34. $\frac{3}{4}x^3$ **35.** a^2d **36.** xy^2

37. $3z^2a$ **38.** $2a^2b^3$ **39.** $\frac{1}{4}x^2y^2$

40. $(2d)^2$ **41.** $a^2 + b^2$ **42.** $b^2 - y^2$

43. $a^2 + b^2 - d^2$ **44.** $x^2 + x$ **45.** $b^2 + 2b$

46. $y^2 - 4y$ **47.** $2b^2 + b$ **48.** $9a - a^2$

49. $x^2 + 3x + 5$ **50.** $y^2 + 2y - 7$ **51.** $2a^2 - 4a + 6$

Grouping Symbols

In 52–72, find the numerical value of the expression. Use $w = 10$, $x = 8$, $y = 5$, and $z = 2$.

52. $2(x + 5)$ **53.** $x(y - 2)$ **54.** $3(2x + z)$

55. $4(2x - 3y)$ **56.** $\frac{x}{2}(y + z)$ **57.** $\frac{1}{2}x(y + z)$

58. $3y - (x - z)$ **59.** $2x + 5(y - 1)$ **60.** $2(x + z) - 5$

61. $3x^2$ **62.** $(3x)^2$ **63.** $y^2 + z^2$

64. $(y + z)^2$ **65.** $w^3 - x^3$ **66.** $(w - x)^3$

67. $3w^2 - 2x^2$ **68.** $(3w - 2x)^2$ **69.** $(3w)^2 - (2x)^2$

70. $(xy)^2$ **71.** $(yw)^2$ **72.** $(w^2)(x^2)$

73. Find the value of $x^2 - 8y$ when $x = 5$ and $y = \frac{1}{2}$.

74. Find the value of $r^2 + 4s$ when $r = 3$ and $s = .5$.

75. Evaluate $\frac{5}{9}(F - 32)$ when the value of F is:

a. 50 **b.** 77 **c.** 86 **d.** 32 **e.** 212

3-5 TRANSLATING VERBAL SENTENCES INTO FORMULAS

A ***formula*** uses mathematical language to express the relationship between two or more variables. Some formulas can be found by the strategy of looking for patterns. For example, how many square units are shown in the rectangle? The rectangle, measuring 5 units in length and 3 units in width, contains a total of 15 square units of area. Several such examples lead to the conclusion that the area of a rectangle is equal to the product of its length and width. This relationship can be expressed by the formula $A = lw$. Here, A, l, and w are variables that represent, respectively, the area, the length, and the width of a rectangle.

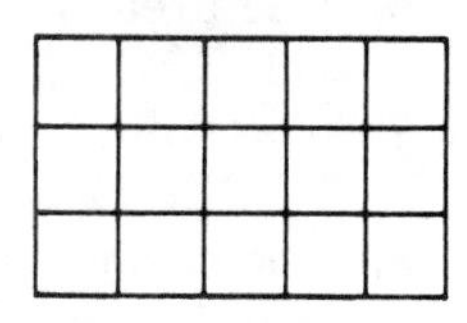

MODEL PROBLEMS

1. Write a formula for each of the following relationships.

a. The perimeter P of a square is equal to 4 times the length of each side s.

Answer: $P = 4s$

b. The cost C of a number of articles is the product of the number of articles n and the price p of each article.

Answer: $C = np$

2. Write a formula that expresses the number of months m that there are in y years.

Solution: Apply the strategy of discovering a pattern.

Since there are 12 months in a year, the number of months m in y years is equal to 12 times the number of years y.

Answer: $m = 12y$

EXERCISES

In 1–15, write a formula that expresses the relationship.

1. The total length l of 10 pieces of lumber, each m meters in length, is 10 times the length of each piece of lumber.

2. An article's selling price s equals its cost c plus the margin m.

3. The perimeter p of a rectangle is equal to the sum of twice its length l and twice its width w.
4. The average M of three numbers, a, b, c, is their sum divided by 3.
5. The area A of a triangle is equal to one-half the length of the base b multiplied by the length of the altitude h.
6. The area A of a square is equal to the square of the length of a side s.
7. The volume V of a cube is equal to the cube of the length of an edge e.
8. The surface area S of a cube is equal to 6 times the square of the length of an edge e.
9. The surface area S of a sphere is equal to the product of 4π and the square of the radius r.
10. The average rate of speed R is equal to the distance that is traveled D divided by the time spent on the trip T.
11. The Fahrenheit temperature F is 32° more than nine-fifths of the Celsius temperature C.
12. The Celsius temperature C is equal to five-ninths of the difference between the Fahrenheit temperature F and 32.
13. The dividend D equals the product of the divisor d and the quotient Q plus the remainder R.
14. A sales tax T that must be paid when an article is purchased is equal to 8% of the value of the article V.
15. A salesman's weekly earnings E is equal to his weekly salary S increased by 2% of his total volume of sales V.

In 16–20, each formula you write will express one of the variables in terms of the others.

16. Write a formula for finding the number of trees n in an orchard containing r rows of t trees each.
17. Write a formula for the total number of seats n in the school auditorium, if it has two sections, each with r rows having s seats in each row.
18. A group of n persons in an automobile crosses a river on a ferry. Write a formula for the total ferry charge c in cents, if the charge is $2.00 for the car and driver and t cents for each additional person.
19. Write a formula for the cost in cents c of a telephone conversation lasting 9 minutes if the charge for the first 3 minutes is x cents and the charge for each additional minute is y cents.
20. Write a formula for the cost in cents c of sending a telegram of 18 words if the cost of sending the first 10 words is a cents and each additional word costs b cents.

21. A gasoline dealer is allowed a profit of 12 cents a gallon for each gallon she sells. If she sells more than 25,000 gallons in a month, she is given an additional profit of 3 cents for every gallon over that number. Assuming that she always sells more than 25,000 gallons a month, express as a formula the number of dollars D in her monthly income in terms of the number N of gallons sold.

3-6 USING FORMULAS FOR PERIMETER, AREA, AND VOLUME

The variable for which a formula is solved is called the ***subject of the formula***. For example, P is the subject of $P = 4s$, the formula for the perimeter P of a square each of whose sides has a length represented by s.

If the values of all the variables of a formula except the subject are known, you can compute its value; that is, you can evaluate the subject of the formula.

PROCEDURE. To evaluate the subject of a formula:

1. Replace the other variables in the formula by their values.
2. Perform the indicated operations.

Evaluating Perimeter Formulas

Recall that the ***perimeter*** of a geometric figure is the sum of the lengths of its sides.

1. If $P = 3s$, find P when $s = 5$ ft.

Solution

$P = 3s$
$P = 3(5)$
$P = 15$

Answer: 15 ft.

2. If $P = 2b + 2h$, find P when $b = 3$ cm and $h = 7$ cm.

Solution

$P = 2b + 2h$
$P = 2(3) + 2(7)$
$P = 6 + 14 = 20$

Answer: 20 cm

To find the perimeter of a figure, you must express all lengths in the same unit of measure, which will be the unit of measure of the perimeter.

EXERCISES

1. The formula for the perimeter of a triangle is $P = a + b + c$. Find P when:
   a. $a = 15$ in., $b = 10$ in., $c = 7$ in.
   b. $a = 4.5$ m, $b = 1.7$ m, $c = 3.8$ m
   c. $a = 9$ ft., $b = 8$ ft., $c = 18$ in.
   d. $a = 7\frac{1}{2}$ ft., $b = 5\frac{3}{4}$ ft., $c = 6\frac{1}{2}$ ft.

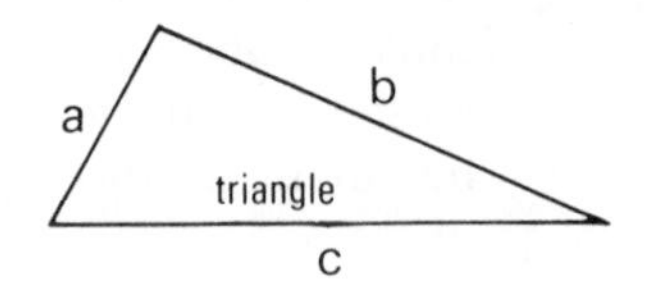

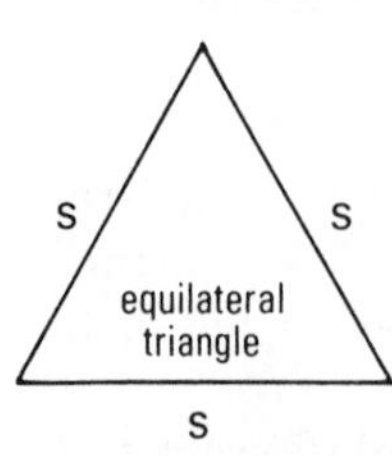

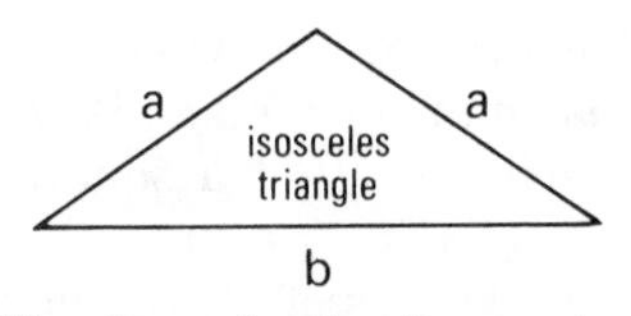

2. The formula for the perimeter of an equilateral triangle is $P = 3s$. Find P when s equals:
   a. 12 cm b. 4.8 m
   c. $9\frac{1}{3}$ ft.

3. The formula for the perimeter of an isosceles triangle is $P = 2a + b$. Find P when:
   a. $a = 6$ m, $b = 4$ m
   b. $a = 7.5$ m, $b = 5.4$ m
   c. $a = 3\frac{1}{2}$ ft., $b = 5$ ft.

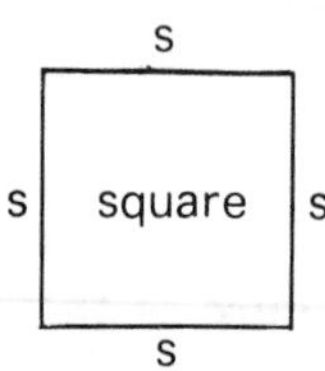

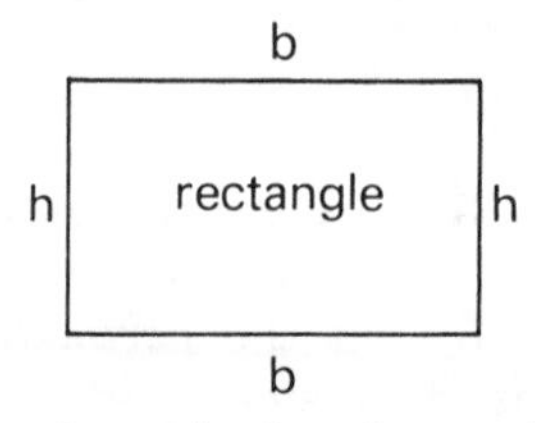

4. The formula for the perimeter of a square is $P = 4s$. Find P when s equals:
   a. 4 cm b. 3.5 m
   c. $8\frac{3}{4}$ in.

5. The formula for the perimeter of a rectangle is $P = 2b + 2h$. Find P when:
   a. $b = 20$ cm, $h = 9$ cm
   b. $b = 7.3$ m, $h = 6.9$ m
   c. $b = 5\frac{1}{2}$ in., $h = 5\frac{1}{4}$ in.

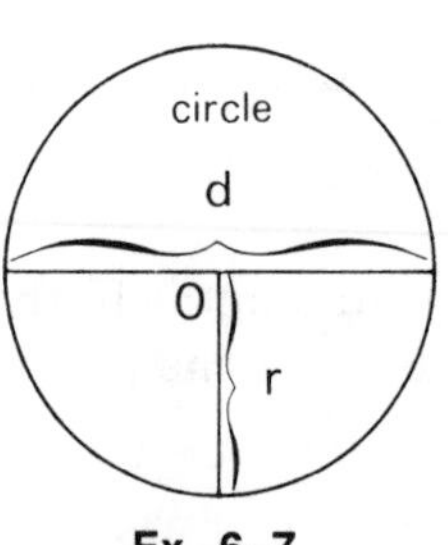

Ex. 6–7

6. The formula for the circumference of a circle is $C = \pi d$. Using $\pi = \frac{22}{7}$, find C when d equals: a. 14 ft. b. $3\frac{1}{2}$ cm c. 3 ft.

7. The formula for the circumference of a circle is $C = 2\pi r$. Using $\pi = 3.14$, find C when r equals: a. 10 ft. b. 13 ft. c. 5.6 in.

Evaluating Area Formulas

Recall that the ***area*** of a region is the number of unit squares it contains.

1. If $A = s^2$, find A when $s = 7$ yd.

Solution

$A = s^2$
$A = (7)^2$
$A = (7)(7) = 49$

Answer: $A = 49$ sq. yd.

2. If $A = \frac{1}{2}h(b + c)$, find A when $h = 3$ m, $b = 4$ m, and $c = 5$ m.

Solution

$A = \frac{1}{2}\,h(b + c)$

$A = \frac{1}{2}\,(3)(4 + 5)$

$A = \frac{1}{2}\,(3)(9) = \frac{1}{2}\,(27) = 13.5$

Answer: $A = 13.5\text{ m}^2$

To find the area of a figure, you must express all lengths in the same unit of measure. Then, the area will be square units of this measure.

EXERCISES

In a polygon, the ***base*** is a side, and the ***height*** (or ***altitude***) is a segment from an opposite vertex, perpendicular to the base. The length of the base is often represented by b and the length of the altitude is represented by h.

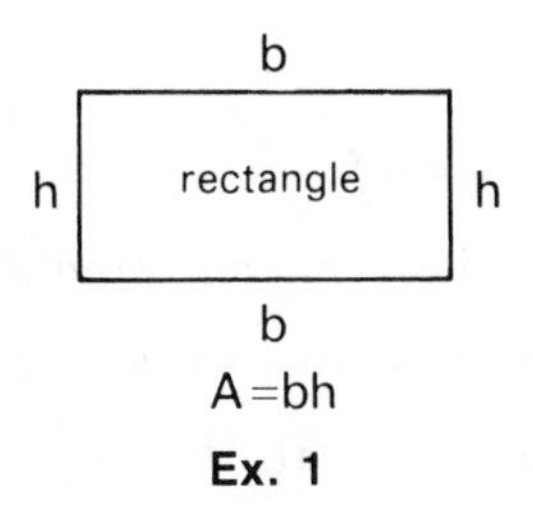

Ex. 1

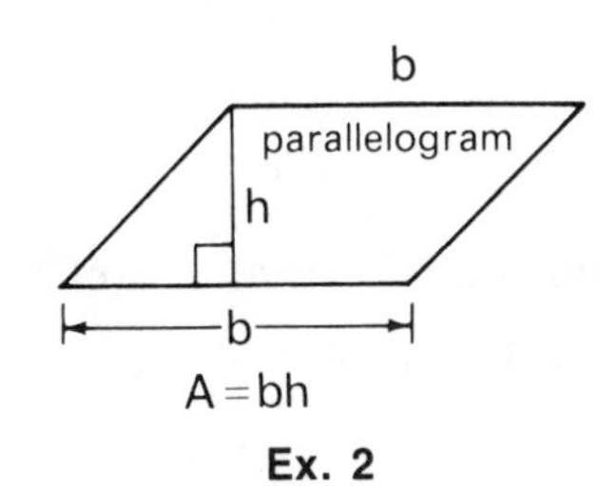

Ex. 2

1. The formula for the area of a rectangle is $A = bh$. Find A when:
a. $b = 15$ ft., $h = 13$ ft. **b.** $b = 7.5$ m, $h = 3.4$ m
c. $b = 8\frac{1}{2}$ ft., $h = 6$ ft. **d.** $b = 1$ m, $h = 40$ cm

2. The formula for the area of a parallelogram is $A = bh$. Find A when:
a. $b = 8$ ft., $h = 12$ ft. **b.** $b = 3.5$ m, $h = 6.4$ m
c. $b = 7\frac{1}{2}$ in., $h = 8$ in. **d.** $b = 1$ m, $h = 10$ cm

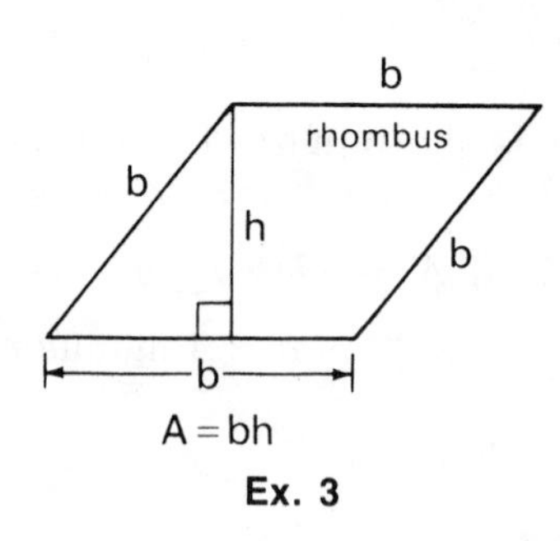

Ex. 3

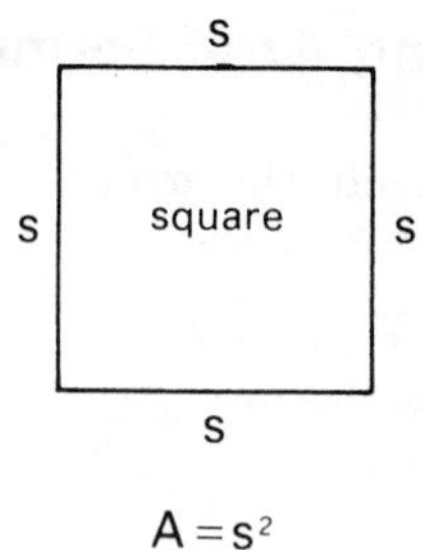

Ex. 4

3. The formula for the area of a rhombus (a parallelogram all of whose sides have the same length) is $A = bh$. Find A when:

a. $b = 5$ m, $h = 3$ m

b. $b = 7$ in., $h = 4.5$ in.

c. $b = 10$ ft., $h = 8\frac{1}{2}$ ft.

d. $b = 14.5$ cm, $h = 11.4$ cm

4. The formula for the area of a square is $A = s^2$. Find A when s equals:

a. 25 in.

b. 9 cm

c. $2\frac{1}{2}$ ft.

d. 6.1 m

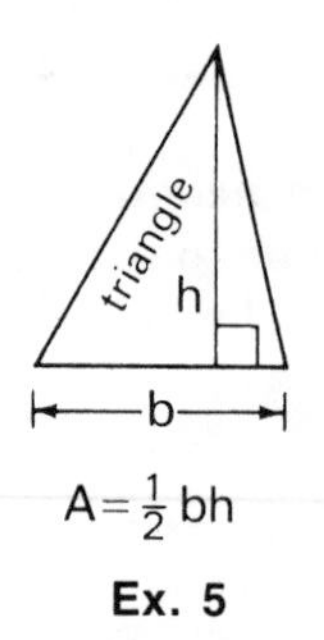

Ex. 5

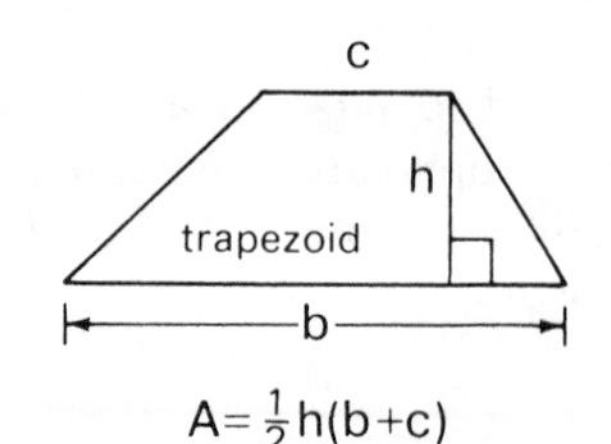

Ex. 6

5. The formula for the area of a triangle is $A = \frac{1}{2}bh$. Find A when:

a. $b = 10$ cm, $h = 6$ cm

b. $b = 10.5$ m, $h = 7.6$ m

c. $b = 3\frac{1}{2}$ in., $h = 8$ in.

d. $b = 1$ ft., $h = 5\frac{1}{2}$ in.

6. The formula for the area of a trapezoid is $A = \frac{1}{2}h(b + c)$. Find A when:

a. $h = 9$ ft., $b = 14$ ft., $c = 8$ ft.

b. $h = 5$ in., $b = 3\frac{1}{4}$ in., $c = \frac{3}{4}$ in.

c. $h = 2$ m, $b = 1.8$ m, $c = 1.1$ m

Evaluating Formulas for Volumes of Solids

Recall that the ***volume*** of a solid is the number of unit cubes it contains. To find the volume of a figure, you must express all lengths in the same unit of measure. The volume will be expressed in cubic units of this measure.

A ***right prism*** is a solid with bases of the same size and shape, and with its height perpendicular to the bases. The volume, V, of a right prism is the product of the area of a base, B, and the height, h, of the prism:

$$V = Bh$$

triangular right prism

h

B

V = Bh

The bases of a right prism are triangles with base 16.2 mm and height 10 mm. Find the volume of the prism if its height is 12 mm.

How to Proceed	*Solution*
(1) Find the value of B, the area of a triangle.	$\frac{1}{2}$(base)(height) $= \frac{1}{2}(16.2)(10)$ $= 81$
(2) Write the volume formula.	$V = Bh$
(3) Substitute into the volume formula.	$V = (81)(12) = 972$

Answer: Volume $= 972$ mm^3

Rectangular solids and cubes are two special types of right prisms. The general volume formula $V = Bh$ is adapted for these special cases, as given in the following exercises.

EXERCISES

1. The formula for the volume of a rectangular solid is $V = lwh$. Find V when:

a. $l = 5$ ft., $w = 4$ ft., $h = 7$ ft.
b. $l = 8$ cm, $w = 7$ cm, $h = 5$ cm
c. $l = 8.5$ m, $w = 4.2$ m, $h = 6.0$ m
d. $l = 2\frac{1}{2}$ in., $w = 8$ in., $h = 5\frac{1}{4}$ in.

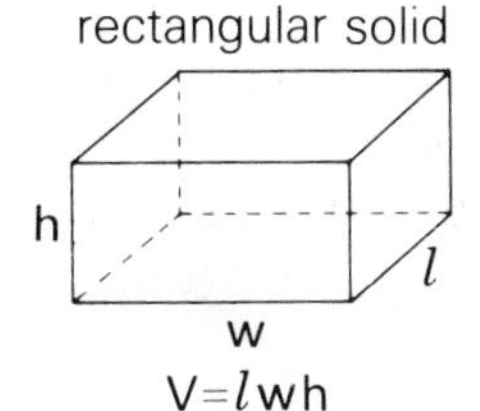

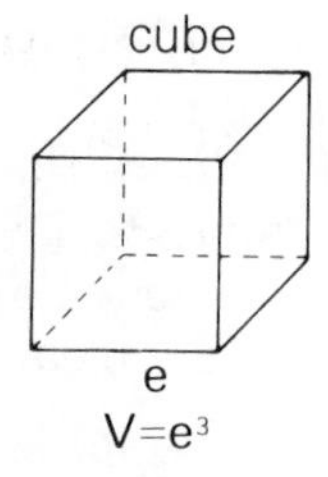

2. The formula for the volume of a cube is $V = e^3$. Find V when e equals:
   a. 2 in. **b.** 3 m **c.** 8 cm
   d. $\frac{1}{3}$ ft. **e.** 1.5 in.

3. Use the formula $V = Bh$ for the volume of the right prisms shown.

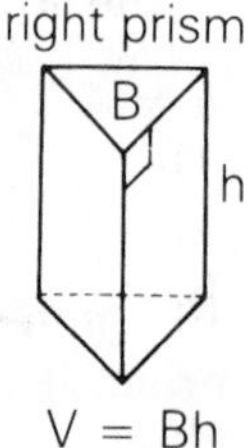

   a. In a triangular right prism, the base and altitude of a triangle have lengths 8 cm and 6 cm, respectively. Find the volume if the height of the prism is 9 cm.

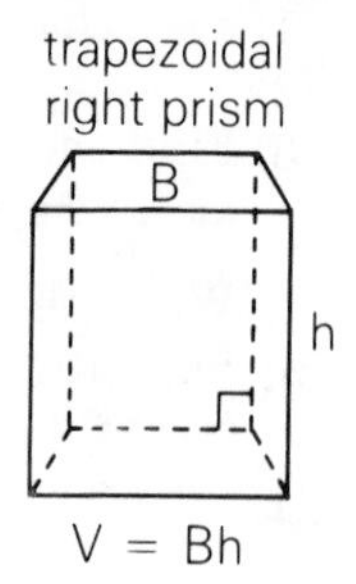

   b. A right prism with trapezoidal bases is 7 inches high. If a trapezoid has bases 3 inches and 5 inches long, and height 4 inches, find the volume of the prism.

4. In the ***pyramid*** and the ***prism*** pictured at the right, the bases are squares whose measures are equal and the measures of the heights are also equal. If we were to fill the pyramid with water three times and each time pour the water into the prism, we would find that the prism would be full.

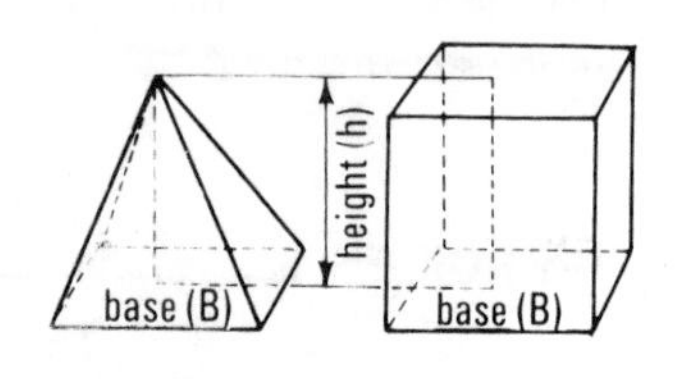

Pyramid $= \frac{1}{3}$ Prism

This means that the volume, or capacity, of the pyramid is $\frac{1}{3}$ of the volume of the prism.

Since the formula for the volume of the prism is $V = Bh$, the formula for the volume of the pyramid is:

$$V = \frac{1}{3} Bh$$

a. Find the volume of a pyramid whose height is 4 inches, and whose base is a rectangle 6 inches long and $3\frac{1}{2}$ inches wide.

b. The base of a pyramid is a parallelogram whose base and height are 5.2 cm and 3 cm, respectively. If the height of the pyramid is 9 cm, find its volume.

Finding Perimeters and Areas of Geometric Figures

The measures of the sides of a triangle are $a = 4.5$ cm, $b = 6.1$ cm, and $c = 5.7$ cm. The altitude h to side b measures 4 cm.
a. Find the perimeter of the triangle.
b. Find the area of the triangle.

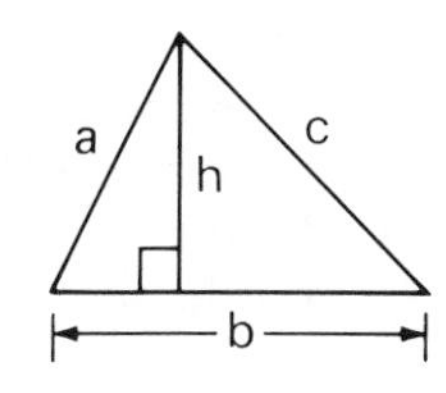

a. *How to Proceed*	*Solution*
(1) Write the proper perimeter formula.	$P = a + b + c$
(2) Substitute the given values of the variables.	$P = 4.5 + 6.1 + 5.7$
(3) Perform the computation.	$P = 16.3$

Answer: Perimeter = 16.3 cm

b. *How to Proceed*	*Solution*
(1) Write the proper area formula.	$A = \frac{1}{2}bh$
(2) Substitute the given values of the variables.	$A = \frac{1}{2}(6.1)(4)$
(3) Perform the computation.	$A = 12.2$

Answer: Area = 12.2 cm^2

EXERCISES

In 1–8: For each figure, find **(a)** its perimeter and **(b)** its area. Recall that: (1) For figures such as the parallelogram and the rectangle, the opposite sides are equal in length. (2) For the square and the rhombus, all sides are equal in length.

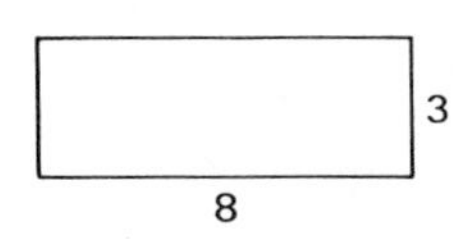

1. Rectangle

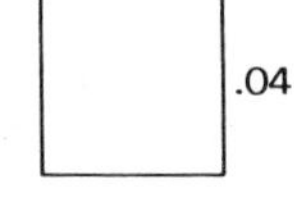

2. Square

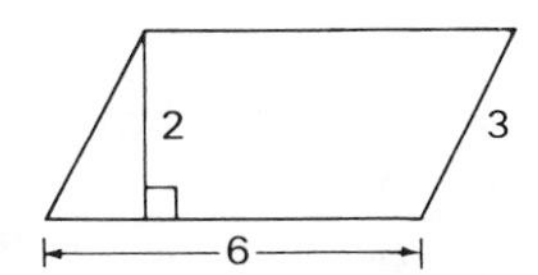

3. Parallelogram

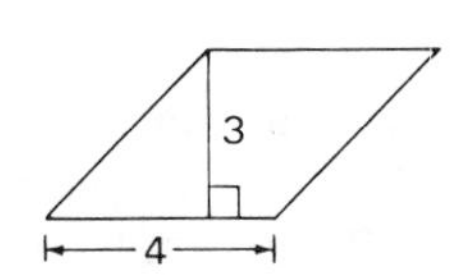

4. Rhombus

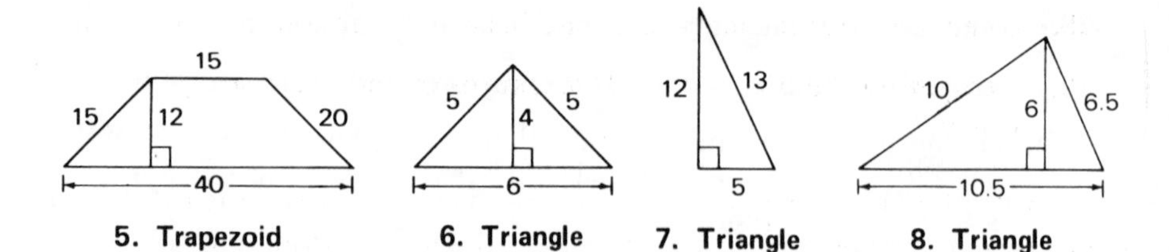

5. Trapezoid **6.** Triangle **7.** Triangle **8.** Triangle

9. Find the area of a rectangle in which the base measures 20 cm and the height is half the length of the base.

10. Find the area of a parallelogram in which the height measures 14 inches and the base measures 6 inches more than the height.

11. Find the area of a triangle in which the base measures 8.5 cm and the height measures 3 cm less than the base.

In 12–16, the given measure represents the perimeter of a square. Find: **(a)** the length of a side of the square; **(b)** the area of the square.

12. 20 cm **13.** 100 mm **14.** 4 ft. **15.** 2 in. **16.** 2.8 m

In 17–29: **a.** Find the measure of each line segment whose algebraic representation is given, when $x = 5$ and $y = 4$. **b.** Using the results found in part **a,** find the area of the geometric figure.

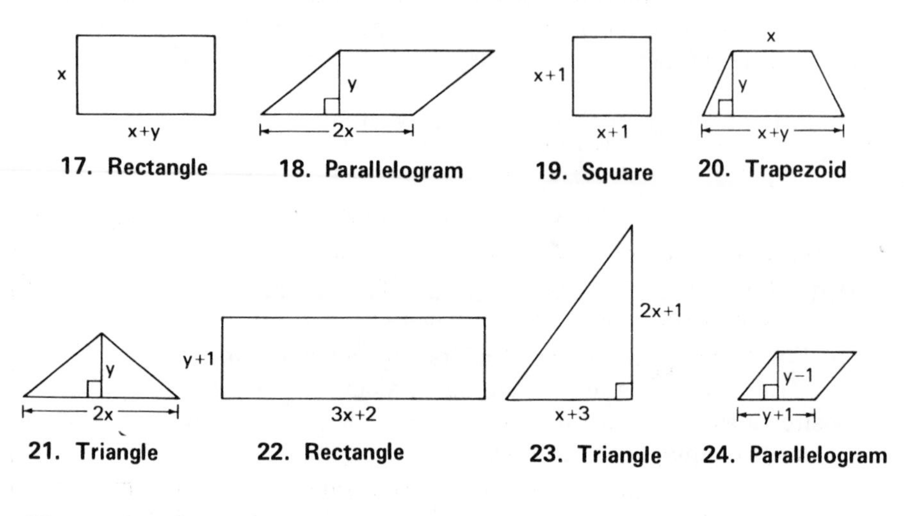

17. Rectangle **18.** Parallelogram **19.** Square **20.** Trapezoid

21. Triangle **22.** Rectangle **23.** Triangle **24.** Parallelogram

25. a triangle: measure of the base $= 3x + 1$; height $= 2x$

26. a rectangle: length $= 2x$; width $= y + 2$

27. a square: the length of a side $= 2x$

28. a parallelogram: measure of the base $= \frac{y}{2}$; height $= x + 7$

29. a trapezoid: height $= x + 3$; measures of the bases are $y + 3$ and $x + y$

In 30–34: When $x = 3$, evaluate the volume of a cube if the length of an edge is represented by the given algebraic expression.

30. x **31.** $x + 2$ **32.** $2x - 5$ **33.** $2(x - 1)$ **34.** $\frac{x}{2}$

In 35 and 36: When $x = 4$, find the volume of a rectangular solid whose measures are represented by the given algebraic expressions.

35. length $= x + 2$, width $= x - 1$, height $= 2x$

36. length $= 2x + 1$, width $= \frac{1}{2}x$, height $= \frac{3}{4}x + 1$

37. A rectangular mall located between two parallel streets is 50 feet wide and 200 yards long. Find the area of the mall.

38. **a.** An area of a lobby floor measures 15 feet by 18 feet. In one corner, an 8-foot-by-7-foot rectangular portion of the floor space is to be covered with tile. The remainder of the floor is to be carpeted. How many square feet of carpet are needed?

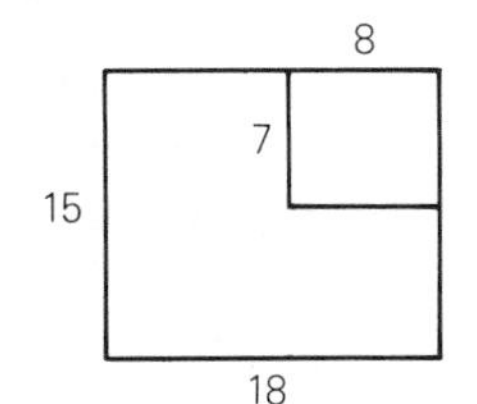

b. When the carpet is laid, a metal strip is used along each edge of the carpet to hold it in place. How many feet of stripping are needed?

c. If the uncarpeted area were in the center along one side as shown in the diagram, would the amount of carpet change? Would the amount of stripping change? Explain.

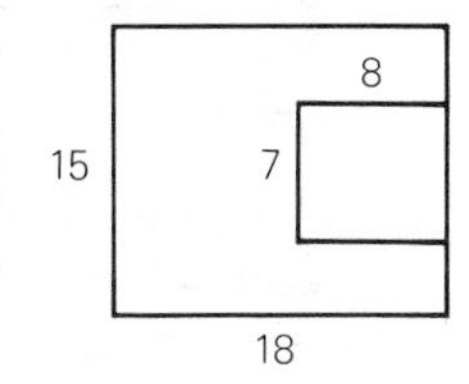

d. If the uncarpeted section were in the center of the area as shown in the diagram, would the amount of carpet change? Would the amount of stripping change? Explain.

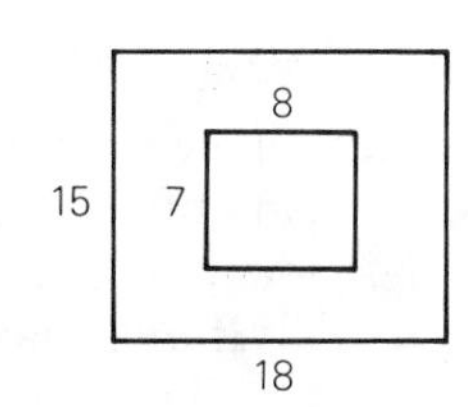

In 39–43, find the area of the shaded region.

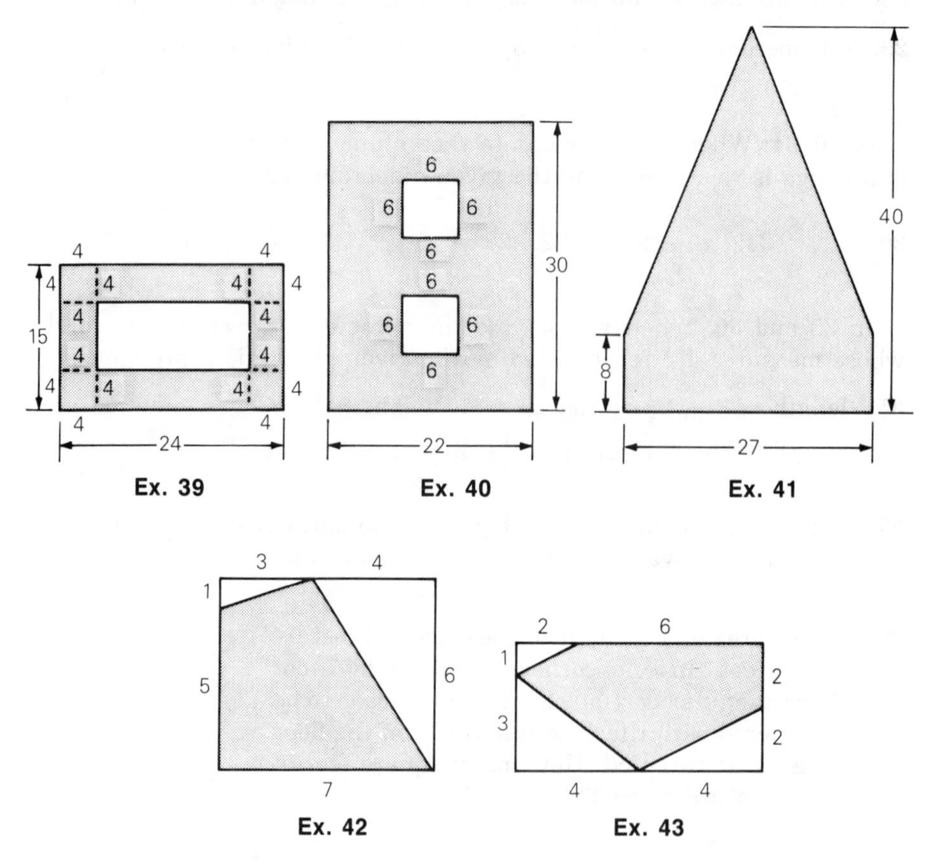

44. Chow Li is planting a border of 48 seedlings that will completely surround a rectangular garden plot. He wants to plant the seedlings 2 feet apart. If the garden plot is to be 30 feet long, how wide should it be?

45. A square is folded in half horizontally. If the perimeter of the resulting figure is 24 feet, what was the area of the original square?

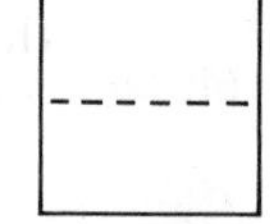

46. Ann wants to make a rectangular pen for her rabbit by using one side of the garage as one long side of the pen. She has 26 feet of fencing to use for the other three sides.

a. If she makes the pen 12 feet long and uses all of the fencing, what are the dimensions of the pen? What is the area?

b. If she makes the pen 10 feet long and uses all of the fencing, what are the dimensions of the pen? Will the rabbit have more room?

c. Using the garage wall and all of the 26 feet of fencing, what are the dimensions of the largest possible pen?

3-7 OPEN SENTENCES AND SOLUTION SETS

If 3 is added to a number, the sum is 7. Using n to represent the number, you can write the sentence algebraically as $n + 3 = 7$. The sentence may be either true or false depending on the number that you use to replace n. Such a sentence is called an ***open sentence***. Assume that the ***domain*** or ***replacement set*** for n is $\{1, 2, 3, 4, 5\}$.

The subset of the domain of the variable consisting of those elements of the domain that make the open sentence true is called the ***solution set***, or the ***truth set***, of the open sentence.

You can substitute values to discover the solution set of the open sentence $n + 3 = 7$ when the domain is $\{1, 2, 3, 4, 5\}$.

$$n + 3 = 7$$

If $n = 1$, then $1 + 3 = 7$ is a false sentence.
If $n = 2$, then $2 + 3 = 7$ is a false sentence.
If $n = 3$, then $3 + 3 = 7$ is a false sentence.
If $n = 4$, then $4 + 3 = 7$ is a *true* sentence.
If $n = 5$, then $5 + 3 = 7$ is a false sentence.

Since 4 is the only element of the domain that makes the open sentence true, the solution set of the open sentence $n + 3 = 7$ is $\{4\}$.

The elements of the domain of a variable are called the ***values*** of the variable. If a variable has only one value, it is called a ***constant***.

If no member of a replacement set will make a sentence true, we say that the solution set is the ***empty set***, or ***null set***, represented by the symbol $\varnothing$.

MODEL PROBLEMS

1. Using the replacement set $\{0, 1, 2, 3\}$, find the solution set for the open sentence $2n > 3$.

   Solution: Replace the variable n in the open sentence $2n > 3$ by each member of the replacement set.

$$2n > 3$$

If $n = 0$, then $2(0) > 3$, or $0 > 3$, is a false sentence.

If $n = 1$, then $2(1) > 3$, or $2 > 3$, is a false sentence.
If $n = 2$, then $2(2) > 3$, or $4 > 3$, is a *true* sentence.
If $n = 3$, then $2(3) > 3$, or $6 > 3$, is a *true* sentence.

Answer: Since $n = 2$ and $n = 3$ are replacements that make $2n > 3$ a true sentence, the solution set is $\{2, 3\}$.

The answer may also be written as follows:

$$\{n \mid 2n > 3\} = \{2, 3\} \text{ when } n \in \{0, 1, 2, 3\}$$

2. Using the replacement set $\{1, 2, 3\}$, find the solution set for the open sentence $y + 5 = 9$.

Solution: Replace the variable y in the open sentence $y + 5 = 9$ by each member of the replacement set.

$y + 5 = 9$
If $y = 1$, then $1 + 5 = 9$ is a false sentence.
If $y = 2$, then $2 + 5 = 9$ is a false sentence.
If $y = 3$, then $3 + 5 = 9$ is a false sentence.

Answer: Since no member of the replacement set makes $y + 5 = 9$ a true sentence, the solution set is the empty set.

This answer may also be written as follows:

$$\{y \mid y + 5 = 9\} = \varnothing \text{ when } y \in \{1, 2, 3\}$$

EXERCISES

In 1–6, tell whether or not the sentences are open sentences.

1. $2 + 3 = 5 + 0$ **2.** $x + 10 = 14$ **3.** $y - 4 = 12$
4. $3 + 2 < 10 \times 0$ **5.** $n > 7$ **6.** $r < 5 + 2$

In 7–9, name the variable.

7. $x + 5 = 9$ **8.** $4y = 20$ **9.** $r - 6 = 12$

In 10–21, use the domain $\{0, 1, 2, 3, 4, 5\}$ to find all the replacements that will change the open sentence to a true sentence. If no replacement will make a true sentence, write *None.*

10. $n + 3 = 7$ **11.** $5 - n = 2$ **12.** $5z = 0$
13. $2m = 7$ **14.** $x - x = 0$ **15.** $n > 2$
16. $n + 3 > 9$ **17.** $2n + 1 < 8$ **18.** $\frac{n + 1}{2} = 2$

19. $\frac{2n + 1}{3} = 4$ 20. $\frac{n}{4} > 1$ 21. $\frac{3x}{2} < x$

In 22–29, using the replacement set {1, 2, 3, 4, 5, 6, 7, 8, 9, 10}, find the solution set.

22. $x + 6 = 9$ 23. $8 - x = 5$ 24. $2x + 1 = 24$
25. $16 = 18 - x$ 26. $y > 9$ 27. $4 < m$
28. $2m > 17$ 29. $2x - 1 > 50$

In 30–37, using the domain $\{2, 2\frac{1}{2}, 3, 3\frac{1}{2}, 4, 4\frac{1}{2}\}$, find the solution set.

30. $x + 2 = 4\frac{1}{2}$ 31. $2x = 7$ 32. $5 - r = \frac{1}{2}$ 33. $\frac{x}{2} = 2.25$
34. $y > 4$ 35. $m < 3$ 36. $2x > 8$ 37. $3a < 4.5$

In 38–41, using the domain {2.1, 2.2, 2.3, 2.4, 2.5}, find the solution set.

38. $x + .1 = 2.4$ 39. $3x - 4 = 2.3$
40. $\frac{y}{2} = 3.6$ 41. $2x + 3 < 6.5$

In 42–47, determine the elements of the set if the domain of the variable is the one indicated.

42. $\{n \mid n + 2 = 5\}$, $n \in \{0, 1, 2, 3, 4, 5\}$
43. $\{x \mid x - 4 = 6\}$, $x \in \{7, 8, 9, 10\}$
44. $\{y \mid y - 1 < 8\}$, $y \in \{7, 8, 9, 10\}$
45. $\{r \mid 2r - 1 > 4\}$, $r \in \{1, 2, 3, 4\}$
46. $\{d \mid 9 - d = 5\}$, $d \in \{0, 1, 2, 3, 4, 5\}$
47. $\{x \mid 3x + 1 < 12\}$, $x \in \{0, 1, 2, 3, 4, 5\}$

3-8 REVIEW EXERCISES

In 1–4, use mathematical symbols to translate the verbal phrases into algebraic language.

1. x divided by b 2. 4 less than r
3. q decreased by d 4. 3 more than twice g

5. Nancy and Gary scored a total of 30 points in a game. If Gary scored x points, represent Nancy's score in terms of x.

In 6–13, find the value of the expression when $a = 10$, $b = 8$, $c = 5$, and $d = 24$.

6. $ac - d$ **7.** $4c^2$ **8.** $3b + c$ **9.** $d - b + c$

10. $\frac{bc}{a}$ **11.** $2a^2 - 2a$ **12.** $(2a)^2 - 2a$ **13.** $a(b - c)$

14. If $p = 5$ and $q = 1$, find the value of $(p - q)^3$.

15. Write a formula to express the number of grams g in k kilograms.

16. If p represents the perimeter of an equilateral triangle, represent the length of one of its sides in terms of p.

17. If the base of a triangle is 8 cm and its height is 12 cm, find the number of square centimeters in the area of the triangle.

18. The lengths, in inches, of the sides of a triangle are represented by y, $y + 3$, and $2y - 2$. Find the perimeter of the triangle when $y = 7$.

19. The length and width, in meters, of a rectangle can be represented by $2x$ and $x - 5$. Find the area of the rectangle when $x = 12$.

In 20–23, using the domain $\{1, 3, 5, 7, 9\}$, find the solution set.

20. $2a < 8$ **21.** $3b > 21$ **22.** $\frac{x}{5} < 2$ **23.** $5 < 12 - y$

In 24–27, select the numeral preceding the expression that best completes the sentence or answers the question.

24. What is the total number of cents in n nickels and q quarters?

(1) $n + q$ (2) $n + 5q$ (3) $5n + 25q$ (4) $30nq$

25. If the perimeter of a square is $4y$, then its area is represented as

(1) $4y$ (2) y^2 (3) $4y^2$ (4) $16y^2$

26. In the term $2xy^3$, the base that is used three times as a factor is

(1) y (2) xy (3) $2xy$ (4) $2x$

27. One member of the solution set of $19 < y \leq 29$ is

(1) 9 (2) 19 (3) 29 (4) 39

28. A bag contains between \$5 and \$6 in pennies, dimes, and quarters. There are three times as many dimes as pennies and twice as many quarters as dimes. How many of each kind of coin are in the bag?

Chapter 4

Simple Equations and Problems

4-1 PREPARING TO SOLVE AN EQUATION

An ***equation*** is a sentence that states that two algebraic expressions are equal. For example, $x + 3 = 9$ is an equation in which $x + 3$ is called the ***left side***, or ***left member***, and 9 is called the ***right side***, or ***right member***.

An equation may be a true sentence such as $5 + 2 = 7$, a false sentence such as $6 - 3 = 4$, or an open sentence such as $x + 3 = 9$. When the variable in an open sentence can be replaced by a number that will make the equation true, that is called a ***root***, or a ***solution***, of the equation. For example, 6 is a root of $x + 3 = 9$.

The set consisting of all the solutions of an equation is called its ***solution set*** or ***truth set***. The solution set of an equation is a subset of the replacement set (the domain) of the variable. This subset consists of the elements of the replacement set that make the open sentence true. Therefore, if the replacement set of x is {numbers of arithmetic}, then the solution set of $x + 3 = 9$ is {6}. Observe that {6} is a subset of {numbers of arithmetic}.

To ***solve an equation*** means to find its solution set.

If every element of the domain satisfies an equation, the equation is called an ***identity***. Thus, $5 + x = x + 5$ is an identity when the domain of x is {numbers of arithmetic} because every element of the domain makes the sentence true.

If not every element of the domain satisfies an equation, the equation is called a ***conditional equation***, or simply an equation. Therefore, $x + 3 = 9$ is a conditional equation.

PROCEDURE. To verify, or check, whether a number is a root of an equation:

1. Replace the variable in the equation by the number.
2. Do the indicated operations to determine whether the resulting statement is true.

MODEL PROBLEM

Is 7 a root of the equation $5x - 10 = 25$?

How to Proceed	*Solution*
(1) Write the equation.	$5x - 10 = 25$
(2) Replace the variable x by 7.	$5(7) - 10 \stackrel{?}{=} 25$
(3) Do the multiplication.	$35 - 10 \stackrel{?}{=} 25$
(4) Do the subtraction.	$25 = 25$ (True)

Answer: Yes

Postulates of Equality

In mathematics, any statement that we accept as being true without proof is called an ***assumption***, a ***postulate***, or an ***axiom***.

At this point, we will state several postulates of equality. These postulates, and others that are stated in later sections, will be used to solve equations in a systematic manner.

● **POSTULATE 1. Reflexive Property of Equality**

The *reflexive property of equality* states that for every number a:

$$a = a$$

● **POSTULATE 2. Symmetric Property of Equality**

The *symmetric property of equality* states that for all numbers a and b:

If $a = b$, then $b = a$.

● **POSTULATE 3. Transitive Property of Equality**

The *transitive property of equality* states that for all numbers a, b, and c:

If $a = b$, and $b = c$, then $a = c$.

Inverse Operations

When solving equations, you will need to be aware of the combined effect of pairs of related operations.

Subtracting a number undoes the effect of having added that number.

$$8 + 2 - 2 = 8 \qquad x + a - a = x$$

Adding a number undoes the effect of having subtracted that number.

$$8 - 2 + 2 = 8 \qquad x - a + a = x$$

Addition and subtraction are called inverse operations.

Dividing by a number undoes the effect of having multiplied by that number.

$$\frac{8 \cdot 2}{2} = 8 \qquad \frac{x \cdot a}{a} = x \text{ where } a \neq 0$$

Multiplying by a number undoes the effect of having divided by that number.

$$\frac{8}{2} \cdot 2 = 8 \qquad \frac{x}{a} \cdot a = x \text{ where } a \neq 0$$

Multiplication and division are called inverse operations, where the numbers are not equal to 0.

Whenever a variable and a number of arithmetic are related by addition, subtraction, multiplication, or division, we can use the inverse operation to obtain the variable itself.

MODEL PROBLEM

State the operation that is the inverse of the operation used in the algebraic expression.

	Answers
a. $5x$	The inverse of multiplication is division.
b. $\frac{d}{6}$	The inverse of division is multiplication.
c. $t + 4$	The inverse of addition is subtraction.
d. $y - 8$	The inverse of subtraction is addition.

EXERCISES

In 1–9, tell whether the number in the parentheses is a root of the given equation.

1. $5x = 50$ (10)
2. $\frac{1}{2}x = 18$ (36)
3. $\frac{1}{3}y = 12$ (4)
4. $x + 5 = 11$ (6)
5. $y + 8 = 14$ (22)
6. $x - 5 = 13$ (8)
7. $m - 4\frac{1}{2} = 9$ $\left(4\frac{1}{2}\right)$
8. $2x + 7 = 21$ (14)
9. $19 = 4x - 1$ (.5)

In 10–18, using the domain {1, 2, 3, 4, 5, 6, 7, 8, 9, 10}, find the solution set of the equation. If the equation has no roots, indicate the solution set as the null set, $\varnothing$.

10. $x + 5 = 7$
11. $y - 3 = 4$
12. $2x + 1 = 9$
13. $6 = 3x - 1.8$
14. $\frac{1}{2}x + 4 = 50$
15. $14 - 2x = 2$
16. $\frac{x + 8}{4} = 3$
17. $3x + .2 = 2.8$
18. $\frac{5x}{4} - 1 = 49$

In 19–24, using the domain {5, 6, 7, 8, 9, 10}, tell whether the equation is a conditional equation or an identity.

19. $x + 3 = 3 + x$
20. $x + 3 = 10$
21. $y + 3 + 4 = 7 + y$
22. $5a = a \cdot 5$
23. $5a = 40$
24. $5 \cdot 2 \cdot a = a \cdot 2 \cdot 5$

In 25–30, name the property of equality that the sentence illustrates.

25. $5 + 2 = 5 + 2$
26. If $6 + 2 = 5 + 3$, and $5 + 3 = 7 + 1$, then $6 + 2 = 7 + 1$.
27. If $4 + 3 = 6 + 1$, then $6 + 1 = 4 + 3$.
28. $x + y = x + y$
29. If $x = y$, then $y = x$.
30. If $m + n = r + s$, and $r + s = x + y$, then $m + n = x + y$.

In 31–50, state the operation that is the inverse of the operation used in the algebraic expression.

31. $8y$
32. $\frac{m}{4}$
33. $d + 3$
34. $x - 6$
35. $\frac{m}{5}$
36. $\frac{1}{5}w$
37. $\frac{1}{3}y$
38. $c + 7$
39. $t - 1$
40. $d + 14$
41. $1\frac{1}{4}m$
42. $\frac{m}{10}$
43. $d + 1\frac{1}{2}$
44. $r - \frac{2}{3}$
45. $c + 2\frac{1}{2}$
46. $.5c$
47. $\frac{r}{100}$
48. $n + .7$
49. $s - .8$
50. $x - 1.5$

4-2 SOLVING SIMPLE EQUATIONS BY USING ADDITION OR SUBTRACTION POSTULATES

● POSTULATE 4. Addition Property of Equality

The *addition property of equality* states that for all numbers a, b, and c:

$$\textbf{If } a = b, \textbf{ then } a + c = b + c.$$

Therefore, we can say: **If the same number is added to both sides of an equality, the equality is retained.**

Study the following examples:

In Arithmetic	*In Algebra*
If $8 = 8$	If $x - 2 = 8$
Then, $8 + 2 = 8 + 2$	Then, $x - 2 + 2 = 8 + 2$ [A_2]
And, $10 = 10$	And, $x = 10$
	A_2 means that 2 has been added to both sides of the equation.

Recall that the substitution principle allows us to replace a quantity by its equal. Using this principle, we can replace x with 10 in the preceding algebraic example.

$$\begin{aligned} x - 2 &= 8 \\ 10 - 2 &\stackrel{?}{=} 8 \\ 8 &= 8 \quad \text{(True)} \end{aligned}$$

Since replacing x with 10 results in a true sentence, we have checked the work done and we have established that 10 is a root of the equation.

Equations that have the same solution set are called ***equivalent equations***. Notice that, previously, when we applied the addition property to the equation $x - 2 = 8$, we obtained the two equivalent equations $x - 2 + 2 = 8 + 2$ and $x = 10$.

When we solve an equation, we transform it into a simpler equivalent equation in which we can easily see the value that can replace the variable and make the resulting sentence true.

In this chapter, when the domain of the variable is not stated, we will assume that it is the set of the numbers of arithmetic.

Now, we will use the addition property of equality in solving equations.

MODEL PROBLEMS

1. Solve and check: $x - 5 = 4$

Solution

$$x - 5 = 4$$
$$x - 5 + 5 = 4 + 5 \quad [A_5]$$
$$x = 9$$

Check

$$x - 5 = 4$$
$$9 - 5 \stackrel{?}{=} 4$$
$$4 = 4 \quad \text{(True)}$$

Answer: $x = 9$

2. Solve and check: $5 = y - 3\frac{1}{2}$

Solution

$$5 = y - 3\frac{1}{2}$$
$$5 + 3\frac{1}{2} = y - 3\frac{1}{2} + 3\frac{1}{2} \quad [A_{3\frac{1}{2}}]$$
$$8\frac{1}{2} = y$$

Check

$$5 = y - 3\frac{1}{2}$$
$$5 \stackrel{?}{=} 8\frac{1}{2} - 3\frac{1}{2}$$
$$5 = 5 \quad \text{(True)}$$

Answer: $y = 8\frac{1}{2}$

● POSTULATE 5. Subtraction Property of Equality

The *subtraction property of equality* states that for all numbers a, b, and c:

If $a = b$, then $a - c = b - c$.

Therefore: If the same number is subtracted from both sides of an equality, the equality is retained.

Study the following examples:

In Arithmetic

If $8 = 8$
Then, $8 - 3 = 8 - 3$
And, $5 = 5$

In Algebra

If $x + 3 = 8$
Then, $x + 3 - 3 = 8 - 3 \quad [S_3]$
And, $x = 5$

S_3 means that 3 has been subtracted from both sides of the equation.

Notice that the application of the subtraction property in the equation $x + 3 = 8$ resulted in the equivalent equations $x + 3 - 3 = 8 - 3$ and $x = 5$. All three equations have the same solution set, $\{5\}$.

Now, we will use the subtraction property of equality in solving equations.

MODEL PROBLEMS

1. Solve and check: $n + 7 = 9$

Solution

$$n + 7 = 9$$
$$n + 7 - 7 = 9 - 7 \quad [S_7]$$
$$n = 2$$

Check

$$n + 7 = 9$$
$$2 + 7 \stackrel{?}{=} 9$$
$$9 = 9 \quad \text{(True)}$$

Answer: $n = 2$

2. Solve and check: $.8 = .3 + t$

Solution

$$.8 = .3 + t$$
$$.8 = t + .3 \quad \text{(Commutative property)}$$
$$.8 - .3 = t + .3 - .3 \quad [S_{.3}]$$
$$.5 = t$$

Check

$$.8 = .3 + t$$
$$.8 \stackrel{?}{=} .3 + .5$$
$$.8 = .8 \quad \text{(True)}$$

Answer: $t = .5$

PROCEDURE. To solve an equation in which a constant is added to or subtracted from a variable:

1. Use the inverse operation, subtracting or adding the constant on both sides of the equation. Obtain an equivalent equation in which only the variable itself is one member of the equation.
2. Check by substitution that the value obtained for the variable makes the given equation true.

EXERCISES

In 1–60, solve and check the equation.

1. $x - 1 = 7$ **2.** $m - 3 = 0$ **3.** $x - 13 = 25$

4. $y - 64 = 77$ **5.** $9 = x - 3$ **6.** $15 = w - 15$

7. $17 = x - 21$ **8.** $89 = b - 73$ **9.** $x - .3 = .4$

10. $r - .07 = .32$ **11.** $.9 = c - .3$ **12.** $7 = d - .7$

13. $y - \frac{1}{2} = \frac{1}{2}$ **14.** $t - 1\frac{1}{3} = 4$ **15.** $4\frac{2}{3} = m - \frac{1}{3}$

16. $12\frac{1}{2} = y - 7\frac{3}{4}$
17. $y + 1 = 8$
18. $t + 8 = 8$
19. $y + 18 = 29$
20. $d + 53 = 81$
21. $9 = x + 3$
22. $17 = t + 17$
23. $39 = e + 21$
24. $98 = f + 39$
25. $x + .7 = .8$
26. $s + .02 = .08$
27. $.9 = m + .3$
28. $12.9 = k + .6$
29. $x + \frac{3}{8} = \frac{7}{8}$
30. $r + \frac{1}{2} = \frac{3}{4}$
31. $3\frac{2}{3} = n + \frac{1}{6}$
32. $18\frac{3}{10} = k + 5\frac{7}{10}$
33. $a - 4 = 9$
34. $a + 5 = 17$
35. $b + 4 = 13$
36. $d + 1 = 12$
37. $18 = a - 3$
38. $16 = r + 2$
39. $25 = s + 11$
40. $54 = t + 39$
41. $y - 13 = 14$
42. $m + 15 = 15$
43. $n - 7 = 3\frac{1}{3}$
44. $y + 5 = 8\frac{1}{4}$
45. $x + \frac{1}{2} = 14\frac{1}{2}$
46. $3\frac{2}{3} = m - \frac{1}{3}$
47. $b - 2\frac{3}{4} = 9$
48. $\frac{7}{8} = y + \frac{3}{4}$
49. $c - 1\frac{1}{4} = 6\frac{1}{2}$
50. $9\frac{1}{4} = d + 3\frac{1}{2}$
51. $d - 5 = 2.3$
52. $m + .7 = 2.9$
53. $8.6 = c - .2$
54. $12 = p + 1.8$
55. $15 = x + 1.5$
56. $3.1 = z - .8$
57. $4 + x = 50$
58. $19 = 7 + y$
59. $1 + x = 7\frac{1}{2}$
60. $.15 + y = 2.25$
61. If $g + 9 = 11$, find the value of $7g$.
62. If $t - .5 = 2.5$, find the value of $t + 7$.
63. If $22 = y + 8$, find the value of $\frac{1}{2}y$.
64. If $c - 1\frac{1}{4} = 2\frac{1}{2}$, find the value of $8c - 2$.
65. If $1.8 + b = 2.7$, find the value of $\frac{1}{3}b - .3$.

4-3 SOLVING SIMPLE EQUATIONS BY USING DIVISION OR MULTIPLICATION POSTULATES

● POSTULATE 6. Division Property of Equality

The *division property of equality* states that for all numbers a, b, and c ($c \neq 0$):

$$\textbf{If } a = b, \textbf{ then } \frac{a}{c} = \frac{b}{c}.$$

Therefore: **If both sides of an equality are divided by the same nonzero number, the equality is retained.**

Study the following examples:

In Arithmetic	*In Algebra*
If $8 = 8$	If $4w = 8$
Then, $\frac{8}{4} = \frac{8}{4}$	Then, $\frac{4w}{4} = \frac{8}{4}$ $[D_4]$
And, $2 = 2$	And, $w = 2$
	D_4 means that both sides of the equation have been divided by 4.

Notice that the application of the division property in the equation $4w = 8$ resulted in the equivalent equations $\frac{4w}{4} = \frac{8}{4}$ and $w = 2$. All three equations have the same solution set, $\{2\}$.

It is important to remember that when we apply the division property of equality, we divide both sides of an equation by the same *nonzero* number. Division by zero is impossible.

Now, we will use the division property of equality in solving equations.

MODEL PROBLEMS

Solve and check:

1. $8y = 56$

Solution

$$8y = 56$$

$$\frac{8y}{8} = \frac{56}{8} \quad [D_8]$$

$$y = 7$$

Check

$$8y = 56$$

$$8(7) \stackrel{?}{=} 56$$

$$56 = 56 \quad \text{(True)}$$

Answer: $y = 7$

2. $22 = 4x$

Solution

$$22 = 4x$$

$$\frac{22}{4} = \frac{4x}{4} \quad [D_4]$$

$$5\tfrac{1}{2} = x$$

Check

$$22 = 4x$$

$$22 \stackrel{?}{=} 4\left(5\tfrac{1}{2}\right)$$

$$22 = 22 \quad \text{(True)}$$

Answer: $x = 5\frac{1}{2}$

3. $.3x = 9$

Solution

$$.3x = 9$$

$$\frac{.3x}{.3} = \frac{9}{.3} \quad [D_{.3}]$$

$$x = 30$$

Check

$$.3x = 9$$

$$.3(30) \stackrel{?}{=} 9$$

$$9 = 9 \quad \text{(True)}$$

Answer: $x = 30$

● POSTULATE 7. Multiplication Property of Equality

The *multiplication property of equality* states that for all numbers a, b, and c:

$$\textbf{If } a = b\textbf{, then } ac = bc.$$

Therefore: **If both sides of an equality are multiplied by the same number, the equality is retained.**

Study the following examples:

In Arithmetic	*In Algebra*
If $8 = 8$	If $\frac{x}{4} = 8$
Then, $4 \cdot 8 = 4 \cdot 8$	Then, $4 \cdot \frac{x}{4} = 4 \cdot 8$ $[M_4]$
And, $32 = 32$	And, $x = 32$
	M_4 means that both sides of the equation have been multiplied by 4.

Notice that the application of the multiplication property in the equation $\frac{x}{4} = 8$ resulted in the equivalent equations $4 \cdot \frac{x}{4} = 4 \cdot 8$ and $x = 32$. All three equations have the same solution set, $\{32\}$.

One might now expect that the application of the multiplication property of equality to an equation always gives an equivalent equation. This is true, *with one important exception.* We may not multiply both sides of an equation by zero to obtain an equivalent equation. For example:

$x = 2$	$0 \cdot x = 0 \cdot 2$ $0 \cdot x = 0$
The solution set is $\{2\}$.	The solution set is {numbers of arithmetic}.

Now, we will use the multiplication property of equality in solving equations.

MODEL PROBLEMS

In 1–3, solve and check.

1. $\frac{n}{3} = 12$

Solution

$$\frac{n}{3} = 12$$

$$3 \cdot \frac{n}{3} = 3 \cdot 12 \quad [M_3]$$

$$n = 36$$

Check

$$\frac{n}{3} = 12$$

$$\frac{36}{3} \stackrel{?}{=} 12$$

$$12 = 12 \quad \text{(True)}$$

Answer: $n = 36$

2. $8 = \frac{1}{2}x$

Solution

$$8 = \frac{x}{2}$$

$$2 \cdot 8 = 2 \cdot \frac{x}{2} \quad [M_2]$$

$$16 = x$$

Check

$$8 = \frac{1}{2}x$$

$$8 \stackrel{?}{=} \frac{1}{2}(16)$$

$$8 = 8 \quad \text{(True)}$$

Answer: $x = 16$

3. $\frac{x}{9} = \frac{4}{3}$

Solution

$$\frac{x}{9} = \frac{4}{3}$$

$$9 \cdot \frac{x}{9} = 9 \cdot \frac{4}{3} \quad [M_9]$$

$$x = 12$$

Check

$$\frac{x}{9} = \frac{4}{3}$$

$$\frac{12}{9} \stackrel{?}{=} \frac{4}{3}$$

$$\frac{4}{3} = \frac{4}{3} \quad \text{(True)}$$

Answer: $x = 12$

PROCEDURE. To solve an equation in which a constant multiplies or divides a variable:

1. Use the inverse operation, multiplying or dividing both sides of the equation by the constant. Obtain an equivalent equation in which only the variable itself is one member of the equation.
2. Check by substitution that the value obtained for the variable makes the given equation true.

EXERCISES

In 1–40, solve and check the equation.

1. $3x = 15$ **2.** $10c = 90$ **3.** $20 = 5d$ **4.** $84 = 7y$
5. $5c = 5$ **6.** $6s = 0$ **7.** $5p = 9$ **8.** $16 = 4y$
9. $8m = 1$ **10.** $1 = 6b$ **11.** $4x = .8$ **12.** $.36 = 6m$

13. $\frac{1}{2}d = 18$ **14.** $20 = \frac{1}{10}r$ **15.** $\frac{2}{3}x = 18$ **16.** $\frac{a}{2} = 3$

17. $3y = 3$ **18.** $2b = 0$ **19.** $81 = 27q$ **20.** $15 = 2y$

21. $6c = 44$ **22.** $6w = 3$ **23.** $\frac{1}{10}x = 0$ **24.** $2a = .6$

25. $.4x = 3.2$ **26.** $1.4x = 5.6$ **27.** $.3c = 1.2$ **28.** $.02x = 25$

29. $.06x = 54$ **30.** $32 = .04z$ **31.** $.06y = 12$ **32.** $.15c = 300$

33. $\frac{1}{8}x = \frac{1}{4}$ **34.** $\frac{1}{3} = \frac{m}{4}$ **35.** $\frac{2}{3}b = 8$ **36.** $3\frac{1}{2}x = 7$

37. $7x = 12\frac{1}{4}$ **38.** $\frac{1}{3}y = \frac{5}{9}$ **39.** $\frac{x}{.5} = 4$ **40.** $\frac{t}{1.4} = 1$

41. If $9x = 36$, find the value of $2x$.

42. If $2x = 64$, find the value of $\frac{1}{4}x$.

43. If $\frac{t}{2} = 12$, find the value of $5t$.

44. If $\frac{2}{3}y = 16$, find the value of $3y + 7$.

45. If $.08y = .96$, find the value of $\frac{1}{2}y - 3$.

4-4 WRITING VERBAL SENTENCES AS EQUATIONS

In algebra, many verbal problems involving number relations are solved by expressing verbal sentences as equations and then solving the equations. At the start, this procedure may seem more difficult than solving the problem by other strategies, such as trial and error. With practice, you will find equations easier to write. Most importantly, as the problems become more complex, you will find that using an algebraic strategy is powerful and efficient.

Verbal Sentence: Four times a number s equals 20.

Equation: $4s = 20$

Verbal Sentence: A number y increased by 6 equals 8.

Equation: $y + 6 = 8$

Verbal Sentence: A number x decreased by 3 equals 5.

Equation: $x - 3 = 5$

Verbal Sentence: A number n divided by 2 equals 4.

Equation: $n \div 2 = 4$

Note that the words of a sentence may need to be replaced by symbols in a different order.

Verbal Sentence: Eight less than a number is 10.
Equation: $x - 8 = 10$

PROCEDURE. To write a verbal sentence as an equation, choose a letter to represent the variable. Then use this letter, with the symbols for arithmetic operations, to express the verbal sentence as an equation.

MODEL PROBLEM

Write the following sentence as an equation: "5 times a number decreased by 7 equals 13."

Solution: Let x represent the number.

5 times a number decreased by 7 equals 13.

$5x - 7 = 13$

Answer: $5x - 7 = 13$

EXERCISES

In 1–9, select the equation that represents, in terms of the given variable, the numerical relationship expressed in the sentence.

1. Three times Harold's height is 108 inches. Let h = Harold's height.
(1) $h + 3 = 108$ (2) $3h = 108$
(3) $h - 3 = 108$ (4) $\frac{1}{3}h = 108$

2. One-half of Mary's weight is 20 kilograms. Let w = Mary's weight.
(1) $\frac{1}{2}w = 20$ (2) $2w = 20$ (3) $w - 2 = 20$ (4) $w + 2 = 20$

3. A number increased by 7 equals 28. Let n = the number.
(1) $7n = 28$ (2) $n + 7 = 28$ (3) $n - 7 = 28$ (4) $\frac{1}{7}n = 28$

4. A number decreased by 5 equals 15. Let x = the number.

(1) $x + 5 = 15$ (2) $5x = 15$ (3) $\frac{1}{5}x = 15$ (4) $x - 5 = 15$

5. If 7 is subtracted from a number, the result is 8. Let x = the number.

(1) $7 - x = 8$ (2) $x - 7 = 8$ (3) $8 - x = 7$ (4) $x + 8 = 7$

6. A movie star bought 15 suits and now has 75 suits. Let s = the number of suits he had originally.

(1) $s + 15 = 75$ (2) $s - 15 = 75$ (3) $15s = 75$ (4) $\frac{s}{15} = 75$

7. In a rectangle, four times the width is 100 cm. Let w = width of the rectangle.

(1) $w + 4 = 100$ (2) $4w = 100$

(3) $w - 4 = 100$ (4) $\frac{1}{4}w = 100$

8. In a parallelogram whose area measures 24 cm^2, the base measures 6 cm. Let h = height of the parallelogram.

(1) $h + 6 = 24$ (2) $6h = 24$ (3) $h \div 6 = 24$ (4) $24h = 6$

9. In a triangle whose area measures 32 m^2, the base measures 4 m. Let h = height of the triangle.

(1) $4h = 32$ (2) $2(4h) = 32$

(3) $\frac{1}{2}(h + 4) = 32$ (4) $\frac{1}{2}(4h) = 32$

In 10–26, write the sentence as an equation. Use n to represent the number.

10. Eight more than a number is 15.
11. Four less than a number is 24.
12. Twelve added to a number is 26.
13. A number decreased by 5 equals 25.
14. A number multiplied by 3 equals 39.
15. A number divided by 4 equals 16.
16. The product of 7 and a number equals 70.
17. One-half of a number, decreased by 7, equals 11.
18. Twice a number, increased by 7, equals 27.
19. Twice a number, decreased by 5, equals 25.
20. The sum of three times a number and 7 is 22.
21. When 9 is subtracted from 5 times a number, the result is 31.
22. The sum of 100 and a number is equal to three times that number.
23. If 3 times a number is increased by 12, the result is the same as twice the number increased by 24.

24. If 8 times a number is decreased by 20, the result is the same as 3 times the number increased by 80.
25. The sum of a number and twice that number equals 45.
26. Three times a number decreased by half of that number equals 40.

4-5 SOLVING PROBLEMS BY USING VARIABLES AND EQUATIONS

Now you are ready to use an important algebraic problem-solving strategy.

PROCEDURE. To solve a verbal problem by using an equation involving one variable:

1. Read the problem carefully.
2. Select a variable that can be used to represent the number to be found.
3. Write an equation using a relationship given in the problem, a previously known relationship, or a formula.
4. Solve the equation.
5. Check the answer by testing it in the given problem.

MODEL PROBLEMS

1. When a number is decreased by 7, the result is 9. Find the number.

How to Proceed	*Solution*
(1) Read carefully and determine what you must find.	You want to find a certain number.
(2) Represent the number by a variable.	Let x = the number.
(3) Write the word statement as an equation.	$x - 7 = 9$
(4) Solve the equation. $[A_7]$	$\underline{\quad + 7 = +7}$ $x = 16$
(5) Check in the original problem.	Is 16 decreased by 7 equal to 9? Yes.

Answer: The number is 16.

2. Ned traveled $\frac{1}{4}$ of the distance that Ben traveled. If Ned traveled 12 kilometers, how far did Ben travel?

How to Proceed	*Solution*
(1) What must you find?	The distance Ben traveled.
(2) Represent the distance Ben traveled by a variable.	Let d = distance Ben traveled.
(3) Using the same variable, represent the distance Ned traveled.	Then $\frac{1}{4}d$ = distance Ned traveled.
(4) Write the word statement as an equation.	$\frac{1}{4}d = 12$
(5) Solve the equation. $[M_4]$	$4 \cdot \frac{1}{4}d = 12 \cdot 4$ $d = 48$
(6) Check in the original problem.	Ben traveled 48 km, Ned traveled 12 km. Did Ned travel $\frac{1}{4}$ of the distance that Ben traveled? Yes.

Answer: Ben traveled 48 kilometers.

EXERCISES

In 1–32, solve the problem using a variable and an equation.

1. A number decreased by 20 equals 36. Find the number.
2. Seven times a number is 63. Find the number.
3. If 7 is subtracted from a number, the result is 46. Find the number.
4. Five times a number is 50. Find the number.
5. What number increased by 25 equals 40?
6. When a number is doubled, the result is 36. Find the number.
7. If 18 is added to a number, the result is 32. Find the number.
8. A number divided by 5 equals 17. Find the number.
9. Ten less than a number is 42. Find the number.
10. A number divided by 4 is $3\frac{1}{2}$. Find the number.
11. One-half of a number is 12. Find the number.
12. Three-fifths of a number is 30. Find the number.
13. The sum of 42 and a number is 96. Find the number.

14. A number multiplied by .3 is 6. Find the number.
15. Four-hundredths of a number is 16. Find the number.
16. After Helen had spent $.25, she had $.85 left. How much money did she have originally?
17. After he had lost 13 kg, Ben weighed 90 kg. Find Ben's original weight.
18. After $2\frac{1}{2}$ feet had been cut from a piece of lumber, there were $9\frac{1}{2}$ feet left. What was the original length of the piece of lumber?
19. After a car had increased its rate of speed by 24 kilometers per hour, it was traveling 76 kilometers per hour. What was its original rate of speed?
20. During a charity drive, the boys in a class contributed $3.75 more than the girls. If the boys contributed $8.25, how much did the girls contribute?
21. The width of a rectangle is 8 feet less than its length. If the width is 9.5 feet, find the length of the rectangle.
22. A high school admitted 1,125 sophomores, which was 78 fewer than the number admitted last year. How many sophomores were admitted last year?
23. How many hours do you have to work to earn $132 if you are paid $5.50 per hour?
24. A man earned $1,000 in 4 weeks. What was his weekly salary?
25. A dealer sold an electric broiler for $39.98. This amount was $12.50 more than the broiler had cost him. How much did the broiler cost the dealer?
26. After using his baseball glove for some time, Charles sold it for $12.25 less than he paid for it. If Charles sold the glove for $3.50, how much did he pay for it originally?
27. Baseballs cost $3.50 each. How many baseballs can be bought for $28.00?
28. The width of a rectangle is $\frac{1}{5}$ of its length. If the width of the rectangle is 6 meters, what is its length?
29. Sue wishes to buy a radio that costs $38. If she has already saved $26 for this purpose, how much must she still save to buy the radio?
30. A merchant bought 8 dozen shirts. If he has sold all but 18 of them, how many shirts has he sold?
31. Mr. Alvarez withdrew from his savings account $25 per week for each of 8 weeks. He then had $1,623 in his account. How much did he have in his account before he made these withdrawals?
32. In a basketball game, the Knicks scored 12 points more than the Celtics. If the Knicks scored 108 points, how many points did the Celtics score?

4-6 SOLVING PERCENT PROBLEMS

A ***percent*** is a way of expressing the quotient of a number divided by 100. When used in computation, a percent is always expressed as an equivalent decimal or common fraction.

In Arithmetic

1. To find 12% of 420, write 12% as .12:

$$12\% \text{ of } 420 = .12(420) = 50.40$$

2. To find 7% of $300, write 7% as $\frac{7}{100}$:

$$7\% \text{ of } \$300 = \frac{7}{100}(300) = \$21$$

In Algebra

1. To express 12% of a number n, write 12% as .12:

$$12\% \text{ of a number } n = .12n$$

2. To express 7% of an amount a, write 7% as $\frac{7}{100}$:

$$7\% \text{ of an amount } a = \frac{7}{100}a$$

MODEL PROBLEM

The Hesters save 8% of the family income. What was the income last week if they saved $34?

How to Proceed	*Solution*
(1) Represent last week's income by a variable.	Let n = last week's income.
(2) Using the same variable, represent last week's savings.	Then, $.08n$ = last week's savings.
(3) Write an equation.	$.08n = 34$
(4) Solve the equation. $[D_{.08}]$	$n = 425$
(5) Check.	Is 8% of 425 equal to 34? Yes, $.08(425) = 34$.

Answer: The Hester family's income was $425.

EXERCISES

In 1–18, use an algebraic equation to solve each problem.

1. 4% of a number is 8. Find the number.
2. 25% of a number is 12. Find the number.
3. 120% of a number is 54. Find the number.

4. 15% of what number is 4.5?
5. $33\frac{1}{3}\%$ of what number is $3\frac{1}{3}$?
6. Marvin saved 25% of his allowance. If he saved $2.50, how much was his allowance?
7. The Giants won 24 games. If that was 60% of all the games they played, how many games did they play?
8. Pearl bought a coat at a sale that offered 40% off. If she saved $48, what was the original price of the coat?
9. Patricia answered 90% of the questions on a test correctly. If she answered 45 questions correctly, how many questions were on the test?
10. At a sale, a radio sold for $20. This amount was 80% of the original price. What was the original price?
11. In a restaurant, Mrs. Green left a tip of $1.20, which was 15% of the amount of the bill. What was the amount of the bill?
12. There are 210 freshmen at East High School. The number of freshmen is 30% of the number of students enrolled in the school. How many students are enrolled in the school?
13. One day 92 students, which was 8% of the total enrollment, were absent. Find the total enrollment.
14. Mr. Hendrick's salary this year is 106% of last year's salary. If this year's salary is $26,341, what was last year's salary?
15. A dealer sold a suit for 150% of the amount he paid for it. If the dealer sold the suit for $120, how much did it cost him?
16. The selling price of an article is 175% of the dealer's cost price. If the dealer sold the article for $28, how much did she pay for it?
17. When Irene bought a dress, she had to pay an 8% sales tax. If the sales tax was $2.40, what was the original price of the dress?
18. The Surevalve Company tested 5% of the valves it had produced. If 350 valves were tested, how many valves had the company produced?

4-7 SOLVING EQUATIONS BY USING SEVERAL OPERATIONS

In the equation $2x + 3 = 15$, there are two operations indicated in the left member: *multiplication and addition.* To solve the equation, we use the inverse operations: *division and subtraction.* As shown on the next page, in Method 1, we first perform subtraction to undo the addition and then perform division to undo the multiplication. In Method 2, we first perform division to undo the multiplication and then perform subtraction to undo the addition. Notice that solution by either method results in the same root.

Method 1

$$2x + 3 = 15$$
$$2x + 3 - 3 = 15 - 3 \quad [S_3]$$
$$2x = 12$$
$$x = 6 \quad [D_2]$$

Check

$$2x + 3 = 15$$
$$2(6) + 3 \stackrel{?}{=} 15$$
$$12 + 3 \stackrel{?}{=} 15$$
$$15 = 15 \quad \text{(True)}$$

Method 2

$$2x + 3 = 15$$
$$\frac{2x + 3}{2} = \frac{15}{2} \quad [D_2]$$
$$\frac{2x}{2} + \frac{3}{2} = \frac{15}{2} \quad \left(\frac{2x}{2} = x\right)$$
$$x = \frac{12}{2} \quad \left[S_{\frac{3}{2}}\right]$$
$$x = 6$$

Check is shown at the left.

Answer: $x = 6$

While both methods yield the correct answer, Method 1 is simpler. In general, it is preferable to perform addition or subtraction first and then to perform multiplication or division.

PROCEDURE. To solve an equation in which several operations are indicated, perform their inverse operations.

MODEL PROBLEMS

1. Solve and check: $7x + 15 = 71$

How to Proceed	*Solution*
(1) Write the equation.	$7x + 15 = 71$
(2) Use the subtraction property.	$-15 = -15 \quad [S_{15}]$
	$7x = 56$
(3) Use the division property.	$\frac{7x}{7} = \frac{56}{7} \quad [D_7]$
	$x = 8$
(4) Check the solution.	$7x + 15 = 71$
	$7(8) + 15 \stackrel{?}{=} 71$
	$56 + 15 \stackrel{?}{=} 71$
	$71 = 71$ (True)

Answer: $x = 8$

2. Find the solution set and check: $\frac{3}{5}x - 6 = 12$

Solution

$$\frac{3x}{5} - 6 = 12$$
$$+6 = +6$$
$$\frac{3x}{5} = 18$$
$$5\left(\frac{3x}{5}\right) = 5(18)$$
$$3x = 90$$
$$x = 30$$

Check

$$\frac{3}{5}x - 6 = 12$$
$$\frac{3(30)}{5} - 6 \stackrel{?}{=} 12$$
$$\frac{90}{5} - 6 \stackrel{?}{=} 12$$
$$18 - 6 \stackrel{?}{=} 12$$
$$12 = 12 \quad \text{(True)}$$

Answer: The solution set is $\{30\}$.

3. If 4 times a number is increased by 7, the result is 43. Find the number.

How to Proceed	*Solution*
(1) Represent the number by a letter.	Let x = the number.
(2) Write the word statement as an equation.	$4x + 7 = 43$
(3) Solve the equation.	$4x = 36$ $x = 9$

Check: Does 4 times 9, increased by 7, give a result of 43?
Yes, $36 + 7 = 43$.

Answer: The number is 9.

EXERCISES

In 1–33, solve and check the equation.

1. $3x + 5 = 35$ **2.** $5a + 17 = 47$ **3.** $4x - 1 = 15$

4. $3y - 5 = 16$ **5.** $55 = 6a + 7$ **6.** $17 = 8c - 4$

7. $15x + 14 = 19$ **8.** $75 = 11 + 16x$ **9.** $14 = 12b + 8$

10. $8 = 18c - 1$ **11.** $11 = 15t + 1$ **12.** $11 = 16d - 1$

13. $\frac{3a}{8} = 12$ **14.** $\frac{4c}{9} = 20$ **15.** $42 = \frac{7d}{8}$

16. $\frac{2}{3}x = 18$

17. $12 = \frac{3}{4}y$

18. $\frac{3}{5}m = 30$

19. $\frac{5t}{4} = \frac{45}{2}$

20. $\frac{7t}{3} = \frac{14}{3}$

21. $1.2 = \frac{4m}{5}$

22. $\frac{x}{3} + 4 = 13$

23. $\frac{a}{4} - 9 = 51$

24. $15 = \frac{b}{7} - 8$

25. $12 = \frac{y}{5} + 3$

26. $.6m + \frac{1}{3} = 18\frac{1}{3}$

27. $9d - \frac{1}{2} = 17\frac{1}{2}$

28. $4a + .2 = 5$

29. $4 = 3t - .2$

30. $\frac{1}{4}x - 5 = 11$

31. $\frac{1}{9}x + 4 = 13$

32. $13 = \frac{1}{3}y - 5$

33. $47 = \frac{4}{5}t + 7$

In 34–50, use an algebraic equation to solve the problem.

34. Ten times a number, increased by 9, is 59. Find the number.
35. The sum of 8 times a number and 5 is 37. Find the number.
36. If six times a number is decreased by 4, the result is 68. Find the number.
37. The difference between 4 times a number and 3 is 25. Find the number.
38. If a number is multiplied by 7, and the product is increased by 2, the result is 100. Find the number.
39. When 12 is subtracted from 3 times a number, the result is 24. Find the number.
40. If 9 is added to 50% of a number, the result is 29. Find the number.
41. If two-thirds of a number is decreased by 4, the result is 56. Find the number.
42. If 38 is added to $\frac{5}{9}$ of a number, the result is 128. Find the number.
43. The sum of 60% of a number and 2.3 is 14.6. Find the number.
44. The larger of two numbers is 12 more than twice the smaller. The larger number is 36. What is the smaller number?
45. The larger of two numbers exceeds six times the smaller by 10. If the larger number is 76, find the smaller number.
46. Find the width of a rectangle if the length is 12 cm and the perimeter is 42 cm.
47. When Rosalie bought 4 packages of paper napkins, the tax was \$.22. If the total cost of the napkins and the tax was \$3.38, what was the cost of a package of napkins?
48. The bookstore ordered boxes of pencils for resale. The cost of the pencils was \$2.25 per box and the shipping cost of the order was \$2.80. If the total cost of the pencils and shipping was \$29.80, how many boxes of pencils were ordered?

49. Dr. Cortez spent 30 minutes driving from home to Seaview Hospital. This was 12 minutes less than twice the time it took him to drive home. How long did it take him to drive home?

50. Last week, Juan worked 7 hours more than 3 times the number of hours that George worked. If Juan worked 40 hours, how long did George work?

4-8 MORE PRACTICE IN SOLVING EQUATIONS

In 1–42, solve and check the equation.

1. $9x = 108$ **2.** $9b + 8 = 8$ **3.** $4y + 2 = 39$

4. $.15x = .06$ **5.** $79 = 5x - 6$ **6.** $5y + 19 = 27$

7. $18 = 4a + 10$ **8.** $14 = n - 5.6$ **9.** $\frac{5}{7}b = 35$

10. $39 = 5x$ **11.** $7x - 2 = 68$ **12.** $48 = 7m - 1$

13. $48 = 5.7b - 3.3$ **14.** $\frac{2x}{5} + 15 = 37$ **15.** $3x - \frac{1}{3} = 5\frac{2}{3}$

16. $87 = 2x - 13$ **17.** $.11a = 44$ **18.** $x - 3\frac{1}{8} = 7\frac{1}{4}$

19. $3y + 19 = 94$ **20.** $\frac{3t}{4} = 84$ **21.** $6x + 3 = 60$

22. $10r = 2$ **23.** $\frac{s}{4} - 5 = 7$ **24.** $.8c - 4 = 3.2$

25. $.01x = 5$ **26.** $.9b = .18$ **27.** $5n - 5 = 30$

28. $8m + 3 = 91$ **29.** $.3x + .2 = 8$ **30.** $\frac{n}{6} + 5\frac{2}{3} = 12\frac{1}{3}$

31. $\frac{4e}{5} - 32 = 28$ **32.** $29 = \frac{1}{3}a + 12$ **33.** $9x - 3 = 95$

34. $\frac{15}{4} = \frac{r}{8}$ **35.** $\frac{7}{10}x = 35$ **36.** $\frac{1}{2}t + \frac{1}{4}t = 12$

37. $\frac{x}{14} = \frac{5}{7}$ **38.** $x + \frac{1}{8} = \frac{7}{8}$ **39.** $\frac{1}{3}c - 1 = 14$

40. $3m - 6\frac{1}{2} = 8\frac{1}{2}$ **41.** $\frac{y}{8} = \frac{9}{2}$ **42.** $18 = 8c - 2$

4-9 REVIEW EXERCISES

In 1–3, using the domain {1, 3, 5, 7, 9}, find the solution set of the equation. The null set, $\varnothing$, is a possible answer.

1. $18 - 3x = 9$ **2.** $0.2y + 3 = 4$ **3.** $4t - 7 = 17$

4. Twelve less than a number is 18. Find the number.

5. If $6k = 48$, find the value of $k + 5$.

6. If $d + \frac{3}{4} = 2\frac{1}{2}$, what is the value of $12d$?

In 7 and 8, select the numeral preceding the correct response.

7. Which of the numbers is a root of the equation $0.2w = 4$?
(1) 0.2 (2) 2 (3) 20 (4) 4

8. Which statement is true if $x = a + b$ and $a + b = 2r$?
(1) $x = r$ (2) $x = 2r$ (3) $x = 4r$ (4) $a = 2r$

In 9–20, solve and check the equation.

9. $t - 9 = 14$ **10.** $23 + y = 31$ **11.** $6m = 3$

12. $w - 0.5 = 9$ **13.** $8 = \frac{1}{4}b$ **14.** $7x - 2 = 61$

15. $5b - 3 = 32$ **16.** $9x + 4 = 76$ **17.** $0.03p = 0.3$

18. $\frac{k}{10} = \frac{5}{2}$ **19.** $3t = \frac{3}{4}$ **20.** $8 + \frac{2}{3}c = 8$

In 21–24: **a.** Write the sentence as an equation where n is used to represent the number. **b.** Solve for n.

21. When twice a number is decreased by 5, the result is 27. Find the number.

22. If one-sixth of a number is $1\frac{1}{3}$, find the number.

23. Five times a number increased by 4 is 40. Find the number.

24. A store sold 5% of a new shipment of toys. If the store sold 16 of these toys, how many toys were in the shipment?

25. Use an algebraic equation to solve the problem. On a quiz show, Georgette won $10 less than three times as much as Charlie. If Georgette won $800, how much did Charlie win?

26. In his will, Mr. Banks left 50% of the value of his property to his wife, 22% to his daughter, 16% to his son, and the remaining $30,000 to charity. What was the value of his property?

Chapter 5

Introducing Logic

5-1 SENTENCES, STATEMENTS, AND TRUTH VALUES

Still another effective approach to problem solving is through a branch of mathematics called logic. ***Logic*** is the study of reasoning. All *reasoning*, whether it is in mathematics or in everyday living, is based on the ways in which we put sentences together.

Sentences That Are True or False

In mathematics, we are concerned with only one type of sentence. A ***mathematical sentence*** must state a fact or contain a complete idea. A mathematical sentence is like a simple declarative sentence in English. It contains a subject and a predicate. Because every mathematical sentence states a fact, we can usually judge such a sentence to be true or false. For example:

1. Every triangle has three sides.	True mathematical sentence
2. $5 + 7 = 12$	True mathematical sentence
3. Chicago is a city.	True mathematical sentence
4. $3 + 5 = 7$	False mathematical sentence
5. Chicago is the capital of Texas.	False mathematical sentence

Nonmathematical Sentences and Phrases

Sentences that ask questions or give commands are not mathematical sentences. We never use these sentences in reasoning because we cannot judge if they are true or false. A ***phrase*** is an expression that is only part of a sentence. Because a phrase does not contain a complete idea, we cannot judge if it is true or false. A phrase is not a mathematical sentence. The following examples are not mathematical sentences:

1. Did you do your homework? — This is not a mathematical sentence because it asks a question.
2. Get home early. — This is not a mathematical sentence because it gives a command.
3. Every triangle — This is not a mathematical sentence because it is a phrase.
4. $4 + 5$ — This is not a mathematical sentence because it is a phrase.

In reasoning, we try to judge whether sentences are true or false. Although some sentences contain complete thoughts, they are true for some people and false for other people. For example:

1. It's hot in this room.
2. That sound is too loud.
3. Liver really tastes good.

Conclusions based on statements like these may also be true for some people and false for others.

Open Sentences

Sometimes it is impossible to tell whether the sentence is true or false because more information is needed. These sentences contain ***variables*** or unknowns. You have seen that a variable may be a symbol such as n or x, but a variable may also be a pronoun like *he* or *it*. Any sentence that contains a variable is called an ***open sentence***. For example:

1. $x + 3 = 7$ — *Open* sentence; the variable is x.
2. She is my sister. — *Open* sentence; the variable is *she*.
3. It's on TV at 8 o'clock. — *Open* sentence; the variable is *it*.

In studying equations, you have seen that a variable is a placeholder for an element from a set of replacements called the ***domain*** or ***replacement set***. The set of all replacements that will change the open sentence into true sentences is called the ***solution set*** or ***truth set***. For example:

Open sentence: $x + 10 = 13$

Variable: x

Domain: $\{1, 2, 3, 4\}$

When $x = 3$, then $3 + 10 = 13$ is a true sentence.

Solution set: $\{3\}$

The concepts that we apply to sentences in algebra are exactly the same as those we apply to sentences that we speak to one another. Of course, we would not use a domain like {1, 2, 3, 4} for the open sentence "She is my sister." Common sense tells us to use a domain of girls' names. The following example compares the sentence with the use of sentences in algebra. Open sentences, variables, domains, and solution sets behave in exactly the same way.

Open sentence: She is my sister.

Variable: She

Domain: {Girls' names}

Replace "she" with "Joanna." If "Joanna is my sister" is a true sentence, then Joanna is an element of the solution set.

Sometimes a solution set contains more than one element. If Mary has two sisters, then for her, the solution set of "She is my sister" may be {Joanna, Jennifer}. Some people have no sisters. For them, the solution set for the open sentence "She is my sister" is the empty set.

Statements

A sentence that can be judged to be true or false is called a ***statement*** or a ***closed sentence***. In a statement, there are no variables. The following diagram shows how different kinds of sentences are related to one another.

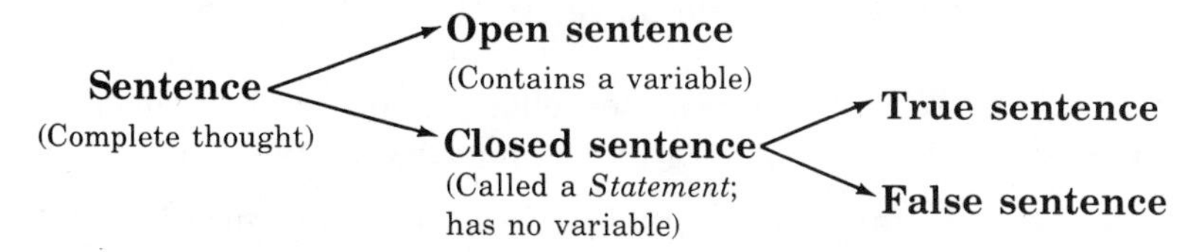

MODEL PROBLEMS

Identify each of the following as a true sentence, false sentence, open sentence, or not a mathematical sentence at all.

	Answers
1. John Wayne was a U.S. President.	**1.** False sentence
2. John Wayne was a movie star.	**2.** True sentence
3. He acted in many Westerns.	**3.** Open sentence (The variable is *he*.)
4. Do you like Westerns?	**4.** Not a mathematical sentence (It asks a question.)
5. Read this book.	**5.** Not a mathematical sentence (It gives a command.)
6. $3x + 5 < 26$	**6.** Open sentence (The variable is x.)
7. $5 + 7 + 8$	**7.** Not a mathematical sentence (It is a phrase.)

EXERCISES

In 1–8, tell whether each is or is not a mathematical sentence.

1. The school day ends at 3:15.
2. Take the bus.
3. Are you going?
4. If John goes.
5. Atlanta is a city in Alaska.
6. $x + 3 = 2x + 1$
7. Two teaspoons, three times a day.
8. Do the next seven problems.

In 9–15, all of the sentences are open sentences. Find the variable in every sentence.

9. She is smart.
10. We elect a president.
11. It is my favorite color.
12. $4y < 20$
13. This is a great country.
14. It is a counting number.
15. He was the most valuable player in the World Series.

In 16–23: **a.** Tell whether the sentence is true, false, or open. **b.** If the sentence is an open sentence, identify the variable.

16. The United States of America declared independence in 1776.
17. They celebrate Independence Day on July 14 every year.

18. San Francisco is a city in New York State.
19. A rectangle is a four-sided polygon.
20. $5x + 2 = 17$
21. $5(10) + 2 = 17$
22. $5(3) + 2 = 17$
23. $2^3 = 3^2$

In 24–28, use the replacement set {New York, Florida, California, Hawaii, Kansas} to find the truth set for each open sentence.

24. It was one of the last two states admitted to the U.S.A.
25. It does not border on or touch an ocean.
26. It is on the east coast of the United States.
27. Its capital is Sacramento.
28. It is one of the states of the United States of America.

In 29–40, use the domain of natural numbers to find the solution set for each open sentence. It is possible that no replacements make true sentences.

29. $x + 5 = 17$
30. $x - 5 = 17$
31. $\frac{2x}{3} = 12$
32. $56 = 3x - 4$
33. $2x + x = 6$
34. $2x - x = 1$
35. $x < 3$
36. $x + 1 < 6$
37. $\frac{x}{8} = \frac{1}{2}$
38. $\frac{2}{3} = \frac{x}{6}$
39. $.2x + 3 = 6$
40. $x + \frac{1}{3} = 3$

In 41–48, use the domain {square, triangle, rectangle, parallelogram, rhombus, trapezoid} to find the truth set for each open sentence.

41. It has three and only three sides.
42. It has two pairs of opposite sides that are parallel.
43. It has four sides that are all equal in measure.
44. It must contain only right angles.
45. It has exactly six sides.
46. It has less than four sides.
47. It has exactly one pair of opposite sides that are parallel.
48. It has four angles that are equal in measure and four sides that are equal in measure.

5-2 NEGATIONS AND SYMBOLS

A sentence that has a ***truth value*** is called a statement. There are two truth values: ***true*** and ***false***, shown by the symbols ***T*** and ***F***. Every statement is either true or false. A statement cannot be both true and false at the same time.

In reasoning, you will learn how to make new statements based upon statements that you already know. One of the simplest examples of this type of reasoning is found in negating a statement.

The ***negation*** of a statement is usually formed by adding the word *not* to the original or given statement. The negation will always have the opposite truth value of the original statement. For example:

1.	Original:	John Kennedy was a U.S. President.	(True)
	Negation:	John Kennedy was *not* a U.S. President.	(False)
2.	Original:	An owl is a fish.	(False)
	Negation:	An owl is *not* a fish.	(True)

There are many ways to place the word "not" into a statement to form its negation. One method starts the negation with the phrase "It is not true that . . . ." For example:

Original:	The post office handles mail.	(True)
Negation:	It is *not* true that the post office handles mail.	(False)
Negation:	The post office does *not* handle mail.	(False)

Both negations here express the same false statement.

The First Symbols in Logic

In logic, we use a single letter to represent a single complete thought. This means that an entire sentence may be replaced by a single letter of the alphabet. Although we can use any letter to represent a simple statement, the letters most often used in logic are p, q, and r. For example, p might represent "John Kennedy was a U.S. President."

To show a negation of a simple statement, place the symbol $\sim$ before the letter for the original or given statement. Then $\sim p$ would represent "John Kennedy was *not* a U.S. President." We read the symbol $\sim p$ as "*not p*."

	Symbol	*Statement in Words*	*Truth Value*
1.	p:	There are twelve months in every year.	(True)
	$\sim p$:	There are *not* twelve months in every year.	(False)
2.	q:	$8 + 9 = 10$	(False)
	$\sim q$:	$8 + 9 \neq 10$	(True)

When p is true, then its negation $\sim p$ is false. When q is false, then its negation $\sim q$ is true.

● **A statement and its negation have opposite truth values.**

However, every sentence is not necessarily a statement with a known truth value. Questions arise in reasoning when we must deal with sentences having an uncertain truth value, such as "The radio is too loud."

The First Truth Table in Logic

To study sentences where we wish to consider all truth values that could be assigned, we use a device called a truth table. A ***truth table*** is a compact way of listing symbols to show all possible truth values for a set of sentences. The truth table for negation is very simple. You will see more complicated truth tables as you continue your study of logic.

Building a Truth Table for *Negation*, $\sim p$

1. Assign a letter to represent the original sentence.
   For example: Let p represent "The radio is too loud."

p:	the original sentence
$\sim p$:	the negation, not p

2. In column 1, list all possible truth values for the original sentence. Here p can be true (shown as T in row 1) or p can be false (shown as F in row 2).

column

	1	**2**
	p	
row 1 →	T	
row 2 →	F	

3. In the heading of column 2, write the negation of the original sentence in symbolic form.

p	$\sim p$
T	
F	

4. Finally, assign truth values for every possible case. In the first row, if p is true, then $\sim p$ must be false. In the second row, if p is false, then $\sim p$ must be true.

p	$\sim p$
T	F
F	T

Truth Table for Negation

As a final note, be aware that many negations can be given in a statement. Each time another negation is included, the truth value of the statement will change. For example:

1. k:	A dime is a coin.	(True)
2. $\sim k$:	A dime is not a coin.	(False)
3. $\sim(\sim k)$:	It is not true that a dime is not a coin.	(True)
4. $\sim(\sim(\sim k))$:	It is not the case that it is not true that a dime is not a coin.	(False)

Of course, we don't usually talk like this because it is too confusing. Since $\sim(\sim k)$ always has the same truth value as k, we can use k in place of $\sim(\sim k)$. Therefore, we can negate a sentence that contains the word *not* by omitting that word.

$\sim k$: A dime is not a coin.
$\sim(\sim k)$: A dime is a coin.

MODEL PROBLEMS

In 1–6: Let q represent "Oatmeal is a cereal."
Let r represent "She has cereal every morning."

For each given sentence: **a.** Write the sentence in symbolic form. **b.** Tell whether the sentence is true, false, or open.

	Answers	
1. Oatmeal is a cereal.	**a.** q	**b.** True
2. Oatmeal is not a cereal.	**a.** $\sim q$	**b.** False
3. It is not true that oatmeal is a cereal.	**a.** $\sim q$	**b.** False
4. She has cereal every morning.	**a.** r	**b.** Open
5. She does not have cereal every morning.	**a.** $\sim r$	**b.** Open
6. It is not true that oatmeal is not a cereal.	**a.** $\sim(\sim q)$	**b.** True

In 7 and 8, symbols are used to represent statements and the truth value of each statement is given.

k: An obtuse triangle contains exactly one obtuse angle. (True)
m: An acute triangle contains exactly one acute angle. (False)

For each sentence given in symbolic form: **a.** Write a complete sentence in words to show what the symbols represent. **b.** Tell if the statements are true or false.

Answers

7. $\sim k$ **a.** An obtuse triangle does *not* contain exactly one obtuse angle. (Or, It is *not* true that an obtuse triangle contains exactly one obtuse angle.) **b.** False

8. $\sim m$ **a.** An acute triangle does *not* contain exactly one acute angle. **b.** True

EXERCISES

In 1–8, write the negation of each sentence.

1. The school has a cafeteria.
2. Georgia is not a city.
3. A school bus is painted yellow.
4. $18 + 20 \div 2 = 28$
5. The measure of a right angle is 90°.
6. $1 + 2 + 3 \neq 4$
7. There are 100 centimeters in a meter.
8. Today is not Saturday.

In 9–18, for each given sentence: **a.** Write the sentence in symbolic form, using the symbols shown below. **b.** Then tell if the sentence is true, false, or open.

Let p represent "A cat is an animal."
Let q represent "A poodle is a cat."
Let r represent "His cat is gray."

9. A cat is an animal.
10. A poodle is a cat.
11. A poodle is not a cat.
12. A cat is not an animal.
13. His cat is gray.
14. His cat is not gray.
15. It is not true that a poodle is a cat.
16. It is not the case that a cat is an animal.
17. It is not the case that a cat is not an animal.
18. It is not the case that a poodle is not a cat.

In 19–22, copy the truth table for negation and fill in all missing symbols.

19.

p	$\sim p$
T	
F	

20.

q	$\sim q$
T	
	T

21.

r	$\sim r$
	F
	T

22.

k	$\sim k$

23. Copy the truth table and fill in all missing symbols.

q	$\sim q$	$\sim(\sim q)$	$\sim(\sim(\sim q))$
T			
F			

24. A truth table is shown for a sentence and its negation.

	q	$\sim q$
row 1 →	T	F
row 2 →	F	T

Let q represent "$x + 3 > 8$."

Tell which row of the truth table shows the correct truth values when:

a. $x = 2$ **b.** $x = 9$ **c.** $x = 0$ **d.** $x = 5$ **e.** $x = \frac{1}{2}(12)$

f. $x =$ the sum of 8 and 3 **g.** $x =$ the difference of 8 and 3

h. $x =$ the product of 8 and 3 **i.** $x =$ the quotient of 8 and 3

In 25–32, the symbols represent sentences.

p: Summer follows spring. r: Baseball is a summer sport.

q: Baseball is a sport. s: He likes baseball.

For each sentence given in symbolic form: **a.** Write a complete sentence in words to show what the symbols represent. **b.** Tell if the sentence is true, false, or open.

25. $\sim p$ **26.** $\sim q$ **27.** $\sim r$ **28.** $\sim s$

29. $\sim(\sim q)$ **30.** $\sim(\sim p)$ **31.** $\sim(\sim r)$ **32.** $\sim(\sim s)$

In 33–36, give a word, a phrase, or a symbol that can be placed in the blank to make the resulting sentence true.

33. When p is true, then $\sim p$ is ____.
34. When p is false, then $\sim p$ is ____.
35 $\sim(\sim p)$ has the same truth value as ____.
36. A sentence that has a truth value is called a ____.

5-3 CONJUNCTIONS

You have seen that a single letter can be used in logic to represent a single complete thought. Sometimes a sentence contains more than one thought. In English, connectives are used to form compound sentences that have two or more thoughts. One of the simplest connectives is the word *and.*

In logic, a ***conjunction*** is a compound sentence formed by combining two simple sentences, using the word ***and***. Each of the simple sentences is called a ***conjunct***. When p and q represent simple sentences, the conjunction p *and* q is written in symbols as $p \wedge q$. For example:

p: There is no school on Saturday.

q: I sleep late.

$p \wedge q$: There is no school on Saturday and I sleep late.

In order for this sentence to be true, both parts must be true: "There is no school on Saturday" must be true and "I sleep late" must be true.

The truth table for a compound sentence contains more than two rows. The first thought p can be true or false, and the second thought q can be true or false. You must consider every possible combination of these true and false statements. The diagram shown below is called a ***tree diagram***. By following its "branches," you can see that there are four possible "true-false" combinations for every two simple statements.

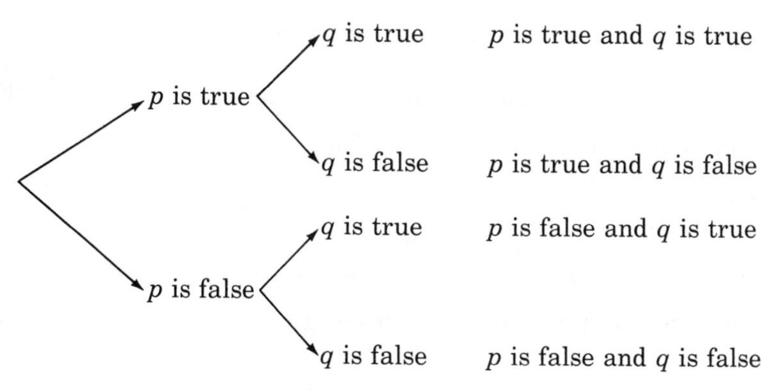

Use the order shown in the table to set up the first two columns in every truth table containing two thoughts p and q. By using the same order all the time, you can find specific cases quickly and you will reduce your chances of making errors.

p	q
T	T
T	F
F	T
F	F

The truth value of every compound sentence depends upon the truth value of the simple sentences used within the compound sentence.

● **The conjunction *p and q* is true only when both parts are true: *p* must be true, and *q* must be true.**

If p is false, or if q is false, or if both are false, then the conjunction *p and q* must be false.

Consider these open sentences:

Let p represent "x is an even number."
Let q represent "$x < 8$."
Then, $p \wedge q$ represents "x is an even number and $x < 8$."

1. Let $x = 6$.
   p: 6 is an even number. (True)
   q: $6 < 8$ (True)
   $p \wedge q$: 6 is an even number and $6 < 8$. (True)

   When both parts of the conjunction are true, then *p and q* is true.

2. Let $x = 10$.
   p: 10 is an even number. (True)
   q: $10 < 8$ (False)
   $p \wedge q$: 10 is an even number and $10 < 8$. (False)

   If any part of the conjunction is false, then *p and q* is false.

3. Let $x = 7$.
   p: 7 is an even number. (False)
   q: $7 < 8$ (True)
   $p \wedge q$: 7 is an even number and $7 < 8$. (False)

   If any part of the conjunction is false, then *p and q* is false.

4. Let $x = 13$.
   p: 13 is an even number. (False)
   q: $13 < 8$ (False)
   $p \wedge q$: 13 is an even number and $13 < 8$. (False)

   When both parts of the conjunction are false, then *p and q* is false.

Building a Truth Table for the *Conjunction* $p \wedge q$

1. Assign two letters, each letter to serve as a symbol for a different simple sentence.

   For example: Let p represent "It is cold."
   Let q represent "It is snowing."

2. List all possible truth values for the simple sentences p and q in the first two columns. Follow the order shown.

p	q
T	T
T	F
F	T
F	F

3. In the heading of the third column, write the conjunction of the sentences in symbolic form.

p	q	$p \wedge q$
T	T	
T	F	
F	T	
F	F	

4. Finally, assign truth values for the conjunction. When both simple sentences p and q are true, then the conjunction p *and* q is true. In all other cases, the conjunction is false.

   $p \wedge q$: "It is cold and it is snowing" will be true only when both parts are true.

p	q	$p \wedge q$
T	T	T
T	F	F
F	T	F
F	F	F

Truth Table for Conjunction

A compound sentence may contain both conjunctions and negations at the same time. The truth table for such a sentence is developed in Model Problem 12 on page 134.

To build any truth table, work from the innermost parentheses first, or from the simplest level of thinking. This is very much like the order of operations we use in arithmetic and in solving equations.

MODEL PROBLEMS

In 1–6: Let p represent "Coffee is a beverage." (True)
Let q represent "Toast is a beverage." (False)
Let r represent "10 is divisible by 2." (True)
Let s represent "10 is divisible by 3." (False)

For each given sentence: **a.** Write the sentence in symbolic form. **b.** Construct that row of the truth table that will tell whether the statement is true or false.

1. Coffee is a beverage and 10 is divisible by 2.

Answer: **a.** $p \wedge r$ **b.** True

p	r	$p \wedge r$
T	T	T

2. Coffee and toast are beverages.

Answer: **a.** $p \wedge q$ **b.** False

p	q	$p \wedge q$
T	F	F

3. Toast is a beverage and 10 is divisible by 2.

Answer: **a.** $q \wedge r$ **b.** False

p	q	$p \wedge q$
F	T	F

4. 10 is divisible by 2 and 10 is not divisible by 3.

Answer: **a.** $r \wedge \sim s$ **b.** True

r	s	$\sim s$	$r \wedge \sim s$
T	F	T	T

5. 10 is divisible by 2 and by 3.

Answer: **a.** $r \wedge s$ **b.** False

r	s	$r \wedge s$
T	F	F

6. Toast is not a beverage and 10 is not divisible by 3.

Answer: **a.** $\sim q \wedge \sim s$ **b.** True

q	s	$\sim q$	$\sim s$	$\sim q \wedge \sim s$
F	F	T	T	T

7. Use the domain $\{1, 2, 3\}$ to find the truth set for the open sentence: $(x > 1) \wedge (x < 4)$.

Solution: Let $x = 1$. $(1 > 1) \wedge (1 < 4)$

$$F \wedge T = \text{False statement.}$$

Since one part is false, the conjunction is false.

Let $x = 2$. $(2 > 1) \wedge (2 < 4)$

$T \wedge T$ = True statement.

Since both parts are true, the conjunction is true.

Let $x = 3$. $(3 > 1) \wedge (3 < 4)$

$T \wedge T$ = True statement.

Since a conjunction is true only when both simple sentences are true, the truth set or solution set is $\{2, 3\}$.

Answer: $\{2, 3\}$

In 8–11, symbols are used to represent statements about a rectangle with length l and width w. For each statement made, its truth value is noted.

Let p represent "Area $= lw$." (True)
Let q represent "Perimeter $= l + w$." (False)
Let r represent "Perimeter $= 2l + 2w$." (True)

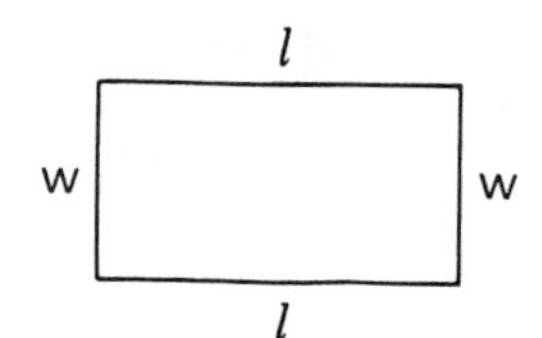

For each sentence given in symbolic form: **a.** Write a complete sentence in words to show what the symbols represent. **b.** Tell if the statement is true or false.

8. $p \wedge q$

Answer: **a.** Area $= lw$ *and* Perimeter $= l + w$.
b. $T \wedge F$ = False statement.

9. $p \wedge r$

Answer: **a.** Area $= lw$ *and* Perimeter $= 2l + 2w$.
b. $T \wedge T$ = True statement.

10. $\sim q \wedge p$

Answer: **a.** Perimeter $\neq l + w$ *and* Area $= lw$.
b. $\sim F \wedge T = T \wedge T$ = True statement.

11. $\sim(q \wedge p)$

Answer: **a.** It is *not* the case that Perimeter $= l + w$ *and* Area $= lw$.
b. $\sim(F \wedge T) = \sim(F)$ = True statement.

Note that "It is not the case that . . ." applies to the *entire* statement that follows it.

12. Build a truth table for the sentence $\sim(\sim p \wedge q)$.

Solution

(1) First, write all possible combinations of true-false statements for p and q.

p	q
T	T
T	F
F	T
F	F

(2) Then, working inside the parentheses, you first need $\sim p$. Write $\sim p$ in column 3 and negate the truth values for p found in column 1.

1	2	3
p	q	~p
T	T	F
T	F	F
F	T	T
F	F	T

(3) Next, to get $\sim p \wedge q$, use the truth values found in columns 2 and 3 to write the truth values for the conjunction. Since q and $\sim p$ are both true in the third row, this is the only conjunction that is true. All other rows are false.

1	2	3	4
p	q	~p	~p ∧ q
T	T	F	F
T	F	F	F
F	T	T	T
F	F	T	F

(4) Finally, negate the truth values for $\sim p \wedge q$ found in column 4 to find the truth values of $\sim(\sim p \wedge q)$ to write in column 5.

1	2	3	4	5
p	q	~p	~p ∧ q	~(~p ∧ q)
T	T	F	F	T
T	F	F	F	T
F	T	T	T	F
F	F	T	F	T

13. Three sentences are written. The truth values are given for the first two sentences. Determine if the third sentence is true, is false, or has an uncertain truth value.

Trudie likes steak and Phil likes fish.	(False)
Trudie likes steak.	(True)
Phil likes fish.	(?)

Solution:

(1) Use symbols to represent the sentences. Indicate their truth values.

$p \wedge q$ (False): Trudie likes steak and Phil likes fish.
p (True): Trudie likes steak.
q (?): Phil likes fish.

(2) Construct a truth table for conjunction as follows:

The first sentence $p \wedge q$ is given as false. In the truth table, you see that $p \wedge q$ is false in rows 2, 3, and 4. Therefore, eliminate the case where $p \wedge q$ is true by crossing out row 1 of the truth table.

p	q	$p \wedge q$
~~T~~	~~T~~	~~T~~
T	F	F
F	T	F
F	F	F

(3) The second sentence p is given as true. In the three rows remaining in our truth table, p is true only in row 2. Eliminate the cases where p is false by crossing out rows 3 and 4 in the truth table.

p	q	$p \wedge q$
~~T~~	~~T~~	~~T~~
T	F	F
~~F~~	~~T~~	~~F~~
~~F~~	~~F~~	~~F~~

(4) There is only one case where $p \wedge q$ is false and where p is true. This occurs in row 2 of the truth table. In row 2, q is false. You can conclude that the sentence q, Phil likes fish, is false.

Answer: "Phil likes fish" is false.

14. Three sentences are given; the truth values are noted for the first two sentences. Determine if the third sentence is true, is false, or has an uncertain truth value.

Jean likes hot weather and Bob likes cold weather. (False)
Jean likes hot weather. (False)
Bob likes cold weather. (?)

Solution:

(1) Use symbols to represent the sentences. Include their truth values.

$p \wedge q$ (False): Jean likes hot weather and Bob likes cold weather.
p (False): Jean likes hot weather.
q (?): Bob likes cold weather.

(2) Construct a truth table for conjunction as follows:

The first sentence $p \wedge q$ is false. Eliminate the case where $p \wedge q$ is true by crossing out row 1 of the truth table.

p	q	$p \wedge q$
~~*T*~~	~~*T*~~	~~*T*~~
T	*F*	*F*
F	*T*	*F*
F	*F*	*F*

(3) The second sentence p is false. This occurs in rows 3 and 4 of the truth table. Eliminate the case where p is true by crossing out row 2 of the truth table.

p	q	$p \wedge q$
~~*T*~~	~~*T*~~	~~*T*~~
~~*T*~~	~~*F*~~	~~*F*~~
F	*T*	*F*
F	*F*	*F*

(4) The remaining rows of the truth table show that there are *two* cases where $p \wedge q$ is false and where p is false: rows 3 and 4. Since q could be true, as in row 3, or q could be false, as in row 4, conclude that q has an uncertain truth value.

Answer: The truth value of "Bob likes cold weather" is uncertain.

EXERCISES

In 1–10, write each sentence in symbolic form, using the given symbols.

Let p represent "It is cold."
Let q represent "It is snowing."
Let r represent "The sun is shining."

1. It is cold and it is snowing.
2. It is cold and the sun is shining.
3. It is not cold.
4. It is not cold and the sun is shining.
5. It is snowing and the sun is not shining.
6. It is not cold and it is not snowing.
7. The sun is not shining and it is not cold.
8. The sun is not shining and it is cold.
9. It is not the case that it is cold and it is snowing.
10. It is not the case that it is snowing and it is not cold.

In 11–20, for each given statement: **a.** Write the statement in symbolic form, using the symbols shown below. **b.** Construct that row of the truth table that will tell whether the statement is true or false.

Let b represent "Water boils at 100°C." (True)
Let f represent "Water freezes at 0°C." (True)
Let t represent "Normal body temperature is 37°C." (True)
Let r represent "Room temperature is 60°C." (False)

11. Normal body temperature is 37°C and water boils at 100°C.
12. Normal body temperature is 37°C and room temperature is 60°C.
13. Water freezes at 0°C and boils at 100°C.
14. Water freezes at 0°C and room temperature is 60°C.
15. Water does not boil at 100°C and water does not freeze at 0°C.
16. Room temperature is not 60°C and water boils at 100°C.
17. Normal body temperature is not 37°C and room temperature is 60°C.
18. It is not the case that water does not boil at 100°C.
19. It is not true that water boils at 100°C and freezes at 0°C.
20. It is not the case that water boils at 100°C and room temperature is 60°C.

In 21–27, tell if the sentence is true, false, or open.

21. France is in South America and Germany is in Asia.
22. Every square contains 4 right angles and every triangle contains 1 right angle.

23. Tuesday follows Monday and $1 + 2 = 3$.

24. It is a math book and it contains problems.

25. Most math books contain problems and $\frac{2}{3}$ of 12 is 8.

26. $x = 28 - 17$ and $x = 11$.

27. The Surgeon General of the U.S. has determined that cigarette smoking is dangerous to your health and warning labels are printed on cigarette packs.

In 28–30, copy the truth table for conjunction and fill in all missing symbols.

28.

p	q	$p \wedge q$
T	T	
T	F	
F	T	
F	F	

29.

m	r	$m \wedge r$
T	T	T
T		F
	T	F
		F

30.

f	g	$f \wedge g$

In 31–36, copy the truth table and fill in all missing symbols. (*Note:* In 32–36, prepare a complete truth table similar to the one shown in Exercise 31.)

31.

p	q	$\sim p$	$\sim q$	$\sim p \wedge \sim q$
T	T			
T	F			
F	T			
F	F			

32.

p	q	$p \wedge q$	$\sim(p \wedge q)$

33.

p	q	$\sim q$	$p \wedge \sim q$

34.

p	q	$\sim p$	$\sim p \wedge q$

35.

p	q	$\sim q$	$q \wedge \sim q$	$\sim(q \wedge \sim q)$

36.

p	q	$\sim q$	$p \wedge \sim q$	$\sim(p \wedge \sim q)$

37. Use the domain of whole numbers to find the truth set for each compound open sentence.

a. $(x > 5) \wedge (x < 8)$ **b.** $(x > 4) \wedge (x \leq 6)$ **c.** $(x \geq 3) \wedge (x < 7)$
d. $(x < 5) \wedge (x < 2)$ **e.** $(x \leq 3) \wedge (x < 4)$ **f.** $(x > 8) \wedge (x > 3)$
g. $(x \geq 3) \wedge (x + 5 < 7)$ **h.** $(x - 8 \leq 2) \wedge (3x > 24)$

In 38–49, symbols are assigned to represent sentences.

Let b represent "A banjo is a stringed instrument."
Let d represent "A drum is a stringed instrument."
Let g represent "A guitar is a stringed instrument."
Let s represent "She plays a guitar."

For each sentence given in symbolic form: **a.** Write a complete sentence in words to show what the symbols represent. **b.** Tell if the sentence is true, false, or open.

38. $b \wedge g$
39. $b \wedge d$
40. $g \wedge s$
41. $b \wedge \sim d$
42. $b \wedge \sim g$
43. $b \wedge \sim s$
44. $g \wedge \sim d$
45. $\sim s \wedge d$
46. $\sim d \wedge \sim b$
47. $\sim(d \wedge b)$
48. $\sim(b \wedge g)$
49. $\sim(g \wedge s)$

In 50–57, give the word, phrase, or symbol that can be placed in the blank to make the resulting sentence true.

50. When p is true and q is true, then $p \wedge q$ is ___.
51. When p is false, then $p \wedge q$ is ___.
52. If p is true, or q is true, but not both, then $p \wedge q$ is ___.
53. When $p \wedge q$ is true, then p is ___ and q is ___.
54. When $p \wedge \sim q$ is true, then p is ___ and q is ___.
55. When $\sim p \wedge q$ is true, then p is ___ and q is ___.
56. When p is false and q is true, then $\sim(p \wedge q)$ is ___.
57. If both p and q are false, then $\sim p \wedge \sim q$ is ___.

In 58–64, three sentences are written. The truth values are given for the first two sentences. Determine if the third sentence is true, is false, or has an uncertain truth value.

58. It is raining and I get wet. (True)
It is raining. (True)
I get wet. (?)

59. I have a headache and I take a nap. (False)
I have a headache. (True)
I take a nap. (?)

60. I have a headache and I take a nap. (False)
I have a headache. (False)
I take a nap. (?)

61. Both a potato and a hurricane have many eyes. (False)
A potato has many eyes. (True)
A hurricane has many eyes. (?)

62. Anna loves TV and Anna loves to stay home. (True)
Anna loves TV. (True)
Anna loves to stay home. (?)

63. Joe and Edna like to stay at home. (False)
Joe likes to stay at home. (True)
Edna likes to stay at home. (?)

64. Juan takes the train and the bus to go to work. (False)
Juan takes the bus to go to work. (False)
Juan takes the train to go to work. (?)

In 65 and 66, a compound sentence is given using a conjunction. After examining the truth value of the compound sentence, determine if the truth value for each sentence that follows is true or false.

65. Most plants need light and water to grow. (True)
- **a.** Most plants need light to grow.
- **b.** Most plants need water to grow.
- **c.** Most plants do not need light to grow.

66. I do not exercise and I know that I should. (True)
- **a.** I do not exercise.
- **b.** I know that I should exercise.
- **c.** I exercise.

5-4 DISJUNCTIONS

Many different connectives are used in everyday conversation. Some connectives that might be heard at the local diner include: bacon and eggs; lettuce and tomato; white bread or rye; tea or coffee; mustard or catsup. In addition to *and*, another common connective in our language is the word *or*.

In logic, a ***disjunction*** is a compound sentence formed by combining two simple sentences using the word ***or***. Each of the simple sentences is called a ***disjunct***. When p and q represent simple sentences, the disjunction ***p or q*** is written in symbols as $p \vee q$. For example:

p: You can use pencil to answer the test.

q: You can use pen to answer the test.

$p \vee q$: You can use pencil or you can use pen to answer the test.

In this example, the compound sentence *p or q* is true in many cases:

1. You answer in pencil. Here p is true and q is false. The disjunction $p \vee q$ is true.
2. You answer in pen. Here p is false while q is true. The disjunction $p \vee q$ is true.

3. You answer in pen but soon you run out of ink. You finish the test in pencil. Here p is true and q is true. Since you obeyed the rules and did nothing false, the disjunction $p \vee q$ is true.

● **The disjunction *p or q* is true when any part of the compound sentence is true: *p* is true, or *q* is true, or both *p* and *q* are true.**

In fact, there is only one case where the disjunction *p or q* is false: when both p and q are false.

Building a Truth Table for the *Disjunction* $p \vee q$

1. Assign two letters; each letter is to serve as a symbol for a different simple sentence.

   For example: Let p represent "The battery is dead."
   Let q represent "We are out of gas."

2. List all possible truth values for the simple sentences p and q in the first two columns. Use the same order that was established for conjunction.

p	q
T	T
T	F
F	T
F	F

3. In the heading of the third column, write the disjunction of the sentences in symbolic form.

p	q	$p \vee q$
T	T	
T	F	
F	T	
F	F	

4. Finally, assign truth values for the disjunction. All cases will be true except for the last row. When both p and q are false, the disjunction is false.

   $p \vee q$: "The battery is dead or we are out of gas" will be true when the battery is dead, or when there is no gas, or when both problems exist.

p	q	$p \vee q$
T	T	T
T	F	T
F	T	T
F	F	F

Truth Table for Disjunction

Two Uses of the Word "Or"

When we use the word *or* to mean that *one or both* of the simple sentences are true, we call this the ***inclusive or***. The truth table we have just developed shows the truth values for the *inclusive or*.

There are times when the word *or* is used in a different way, as in "He is in grade 9 or he is in grade 10." Here it is not possible for both simple sentences to be true at the same time. When we use the word *or* to mean that *one and only one* of the simple sentences is true, we call this the ***exclusive or***. The truth table for the *exclusive or* will be different from the table shown for disjunction. In the *exclusive or*, the disjunction p *or* q will be true when p is true, or when q is true, but not both.

In everyday conversation, it is often evident from the context which of these uses of *or* is intended. In legal documents or when ambiguity can cause difficulties, the *inclusive or* is sometimes written as *and/or*.

● **We will use only the *inclusive or* in this book. Whenever we speak of disjunction, *p or q* will be true when *p* is true, or when *q* is true, or when *both p* and *q* are true.**

Logic and Sets

Some people believe that logic is not like any other kind of mathematics that they have ever seen. This is not true. Conjunction and disjunction behave in exactly the same way as operations you have seen with sets.

The truth sets or solution sets of open sentences that use the connectives *not*, *and*, and *or* can be compared to the *complement*, *intersection*, and *union* of sets.

Let p represent $x < 4$.
Let q represent $x > 1$.
Let the domain be $\{1, 2, 3, 4, 5\}$.

In order to find the truth sets of $\sim p$, $\sim q$, $p \wedge q$, and $p \vee q$, we will first find the truth sets of p and of q.

The truth set of p is $\{1, 2, 3\}$. Therefore, the truth set of $\sim p$ is $\{4, 5\}$, the complement of the truth set of p.

The truth set of q is $\{2, 3, 4, 5\}$. Therefore, the truth set of $\sim q$ is $\{1\}$, the complement of the truth set of q.

The truth set of $p \wedge q$ is $\{2, 3\}$, the intersection of the truth set of p and the truth set of q.

The truth set of $p \vee q$ is $\{1, 2, 3, 4, 5\}$, the union of the truth set of p and the truth set of q.

Let P be the truth set of p.
Let Q be the truth set of q.

The truth set of $\sim p$ is $\overline{P}$.
The truth set of $\sim q$ is $\overline{Q}$.
The truth set of $p \wedge q$ is $P \cap Q$.
The truth set of $p \vee q$ is $P \cup Q$.

MODEL PROBLEMS

In 1–6: Let k represent "Kevin won the play-off."
Let a represent "Alexis won the play-off."
Let n represent "Nobody won."

Write each given sentence in symbolic form. *Answers*

1. Kevin or Alexis won the play-off. $k \vee a$
2. Kevin won the play-off or nobody won. $k \vee n$
3. Alexis won the play-off or Alexis didn't win. $a \vee \sim a$
4. It is not true that Kevin or Alexis won the play-off. $\sim(k \vee a)$
5. Either Kevin did not win the play-off or Alexis did not win. $\sim k \vee \sim a$
6. It's not the case that Alexis and Kevin won the play-off. $\sim(a \wedge k)$

In 7–10, symbols are used to represent the statements as shown. For each statement, its truth value is noted.

Let k represent "Every square is a rhombus." (True)
Let m represent "Every rhombus is a square." (False)
Let q represent "Every square is a parallelogram." (True)

For each sentence given in symbolic form: **a.** Write a complete sentence in words to show what the symbols represent. **b.** Tell if the statement is true or false.

Answers

7. $k \vee q$
a. Every square is a rhombus or every square is a parallelogram.
b. $T \vee T$ is a true disjunction.

8. $k \vee m$
a. Every square is a rhombus or every rhombus is a square.
b. $T \vee F$ is a true disjunction.

9. $m \vee \sim q$ **a.** Every rhombus is a square or every square is not a parallelogram.
b. $F \vee \sim T = F \vee F =$ a false disjunction.

10. $\sim(m \vee q)$ **a.** It is not the case that every rhombus is a square or every square is a parallelogram.
b. $\sim(F \vee T) = \sim(T) =$ a false statement.

In 11–13, use the domain {0, 1, 2, 3, 4, 5, 6} to find the truth set of each compound sentence.

11. $(x > 4) \vee (x \leq 2)$
$\{5, 6\} \cup \{0, 1, 2\} = \{0, 1, 2, 5, 6\}$ *Ans.*

12. $(x \not> 4) \wedge (x \not\leq 2)$
$\{0, 1, 2, 3, 4\} \cap \{3, 4, 5, 6\} = \{3, 4\}$ *Ans.*

13. $(x \geq 5) \wedge (x \leq 3)$
$\{5, 6\} \cap \{0, 1, 2, 3\} = \varnothing$ *Ans.*

EXERCISES

In 1–10, write each sentence in symbolic form, using the given symbols.

Let s represent "I will study."
Let p represent "I will pass the test."
Let f represent "I am foolish."

1. I will study or I am foolish.
2. I will study or I will not pass the test.
3. I will study and I will pass the test.
4. I will pass the test or I am foolish.
5. I am not foolish and I will pass the test.
6. I will not study or I am foolish.
7. I will study or I will not study.
8. I will study and I will pass the test, or I am foolish.
9. It is not true that I will study or I am foolish.
10. It is not the case that I will not study or I am not foolish.

In 11–20, for each given statement: **a.** Write the statement in symbolic form, using the symbols given below. **b.** Tell whether the statement is true or false.

Let c represent "A meter contains 100 centimeters." (True)
Let m represent "A meter contains 1,000 millimeters." (True)
Let k represent "A kilometer is 1,000 meters." (True)
Let l represent "A meter is a liquid measure." (False)

11. A meter contains 1,000 millimeters or a kilometer is 1,000 meters.
12. A meter contains 100 centimeters or a meter is a liquid measure.
13. A meter contains 100 centimeters or 1,000 millimeters.
14. A kilometer is not 1,000 meters or a meter does not contain 100 centimeters.
15. A meter is a liquid measure or a kilometer is 1,000 meters.
16. A meter is a liquid measure and a meter contains 100 centimeters.
17. It is not the case that a meter contains 100 centimeters or 1,000 millimeters.
18. It is false that a kilometer is not 1,000 meters or a meter is a liquid measure.
19. A meter contains 100 centimeters and 1,000 millimeters or a meter is a liquid measure.
20. It is not true that a meter contains 100 centimeters or a meter is a liquid measure.

In 21–23, copy the truth table for disjunction and fill in the missing symbols.

21.

p	q	$p \vee q$
T	T	
T	F	
F	T	
F	F	

22.

k	t	$k \vee t$
T		
T		
F		
F	F	F

23.

p	r	$p \vee r$

In 24–29, copy the truth table and fill in all missing symbols. (*Note:* In 24–29, prepare a complete truth table similar to the one shown in Exercise 21.)

24.

p	q	$p \vee q$	$\sim(p \vee q)$

25.

p	q	$\sim p$	$\sim p \vee q$	$\sim(\sim p \vee q)$

26.

p	q	$\sim p$	$\sim q$	$\sim p \vee \sim q$

27.

p	q	$\sim q$	$p \vee \sim q$	$q \vee (p \vee \sim q)$

28.

p	q	$\sim q$	$q \vee \sim q$	$p \vee (q \vee \sim q)$

29.

p	q	$p \vee q$	$p \wedge q$	$(p \vee q) \vee (p \wedge q)$

30. Use the domain of one-digit whole numbers {0, 1, 2, . . . , 9} to find the truth set of each compound open sentence.

a. $(x < 3) \vee (x < 2)$ **b.** $(x \geq 8) \vee (x < 1)$ **c.** $(x > 9) \vee (x \leq 3)$

d. $(x > 2) \wedge (x < 7)$ **e.** $(x < 5) \vee (x > 9)$ **f.** $(x < 5) \wedge (x > 8)$

In 31–42, symbols are assigned to represent sentences.

Let b represent "Biology is a science."
Let s represent "Spanish is a language."
Let h represent "Homemaking is a language."
Let d represent "It's a difficult course."

For each sentence given in symbolic form: **a.** Write a complete sentence in words to show what the symbols represent. **b.** Tell if the sentence is true, false, or open.

31. $s \vee h$ **32.** $b \vee s$ **33.** d **34.** $\sim s \vee h$

35. $\sim d$ **36.** $b \vee \sim h$ **37.** $\sim b \vee \sim s$ **38.** $\sim(s \vee h)$

39. $s \wedge h$ **40.** $s \wedge b$ **41.** $\sim(\sim b \vee s)$ **42.** $\sim(\sim s \vee h)$

In 43–49, give the word, phrase, or symbol that can be placed in the blank to make the resulting sentence true.

43. When p is true, then $p \vee q$ is ____.
44. When q is true, then $p \vee q$ is ____.
45. When p is false and q is false, then $p \vee q$ is ____.
46. When $p \vee \sim q$ is false, then p is ____ and q is ____.
47. When $\sim p \vee q$ is false, then p is ____ and q is ____.
48. When p is false and q is true, then $\sim(p \vee q)$ is ____.
49. When p is false and q is true, then $\sim p \vee \sim q$ is ____.

In 50–54, three sentences are written. The truth values are given for the first two sentences. Determine whether the third sentence is true, is false, or has an uncertain truth value.

50. She will sink. (False)
She will swim. (True)
She will sink or she will swim. (?)

51. I will work after school or I will study more. (True)
I will work after school. (False)
I will study more. (?)

52. Michael cannot swim or Michael cannot skate. (False)
Michael cannot swim. (False)
Michael cannot skate. (?)

53. Nicolette is my friend or I have two left feet. (True)
Nicolette is my friend. (True)
I have two left feet. (?)

54. Jennifer draws well or plays the cello. (True)
Jennifer does not draw well. (False)
Jennifer plays the cello. (?)

5-5 CONDITIONALS

The connective used most often in reasoning can be found in the following sentence:

If the fever continues, then he should see a doctor.

To find the connective, first list the simple sentences using p and q.

p: The fever continues. q: He should see a doctor.

The connective is the words that remain, If . . . then.

In English, such a sentence is called a *complex sentence*. In mathematics, however, all sentences formed by connectives are called *compound sentences*.

In logic, a ***conditional*** is a compound sentence formed by combining two simple sentences using the words ***if . . . then***. When p and q represent simple sentences, the conditional ***if p then q*** is written in symbols as $\boldsymbol{p \rightarrow q}$.

Since a conditional is sometimes called an ***implication***, the symbols for the conditional $\boldsymbol{p \rightarrow q}$ can be read ***p implies q***. Let us look at another example.

p: It is snowing.
q: The temperature is below freezing.
$p \rightarrow q$: *If* it is snowing, *then* the temperature is below freezing.

or

$p \rightarrow q$: It is snowing *implies* that the temperature is below freezing.

Certainly we would agree that the compound sentence *if p then q* is true for this example: "If it is snowing, then the temperature must be below freezing." However, if we reverse the order of the simple sentences to form the new conditional *if q then p*, we will get a sentence with a completely different meaning:

$q \rightarrow p$: *If* the temperature is below freezing, *then* it is snowing.

When the temperature is below freezing, it does not necessarily follow that it must be snowing. The conditional *if q then p* is not necessarily a true statement. Changing the ***order*** in which we connect two simple sentences can sometimes result in forming two conditionals with *different truth values.*

Parts of a Conditional

The parts of the conditional *if p then q* can be identified by name:

p is called the ***premise***, the ***hypothesis***, or the ***antecedent***. It is an assertion or a sentence that begins our argument. The antecedent usually follows the word *if.*

q is called the ***conclusion*** or the ***consequent***. It is an ending or a sentence that closes our argument. The consequent usually follows the word *then.*

There are different ways to write the conditional *if p then q.* Notice that the antecedent p is connected to the word *if* in the examples shown:

$p \rightarrow q$: If Alice scores one more point, then our team will win.
(antecedent or hypothesis: "Alice scores one more point"; consequent or conclusion: "our team will win")

$p \rightarrow q$: Our team will win if Alice scores one more point.
(consequent: "Our team will win"; antecedent: "Alice scores one more point")

Both sentences say the same thing: We hypothesize or hope that Alice scores one more point. When that happens, we can conclude that our team will win. Although the word order of a conditional may vary, the antecedent is always written first when using symbols.

Building a Truth Table for the *Conditional* $p \rightarrow q$

1. Let p serve as the symbol for the antecedent and let q serve as the symbol for the consequent.

   For example: Dr. Cathy Russo told her patient, Bill, "If you take the medicine, then you'll feel better in 24 hours."

   p: You take the medicine.
   q: You'll feel better in 24 hours.

2. List all possible truth values for the antecedent p and the consequent q in the first two columns. Use the same order established for the other connectives.

p	q
T	T
T	F
F	T
F	F

3. In the heading of the third column, write the conditional *if p then q* in symbolic form.

p	q	$p \to q$
T	T	
T	F	
F	T	
F	F	

4. Assign truth values to the conditional by considering the truth values for p and q in each row.

Row 1:
Bill does take the medicine. (p is true)
He does feel better in 24 hours. (q is true)
Dr. Russo has told Bill the truth.
The conditional statement is true.

p	q	$p \to q$
T	T	T
T	F	
F	T	
F	F	

Row 2:
Bill does take the medicine. (p is true)
He does not feel better in 24 hours. (q is false)
Dr. Russo did not tell Bill the truth.
The conditional statement is false.

p	q	$p \to q$
T	T	T
T	F	F
F	T	
F	F	

Rows 3 and 4:
Bill does not take the medicine. (p is false)
It is possible that Bill feels better, (Row 3, q is true)
or that he does not feel better. (Row 4, q is false)

p	q	$p \to q$
T	T	T
T	F	F
F	T	?
F	F	?

In both cases, the doctor did not tell Bill a lie because she told him only what would happen if he did take the medicine. Since the doctor did not make a false statement in these two cases, we will assign *true* to the conditional statement.

5. Assign the truth values to complete the table.

$p \to q$: "If you take the medicine, then you'll feel better in 24 hours" will be true in all cases except one: when the medicine is taken and Bill does not feel better in 24 hours.

p	q	$p \to q$
T	T	T
T	F	F
F	T	T
F	F	T

Truth Table for Conditional

● **A conditional is false when a true hypothesis leads to a false conclusion.** In all other cases, the conditional is true.

Hidden Conditionals

We constantly use conditionals in our everyday lives. Often the words *if. . . then* do not appear in a sentence. In such a case, we say that the sentence has a ***hidden conditional***. We can still understand that the sentence is a conditional because it contains an antecedent p and a consequent q. We can rewrite the words in the sentence so that the conditional form *if* p *then* q becomes more obvious. For example:

1. "When this assignment has been written, you should hand it in" becomes:

   $p \to q$: *If* the assignment has been written, *then* you should hand it in.

2. "For good health, exercise regularly" becomes:

   $p \to q$: *If* you want to have good health, *then* you should exercise regularly.

3. "Vote for me and I'll whip unemployment" becomes:

   $p \rightarrow q$: *If* you vote for me, *then* I'll whip unemployment.

4. "$x + 3 = 10$, therefore $x = 7$" becomes:

   $p \rightarrow q$: *If* $x + 3 = 10$, *then* $x = 7$.

MODEL PROBLEMS

In 1–3, for each given sentence: **a.** Identify the hypothesis p. **b.** Identify the conclusion q.

1. If Mrs. Garbowski assigns homework, then you'd better do it.

Answer: **a.** p: Mrs. Garbowski assigns homework.
b. q: You'd better do the homework.

2. You can assemble the bicycle if you follow these easy directions.

Answer: **a.** p: You follow these easy directions.
b. q: You can assemble the bicycle.

3. I carry an umbrella when the forecast predicts rain and the weather looks threatening.

Answer: **a.** p: The forecast predicts rain and the weather looks threatening.
b. q: I carry an umbrella.

In 4–7, identify the truth value to be assigned to each conditional statement.

4. If $2^2 = 4$, then $2^3 = 8$.

Solution: The hypothesis p is "$2^2 = 4$," which is true.
The conclusion q is "$2^3 = 8$," which is true.
The conditional $p \rightarrow q$ is true. *Ans.*

5. If 9 is an odd number, then 9 is prime.

Solution: The hypothesis p is "9 is an odd number," which is true.
The conclusion q is "9 is prime," which is false because 9 is divisible by 3.
The conditional $p \rightarrow q$ is false. *Ans.*

6. If a square has 5 sides, then $5 + 5 = 10$.

Solution: The antecedent p is "A square has 5 sides," which is false.
The consequent q is "$5 + 5 = 10$," which is true.
The conditional $p \rightarrow q$ is true. *Ans.*

7. If time goes backward, then I'll get younger every day.

Solution: The antecedent p is "Time goes backward," which is false.
The consequent q is "I'll get younger every day," which is false.
The conditional $p \rightarrow q$ is true. *Ans.*

In 8–12, for each given statement: **a.** Write the statement in symbolic form, using the symbols given below. **b.** Tell whether the statement is true or false.

Let m represent "Tuesday follows Monday." (True)
Let w represent "There are 7 days in one week." (True)
Let h represent "There are 40 hours in a day." (False)

	Answers
8. If Tuesday follows Monday, then there are 7 days in one week.	**a.** $m \rightarrow w$ **b.** $T \rightarrow T$ is true.
9. If there are 7 days in one week, then there are 40 hours in a day.	**a.** $w \rightarrow h$ **b.** $T \rightarrow F$ is false.
10. If Tuesday does not follow Monday, then there are not 7 days in one week.	**a.** $\sim m \rightarrow \sim w$ **b.** $F \rightarrow F$ is true.
11. Tuesday follows Monday if there are not 7 days in one week.	**a.** $\sim w \rightarrow m$ **b.** $F \rightarrow T$ is true.
12. If Tuesday follows Monday and there are 7 days in one week, then there are 40 hours in a day.	**a.** $(m \wedge w) \rightarrow h$ **b.** $(T \wedge T) \rightarrow F$ $T \rightarrow F$ is false.

EXERCISES

In 1–8, for each given sentence: **a.** Identify the hypothesis p. **b.** Identify the conclusion q.

1. If it rains, then the game is cancelled.
2. If it is 9:05 A.M., then I'm late to class.
3. When it rains, then I do not have to water the lawn.
4. You can get to the stadium if you take the Third Avenue bus.
5. The perimeter of a square is $4x + 8$ if one side of the square is $x + 2$.
6. If a polygon has exactly three sides, it is a triangle.

7. If the shoe fits, wear it.
8. When you have a headache, you should take time out and get some rest.

In 9–14, write each sentence in symbolic form, using the given symbols.

p: The test is easy. *q*: Sam studies. *r*: Sam passes the test.

9. If the test is easy, then Sam will pass the test.
10. If Sam studies, then Sam will pass the test.
11. If the test is not easy, then Sam will not pass the test.
12. The test is easy if Sam studies.
13. Sam will not pass the test if Sam doesn't study.
14. Sam passes the test if the test is easy.

In 15–22, for each given statement: **a.** Write the statement in symbolic form, using the symbols given below. **b.** Tell whether the conditional statement is true or false, based upon the truth values given.

r: The race is difficult. (True)
p: Karen practices. (False)
w: Karen wins the race. (True)

15. If Karen practices, then Karen will win the race.
16. If Karen wins the race, then Karen has practiced.
17. If Karen wins the race, the race is difficult.
18. Karen wins the race if the race is not difficult.
19. Karen will not win the race if Karen does not practice.
20. Karen practices if the race is difficult.
21. If the race is not difficult and Karen practices, then Karen will win the race.
22. If the race is difficult and Karen does not practice, then Karen will not win the race.

In 23–25, copy the truth table for the conditional and fill in the missing symbols.

23.

r	t	$r \rightarrow t$
T	*T*	*T*
T	*F*	*F*
F	*T*	
F	*F*	

24.

k	m	$k \rightarrow m$
T	*T*	*T*
F	*T*	*T*
F	*F*	*T*

25.

q	r	$q \rightarrow r$

In 26–33, find the truth value to be assigned to the conditional statement.

26. If $5 + 7 = 12$, then $7 + 5 = 12$.
27. If $3 > 10$, then $10 > 13$.
28. If $1 \cdot 1 = 1$, then $1 \cdot 1 \cdot 1 = 1$.
29. $12 \div 3 = 9$ if $12 \div 9 = 3$.
30. $2 + 2 = 2^2$ if $3 + 3 = 3^2$.
31. $2^3 = 3^2$ if $2^4 = 4^2$.
32. If $1 + 2 + 3 = 1 \cdot 2 \cdot 3$, then $1 + 2 = 1 \cdot 2$.
33. If every square is a rectangle, then every rectangle is a square.

In 34–49, symbols are assigned to represent sentences and truth values are assigned to these sentences.

Let j represent "I jog."	(True)
Let d represent "I diet."	(False)
Let g represent "I feel well."	(True)
Let h represent "I get hungry."	(True)

For the compound sentences in symbolic form: **a.** Write a complete sentence in words to show what the symbols represent. **b.** Tell whether the compound sentence is true or false.

34. $j \rightarrow g$
35. $d \rightarrow h$
36. $h \rightarrow d$
37. $\sim g \rightarrow j$
38. $\sim g \rightarrow \sim j$
39. $g \rightarrow \sim h$
40. $h \rightarrow \sim d$
41. $\sim d \rightarrow \sim h$
42. $\sim j \rightarrow \sim g$
43. $j \rightarrow d$
44. $(j \wedge h) \rightarrow g$
45. $j \rightarrow (h \wedge g)$
46. $(j \vee d) \rightarrow h$
47. $d \rightarrow (h \wedge \sim g)$
48. $\sim j \rightarrow (d \wedge h)$
49. $(j \wedge h) \rightarrow d$

In 50–55, state the word, phrase, or symbol that can be placed in the blank to make the resulting sentence true.

50. When p and q represent two simple sentences, the conditional *if p then q* is written symbolically as ____.
51. The conditional *if q then p* is written symbolically as ____.
52. The conditional $p \rightarrow q$ is false only when p is ____ and q is ____.
53. When the conclusion q is true, then $p \rightarrow q$ must be ____.
54. When the hypothesis p is false, then $p \rightarrow q$ must be ____.
55. If the hypothesis p is true and the conditional $p \rightarrow q$ is true, then the conclusion q must be ____.

In 56–60, three sentences are written. The truth values are given for the first two sentences. Determine if the third sentence is true, is false, or has an uncertain truth value.

56. If you read in dim light, then you can strain your eyes. (True)
You read in dim light. (True)
You can strain your eyes. (?)

57. If the quadrilateral has 4 right angles, then the quadrilateral is a square. (False)
The quadrilateral has 4 right angles. (True)
The quadrilateral is a square. (?)

58. If n is an odd number, then $2 \cdot n$ is an even number. (True)
$2 \cdot n$ is an even number. (True)
n is an odd number. (?)

59. If the report is late, then you will not get an A. (True)
The report is late. (False)
You will not get an A. (?)

60. Area $= \frac{1}{2}bh$, if the polygon is a triangle. (True)
The polygon is a triangle. (True)
Area $= \frac{1}{2}bh$ (?)

5-6 REVIEW EXERCISES

In 1–5, for each given statement: **a.** Write the statement in symbolic form, using the symbols given below. **b.** Tell whether the statement is true or false.

Let j represent "July follows June." (True)
Let a represent "August follows July." (True)
Let w represent "July is a winter month." (False)

1. If July follows June, then August follows July.
2. July follows June and July is a winter month.
3. July is a winter month or July follows June.
4. If August follows July, then July does not follow June.
5. August does not follow July and July is not a winter month.

In 6 and 7: **a.** Identify the hypothesis p. **b.** Identify the conclusion q.

6. If at first you don't succeed, then you should try again.
7. You will get a detention if you are late one more time.

8. Let p represent "I sleep" and let q represent "I'm grouchy." Using p and q, write in symbolic form: "If I don't sleep, I'm grouchy."
9. Let p represent "A parallelogram is a rhombus" and let q represent "Its sides are congruent." Using p and q, write in symbolic form: "A parallelogram is a rhombus if its sides are congruent."
10. Which whole number, when substituted for y, will make the following sentence true? $(y + 3 > 8) \wedge (y < 7)$

In 11–14, write the word, phrase, or symbol that can be placed in each blank to make the resulting statement true.

11. $\sim(\sim p)$ has the same truth value as ____.
12. When p is true and q is false, then $p \wedge \sim q$ is ____.
13. When $p \vee \sim q$ is false, then p is ____ and q is ____.
14. If the conclusion q is true, then $p \rightarrow q$ must be ____.

In 15–19, find the truth value of the sentence when a, b, and c are all true.

15. $\sim a$ **16.** $\sim b \wedge c$ **17.** $b \rightarrow \sim c$ **18.** $a \vee \sim b$ **19.** $\sim a \rightarrow \sim b$

20. Let the domain be $\{1, 2, 3, 4, \ldots, 10\}$.
Let p represent "$x > 5$."
Let q represent "x is prime."
Find the solution set for each of the following:
a. p **b.** $\sim p$ **c.** q **d.** $\sim q$ **e.** $p \vee q$ **f.** $p \wedge q$ **g.** $\sim p \wedge q$

21. Peter, Carl, and Ralph play different musical instruments and different sports. The instruments that the boys play are violin, cello, and flute. The sports that the boys play are baseball, tennis, and soccer. From the clues given below, determine what instrument and what sport each boy plays.
1. The violinist plays tennis.
2. The boy who plays the flute does not play soccer.
3. Peter does not play the cello.
4. Ralph plays baseball.

Chapter **6**

Using Logic

6-1 COMPOUND STATEMENTS AND TRUTH VALUES

Your study of logic to this point has included negation, conjunction, disjunction, and conditional, as shown in the following truth tables:

		Negations		Conjunction	Disjunction	Conditional
p	q	$\sim p$	$\sim q$	$p \wedge q$	$p \vee q$	$p \rightarrow q$
T	*T*	*F*	*F*	*T*	*T*	*T*
T	*F*	*F*	*T*	*F*	*T*	*F*
F	*T*	*T*	*F*	*F*	*T*	*T*
F	*F*	*T*	*T*	*F*	*F*	*T*

A compound sentence often contains more than one connective. To judge the truth value of any compound sentence, we examine the truth values of its component parts. When the truth value of every simple sentence is known within the compound being formed, we have a ***compound statement.***

For example: 8 is an even number *and* 8 is *not* prime.

We may express this compound statement in symbolic form, assigning a letter to represent every *simple "positive" statement.*

Let e represent "8 is an even number."
Let p represent "8 is prime."
Then $\sim p$ represents "8 is *not* prime."
While $e \wedge \sim p$ represents "8 is an even number *and* 8 is *not* prime."

Substituting truth values directly into the statement $e \wedge \sim p$, we get $T \wedge \sim F$, which reduces to $T \wedge T$, or simply T. The compound statement is true. Other examples appear in the model problems.

PROCEDURE. **To find the truth value of a compound statement:**

1. **Simplify the truth values within parentheses or other groupings, always working from the innermost group first.**
2. **Simplify negations.**
3. **Simplify other connectives, working from left to right.**

MODEL PROBLEMS

1. Let p represent "$7^2 = 49$" and let q represent "a rectangle is a parallelogram."

a. Write in symbolic form using p and q: $7^2 = 49$ *or* a rectangle is *not* a parallelogram.
b. Find the truth value of this compound statement.

Solution:

a. $p \vee \sim q$
b. Substitute the truth values for p and q: $T \vee \sim T$
Simplify the negation: $T \vee F$
Simplify the disjunction: T *Ans.*

In 2 and 3, find the truth value of the compound sentence when p, q, and r are all true.

2. $(\sim p \wedge r) \rightarrow q$

Solution: Substitute the truth values: $(\sim T \wedge T) \rightarrow T$
Within parentheses, negate: $(F \wedge T) \rightarrow T$
Simplify parentheses: $F \rightarrow T$
Simplify the conditional: T *Ans.*

3. $(p \rightarrow \sim q) \vee \sim r$

Solution: Substitute the truth values: $(T \rightarrow \sim T) \vee \sim T$
Within parentheses, negate: $(T \rightarrow F) \vee \sim T$
Simplify parentheses: $F \vee \sim T$
Simplify the negation: $F \vee F$
Simplify the disjunction: F *Ans.*

EXERCISES

In 1–4, write a statement in symbolic form to show the correct heading for column 3.

1.

p	q	Col. 3
T	T	T
T	F	F
F	T	F
F	F	F

2.

p	q	Col. 3
T	T	T
T	F	F
F	T	T
F	F	T

3.

p	q	Col. 3
T	T	T
T	F	T
F	T	T
F	F	F

4.

p	q	Col. 3
T	T	F
T	F	T
F	T	F
F	F	T

In 5–12, select the numeral preceding the word or expression that best completes the statement or answers the question.

5. When $p \vee q$ is false, then:
(1) p is true and q is false. (2) p is false and q is true.
(3) p and q are both false. (4) p and q are both true.

6. When $p \rightarrow q$ is false, then:
(1) p is true and q is false. (2) p is false and q is true.
(3) p and q are both false. (4) p and q are both true.

7. If p represents "x is divisible by 2" and q represents "x is divisible by 5," then which is true when $x = 14$?
(1) $p \wedge q$ (2) $p \vee q$ (3) $\sim p$ (4) q

8. If p represents "the polygon has 4 sides" and q represents "opposite sides of the polygon are parallel," then which is true when the polygon is a triangle?
(1) $p \wedge q$ (2) $p \vee q$ (3) $\sim p$ (4) q

9. If p represents "x is a prime number" and q represents "x is divisible by 3," then which is true when $x = 9$?
(1) $p \wedge q$ (2) $p \vee q$ (3) $\sim q$ (4) p

10. If $p \wedge q$ is true, then:
(1) $p \vee q$ is false. (2) $p \rightarrow q$ is false.
(3) $p \vee q$ is true. (4) $\sim(p \wedge q)$ is true.

11. When p is true and q is false, then:
(1) $p \wedge q$ is true. (2) $p \rightarrow q$ is false.
(3) $p \vee q$ is false. (4) $\sim p \wedge \sim q$ is true.

12. Let p represent "$3x + 1 = 13$" and let q represent "$2x + 3x = 25$." Which is true when $x = 4$?
(1) $p \wedge q$ (2) $p \rightarrow q$ (3) $p \vee q$ (4) $\sim p$

In 13–24: Let m represent "28 is a multiple of 7." (True)
Let s represent "28 is the square of an integer." (False)
Let f represent "7 is a factor of 28." (True)

a. Write a correct translation in words for the statement given in symbolic form.
b. Give the truth value of the compound statement.

13. $m \wedge \sim s$ **14.** $f \vee \sim m$ **15.** $\sim m \rightarrow s$ **16.** $\sim(f \wedge s)$
17. $s \vee \sim f$ **18.** $\sim(m \vee s)$ **19.** $\sim f \rightarrow \sim m$ **20.** $s \rightarrow \sim m$
21. $(m \wedge f) \rightarrow s$ **22.** $m \rightarrow (s \vee f)$ **23.** $(m \vee s) \rightarrow f$ **24.** $f \rightarrow (s \vee m)$

In 25–32, find the truth value for the given statement.

25. If $2 + 3 = 5$ and $3 + 5 = 8$, then $5 + 8 = 13$.
26. If $2^2 = 4$ and $2 + 2 = 4$, then $3^2 = 3 + 3$.
27. If $2^2 = 4$, then $3^2 = 6$ and $4^2 = 8$.
28. If $4 + 8 = 10$ or $4 + 8 = 32$, then $10 - 8 = 4$ or $32 - 8 = 4$.
29. If $2(3) = 5$ and $3(2) = 5$, then $2(3) = 3(2)$.
30. $2^3 = 3^2$ if $2^3 = 6$ and $3^2 = 6$.
31. It is not true that $3 + 10 = 13$ and $3 - 10 = 7$.
32. It is not true that 12 is even and prime.

In 33–44, find the truth value of the compound sentence when p, q, and r are all true.

33. $p \rightarrow \sim q$ **34.** $p \wedge \sim r$ **35.** $q \vee \sim p$ **36.** $p \wedge \sim p$
37. $\sim(p \vee q)$ **38.** $\sim q \rightarrow r$ **39.** $\sim r \rightarrow \sim p$ **40.** $p \vee \sim p$
41. $\sim p \vee \sim q$ **42.** $(p \wedge q) \rightarrow r$ **43.** $(p \vee q) \vee r$ **44.** $(p \wedge q) \vee \sim r$

45. Elmer Megabucks does not believe that girls should marry before the age of 21 and he disapproves of smoking. Therefore, he put the following provision in his will: I leave $100,000 to each of my nieces who, at the time of my death, is over 21 or unmarried, and does not smoke.

Each of his nieces is described below at the time of Elmer's death. Which nieces will inherit $100,000?

Judy is 24, married, and smokes.
Diane is 20, married, and does not smoke.
Janice is 26, unmarried, and does not smoke.
Peg is 19, unmarried, and smokes.
Sue is 30, unmarried, and smokes.
Sarah is 18, unmarried, and does not smoke.
Laurie is 28, married, and does not smoke.
Pam is 19, married, and smokes.

46. Some years after Elmer Megabucks prepared his will, he amended the conditions, by moving a comma, to read: I leave $100,000 to each of my nieces who, at the time of my death, is over 21, or unmarried and does not smoke. Which nieces, described in Exercise 45, will now inherit $100,000?

47. Let p represent "x is divisible by 6."
Let q represent "x is divisible by 2."

a. If possible, find a value of x that will:
(1) make p true and q true.
(2) make p true and q false.
(3) make p false and q true.
(4) make p false and q false.

b. What conclusion can be drawn about the truth value of the compound sentence $p \rightarrow q$?

6-2 COMPOUND SENTENCES AND TRUTH TABLES

A truth table shows all possible truth values of a logic statement. To construct a truth table for a compound sentence that uses two or more logical connectives, we consider one logical connective at a time.

MODEL PROBLEM

Construct a truth table for the sentence $q \rightarrow \sim(p \vee \sim q)$.

Solution:

(1) List truth values for p and q in the first two columns.

p	q
T	T
T	F
F	T
F	F

(2) Consider the expression within parentheses, $(p \vee \sim q)$. First get the negation of q by negating column 2.

p	q	$\sim q$
T	T	F
T	F	T
F	T	F
F	F	T

(3) Then, find the truth values for $(p \vee \sim q)$ by combining columns 1 and 3 under disjunction.

p	q	$\sim q$	$p \vee \sim q$
T	T	F	T
T	F	T	T
F	T	F	F
F	F	T	T

(4) Negate column 4 to find the truth values for $\sim(p \vee \sim q)$.

p	q	$\sim q$	$p \vee \sim q$	$\sim(p \vee \sim q)$
T	T	F	T	F
T	F	T	T	F
F	T	F	F	T
F	F	T	T	F

(5) Use column 2 as the hypothesis and column 5 as the conclusion to arrive at the truth values for the conditional sentence.

p	q	$\sim q$	$p \vee \sim q$	$\sim(p \vee \sim q)$	$q \rightarrow \sim(p \vee \sim q)$
T	T	F	T	F	F
T	F	T	T	F	T
F	T	F	F	T	T
F	F	T	T	F	T

Observe that the compound sentence, $q \rightarrow \sim(p \vee \sim q)$, is sometimes true and sometimes false, depending upon the truth values of p and q.

EXERCISES

In 1–8, copy and complete the truth table for the given sentence which is the last column head on the right. (In 2–8, prepare a truth table similar to the one shown in Exercise 1.)

1.

p	q	$p \vee q$	$\sim q$	$(p \vee q) \rightarrow \sim q$
T	T			
T	F			
F	T			
F	F			

2.

p	q	$p \wedge q$	$p \vee q$	$(p \wedge q) \rightarrow (p \vee q)$

3.

p	q	$\sim p$	$q \rightarrow \sim p$	$(q \rightarrow \sim p) \wedge p$

4.

p	q	$\sim p$	$\sim p \wedge q$	$p \vee (\sim p \wedge q)$

5.

p	q	$\sim p$	$p \vee \sim p$	$q \rightarrow (p \vee \sim p)$

6.

p	q	$p \rightarrow q$	$\sim q$	$(p \rightarrow q) \rightarrow \sim q$

7.

p	q	$\sim p$	$\sim p \vee q$	$p \wedge q$	$(\sim p \vee q) \rightarrow (p \wedge q)$

8.

p	q	$p \wedge q$	$\sim(p \wedge q)$	$\sim p$	$\sim q$	$\sim p \wedge \sim q$	$\sim(p \wedge q) \rightarrow (\sim p \wedge \sim q)$

In 9–14, construct a truth table for the given sentence.

9. $\sim q \rightarrow (\sim q \wedge p)$ **10.** $p \rightarrow \sim(p \vee q)$ **11.** $\sim(p \wedge q) \vee p$

12. $(\sim p \wedge q) \vee p$ **13.** $(p \rightarrow \sim q) \wedge q$ **14.** $(\sim p \wedge \sim q) \vee (p \wedge q)$

15. Given the sentence: "*If* a number is *not* even, *then* it is even *or* it is prime."

Let e represent "A number is even."
Let p represent "A number is prime."

a. Write the compound sentence in symbolic form, using e and p.
b. Construct a truth table for the compound sentence.
c. For any case where the compound sentence is false, give the truth values of e and of p.
d. Find a whole number that fits the truth values listed in part **c**.

6-3 BICONDITIONALS

The truth table for the **conditional $p \to q$** is shown below. The conditional $p \to q$ is false when the hypothesis p is true and the conclusion q is false as shown in row 2. In all other cases, the conditional $p \to q$ is true.

When we reverse the order of the sentences p and q, we form a new **conditional $q \to p$**. The conditional $q \to p$ is false when the hypothesis q is true and the conclusion p is false, as shown in row 3. In all other cases, the conditional $q \to p$ is true.

Hypothesis (p) **Conclusion** (q)

p	q	p → q
T	*T*	*T*
T	*F*	*F*
F	*T*	*T*
F	*F*	*T*

Conclusion (p) **Hypothesis** (q)

p	q	q → p
T	*T*	*T*
T	*F*	*T*
F	*T*	*F*
F	*F*	*T*

The prefix *bi-* means *two* as in bicycle, binary, and bifocals. The biconditional consists of two conditionals. In logic, the ***biconditional*** is a compound sentence formed by the conjunction of the two conditionals $p \to q$ and $q \to p$. To find the truth value of the biconditional, we can construct a truth table for the sentence $(p \to q) \wedge (q \to p)$.

p	q	p → q	q → p	(p → q) ∧ (q → p)
T	*T*	*T*	*T*	*T*
T	*F*	*F*	*T*	*F*
F	*T*	*T*	*F*	*F*
F	*F*	*T*	*T*	*T*

The compound sentence formed in the truth table is both lengthy to read and to write.

$(p \to q) \wedge (q \to p)$: **If p then q, and if q then p**

or

p implies q, and q implies p.

We shorten the writing of the biconditional by introducing the symbol, $p \leftrightarrow q$, to replace the compound sentence. We shorten the reading of the biconditional by using the words **p if and only if q**. These abbreviated versions, as well as the lengthy ones, are all acceptable ways of indicating the biconditional. From the truth table, we observe that:

● **The biconditional *p if and only if q* is true when p and q are both true or both false.**

In other words, $p \leftrightarrow q$ is true when p and q have the **same truth value**. When p and q have different truth values, the biconditional is false.

Building the Truth Table for the *Biconditional* $p \leftrightarrow q$

1. Assign two letters p and q, each to serve as a symbol for a different simple sentence.
2. List all possible truth values for p and q in columns 1 and 2.
3. In the third column, write the biconditional in symbolic form.

p	q	$p \leftrightarrow q$
T	T	
T	F	
F	T	
F	F	

4. Finally, assign truth values for the biconditional. When p and q have the same truth value, as in rows 1 and 4, the biconditional is true. When p and q have different truth values, as in rows 2 and 3, the biconditional is false.

p	q	$p \leftrightarrow q$
T	T	T
T	F	F
F	T	F
F	F	T

Truth Table for Biconditional

Applications of the Biconditional

The truth table shows that the biconditional is not true in all cases. For example:

$p \leftrightarrow q$: $x > 5$ *if and only if* $x > 3$.

or

$p \leftrightarrow q$: If $x > 5$, then $x > 3$, *and* if $x > 3$, then $x > 5$.

This biconditional is *false* because it fails for certain numbers. Let $x = 4$. Then p, $x > 5$, is false while q, $x > 3$, is true. Since the truth values are not the same, the biconditional is not true.

However, there are many examples where the biconditional is always true.

EXAMPLE 1. Definitions

Two conditionals are stated.

$p \rightarrow t$: If a polygon has exactly 3 sides, then it is a triangle.

$t \rightarrow p$: If a polygon is a triangle, then it has exactly 3 sides.

The conjunction $(p \rightarrow t) \wedge (t \rightarrow p)$ is the biconditional $(p \leftrightarrow t)$. When p is true, t is true. When p is false, t is false. Since the truth values are the same, the biconditional is always true. *We use the biconditional to serve as a definition.*

$p \leftrightarrow t$: A polygon is a triangle *if and only if* it has exactly three sides.

Precise definitions in mathematics are often stated using the words *if and only if*. The *if* tells us what we must include, as in "exactly three sides." The *only if* tells us what to exclude, as in "all cases that do not have exactly three sides."

EXAMPLE 2. Equations

$$\begin{aligned} 2x + 1 &= 17 \\ 2x &= 16 \\ x &= 8 \end{aligned}$$

In simplifying equations, we follow a process, as shown at the left. Since $x = 8$ will be true in every equation listed here and any other value will be false, we can reverse this process.

$a \rightarrow b$: $2x + 1 = 17$ implies $2x = 16$.
$b \rightarrow a$: $2x = 16$ implies $2x + 1 = 17$.

Equations of this type form biconditional statements.

$a \leftrightarrow b$: $2x + 1 = 17$ *if and only if* $2x = 16$.
$b \leftrightarrow c$: $2x = 16$ *if and only if* $x = 8$.

EXAMPLE 3. Equivalences

When any two statements always have the same truth value, we can substitute one statement for the other. The statements are said to be ***logically equivalent***. You will see examples of equivalences in the next section.

MODEL PROBLEMS

In 1 and 2, identify the truth value to be assigned to each biconditional.

1. Cars stop if and only if there is a red light.

Solution: Let p represent "Cars stop."
Let q represent "There is a red light."

When p is true, it does not follow that q must be true. Cars stop at stop signs, railroad crossings, or when parking. Since p and q do not have the same truth value, the biconditional $p \leftrightarrow q$ is false.

Answer: False

2. $x + 2 = 7$ if and only if $x = 5$.

Solution: Let p represent "$x + 2 = 7$."
Let q represent "$x = 5$."

When $x = 5$, both p and q are true. When $x \neq 5$, both p and q are false. In any event, p and q have the same truth value. Thus, the biconditional $p \leftrightarrow q$ is true.

Answer: True

EXERCISES

In 1–5, write each biconditional in symbolic form, using the symbols given.

t: The triangle is a right triangle.
r: The triangle contains a right angle.
n: The triangle contains a 90° angle.

1. A triangle is a right triangle if and only if it contains a right angle.
2. A triangle contains a 90° angle if and only if it is a right triangle.
3. A triangle contains a 90° angle if and only if it contains a right angle.
4. If a triangle is a right triangle, then it contains a 90° angle, and if the triangle contains a 90° angle, then it is a right triangle.
5. A triangle is not a right triangle if and only if it does not contain a 90° angle.

In 6–8, copy the truth table for the biconditional and fill in the missing symbols.

6.

p	q	$p \leftrightarrow q$
T	T	
T	F	
F	T	
F	F	

7.

r	s	$r \leftrightarrow s$
T	T	
T	F	F
F	T	F
F	F	

8.

d	k	$d \leftrightarrow k$

In 9–15, complete the truth tables, filling in all missing symbols. (In 10–15, prepare a truth table similar to the one shown in Exercise 9.)

9.

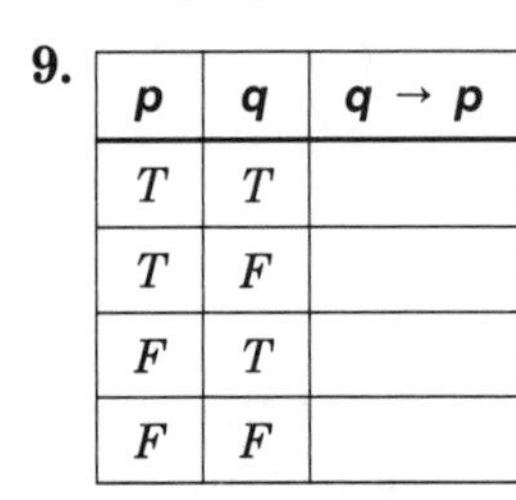

p	q	$q \rightarrow p$
T	T	
T	F	
F	T	
F	F	

10.

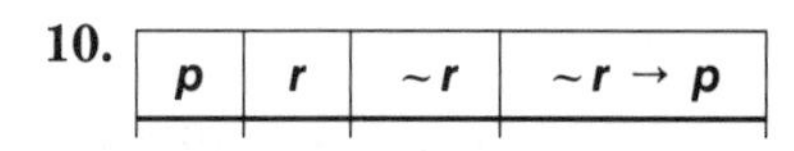

p	r	$\sim r$	$\sim r \rightarrow p$

11.

t	v	$t \wedge v$	$(t \wedge v) \rightarrow t$

12.

b	c	$b \vee c$	$(b \vee c) \leftrightarrow c$

13.

p	t	$t \rightarrow p$	$p \rightarrow t$	$(t \rightarrow p) \wedge (p \rightarrow t)$

14.

p	q	$q \rightarrow p$	$(q \rightarrow p) \rightarrow q$

15.

r	s	$\sim r$	$\sim r \rightarrow s$	$(\sim r \rightarrow s) \leftrightarrow r$

In 16–25, identify the truth value to be assigned to each biconditional.

16. $x + 3 = 30$ if and only if $x = 27$.
17. A polygon is a pentagon if and only if it has exactly 3 sides.
18. $x + 4 = 12$ if and only if $x + 6 = 14$.
19. A rectangle is a square if and only if the rectangle has all sides of equal length.
20. The refrigerator runs if and only if the electricity is on.
21. An angle is right if and only if it contains 90°.
22. A set is empty if and only if it contains no elements.
23. A number is even if and only if it is exactly divisible by 2.

24. Two angles have the same measure if and only if they are right angles.

25. A sentence is a statement if and only if it has truth values.

26. Using the symbols p and q, write each compound sentence in symbolic form.

a. p if and only if q.
b. if p then q and if q then p.
c. p implies q and q implies p.
d. q if and only if p.
e. if p then q or if q then p.

27. Of the five responses given in Exercise 26, four of the answers name the biconditional, $p \leftrightarrow q$. Which choice does *not* represent the biconditional?

6-4 TAUTOLOGIES

We have seen many compound sentences that are sometimes true and sometimes false.

In logic, a ***tautology*** is a compound sentence that is *always true*, regardless of the truth values assigned to the simple sentences within the compound. For example, $(p \wedge q) \rightarrow p$ is a tautology.

To demonstrate that the compound $(p \wedge q) \rightarrow p$ will always be true, we build a truth table. No matter what truth values p and q have, every element in the last column is true. This tells us that we have a tautology, or a basic truth in logic.

p	q	$p \wedge q$	$(p \wedge q) \rightarrow p$
T	T	T	T
T	F	F	T
F	T	F	T
F	F	F	T

tautology

Examples of Tautologies

EXAMPLE 1. The *simplest tautology* is seen in the sentence $p \vee \sim p$. Its truth table is shown at the right. We may replace the symbol p with any simple sentence, whether its truth values are known or not.

p	$\sim p$	$p \vee \sim p$
T	F	T
F	T	T

tautology

$p \vee \sim p$: A number is odd or a number is not odd.
$p \vee \sim p$: A square is a triangle or a square is not a triangle.
$p \vee \sim p$: It will rain or it will not rain.
$p \vee \sim p$: $x + 3 = 15$ or $x + 3 \neq 15$.
$p \vee \sim p$: A statement is true or a statement is not true.

This last interpretation of $p \vee \sim p$ can also be read as "A statement is true or false."

EXAMPLE 2. Tautologies are used to develop strong arguments. For example, these two statements are made:

$r \rightarrow m$: If it rains, then we'll go to the movies.
r: It rains.

What can we conclude?

m: We'll go to the movies.

Combine the first two sentences as a conjunction. This compound sentence will be used as the hypothesis of a conditional.

The reasoning becomes: *If* $r \rightarrow m$ *and* r, then m.

Written symbolically: $[(r \rightarrow m) \wedge r] \rightarrow m$

Test the reasoning in a truth table, working from the innermost parentheses first. Since, as shown in the last column, the statement is always true, this is a tautology.

r	m	$r \rightarrow m$	$(r \rightarrow m) \wedge r$	$[(r \rightarrow m) \wedge r] \rightarrow m$
T	T	T	T	T
T	F	F	F	T
F	T	T	F	T
F	F	T	F	T

tautology

EXAMPLE 3. When two statements always have the same truth values, the two statements are ***logically equivalent***.

To show that an *equivalence* exists, compare the two statements, using the biconditional. If the statements have the same truth values, then the biconditional will be true in every case. The biconditional will be a tautology.

Let p represent "I study."
Let q represent "I fail."

The two statements being tested for an equivalence follow.

$\sim p \rightarrow q$: If I don't study, then I'll fail.
$p \vee q$: I study or I fail.

Use the biconditional to test the statement:

$$(\sim p \rightarrow q) \leftrightarrow (p \vee q)$$

In the truth table, work within parentheses first to find $\sim p$, then $\sim p \rightarrow q$. Notice that the truth values of $\sim p \rightarrow q$ in column 4 match exactly the truth values of $p \vee q$ in column 5. The last column shows that the biconditional will always be true.

p	q	$\sim p$	$\sim p \rightarrow q$	$p \vee q$	$(\sim p \rightarrow q) \leftrightarrow (p \vee q)$
T	T	F	T	T	T
T	F	F	T	T	T
F	T	T	T	T	T
F	F	T	F	F	T

Therefore, $p \vee q$ and $\sim p \rightarrow q$ are called ***equivalent statements***; they say the same thing in two different ways. This is verified by their matching truth values.

MODEL PROBLEMS

1. a. On your paper, copy and complete the truth table for the statement:

$$(p \rightarrow \sim q) \leftrightarrow (\sim p \vee \sim q)$$

p	q	$\sim p$	$\sim q$	$p \rightarrow \sim q$	$\sim p \vee \sim q$	$(p \rightarrow \sim q) \leftrightarrow (\sim p \vee \sim q)$

b. Is $(p \rightarrow \sim q) \leftrightarrow (\sim p \vee \sim q)$ a tautology?

c. Let p represent "I get a job."
Let q represent "I go to the dance."
Which sentence is equivalent to $(p \rightarrow \sim q)$?

(1) If I don't get a job, then I won't go to the dance.
(2) I get a job or I don't go to the dance.
(3) I don't get a job or I don't go to the dance.
(4) If I get a job, then I'll go to the dance.

Solution

a.

p	q	$\sim p$	$\sim q$	$p \to \sim q$	$\sim p \lor \sim q$	$(p \to \sim q) \leftrightarrow (\sim p \lor \sim q)$
T	T	F	F	F	F	T
T	F	F	T	T	T	T
F	T	T	F	T	T	T
F	F	T	T	T	T	T

b. Since $(p \to \sim q) \leftrightarrow (\sim p \lor \sim q)$ is always true, this statement is a tautology. *Answer:* Yes

c. Because the compound sentence $(p \to \sim q) \leftrightarrow (\sim p \lor \sim q)$ is a tautology, $(p \to \sim q)$ and $(\sim p \lor \sim q)$ are logically equivalent statements. Given $(p \to \sim q)$, look for $(\sim p \lor \sim q)$. In choice 3, $(\sim p \lor \sim q)$: I don't get a job or I don't go to the dance.
Answer: (3)

2. **a.** Construct the truth tables for each of the following statements:
1. $\sim q \to p$ 2. $\sim(p \to q)$ 3. $p \lor q$

b. Which, if any, of the three statements in part **a** are logically equivalent? State a tautology to show this equivalence or give a reason why there is no equivalence.

Solution

a. 1. $\sim q \to p$

p	q	$\sim q$	$\sim q \to p$
T	T	F	T
T	F	T	T
F	T	F	T
F	F	T	F

a. 2. $\sim(p \to q)$

p	q	$p \to q$	$\sim(p \to q)$
T	T	T	F
T	F	F	T
F	T	T	F
F	F	T	F

a. 3. $p \lor q$

p	q	$p \lor q$
T	T	T
T	F	T
F	T	T
F	F	F

b. Because the truth tables for $(\sim q \to p)$ and $(p \lor q)$ have the same truth values, statement 1 and statement 3 are logically equivalent. We can write the tautology:

$$(\sim q \to p) \leftrightarrow (p \lor q)$$

3. Is $[(p \wedge q) \vee \sim q] \rightarrow \sim p$ a tautology? Give a reason why.

Solution

p	q	$p \wedge q$	$\sim q$	$(p \wedge q) \vee \sim q$	$\sim p$	$[(p \wedge q) \vee \sim q] \rightarrow \sim p$
T	*T*	*T*	*F*	*T*	*F*	*F*
T	*F*	*F*	*T*	*T*	*F*	*F*
F	*T*	*F*	*F*	*F*	*T*	*T*
F	*F*	*F*	*T*	*T*	*T*	*T*

Answer: No. This is not a tautology because the statement is false in rows 1 and 2.

EXERCISES

In 1–10: **a.** Copy and complete the truth table for the given statement. (*Note:* In 6–10, prepare a complete truth table similar to the one in Exercise 5.) **b.** Indicate if the statement is or is not a tautology.

1.

p	$\sim p$	$\sim(\sim p)$	$p \leftrightarrow \sim(\sim p)$
T			
F			

2.

p	$\sim p$	$\sim p \wedge p$	$\sim(\sim p \wedge p)$
T			
F			

3.

p	$\sim p$	$p \rightarrow \sim p$	$(p \rightarrow \sim p) \leftrightarrow \sim p$
T			
F			

4.

q	$\sim q$	$\sim q \rightarrow q$
T		
F		

5.

p	q	$\sim p$	$\sim p \vee q$	$p \vee (\sim p \vee q)$
T	*T*			
T	*F*			
F	*T*			
F	*F*			

6.

p	q	$p \wedge q$	$p \vee q$	$(p \wedge q) \rightarrow (p \vee q)$

7.

p	q	$p \vee q$	$p \wedge q$	$(p \vee q) \rightarrow (p \wedge q)$

8.

p	q	$p \vee q$	$p \to (p \vee q)$

9.

p	q	$\sim p$	$\sim p \wedge q$	$p \wedge (\sim p \wedge q)$	$\sim[p \wedge (\sim p \wedge q)]$

10.

p	q	$\sim q$	$p \to \sim q$	$\sim(p \to \sim q)$	$p \wedge q$	$\sim(p \to \sim q) \leftrightarrow (p \wedge q)$

In 11–20, construct truth tables for the given tautologies.

11. $p \to (p \vee q)$
12. $(p \wedge q) \leftrightarrow (q \wedge p)$
13. $(p \vee p) \to p$
14. $(p \wedge q) \to q$
15. $[p \vee (p \wedge q)] \leftrightarrow p$
16. $(\sim p \vee q) \to (p \to q)$
17. $[(p \to q) \wedge \sim q] \to \sim p$
18. $\sim[(\sim p \vee q) \vee \sim q] \to (p \wedge q)$
19. $\sim(p \vee q) \leftrightarrow (\sim p \wedge \sim q)$
20. $[p \wedge (\sim p \vee q)] \leftrightarrow (p \wedge q)$

In 21–25: **a.** Construct a truth table for each of the three statements in the row. **b.** Using the results from part **a**, either write a tautology that states that two of the three statements are logically equivalent, or tell why no tautology exists.

21. (1) $q \to \sim p$ (2) $\sim p \vee \sim q$ (3) $\sim q \to \sim p$
22. (1) $q \vee \sim p$ (2) $p \vee \sim q$ (3) $p \to q$
23. (1) $\sim p \to q$ (2) $q \wedge \sim p$ (3) $\sim q \to p$
24. (1) $p \wedge q$ (2) $p \leftrightarrow q$ (3) $\sim p \leftrightarrow \sim q$
25. (1) $p \to \sim q$ (2) $\sim(p \wedge q)$ (3) $p \leftrightarrow \sim q$

26. **a.** Construct a truth table for the statement $(p \to q) \leftrightarrow (\sim p \vee q)$.
b. p: I like baseball. q: I join the team.
Which sentence, if any, is logically equivalent to $(p \to q)$?
(1) I like baseball or I join the team.
(2) I don't like baseball or I join the team.
(3) If I like baseball, then I don't join the team.
(4) If I don't like baseball, then I join the team.

27. **a.** Construct a truth table for the statement $\sim(p \vee \sim q) \leftrightarrow (\sim p \wedge q)$.
b. p: I save money. q: I work.
Which sentence, if any, is logically equivalent to $\sim(p \vee \sim q)$?
(1) I don't save money and I work.
(2) If I don't save money, then I don't work.
(3) I save money or I don't work.
(4) I don't work and I save money.

28. **a.** Construct a truth table for the statement $(\sim p \rightarrow q) \leftrightarrow (p \vee q)$.

b. Let p represent "I get home late."
Let q represent "We'll go out."
Tell which sentence, if any, is logically equivalent to "If I don't get home late, then we'll go out."
(1) If I get home late, then we'll go out.
(2) If I get home late, then we won't go out.
(3) I get home late and we go out.
(4) I get home late or we'll go out.

29. **a.** Construct a truth table for the statement
$[p \vee (q \wedge \sim p)] \leftrightarrow (p \vee q)$.

b. Let p represent "I'll go to college."
Let q represent "I'll go to work."
Which sentence, if any, is logically equivalent to $(p \vee q)$?
(1) I'll go to college or I won't go to work.
(2) I'll go to work or I won't go to college.
(3) I'll go to college or I'll go to work and not college.
(4) I'll go to work or I'll go to college and work.

30. l: Mark stays up late at night.
t: Mark is tired in the morning.

a. Write the symbolic form of the sentence "If Mark stays up late at night, then he is tired in the morning."

b. Write the symbolic form of the sentence "If Mark is not tired in the morning, then he did not stay up late at night."

c. Test to see if these sentences in **a** and **b** are equivalent by constructing a truth table, using the biconditional.

31. George made two statements, shown symbolically below.
$f \vee s$: I will see you on Friday or Saturday.
$\sim f$: I won't see you on Friday.
Can we conclude that s is true? (*Hint:* Let s represent "I will see you on Saturday.") Answer the question by constructing a truth table for the compound sentence $[(f \vee s) \wedge \sim f] \rightarrow s$.

6-5 INVERSES, CONVERSES, AND CONTRAPOSITIVES

The conditional *if p then q* is the connective most often used in reasoning. Too often, people attempt to win an argument or to make a point by "twisting words around." In order to help you avoid becoming the victim of this kind of argument, you will study three new conditionals, each of which is formed by making some changes in an original conditional statement. These new conditionals are called the *inverse*, the *converse*, and the *contrapositive*.

The Inverse

Starting with an original conditional ($p \rightarrow q$), we form the ***inverse*** ($\sim p \rightarrow \sim q$) by negating the hypothesis and negating the conclusion. The symbols for the inverse may be read as *not p implies not q*, or *if not p, then not q*.

				Conditional	Inverse
p	q	$\sim p$	$\sim q$	$p \rightarrow q$	$\sim p \rightarrow \sim q$
T	T	F	F	T	T
T	F	F	T	F	T
F	T	T	F	T	F
F	F	T	T	T	T

Note that a conditional and its inverse are *not equivalent* statements because they do not have matching truth values in every case.

EXAMPLE 1. In advertising a product, the manufacturer sometimes would like the reader to assume the truth of the inverse of a true conditional statement. For example, the Spritz Company makes only root beer. We can agree that the conditional "If you have a Spritz, then you have a root beer" is true. Would it follow that the company's advertisement, "When you're out of Spritz, then you're out of root beer," is also true? Let us see.

Conditional ($p \rightarrow q$): If you have a Spritz, then you have a root beer.

Inverse ($\sim p \rightarrow \sim q$): If you do *not* have a Spritz, then you do *not* have a root beer.

Notice that the advertisement is a "hidden" form of the inverse.

Inverse ($\sim p \rightarrow \sim q$): When you're out of Spritz, then you're out of root beer.

Suppose we do not have a Spritz, but we have a root beer from another manufacturer. Then p is false and q is true. We see that in this case the conditional has a false hypothesis and a true conclusion and is true, and the inverse has a true hypothesis and a false conclusion and is false.

This conditional and its inverse do not always have the same truth value.

EXAMPLE 2. s: It is spring. m: The month is May.

Conditional ($s \to m$): If it is spring, then the month is May.

Inverse ($\sim s \to \sim m$): If it is not spring, then the month is not May.

The calendar time for spring begins in March and ends in June.

To compare the truth values of the conditional and the inverse, suppose the month is April. Then, s is true and m is false. The conditional has a true hypothesis and a false conclusion and is false. The inverse has a false hypothesis and a true conclusion and is true.

This conditional and its inverse do not always have the same truth value.

EXAMPLE 3. There are instances in which a conditional and its inverse do have the same truth value.

Conditional ($e \to f$): If $x + 3 = 7$, then $x = 4$.

Inverse ($\sim e \to \sim f$): If $x + 3 \neq 7$, then $x \neq 4$.

When $x = 4$, e and f are both true, as in row 1 of the truth table.

When $x \neq 4$, e and f are both false, as in row 4 of the truth table.

There are no values of x for which e is true and f is false, as in row 2, or for which e is false and f is true, as in row 3.

Therefore, this conditional and its inverse always have the same truth value.

Hence, we must judge the truth value of each inverse on its own merits.

● **Conclusion: When a conditional ($p \to q$) is true, its inverse ($\sim p \to \sim q$) may be true or it may be false.**

The Converse

Starting with an original conditional ($p \to q$), we form the ***converse*** ($q \to p$) by interchanging the hypothesis and the conclusion. The symbols for the converse may be read as *q implies p* or *if q, then p.*

Does the converse have the same truth value as the conditional? To answer this question, we will construct a truth table. Note that the conditional and its converse are *not equivalent* statements because they do not have matching truth values in every case.

		Conditional	Converse
p	q	$p \to q$	$q \to p$
T	T	T	T
T	F	F	T
F	T	T	F
F	F	T	T

EXAMPLE 1. p: It rained. q: The ground got wet.

Conditional ($p \to q$): If it rained, then the ground got wet.

Converse ($q \to p$): If the ground got wet, then it rained.

In rows 1, 3, and 4, the conditional ($p \to q$) is true. If it rained, then the ground got wet. However, we notice that in row 3, the converse ($q \to p$) is false. Suppose that the ground got wet, but the conclusion, it rained, is false. A sprinkler or a burst pipe or snow could have caused the ground to become wet.

Since we can find instances where the converse is not true when the conditional is true, then the conditional and its converse do not always have the same truth value.

EXAMPLE 2. A television commercial shows a series of beautiful women, all of whom use Cleanse soap. Assuming that these models really do use the product, let us also assume that the following conditional is true.

Conditional ($b \to c$): If you are beautiful, then you use Cleanse soap. Of course, what the advertiser wants you to believe is that the converse is true. This is not necessarily so.

Converse ($c \to b$): If you use Cleanse soap, then you will be beautiful. This converse is false because using the soap will not guarantee that you will become beautiful.

EXAMPLE 3. There are instances in which a conditional and its converse are both true.

Conditional ($h \to s$): If a polygon is a hexagon, then the polygon has exactly six sides.

Converse ($s \rightarrow h$): If a polygon has exactly six sides, then the polygon is a hexagon.

We must judge the truth value of each converse on its own merits.

● **Conclusion: When a conditional ($p \rightarrow q$) is true, its converse ($q \rightarrow p$) may be true or it may be false.**

Remember that we normally start with a true conditional when reasoning. Observe that when we start with a false conditional, as is shown in the second row of the truth table on page 178, the converse is true.

The Contrapositive

Starting with a conditional ($p \rightarrow q$), we form the ***contrapositive*** ($\sim q \rightarrow \sim p$) by negating both the hypothesis and the conclusion, and then interchanging the resulting negations. The symbols for the contrapositive may be read as *not q implies not p* or *if not q, then not p.*

Does the contrapositive have the same truth value as the conditional? To answer this question, we will construct a truth table.

				Conditional	Contrapositive
p	q	$\sim q$	$\sim p$	$p \rightarrow q$	$\sim q \rightarrow \sim p$
T	T	F	F	T	T
T	F	T	F	F	F
F	T	F	T	T	T
F	F	T	T	T	T

From the table, we see that **the conditional and its contrapositive are logically equivalent statements.**

EXAMPLE 1. p: It rained. q: The ground got wet.

Conditional ($p \rightarrow q$): If it rained, then the ground got wet.

Contrapositive ($\sim q \rightarrow \sim p$): If the ground did not get wet, then it did not rain.

Because the conditional and its contrapositive have matching truth values in every case, we can write a *tautology*:

$$(p \rightarrow q) \leftrightarrow (\sim q \rightarrow \sim p)$$

Therefore, this conditional and its contrapositive always have the same truth value.

EXAMPLE 2. Let us consider an example in which we start with a *false* conditional. We will see that the contrapositive, as in the second row of the previous truth table, must also be *false*.

Conditional ($o \rightarrow r$): If 15 is an odd number, then 15 is a prime number. Here, ($o \rightarrow r$) is false because ($T \rightarrow F$) is false.

Contrapositive ($\sim r \rightarrow \sim o$): If 15 is not a prime number, then 15 is not an odd number. Here, ($\sim r \rightarrow \sim o$) is false because ($T \rightarrow F$) is false.

● **Conclusion: When a conditional ($p \rightarrow q$) is true, its contrapositive ($\sim q \rightarrow \sim p$) must also be true.**
When a conditional is false, its contrapositive must be false.

Logical Equivalents

You have seen that a *conditional* and its *contrapositive* are ***logical equivalents*** because they have the same truth value.

Write the converse of a conditional and the inverse of that same conditional. Notice that one is the contrapositive of the other. For example:

Conditional ($p \rightarrow q$): If it snows, then it is cold.
Converse ($q \rightarrow p$): If it is cold, then it snows.
Inverse ($\sim p \rightarrow \sim q$): If it does not snow, then it is not cold.

Since the contrapositive of ($q \rightarrow p$) is ($\sim p \rightarrow \sim q$), we can say that a *converse* of a statement and an *inverse* of that same statement are *logical equivalents*. We may also verify that the converse and inverse are logically equivalent by examining their truth tables.

				Conditional	Inverse	Converse	Contrapositive
p	q	$\sim p$	$\sim q$	$p \rightarrow q$	$\sim p \rightarrow \sim q$	$q \rightarrow p$	$\sim q \rightarrow \sim p$
T	T	F	F	T	T	T	T
T	F	F	T	F	T	T	F
F	T	T	F	T	F	F	T
F	F	T	T	T	T	T	T

● **The truth table shows that the conditional and the contrapositive are logically equivalent and that the inverse and converse are logically equivalent.**

Forming Related Statements

A conditional statement is not always formed by using a "positive" hypothesis and a "positive" conclusion. For example, "If it does not rain, then we'll go to the park." In this case,

r: It rains.

$\sim r$: It does *not* rain.

k: We'll go to the park.

$\sim r \rightarrow k$: If it does *not* rain, then we'll go to the park.

Using $(\sim r \rightarrow k)$ as the original conditional statement, we can form its inverse, its converse, and its contrapositive by following the rules established within each definition.

The General Case and Rules	A Specific Case
Conditional: $p \rightarrow q$ The hypothesis is p; the conclusion is q.	**Conditional:** $\sim r \rightarrow k$ If it does not rain, then we'll go to the park.
Inverse: $\sim p \rightarrow \sim q$ Negate the hypothesis; negate the conclusion.	**Inverse:** $r \rightarrow \sim k$ If it rains, then we will not go to the park.
Converse: $q \rightarrow p$ Interchange hypothesis and conclusion.	**Converse:** $k \rightarrow \sim r$ If we go to the park, then it does not rain.
Contrapositive: $\sim q \rightarrow \sim p$ Negate hypothesis; negate conclusion; interchange these negations.	**Contrapositive:** $\sim k \rightarrow r$ If we do not go to the park, then it rained.

Since the conditional $(\sim r \rightarrow k)$ and the contrapositive $(\sim k \rightarrow r)$ are logically equivalent, we can write the tautology:

$$(\sim r \rightarrow k) \leftrightarrow (\sim k \rightarrow r)$$

Since the inverse $(r \rightarrow \sim k)$ and the converse $(k \rightarrow \sim r)$ are logically equivalent, we can write the tautology:

$$(r \rightarrow \sim k) \leftrightarrow (k \rightarrow \sim r)$$

MODEL PROBLEMS

In 1–6, write the required statements in symbolic form.

		Answers
1.	the inverse of $k \rightarrow t$	$\sim k \rightarrow \sim t$
2.	the inverse of $\sim m \rightarrow r$	$m \rightarrow \sim r$
3.	the converse of $v \rightarrow y$	$y \rightarrow v$
4.	the converse of $s \rightarrow \sim t$	$\sim t \rightarrow s$
5.	the contrapositive of $l \rightarrow m$	$\sim m \rightarrow \sim l$
6.	the contrapositive of $\sim c \rightarrow d$	$\sim d \rightarrow c$

7. Given the true statement: If the polygon is a rectangle, then it has four sides. Which statement must also be true?
(1) If the polygon has four sides, then it is a rectangle.
(2) If the polygon is not a rectangle, then it does not have four sides.
(3) If the polygon does not have four sides, then it is not a rectangle.
(4) If the polygon has four sides, then it is not a rectangle.

Solution: A conditional and its contrapositive always have the same truth value. When the conditional states "rectangle → four sides," the contrapositive is "not four sides → not rectangle."

Answer: (3)

EXERCISES

In 1–4, write the inverse of the statement in symbolic form.

1. $p \rightarrow q$ **2.** $t \rightarrow \sim w$ **3.** $\sim m \rightarrow p$ **4.** $\sim p \rightarrow \sim q$

In 5–8, write the converse of the statement in symbolic form.

5. $p \rightarrow q$ **6.** $t \rightarrow \sim w$ **7.** $\sim m \rightarrow p$ **8.** $q \rightarrow p$

In 9–12, write the contrapositive of the statement in symbolic form.

9. $p \rightarrow q$ **10.** $t \rightarrow \sim w$ **11.** $\sim m \rightarrow p$ **12.** $\sim q \rightarrow \sim p$

In 13–16, write the inverse of the statement in words.

13. If you use Charm face powder, then you will be beautiful.
14. If you buy Goal toothpaste, then your children will brush longer.
15. When you serve imported sparkling water, it shows that you have good taste.
16. The man who wears Cutrite clothes is well dressed.

In 17–20, **(a)** write the inverse of the conditional statement in words, **(b)** give the truth value of the conditional, and **(c)** give the truth value of the inverse.

17. If a polygon is a triangle, then the polygon has exactly three sides.
18. If a polygon is a trapezoid, then the polygon has exactly four sides.
19. If $2 \cdot 2 = 4$, then $2 \cdot 3 = 6$.
20. If $2^2 = 4$, then $3^2 = 6$.

In 21–24, write the converse of the statement in words.

21. If you live to an old age, then you eat Nano yogurt.
22. If you take pictures of your family with a Blinko camera, then you care about your family.
23. If you drive a Superb car, then you'll get good mileage.
24. If you use Dust and Roast, you'll make a better chicken dinner.

In 25–28, **(a)** write the converse of the conditional statement in words, **(b)** give the truth value of the conditional, and **(c)** give the truth value of the converse.

25. If a number is even, then the number is exactly divisible by 2.
26. If two segments are 5 cm each, then the two segments are equal in measure.
27. If $5 = 1 + 4$, then $5^2 = 1^2 + 4^2$.
28. If $2(5) + 3 = 10 + 3$, then $2(5) + 3 = 13$.

In 29–32, write the contrapositive of the statement in words.

29. If you care enough to send the best, then you send Trademark cards.
30. If you use Trickle deodorant, then you won't have body odor.
31. If you brush with Brite, then your teeth will be pearly white.
32. If you want a good job, then you'll get a high school diploma.

In 33–37, **(a)** write the contrapositive of the conditional statement in words, **(b)** give the truth value of the conditional, and **(c)** give the truth value of the contrapositive.

33. If opposite sides of a quadrilateral are parallel, then the quadrilateral is a parallelogram.
34. If two segments are 8 cm each, then the two segments are equal in measure.
35. If $1 + 2 = 3$, then $2 + 3 = 4$.
36. If all angles of a quadrilateral are equal in measure, then the quadrilateral is a rectangle.
37. If a number is prime, then it is not an even number.

In 38–42, write the numeral preceding the word or expression that best completes the statement or answers the question.

38. When $p \to q$ is true, which related conditional must be true?
(1) $q \to p$ (2) $\sim p \to \sim q$ (3) $p \to \sim q$ (4) $\sim q \to \sim p$

39. Which is the contrapositive of "If winter comes, then spring is not far behind"?
(1) If spring is not far behind, then winter comes.
(2) If spring is far behind, then winter comes.
(3) If spring is not far behind, then winter does not come.
(4) If spring is far behind, then winter does not come.

40. Which is the converse of "If a polygon is a trapezoid, its area = $\frac{1}{2}(b + c) \cdot h$"?
(1) If a polygon is not a trapezoid, then its area $\neq \frac{1}{2}(b + c) \cdot h$.
(2) If a polygon has area $= \frac{1}{2}(b + c) \cdot h$, then it is a trapezoid.
(3) If a polygon has area $\neq \frac{1}{2}(b + c) \cdot h$, then it is not a trapezoid.
(4) If the area of a trapezoid is $\frac{1}{2}(b + c) \cdot h$, then it is a polygon.

41. Which is the inverse of "If $x = 2$, then $x + 3 \neq 9$"?
(1) If $x + 3 \neq 9$, then $x = 2$. (2) If $x + 3 = 9$, then $x \neq 2$.
(3) If $x \neq 2$, then $x + 3 = 9$. (4) If $x \neq 2$, then $x + 3 \neq 9$.

42. Which is the contrapositive of "If $x > 2$, then $3x + 5x \neq 16$"?
(1) If $3x + 5x \neq 16$, then $x > 2$.
(2) If $3x + 5x = 16$, then $x \not> 2$.
(3) If $3x + 5x \neq 16$, then $x \not> 2$.
(4) If $3x + 5x = 16$, then $x > 2$.

43. For a conditional statement $(p \to q)$:
a. Write its inverse in symbolic form.
b. Write the converse of this inverse in symbolic form.
c. What is the relationship between the conditional statement $(p \to q)$ and the converse of the inverse of that statement?
d. What might have been another way to define the contrapositive of a conditional statement?

In 44–51, a conditional statement is given. Write in words **(a)** the inverse, **(b)** the converse, and **(c)** the contrapositive of that conditional.

44. If today is Friday, then tomorrow is Saturday.
45. If Douglas does well in college, then he will apply to medical school.
46. Arlette will get a role in the play if she auditions.
47. Dorothea will graduate from law school in January if she takes courses this summer.

48. If John is accepted at the Culinary Institute, then he has a chance of earning a high salary as a chef.
49. If a man is honest, he does not steal.
50. If Julia doesn't water the plants, then the plants will die.
51. Rachel will not get her allowance if she forgets to do her chores.

In 52–54, if the given statement is assumed to be true, which of the four statements that follow must also be true?

52. If a figure is a square, then it is a polygon.
(1) If the figure is a polygon, then it is a square.
(2) If the figure is not a square, then it is not a polygon.
(3) If the figure is not a polygon, then it is not a square.
(4) If the figure is not a square, then it is a polygon.
53. If $a = b$, then $b \neq d$.
(1) If $a \neq b$, then $b \neq d$.
(2) If $b \neq d$, then $a = b$.
(3) If $b = d$, then $a \neq b$.
(4) If $a \neq b$, then $b = d$.
54. I'll get into shape if I exercise.
(1) If I don't exercise, then I won't get into shape.
(2) If I don't get into shape, then I do not exercise.
(3) If I get into shape, then I exercise.
(4) I exercise if I get into shape.

In 55–60, assume that the conditional statement is true. Then:
a. Write its converse in words.
b. Is the converse always true, always false, or uncertain?
c. Write its inverse in words.
d. Is the inverse always true, always false, or uncertain?
e. Write its contrapositive in words.
f. Is the contrapositive always true, always false, or uncertain?

55. If Eddie lives in San Francisco, then he lives in California.
56. If a number is divisible by 12, then it is divisible by 3.
57. If I have the flu, then I am ill.
58. If one pen costs 29 cents, then three pens cost 87 cents.
59. If a quadrilateral is a rhombus, then it has four congruent sides.
60. If Alex loves computers, then he will learn how to write programs.

6-6 DRAWING CONCLUSIONS

In the process of reasoning, we use statements that we believe to be true and the rules of logic in order to draw conclusions, that is, to determine other statements that are true. When people disagree on the truth of the conclusions that have been drawn, it is often because they disagree on the truth value of the initial statements to which the laws of logic are applied.

MODEL PROBLEMS

1. What conclusions can be drawn when the following statements are true?

Today is Monday or I have gym.
Today is not Monday.

Solution:

(1) Use symbols to represent the sentences that we know to be true.

$p \vee q$: Today is Monday or I have gym.
$\sim p$: Today is not Monday.

(2) Construct a truth table for $p \vee q$. Since $p \vee q$ is given true, eliminate the case where $p \vee q$ is false by crossing out row 4.

p	q	$p \vee q$
T	T	T
T	F	T
F	T	T
~~F~~	~~F~~	~~F~~

Since $\sim p$ is given true, p is false. Eliminate the cases where p is true by crossing out rows 1 and 2.

p	q	$p \vee q$
~~T~~	~~T~~	~~T~~
~~T~~	~~F~~	~~T~~
F	T	T
~~F~~	~~F~~	~~F~~

(3) There is only one case that remains. Row 3 tells us that q is true. Therefore, we may conclude that "I have gym" is a true statement.

Answer: I have gym.

Note: In most cases, there will be many possible conclusions. Any true statement could be considered to be a conclusion. The following statements could all be shown to be true when the given statements are true.

Today is not Monday and I have gym.
If today is not Monday, then I have gym.
If I do not have gym, then today is Monday.

2. Draw a conclusion based on the following true statements.

If I do not finish my English essay, I will get a *C* in English.
I do not get a *C* in English.

Solution:

(1) Use symbols to represent the sentences that we know to be true.

$\sim p \rightarrow q$: If I do not finish my English essay, I will get a *C* in English.
$\sim q$: I do not get a *C* in English.

(2) Construct a truth table for $\sim p \rightarrow q$.

Since $\sim p \rightarrow q$ is true, eliminate the case where it is false by crossing out row 4.

p	q	$\sim p$	$\sim p \rightarrow q$
T	*T*	*F*	*T*
T	*F*	*F*	*T*
F	*T*	*T*	*T*
~~*F*~~	~~*F*~~	~~*T*~~	~~*F*~~

Since $\sim q$ is true, q is false. Eliminate the cases where q is true by crossing out rows 1 and 3.

p	q	$\sim p$	$\sim p \rightarrow q$
~~*T*~~	~~*T*~~	~~*F*~~	~~*T*~~
T	*F*	*F*	*T*
~~*F*~~	~~*T*~~	~~*T*~~	~~*T*~~
~~*F*~~	~~*F*~~	~~*T*~~	~~*F*~~

(3) There is only one case that remains. Row 2 tells us that p is true. Therefore, we may conclude that "I did finish my English essay" is a true statement.

Answer: I did finish my English essay.

3. What conclusions can be drawn from the following true statement?

Today is Monday and I have gym.

Solution: When a conjunction is true, each of the simple statements from which it is constructed must be true. Therefore, each of the following statements is true.

Today is Monday.
I have gym.

4. What conclusions, if any, can be drawn from the following true statements?

If today is Monday, then I have gym.
Today is not Monday.

Solution:

(1) Use symbols to represent the sentences that we know to be true.

$p \rightarrow q$: If today is Monday, then I have gym.
$\sim p$: Today is not Monday.

(2) Construct a truth table for $p \rightarrow q$.

Since $p \rightarrow q$ is true, eliminate the case where $p \rightarrow q$ is false by crossing out row 2.

p	q	$p \rightarrow q$
T	*T*	*T*
~~*T*~~	~~*F*~~	~~*F*~~
F	*T*	*T*
F	*F*	*T*

Since $\sim p$ is true, p is false. Eliminate the cases where p is true. Row 2 has already been eliminated; cross out row 1.

p	q	$p \rightarrow q$
~~*T*~~	~~*T*~~	~~*T*~~
~~*T*~~	~~*F*~~	~~*F*~~
F	*T*	*T*
F	*F*	*T*

(3) There are two cases that remain. Row 3 tells us that q is true. Row 4 tell us that q is false. Therefore, we can draw no conclusion about the truth value of q.

Answer: There is no simple sentence that is a conclusion.

Note: It may, of course, be possible to write some conclusions that do not depend on the truth value of q. For example, the contrapositive of a conditional always has the same truth value as the conditional. Therefore, when "If today is Monday, then I have gym" is true, the contrapositive "If I do not have gym, then today is not Monday" is also true.

EXERCISES

In 1–10, assume that the first two sentences are true. Determine whether the third sentence is true, false, or cannot be determined to be true or false.

1. I study hard or I do not pass.
I pass.
I study hard.

2. If Toy is a dog, then Toy is an animal.
Toy is a dog.
Toy is an animal.

3. If I am late for dinner, then my dinner will be cold.
I am late for dinner.
My dinner is cold.

4. If I am late for dinner, then my dinner will be cold.
I am not late for dinner.
My dinner is not cold.

5. I like skating or skiing.
I like skating.
I like skiing.

6. I like skating or skiing.
I do not like skating.
I like skiing.

7. I will go to college if and only if I work this summer.
I do not work this summer.
I will go to college.

8. I live in New York State if I live in Albany.
I do not live in Albany.
I live in New York State.

9. $x > 10$ if $x = 15$.
$x = 15$.
$x > 10$.

10. The average of two numbers is 10 when the numbers are 7 and 13.
The average of two numbers is 10.
The two numbers are 7 and 13.

In 11–25, assume that the given sentences are true. Write a conclusion that is a simple sentence, if possible. If no conclusion is possible, write "no conclusion."

11. If I do my assignments, I pass the course.
I do my assignments.

12. On Saturday, we go skiing or we play hockey.
Last Saturday, we did not go skiing.

13. If x is a prime, then $x \neq 9$.
$x = 9$.

14. If a parallelogram contains a right angle, that parallelogram is a rectangle.
Parallelogram *ABCD* is not a rectangle.

15. $x \geq 10$ or $x < 10$.
$x \nless 10$.

16. If $2x + 5 = 7$, then $2x = 2$.
If $2x = 2$, then $x = 1$.
$2x + 5 = 7$.

17. 2 is a prime if and only if 2 has exactly two factors.
2 has exactly two factors.

18. 3 is a prime if and only if 3 has exactly two factors.
3 is a prime.

19. If x is divisible by 4, then x is divisible by 2.
x is divisible by 2.

20. x is prime or x is even.
x is prime.

21. If it is July, then it is summer.
It is not summer.

22. It is October or November, and it is not spring.
It is not November.

23. I study math, and I study French or Spanish.
I do not study French.

24. x is even and a prime if and only if $x = 2$.
$x = 2$.

25. x is even and a prime if and only if $x = 2$.
$x \neq 2$.

6-7 REVIEW EXERCISES

In 1–4, for each given statement: **a.** Write the symbolic statement in words. **b.** Tell whether the statement is true or false.

Let p represent "2 is a prime." (true)
Let q represent "There is one even prime." (true)
Let r represent "4 is a prime." (false)

1. $q \rightarrow \sim p$ **2.** $r \rightarrow (p \wedge q)$ **3.** $\sim p \vee \sim r$ **4.** $q \rightarrow (p \vee r)$

5. Write in symbolic form:
a. the converse of $p \rightarrow q$
b. the inverse of $p \rightarrow q$
c. the contrapositive of $p \rightarrow q$

In 6–13, write the numeral preceding the expression that best answers the question or completes the statement.

6. Let p represent "x is prime." Let q represent "$x > 25$." When x is 21, which sentence is true?
(1) p (2) q (3) $p \wedge q$ (4) $p \rightarrow q$

7. Let e represent "x is even." Let d represent "x is divisible by 6." When x is 46, which sentence is true?
(1) $e \wedge d$ (2) $e \vee d$ (3) $e \rightarrow d$ (4) $\sim e \wedge d$

8. The sentence $m \vee r$ is false if and only if
(1) m is false and r is true. (2) both m and r are false.
(3) m is true and r is false. (4) both m and r are true.

9. If $p \leftrightarrow q$ is true, which sentence is also true?
(1) $p \rightarrow q$ (2) $p \wedge q$ (3) $p \vee q$ (4) $\sim p \wedge q$

10. What is the inverse of the sentence "If $8x = 24$, then $x = 3$"?
(1) If $x = 3$, then $8x = 24$. (2) If $8x \neq 24$, then $x \neq 3$.
(3) If $8x = 24$, then $x \neq 3$. (4) If $x \neq 3$, then $8x \neq 24$.

11. Which is the converse of the sentence $\sim p \rightarrow q$?
(1) $\sim q \rightarrow p$ (2) $q \rightarrow \sim p$ (3) $p \rightarrow \sim q$ (4) $q \rightarrow p$

12. Which has the same truth value as the sentence "If Will won, then Lon lost"?
(1) If Lon lost, then Will won.
(2) If Will did not win, then Lon did not lose.
(3) If Lon lost, then Will did not win.
(4) If Lon did not lose, then Will did not win.

13. Which sentence is always false?
(1) $p \vee \sim p$ (2) $q \wedge \sim q$ (3) $p \wedge \sim q$ (4) $q \vee \sim p$

14. **a.** Complete the truth table for the sentence: $[(p \rightarrow \sim q) \wedge p] \rightarrow \sim q$

p	q	$\sim q$	$p \rightarrow \sim q$	$(p \rightarrow \sim q) \wedge p$	$[(p \rightarrow \sim q) \wedge p] \rightarrow \sim q$

b. Is $[(p \rightarrow \sim q) \wedge p] \rightarrow \sim q$ a tautology?
c. Justify the answer given for part **b.**

15. **a.** Construct a truth table for the sentence $\sim(p \wedge q) \leftrightarrow (\sim p \vee \sim q)$.
b. Is the sentence a tautology? Why?

16. **a.** Complete the truth table for the sentence: $p \leftrightarrow [\sim(p \rightarrow q) \vee q]$

p	q	$p \rightarrow q$	$\sim(p \rightarrow q)$	$\sim(p \rightarrow q) \vee q$	$p \leftrightarrow [\sim(p \rightarrow q) \vee q]$

b. Is the sentence a tautology? Why?

17. Which of the following sentences is equivalent to $\sim(p \vee q)$?
(1) $\sim p \vee \sim q$ (2) $\sim p \vee q$ (3) $\sim p \wedge \sim q$ (4) $\sim p \rightarrow q$

18. Assume that the given sentences are true. Write a simple sentence that is a conclusion.

If $\triangle ABC$ is isosceles, then $AB = AC$.
$AB \neq AC$

19. At a track meet, Janice, Kay, and Virginia were the first three finishers of a 50-meter dash. Virginia did not come in second. Kay did not come in third. Virginia came in ahead of Janice. In what order did they finish the race?

Chapter **7**

Signed Numbers

7-1 EXTENDING THE NUMBER LINE

Until now you have used only zero and numbers greater than zero. But you know that you often want to use a number that is less than zero. For example, the temperature on a winter day may be below 0° Celsius. In order to have numbers that are smaller than zero as well as larger than or equal to zero, we will consider a new set of numbers called the ***real numbers***. In the set of real numbers, numbers larger than zero are *positive real numbers* and numbers smaller than zero are *negative real numbers*. Like the numbers of arithmetic, the real numbers can be associated with points on a number line, which we now extend to include the negative numbers.

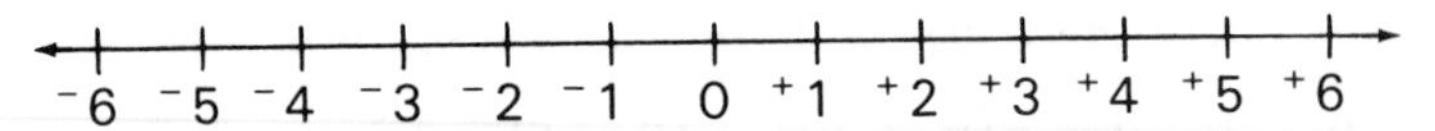

The + sign is part of the numeral for a positive number, the − sign is part of the numeral for a negative number, and 0 is not written with a sign. These numbers are called ***signed numbers*** or ***directed numbers***.

We will make frequent use of an important subset of the set of signed numbers

$$\{\ldots, {}^{-}4, {}^{-}3, {}^{-}2, {}^{-}1, 0, {}^{+}1, {}^{+}2, {}^{+}3, {}^{+}4, \ldots\}$$

called the set of ***integers***. The numbers ${}^{+}1$, ${}^{+}2$, ${}^{+}3$, ${}^{+}4$, . . . are called *positive integers*; the numbers ${}^{-}1$, ${}^{-}2$, ${}^{-}3$, ${}^{-}4$, . . . are called *negative integers*. We think of the positive integers as the same set of numbers as the natural numbers.

Note that the real number line contains more than just the integers. Included between any two integers are fractions and decimals, and still other real numbers that you will study in Chapter 19.

Ordering Signed Numbers

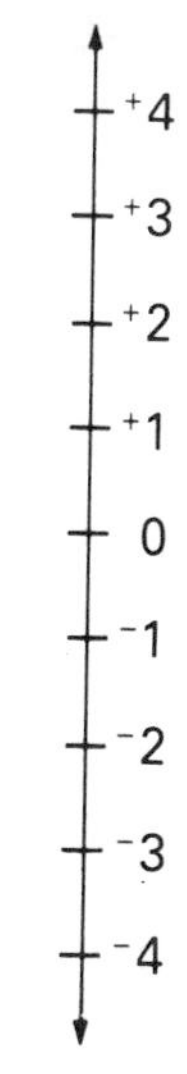

A vertical number line is used in thermometers. A temperature of $^{-}1°$ is lower than a temperature of $^{+}4°$ and higher than a temperature of $^{-}3°$. That is, $^{-}1 < {}^{+}4$ and $^{-}1 > {}^{-}3$.

On a horizontal number line, $^{+}4$ is to the right of $^{-}1$, and $^{-}3$ is to the left of $^{-}1$.

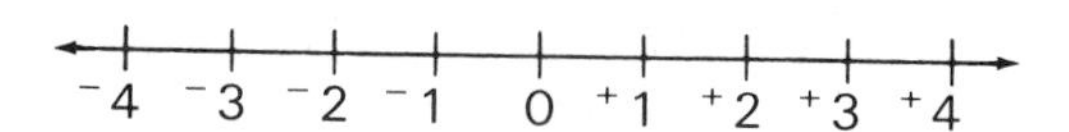

All signed numbers are ordered on the real number line. In this ordering, any number is greater than every number to its left and is less than every number to its right.

Since $^{-}1$ is less than $^{+}4$ and greater than $^{-}3$, $^{-}1$ is *between* $^{+}4$ and $^{-}3$. That is, $^{+}4 > {}^{-}1 > {}^{-}3$ or $^{-}3 < {}^{-}1 < {}^{+}4$. Using the conjunction symbol of logic, we can write:

$$({}^{+}4 > {}^{-}1) \wedge ({}^{-}1 > {}^{-}3) \text{ or } ({}^{-}3 < {}^{-}1) \wedge ({}^{-}1 < {}^{+}4)$$

MODEL PROBLEMS

1. State whether each of the sentences is true or false and give a reason for your answer:

Answers

a. $^{+}7 > {}^{-}2$ True because $^{+}7$ is to the right of $^{-}2$ on a number line.

b. $^{-}5 > {}^{-}3$ False because $^{-}5$ is to the left of $^{-}3$ on a number line.

2. Use the conjunction symbol $\wedge$ to rewrite each sentence.

Answers

a. $^{+}9 > {}^{+}4 > {}^{+}1$ $\quad ({}^{+}9 > {}^{+}4) \wedge ({}^{+}4 > {}^{+}1)$

b. $^{-}2 < 0 < {}^{+}3$ $\quad ({}^{-}2 < 0) \wedge (0 < {}^{+}3)$

EXERCISES

In 1–3, draw a real number line. Then locate the points whose coordinates are given.

1. $^{+}4, {}^{-}2, 0, {}^{-}5, {}^{+}3$ **2.** $^{+}\frac{1}{2}, 0, {}^{-}1\frac{1}{2}, {}^{+}3\frac{1}{2}, \frac{-3}{4}$ **3.** $^{+}1, \frac{-1}{3}, 0, {}^{+}1\frac{2}{3}, \frac{-5}{6}$

In 4–7, state whether the sentence is true or false. Give a reason for your answer.

4. $^{+}5 > {}^{+}2$ **5.** $^{-}3 < 0$ **6.** $^{-}7 > {}^{-}1$ **7.** $^{-}1\frac{3}{4} > {}^{-}1\frac{7}{8}$

In 8–13, use the symbol < to order the numbers.

8. $^{-}4, {}^{+}8$ **9.** $^{-}3, {}^{-}6$ **10.** $^{+}1\frac{1}{2}, {}^{-}1\frac{1}{2}$

11. $^{+}3, {}^{-}2, {}^{-}4$ **12.** $^{-}2, {}^{+}8, 0$ **13.** $^{-}3\frac{1}{2}, {}^{+}6, {}^{+}2\frac{1}{2}$

In 14–19, use the symbol > to order the numbers.

14. $^{+}7, {}^{-}4$ **15.** $^{-}12, {}^{+}12$ **16.** $^{-}1\frac{1}{2}, 0$

17. $^{+}3, {}^{-}3, {}^{+}5$ **18.** $^{-}5, {}^{-}1, 0$ **19.** $^{-}1.5, {}^{+}3\frac{1}{2}, {}^{-}2\frac{1}{2}$

In 20–22, state which number is between the other two.

20. $^{+}8, {}^{-}2, {}^{+}2$ **21.** $^{+}9, {}^{-}9, {}^{-}4$ **22.** $^{+}.6, {}^{+}1.1, {}^{-}.8$

In 23–28, state whether the sentence is true or false.

23. $^{+}5 \geq {}^{+}2$ **24.** $^{-}2 \leq {}^{+}5$ **25.** $^{-}9 \leq {}^{-}12$

26. $^{-}3 \geq {}^{+}3$ **27.** $^{+}6 > {}^{-}1 > {}^{-}6$ **28.** $^{-}10 > {}^{-}2 > 0$

29. Use the conjunction symbol $\wedge$ to rewrite each sentence.
a. $^{+}2 < {}^{+}5 < {}^{+}7$ **b.** $^{-}3 < 0 < {}^{+}9$ **c.** $^{-}8 > {}^{-}10 > {}^{-}12$

30. If x is a positive number and y is a negative number ($x > 0$ and $y < 0$), tell whether each statement is true or false.
a. $x = y$ **b.** $x > y$ **c.** $y > x$ **d.** $x < 0$ **e.** $y < 0 < x$

31. What is the smallest positive integer?
32. What is the greatest negative integer?
33. **a.** Is there a greatest positive integer? **b.** Why?
34. **a.** Is there a least negative integer? **b.** Why?

7-2 GRAPHING THE SOLUTION SET OF AN OPEN SENTENCE INVOLVING ONE VARIABLE ON A NUMBER LINE

PROCEDURE. To graph the solution set of an open sentence involving one variable on a number line:

1. Find the set of numbers that are solutions of the open sentence.
2. Graph the solution set on a number line.

MODEL PROBLEMS

1. If the domain of x is $\{^-3, ^-2, ^-1, 0, ^+1, ^+2, ^+3\}$, graph the solution set for the sentence $x \geq {}^-2$.

Solution: When any element of the domain except $^-3$ replaces x in the sentence $x \geq {}^-2$ (which means $x > {}^-2$ or $x = {}^-2$), the resulting statement is true. Therefore, the solution set of $x \geq {}^-2$ is $\{^-2, ^-1, 0, ^+1, ^+2, ^+3\}$.

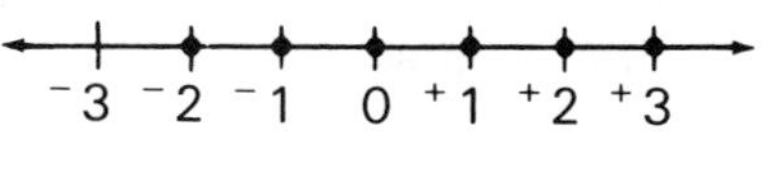

We are able to list the solution set of this inequality because it is a finite set.

2. Using the set of signed numbers as the replacement set, graph the solution set of each of the following sentences:

a. $y > {}^-4$ **b.** $m \leq {}^-2$ **c.** $^-3 \leq t < {}^+2$

Solution:

a. The graph of $y > {}^-4$ consists of all points to the right of $^-4$. The nondarkened circle shows that $^-4$ is not included.

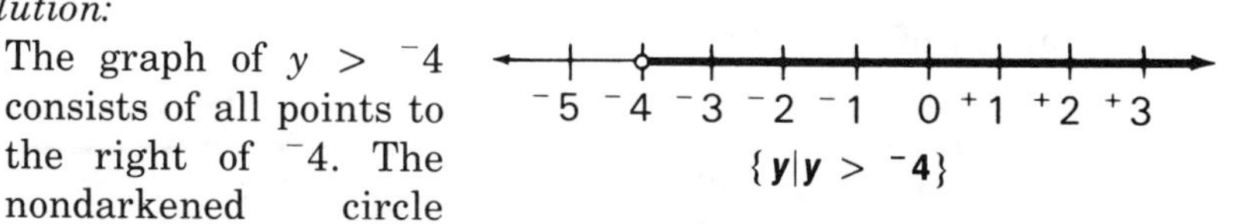

b. The graph of $m \leq {}^-2$ consists of the point $^-2$ and all points to the left of $^-2$. The darkened circle shows that $^-2$ is included.

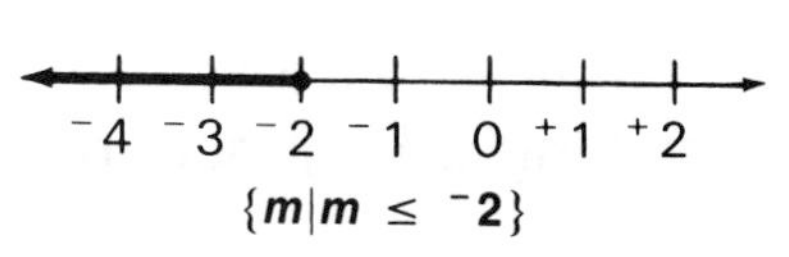

c. The graph of $^-3 \leq t < {}^+2$ consists of the point $^-3$ and all points between $^-3$ and $^+2$. The point $^+2$ is not included.

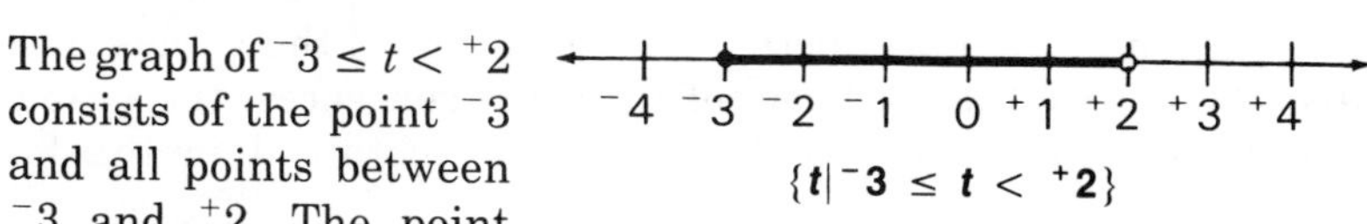

Notice that we are not able to list these solution sets because they are infinite sets. Even in part **c**, where the graph does not go on indefinitely, the set of points between $^{-}3$ and $^{+}2$ is an infinite set.

Graphing a Conjunction

The algebraic sentence $^{+}3 < x < {}^{+}6$ is equivalent to the conjunction $(^{+}3 < x) \wedge (x < {}^{+}6)$. Since the solution set of a conjunction must contain all values of the variable that satisfy both open sentences, the graph of the conjunction $(^{+}3 < x) \wedge (x < {}^{+}6)$ can be obtained in the following manner:

Steps | *Solution*

(1) Graph the solution set of the first open sentence, $^{+}3 < x$.

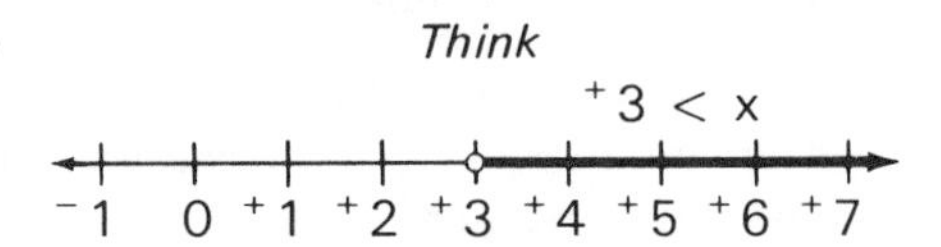

(2) Graph on the same number line the solution set of the second open sentence, $x < {}^{+}6$.

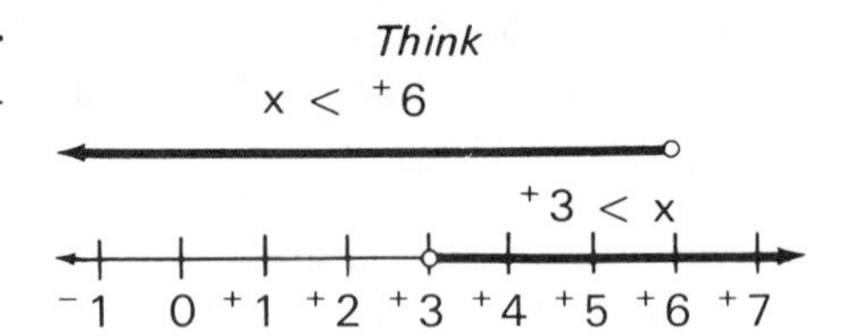

(3) The graph of the solution set of the conjunction is the set of points common to the graphs made in steps (1) and (2), that is, the *intersection* of the two sets of points graphed in steps (1) and (2).

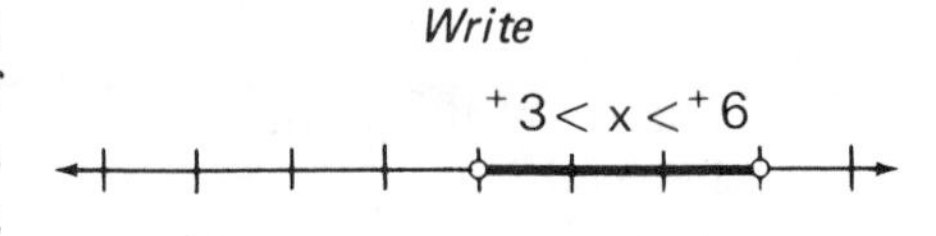

Graphing a Disjunction

The disjunction $(^{+}3 < x) \vee (x > {}^{+}6)$ means $^{+}3 < x$ or $x > {}^{+}6$. Since the solution set of a disjunction must contain all the values of the variable that satisfy at least one of the open sentences, the graph of the disjunction $(^{+}3 < x) \vee (x > {}^{+}6)$ can be obtained in the following manner:

Steps	*Solution*
(1) Graph the solution set of the first open sentence, $^{+}3 < x$.	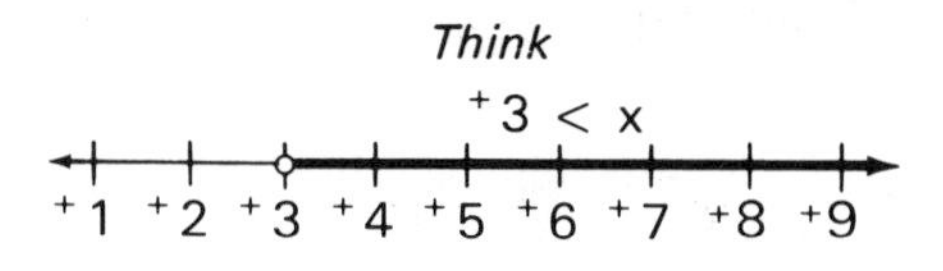
(2) Graph on the same number line the solution set of the second open sentence, $x > {}^{+}6$.	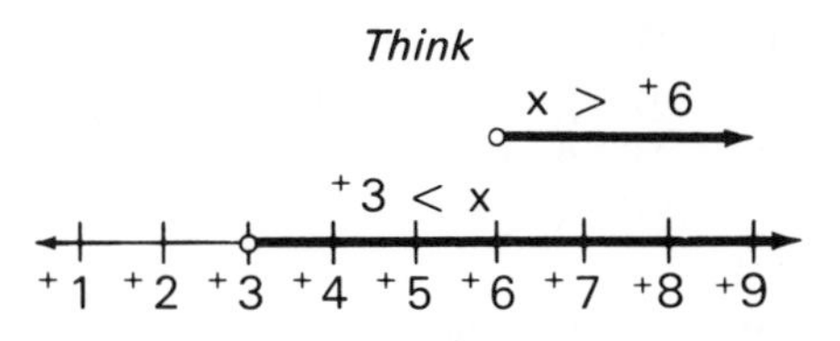
(3) The graph of the solution set of the disjunction is the set of points that are in at least one of the graphs made in step (1) and step (2), that is, the *union* of the two sets of points graphed in step (1) and step (2).	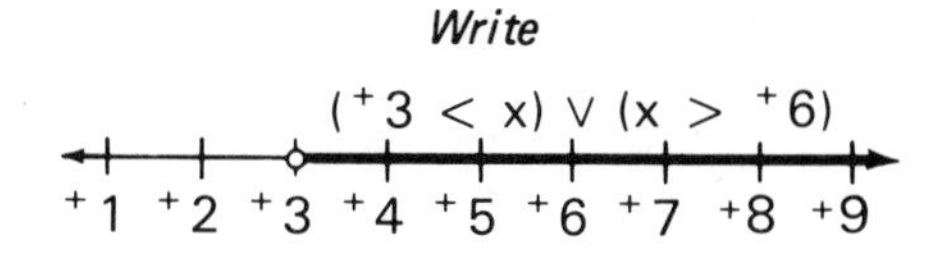

It is possible that two open sentences that are involved in a disjunction have no elements in common. The graph of the disjunction still shows the union of both graphs. For example, the graph of the disjunction $(x < {}^{+}2) \vee (x \geq {}^{+}5)$ is:

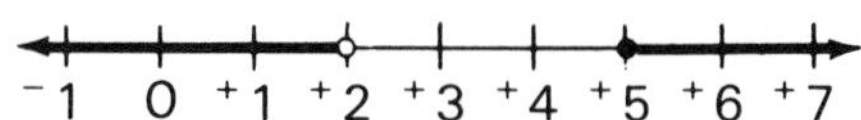

EXERCISES

In 1–9, if the replacement set is $\{^{-}4, {}^{-}3, {}^{-}2, {}^{-}1, 0, {}^{+}1, {}^{+}2, {}^{+}3, {}^{+}4\}$, graph the solution set of the open sentence.

1. $x > 0$ **2.** $x \leq 0$ **3.** $y > {}^{-}1$

4. $t < {}^{+}2$ **5.** $m \geq {}^{+}1$ **6.** $^{-}1 < d < {}^{+}3$

7. $^{-}1 \leq x < {}^{+}2$ **8.** $^{-}1 < y \leq {}^{+}2$ **9.** $^{-}3 \leq t \leq {}^{+}3$

In 10–20, if the domain is the set of signed numbers, graph the solution set of the sentence.

10. $x > {}^{+}6$ **11.** $y < {}^{-}3$ **12.** $s \leq 0$

13. $x \neq {}^{+}2$ **14.** $z \geq {}^{+}2\frac{1}{2}$ **15.** $^{-}3 < x < {}^{+}2$

16. $^{-}2 \le x < {}^{+}4$ **17.** $^{-}1 < y \le {}^{+}3$ **18.** $^{-}3 \le m \le {}^{+}3$

19. $(t \ge {}^{-}3) \vee (t > {}^{+}1)$ **20.** $(w \le {}^{-}1) \vee (w \ge {}^{+}2)$

In 21–26, choose the inequality that is represented by the accompanying graph.

21. (1) $^{-}3 < x \le {}^{+}2$
(2) $^{-}3 \le x < {}^{+}2$
(3) $^{-}3 \le x \le {}^{+}2$
(4) $^{-}3 < x < {}^{+}2$

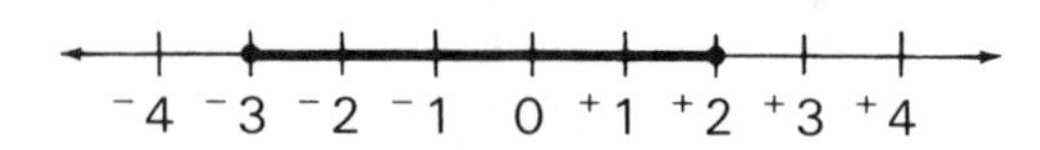

22. (1) $^{-}2 \le x \le {}^{+}4$
(2) $^{-}2 < x \le {}^{+}4$
(3) $^{-}2 < x < {}^{+}4$
(4) $^{-}2 \le x < {}^{+}4$

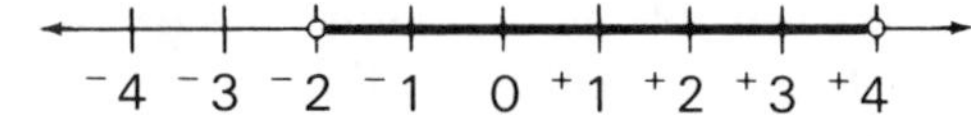

23. (1) $^{-}4 \le x \le {}^{+}3$
(2) $^{-}4 < x < {}^{+}3$
(3) $^{-}4 < x \le {}^{+}3$
(4) $^{-}4 \le x < {}^{+}3$

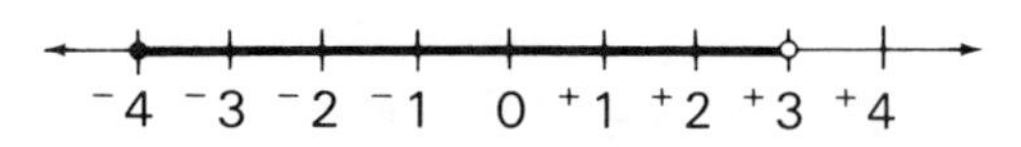

24. (1) $m \le 0$
(2) $m > 0$
(3) $m < 0$
(4) $m \ge 0$

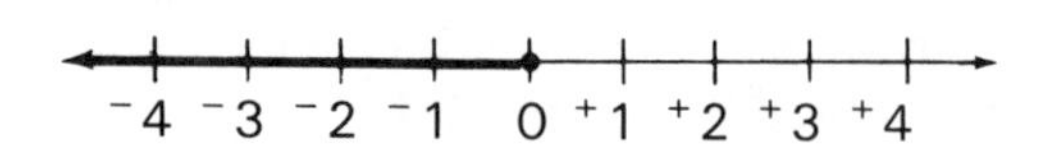

25. (1) $({}^{-}2 \ge y) \wedge (y > {}^{+}1)$
(2) $({}^{-}2 \le y) \wedge (y \le {}^{+}1)$
(3) $({}^{-}2 < y) \wedge (y \le {}^{+}1)$
(4) $({}^{-}2 \le y) \wedge (y < {}^{+}1)$

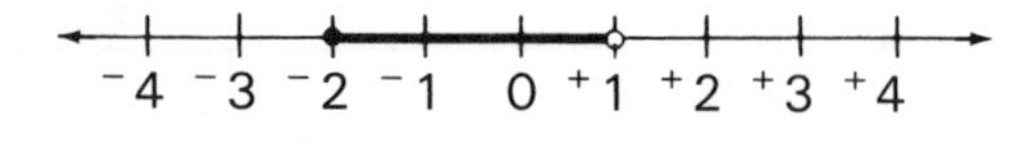

26. (1) $(x \le {}^{-}1) \vee (x \ge {}^{+}2)$
(2) $(x < {}^{-}1) \vee (x \ge {}^{+}2)$
(3) $(x > {}^{-}1) \vee (x > {}^{+}2)$
(4) $(x \le {}^{-}1) \vee (x > {}^{+}2)$

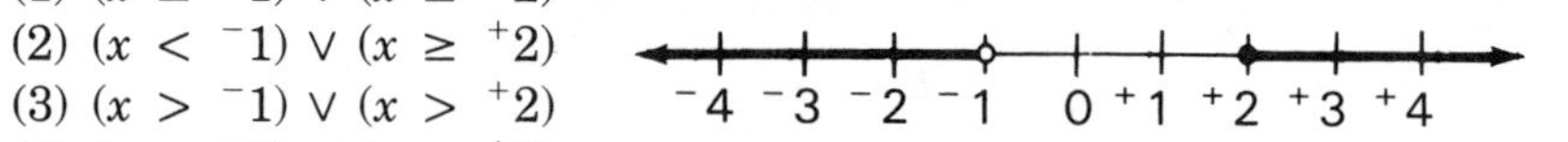

27. Select the graph of the solution set of $^{-}1 < x \le {}^{+}2$.

(1) 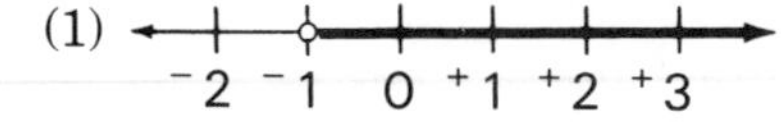

(2) 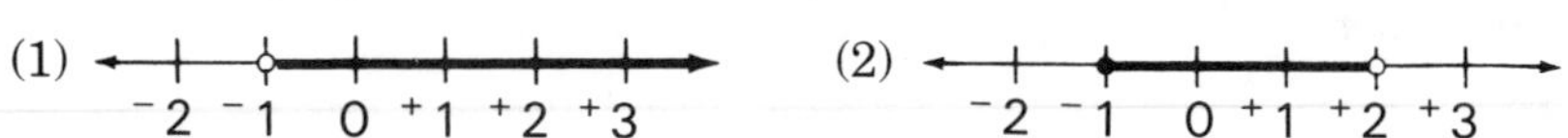

(3) 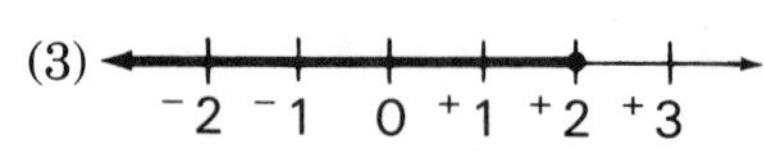

(4) 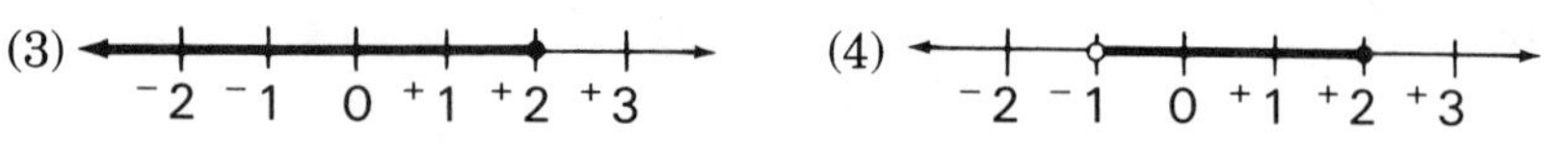

28. Select the graph which shows the solution set of $^{-}1 \le y \le {}^{+}3$.

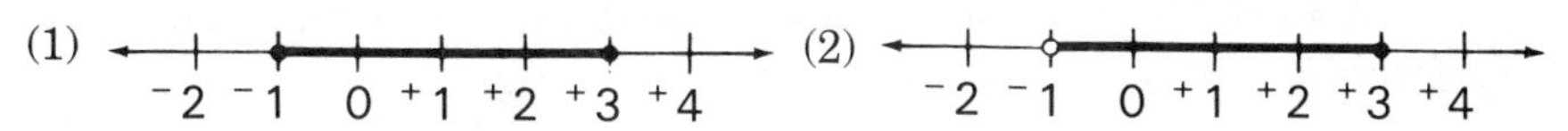

(3) $^{-}2 \quad ^{-}1 \quad 0 \quad ^{+}1 \quad ^{+}2 \quad ^{+}3 \quad ^{+}4$ (4) $^{-}2 \quad ^{-}1 \quad 0 \quad ^{+}1 \quad ^{+}2 \quad ^{+}3 \quad ^{+}4$

29. Select the graph of the solution set of $(w \le 0) \vee (w > {}^{+}3)$.

(1) $^{-}2 \quad ^{-}1 \quad 0 \quad ^{+}1 \quad ^{+}2 \quad ^{+}3 \quad ^{+}4$ (2) $^{-}2 \quad ^{-}1 \quad 0 \quad ^{+}1 \quad ^{+}2 \quad ^{+}3 \quad ^{+}4$

(3) $^{-}2 \quad ^{-}1 \quad 0 \quad ^{+}1 \quad ^{+}2 \quad ^{+}3 \quad ^{+}4$ (4) $^{-}2 \quad ^{-}1 \quad 0 \quad ^{+}1 \quad ^{+}2 \quad ^{+}3 \quad ^{+}4$

In 30–37, graph the given set.

30. $\{x|(x < {}^{+}2) \wedge (x > {}^{-}3)\}$
31. $\{x|(x \ge {}^{-}1) \wedge (x \le {}^{+}4)\}$
32. $\{x|(x \ge {}^{-}4) \wedge (x < {}^{+}1)\}$
33. $\{x|(x > 0) \wedge (x \le {}^{+}3)\}$
34. $\{x|(x \le {}^{-}1) \vee (x \ge {}^{+}2)\}$
35. $\{x|(x < {}^{-}2) \vee (x \ge {}^{+}5)\}$
36. $\{x|(x \le {}^{-}2) \vee (x < {}^{+}4)\}$
37. $\{x|(x \le {}^{-}3) \vee (x \ge {}^{-}3)\}$

7-3 THE OPPOSITE OF A DIRECTED NUMBER

On a real number line any number can be paired with another number that is the same distance from 0 and on the opposite side of 0. We call such a pair of numbers ***opposites***.

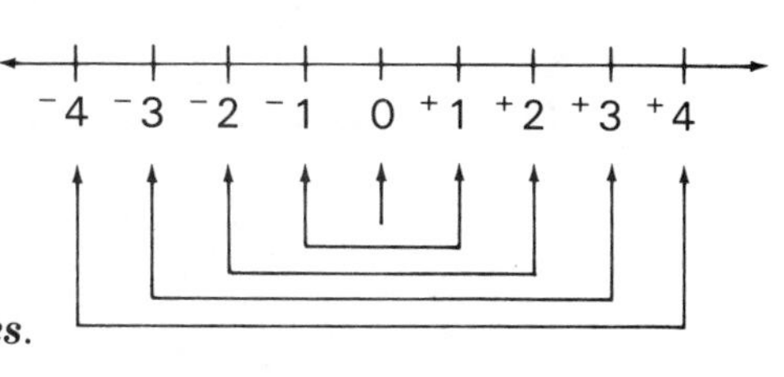

$-({}^{+}1) = {}^{-}1$ is read, "the opposite of $^{+}1$ is $^{-}1$"
$-({}^{-}2) = {}^{+}2$ is read, "the opposite of $^{-}2$ is $^{+}2$"

We write the opposite of 4 as -4.

We know that the opposite of 4 is $^{-}4$.

Since -4 and $^{-}4$ represent the same number, we will not use the raised sign when writing negative numbers.

$$-4 \text{ can mean} \begin{cases} \text{negative } 4 \\ \quad \text{or} \\ \text{the opposite of } 4 \end{cases}$$

We will write positive numbers with a lowered plus sign or with no sign. A variable, whether or not it has a sign, can represent either a positive or a negative number, or zero.

Number	*Opposite*	*Opposite of the Opposite*
+5 or 5	−5	+5 or 5
−6	6	−6
y	$-y$	$-(-y) = y$

The opposite of the opposite of a number is always the number itself.

Notice the following similarity.
If p is a sentence, then $\sim(\sim p) \leftrightarrow p$.
If y is a signed number, then $-(-y) = y$.

MODEL PROBLEMS

In 1–4, write the simplest symbol that represents the opposite of the number.

	Opposites		*Opposites*
1. 15	−15	**2.** −10	10 *or* +10
3. (4 + 8)	−12	**4.** $-[-(9 - 3)]$	−6

EXERCISES

In 1–12, write the simplest symbol that represents the opposite of the number.

1. 8 **2.** −8 **3.** $+3\frac{1}{2}$ **4.** −6.5

5. (10 + 9) **6.** (24 − 10) **7.** (9 − 9) **8.** 8×0

9. $-(-7)$ **10.** $-\left(-\frac{3}{4}\right)$ **11.** $-[-(+5)]$ **12.** $-[-(6 + 8)]$

In 13–16, select the greater of the two numbers.

13. 10, −5 **14.** −1, 7 **15.** −8, −4 **16.** −12, 0

In 17–26, tell whether the statement is true or false.

17. If a is a real number, then $-a$ is always a negative number.
18. If a is a negative number, then $-a$ is always a positive number.
19. The opposite of a number is always a different number.

20. On a number line, the opposite of a positive number is to the left of the number.
21. On a number line, the opposite of any number is always to the left of the number.
22. If x is a positive number, then x is greater than its opposite.
23. The opposite of the opposite of a number is that number itself.
24. If $x > 0$, then $-x < 0$.
25. If $x < 0$, then $-x > 0$.
26. $x > 0$ if and only if $-x < 0$.

7-4 THE ABSOLUTE VALUE OF A NUMBER

In every pair of nonzero opposite numbers, the positive number is the greater. On a number line, the positive number is always to the right of the negative number that is its opposite. For example, 10 is greater than its opposite -10; on a number line, 10 is to the right of -10.

The greater of a nonzero number and its opposite is called the ***absolute value*** of the number. The absolute value of 0 is 0.

The absolute value of a number, a, is symbolized by $|a|$. Since 10 is the greater of the two numbers 10 and -10, the absolute value of 10 is 10 and the absolute value of -10 is 10.

$$|10| = 10$$

$$|-10| = 10$$

$$|10| = |-10|$$

Notice that the absolute value of a positive number is the number itself; the absolute value of a negative number is its opposite.

The absolute value of a number can also be considered as the distance between 0 and the graph of that number on the real number line. For example, $|3| = 3$, the distance from 0 to P, the graph of 3 on the real number line; $|-3| = 3$, the distance from 0 to S, the graph of -3 on the real number line.

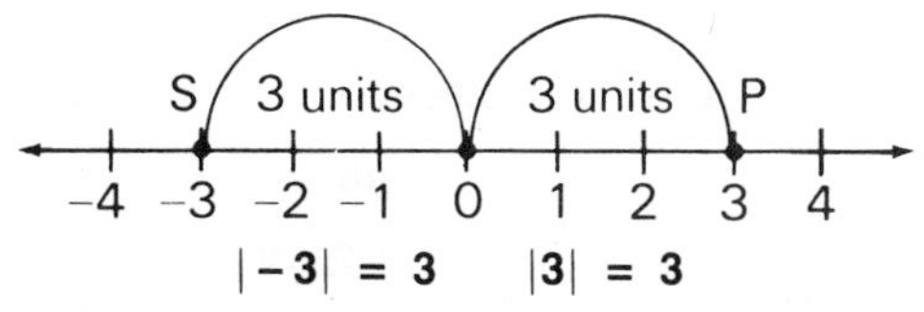

Observe, too, that the absolute value of any real number x is the maximum of the number and its opposite, symbolized as follows:

$$|\boldsymbol{x}| = \boldsymbol{x} \textbf{ max } (-\boldsymbol{x})$$

MODEL PROBLEMS

1. Find the value of the expression $|12| + |-3|$.

Solution:
Since $|12| = 12$, and $|-3| = 3$, $|12| + |-3| = 12 + 3 = 15$ *Ans.*

2. Find the value of the expression $|12 - 3|$.

Solution: First, evaluate the expression inside the absolute value symbol. Then, find the absolute value.

$|12 - 3| = |9|$, and $|9| = 9$ *Ans.*

EXERCISES

In 1–10: **a.** Give the absolute value of the given number. **b.** Give another number that has the same absolute value.

1. 3 **2.** -5 **3.** $+18$ **4.** -13 **5.** -20

6. $1\frac{1}{2}$ **7.** $-3\frac{3}{4}$ **8.** $-1\frac{1}{2}$ **9.** $+2.7$ **10.** -1.4

In 11–18, state whether the sentence is true or false.

11. $|20| = 20$ **12.** $|-13| = 13$ **13.** $|-15| = -15$ **14.** $|-9| = |9|$

15. $|-7| < |7|$ **16.** $|-10| > |3|$ **17.** $|8| < |-19|$ **18.** $|-21| > 21$

In 19–33, find the value of the number expression.

19. $|9| + |3|$ **20.** $|+8| - |+2|$ **21.** $|-6| + |4|$

22. $|-10| - |-5|$ **23.** $|4.5| - |4.5|$ **24.** $|+6| + |-4|$

25. $|8 + 6|$ **26.** $|7 - 2|$ **27.** $|15 - 15|$

28. $|9| + |-3| - |-4|$ **29.** $|-8| - |-2| + |-3|$ **30.** $|10| - |-6| - |4|$

31. $|(8 - 4)| + |-3|$ **32.** $-(|-9| - |7|)$ **33.** $-(|-8| - 2)$

In 34–41, state whether the sentence is true or false.

34. $|+5| - |-5| = 0$ **35.** $|+9| + |-9| = 0$

36. $|3| \cdot |-3| = -9$ **37.** $2 \cdot |-4| = |-2| \cdot |-4|$

38. $\frac{|-8|}{|-4|} = -|+2|$ **39.** $|4| \cdot |-2| - \frac{|-16|}{|2|} = 0$

40. $|-6| \cdot |4| > 0$ **41.** $|6| + |-4| < 6 - 4$

7-5 ADDING SIGNED NUMBERS ON A NUMBER LINE

Addition of signed numbers may be shown on a number line. To add a positive number, move to the right. To add a negative number, move to the left.

MODEL PROBLEMS

1. Add +3 and +2.

Solution: Start at 0 and move 3 units to the right to +3; then, move 2 more units to the right, arriving at +5.

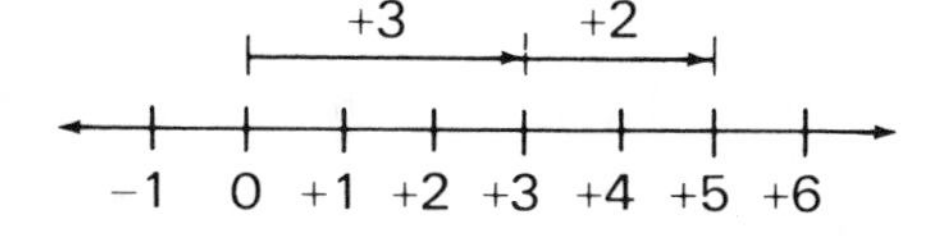

Answer: $(+3) + (+2) = +5$

2. Add −3 and −2.

Solution: Start at 0 and move 3 units to the left to −3; then, move 2 more units to the left, arriving at −5.

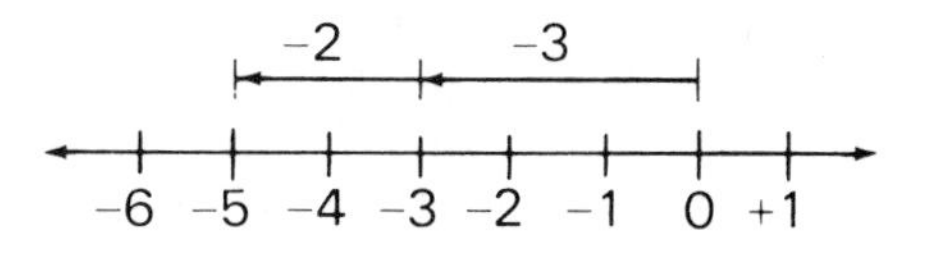

Answer: $(-3) + (-2) = -5$

3. Add +3 and −2.

Solution: Start at 0 and move 3 units to the right to +3; then, move 2 units to the left, arriving at +1.

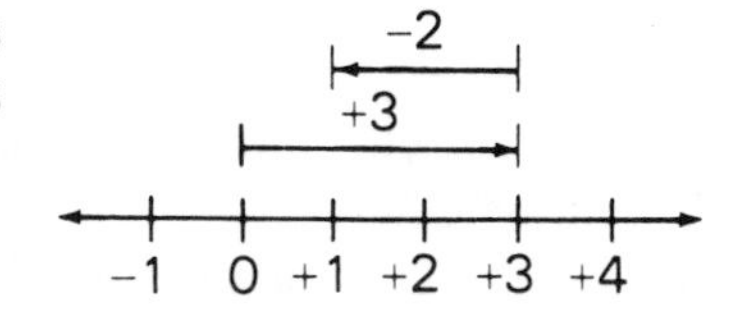

Answer: $(+3) + (-2) = +1$

4. Add −3 and +2.

Solution: Start at 0 and move 3 units to the left to −3; then, move 2 units to the right, arriving at −1.

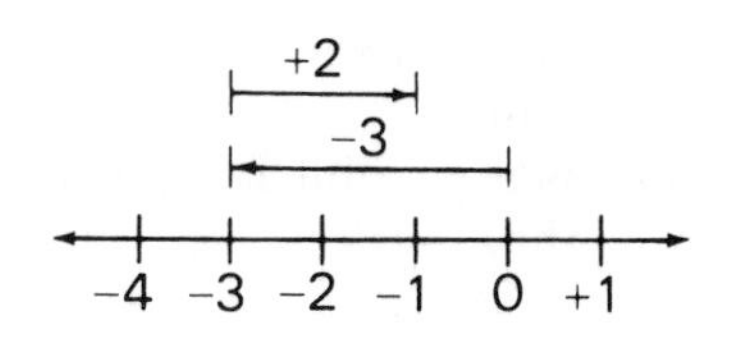

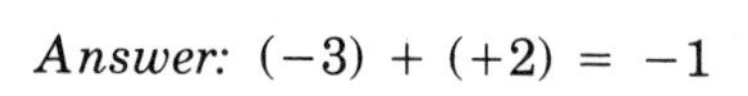

Answer: $(-3) + (+2) = -1$

EXERCISES

In 1–19, use a number line to find the sum of the signed numbers.

1. $(+3) + (+4)$ **2.** $(+6) + (+8)$ **3.** $(-2) + (-4)$
4. $(-5) + (-3)$ **5.** $(+7) + (-4)$ **6.** $(+4) + (-1)$
7. $(-6) + (+5)$ **8.** $(-8) + (+10)$ **9.** $(+4) + (-4)$
10. $(-7) + (+7)$ **11.** $(-2) + (+2)$ **12.** $(0) + (+4)$
13. $(0) + (-6)$ **14.** $(+6) + (0)$ **15.** $(-8) + (0)$
16. $[(+3) + (+4)] + (+2)$ **17.** $[(+8) + (-4)] + (-6)$
18. $[(-7) + (-3)] + (+6)$ **19.** $[(-5) + (+2)] + (+3)$

In 20–24, use signed numbers to solve the problem.

20. In one hour, the Celsius temperature rose 4° and in the next hour, the Celsius temperature rose 3°. What was the net change in temperature during the two-hour period?

21. An elevator started on the ground floor and rose 30 floors. Then, it came down 12 floors. At which floor was it at that time?

22. A football team gained 7 yards on the first play, lost 2 yards on the second, and lost 8 yards on the third. What was the net result of the three plays?

23. Fay deposited $250 in a bank. During the next month, she made a deposit of $60 and a withdrawal of $80. How much money did Fay have in the bank at that time?

24. During a four-day period, the dollar value of a share of stock rose $1\frac{1}{2}$ on the first day, dropped $\frac{5}{8}$ on the second day, rose $\frac{1}{8}$ on the third day, and dropped $1\frac{3}{4}$ on the fourth day. What was the net change in the stock during this period?

25. What type of number does the sum of two positive numbers always appear to be?

26. What type of number does the sum of two negative numbers always appear to be?

27. Is it possible for the sum of a positive and a negative number to be **(a)** a positive number? **(b)** a negative number?

28. If two given signed numbers (not opposites) are to be added, how can you tell whether the sum is a positive number or a negative number?

29. Is it possible for the sum of two numbers of arithmetic to be smaller than either of the numbers?

30. **a.** Is it possible for the sum of two signed numbers to be smaller than either of the numbers?
b. If your answer in part **a** is yes, give an example.

7-6 ADDITION OF SIGNED NUMBERS

In Chapter 1, you learned that the numbers of arithmetic have various properties of addition. Now, we will define the operation of addition so that these properties will also be true for signed numbers. By doing this, you will be able to add signed numbers without the use of a number line.

Addition of Two Positive Numbers

If +2 and +4 are added on a number line, the sum is +6. We write $(+2) + (+4) = +6$, or $2 + 4 = 6$. This example illustrates that you can add positive numbers in the same manner that you added the numbers of arithmetic.

The sum +6 is a positive number whose absolute value 6 is the sum of 2 and 4, the absolute values of +2 and +4.

+2	The absolute value of +2 is 2.
+4	The absolute value of +4 is 4.
+6	The sum of the absolute values is 6.

● **Rule 1.** The sum of two positive numbers is a positive number whose absolute value is found by adding the absolute values of the numbers.

In general, if both a and b are positive numbers:

$$\boldsymbol{a + b = |a| + |b|}$$

MODEL PROBLEMS

In 1–4, add the two numbers.

1. $\begin{array}{r} +8 \\ +10 \\ \hline +18 \end{array}$ **2.** $\begin{array}{r} +9.1 \\ +7.5 \\ \hline +16.6 \end{array}$ **3.** $(+7) + (+5) = +12$ **4.** $\frac{3}{8} + \frac{4}{8} = \frac{7}{8}$

Addition of Two Negative Numbers

If -2 and -4 are added on a number line, the sum is -6. We can write $(-2) + (-4) = -6$.

The sum -6 is a negative number whose absolute value 6 is the sum of 2 and 4, the absolute values of -2 and -4.

-2	The absolute value of -2 is 2.
$\underline{-4}$	The absolute value of -4 is 4.
-6	The sum of the absolute values is 6.

● **Rule 2.** The sum of two negative numbers is a negative number whose absolute value is found by adding the absolute values of the numbers.

In general, if both a and b are negative numbers:

$$\boldsymbol{a + b = -(|a| + |b|)}$$

Observe that in effect Rule 1 and Rule 2 tell us that:

● **To add two numbers that have the same sign, find the sum of their absolute values and give to this sum the common sign.**

MODEL PROBLEMS

In 1–6, add.

1. $\begin{array}{r} -4 \\ \underline{-3} \\ -7 \end{array}$ 2. $\begin{array}{r} -10 \\ \underline{-\ 8} \\ -18 \end{array}$ 3. $\begin{array}{r} -4.2 \\ \underline{-3.6} \\ -7.8 \end{array}$ 4. $\begin{array}{r} -\ 7.4 \\ \underline{-\ 8.7} \\ -16.1 \end{array}$

5. $\left(-\frac{3}{10}\right) + \left(-\frac{4}{10}\right) = -\frac{7}{10}$ 6. $\left(-5\frac{1}{8}\right) + \left(-5\frac{7}{8}\right) = -11$

Addition of a Positive Number and a Negative Number

If $+5$ and -3 are added on a number line, the sum is $+2$. We can write this as $(+5) + (-3) = +2$.

$+5$	The absolute value of $+5$ is 5.
$\underline{-3}$	The absolute value of -3 is 3.
$+2$	The difference of the absolute values is 2.

If -5 and $+3$ are added on a number line, the sum is -2. We can write this as $(-5) + (+3) = -2$.

-5	The absolute value of -5 is 5.
$+3$	The absolute value of $+3$ is 3.
-2	The difference of the absolute values is 2.

In both examples, the sum has the sign of the number with the greater absolute value. The absolute value of the sum is the difference of the absolute values of the numbers to be added.

If -5 and $+5$ are added on the number line, the sum is 0. We can write this as $(-5) + (+5) = 0$.

● **Rule 3.** To find the sum of two numbers, one of which is positive or 0 and the other negative, find a number whose absolute value is the difference of the absolute values of the numbers.

The sum is positive when the positive number has the greater absolute value; the sum is negative when the negative number has the greater absolute value; the sum is 0 if both numbers have the same absolute value.

In general, if a is a positive number or 0, that is, $a \geq 0$; and if b is a negative number, that is, $b < 0$, then:

$$\textbf{if } |a| > |b|, \textbf{ then } a + b = |a| - |b|$$

and

$$\textbf{if } |b| > |a|, \textbf{ then } a + b = -(|b| - |a|)$$

and

$$\textbf{if } |a| = |b|, \textbf{ then } a + b = 0$$

Observe that in effect Rule 3 tells us that:

● **To add two signed numbers that have different signs, find the difference of the absolute values of these numbers and give to this difference the sign of the number that has the greater absolute value; the sum is 0 if both numbers have the same absolute value.**

MODEL PROBLEMS

In 1–6, add the numbers.

1.	2.	3.	4.	5.	6.
$+9$	-8	-6	-6.7	-1.8	$-7\frac{3}{4}$
-2	$+3$	$+8$	$+4.2$	7.2	$5\frac{1}{4}$
$+7$	-5	$+2$	-2.5	5.4	$-2\frac{1}{2}$

Closure Property of Addition

The sum of two signed numbers is always a unique member of the set of signed numbers. Thus, addition is a binary operation for the set of signed numbers.

In general, for all signed numbers a and b:

$$a + b \text{ is a unique signed number}$$

Commutative Property of Addition

In general, for all signed numbers a and b:

$$a + b = b + a$$

Associative Property of Addition

In general, for all signed numbers a, b and c:

$$(a + b) + c = a + (b + c)$$

Addition Property of Zero

Since the sum of 0 and a signed number is that number itself, 0 is called the ***identity element of addition***, or the ***additive identity***.

In general, for every signed number a:

$$a + 0 = a \quad \text{and} \quad 0 + a = a$$

Other properties of signed numbers which we will assume are:

Addition Property of Opposites

Every signed number a has an ***opposite*** $-a$, such that their sum is 0. The opposite of a number is called the ***additive inverse*** of the number.

In general, for every signed number a and its opposite $-a$:

$$a + (-a) = 0$$

Property of the Opposite of a Sum

The opposite of the sum of two signed numbers is equal to the sum of the opposites.

In general, for all signed numbers a and b:

$$-(a + b) = (-a) + (-b)$$

Note: When adding more than two signed numbers, the commutative and associative properties allow us to arrange them in any order and group them in any way. It may prove helpful to add the positive numbers first, to add the negative numbers second, and then to add these two results.

MODEL PROBLEM

Add: $(+6) + (-2) + (+7) + (-4)$

How to Proceed	*Solution*
(1) Write the expression.	$(+6) + (-2) + (+7) + (-4)$
(2) Use the commutative and associative properties and add the positive and negative numbers separately.	$[(+6) + (+7)] + [(-2) + (-4)]$ $+6 \quad\quad -2$ $\underline{+7} \quad\quad \underline{-4}$ $+13 \quad\quad -6$
(3) Add the positive and negative sums.	$(+13) + (-6) = +7$ *Ans.*

Note that this sum could have been written as $6 - 2 + 7 - 4$. We will agree that this expression will mean the sum of the four signed numbers $+6$, -2, $+7$, and -4.

EXERCISES

In 1–51, add the numbers.

1. $\begin{array}{r} +6 \\ \underline{+4} \end{array}$ **2.** $\begin{array}{r} +7 \\ \underline{+6} \end{array}$ **3.** $\begin{array}{r} -14 \\ \underline{-23} \end{array}$ **4.** $\begin{array}{r} -17 \\ \underline{-28} \end{array}$ **5.** $\begin{array}{r} +8 \\ \underline{-6} \end{array}$

6. $\begin{array}{r} -9 \\ \underline{+7} \end{array}$ **7.** $\begin{array}{r} +8 \\ \underline{-4} \end{array}$ **8.** $\begin{array}{r} +2 \\ \underline{-9} \end{array}$ **9.** $\begin{array}{r} +23 \\ \underline{-35} \end{array}$ **10.** $\begin{array}{r} +6 \\ \underline{0} \end{array}$

11. $\begin{array}{r} -5 \\ 0 \\ \hline \end{array}$ 12. $\begin{array}{r} 0 \\ +4 \\ \hline \end{array}$ 13. $\begin{array}{r} 0 \\ -9 \\ \hline \end{array}$ 14. $\begin{array}{r} +5 \\ -5 \\ \hline \end{array}$ 15. $\begin{array}{r} -9 \\ +9 \\ \hline \end{array}$

16. $\begin{array}{r} +15 \\ +9 \\ \hline \end{array}$ 17. $\begin{array}{r} -28 \\ -38 \\ \hline \end{array}$ 18. $\begin{array}{r} -15 \\ -15 \\ \hline \end{array}$ 19. $\begin{array}{r} +6\frac{2}{3} \\ +1\frac{1}{3} \\ \hline \end{array}$ 20. $\begin{array}{r} -5\frac{1}{2} \\ -3\frac{1}{2} \\ \hline \end{array}$

21. $\begin{array}{r} 9\frac{1}{2} \\ 8\frac{3}{4} \\ \hline \end{array}$ 22. $\begin{array}{r} -6\frac{5}{6} \\ -1\frac{2}{3} \\ \hline \end{array}$ 23. $\begin{array}{r} -33\frac{1}{3} \\ +19\frac{2}{3} \\ \hline \end{array}$ 24. $\begin{array}{r} -5\frac{3}{4} \\ 8\frac{1}{2} \\ \hline \end{array}$ 25. $\begin{array}{r} -5.6 \\ -2.2 \\ \hline \end{array}$

26. $\begin{array}{r} +5.4 \\ +2.9 \\ \hline \end{array}$ 27. $\begin{array}{r} -8.8 \\ -7.5 \\ \hline \end{array}$ 28. $\begin{array}{r} 7.9 \\ -5.6 \\ \hline \end{array}$ 29. $\begin{array}{r} -6.9 \\ 9.4 \\ \hline \end{array}$ 30. $\begin{array}{r} +7 \\ -8\frac{3}{4} \\ \hline \end{array}$

31. $(+8) + (-14)$
32. $(-12) + (+37)$
33. $(+40) + (-17)$
34. $(-18) + 0$
35. $0 + (-28)$
36. $(+15) + (-15)$
37. $|-34| + |+20|$
38. $-|7| + (-10)$
39. $|15| + (-|-15|)$

40. $\begin{array}{r} +27 \\ -9 \\ -12 \\ \hline \end{array}$ 41. $\begin{array}{r} -45 \\ +12 \\ +13 \\ \hline \end{array}$ 42. $\begin{array}{r} 15 \\ -28 \\ 13 \\ \hline \end{array}$ 43. $\begin{array}{r} +20 \\ -12 \\ -8 \\ \hline \end{array}$ 44. $\begin{array}{r} -1.5 \\ +3.7 \\ -8.3 \\ \hline \end{array}$ 45. $\begin{array}{r} 8\frac{1}{2} \\ -4\frac{1}{4} \\ 7\frac{3}{4} \\ \hline \end{array}$

46. $+18 - 15 + 9$
47. $30 - 18 - 12$
48. $-19 + 8 - 15$
49. $-17 - 19 + 40$
50. $48 - 32 + 19 - 41$
51. $-4\frac{1}{3} + 7 + 8\frac{1}{3} - 11$

In 52–56, name the signed number that represents the sum of the quantities. Represent a rise or a profit by a positive number.

52. a rise of 4 meters and a rise of 6 meters
53. a loss of 6 yards and a loss of 2 yards
54. a rise of 7 meters and a fall of 5 meters
55. a loss of $20 and a profit of $20
56. a rise of 4°, a drop of 3°, and a drop of 5°

In 57–68, give a replacement for the question mark that will make the resulting sentence true.

57. $(+4) + (?) = 0$ **58.** $(-2) + (?) = 0$

59. $(0) + (?) = 0$ **60.** $(12) + (?) = 0$

61. $(b) + (?) = 0$ **62.** $(-y) + (?) = 0$

63. $(+8) + (?) = (+12)$ **64.** $(+10) + (?) = 7$

65. $(6) + (?) = -4$ **66.** $\left(-4\frac{1}{2}\right) + (?) = -9\frac{1}{2}$

67. $\left(-\frac{6}{7}\right) + (?) = \left(+\frac{2}{7}\right)$ **68.** $(-3.75) + (?) = -3.75$

In 69–74, give a replacement for the variable that will make the resulting sentence true.

69. $9 + y = 0$ **70.** $x + (-12) = 0$ **71.** $5 + c = 1$

72. $x + 4 = -2$ **73.** $x + (-6) = 8$ **74.** $d + (-5) = -3$

In 75–79, name the addition property that makes each sentence true.

75. $(-3) + (+8) = (+8) + (-3)$ **76.** $(+50) + (-50) = 0$

77. $(-8) + 0 = -8$ **78.** $-(8 + 9) = (-8) + (-9)$

79. $(-6) + [(-4) + (+2)] = [(-6) + (-4)] + (+2)$

In 80–83, state whether the sentence is true or false.

80. $|x| + |-x| = 0 \ (x \neq 0)$ **81.** $(-c) + (-d) = -(c + d)$

82. $-(-b) = b$ **83.** $(a + b) + [-(a + b)] = 0$

7-7 SUBTRACTION OF SIGNED NUMBERS

In arithmetic, to subtract 3 from 7, we find a number which, when added to 3, will give 7. That number is 4. We know that $7 - 3 = 4$ because $3 + 4 = 7$.

Subtraction in the set of numbers of arithmetic is defined as the inverse operation of addition.

● **In general, for every number c and every number b, the expression $c - b$ means to find a number a such that $b + a = c$.**

We use the same definition of subtraction in the set of signed numbers. To subtract (-2) from $(+3)$, written $(+3) - (-2)$, we must find a number which, when added to -2, will give $+3$. We write:

$$(-2) + (?) = +3$$

We can use a number line to find the answer to $(-2) + (?) = +3$. Think as follows: From a point 2 units to the left of 0, what motion must be made to arrive at a point 3 units to the right of 0? We must move 5 units to the right. This motion is represented by $+5$.

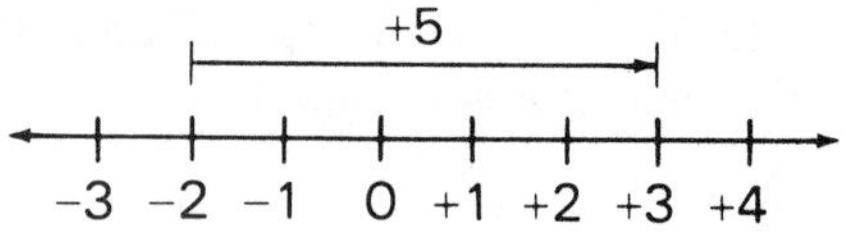

Therefore, $(+3) - (-2) = +5$ because $(-2) + (+5) = +3$. We can write $(+3) - (-2) = +5$ vertically as follows:

$(+3)$	*or*	*Subtract:*	$(+3)$	minuend
$-(-2)$			(-2)	subtrahend
$+5$			$+5$	difference

Check each of the following examples by using a number line to answer the related question: subtrahend + (?) = minuend.

Subtract:				
	$+9$	-7	$+5$	-3
	$+6$	-2	-2	$+1$
	$+3$	-5	$+7$	-4

Now, we will consider another way in which addition and subtraction are related. In each of the following examples, compare the result obtained when subtracting the signed number with the result obtained when adding the opposite of that signed number:

Subtract	*Add*	*Subtract*	*Add*	*Subtract*	*Add*	*Subtract*	*Add*
$+9$	$+9$	-7	-7	$+5$	$+5$	-3	-3
$+6$	-6	-2	$+2$	-2	$+2$	$+1$	-1
$+3$	$+3$	-5	-5	$+7$	$+7$	-4	-4

Observe that in each example, adding the opposite (the additive inverse) of a signed number gives the same result as subtracting that signed number. It therefore seems reasonable to define subtraction as follows:

If a is any signed number and b is any signed number, then:

$$a - b = a + (-b)$$

PROCEDURE. To subtract one signed number from another, add the opposite (additive inverse) of the subtrahend to the minuend.

It is always possible to subtract one signed number from another and obtain a unique signed number as the result. Therefore, the set of signed numbers is closed with respect to subtraction and, for the set of signed numbers, subtraction is a binary operation.

Uses of the Symbol −

There are three ways in which we have used the symbol −.

To indicate that a number is negative:

Example: -2 negative 2

To indicate the opposite of a number:

Example: $-(-4)$ opposite of negative 4
$-a$ opposite of a

To indicate subtraction:

Example: $4 - (-3)$ the difference between 4 and −3

MODEL PROBLEMS

1. Perform the indicated subtractions.
a. $(+30) - (+12) = (+30) + (-12) = +18$
b. $(-19) - (-7) = (-19) + (+7) = -12$
c. $(-4) - (0) = (-4) + (0) = -4$
d. $0 - 8 = 0 + (-8) = -8$

2. Subtract the lower number from the upper number.

a.	**b.**	**c.**	**d.**	**e.**	**f.**	**g.**
+45	−19	−19	+25	0	−8	−8
+20	−14	+17	−18	−6	−8	+8
+25	−5	−36	+43	+6	0	−16

Note: In each problem, the signed number is subtracted by adding its opposite to the minuend.

3. How much greater than −3 is 9?

Solution:

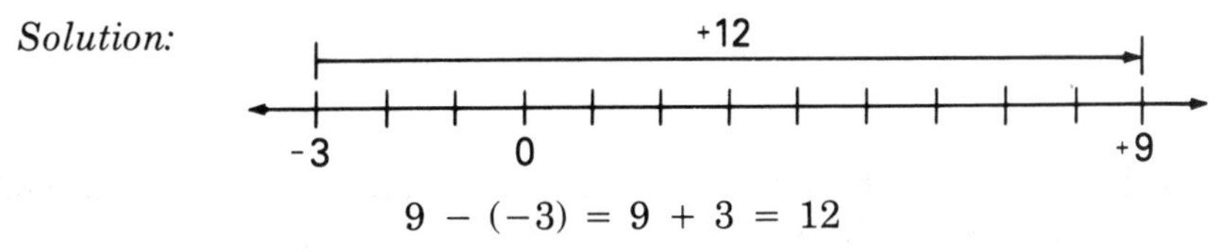

$$9 - (-3) = 9 + 3 = 12$$

Answer: 9 is 12 greater than −3.

EXERCISES

In 1–36, subtract the lower number from the upper number.

1. $\begin{array}{r}+9\\ \underline{+3}\end{array}$ 2. $\begin{array}{r}+25\\ \underline{+18}\end{array}$ 3. $\begin{array}{r}+6\\ \underline{+8}\end{array}$ 4. $\begin{array}{r}+16\\ \underline{+24}\end{array}$ 5. $\begin{array}{r}-7\\ \underline{-3}\end{array}$ 6. $\begin{array}{r}-26\\ \underline{-12}\end{array}$

7. $\begin{array}{r}-2\\ \underline{-6}\end{array}$ 8. $\begin{array}{r}-34\\ \underline{-50}\end{array}$ 9. $\begin{array}{r}+5\\ \underline{-1}\end{array}$ 10. $\begin{array}{r}+26\\ \underline{-19}\end{array}$ 11. $\begin{array}{r}+4\\ \underline{-9}\end{array}$ 12. $\begin{array}{r}+65\\ \underline{-75}\end{array}$

13. $\begin{array}{r}-6\\ \underline{+2}\end{array}$ 14. $\begin{array}{r}-63\\ \underline{+29}\end{array}$ 15. $\begin{array}{r}-5\\ \underline{+7}\end{array}$ 16. $\begin{array}{r}-36\\ \underline{+50}\end{array}$ 17. $\begin{array}{r}+10\\ \underline{0}\end{array}$ 18. $\begin{array}{r}0\\ \underline{+7}\end{array}$

19. $\begin{array}{r}0\\ \underline{-20}\end{array}$ 20. $\begin{array}{r}-19\\ \underline{-19}\end{array}$ 21. $\begin{array}{r}+18\\ \underline{+29}\end{array}$ 22. $\begin{array}{r}+15\\ \underline{+15}\end{array}$ 23. $\begin{array}{r}+36\\ \underline{-15}\end{array}$ 24. $\begin{array}{r}-39\\ \underline{+15}\end{array}$

25. $\begin{array}{r}-45\\ \underline{+17}\end{array}$ 26. $\begin{array}{r}-6\\ \underline{+6}\end{array}$ 27. $\begin{array}{r}-8\\ \underline{-8}\end{array}$ 28. $\begin{array}{r}0\\ \underline{-15}\end{array}$ 29. $\begin{array}{r}+8.7\\ \underline{+6.5}\end{array}$ 30. $\begin{array}{r}+8.3\\ \underline{-6.2}\end{array}$

31. $\begin{array}{r}-6.9\\ \underline{+3.7}\end{array}$ 32. $\begin{array}{r}5.9\\ \underline{7.2}\end{array}$ 33. $\begin{array}{r}+9\frac{1}{2}\\ \underline{+6\frac{1}{2}}\end{array}$ 34. $\begin{array}{r}-3\frac{1}{4}\\ \underline{-7\frac{3}{4}}\end{array}$ 35. $\begin{array}{r}7\frac{3}{4}\\ \underline{-2\frac{1}{4}}\end{array}$ 36. $\begin{array}{r}-6\frac{5}{6}\\ \underline{+3\frac{1}{3}}\end{array}$

In 37–42, perform the indicated subtraction.

37. $(+19) - (+30)$ 38. $(-12) - (-25)$ 39. $22 - (-8)$

40. $(+6.4) - (+8.1)$ 41. $(-3.7) - (-5.2)$ 42. $(-9.2) - 8.3$

43. How much is 18 decreased by -7?
44. How much greater than -15 is 12?
45. How much greater than -4 is -1?
46. How much less than 6 is -3?
47. What number is 6 less than -6?
48. From the sum of 25 and -10, subtract -4.
49. Subtract 8 from the sum of -6 and -12.

In 50–55, state the number that must be added to the given number to make the result equal to 0.

50. $+5$ 51. -3 52. $+8.5$ 53. -3.7 54. $+1\frac{7}{8}$ 55. $-\frac{9}{2}$

In 56–61, find the value of the given expression.

56. $(+7) + (+9) - (-4)$

57. $(-12) - (+9) + (-20)$

58. $32 - 49 - 21 + 10$

59. $-15 + 8 - 5 + 12$

60. $6\frac{1}{4} - 5 + 7\frac{3}{4} - 1\frac{1}{2}$

61. $-5\frac{1}{3} + 8 + 9\frac{1}{3} - 12$

In 62–70, use signed numbers to do the problem.

62. Find the change when the Celsius temperature changes from:
 a. $+5°$ to $+8°$ **b.** $-10°$ to $+18°$
 c. $-6°$ to $-18°$ **d.** $+12°$ to $-4°$

63. Find the change in altitude when you go from a place that is 15 meters below sea level to a place that is 95 meters above sea level.

64. In a game, Sid was 35 points "in the hole." How many points must he make in order to have a score of 150 points?

65. The record high Fahrenheit temperature in New City is 105°; the record low is $-9°$. Find the difference between these temperatures.

66. At one point, the Pacific Ocean is .5 kilometer in depth; at another point it is .25 kilometer in depth. Find the difference between these depths.

67. State whether the following sentences are true or false:
 a. $(+5) - (-3) = (-3) - (+5)$ **b.** $(-7) - (-4) = (-4) - (-7)$

68. If x and y represent numbers:
 a. Does $x - y = y - x$ for all replacements of x and y?
 b. Does $x - y = y - x$ for any replacements of x and y? For which values of x and y?
 c. What is the relation between $x - y$ and $y - x$ for all replacements of x and y?
 d. Is the operation of subtraction commutative? That is, for all signed numbers x and y, does $x - y = y - x$?

69. State whether the following sentences are true or false:
 a. $(15 - 9) - 6 = 15 - (9 - 6)$
 b. $[(-10) - (+4)] - (+8) = (-10) - [(+4) - (+8)]$

70. Is the operation of subtraction associative? That is, for all signed numbers x, y, and z, does $(x - y) - z = x - (y - z)$?

7-8 MULTIPLICATION OF SIGNED NUMBERS

We will use a common experience to illustrate the various cases that can arise in the multiplication of signed numbers. Represent a gain of weight by a positive number and a loss of weight by a negative number, and represent a number of weeks in the future by a positive number and a number of weeks in the past by a negative number.

Case 1. Multiplying a Positive Number by a Positive Number

If a girl gains 1 kilogram each week, 4 weeks from now she will be 4 kilograms heavier. Using signed numbers, we may write:

$$(+4) \cdot (+1) = +4$$

The product of the two positive numbers is a positive number.

Case 2. Multiplying a Negative Number by a Positive Number

If a girl loses 1 kilogram each week, 4 weeks from now she will be 4 kilograms lighter than she is now. Using signed numbers, we may write:

$$(+4) \cdot (-1) = -4$$

The product of the negative number and the positive number is a negative number.

Case 3. Multiplying a Positive Number by a Negative Number

If a girl has gained 1 kilogram each week, 4 weeks ago she was 4 kilograms lighter than she is now. Using signed numbers, we may write:

$$(-4) \cdot (+1) = -4$$

The product of the positive number and the negative number is a negative number.

Case 4. Multiplying a Negative Number by a Negative Number

If a girl has lost 1 kilogram each week, 4 weeks ago she was 4 kilograms heavier than she is now. Using signed numbers, we may write:

$$(-4) \cdot (-1) = +4$$

The product of the two negative numbers is a positive number.

In all four cases, the absolute value of the product, 4, is equal to the product of the absolute values of the factors, 4 and 1.

These examples illustrate the reasonableness of the following rules:

Rules for Multiplying Signed Numbers

● **Rule 1.** The product of two positive numbers or of two negative numbers is a positive number whose absolute value is the product of the absolute values of the numbers.

● **Rule 2.** The product of a positive number and a negative number is a negative number whose absolute value is the product of the absolute values of the numbers.

In general, if a and b are both positive or are both negative, then:

$$\boldsymbol{ab = |a| \cdot |b|}$$

If one of the numbers a and b is positive and the other is negative, then:

$$ab = -(|a| \cdot |b|)$$

In effect, these rules tell us that:

● **To multiply two signed numbers, multiply their absolute values. Then, write a plus sign before this product when the two numbers have the same sign, or, write a minus sign before this product when the two numbers have different signs.**

MODEL PROBLEMS

In 1–6, multiply the two numbers.

1. $\begin{array}{r} +12 \\ +4 \\ \hline +48 \end{array}$ **2.** $\begin{array}{r} -13 \\ -5 \\ \hline +65 \end{array}$ **3.** $\begin{array}{r} +18 \\ -3 \\ \hline -54 \end{array}$ **4.** $\begin{array}{r} -15 \\ 6 \\ \hline -90 \end{array}$ **5.** $\begin{array}{r} +3.4 \\ -3 \\ \hline -10.2 \end{array}$ **6.** $\begin{array}{r} -7\frac{1}{8} \\ -3 \\ \hline +21\frac{3}{8} \end{array}$

Closure Property of Multiplication

The product of two signed numbers is always a unique member of the set of signed numbers. Thus, multiplication is a binary operation for the set of signed numbers.

In general, for all signed numbers a and b:

ab is a unique signed number

Commutative Property of Multiplication

In general, for all signed numbers a and b:

$$ab = ba$$

Associative Property of Multiplication

In general, for all signed numbers a, b, and c:

$$(ab)c = a(bc)$$

Distributive Property of Multiplication Over Addition

In general, for all signed numbers a, b, and c:

$$\boldsymbol{a(b + c) = ab + ac} \quad \text{and} \quad \boldsymbol{ab + ac = a(b + c)}$$

Multiplication Property of One

The product of 1 and a signed number is that number itself. For this reason, 1 is called the ***identity element of multiplication***, or the ***multiplicative identity***.

In general, for every signed number a:

$$\boldsymbol{a \cdot 1 = a} \quad \text{and} \quad \boldsymbol{1 \cdot a = a}$$

Multiplication Property of Zero

In general, for every signed number a:

$$\boldsymbol{a \cdot 0 = 0} \quad \text{and} \quad \boldsymbol{0 \cdot a = 0}$$

We may also note one new property that we will assume for signed numbers:

Multiplication Property of Negative One

In general, for every signed number a:

$$\boldsymbol{(a) \cdot (-1) = -a} \quad \text{and} \quad \boldsymbol{(-1) \cdot (a) = -a}$$

MODEL PROBLEMS

1. Find the value of $(-2)^3$.

Solution: $(-2)^3 = (-2)(-2)(-2) = -8$ *Ans.*

Note: The answer is negative because there is an odd number of negative factors (3 negative factors).

2. Find the value of $(-3)^4$.

Solution: $(-3)^4 = (-3)(-3)(-3)(-3) = +81$ *Ans.*

Note: The answer is positive because there is an even number of negative factors (4 negative factors).

EXERCISES

In 1–27, find the product of the numbers.

1. $\begin{array}{r}+6\\ \underline{+4}\end{array}$ 2. $\begin{array}{r}-7\\ \underline{-1}\end{array}$ 3. $\begin{array}{r}-15\\ \underline{-7}\end{array}$ 4. $\begin{array}{r}-8\\ \underline{+4}\end{array}$ 5. $\begin{array}{r}+15\\ \underline{-8}\end{array}$ 6. $\begin{array}{r}24\\ \underline{-6}\end{array}$

7. $\begin{array}{r}-12\\ \underline{0}\end{array}$ 8. $\begin{array}{r}-25\\ \underline{-4}\end{array}$ 9. $\begin{array}{r}-24\\ \underline{+8}\end{array}$ 10. $\begin{array}{r}0\\ \underline{-5}\end{array}$ 11. $\begin{array}{r}-75\\ \underline{-3}\end{array}$ 12. $\begin{array}{r}15\\ \underline{-9}\end{array}$

13. $\begin{array}{r}+1.5\\ \underline{-2.4}\end{array}$ 14. $\begin{array}{r}-.25\\ \underline{80}\end{array}$ 15. $\begin{array}{r}+8\\ \underline{+\frac{1}{2}}\end{array}$ 16. $\begin{array}{r}-15\\ \underline{+\frac{3}{5}}\end{array}$ 17. $\begin{array}{r}-\frac{1}{2}\\ \underline{-\frac{1}{3}}\end{array}$ 18. $\begin{array}{r}+16\\ \underline{-2\frac{1}{4}}\end{array}$

19. $(+8)\left(+\frac{1}{4}\right)$
20. $\left(-\frac{3}{5}\right)(-20)$
21. $(+2)\left(-\frac{1}{2}\right)$
22. $(+4)(+3)(+2)$
23. $(-1)(-7)(-8)$
24. $(-3)(-5)(+4)(-1)$
25. $(-7)(+2)(0)$
26. $|+10| \cdot |-3| \cdot (-4)$
27. $(+8)(-9)(0)(-10)$

In 28–39, find the value of the expression.

28. $(+4)^2$
29. $(-3)^2$
30. $(+5)^3$
31. $(-4)^3$
32. $(-5)^3$
33. $(-1)^4$
34. $\left(+\frac{1}{2}\right)^2$
35. $\left(-\frac{1}{2}\right)^2$
36. $\left(+\frac{2}{3}\right)^3$
37. $\left(-\frac{3}{5}\right)^3$
38. $\left(-\frac{1}{4}\right)^3$
39. $\left(-\frac{1}{5}\right)^4$

In 40–44, fill in the blanks so that the resulting sentence is an illustration of the distributive property.

40. $5(9 + 7) =$ ____
41. $-4(x + y) =$ ____
42. ____ $= 6(-3) + 6(-5)$
43. ____ $= 7a + 7b$
44. $8($____$) = ($____$)(5) + ($____$)(-3)$

In 45–48, name the multiplication property illustrated.

45. $(-6) \cdot (-5) = (-5) \cdot (-6)$
46. $[(-3) \cdot 4] \cdot 7 = (-3) \cdot [4 \cdot 7]$
47. $-8 \cdot [4 + (-1)] = (-8) \cdot (4) + (-8) \cdot (-1)$
48. $5x + 5 \cdot (-y) = 5 \cdot [x + (-y)]$

49. Name the property that justifies each of statements **a** through **e**.
 a. $2(1 + 3) = 2(1) + 2(3)$
 b. $(-2) + (-1) = (-1) + (-2)$
 c. $(1 + 2) + 3 = 1 + (2 + 3)$
 d. $2 + 0 = 2$
 e. $(-5) \times 7 = 7 \times (-5)$

50. State whether the following sentences are true or false:
a. $5(7 - 3) = 5(7) - 5(3)$
b. $8[(+4) - (-2)] = 8(+4) - 8(-2)$

51. Is the operation of multiplication distributive over subtraction? That is, does $x(y - z) = xy - xz$ for all signed numbers x, y, and z?

52. **a.** Complete the first table, showing the product ab for each pair of numbers a and b.
b. In the second table, what is the heading for the third column, written in symbolic logic?
c. What similarities, if any, can you find when comparing the two tables?

a	b	ab
+5	+2	
+5	−2	
−5	+2	
−5	−2	

Table 1

p	q	
T	T	T
T	F	F
F	T	F
F	F	T

Table 2

7-9 DIVISION OF SIGNED NUMBERS

Using the Inverse Operation in Dividing Signed Numbers

Division may be defined as the inverse operation of multiplication, just as subtraction is defined as the inverse operation of addition. To divide 6 by 2 means to find a number which, when multiplied by 2, gives 6. The number is 3 because $3(2) = 6$. Thus, $\frac{6}{2} = 3$, or $6 \div 2 = 3$. The number 6 is the ***dividend***, 2 is the ***divisor***, and 3 is the ***quotient***.

It is impossible to divide a signed number by 0. That is, division by 0 is undefined. For example, to solve $(-9) \div 0 = ?$, we would have to find a number which, when multiplied by 0, would give -9. There is no such number since the product of any signed number and 0 is 0.

In general, for all signed numbers a and b $(b \neq 0)$:

$a \div b$, or $\frac{a}{b}$, means to find a unique number c such that $cb = a$

In dividing nonzero signed numbers, there are four possible cases. Consider the following examples:

Case 1. $\frac{+6}{+3} = ?$ implies $(?)(+3) = +6$. *Answer:* $\frac{+6}{+3} = +2$

Case 2. $\frac{-6}{-3} = ?$ implies $(?)(-3) = -6$. *Answer*: $\frac{-6}{-3} = +2$

Case 3. $\frac{-6}{+3} = ?$ implies $(?)(+3) = -6$. *Answer*: $\frac{-6}{+3} = -2$

Case 4. $\frac{+6}{-3} = ?$ implies $(?)(-3) = +6$. *Answer*: $\frac{+6}{-3} = -2$

In the preceding examples, observe:

1. When the dividend and divisor are both positive or both negative, the quotient is positive.
2. When the dividend and divisor have opposite signs, the quotient is negative.
3. In all cases, the absolute value of the quotient is the absolute value of the dividend divided by the absolute value of the divisor.

The previous examples illustrate the following rules of division:

Rules for Dividing Signed Numbers

● **Rule 1.** The quotient of two positive numbers, or of two negative numbers, is a positive number whose absolute value is the absolute value of the dividend divided by the absolute value of the divisor.

● **Rule 2.** The quotient of a positive number and a negative number is a negative number whose absolute value is the absolute value of the dividend divided by the absolute value of the divisor.

In general, if a and b are both positive or are both negative, then:

$$\frac{a}{b} = \frac{|a|}{|b|}$$

If one of the numbers a and b is positive and the other number is negative, then:

$$\frac{a}{b} = -\left(\frac{|a|}{|b|}\right)$$

Rule for Dividing Zero by a Nonzero Number

If the expression $\frac{0}{-5} = ?$, then $(?)(-5) = 0$. Since 0 is the only number that can replace ? and result in a true statement, $\frac{0}{-5} = 0$. This illustrates that zero divided by any nonzero number is zero.

In general, if a is a nonzero number ($a \neq 0$):

$$\frac{\mathbf{0}}{a} = \mathbf{0}$$

MODEL PROBLEMS

In 1–6, perform the indicated division, if possible.

1. $\frac{+60}{+15} = +\left(\frac{60}{15}\right) = +4$

2. $\frac{+10}{-90} = -\left(\frac{10}{90}\right) = -\frac{1}{9}$

3. $\frac{-27}{-3} = +\left(\frac{27}{3}\right) = +9$

4. $(-45) \div 9 = -(45 \div 9) = -5$

5. $0 \div 9 = 0$

6. $-3 \div (0)$ is undefined

Using the Reciprocal in Dividing Signed Numbers

When the product of two numbers is 1, one number is called the ***reciprocal*** or ***multiplicative inverse*** of the other. For example, since $(+8)\left(+\frac{1}{8}\right) = 1$, we say $+\frac{1}{8}$ is the reciprocal or multiplicative inverse of $+8$ and $+8$ is the reciprocal or multiplicative inverse of $+\frac{1}{8}$.

Since $\left(\frac{3}{5}\right)\left(\frac{5}{3}\right) = 1$, we say $\frac{5}{3}$ is the reciprocal or multiplicative inverse of $\frac{3}{5}$ and $\frac{3}{5}$ is the reciprocal or multiplicative inverse of $\frac{5}{3}$.

Since $\left(-\frac{1}{2}\right)(-2) = 1$, we say -2 is the reciprocal or multiplicative inverse of $-\frac{1}{2}$ and $-\frac{1}{2}$ is the reciprocal or multiplicative inverse of -2.

Since there is no number which, when multiplied by 0, gives 1, the number 0 has no reciprocal.

In general, for every nonzero signed number a, there is a unique signed number $\frac{1}{a}$ such that:

$$\boldsymbol{a \cdot \frac{1}{a} = 1}$$

Using the reciprocal of a number, we can define division in terms of multiplication as follows:

For all signed numbers a and b $(b \neq 0)$:

$$\boldsymbol{a \div b = \frac{a}{b} = a \cdot \frac{1}{b} \quad (b \neq 0)}$$

PROCEDURE. To divide a signed number by a nonzero signed number, multiply the dividend by the reciprocal of the divisor.

Notice that if we exclude division by 0, the set of signed numbers is closed with respect to division because every nonzero signed number has a unique reciprocal, and multiplication by this reciprocal is always possible. This is to say that if division by zero is excluded, division is a binary operation for the set of signed numbers.

MODEL PROBLEMS

In 1–5, perform the indicated division.

1. $\frac{+10}{+2} = (+10)\left(+\frac{1}{2}\right) = +5$

2. $\frac{-12}{+8} = (-12)\left(+\frac{1}{8}\right) = -\frac{3}{2}$

3. $\frac{-28}{-7} = (-28)\left(-\frac{1}{7}\right) = +4$

4. $\frac{0}{-3} = (0)\left(-\frac{1}{3}\right) = 0$

5. $(+18) \div \left(-\frac{1}{2}\right) = (+18)(-2) = -36$

EXERCISES

In 1–12, name the reciprocal (the multiplicative inverse) of the given number.

1. 6
2. −5
3. 9
4. −7
5. 1
6. −1
7. $\frac{1}{5}$
8. $-\frac{1}{10}$
9. $\frac{3}{4}$
10. $-\frac{2}{3}$
11. $x\ (x \neq 0)$
12. $-x\ (x \neq 0)$

In 13–48, find the indicated quotients or write "undefined" if no quotient exists.

13. $\frac{+10}{+2}$
14. $\frac{-63}{-9}$
15. $\frac{-8}{+4}$
16. $\frac{-48}{+16}$
17. $\frac{+25}{-1}$
18. $\frac{+84}{-14}$
19. $\frac{-1}{-1}$
20. $\frac{-10}{+10}$
21. $\frac{0}{+6}$
22. $\frac{+52}{-4}$
23. $\frac{+84}{-12}$
24. $\frac{-30}{-6}$
25. $\frac{+100}{-25}$
26. $\frac{-108}{0}$
27. $\frac{-65}{+5}$
28. $\frac{0}{3}$
29. $\frac{+4}{-8}$
30. $\frac{-6}{-9}$
31. $\frac{-15}{-12}$
32. $\frac{+18}{-4}$
33. $\frac{-16}{+6}$
34. $\frac{-34}{4}$
35. $\frac{+100}{-8}$
36. $\frac{-36}{-8}$
37. $\frac{0}{-4}$
38. $\frac{-5}{-9}$
39. $\frac{3}{-7}$
40. $\frac{8.4}{-4}$
41. $\frac{-9.6}{-.3}$
42. $\frac{-3.6}{1.2}$
43. $(+48) \div (-6)$
44. $(-75) \div (-15)$
45. $(-50) \div (0)$
46. $(+12) \div \left(-\frac{1}{3}\right)$
47. $\left(-\frac{3}{4}\right) \div (+6)$
48. $\left(-\frac{3}{4}\right) \div \left(-\frac{2}{3}\right)$

49. **a.** Find the value of x for which the denominator of the fraction $\frac{1}{x - 2}$ has a value of 0.
    b. State the value of x for which the multiplicative inverse of $(x - 2)$ is not defined.

In 50–53, give the multiplicative inverse of the expression and state the value of x for which the multiplicative inverse is not defined.

50. $x - 5$
51. $x + 3$
52. $2x - 1$
53. $3x + 1$

54. State whether the following sentences are true or false:
    a. $(+10) \div (-5) = (-5) \div (+10)$
    b. $(-16) \div (-2) = (-2) \div (-16)$

55. If x and y represent signed numbers:
   a. Does $x \div y = y \div x$ for all replacements of x and y?
   b. Does $x \div y = y \div x$ for any replacements of x and y? If your answer is yes, give an example.
   c. What is the relation between $x \div y$ and $y \div x$ when $x \neq 0$ and $y \neq 0$?
   d. Is division commutative? That is, does $x \div y = y \div x$ for every nonzero signed number x and every nonzero signed number y?

56. State whether the following sentences are true or false:
   a. $[(+16) \div (+4)] \div (+2) = (+16) \div [(+4) \div (+2)]$
   b. $[(-36) \div (+6)] \div (-2) = (-36) \div [(+6) \div (-2)]$

57. Is division associative? That is, does $(x \div y) \div z = x \div (y \div z)$ for every signed number x, y, and z, when $y \neq 0$ and $z \neq 0$?

58. State whether the following sentences are true or false:
   a. $(12 + 6) \div 2 = 12 \div 2 + 6 \div 2$
   b. $[(+25) - (-10)] \div (+5) = (+25) \div (+5) - (-10) \div (+5)$

59. Does it appear that the operation of division is distributive over addition? That is, does $(x + y) \div z = x \div z + y \div z$ for every signed number x, y, and z when $z \neq 0$?

60. **a.** What is the additive identity for the set of signed numbers?
   b. What is the multiplicative identity for the set of signed numbers?
   c. What is the additive inverse of 3?
   d. What is the multiplicative inverse of -6?
   e. What number is its own multiplicative inverse?

7-10 USING SIGNED NUMBERS IN EVALUATING ALGEBRAIC EXPRESSIONS

When you evaluate an algebraic expression by replacing the variables by signed numbers, you follow the same procedure that you used when you evaluated algebraic expressions by replacing the variables by the numbers of arithmetic.

MODEL PROBLEMS

1. Find the value of $-3x^2y^3$ when $x = +2$ and $y = -1$.

How to Proceed	*Solution*
(1) Write the expression.	$-3x^2y^3$
(2) Replace the variables by the given values.	$= -3(+2)^2(-1)^3$
(3) Evaluate the powers.	$= -3(+4)(-1)$
(4) Multiply the signed numbers.	$= +12$ *Ans.*

2. Find the value of $x^2 - 3x - 54$ when $x = -5$.

How to Proceed	*Solution*
(1) Write the expression.	$x^2 - 3x - 54$
(2) Replace the variable by its given value.	$= (-5)^2 - 3(-5) - 54$
(3) Evaluate the power.	$= 25 - 3(-5) - 54$
(4) Do the multiplication.	$= 25 + 15 - 54$
(5) Do the addition.	$= -14$ *Ans.*

Note that the algebraic expression may be thought of as the sum of three terms: $x^2 + (-3x) + (-54)$

EXERCISES

In 1–52, find the numerical value of the expression. Use $a = -8$, $b = +6$, $d = -3$, $x = -4$, $y = 5$, and $z = -1$.

1. $6a$ **2.** $-5b$ **3.** ab **4.** $2xy$

5. $-4bz$ **6.** $\frac{1}{3}d$ **7.** $-\frac{2}{3}b$ **8.** $\frac{3}{8}a$

9. $\frac{1}{2}xy$ **10.** $-\frac{3}{4}ab$ **11.** a^2 **12.** d^3

13. $-y^2$ **14.** $-d^2$ **15.** $-z^3$ **16.** $2x^2$

17. $-3y^2$ **18.** $-3b^2$ **19.** $4d^2$ **20.** $-2z^3$

21. xy^2 **22.** a^2b **23.** $2d^2y^2$ **24.** $\frac{1}{2}db^2$

25. $-2d^3z^2$ **26.** $a + b$ **27.** $a - x$ **28.** $2x + z$

29. $3y - b$ **30.** $a - 2d$ **31.** $b - 4d$ **32.** $5x + 2y$

33. $7b - 5x$ **34.** $x^2 + x$ **35.** $2b^2 + b$ **36.** $y^2 - y$

37. $2d^2 - d$ **38.** $2a + 5d + 3x$ **39.** $8y + 5b - 6d$

40. $9b - 3z - 2x$ **41.** $x^2 + 3x + 5$ **42.** $z^2 + 2z - 7$

43. $a^2 - 5a - 6$ **44.** $d^2 - 4d + 6$ **45.** $2x^2 - 3x + 5$

46. $15 + 5z - z^2$ **47.** $2(a + b)$ **48.** $3(2x - 1) + 6$

49. $10 - 3(x - 4)$ **50.** $(x + 2)(x - 1)$ **51.** $(a - b)(a + b)$

52. $(x + d)(x - 4z)$

53. Find the value of $a^2 - 9b$ when $a = 4$ and $b = \frac{1}{3}$.

54. Find the value of $9x^2 - 4y^2$ when $x = \frac{1}{3}$ and $y = \frac{1}{2}$.

55. If $x = 4$, find the value of **(a)** $2x^2$ and **(b)** $(2x)^2$.

56. If $y = -2$, find the value of **(a)** $3y^2$ and **(b)** $(3y)^2$.

57. If $z = -\frac{1}{2}$, find the value of **(a)** $4z^2$ and **(b)** $(4z)^2$.

In 58–62, find the value of the expression. Use $a = -12$, $b = +6$, and $c = -1$.

58. $\frac{ac}{-3b}$ **59.** $\frac{b^2c}{a}$ **60.** $\frac{3a^2c^3}{b^3}$ **61.** $\frac{a - b^2}{-2c^2}$ **62.** $\frac{b^2 - a^2}{b^2 + a^2}$

7-11 USING THE ADDITIVE INVERSE IN SOLVING EQUATIONS

When we were dealing with the set of numbers of arithmetic, we used the addition property of equality to solve the first-degree equation $x - 4 = 7$ and the subtraction property to solve $x + 5 = 9$.

$$\begin{aligned} x - 4 &= 7 \\ x - 4 + 4 &= 7 + 4 \\ x &= 11 \end{aligned} \qquad \begin{aligned} x + 5 &= 9 \\ x + 5 - 5 &= 9 - 5 \\ x &= 4 \end{aligned}$$

The addition and subtraction properties of equality also hold true in the set of signed numbers. When the same signed number is added to both members of an equation or subtracted from both members of an equation, the equality is retained. However, using signed numbers, we can solve equations such as $x - 4 = 7$ and $x + 5 = 9$ by using only the addition property of equality.

Remember that the sum of a number and its additive inverse (opposite) is 0; that is, $n + (-n) = 0$. The following model problems illustrate how the additive inverse is used in solving equations:

MODEL PROBLEMS

1. Solve and check: $x - 4 = 7$

How to Proceed	*Solution*	*Check*
Add +4, the additive inverse (opposite) of −4, to both members of the equation.	$x - 4 = 7$	$x - 4 = 7$
	$x - 4 + (+4) = 7 + (+4)$	$11 - 4 \stackrel{?}{=} 7$
	$x + 0 = 11$	$7 = 7$
	$x = 11$	(True)

Answer: $x = 11$ *or* the solution set is $\{11\}$.

2. Solve and check: $x + 5 = 3$

How to Proceed	*Solution*	*Check*
Add −5, the additive inverse (opposite) of +5, to both members of the equation.	$x + 5 = 3$ $x + 5 + (-5) = 3 + (-5)$ $x + 0 = -2$ $x = -2$	$x + 5 = 3$ $(-2) + 5 \stackrel{?}{=} 3$ $3 = 3$ (True)

Answer: $x = -2$

EXERCISES

In 1–20, solve for the variable and check.

1. $x - 5 = 13$ **2.** $y + 8 = 12$ **3.** $17 = t - 9$
4. $36 = c + 20$ **5.** $x + 6 = 4$ **6.** $x - 5 = -9$
7. $n + 7 = 4$ **8.** $3 = y + 12$ **9.** $5 + r = -9$
10. $-5 = -7 + c$ **11.** $-4 = d - 8$ **12.** $s + 12 = 8$
13. $x + .9 = .5$ **14.** $w - 1.6 = .3$ **15.** $.6 + y = .2$
16. $-.3 = s + .7$ **17.** $n + 3\frac{1}{2} = 2$ **18.** $x - 2\frac{1}{3} = -5$
19. $-\frac{1}{2} = n - 1\frac{3}{4}$ **20.** $3\frac{1}{4} = y + 6\frac{1}{2}$

In 21–23, determine the element(s) of the set if $x \in$ {signed numbers}.

21. $\{x \mid x + 7 = 2\}$ **22.** $\{x \mid 9 = x + 15\}$ **23.** $\{x \mid 10 = x + 10\}$

7-12 USING THE MULTIPLICATIVE INVERSE IN SOLVING EQUATIONS

The multiplication and division properties of equality are true in the set of signed numbers as well as in the set of numbers of arithmetic. When both members of an equation are multiplied or divided by the same nonzero signed number, the equality is retained. However, using signed numbers, we can solve equations such as $5x = -20$ and $\frac{x}{3} = -2$ by using only the multiplication property of equality.

Remember that the product of a number and its multiplicative inverse (reciprocal) is 1; that is, $(n)\left(\frac{1}{n}\right) = 1$. The following problems illustrate how the multiplicative inverse is used in solving equations.

MODEL PROBLEMS

1. Solve and check: $5x = -20$

How to Proceed	*Solution*	*Check*
Multiply both members of the equation by $\frac{1}{5}$, the multiplicative inverse (reciprocal) of the coefficient 5.	$5x = -20$ $\frac{1}{5}(5x) = \frac{1}{5}(-20)$ $1 \cdot x = -4$ $x = -4$	$5x = -20$ $5(-4) \stackrel{?}{=} -20$ $-20 = -20$ (True)

Answer: $x = -4$

The equation could have been solved by dividing both members by 5.

2. Solve and check: $-\frac{2}{3}y = 18$

How to Proceed	*Solution*	*Check*
Multiply both members of the equation by $-\frac{3}{2}$, the multiplicative inverse (reciprocal) of the coefficient $-\frac{2}{3}$.	$-\frac{2}{3}y = 18$ $\left(-\frac{3}{2}\right)\left(-\frac{2}{3}y\right) = \left(-\frac{3}{2}\right)(18)$ $1 \cdot y = -27$ $y = -27$	$-\frac{2}{3}y = 18$ $\left(-\frac{2}{3}\right)(-27) \stackrel{?}{=} 18$ $18 = 18$ (True)

Answer: $y = -27$

EXERCISES

In 1–20, find the solution set of the sentence and check.

1. $3m = 15$ **2.** $15x = -45$ **3.** $-77 = 11k$

4. $-5 = 2y$ **5.** $-13a = 65$ **6.** $-8k = 8.8$

7. $-5m = -35$ **8.** $2x = -\frac{1}{9}$ **9.** $-x = 18$

10. $\frac{1}{3}z = 6$ **11.** $-20 = \frac{2}{5}d$ **12.** $\frac{5}{8}x = -10$

13. $\frac{1}{3}x = -1.8$ **14.** $\frac{3}{5}y = \frac{6}{8}$ **15.** $\frac{12}{9} = \frac{-4}{3}c$

16. $-\frac{3}{2}x = 1\frac{1}{2}$ **17.** $\frac{-25}{9} = 8\frac{1}{3}t$ **18.** $\frac{y}{3} = -15$

19. $\frac{c}{9} = -\frac{2}{3}$ **20.** $\frac{2x}{3} = \frac{4}{9}$

In 21–23, determine the elements of the set if $x \in$ {signed numbers}.

21. $\left\{x \middle| \frac{x}{5} = -20\right\}$ **22.** $\left\{x \middle| -\frac{4}{5}x = 40\right\}$ **23.** $\left\{x \middle| \frac{3}{10}x = 0\right\}$

7-13 REVIEW EXERCISES

In 1–3, find the sum.

1. $\begin{array}{r} +9 \\ \underline{-2} \end{array}$ **2.** $\begin{array}{r} -17 \\ \underline{+4} \end{array}$ **3.** $\begin{array}{r} -18 \\ \underline{-36} \end{array}$

In 4–6, find the difference.

4. $\begin{array}{r} -8 \\ \underline{+6} \end{array}$ **5.** $\begin{array}{r} +19 \\ \underline{-43} \end{array}$ **6.** $\begin{array}{r} -8.3 \\ \underline{-9.6} \end{array}$

In 7–9, find the product.

7. $\begin{array}{r} -7 \\ \underline{+6} \end{array}$ **8.** $\begin{array}{r} +26 \\ \underline{-4} \end{array}$ **9.** $\begin{array}{r} -1.5 \\ \underline{-.6} \end{array}$

In 10–12, find the quotient.

10. $\frac{-6}{+2}$ **11.** $\frac{-56}{-8}$ **12.** $\frac{+2}{-8}$

In 13–18, state whether the sentence is true or false.

13. $-7 > -2$ **14.** $|-7| > |-2|$ **15.** $|-9| = -9$

16. $|-4| + |+4| = 0$ **17.** $-3.9 < -3.2$ **18.** $\frac{|-24|}{|+4|} = -6$

19. How much greater than -8 is $+14$?
20. Name the greatest negative integer.
21. From the sum of 15 and -18, subtract -4.

In 22–30, find the value of the expression.

22. $-7 + (-9)$ **23.** $(-3)(+17)$ **24.** $(-75) \div (-5)$

25. $-28 - (+4)$ **26.** $-1.7 - (-2.3)$ **27.** $(-3)(-10)(-8)$

28. $(-14)\left(-\frac{3}{7}\right)$ **29.** $\frac{0}{-12}$ **30.** $\left(-\frac{1}{3}\right)^2$

In 31–36, graph the solution set of the open sentence using **(a)** a replacement set of $\{-3, -2, -1, 0, 1, 2, 3\}$ and **(b)** a replacement set of all real numbers.

31. $x \geq 2$ **32.** $x < 1$ **33.** $-2 < x \leq 2$

34. $(x \leq -1) \lor (x > 1)$ **35.** $(x > -2) \land (x \leq 2)$

36. $(x > 0) \lor (x \leq 3)$

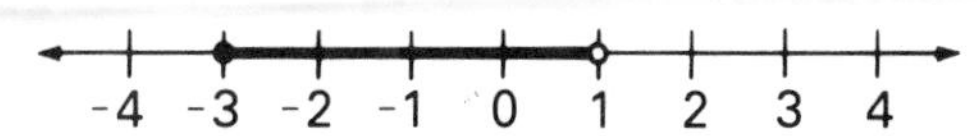

37. The solution set of which inequality is shown in the graph above?

(1) $(x \leq -3) \wedge (x > 1)$ (2) $(x \geq -3) \wedge (x > 1)$
(3) $(x \geq -3) \wedge (x < 1)$ (4) $(x \geq -3) \vee (x < 1)$

In 38–49, find the numerical value of the expression. Use $a = -3$, $b = +2$, $c = -4$, $x = -1$, and $y = -5$.

38. $-2x$ **39.** $-2 + x$ **40.** $4a^2$
41. $(4a)^2$ **42.** cx^2 **43.** $y^2 + y$
44. $b^2 + by$ **45.** $abcx$ **46.** $a - b + c$
47. $(a + b)(a - b)$ **48.** $(c - 5)(c + 8)$ **49.** $|-bcx - y|$

50. For what replacement of x will $12 + x = -3$ be a true statement?
51. What is the opposite of $-(-2)$?

In 52–59, solve for the variable and check.

52. $x + 10 = 21$ **53.** $y - 11 = 4$ **54.** $z + 3 = -5$
55. $7 = r - 3$ **56.** $3a = 42$ **57.** $12 = -3x$
58. $\frac{3}{5}q = 15$ **59.** $-8 = \frac{b}{2}$

60. Maurice answered all of the 60 questions on a multiple-choice test. To penalize for guessing, the test was scored by using the formula $S = R - \frac{W}{4}$, where S is the score on the test, R is the number he got right, and W is the number he got wrong.

a. What is the lowest possible score?
b. How many answers did Maurice get right if his score was -5?
c. Is it possible to get a score of -4?

Chapter **8**

Operations With Monomials

8-1 ADDING LIKE MONOMIALS

An algebraic expression consisting of one term that is a constant, a variable, or the product of constants and variables is called a ***monomial***. Examples of monomials are:

$$5 \qquad x \qquad 8z \qquad -4y^2 \qquad 7a^2b^3$$

Two or more terms containing the same variables, with corresponding variables having the same exponents, are called ***like terms***. For example, pairs of like terms include:

$$6k \text{ and } k \qquad 5x^2 \text{ and } -7x^2 \qquad 9ab \text{ and } 2ab \qquad 9x^2y^3 \text{ and } -11x^2y^3$$

Two or more terms are called ***unlike terms*** when they are not like terms. For example, pairs of unlike terms include:

$$3x \text{ and } 4y \qquad 5x^2 \text{ and } 5x^3 \qquad 9ab \text{ and } 2a \qquad 8x^3y^2 \text{ and } 4x^2y^3$$

To add like terms or like monomials, we use the distributive property of multiplication over addition.

$$9x + 2x = (9 + 2)x = 11x$$
$$-16cd + 3cd = (-16 + 3)cd = -13cd$$
$$18y^2 + (-y^2) = [(18) + (-1)]y^2 = 17y^2$$

Since the distributive property is true for any number of terms, we may add any number of monomials.

$$-3ab^2 + 4ab^2 - 2ab^2 = (-3 + 4 - 2)ab^2 = -1ab^2 \text{ or } -ab^2$$
$$x^3 + 11x^3 - 8x^3 - 4x^3 = (1 + 11 - 8 - 4)x^3 = 0x^3 \text{ or } 0$$

Note in the above examples that when like terms are added:

1. The result has the same variable factor as the original terms.
2. The numerical coefficient is the sum of the numerical coefficients of the terms.

The sum of unlike terms cannot be expressed as a single term. For example, $2x + 3y$ cannot be written as a monomial. An algebraic expression in simplest form contains no like terms.

PROCEDURE. To add like monomials, use the distributive property of multiplication; or find the sum of the numerical coefficients and multiply this sum by the common variable factors.

We often represent the measures of sides and the measures of angles of geometric figures in algebraic terms. For example, if the length of each side of a square is represented by s, the perimeter of the square will be represented by $s + s + s + s$, or $4s$. Since the length of the side of a geometric figure is always a positive number, the variable s cannot have a nonpositive value.

When an algebraic expression involving a variable is used to represent the measure of a line segment or the measure of an angle of a geometric figure, the domain of the variable must be restricted to such values that result in a positive value for the measure involved.

MODEL PROBLEMS

1. Add.

a.	**b.**	**c.**	**d.**	**e.**	**f.**
$+7x$	$-3y^2$	$-15abc$	$+8x^2y$	$-9y$	$+2(a + b)$
$-3x$	$-5y^2$	$+6abc$	$-x^2y$	$+9y$	$+6(a + b)$
$+4x$	$-8y^2$	$-9abc$	$+7x^2y$	0	$+8(a + b)$

2. José has twice as many nickels in his pocket as does Ralph. Ralph has twice as many nickels as does Lucy. If Lucy has n nickels, express in simplest form the total number of nickels these three people have.

Solution:

Since Lucy has n nickels,
then Ralph has $2n$,
and Jose has $2(2n)$, or $4n$.

The sum of the numbers of nickels is $n + 2n + 4n$, or $7n$. *Ans.*

EXERCISES

In 1–6, simplify the expression by adding the monomials.

1. $(+8c) + (+7c)$
2. $(+10t) + (-3t)$
3. $(-4a) + (-6a)$
4. $(-20r) + (5r)$
5. $(-7w) + (+7w)$
6. $(5ab) + (-9ab)$

In 7–31, add:

7. $\begin{array}{r} +7c \\ \underline{+8c} \end{array}$
8. $\begin{array}{r} -39r \\ \underline{-22r} \end{array}$
9. $\begin{array}{r} -19t \\ \underline{+6t} \end{array}$
10. $\begin{array}{r} +14c \\ \underline{-c} \end{array}$
11. $\begin{array}{r} -1.5m \\ \underline{+1.2m} \end{array}$
12. $\begin{array}{r} +3e \\ \underline{-3e} \end{array}$
13. $\begin{array}{r} +2x^2 \\ \underline{+9x^2} \end{array}$
14. $\begin{array}{r} -48y^2 \\ \underline{-13y^2} \end{array}$
15. $\begin{array}{r} -d^2 \\ \underline{+7d^2} \end{array}$
16. $\begin{array}{r} .5y^3 \\ \underline{.8y^3} \end{array}$
17. $\begin{array}{r} +\frac{5}{3}c^4 \\ \underline{-\frac{7}{3}c^4} \end{array}$
18. $\begin{array}{r} -10r^3 \\ \underline{10r^3} \end{array}$
19. $\begin{array}{r} 8rs \\ \underline{6rs} \end{array}$
20. $\begin{array}{r} -6mn \\ \underline{-mn} \end{array}$
21. $\begin{array}{r} -4xyz \\ \underline{+5xyz} \end{array}$
22. $\begin{array}{r} +.4cd \\ \underline{-.8cd} \end{array}$
23. $\begin{array}{r} -8xy \\ \underline{+8xy} \end{array}$
24. $\begin{array}{r} +3(x+y) \\ \underline{+9(x+y)} \end{array}$
25. $\begin{array}{r} +6a^2b \\ \underline{+7a^2b} \end{array}$
26. $\begin{array}{r} -xy^2 \\ \underline{-3xy^2} \end{array}$
27. $\begin{array}{r} -16x^2 \\ -x^2 \\ \underline{+15x^2} \end{array}$
28. $\begin{array}{r} -4rst \\ +8rst \\ \underline{+9rst} \end{array}$
29. $\begin{array}{r} -6xy^2 \\ +9xy^2 \\ \underline{-3xy^2} \end{array}$
30. $\begin{array}{r} 9c^2d^2 \\ 3c^2d^2 \\ \underline{-7c^2d^2} \end{array}$
31. $\begin{array}{r} +5(r+s) \\ -6(r+s) \\ \underline{+(r+s)} \end{array}$

In 32–37, simplify the expression by combining like terms.

32. $(+6x) + (-4x) + (-5x) + (+10x)$
33. $-5y + 6y + 9y - 14y$
34. $(+7c) + (-15c) + (+2c) + (+12c)$
35. $4m + 9m - 12m - m$
36. $(+8x^2) + (-x^2) + (-12x^2) + (+2x^2)$
37. $13y^2 - 15y^2 - y^2 + 8y^2$

38. Express in simplest form the perimeter of a rectangle whose width is represented by $2y$ and whose length is represented by

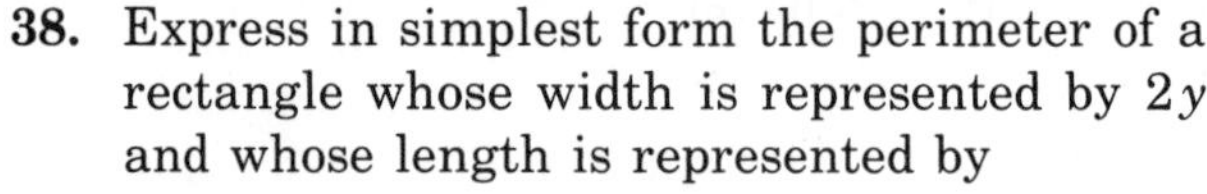

a. $6y$ b. $10y$ c. $7.5y$

d. $1.8y$ e. $2\frac{1}{4}y$ f. $7\frac{3}{4}y$

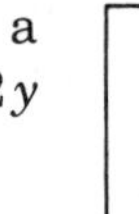

39. Express the perimeter of each of the following figures and simplify the result by combining like terms.

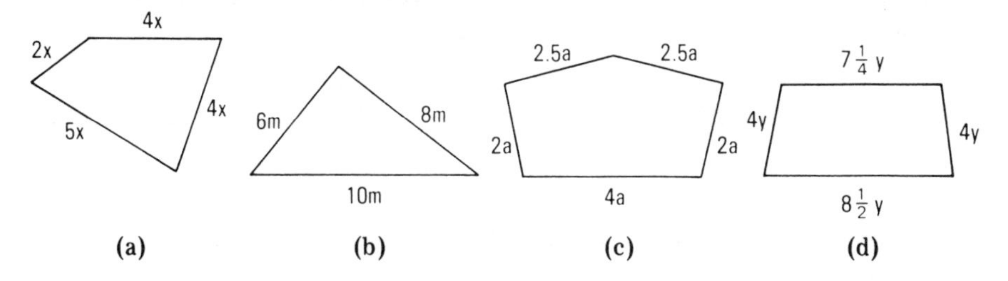

40. Sharon is twice as old as Amanda. If Amanda's age is represented by a, express the sum of the ages of Sharon and Amanda in simplest form.

41. Ruth earned three times as much as Bernadette. If Bernadette earned x dollars, express the sum of their earnings in simplest form.

42. A cheeseburger costs three times as much as a soft drink, and an order of fries costs twice as much as a soft drink. If a soft drink costs s cents, express in simplest form the total cost of a cheeseburger, an order of fries, and a soft drink.

43. Jack deposited some money in his savings account in September. In October, he deposited twice as much as in September, and in November he deposited one-half as much as in September. If x represents the amount of money deposited in September, represent in simplest form the total amount deposited in the three months.

44. On Tuesday, Melita read three times as many pages as she did on Monday. On Wednesday, she read one-and-one-half times as many pages as she did on Monday, and on Thursday, she read half as many pages as she did on Monday. If Melita read p pages on Monday, represent in simplest form the total number of pages read in the four days.

8-2 SUBTRACTING LIKE MONOMIALS

We can subtract like monomials by using the same method that we used to subtract signed numbers; we add the opposite (additive inverse) of the subtrahend to the minuend.

$$(+7) - (-3) = (+7) + (+3) = +10$$
$$(+7x) - (-3x) = (+7x) + (+3x) = +10x$$

> **PROCEDURE.** To subtract one monomial from another like monomial, add the opposite (additive inverse) of the subtrahend to the minuend.

MODEL PROBLEMS

1. Subtract.

a.	**b.**	**c.**	**d.**	**e.**	**f.**
$+8y$	$-5x^2$	$+15rst$	0	$-8m$	$+4(m+n)$
$+3y$	$-3x^2$	$+14rst$	$-5t$	$-8m$	$-5(m+n)$
$+5y$	$-2x^2$	$+rst$	$+5t$	0	$+9(m+n)$

2. What must be subtracted from $-8x$ to give the result $+5x$?

Solution: Reword the problem as an addition: What must be added to $+5x$ to give the result $-8x$?

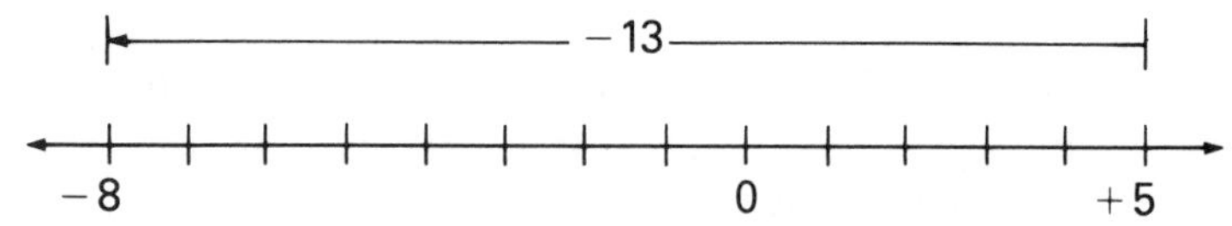

Answer: $-13x$ must be subtracted from $-8x$ to give the result $+5x$.

EXERCISES

In 1–30, subtract.

1.	**2.**	**3.**	**4.**	**5.**
$+9x$	$+12c$	$+7ab$	$+12cd^2$	$-7x$
$+3x$	$+9c$	$+8ab$	$+8cd^2$	$-9x$

6.	**7.**	**8.**	**9.**	**10.**
$-12y$	$-6xy$	$-5a^2b^2$	$+15c$	$-8x$
$-8y$	$-4xy$	$-a^2b^2$	$-3c$	$+3x$

11.	**12.**	**13.**	**14.**	**15.**
$-7xyz$	$+7z$	0	$3m$	$-5m$
$+9xyz$	0	$-5x^2$	$-3m$	$-5m$

16.	**17.**	**18.**	**19.**	**20.**
$+7d$	$-5.1x$	$-7r$	$7d^2$	$-8t^3$
$-d$	$+2.3x$	$-7r$	$-3d^2$	$+t^3$

21.	**22.**	**23.**	**24.**	**25.**
$-1.5y^3$	$+9(m+n)$	$+7cd$	$-8mn$	$-6rs$
$+.7y^3$	$+5(m+n)$	$+9cd$	$-9mn$	$+5rs$

26.	**27.**	**28.**	**29.**	**30.**
$-3ab$	$.4cd$	$-5(x+y)$	$+3y^2z^2$	$-5xy^2$
$7ab$	$-.9cd$	$-3(x+y)$	$+2y^2z^2$	$+2xy^2$

In 31–42, simplify the expression by subtracting the monomials.

31. $(+9r) - (+2r)$
32. $(+15s) - (-5s)$
33. $(-17n) - (11n)$
34. $(-15t) - (-15t)$
35. $(+8x) - (0)$
36. $(0) - (+8x)$
37. $(+9x^2) - (-3x^2)$
38. $(-3y^2) - (+7y^2)$
39. $(+5ab) - (+2ab)$
40. $(-12xy) - (+3xy)$
41. $(-4rs^2) - (2rs^2)$
42. $(+7x^2y) - (9x^2y)$

43. Subtract $-2x$ from $-8x$.
44. From 0 take $-5x$.
45. From $+3xy^2$ subtract 0.
46. Take $-7x$ from 0.
47. What must be added to $+6x$ to give the result $+10x$?
48. What must be subtracted from $+9d$ to give the result $+5d$?
49. What must be subtracted from $-8z$ to give the result $+3z$?
50. From the sum of $-5xy$ and $+12xy$ subtract the sum of $+9xy$ and $-15xy$.

In 51–54, the representations of the perimeter of triangle ABC and the lengths of $\overline{AB}$ and $\overline{BC}$ are given. Find the representation of the length of $\overline{AC}$.

51. Perimeter $= 18x$, $AB = 6x$, $BC = 4x$
52. Perimeter $= 15y$, $AB = y$, $BC = 7y$
53. Perimeter $= 35k$, $AB = 3k$, $BC = 15k$
54. Perimeter $= x$, $AB = \frac{1}{3}x$, $BC = \frac{1}{3}x$

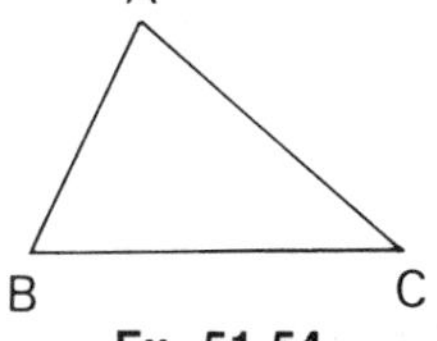

Ex. 51-54

In 55–58, in triangle ABC, two sides have equal measures. The representations of the perimeter of triangle ABC and of the measure of each of the equal sides are given. Find the representation of the third side.

55. Perimeter $= 13c$, side $= 4c$
56. Perimeter $= 20bc$, side $= 6bc$
57. Perimeter $= 12.8d$, side $= 5.2d$
58. Perimeter $= 10x$, side $= 3\frac{1}{4}x$

59. Of three sisters, Ellen weighs twice as much as the baby, Susan. Lisa weighs six times as much as Susan. If Susan weighs p pounds, express in simplest form how much more Lisa weighs than Ellen.
60. Kelly has five times as many dimes in her pocket as does Jill. Jill has twice as many dimes as does Leon. If Leon has d dimes, express in simplest form the difference between the numbers of dimes that Kelly and Jill have.

8-3 MULTIPLYING POWERS OF THE SAME BASE

Finding the Product of Powers

We know that y^2 means $y \cdot y$ and y^3 means $y \cdot y \cdot y$. Therefore:

$$y^2 \cdot y^3 = \overbrace{(y \cdot y)}^{2} \cdot \overbrace{(y \cdot y \cdot y)}^{3} = \overbrace{y \cdot y \cdot y \cdot y \cdot y}^{5} = y^5$$

Similarly, $c^2 \cdot c^4 = \overbrace{(c \cdot c)}^{2} \cdot \overbrace{(c \cdot c \cdot c \cdot c)}^{4} = \overbrace{c \cdot c \cdot c \cdot c \cdot c \cdot c}^{6} = c^6$

$$x \cdot x^3 = \overbrace{(x)}^{1} \cdot \overbrace{(x \cdot x \cdot x)}^{3} = x^4$$

The exponent in each product is the sum of the exponents in the factors as shown in these examples.

In general, when x is a signed number and a and b are positive integers:

$$x^a \cdot x^b = x^{a+b}$$

PROCEDURE. To multiply powers of the same base, find the exponent of the product by adding the exponents of the factors. The base of the product is the same as the base of the factors.

Note that this procedure does not apply to the product of powers that have different bases. For example, $c^2 \cdot d^3$ cannot be simplified because $c^2 \cdot d^3 = c \cdot c \cdot d \cdot d \cdot d$, an expression that does not have 5 factors that are the same.

Finding a Power of a Power

Since $(x^3)^4 = x^3 \cdot x^3 \cdot x^3 \cdot x^3$, then $(x^3)^4 = x^{12}$. The exponent 12 can be obtained by addition, $3 + 3 + 3 + 3 = 12$, or by multiplication, $4 \times 3 = 12$.

In general, when x is a signed number and a and c are positive integers:

$$(x^a)^c = x^{ac}$$

An expression such as $(x^5y^2)^3$ can be simplified by using the commutative and associative properties:

$$\begin{aligned}(x^5y^2)^3 &= (x^5y^2)(x^5y^2)(x^5y^2)\\ &= (x^5 \cdot x^5 \cdot x^5)(y^2 \cdot y^2 \cdot y^2)\\ &= x^{15}y^6\end{aligned}$$

When the base is a product of two or more factors, the power can be simplified by applying the rule to each factor:

$$(x^5y^2)^3 = (x^5)^3(y^2)^3 = x^{5\cdot3}y^{2\cdot3} = x^{15}y^6$$

MODEL PROBLEMS

In 1–5, simplify the expression by multiplying.

1. $x^5 \cdot x^4 = x^{5+4} = x^9$ **2.** $m^6 \cdot m = m^6 \cdot m^1 = m^{6+1} = m^7$

3. $10^3 \cdot 10^2 = 10^{3+2} = 10^5$ **4.** $m^{4a} \cdot m^{3a} = m^{4a+3a} = m^{7a}$

5. $(a^2)^3 = a^2 \cdot a^2 \cdot a^2 = a^{2+2+2} = a^6$ or $(a^2)^3 = a^{2\cdot3} = a^6$

EXERCISES

In 1–35, multiply.

1. $a^2 \cdot a^3$ **2.** $b^3 \cdot b^4$ **3.** $c^2 \cdot c^5$ **4.** $d^4 \cdot d^6$ **5.** $r^2 \cdot r^4 \cdot r^5$

6. $t^2 \cdot t^2$ **7.** $r^3 \cdot r^3$ **8.** $s^4 \cdot s^4$ **9.** $e^5 \cdot e^5$ **10.** $z^3 \cdot z^3 \cdot z^5$

11. $x^3 \cdot x^2$ **12.** $a^5 \cdot a^2$ **13.** $s^6 \cdot s^3$ **14.** $y^4 \cdot y^2$ **15.** $t^8 \cdot t^4 \cdot t^2$

16. $x \cdot x$ **17.** $a^2 \cdot a$ **18.** $b^4 \cdot b$ **19.** $c \cdot c^5$ **20.** $e^4 \cdot e \cdot e^5$

21. $2^3 \cdot 2^2$ **22.** $3^4 \cdot 3^3$ **23.** $5^2 \cdot 5^4$ **24.** $4^3 \cdot 4$ **25.** $2^4 \cdot 2^5 \cdot 2$

26. $(x^3)^2$ **27.** $(a^4)^2$ **28.** $(y^2)^4$ **29.** $(y^5)^2$ **30.** $(z^3)^2 \cdot (z^4)^2$

31. $(x^2y^3)^2$ **32.** $(ab^2)^4$ **33.** $(rs)^3$ **34.** $(2^2 \cdot 3^2)^3$ **35.** $(5 \cdot 2^3)^4$

In 36–40, multiply. (The exponents in each exercise are positive integers.)

36. $x^a \cdot x^{2a}$ **37.** $y^c \cdot y^2$ **38.** $c^r \cdot c^2$ **39.** $x^m \cdot x$ **40.** $(3y)^a \cdot (3y)^b$

In 41–48, state whether the sentence is true or false.

41. $10^4 \cdot 10^3 = 10^7$ **42.** $2^4 \cdot 2^2 = 2^8$ **43.** $3^3 \cdot 2^2 = 6^5$

44. $3^3 \cdot 2^2 = 6^6$ **45.** $5^4 \cdot 5 = 5^5$ **46.** $2^2 + 2^2 = 2^3$

47. $(2^2)^3 = 2^5$ **48.** $(2^3)^5 = 2^{15}$

In 49–51, the representations of the length, l, width, w, and height, h, of a rectangular solid are given. Represent the volume, V, of the rectangular solid. (Remember: $V = l \cdot w \cdot h$)

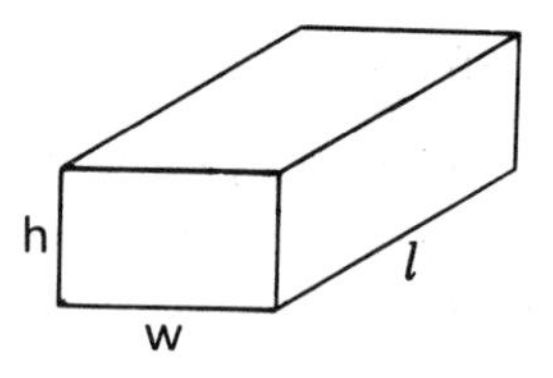

Ex. 49-51

49. $l = x$, $w = x$, $h = x$
50. $l = y$, $w = x$, $h = x$
51. $l = z$, $w = z$, $h = y$

8-4 MULTIPLYING A MONOMIAL BY A MONOMIAL

We know that the commutative property of multiplication makes it possible to arrange the factors of a product in any order and that the associative property of multiplication makes it possible to group the factors in any combination. For example:

$$(5x)(6y) = (5)(6)(x)(y) = (5 \cdot 6)(x \cdot y) = 30xy$$

$$(3x)(7x) = (3)(7)(x)(x) = (3 \cdot 7)(x \cdot x) = 21x^2$$

$$(-2x^2)(+5x^4) = (-2)(x^2)(+5)(x^4) = [(-2)(+5)][(x^2)(x^4)] = -10x^6$$

$$(-3a^2b^3)(-4a^4b) = (-3)(a^2)(b^3)(-4)(a^4)(b)$$

$$= [(-3)(-4)][(a^2)(a^4)][(b^3)(b)] = +12a^6b^4$$

In the preceding examples, the factors may be rearranged and grouped mentally.

PROCEDURE. To multiply monomials:

1. Use the commutative and associative properties to rearrange and group the factors. This may be done mentally.
2. Multiply the numerical coefficients.
3. When variable factors are powers with the same base, multiply by adding exponents.
4. Multiply the products obtained in steps 2 and 3.

MODEL PROBLEMS

1. Multiply.

a. $(+8xy)(+3z) = +24xyz$ **b.** $(-4a^3)(-5a^5) = +20a^8$
c. $(-6y^3)(y) = -6y^4$ **d.** $(+3a^2b^3)(+4a^3b^4) = +12a^5b^7$

e. $(-5x^2y^3)(-2xy^2) = +10x^3y^5$

f. $(+6c^2d^3)\left(-\frac{1}{2}d\right) = -3c^2d^4$

g. $(-3x^2)^3 = (-3x^2)(-3x^2)(-3x^2) = -27x^6$ or
$(-3x^2)^3 = (-3)^3(x^2)^3 = -27x^6$

2. Represent the area of a rectangle whose length is represented by $3x$ and whose width is represented by $2x$.

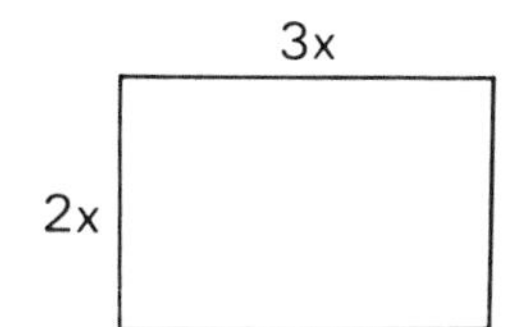

How to Proceed	*Solution*
(1) Write the area formula.	$A = l \cdot w$
(2) Substitute the values of l and w.	$A = (3x) \cdot (2x)$
(3) Perform the multiplication.	$A = (3 \cdot 2) \cdot (x \cdot x)$
	$A = 6x^2$ *Ans.*

Notice that the same answer can be obtained by using the following set of geometric models:

<table>
<tr><td></td><td>3x</td><td></td><td></td><td>x + x + x</td><td></td><td></td><td>x + x + x</td><td></td><td></td><td>3x</td></tr>
<tr><td>2x</td><td></td><td>=</td><td>x
+
x</td><td></td><td>=</td><td>x
+
x</td><td>x² x² x²
x² x² x²</td><td>=</td><td>2x</td><td>6x²</td></tr>
</table>

EXERCISES

In 1–6, multiply.

1. $\begin{array}{r} 4x^2 \\ \underline{3x^3} \end{array}$ **2.** $\begin{array}{r} 5w \\ \underline{3w} \end{array}$ **3.** $\begin{array}{r} -3t^2 \\ \underline{-t} \end{array}$ **4.** $\begin{array}{r} -5d^3 \\ \underline{+5d^3} \end{array}$ **5.** $\begin{array}{r} 6d^5 \\ \underline{-3d} \end{array}$ **6.** $\begin{array}{r} -9c \\ \underline{12c} \end{array}$

In 7–42, find the product

7. $(+6)(-2a)$ **8.** $(-4)(-6b)$ **9.** $(+5)(-2)(-3y)$

10. $(4a)(5b)$ **11.** $(-8r)(-2s)$ **12.** $(+7x)(-2y)(3z)$

13. $(+6x)\left(-\frac{1}{2}y\right)$ **14.** $\left(-\frac{3}{4}a\right)(+8b)$ **15.** $(-6x)\left(\frac{1}{2}y\right)\left(-\frac{1}{3}z\right)$

16. $(+5ab)(-3c)$ **17.** $(-7r)(5st)$ **18.** $(-2)(+6cd)(-e)$

19. $(+9xy)(-2cd)$ **20.** $(3s)(-4m)(5cd)$ **21.** $(+5a^2)(-4a^2)$

22. $(-6x^4)(-3x^3)$ **23.** $(20y^3)(-7y^2)$ **24.** $(18r^5)(-5r^2)$

25. $(+3z^2)(+4z)$ **26.** $(-8y^5)(+5y)$ **27.** $(-9z)(8z^4)(z^3)$

28. $(+6x^2y^3)(-4x^4y^2)$ **29.** $(-7a^3b)(+5a^2b^2)$ **30.** $(+4ab^2)(-2a^2b^3)$

31. $(-2r^4s)(+8rs)$ **32.** $(-9c)(+8cd^2)$ **33.** $(-3y)(5xy)(15xy^2)$

34. $\left(+\frac{2}{3}x^2\right)(-6x)$ **35.** $(+7a)^2$ **36.** $(-.5x)^2$

37. $(-2x^2)^3$ **38.** $\left(-\frac{2}{5}r^2s^2\right)^2$ **39.** $(+2x)^2(+3y)^2$

40. $\left(\frac{1}{2}x^2\right)^3(-4y^3)^2$ **41.** $5x(4x)^2$ **42.** $10(2x)^2(-y^2)^3$

In 43–45, represent the area of the rectangle whose length and width are given.

43. $l = 5y$, $w = 3y$ **44.** $l = 3x$, $w = 5y$ **45.** $l = 3cd^2$, $w = 8cd$

In 46–49, represent the area of the square the length of each of whose sides is given.

46. $2x$ **47.** $\frac{1}{2}y$ **48.** $3xy$ **49.** $5x^2$

In 50–53, represent the volume of the rectangular solid whose length, l, width, w, and height, h, are given.

50. $l = 4x$, $w = 2x$, $h = 5x$ **51.** $l = 4x$, $w = 2x$, $h = 5$

52. $l = 4x$, $w = 2y$, $h = 5z$ **53.** $l = b$, $w = 2a$, $h = d$

In 54–57, represent the volume of a cube the length of each of whose edges, e, is given. [Remember: $V = e^3$]

54. $e = w$ **55.** $e = 3x$ **56.** $e = 5y$ **57.** $e = \frac{1}{2}k$

In 58–60, select the correct answer.

58. The product of $(5x^3)$ and $(3x^5)$ is:
(1) $8x^{15}$ (2) $15x^{15}$ (3) $15x^8$ (4) $8x^8$

59. The product of $(-3a^3)$ and $(4a^2)$ is:
(1) $12a^5$ (2) $-12a^6$ (3) $-12a^5$ (4) $7a^5$

60. When $4rs^2$ is multiplied by $3r^3s^5$, the product is:
(1) $12r^3s^{10}$ (2) $7r^4s^{10}$ (3) $12r^4s^7$ (4) $7r^3s^7$

61. If one pound of grass seed costs $25x$ cents, represent the cost of $7y$ pounds of seed in simplest form.

62. If a bus travels at the rate of $10z$ miles per hour for $4z$ hours, represent the distance traveled in simplest form.

63. If Lois has $2n$ nickels, represent the number of cents she has in simplest form.

8-5 DIVIDING POWERS OF THE SAME BASE

You know that division and multiplication are inverse operations.

Since $x^2 \cdot x^3 = x^5$, then $x^5 \div x^3 = x^2$.
Since $y^5 \cdot y^4 = y^9$, then $y^9 \div y^4 = y^5$.
Since $c^4 \cdot c = c^5$, then $c^5 \div c = c^4$. (Remember: c means c^1)

Observe that the exponent in each quotient is the difference between the exponent of the dividend and the exponent of the divisor.

In general, when $x \neq 0$ and a and b are positive integers with $a > b$:

$$x^a \div x^b = x^{a-b}$$

PROCEDURE. To divide powers of the same base, find the exponent of the quotient by subtracting the exponent of the divisor from the exponent of the dividend. The base of the quotient is the same as the base of the dividend and the base of the divisor.

You know that any nonzero number divided by itself is 1. Therefore, $x \div x = 1$ and $y^3 \div y^3 = 1$.

In general, when $x \neq 0$ and a is a positive integer:

$$x^a \div x^a = 1$$

MODEL PROBLEMS

1. Simplify by performing the indicated division.

a. $x^9 \div x^5 = x^{9-5} = x^4$

b. $y^5 \div y = y^{5-1} = y^4$

c. $c^5 \div c^5 = 1$

d. $10^5 \div 10^3 = 10^{5-3} = 10^2$

e. $y^{6b} \div y^{4b} = y^{6b-4b} = y^{2b}$
(b is a positive integer.)

2. Write in simplest form: $\dfrac{10^7 \cdot 10^4}{10^8}$

Solution: First, simplify the numerator. Then, apply the rule for division of powers with the same base.

$$\frac{10^7 \cdot 10^4}{10^8} = \frac{10^{7+4}}{10^8} = \frac{10^{11}}{10^8} = 10^{11-8} = 10^3 \quad \textit{Ans.}$$

EXERCISES

In 1–20, divide.

1. $x^8 \div x^2$ **2.** $a^{10} \div a^5$ **3.** $b^7 \div b^3$ **4.** $c^5 \div c^4$ **5.** $d^7 \div d^7$

6. $\frac{d^4}{d^2}$ **7.** $\frac{e^9}{e^3}$ **8.** $\frac{m^{12}}{m^4}$ **9.** $\frac{n^{10}}{n^9}$ **10.** $\frac{r^6}{r^6}$

11. $x^8 \div x$ **12.** $y^7 \div y$ **13.** $z^{10} \div z$ **14.** $t^5 \div t$

15. $m \div m$ **16.** $2^5 \div 2^2$ **17.** $10^6 \div 10^4$ **18.** $3^4 \div 3^2$

19. $5^3 \div 5$ **20.** $10^4 \div 10$

In 21–26, divide. (Exponents in each exercise are positive integers.)

21. $x^{5a} \div x^{2a}$ **22.** $y^{10b} \div y^{2b}$ **23.** $r^c \div r^d$ $(c > d)$

24. $s^x \div s^2$ $(x > 2)$ **25.** $a^b \div a^b$ **26.** $2^a \div 2^b$ $(a > b)$

In 27–32, simplify the expression.

27. $\frac{2^3 \cdot 2^4}{2^2}$ **28.** $\frac{5^8}{5^4 \cdot 5}$ **29.** $\frac{10^2 \cdot 10^3}{10^4}$

30. $\frac{10^6}{10^2 \cdot 10^4}$ **31.** $\frac{10^8 \cdot 10^2}{(10^5)^2}$ **32.** $\frac{6^4 \cdot 6^9}{6^2 \cdot 6^3}$

In 33–35, tell whether the sentence is true or false.

33. $4^5 \div 2^3 = 2^2$ **34.** $5^6 \div 5^2 = 5^4$ **35.** $3^8 \div 3^4 = 1^4$

8-6 DIVIDING A MONOMIAL BY A MONOMIAL

You know that $\frac{a}{b} \cdot \frac{c}{d} = \frac{ac}{bd}$.

Therefore, by the reflexive property, $\frac{ac}{bd} = \frac{a}{b} \cdot \frac{c}{d}$.

Using this relationship, we can write:

$$\frac{-30x^6}{2x^4} = \frac{-30}{2} \cdot \frac{x^6}{x^4} = -15x^2$$

$$\frac{-21a^5b^4}{-3a^4b} = \frac{-21}{-3} \cdot \frac{a^5}{a^4} \cdot \frac{b^4}{b} = +7a^1b^3 \text{ or } 7ab^3$$

$$\frac{12y^2z^2}{4y^2z} = \frac{12}{4} \cdot \frac{y^2}{y^2} \cdot \frac{z^2}{z} = 3 \cdot 1 \cdot z^1 \text{ or } 3z$$

PROCEDURE. To divide monomials:

1. Divide their numerical coefficients.
2. When variable factors are powers with the same base, divide by subtracting exponents.
3. Multiply the quotients obtained in steps 1 and 2.

If the area of a rectangle is 42 and its length is 6, we can find its width by dividing the area 42 by the length 6. Thus, $42 \div 6 = 7$, which is the width. Similarly, if the area of a rectangle is represented by $42x^2$ and its length is represented by $6x$, we can represent the width by dividing the area $42x^2$ by the length $6x$. We get $(42x^2) \div (6x) = 7x$, which represents the width.

MODEL PROBLEMS

In 1–7, divide:

1. $(+24a^5) \div (+3a^2) = +8a^3$ **2.** $(-15x^6y^5) \div (-3x^3y^2) = +5x^3y^3$

3. $\dfrac{-18x^3y^2}{+6x^2y} = -3xy$ **4.** $\dfrac{+20a^3c^4d^2}{-5a^3c^3} = -4cd^2$ **5.** $\dfrac{3r^2s^3}{3r^2s^3} = 1$

6. $\dfrac{10ab^4}{5b^3} = 2ab$ **7.** $\dfrac{24(a + b)^5}{4(a + b)^2} = 6(a + b)^3$

EXERCISES

In 1–19, divide.

1. $18x$ by 2 **2.** $14x^2y^2$ by -7 **3.** $-36y^{10}$ by $+6y^2$

4. $40a^4$ by $-4a$ **5.** $24a^2b^2$ by $-8b^2$ **6.** $15c^4d$ by $-5c^3d$

7. $7r^4c$ by $-7r^4c$ **8.** $-28c^2d$ by $7cd$ **9.** $30de^3$ by $5de^2$

10. $(+8cd) \div (-4c)$ **11.** $(-14xy^3) \div (-7xy^3)$

12. $\dfrac{18x^6}{2x^2}$ **13.** $\dfrac{-8c^3}{2c}$ **14.** $\dfrac{5x^2y^3}{-5y^3}$ **15.** $\dfrac{-49c^4b^3}{7c^2b^2}$

16. $\dfrac{-24x^2y}{-3xy}$ **17.** $\dfrac{-27xyz}{9xz}$ **18.** $\dfrac{-56abc}{8abc}$ **19.** $\dfrac{15(c - d)}{-5(c - d)}$

20. If 5 oranges cost $15y$ cents, represent the cost of one orange in simplest form.

21. If a car traveled $300x$ miles in $6x$ hours, find the rate at which the car traveled.
22. Represent in simplest form the time needed for a train to travel $150ab$ miles at $15a$ miles per hour.
23. If $3y$ pens cost $12y^3$ dollars, represent the cost of one pen.
24. If the area of a rectangle is $35x^4$ and the length is $7x^2$, represent the width in simplest form.

In 25–27, represent the width of a rectangle whose area is represented by $24x^3y^2$ and whose length is given.

25. length $= 8x^3$ 26. length $= 24y^2$ 27. length $= 2x^2y$

8-7 NONPOSITIVE INTEGRAL EXPONENTS

Now you will see that nonpositive integers such as 0, -1, and -2 can also be used as exponents. We will define powers having zero and negative integral exponents in such a way that the properties that were valid for positive integral exponents will also be valid for nonpositive integral exponents. Hence, the following properties will be valid for all integral exponents:

1. $x^a \cdot x^b = x^{a+b}$ 2. $x^a \div x^b = x^{a-b}$ $(x \neq 0)$ 3. $(x^a)^b = x^{a \cdot b}$

The Zero Exponent

You know that $\frac{x^3}{x^3} = 1$ when $x \neq 0$. If we wish $\frac{x^3}{x^3} = x^{3-3}$ (that is, $\frac{x^3}{x^3} = x^0$) to be a true meaningful statement, then we must say that $x^0 = 1$, since x^0 and 1 are each equal to $\frac{x^3}{x^3}$. This leads us to make the following definition:

$x^0 = 1$ if x is a number such that $x \neq 0$

It can be shown that all the laws of exponents remain valid when x^0 is defined as 1. For example:

Using the definition $10^0 = 1$, $10^3 \cdot 10^0 = 10^3 \cdot 1 = 10^3$.

Using the law of exponents, $10^3 \cdot 10^0 = 10^{3+0} = 10^3$. Each procedure results in the same product.

The definition $x^0 = 1$ $(x \neq 0)$ permits us to say that the zero power of any number except zero equals 1.

$$4^0 = 1 \quad (-4)^0 = 1 \quad (4x)^0 = 1 \quad (-4x)^0 = 1$$

Note that $4x^0 = 4^1 \cdot x^0 = 4 \cdot 1 = 4$.
But, $(4x)^0 = 4^0 \cdot x^0 = 1 \cdot 1 = 1$.

The Negative Integral Exponent

We know that, for $x \neq 0$, $\frac{x^3}{x^5} = \frac{x \cdot x \cdot x}{x \cdot x \cdot x \cdot x \cdot x} = \frac{1}{x \cdot x} = \frac{1}{x^2}$. If we wish $\frac{x^3}{x^5} = x^{3-5}$ $\left(\text{that is, } \frac{x^3}{x^5} = x^{-2}\right)$ to be a true meaningful statement, then we must say that $x^{-2} = \frac{1}{x^2}$, since x^{-2} and $\frac{1}{x^2}$ are each equal to $\frac{x^3}{x^5}$. This leads us to make the following definition:

$$\boldsymbol{x^{-n} = \frac{1}{x^n}} \textbf{ if } \boldsymbol{x} \textbf{ is a number such that } \boldsymbol{x \neq 0}$$

Now we can say that for all integral values of a and b,

$$\boldsymbol{\frac{x^a}{x^b} = x^{a-b} \quad (x \neq 0)}$$

It can be shown that all the laws of exponents remain valid if x^{-n} is defined as $\frac{1}{x^n}$. For example:

Using the definition $2^{-4} = \frac{1}{2^4}$, $2^2 \cdot 2^{-4} = 2^2 \cdot \frac{1}{2^4} = \frac{2^2}{2^4} = \frac{1}{2^2} = 2^{-2}$

Using the law of exponents, $2^2 \cdot 2^{-4} = 2^{2+(-4)} = 2^{-2}$, the same as the previous result.

MODEL PROBLEMS

1. Transform the given expression into an equivalent expression involving a positive exponent.

a. $4^{-3} = \frac{1}{4^3}$ **b.** $10^{-1} = \frac{1}{10^1}$ **c.** $\left(-\frac{1}{2}\right)^{-4} = \frac{1}{\left(-\frac{1}{2}\right)^4}$

d. $\frac{1}{2^{-5}} = \frac{1}{\frac{1}{2^5}} = 1 \div \frac{1}{2^5} = 1 \times \frac{2^5}{1} = 2^5$

2. Compute the value of the expression.

a. $3^0 = 1$ **b.** $10^{-2} = \frac{1}{10^2} = \frac{1}{100}$

c. $(-5)^0 + 2^{-4} = 1 + \frac{1}{2^4} = 1 + \frac{1}{16} = 1\frac{1}{16}$

d. $6(3^{-3}) = 6\left(\frac{1}{3^3}\right) = \frac{6}{27} = \frac{2}{9}$

3. Use the laws of exponents to perform the indicated operation.

a. $2^7 \cdot 2^{-3} = (2)^{7+(-3)} = 2^4$

b. $3^{-6} \div 3^{-2} = (3)^{-6-(-2)} = 3^{-4}$

c. $(x^4)^{-3} = x^{(4)(-3)} = x^{-12}$

d. $(y^{-2})^{-4} = y^{(-2)(-4)} = y^8$

EXERCISES

In 1–5, transform the given expression into an equivalent expression involving a positive exponent.

1. 10^{-4} **2.** 2^{-1} **3.** $\left(\frac{2}{3}\right)^{-2}$ **4.** m^{-6} **5.** r^{-3}

In 6–21, compute the value of the expression.

6. 10^0 **7.** $(-4)^0$ **8.** y^0 **9.** $(2K)^0$

10. 3^{-2} **11.** 2^{-4} **12.** $(-6)^{-1}$ **13.** $(-1)^{-5}$

14. 10^{-1} **15.** 10^{-2} **16.** 10^{-3} **17.** 10^{-4}

18. $4(10)^{-2}$ **19.** $1.5(10)^{-3}$ **20.** $7^0 + 6^{-2}$ **21.** $\left(\frac{1}{2}\right)^0 + 3^{-3}$

In 22–33, use the laws of exponents to perform the indicated operation.

22. $10^{-2} \cdot 10^5$ **23.** $3^{-4} \cdot 3^{-2}$ **24.** $10^{-3} \div 10^{-5}$ **25.** $3^4 \div 3^0$

26. $(4^{-1})^2$ **27.** $(3^{-3})^{-2}$ **28.** $a^0 \cdot a^4$ **29.** $x^{-5} \cdot x$

30. $m^2 \div m^7$ **31.** $t^{-6} \div t^2$ **32.** $(a^{-4})^3$ **33.** $(x^{-2})^0$

34. Find the value of $7x^0 - (6x)^0$.

35. Find the value of $5x^0 + 2x^{-1}$ when $x = 4$.

8-8 EXPRESSING LARGE NUMBERS IN SCIENTIFIC NOTATION

Scientists frequently deal with very large numbers that have been rounded. In order to be able to write and to compute with such numbers more easily, the ***scientific*** (or ***standard***) ***notation*** system was developed.

Scientific notation of a number is defined as follows:

- **The number is expressed as a product of two quantities: The first is a number equal to or greater than 1, but less than 10, and the second is an integral power of 10.**

The table at the right lists positive integral powers of 10.

Notice that the number of zeros following the 1 in the value of an integral power of 10 is equal to the exponent of 10. For example, in $10^3 = 1{,}000$, the exponent is 3 and there are 3 zeros following 1 in 1,000.

Positive Integral Powers of 10	
$10^1 = 10$	$10^5 = 100{,}000$
$10^2 = 100$	$10^6 = 1{,}000{,}000$
$10^3 = 1{,}000$	$10^7 = 10{,}000{,}000$
$10^4 = 10{,}000$	$10^8 = 100{,}000{,}000$

To express 3,000,000,000 in scientific notation, express the number as the product of two numbers that fit the definition. Here,

$$3{,}000{,}000{,}000 = 3 \times 1{,}000{,}000{,}000$$
$$3{,}000{,}000{,}000 = 3 \times 10^9$$

MODEL PROBLEMS

1. The distance from the earth to the sun is approximately 93,000,000 miles. Write this number in scientific notation.

How to Proceed	*Solution*
(1) Express the number as a product of two factors that fit the definition. (Here, 9.3 is between 1 and 10.)	$93{,}000{,}000 = 9.3 \times 10{,}000{,}000$
(2) Express the second factor as a power of 10, using an exponent.	$93{,}000{,}000 = 9.3 \times 10^7$ *Ans.*

2. The approximate population of the United States in 1970 was 2.03×10^8. Find the approximate number of people in the United States at that time.

How to Proceed	*Solution*
(1) Evaluate the second factor, which is a power of 10.	$2.03 \times 10^8 = 2.03 \times 100{,}000{,}000$
(2) Multiply the factors.	$2.03 \times 10^8 = 203{,}000{,}000$ *Ans.*

Notice that we could have multiplied 2.03 by 10^8 quickly by moving the decimal point in 2.03 eight places to the right.

3. Express $(2 \times 10^4)(3.5 \times 10^5)$ **(a)** in scientific notation and **(b)** as an integer.

a. Use the commutative and associative properties.

$$(2 \times 10^4)(3.5 \times 10^5) = (2 \times 3.5)(10^4 \times 10^5)$$
$$= 7 \times 10^9$$

b. $7 \times 10^9 = 7{,}000{,}000{,}000$

EXERCISES

In 1–3, write the number as a power of 10 using a positive exponent.

1. 100,000 **2.** 1,000,000,000 **3.** 1,000,000,000,000

In 4–15, find the number that is expressed by the numeral.

4. 10^7 **5.** 10^{10} **6.** 10^{13}
7. 10^{15} **8.** 3×10^5 **9.** 4×10^8
10. 6×10^{14} **11.** 9×10^9 **12** 1.3×10^4
13. 8.3×10^{12} **14.** 1.27×10^3 **15.** 6.14×10^{10}

In 16–23, express the number in scientific notation.

16. 400 **17.** 6,000 **18.** 30,000 **19.** 400,000 **20.** 7,000,000
21. 300,000,000 **22.** 80,000,000 **23.** 20,000,000,000

In 24–31, find the number that can replace the question mark and make the resulting statement true.

24. $120 = 1.2 \times 10^?$ **25.** $760 = 7.6 \times 10^?$
26. $9{,}300 = 9.3 \times 10^?$ **27.** $52{,}000 = 5.2 \times 10^?$
28. $5{,}280 = 5.28 \times 10^?$ **29.** $375{,}000 = 3.75 \times 10^?$
30. $1{,}610{,}000 = 1.61 \times 10^?$ **31.** $872{,}000{,}000 = 8.72 \times 10^?$

In 32–39, express the number in scientific notation.

32. 8,400 **33.** 27,000 **34.** 54,000,000 **35.** 320,000,000
36. 6,750 **37.** 81,600 **38.** 453,000 **39.** 375,000,000

In 40–43, compute the result applying the laws of exponents. Represent the result **(a)** in scientific notation and **(b)** as an integer.

40. $(3 \times 10^4)(2 \times 10^3)$ **41.** $(1.5 \times 10^6)(8 \times 10^7)$
42. $(8 \times 10^{15}) \div (4 \times 10^5)$ **43.** $(9.3 \times 10^8) \div (3.1 \times 10^2)$

In 44–48, express the number in scientific notation.

44. The planet Uranus is about 2,000,000,000 miles from the sun.
45. The velocity of light is 30,000,000,000 centimeters per second.
46. The distance between the earth and its nearest star, other than the sun, is 26,000,000,000,000 miles.
47. A light year, which is the distance light travels in one year, is approximately 9,500,000,000,000 kilometers.
48. A star that is about 12,000,000,000,000,000,000 miles away can be revealed by the Palomar telescope.

In 49–52, express the number as an integer.

49. The diameter of the universe is 2×10^9 light years.
50. The distance from the earth to the moon is 2.4×10^5 miles.
51. The sun weighs about 1.8×10^{27} tons.
52. The mass of the earth is approximately 5.9×10^{24} kilograms.

8-9 EXPRESSING SMALL NUMBERS IN SCIENTIFIC NOTATION

Scientists also deal with very small numbers that have been rounded. We can express such numbers in scientific notation by using negative integral powers of 10, as listed in the table below.

Negative Integral Powers of 10	
$.1 = \frac{1}{10^1} = 10^{-1}$	$.00001 = \frac{1}{10^5} = 10^{-5}$
$.01 = \frac{1}{10^2} = 10^{-2}$	$.000001 = \frac{1}{10^6} = 10^{-6}$
$.001 = \frac{1}{10^3} = 10^{-3}$	$.0000001 = \frac{1}{10^7} = 10^{-7}$
$.0001 = \frac{1}{10^4} = 10^{-4}$	$.00000001 = \frac{1}{10^8} = 10^{-8}$

Notice that in each case the absolute value of the negative exponent of 10 is equal to the number of places to the right of the decimal point.

If we wish to express .0003 in scientific notation, we must first express the number as a product of two numbers to fit the definition. Here,

$$.0003 = 3. \times .0001$$
$$.0003 = 3 \times 10^{-4}$$

MODEL PROBLEMS

1. Express .0000029 in scientific notation.

How to Proceed	*Solution*
(1) Express the number as the product of two factors that fit the definition. (Here 2.9 is between 1 and 10.)	$.0000029 = 2.9 \times .000001$
(2) Express the second factor as a power of 10, using exponents.	$.0000029 = 2.9 \times 10^{-6}$ *Ans.*

2. The diameter of a red blood corpuscle expressed in scientific notation is $7.5 \times (10^{-4})$ centimeter. Write the number of centimeters in the diameter as a decimal fraction.

How to Proceed	*Solution*
(1) Evaluate the second factor, which is a power of 10.	$7.5 \times 10^{-4} = 7.5 \times .0001$
(2) Multiply the factors.	$7.5 \times 10^{-4} = .00075$ *Ans.*

Notice that we could have multiplied 7.5 by 10^{-4} quickly by moving the decimal point in 7.5 four places to the left.

EXERCISES

In 1–4, write the number as a power of 10 involving a negative exponent.

1. .01 **2.** .00001 **3.** .00000001 **4.** .0000000001

In 5–16, express the given symbol as a decimal fraction.

5. 10^{-6} **6.** 10^{-8} **7.** 10^{-15}

8. 10^{-18} **9.** 4×10^{-3} **10.** $7 \times (10^{-9})$
11. 8×10^{-10} **12.** $9 \times (10^{-13})$ **13.** 1.2×10^{-4}
14. $3.6 \times (10^{-5})$ **15.** 7.4×10^{-11} **16.** $3.14 \times (10^{-14})$

In 17–20, express the number in scientific notation.

17. .002 **18.** .0005 **19.** .000003 **20.** .00000009

In 21–26, state the number that can replace the question mark to make the statement true.

21. $.023 = 2.3 \times 10^{?}$ **22.** $.000086 = 8.6 \times 10^{?}$
23. $.000000019 = 1.9 \times 10^{?}$ **24.** $.000000000041 = 4.1 \times 10^{?}$
25. $.00156 = 1.56 \times 10^{?}$ **26.** $.000000873 = 8.73 \times 10^{?}$

In 27–34, express the number in scientific notation.

27. .0052 **28.** .00061 **29.** .0000039 **30.** .000000014
31. .156 **32.** .00381 **33.** .0000763 **34.** .000000917

In 35–38, compute the result applying the laws of exponents. Represent the result **(a)** in scientific notation and **(b)** as a decimal.

35. $(2 \times 10^{-5})(3 \times 10^{2})$ **36.** $(2.5 \times 10^{-2})(3 \times 10^{-3})$
37. $(7.5 \times 10^{-4}) \div (2.5 \times 10^{3})$ **38.** $(6.8 \times 10^{-5}) \div (3.4 \times 10^{-8})$

In 39–44, express each decimal fraction in scientific notation.

39. The approximate diameter of the smallest particle visible to the naked eye is .004 inch.
40. A microampere is .000001 of an ampere.
41. The radius of an electron is about .0000000000005 centimeter.
42. The diameter of some white blood corpuscles is approximately .0008 inch.
43. The wavelength of red light is .000065 centimeter.
44. The mass of a hydrogen atom is approximately .00000000000000000000000167 gram.

In 45–48, express the number as a decimal.

45. In a motion-picture film, the image of each picture remains on the screen approximately 6×10^{-2} second.
46. It takes light about 2×10^{-8} second to cross a room.
47. The density of dry air is approximately 1.3×10^{-3} gram per centimeter.
48. An atomic mass unit is 1.66×10^{-24} gram.

8-10 REVIEW EXERCISES

In 1–15, simplify the expression.

1. $3r^2 + 2r^2$ **2.** $5bc - bc$ **3.** $13w - 8w + 5w$

4. $(-6k)^2$ **5.** $(-am^2)^3$ **6.** $(8mg)(-3g)$

7. $-5t - (-8t)$ **8.** $(-5t)(-8t)$ **9.** $(-5t) \div (-8t)$

10. $(3x^3y)(-4xy^4)$ **11.** $-4y^2 - 5y^2$ **12.** $-4y^2(-5y^2)$

13. $\dfrac{40b^3x^6}{-8b^2x}$ **14.** $x^b \cdot x^4$ **15.** $(-4c)^3\left(\dfrac{1}{4c}\right)^2$

In 16–20, write the expression in an equivalent form, using positive exponents only. Where possible, compute the answer.

16. 8^{-2} **17.** k^{-6} **18.** 10^{-3} **19.** $\left(\dfrac{2}{5}\right)^{-2}$ **20.** $2x^{-4}$

In 21–24, use the laws of exponents to perform the operations, and simplify.

21. $3^{-5} \cdot 3^4$ **22.** $(7^{-3})^0$ **23.** $(10^{-2})^{-1}$ **24.** $12^{-3} \cdot 12$

25. If the length of a rectangle is represented by $8x$, and its width by $3x$, find the representation of **(a)** the perimeter of the rectangle and **(b)** the area.

26. If the length of one side of a square is represented by $\frac{1}{2}h$, find the representation of **(a)** the perimeter of the square and **(b)** the area.

27. The perimeter of a triangle is represented by $41px$. If the lengths of two sides are represented by $18px$ and $7px$, represent the length of the third side.

28. If $4x$ tickets cost $16x^4$ dollars, what is the cost of one ticket?

29. If there are $8w$ seats in every row, and there are $2w^2$ rows in an auditorium, how many seats are in the auditorium?

30. Find the value of $8x^0 + 4x^{-1}$ when $x = 2$.

31. What must be subtracted from $5py$ to give the result $9py$?

In 32–35, express the number in scientific notation.

32. 5,800 **33.** 14,200,000 **34.** .00006 **35.** .00000277

In 36–39, find the decimal number that is expressed by the given numeral.

36. 4×10^4 **37.** 3×10^{-3} **38.** 3.9×10^8 **39.** 1.03×10^{-4}

In 40–47, select the numeral preceding the expression that best completes the statement or answers the question.

40. When $-24x^{12}$ is divided by $-3x^3$, the quotient is
(1) $8x^4$ (2) $-8x^4$ (3) $8x^9$ (4) $-8x^9$

41. The product $(5aw^4)(4a^3w^2)$ is
(1) $20a^3w^8$ (2) $20a^4w^6$ (3) $9a^3w^8$ (4) $9a^4w^6$

42. The expression $\frac{10^7 \cdot 10}{(10^3)^2}$ is equivalent to
(1) 1 (2) 10 (3) 100 (4) 1,000

43. The expression $-8t\,(-8t^4)$ is equivalent to
(1) $64t^4$ (2) $64t^5$ (3) $-64t^5$ (4) $-16t^5$

44. The expression $x^3 + x^3$ is equivalent to
(1) x^6 (2) x^9 (3) $2x^3$ (4) $2x^6$

45. The expression $(2a^2)^3$ is equivalent to
(1) $6a^5$ (2) $6a^6$ (3) $8a^5$ (4) $8a^6$

46. Which of the given statements is true?
(1) $20^6 \div 20^3 = 1^3$ (2) $(30^2)^5 = 30^7$
(3) $19^2 \cdot 19^5 = 19^7$ (4) $2^4 \cdot 4^2 = 8^6$

47. The length of a rectangle is represented by $6d^2y^2$. If the area of the rectangle is represented by $18d^2y^6$, then the width is represented by
(1) $3y^4$ (2) $3d^2y^4$ (3) $12y^4$ (4) $3y^3$

48. A freight train one mile long is traveling at a steady speed of 20 miles per hour. It enters a tunnel one mile long at 1 P.M. At what time does the end of the train emerge from the tunnel?

Chapter 9

Operations With Polynomials

9-1 ADDING POLYNOMIALS

You have already learned that terms like 5, x, z^2, and $4y^3$ are called *monomials*. The ***degree*** of a monomial in one variable is the exponent of the variable.

The degree of x or x^1 is 1; the degree of z^2 is 2; the degree of $4y^3$ is 3.

Since a nonzero constant such as 5 is equivalent to $5x^0$, the degree of a nonzero constant is zero. The monomial 0 has no degree.

- **A polynomial is a sum of monomials.**

A monomial such as $4x^2$ may be considered to be a polynomial of one term. (*Mono* means *one*, *poly* means *many*.)

A polynomial of two unlike terms such as $10a + 12b$ is called a ***binomial***. (*Bi* means *two*.)

A polynomial of three unlike terms such as $x^2 + 3x + 2$ is called a ***trinomial***. (*Tri* means *three*.)

A polynomial such as $5x^2 + (-2x) + (-4)$ is usually written as $5x^2 - 2x - 4$.

A polynomial has been simplified or is in ***simplest form*** when it contains no like terms. For example, $5x^3 + 8x^2 - 5x^3 + 7$, when expressed in simplest form, becomes $8x^2 + 7$.

The ***degree of a polynomial*** is the greatest of the degrees of its terms after the polynomial has been simplified. For example, the degree of $5x^3 + 8x^2 - 5x^3 + 7$, which when simplified becomes $8x^2 + 7$, is 2 (not 3).

A polynomial is arranged in ***descending order*** when the exponents of a particular variable decrease as we move from left to right. Thus, $x^3 - 3x^2 + 5x - 7$ is arranged in descending order of x.

A polynomial is arranged in ***ascending order*** when the exponents of a particular variable increase as we move from left to right. Thus, $y - 4y^2 + 5y^3$ is arranged in ascending order of y. The polynomial $x^2 + 2xy + y^2$ is arranged in descending order of x, whereas it is arranged in ascending order of y.

A polynomial in one variable is written in ***standard form*** when its terms are arranged in descending order. The polynomial $3x^2 - 7x + 3$ is in standard form, whereas the polynomial $4z^2 - 2z + 5z^3$ is not in standard form.

To add two polynomials, we use the commutative, associative, and distributive properties to combine like terms. For example, we can add $3x^2 + 5$ and $6x^2 + 8$ as follows:

Step	*Reason*
(1) $(3x^2 + 5) + (6x^2 + 8) = (3x^2 + 6x^2) + (5 + 8)$	(1) Commutative and associative properties
(2) $= (3 + 6)x^2 + (5 + 8)$	(2) Distributive property
(3) $= 9x^2 + 13$	(3) Substitution principle

To find the sum of the polynomials $4x^2 + 3x - 5$, $3x^2 - 6 - 5x$, and $-x + 3 - 2x^2$, we can write the polynomials vertically, first arranging them in descending (or ascending) order. Then, we can add the like terms in each column. As shown at the right, the sum is $5x^2 - 3x - 8$.

$$\begin{array}{rrr} 4x^2 & + 3x & - 5 \\ 3x^2 & - 5x & - 6 \\ -2x^2 & - x & + 3 \\ \hline 5x^2 & - 3x & - 8 \end{array}$$

PROCEDURE. To add polynomials, combine like terms by adding their numerical coefficients.

Addition can be checked by adding again in the opposite direction. Addition can also be checked by substituting convenient values for the variables and evaluating the polynomials and the sum. The sum of the values of the polynomials should be equal to the value of the polynomial that is the sum of the polynomials.

● Do not use 0 or 1 as values for checking the addition of polynomials.

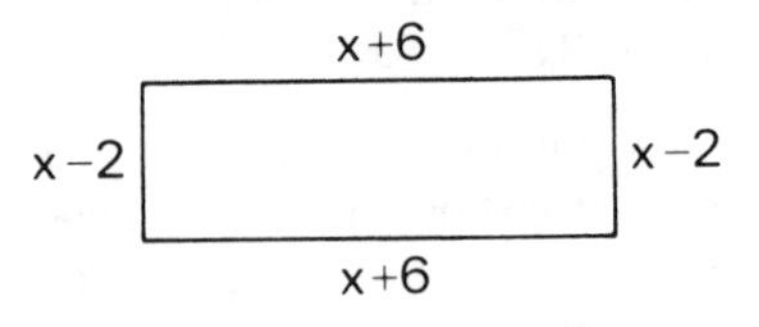

To represent the perimeter of a rectangle whose length is represented by $x + 6$ and whose width is represented by $x - 2$, we would add the measures of the four sides.

$$P = (x + 6) + (x - 2) + (x + 6) + (x - 2) = 4x + 8$$

Since the width $x - 2$ must be a positive number, the value of the variable x cannot be 2 or less than 2.

MODEL PROBLEMS

1. Add and check: $4x + 3y - 5z$, $3x - 5y - 6z$, $-2x - y + 3z$

Solution:

$$\begin{array}{r} 4x + 3y - 5z \\ 3x - 5y - 6z \\ \underline{-2x - y + 3z} \\ 5x - 3y - 8z \end{array}$$

Check: Let $x = 4$, $y = 3$, $z = 2$.

$$\begin{array}{rcr} 16 + 9 - 10 & & = 15 \\ 12 - 15 - 12 & & = -15 \\ \underline{-8 - 3 + 6} & & = \underline{-5} \\ 20 - 9 - 16 = -5 & \longleftrightarrow & -5 \end{array}$$

Answer: $5x - 3y - 8z$

2. Add: $+7x^2 - 5xy + 4y^2$, $+3xy - x^2$, $-9y^2 + 2xy$

How to Proceed

(1) Arrange in descending order of x.
(2) Arrange like terms in the same column.
(3) Add like terms in each column.

Solution

$$\begin{array}{rrr} +7x^2 & - 5xy & + 4y^2 \\ - x^2 & + 3xy & \\ & + 2xy & - 9y^2 \\ \hline +6x^2 & + 0 & - 5y^2 \end{array}$$

Answer: $6x^2 - 5y^2$. Check by adding in the opposite direction.

3. Simplify: $6a + [5a + (6 - 3a)]$

Solution: When one grouping symbol appears within another grouping symbol, first perform the operation involving the algebraic expression within the innermost grouping symbol.

$$\begin{aligned} 6a + [5a + (6 - 3a)] &= 6a + [5a + 6 - 3a] \\ &= 6a + [2a + 6] \\ &= 6a + 2a + 6 \\ &= 8a + 6 \quad \textit{Ans.} \end{aligned}$$

EXERCISES

In 1–5, state the degree of the monomial.

1. x^2 **2.** $4y^5$ **3.** $\frac{1}{2}c^3$ **4.** $5x$ **5.** 0

In 6–10, state the degree of the polynomial.

6. $2x^3 + 7x - 4$ **7.** $\frac{1}{2} + 5x^2$ **8.** $x^3 - 4$

9. $a^2 + 7a^4 - 9$ **10.** 8

In 11–14, state whether the expression is a monomial, a binomial, a trinomial, or none of these.

11. $8x + 3$ **12.** $7y$ **13.** $-2a^2 + 3a - 6$ **14.** $x^3 + 2x^2 + x - 7$

In 15–18: **a.** Arrange the polynomial in descending order (standard form). **b.** Arrange the polynomial in ascending order. **c.** State the degree of the polynomial.

15. $5 + 2x^2 - 3x$ **16.** $y^4 - 9 + y^3$

17. $6 + x^4 - \frac{1}{2}x^3$ **18.** $2a^2 - 3a + a^4$

In 19–26, simplify the polynomials.

19. $5c + 3d + 2c + 8d$ **20.** $9y + 6w + 3w + y$

21. $8x + 9y - 3x - 6y$ **22.** $-4a + 6b + 3a - b$

23. $3r + 2s + 9t + 4r - 5s + t$

24. $-5m + 6n + 8p - 6n + 3m$

25. $3x^2 - 5x + 7 + 2x^2 + 3x - 9$

26. $2x + 4x^2 - 7 - x^2 + 7 - 8x$

In 27–36, add and check the result.

27. $\begin{array}{r} 5x + 3y \\ \underline{6x + 9y} \end{array}$ **28.** $\begin{array}{r} 4a - 6b \\ \underline{9a + 3b} \end{array}$ **29.** $\begin{array}{r} -6m + n \\ \underline{-4m - 5n} \end{array}$ **30.** $\begin{array}{r} -9ab + 8cd \\ \underline{3ab - 8cd} \end{array}$

31. $\begin{array}{r} 15x - 26y + 8z \\ \underline{3x - 14y - 3z} \end{array}$ **32.** $\begin{array}{r} x^2 - 33x + 15 \\ \underline{-4x^2 + 18x - 36} \end{array}$ **33.** $\begin{array}{r} -5a^2 - 6ab - 4b^2 \\ \underline{+7a^2 + 6ab - 3b^2} \end{array}$

34. $\begin{array}{r} x^2 + 3x + 5 \\ 2x^2 - 4x - 1 \\ \underline{-5x^2 + 2x + 4} \end{array}$ **35.** $\begin{array}{r} 5c^2 - 4cd + 6d^2 \\ -c^2 + 3cd + 2d^2 \\ \underline{-3c^2 + cd - 8d^2} \end{array}$ **36.** $\begin{array}{r} 2.1 + .9z + z^2 \\ -.7z - .2z^2 \\ \underline{-.9 + .2z} \end{array}$

In 37–52, simplify the expression.

37. $4a + (9a + 3)$ **38.** $7b + (4b - 6)$ **39.** $8c + (7 - 9c)$

40. $(-6x - 4) + 6x$ **41.** $r + (s + 2r)$ **42.** $8d^2 + (6d^2 - 4d)$

43. $(5x + 3) + (6x - 5)$ **44.** $-6y + [7 + (6y - 7)]$

45. $(5 - 6y) + (-9y + 2)$ **46.** $5a + [3b + (-2a + 4b)]$

47. $(5x^2 + 4) + (-3x^2 - 4)$ **48.** $3y^2 + [6y^2 + (3y - 4)]$

49. $(x^3 + 3x^2) + (-2x^2 - 9)$ **50.** $-d^2 + [9d + (2 - 4d^2)]$

51. $(x^2 + 5x - 24) + (-x^2 - 4x + 9)$

52. $(x^3 + 9x - 5) + (-4x^2 - 12x + 5)$

53. Add: $3y^2 + 7 - 5y$ and $9 + 4y - 5y^2$

54. Add: $2c^2 + 5c - 3$, $4c^2 - 5$, $6 - 5c$

55. Add: $x^2 - 7xy + 3y^2$, $-2y^2 + 3x^2 - 4xy$, $xy - 2x^2 - 4y^2$

In 56–61: **a.** Express the perimeter of the rectangle in simplest form. **b.** Name one possible value for x. **c.** Name one value for x that is not possible.

56. length $= 2x$, width $= x$ **57.** length $= 3x$, width $= x + 2$

58. length $= x$, width $= x - 3$ **59.** length $= x + 3$, width $= x - 8$

60. length $= x$, width $= 3x - 4$

61. length $= x - 4$, width $= x - 6$

In 62–65: **a.** Express the perimeter of the figure as a polynomial in simplest form. **b.** Name one value for x that is not possible.

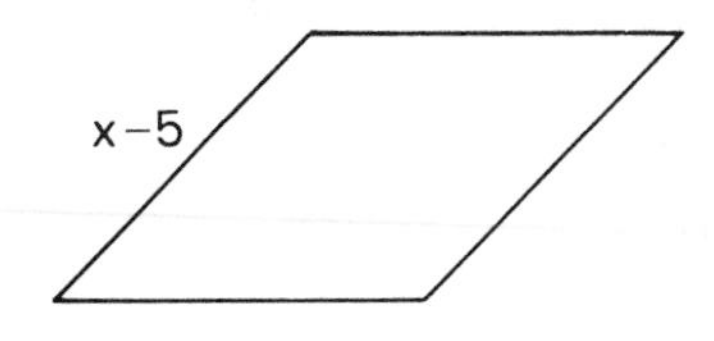

Ex. 62. Rhombus

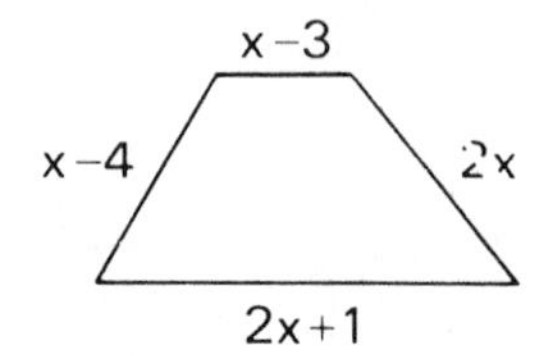

Ex. 63. Trapezoid

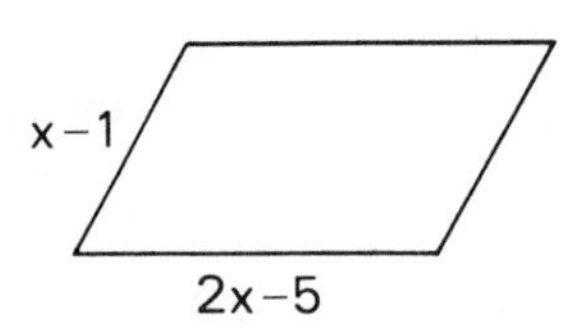

Ex. 64. Parallelogram

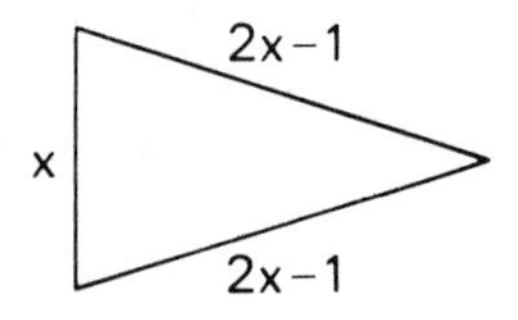

Ex. 65. Triangle

66. Represent the perimeter of a square each of whose sides is represented by:
a. $3x + 5$ **b.** $4x - 1$ **c.** $x^2 + 4x - 3$ **d.** $x^2 + 2xy + y^2$

In 67–72, write each answer as a polynomial in simplest form.

67. The cost of 12 gallons of gas is represented by $12x$ and the cost of a quart of oil is represented by $2x - 30$. Represent the cost of 12 gallons of gas and a quart of oil.

68. In their last basketball game, Tom scored $2x$ points, Tony scored $x + 5$ points, Walt scored $3x + 1$ points, Dick scored $4x - 7$ points, and Dan scored $2x - 2$ points. Represent the total points scored by these five players.

69. The length of a rectangle is 4 more than the width. If x represents the width of the rectangle, represent the perimeter in terms of x.

70. The cost of a chocolate shake is 40 cents less than the cost of a hamburger. If h represents the cost of a hamburger in cents, represent the cost of a hamburger and a chocolate shake in terms of h.

71. Ann spent 12 dollars more for fabric for a new dress than she did for buttons, and 1 dollar less for thread than she did for buttons. If b represents the cost of the buttons in dollars, represent the total cost of the materials needed for the dress in terms of b.

72. Last week, Greg spent twice as much on bus fare as he did on lunch, and 3 dollars less on entertainment than he did on bus fare. If x represents the amount spent on lunch in dollars, express the total amount spent on lunch, bus fare, and entertainment in terms of x.

9-2 SUBTRACTING POLYNOMIALS

To subtract one polynomial from another, we use a procedure similar to that used to subtract like terms: we add the opposite of the subtrahend to the minuend.

We can write the opposite of a polynomial using the symbol $-$. For example, the opposite of $2x^2 - 5x - 3$ can be written $-(2x^2 - 5x - 3)$.

We can also write the opposite of a polynomial by forming a polynomial each of whose terms is the opposite of the corresponding term of the original polynomial. For example:

$$\text{If the original polynomial is } \underbrace{2x^2}_{\downarrow}\ \underbrace{-\ 5x}_{\downarrow}\ \underbrace{-\ 3}_{\downarrow},$$
$$\text{the opposite of this polynomial is } -2x^2 + 5x + 3.$$

The sum of an algebraic expression and its opposite is 0.

In general, for any x, y, and z,

$$-(x + y + z) = (-x) + (-y) + (-z)$$

Now let us subtract $2x^2 - 5x - 3$ from $5x^2 + 8x - 7$ in the following manner, first changing subtraction to addition of the opposite:

$$\begin{aligned}(5x^2 + 8x - 7) - (2x^2 - 5x - 3) &= (5x^2 + 8x - 7) + (-2x^2 + 5x + 3)\\ &= 5x^2 + 8x - 7 - 2x^2 + 5x + 3\\ &= (5 - 2)x^2 + (8 + 5)x + (-7 + 3)\\ &= 3x^2 + 13x - 4\end{aligned}$$

The solution of a subtraction example can also be arranged vertically as shown at the right. We mentally add the opposite of each term of the subtrahend to the corresponding term of the minuend.

$$\begin{array}{rrr} 5x^2 & + \;\; 8x & - 7 \\ 2x^2 & - \;\; 5x & - 3 \\ \hline 3x^2 & + 13x & - 4 \end{array}$$

PROCEDURE. To subtract one polynomial from another, add the opposite (additive inverse) of the subtrahend to the minuend.

Subtraction can be checked by adding the subtrahend and the difference. The result should equal the minuend. Subtraction can also be checked by substituting values for the variables.

MODEL PROBLEMS

1. Subtract and check: $(5x^2 - 6x + 3) - (2x^2 - 9x - 6)$

Solution: Subtract.

$$\begin{array}{rl} 5x^2 - 6x + 3 & \text{Minuend} \\ 2x^2 - 9x - 6 & \text{Subtrahend} \\ \hline 3x^2 + 3x + 9 & \text{Difference} \end{array}$$

Check: Add.

$$\begin{array}{rl} 2x^2 - 9x - 6 & \text{Subtrahend} \\ 3x^2 + 3x + 9 & \text{Difference} \\ \hline 5x^2 - 6x + 3 & \text{Minuend} \end{array}$$

Answer: $3x^2 + 3x + 9$

2. Simplify the expression $9x - [7 - (4 - 2x)]$.

Solution:

$$\begin{aligned}9x - [7 - (4 - 2x)] &= 9x - [7 + (-4 + 2x)]\\ &= 9x - [7 - 4 + 2x]\\ &= 9x - [3 + 2x]\\ &= 9x + [-3 - 2x]\\ &= 9x - 3 - 2x\\ &= 7x - 3\end{aligned}$$

(First, perform the subtraction involving the expression within the innermost grouping symbol.)

Answer: $7x - 3$

EXERCISES

In 1–6, write the opposite (additive inverse) of the expression.

1. $9x + 6$ **2.** $-5x + 3$ **3.** $-6x - 6y$
4. $2x^2 - 3x + 2$ **5.** $-y^2 + 5y - 4$ **6.** $7ab - 3bc$

In 7–20, subtract and check the result.

7. $\begin{array}{r} 10a + 8b \\ \underline{4a + 5b} \end{array}$ **8.** $\begin{array}{r} 5b + 3c \\ \underline{4b + c} \end{array}$ **9.** $\begin{array}{r} 6d + 6e \\ \underline{9d - 8e} \end{array}$ **10.** $\begin{array}{r} 8x - 3y \\ \underline{-4x + 8y} \end{array}$

11. $\begin{array}{r} 4r - 7s \\ \underline{5r - 7s} \end{array}$ **12.** $\begin{array}{l} 0 \\ \underline{8a - 6b} \end{array}$ **13.** $\begin{array}{r} 6rs - 7bc \\ \underline{9rs - 7bc} \end{array}$ **14.** $\begin{array}{r} 5xy - 9cd \\ \underline{-3xy + cd} \end{array}$

15. $\begin{array}{r} x^2 - 6x + 5 \\ \underline{3x^2 - 2x - 2} \end{array}$ **16.** $\begin{array}{r} 3y^2 - 2y - 1 \\ \underline{-5y^2 - 2y + 6} \end{array}$ **17.** $\begin{array}{r} 3a^2 - 2ab + 3b^2 \\ \underline{-a^2 - 5ab + 3b^2} \end{array}$

18. $\begin{array}{l} 7a + 6b - 9c \\ \underline{3a \qquad - 6c} \end{array}$ **19.** $\begin{array}{r} x^2 \qquad - 9 \\ \underline{-2x^2 + 5x - 3} \end{array}$ **20.** $\begin{array}{r} 5 - 6d - d^2 \\ \underline{- 4d - d^2} \end{array}$

In 21–46, simplify the expression.

21. $(3y - 6) - (8 - 9y)$ **22.** $(-4x + 7) - (3x - 7)$
23. $(4a - 3b) - (5a - 2b)$ **24.** $(2c + 3d) - (-6d - 5c)$
25. $(5x^2 + 6x - 9) - (x^2 - 3x + 7)$
26. $(2x^2 - 3x - 1) - (2x^2 + 5x)$
27. $5x - (2x + 5)$ **28.** $3y - (5y - 4)$ **29.** $4z - (6z - 2)$
30. $9m - (6 + 6m)$ **31.** $m - (m - n)$ **32.** $4d - (5c + 4d)$
33. $5c - (4c - 6c^2)$ **34.** $8r - (-6s - 8r)$ **35.** $-2x - (5x + 8)$
36. $-9d - (2c - 4d) + 4c$ **37.** $(3y + z) + (z - 5y) - (2z - 2y)$
38. $(a - b) - (a + b) - (-a - b)$
39. $(x^2 - 3x) + (5 - 9x) - (5x^2 - 7)$
40. $5c - [8c - (6 - 3c)]$ **41.** $12 - [-3 + (6x - 9)]$
42. $10x + [3x - (5x - 4)]$ **43.** $x^2 - [-3x + (4 - 7x)]$
44. $3x^2 - [7x - (4x - x^2) + 3]$ **45.** $9a - [5a^2 - (7 + 9a - 2a^2)]$
46. $4y^2 - \{4y + [3y^2 - (6y + 2) + 6]\}$

47. From $4x + 2y$, subtract $x - 4y$.
48. Subtract $5a - 7b$ from $3a - 9b$.
49. From $5x^2 + 5x - 4$, subtract $x^2 - 3x + 5$.
50. Subtract $7r^2 + 3r - 8$ from $10r^2 - 3r - 7$.
51. From $m^2 + 5m - 7$, subtract $m^2 - 3m - 4$.
52. From $12x - 6y + 9z$, subtract $-x - 3z + 6y$.
53. Subtract $2x^2 - 3x + 7$ from $x^2 + 6x - 12$.

54. The sum of two binomials is $6y^2 + 9y$. One of the binomials is $4y^2 + 5y$. What is the other binomial?
55. The sum of two trinomials is $15x^2 - 7x + 3$. One of the trinomials is $8x^2 + 9x - 7$. What is the other trinomial?
56. Subtract $2c^2 + 3c - 4$ from 0.
57. How much greater than $a^2 + 3ab + b^2$ is $4a^2 + 9ab - 2b^2$?
58. **a.** How much less than 25 is 15?
b. How much less than $5x + 3y$ is $2x + y$?
59. How much less than $4x^2 - 5$ is $3x^2 + 2$?
60. **a.** By how much does 13 exceed 10?
b. By how much does $7x + 5$ exceed $4x - 3$?
61. By how much does $a + b + c$ exceed $a + b - c$?
62. What algebraic expression must be added to $2x^2 + 5x + 7$ to give $8x^2 - 4x - 5$ as the result?
63. What algebraic expression must be added to $4x^2 - 8$ to make the result equal to 0?
64. What algebraic expression must be added to $-3x^2 + 7x - 5$ to give 0 as the result?
65. From the sum of $y^2 + 2y - 7$ and $2y^2 - 4y + 3$, subtract $3y^2 - 8y - 10$.
66. Subtract the sum of $c^2 - 5$ and $-2c^2 + 3c$ from $4c^2 - 6c + 7$.

9-3 MULTIPLYING A POLYNOMIAL BY A MONOMIAL

You know that the distributive property of multiplication over addition states:

$$a(b + c) = ab + ac$$

Therefore,
$$x(4x + 3) = x(4x) + x(3)$$
$$x(4x + 3) = 4x^2 + 3x$$

This result can be illustrated geometrically. Let us separate a rectangle whose length is $4x + 3$ and whose width is x into two smaller rectangles such that the length of one rectangle is $4x$ and the length of the other is 3.

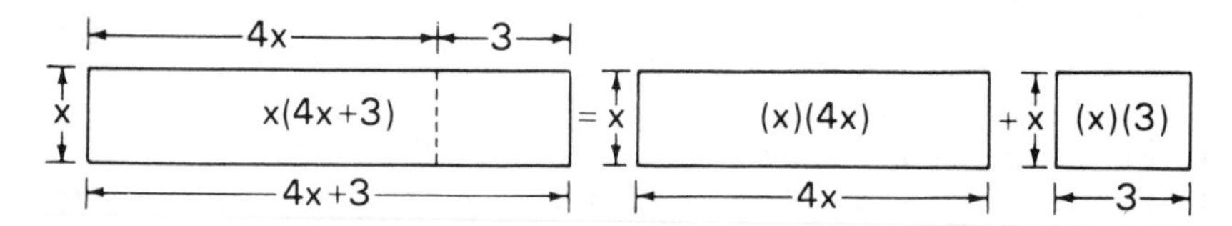

Since the area of the largest rectangle is equal to the sum of the areas of the two smaller rectangles:

$$x(4x + 3) = x(4x) + x(3) = 4x^2 + 3x$$

PROCEDURE. To multiply a polynomial by a monomial, use the distributive property: multiply each term of the polynomial by the monomial and write the result as the sum of these products.

MODEL PROBLEMS

In 1–4, multiply.

1. $5(r - 7) = 5r - 35$

2. $8(3x - 2y + 4z) = 24x - 16y + 32z$

3. $-5x(x^2 - 2x + 4) = -5x^3 + 10x^2 - 20x$

4. $-3a^2b^2(4ab^2 - 3b^2) = -12a^3b^4 + 9a^2b^4$

EXERCISES

In 1–18, multiply.

1. $3(6c + 3d)$ **2.** $-5(4m - 6n)$ **3.** $-2(8a + 6b)$

4. $10\left(2x - \frac{1}{5}y\right)$ **5.** $12\left(\frac{2}{3}m - 4n\right)$ **6.** $-8\left(4r - \frac{1}{4}s\right)$

7. $-16\left(\frac{3}{4}c - \frac{5}{8}d\right)$ **8.** $4x(5x + 6)$ **9.** $5d(d^2 - 3d)$

10. $-5c^2(15c - 4c^2)$ **11.** $mn(m + n)$ **12.** $-ab(a - b)$

13. $3ab(5a^2 - 7b^2)$ **14.** $-r^3s^3(6r^4s - 3s^4)$

15. $10d(2a - 3c + 4b)$ **16.** $-8(2x^2 - 3x - 5)$

17. $3xy(x^2 + xy + y^2)$ **18.** $5r^2s^2(-2r^2 + 3rs - 4s^2)$

In 19–24, write each answer as a polynomial in simplest form.

19. If the cost of a notebook is represented by $2x - 3$, express the cost of 5 notebooks.

20. If the length of a rectangle is represented by $5y - 7$ and the width by $3y$, represent the area of the rectangle.

21. If the measure of the base of a triangle is represented by $3b + 2$ and the height by $4b$, represent the area of the triangle.

22. Represent the distance traveled in 3 hours by a car traveling at $3x - 7$ miles per hour.

23. Represent in terms of x and y the amount saved in $3y$ weeks if $x - 2$ dollars are saved each week.

24. The length of a rectangular skating rink is 2 less than 3 times the width. If w represents the width of the rink, represent the area in terms of w.

9-4 USING MULTIPLICATION TO SIMPLIFY ALGEBRAIC EXPRESSIONS CONTAINING SYMBOLS OF GROUPING

To simplify the expression $3x + 7(2x + 3)$, we use the distributive property and then combine like terms. Thus:

$$3x + 7(2x + 3) = 3x + 7(2x) + 7(3) = 3x + 14x + 21 = 17x + 21$$

We can use the identity $a = 1a$ to simplify the following expression:

$$5 + (2x - 3) = 5 + 1(2x - 3) = 5 + 2x - 3 = 2x + 2$$

Likewise, we can simplify $5y - (2 - 7y)$ by using the identity $-a = -1a$ and then multiplying:

$$5y - (2 - 7y) = 5y - 1(2 - 7y) = 5y - 2 + 7y = 12y - 2$$

MODEL PROBLEMS

In 1–3, simplify the expression by using the distributive property of multiplication and combining like terms.

1. $3(x + 5) - 10$ **2.** $2c + (7c - 4)$ **3.** $-2(3 - 2x^2) - (6 - 5x)$

Solution	*Solution*	*Solution*
$3(x + 5) - 10$	$2c + (7c - 4)$	$-2(3 - 2x^2) - (6 - 5x)$
$= 3x + 15 - 10$	$= 2c + 1(7c - 4)$	$= -2(3 - 2x^2) - 1(6 - 5x)$
$= 3x + 5$ *Ans.*	$= 2c + 7c - 4$	$= -6 + 4x^2 - 6 + 5x$
	$= 9c - 4$ *Ans.*	$= 4x^2 + 5x - 12$ *Ans.*

EXERCISES

In 1–30, simplify the expression.

1. $5(d + 3) - 10$ **2.** $3(2 - 3c) + 5c$ **3.** $7 + 2(7x - 5)$

4. $-2(x - 1) + 6$ **5.** $-4(3 - 6a) - 7a$ **6.** $5 - 4(3e - 5)$

7. $8 + (4e - 2)$ **8.** $a + (b - a)$ **9.** $(6b + 4) - 2b$

10. $9 - (5t + 6)$ **11.** $4 - (2 - 8s)$ **12.** $-(6x - 7) + 14$

13. $5x(2x - 3) + 9x$ **14.** $12y - 3y(2y - 4)$

15. $7x + 3(2x - 1) - 8$ **16.** $7c - 4d - 2(4c - 3d)$

17. $3a - 2a(5a - a) + a^2$ **18.** $(a + 3b) - (a - 3b)$

19. $4(2x + 5) - 3(2 - 7x)$ **20.** $3(x + y) + 2(x - 3y)$

21. $5x(2 - 3x) - x(3x - 1)$ **22.** $y(y + 4) - y(y - 3) - 9y$

23. $7x(x + 3y) - 4y(-4x - y)$

24. $-2c(c + 2d) + 4d(2c - 3d)$

25. $ab(7a - 3c) - bc(2a - b)$

26. $mn(4m^2 - 2n^2) - 2mn(2m^2 - n^2)$

27. $7[5x + 2(x - 3) + 4]$

28. $-4[8y - 7 - 3(2y - 1)]$

29. $4x[2x^2 - 2x(x + 3) - 5]$

30. $x^2z - x[xy - x(y - z)]$

31. In terms of x:

a. express the area of the outer rectangle pictured at the right.

b. express the area of the inner rectangle pictured at the right.

c. express as a polynomial in simplest form the area of the shaded region.

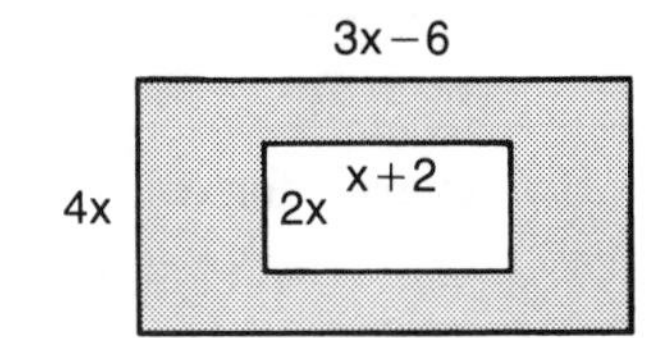

In 32–35, express the answer as a polynomial in simplest form.

32. A mail order catalog lists books for $3x - 5$ dollars each. The cost of postage is 2 dollars for 5 books or less. Represent the cost of an order for 4 books.

33. A store advertises skirts for $x - 5$ dollars and allows an additional 10-dollar reduction on the total purchase if 3 or more skirts are purchased. Represent the cost of 5 skirts.

34. A store advertises skirts for $x - 5$ dollars and allows an additional 2-dollar reduction on each shirt if 3 or more skirts are purchased. Represent the cost of 5 skirts.

35. A store advertises skirts for $x - 5$ dollars and tops for $2x - 3$ dollars. Represent the cost of 2 skirts and 3 tops.

9-5 MULTIPLYING A POLYNOMIAL BY A POLYNOMIAL

To find the product $(x + 4)(a)$, we used the distributive property of multiplication over addition:

$$(x + 4)(a) = x(a) + 4(a)$$

Now, let us use this property to find the product of the two polynomials $(x + 4)(x + 3)$.

$$\begin{aligned} (x + 4)(a) &= x(a) + 4(a) \\ (x + 4)(x + 3) &= x(x + 3) + 4(x + 3) \\ &= x^2 + 3x + 4x + 12 \\ &= x^2 + 7x + 12 \end{aligned}$$

This result can also be illustrated geometrically.

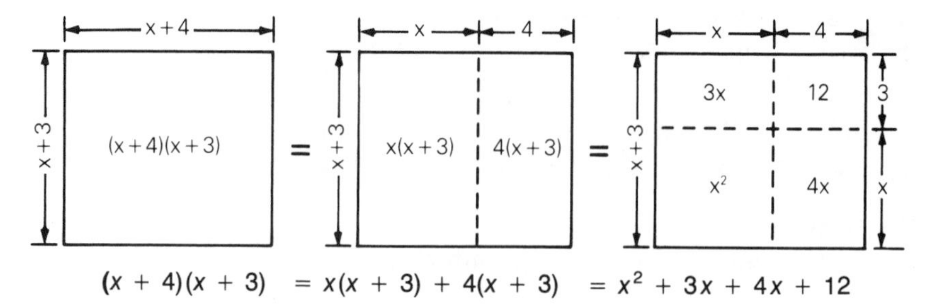

$(x + 4)(x + 3) \quad = x(x + 3) + 4(x + 3) \quad = x^2 + 3x + 4x + 12$

In general, for all a, b, c, and d:

$$\begin{aligned}(\boldsymbol{a} + \boldsymbol{b})(\boldsymbol{c} + \boldsymbol{d}) &= a(c + d) + b(c + d)\\ &= \boldsymbol{ac} + \boldsymbol{ad} + \boldsymbol{bc} + \boldsymbol{bd}\end{aligned}$$

Notice that each term of the first polynomial is multiplying each term of the second.

At the right, you see a convenient vertical arrangement of the previous multiplication, similar to the arrangement used in arithmetic multiplication. Multiply from left to right.

$$\begin{array}{rl} & x + 4 \\ & \underline{x + 3} \\ (x+4)x \rightarrow & x^2 + 4x \\ (x+4)3 \rightarrow & \underline{\quad + 3x + 12} \\ \text{Add like terms: } & x^2 + 7x + 12 \end{array}$$

PROCEDURE. To multiply a polynomial by a polynomial, first arrange the multiplicand and multiplier according to descending or ascending powers of a common variable. Then, use the distributive property: multiply each term of the multiplicand by each term of the multiplier. Finally, combine like terms.

Multiplication can be checked by interchanging the multiplicand and the multiplier and multiplying again. The product should be the same.

MODEL PROBLEMS

In 1 and 2, express the product as a polynomial in simplest form.

1. $(3x - 4)(4x + 5)$

Solution:

$$\begin{aligned}(3x - 4)(4x + 5) &= 3x(4x + 5) - 4(4x + 5)\\ &= 12x^2 + 15x - 16x - 20\\ &= 12x^2 - x - 20\end{aligned}$$

Answer: $12x^2 - x - 20$

Alternate Solution:

$$\begin{array}{r} 3x - 4 \\ \underline{4x + 5} \\ 12x^2 - 16x \\ \underline{+ 15x - 20} \\ 12x^2 - x - 20 \end{array}$$

2. $(x^2 + 3xy + 9y^2)(x - 3y)$

Solution:

$$\begin{aligned} &(x^2 + 3xy + 9y^2)(x - 3y) \\ &= x^2(x - 3y) + 3xy(x - 3y) + 9y^2(x - 3y) \\ &= x^3 - 3x^2y + 3x^2y - 9xy^2 + 9xy^2 - 27y^3 \\ &= x^3 \quad + 0x^2y \quad + 0xy^2 \quad - 27y^3 \end{aligned}$$

Answer: $x^3 - 27y^3$

EXERCISES

In 1–52, multiply.

1. $(a + 2)(a + 3)$ **2.** $(c + 6)(c + 1)$ **3.** $(x - 5)(x - 3)$
4. $(d - 6)(d - 5)$ **5.** $(d + 9)(d - 3)$ **6.** $(x - 7)(x + 2)$
7. $(m + 3)(m - 7)$ **8.** $(z - 5)(z + 8)$ **9.** $(f + 10)(f - 8)$
10. $(t + 15)(t - 6)$ **11.** $(b - 8)(b - 10)$ **12.** $(w - 13)(w + 7)$
13. $(6 + y)(5 + y)$ **14.** $(8 - e)(6 - e)$ **15.** $(12 - r)(6 + r)$
16. $(x + 5)(x - 5)$ **17.** $(y + 7)(y - 7)$ **18.** $(a + 9)(a - 9)$
19. $(2x + 1)(x - 6)$ **20.** $(c - 5)(2c - 4)$ **21.** $(2a + 9)(3a + 1)$
22. $(5y - 2)(3y - 1)$ **23.** $(2x + 3)(2x - 3)$ **24.** $(3d + 8)(3d - 8)$
25. $(x + y)(x + y)$ **26.** $(a - b)(a - b)$ **27.** $(a + b)(a - b)$
28. $(a + 2b)(a + 3b)$ **29.** $(2c - d)(3c + d)$ **30.** $(x - 4y)(x + 4y)$
31. $(2z + 5w)(3z - 4w)$ **32.** $(9x - 5y)(2x + 3y)$
33. $(5k + 2m)(3r + 4s)$ **34.** $(3x + 4y)(3x - 4y)$
35. $(r^2 + 5)(r^2 - 2)$ **36.** $(x^2 - y^2)(x^2 + y^2)$
37. $(x^2 + 3x + 5)(x + 2)$ **38.** $(2c^2 - 3c - 1)(2c + 1)$
39. $(3 - 2d - d^2)(5 - 2d)$ **40.** $(c^2 - 2c + 4)(c + 2)$
41. $(2x^2 - 3x + 1)(3x - 2)$ **42.** $(3x^2 - 4xy + y^2)(4x + 3y)$
43. $(x^3 - 3x^2 + 2x - 4)(3x - 1)$ **44.** $(2x + 1)(3x - 4)(x + 3)$
45. $(x^2 - 4x + 1)(x^2 + 5x - 2)$ **46.** $(x + 4)(x + 4)(x + 4)$
47. $(a + 5)^3$ **48.** $(x - y)^3$
49. $(5 + x^2 - 2x)(2x - 3)$ **50.** $(5x - 4 + 2x^2)(3 + 4x)$
51. $(2xy + x^2 + y^2)(x + y)$ **52.** $a(a + b)(a - b)$

In 53–60, simplify the expression.

53. $(x + 7)(x - 2) - x^2$ **54.** $2(3x + 1)(2x - 3) + 14x$
55. $8x^2 - (4x + 3)(2x - 1)$ **56.** $(x + 4)(x + 3) - (x - 2)(x - 5)$
57. $(3y + 5)(2y - 3) - (y + 7)(5y - 1)$ **58.** $(y + 4)^2 - (y - 3)^2$
59. $r(r - 2s) - (r - s)$ **60.** $a[(a + 2)(a - 2) - 4]$

In 61–63, use symbols of grouping to write an algebraic expression that represents the answer. Then, express the answer as a polynomial in simplest form.

61. The length of a rectangle is $2x - 5$ and its width is $x + 7$. Express the area of the rectangle.
62. The dimensions of a rectangle are represented by $11x - 8$ and $3x + 5$. Represent the area of the rectangle.
63. A plane travels at a rate represented by $(x + 100)$ kilometers per hour. Represent the distance it can travel in $(2x + 3)$ hours.

In 64–69: **a.** Find the area of the rectangle whose length and width are given. **b.** Check the result found in part **a** as follows:
(1) Copy the rectangle that is shown.
(2) Represent the length and width of each of the four small rectangles.
(3) Represent the area of each of the four small rectangles.
(4) Add the four areas found in step (3) to find the area of the original given rectangle.

64. length $= x + 3$, width $= x + 2$
65. length $= x + 6$, width $= x + 5$
66. length $= 2x + 5$, width $= 5x + 3$
67. length $= 3x + 1$, width $= 3x + 1$
68. length $= 2x + 3y$, width $= x + y$
69. length $= 2x + 5y$, width $= 3x + 4y$

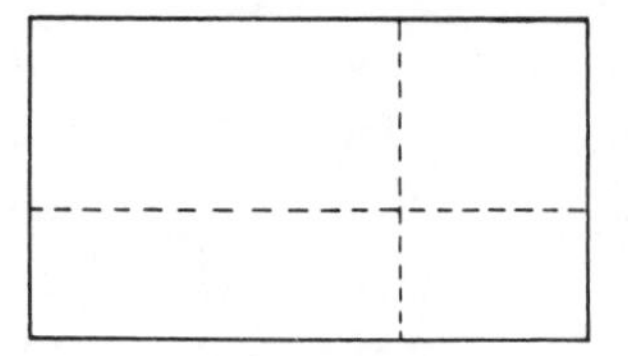

Ex. 64–69

9-6 DIVIDING A POLYNOMIAL BY A MONOMIAL

We know that: $\frac{a}{b} + \frac{c}{b} = \frac{a + c}{b}$

Therefore, using the symmetric property of equality, we can write:

$$\frac{a + c}{b} = \frac{a}{b} + \frac{c}{b}$$

Similarly, $$\frac{2x + 2y}{2} = \frac{2x}{2} + \frac{2y}{2} = x + y$$

and $$\frac{15y^4 - 12xy^3}{3y^2} = \frac{15y^4}{3y^2} - \frac{12xy^3}{3y^2} = 5y^2 - 4xy$$

Usually, we do not need to write the middle expression in the equality.

$$\frac{21a^2b - 3ab}{3ab} = 7a - 1$$

PROCEDURE. To divide a polynomial by a monomial, divide each term of the polynomial by the monomial.

MODEL PROBLEMS

In 1 and 2, divide.

1. $(8a^5 - 6a^4) \div 2a^2 = 4a^3 - 3a^2$
2. $\dfrac{24x^3y^4 - 18x^2y^2 - 6xy}{-6xy} = -4x^2y^3 + 3xy + 1$

EXERCISES

In 1–31, divide.

1. $(10x + 20y) \div 5$
2. $(18r - 27s) \div 9$
3. $(14x + 7) \div 7$
4. $(cm + cn) \div c$
5. $(tr - r) \div r$
6. $\dfrac{12a - 6b}{-2}$
7. $\dfrac{8c^2 - 12d^2}{-4}$
8. $\dfrac{m^2 + 8m}{m}$
9. $\dfrac{p + prt}{p}$
10. $\dfrac{y^2 - 5y}{-y}$
11. $\dfrac{18d^3 + 12d^2}{6d}$
12. $\dfrac{20x^2 + 15x}{5x}$
13. $\dfrac{18r^5 + 12r^3}{6r^2}$
14. $\dfrac{16t^5 - 8t^4}{4t^2}$
15. $\dfrac{9y^9 - 6y^6}{-3y^3}$
16. $\dfrac{8a^3 - 4a^2}{-4a^2}$
17. $\dfrac{3ab^2 - 4a^2b}{ab}$
18. $\dfrac{4c^2d - 12cd^2}{4cd}$
19. $\dfrac{2\pi r^2 + 2\pi rh}{2\pi r}$
20. $\dfrac{-6a^2b - 12ab^2}{-2ab}$
21. $\dfrac{36a^4b^2 - 18a^2b^2}{-18a^2b^2}$
22. $\dfrac{-5y^5 + 15y - 25}{-5}$
23. $\dfrac{-2a^2 - 3a + 1}{-1}$
24. $\dfrac{2.4y^5 + 1.2y^4 - .6y^3}{-.6y^3}$

25. $\dfrac{15r^4s^4 + 20r^3s^3 - 5r^2s^2}{-5r^2s^2}$

26. $(8y^2 + 6y) \div 2y$

27. $(9c^3 - 6c^2 + 3c) \div 3c$

28. $(5ab - 10ac) \div 5a$

29. $(y^3 - y^2 + y) \div -y$

30. Divide $12x^3 - 8x^2 + 4x$ by $4x$.

31. Divide $8d^3 - 6d^2 + 4d$ by $4d$.

9-7 DIVIDING A POLYNOMIAL BY A POLYNOMIAL

To divide one polynomial by another, use a procedure similar to the one used when dividing one arithmetic number by another. When we divide 736 by 32, we see through repeated subtractions how many times 32 is contained in 736. Likewise, when we divide $x^2 + 6x + 8$ by $x + 2$, we see through repeated subtractions how many times $x + 2$ is contained in $x^2 + 6x + 8$.

See how dividing $x^2 + 6x + 8$ by $x + 2$ follows the same pattern as dividing 736 by 32:

How to Proceed	*Solution 1*	*Solution 2*
(1) Write the usual division form.	$32\overline{)736}$	$x + 2\overline{)x^2 + 6x + 8}$
(2) Divide the left number of the dividend by the left number of the divisor to obtain the first number of the quotient.	$\begin{array}{r} 2 \\ 32\overline{)736} \end{array}$	$\begin{array}{r} x \qquad\quad \\ x + 2\overline{)x^2 + 6x + 8} \end{array}$
(3) Multiply the whole divisor by the first number of the quotient.	$\begin{array}{r} 2 \\ 32\overline{)736} \\ \underline{64}\; \end{array}$	$\begin{array}{r} x \qquad\quad \\ x + 2\overline{)x^2 + 6x + 8} \\ \underline{x^2 + 2x} \quad\; \end{array}$
(4) Subtract this product from the dividend and bring down the next number of the dividend to obtain the new dividend.	$\begin{array}{r} 2 \\ 32\overline{)736} \\ \underline{64}\; \\ 96 \end{array}$	$\begin{array}{r} x \qquad\quad \\ x + 2\overline{)x^2 + 6x + 8} \\ \underline{x^2 + 2x} \quad\; \\ 4x + 8 \end{array}$
(5) Divide the left number of the new dividend by the left number of the divisor to obtain the next number of the quotient.	$\begin{array}{r} 23 \\ 32\overline{)736} \\ \underline{64}\; \\ 96 \end{array}$	$\begin{array}{r} x + 4 \qquad \\ x + 2\overline{)x^2 + 6x + 8} \\ \underline{x^2 + 2x} \quad\; \\ 4x + 8 \end{array}$

(6) Repeat steps (3) and (4), multiplying the whole divisor by the second number of the quotient. Subtract the result from the new dividend. The last remainder is 0.

$$\begin{array}{r} 23 \\ 32\overline{)736} \\ \underline{64} \\ 96 \\ \underline{96} \end{array} \qquad \begin{array}{r} x + 4 \\ x + 2\overline{)x^2 + 6x + 8} \\ \underline{x^2 + 2x} \\ 4x + 8 \\ \underline{4x + 8} \end{array}$$

23 *Ans.* $\qquad x + 4$ *Ans.*

Note: The division process comes to an end when the remainder is 0, or the degree of the remainder is less than the degree of the divisor.

To check the division, use the relationship:

quotient × divisor + remainder = dividend

Check 1	*Check 2*
23	$x + 4$
×32	$x + 2$
46	$x^2 + 4x$
69	$+ 2x + 8$
736	$x^2 + 6x + 8$

Division becomes more convenient if the terms of both the divisor and the dividend are arranged in descending or ascending order.

For example, if $3x - 1 + x^3 - 3x^2$ is to be divided by $x - 1$, write:

$$x - 1\overline{)x^3 - 3x^2 + 3x - 1}$$

MODEL PROBLEM

Divide $5s + 6s^2 - 15$ by $2s + 3$. Check.

Solution:

Arrange terms of the dividend in descending order of s.

$$\begin{array}{r} 3s - 2 \\ 2s + 3\overline{)6s^2 + 5s - 15} \\ \underline{6s^2 + 9s} \\ -4s - 15 \\ \underline{-4s - 6} \\ -9 \end{array}$$

Answer: $3s - 2 + \dfrac{-9}{2s + 3}$

Check:

$$\begin{array}{rl} 2s + 3 & \text{Divisor} \\ \underline{3s - 2} & \text{Quotient} \\ 6s^2 + 9s & \\ \underline{-4s - 6} & \\ 6s^2 + 5s - 6 & \\ \underline{-9} & \text{Remainder} \\ 6s^2 + 5s - 15 & \text{Dividend} \end{array}$$

EXERCISES

In 1–14, divide and check.

1. $b^2 + 5b + 6$ by $b + 3$
2. $y^2 + 3y + 2$ by $y + 2$
3. $m^2 - 8m + 7$ by $m - 1$
4. $w^2 + 2w - 15$ by $w + 5$
5. $y^2 + 20y + 61$ by $y + 17$
6. $m^2 + 7m - 27$ by $m - 5$
7. $3x^2 - 8x + 4$ by $3x - 2$
8. $15t^2 - 19t - 56$ by $5t + 7$
9. $10r^2 - r - 24$ by $2r + 3$
10. $12c^2 - 22c + 8$ by $4c - 2$
11. $66 + 17x + x^2$ by $6 + x$
12. $30 - m - m^2$ by $5 - m$
13. $x^2 - 64$ by $x - 8$
14. $4y^2 - 49$ by $2y + 7$
15. One factor of $x^2 - 4x - 21$ is $x - 7$. Find the other factor.
16. The area of a rectangle is represented by $x^2 - 8x - 9$. If its length is represented by $x + 1$, how can its width be represented?
17. The area of a rectangle is represented by $3y^2 + 8y + 4$. If its width is represented by $3y + 2$, how can its length be represented?

9-8 REVIEW EXERCISES

1. **a.** Add the polynomials at the right.
   b. Let $x = 2$ and $b = -3$. Check the sum by substitution.

$$\begin{array}{r} 3x^2 - 3bx - b \\ \underline{x^2 + 5bx + b} \end{array}$$

2. Find the sum of $3c - 7d$, $-4c + 3d$, $5c - 8d$, and $d - c$.

In 3–5, subtract the polynomials.

3. $$\begin{array}{r} 3y^2 - 2y + 6 \\ \underline{y^2 - 8y - 2} \end{array}$$
4. $$\begin{array}{r} 3r + 2s - 7t \\ \underline{4r \qquad + 9t} \end{array}$$
5. $$\begin{array}{r} m^2 \qquad - 16 \\ \underline{-m^2 + m + 5} \end{array}$$

6. The lengths of the sides of a triangle are represented by $k + 6r$, $k - 8r$, and $3k + r$. Express the perimeter of the triangle as a binomial.
7. Subtract $2p^2 - 8p + 9$ from $p^2 - 8p + 2$.

In 8–17, simplify the expression.

8. $2x - (x - 2)$
9. $x + 2(x - 2)$
10. $(x + 2)(x - 2)$
11. $(b - 3)(b - 6)$
12. $(2h + 1)(h - 5)$
13. $(y + 4)^2$
14. $5(m + 1) + 3(1 - m)$
15. $-3ax(a^2 - 4ax - 5x^2)$
16. $\dfrac{6w^3 - 8w^2 + 2w}{2w}$
17. $ab - [a^2 - a(b + a)]$

18. Express the product of $(2y + 3)(2y - 1)$ as a trinomial.

19. If one side of a square is represented by $x + 5$, write the polynomial that represents (**a**) its perimeter and (**b**) its area.

20. Divide $x^2 - 10x - 24$ by $x + 2$, and check.

21. Divide $9y^2 - 25$ by $3y - 5$, and check.

In 22–24, select the numeral preceding the correct answer.

22. The length of a rectangle is represented by $x + 6$, and its width by $x - 9$. What is the area of the rectangle in terms of x?
(1) $x^2 - 54$ (2) $x^2 - 3x - 54$ (3) $x^2 + 3x - 54$ (4) $4x - 6$

23. The area of a rectangle is represented by $y^2 + 2y - 8$, and its width by $y - 2$. What is the length of the rectangle in terms of y?
(1) $y + 4$ (2) $y - 4$ (3) $y - 6$ (4) $y - 8$

24. A regular hexagon has 6 sides of equal length. If the perimeter of a regular hexagon is represented as $12x - 24$, what is the length of one of its sides in terms of x?
(1) $x - 2$ (2) $2x - 4$ (3) $3x - 6$ (4) $72x - 144$

25. Express in simplest form the distance traveled in $2k$ hours at $(3k - 4)$ miles per hour.

26. Express in simplest form the cost of one pencil if $5x$ pencils cost $25x^2 - 5x$ cents.

27. The cost of a pizza is 20 cents less than 9 times the cost of a soft drink. If x represents the cost of a soft drink in cents, express in simplest form the cost of 2 pizzas and 6 soft drinks.

28. What is the units digit of $(27)^{27}$?

29. If $(a + 3)^2 = 25 + 6a$, what is the value of a?

Chapter 10

First-Degree Equations and Inequalities in One Variable

10-1 USING BOTH THE ADDITIVE AND MULTIPLICATIVE INVERSES IN SOLVING EQUATIONS

When the solution of an equation requires the use of both the additive and the multiplicative inverses, either inverse may be used first. However, the solution is usually easier when the additive inverse is used first.

MODEL PROBLEMS

1. Solve and check: $2x + 3x + 4 = -6$

How to Proceed	*Solution*
(1) Write the equation.	$2x + 3x + 4 = -6$
(2) Combine like terms.	$5x + 4 = -6$
(3) Add -4, the additive inverse of $+4$.	$\underline{\quad -4 = -4}$
	$5x = -10$
(4) Multiply by $\frac{1}{5}$, the multiplicative inverse of 5.	$\frac{1}{5}(5x) = \frac{1}{5}(-10)$
	$x = -2$

Check:

$$2x + 3x + 4 = -6$$
$$2(-2) + 3(-2) + 4 \stackrel{?}{=} -6$$
$$(-4) + (-6) + 4 \stackrel{?}{=} -6$$
$$-6 = -6 \quad \text{(True)}$$

Answer: $x = -2$

2. Solve and check: $\frac{3}{4}x - 4 = 17$

How to Proceed	*Solution*	*Check*
(1) Write the equation.	$\frac{3}{4}x - 4 = 17$	$\frac{3}{4}x - 4 = 17$
(2) Add +4, the additive inverse of −4.	$\underline{\quad +4 = +4}$ $\frac{3}{4}x = 21$	$\frac{3}{4}(28) - 4 \stackrel{?}{=} 17$ $21 - 4 \stackrel{?}{=} 17$
(3) Multiply by $\frac{4}{3}$, the multiplicative inverse of $\frac{3}{4}$.	$\frac{4}{3}\left(\frac{3}{4}x\right) = \frac{4}{3}(21)$ $x = 28$	$17 = 17$ (True)

Answer: $x = 28$

3. If 4 times a number is increased by 5, the result is 41. Find the number.

How to Proceed	*Solution*
(1) Represent the number by a letter.	Let x = the number.
(2) Write the word statement as an equation.	$4x + 5 = 41$
(3) Solve the equation.	$4x = 36$ $x = 9$

Check: Does 4 times 9, increased by 5, give a result of 41? Yes.

Answer: The number is 9.

Note: We can arrive at the same result geometrically, making use of the following models:

1. Draw two models to represent the equation.

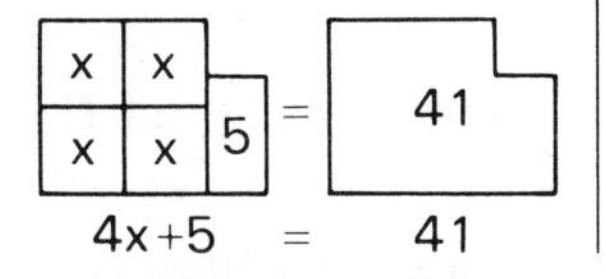

2. Subtract 5 from each model.

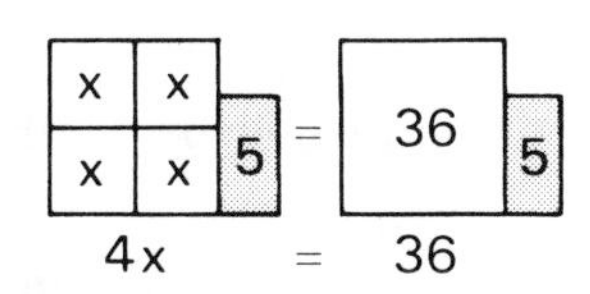

3. Divide by 4 to find each of the equal parts.

x x | 9 9
x x | 9 9
x = 9

EXERCISES

In 1–21, solve the equation and check.

1. $3x + 4 = 16$	**2.** $35 = 21y - 7$	**3.** $2c + 1 = -31$
4. $2y + 18 = 8$	**5.** $2x + 9 = 37$	**6.** $4x + 2 = -34$
7. $-42 = 5x + 28$	**8.** $-34 = 2 - 6t$	**9.** $13 = 8x - 7$

10. $-32 = 24y - 20$ **11.** $\frac{1}{2}z + 6 = 15$ **12.** $\frac{1}{5}y - 3 = -4$

13. $\frac{2}{3}m + 7 = 29$ **14.** $\frac{2}{5}r - 9 = -19$ **15.** $\frac{3}{2}x - 14 = 16$

16. $-25 = \frac{7}{3}r - 11$ **17.** $-5.4 = 2.6 + 2x$

18. $9x - 5x + 9 = 1$ **19.** $5x + 2x - 17 = 53$

20. $2y - 8y + 29 = 5$ **21.** $8x - 21 - 5x = -15$

In 22–25, determine the element(s) of the set if $x \in$ {signed numbers}.

22. $\{x \mid 5x + 30 = 10\}$ **23.** $\{x \mid 15 = 6x - 15\}$

24. $\{x \mid \frac{2}{3}x - 12 = 60\}$ **25.** $\{x \mid 5x + 5 - 9x = -11\}$

In 26–31, represent in terms of x:

26. twice the number represented by x, increased by 8
27. three times the number represented by x, decreased by 12
28. four times the number that is 3 more than x
29. three times the number that exceeds x by 5
30. twice the sum of the number represented by x and 5
31. ten times the number obtained when twice x is decreased by 10

32. If the smaller of two numbers is represented by x, represent the greater when their sum is:
a. 10 **b.** 25 **c.** 36 **d.** 50 **e.** 100 **f.** 3,000
33. If the sum of two numbers is represented by S, and the smaller number is represented by x, represent the greater in terms of S and x.
34. If the sum of two numbers is represented by S, and the greater number is represented by l, represent the smaller in terms of S and l.
35. The sum of 7 times a number and 5 is -51. Find the number.
36. If 10 times a number is decreased by 6, the result is 104. Find the number.
37. If $\frac{2}{3}$ of a number is diminished by 8, the result is 32. Find the number.
38. The Tigers played 78 games. If they won 8 games more than they lost, how many games did they lose?
39. A salesman receives a weekly salary of \$150. He also receives \$4 for each tire he sells. One week, he earned \$330. How many tires did he sell that week?
40. Rosita wishes to make a long-distance telephone call. The charges are \$.75 for the first three minutes and 14 cents for each additional minute. If Rosita has \$2.15, how many minutes can her call last?

41. Ray's father pays him \$5 for cutting the grass and \$3 an hour for cutting the hedges. One day, Ray earned \$15.50 for cutting the grass and cutting the hedges. How many hours did Ray spend cutting the hedges?

10-2 SOLVING EQUATIONS THAT HAVE THE VARIABLE IN BOTH MEMBERS

A variable represents a number. As you know, any number may be added to both members of an equation without changing the solution set. Therefore, the same variable (or the same multiple of the same variable) may be added to or subtracted from both members of an equation without changing the solution set.

To solve $8x = 30 + 5x$, first eliminate $5x$ from the right member of the equation.

Add $-5x$ to both members of the equation.

Method 1

$$\begin{aligned} 8x &= 30 + 5x \\ -5x &= \ - 5x \\ \hline 3x &= 30 \\ x &= 10 \end{aligned}$$

Method 2

$$\begin{aligned} 8x &= 30 + 5x \\ 8x + (-5x) &= 30 + 5x + (-5x) \\ 3x &= 30 \\ x &= 10 \end{aligned}$$

The check is left to the student.

Answer: $x = 10$

PROCEDURE. To solve an equation that has the variable in both members, transform it into an equivalent equation in which the variable appears only in one member. Then, solve this equation.

MODEL PROBLEMS

1. Solve and check: $7x = 63 - 2x$

How to Proceed	*Solution*	*Check*
(1) Write the equation.	$7x = 63 - 2x$	$7x = 63 - 2x$
(2) Add $2x$ to each member of the equation.	$+2x = \quad + 2x$	$7(7) \stackrel{?}{=} 63 - 2(7)$
	$9x = 63$	$49 \stackrel{?}{=} 63 - 14$
(3) Divide each member of the equation by 9.	$x = 7$	$49 = 49$ (True)

Answer: $x = 7$

2. The larger of two numbers is 4 times the smaller. If the larger number exceeds the smaller number by 15, find the number.

> *Note:* When s represents the smaller number and $4s$ represents the larger number, *the larger exceeds the smaller by 15* has the following meanings. Use any one of them.
>
> 1. The larger equals 15 more than the smaller, written $4s = s + 15$.
> 2. The larger decreased by 15 equals the smaller, written $4s - 15 = s$.
> 3. The larger decreased by the smaller equals 15, written $4s - s = 15$.

Solution:

Let s = the smaller number.
Then, $4s$ = the larger number.

$$\underbrace{\text{The larger}}_{4s}\ \underbrace{\text{is}}_{=}\ \underbrace{\text{15 more than the smaller,}}_{s + 15}$$

$$\begin{aligned} 4s &= s + 15 \\ -s &= -s \\ \hline 3s &= 15 \\ s &= 5 \\ 4s &= 20 \end{aligned}$$

Check: The larger number, 20, is 4 times the smaller number, 5. The larger number, 20, exceeds the smaller number, 5, by 15.

Answer: The larger number is 20; the smaller number is 5.

3. The greater of two numbers is twice the smaller. If the greater is decreased by 10, the result is 5 more than the smaller. Find the numbers.

How to Proceed	*Solution*
(1) Represent the two numbers in terms of the same variable.	Let x = the smaller number. Then, $2x$ = the greater number.

(2) Write an equation that symbolizes the relationships stated in the problem.

$$\underbrace{\textit{The greater decreased by 10}}_{2x - 10} \underbrace{\textit{is}}_{=} \underbrace{\textit{5 more than the smaller.}}_{x + 5}$$

(3) Solve the equation.

(*a*) Add 10 to each member.

$$2x - 10 + 10 = x + 5 + 10$$
$$2x = x + 15$$

(*b*) Add $-x$ to each member.

$$2x + (-x) = x + 15 + (-x)$$
$$x = 15$$
$$2x = 30$$

Check: The greater decreased by 10 is $30 - 10 = 20$. 5 more than the smaller is $15 + 5 = 20$. The results are the same, 20.

Answer: The smaller number is 15; the greater number is 30.

EXERCISES

In 1–38, solve the equation and check.

1. $7x = 10 + 2x$ **2.** $9x = 44 - 2x$ **3.** $5c = 28 + c$

4. $y = 4y + 30$ **5.** $2d = 36 + 5d$ **6.** $2\frac{1}{4}y = 1\frac{1}{4}y - 8$

7. $.8m = .2m + 24$ **8.** $8y = 90 - 2y$ **9.** $2.3x + 36 = .3x$

10. $2\frac{3}{4}x + 24 = 3x$ **11.** $5a - 40 = 3a$ **12.** $5c = 2c - 81$

13. $x = 9x - 72$ **14.** $.5m - 30 = 1.1m$ **15.** $4\frac{1}{4}c = 9\frac{3}{4}c + 44$

16. $7r + 10 = 3r + 50$ **17.** $4y + 20 = 5y + 9$ **18.** $7x + 8 = 6x + 1$

19. $x + 4 = 9x + 4$ **20.** $9x - 3 = 2x + 46$

21. $y + 30 = 12y - 14$ **22.** $c + 20 = 55 - 4c$

23. $2d + 36 = -3d - 54$ **24.** $7y - 5 = 9y + 29$

25. $2m - 1 = 6m + 1$ **26.** $4x - 3 = 47 - x$

27. $3b - 8 = 14 - 8b$ **28.** $\frac{2}{3}t - 11 = 64 - 4\frac{1}{3}t$

29. $18 - 4n = 6 - 16n$ **30.** $-2y - 39 = 5y - 18$

31. $7x - 4 = 5x - x + 35$ **32.** $10 - x - 3x = 7x - 23$

33. $8a - 15 - 6a = 85 - 3a$ **34.** $8c + 1 = 7c - 14 - 2c$

35. $12x - 5 = 8x - x + 50$ **36.** $6d - 12 - d = 9d + 53 + d$

37. $3m - 5m - 12 = 7m - 88 - 5$ **38.** $5 - 3z - 18 = z - 1 + 8z$

39. Eight times a number equals 35 more than the number. Find the number.

40. Six times a number equals 3 times the number, increased by 24. Find the number.

41. Twice a number is equal to 35 more than 7 times the number. Find the number.
42. If a number is multiplied by 7, the result is the same as when 25 is added to twice the number. Find the number.
43. If twice a number is subtracted from 132, the result equals four times the number. Find the number.
44. If 3 is added to 5 times a number, the result is the same as when 15 is added to twice the number. Find the number.
45. If 4 times a number is decreased by 9, the result is the same as when 3 times the number is decreased by 1. Find the number.
46. If 3 times a number is increased by 5, the result is the same as when 77 is decreased by 9 times the number. Find the number.
47. If 5 times a number is increased by 50, the result is the same as when 200 is decreased by the number. Find the number.
48. If 10 times a certain number is increased by 4, the result is 12 more than 9 times the number. Find the number.
49. If 3 times a number is increased by 22, the result is 14 less than 7 times the number. Find the number.

10-3 SOLVING EQUATIONS CONTAINING PARENTHESES

PROCEDURE. To solve an equation containing parentheses, transform it into an equivalent equation that does not contain parentheses. Do this by performing the indicated operation on the numbers and variables contained within the parentheses. Combine like terms in each member of the equation before solving the equation.

MODEL PROBLEMS

1. Solve and check: $9t - (2t - 4) = 25$

Note: $9t - (2t - 4)$ means from $9t$ subtract $(2t - 4)$.

How to Proceed	*Solution*
(1) Write the equation.	$9t - (2t - 4) = 25$
(2) To subtract $(2t - 4)$, add its opposite.	$9t + (-2t + 4) = 25$
	$9t - 2t + 4 = 25$
(3) Combine like terms.	$7t + 4 = 25$
(4) Solve the equation.	$7t = 21$
	$t = 3$

Check:

$$9t - (2t - 4) = 25$$
$$9(3) - [2(3) - 4] \stackrel{?}{=} 25$$
$$9(3) - [6 - 4] \stackrel{?}{=} 25$$
$$27 - [2] \stackrel{?}{=} 25$$
$$25 = 25 \quad \text{(True)}$$

Answer: $t = 3$

2. The larger of two numbers is 5 more than the smaller. Twice the larger is 40 more than the smaller. Find the numbers.

Solution: Let x = the smaller number.
Then, $x + 5$ = the larger number.

$$\underbrace{\text{Twice the larger}}_{2(x + 5)} \; \underbrace{\text{is}}_{=} \; \underbrace{\text{40 more than the smaller}}_{x + 40}$$

$$\begin{aligned} 2x + 10 &= x + 40 \\ -x &= -x \\ \hline x + 10 &= 40 \\ -10 &= -10 \\ \hline x &= 30 \\ x + 5 &= 35 \end{aligned}$$

Check: Twice the larger $= 2(35) = 70$
40 more than the smaller $= 30 + 40 = 70$

Answer: The smaller number is 30 and the larger number is 35.

3. Solve and check: $27x - 3(x - 6) = 6$

Note: Since $-3(x - 6)$ means that $(x - 6)$ is to be multiplied by -3, we will use the distributive property.

How to Proceed	*Solution*
(1) Write the equation.	$27x - 3(x - 6) = 6$
(2) Use the distributive property.	$27x - 3x + 18 = 6$
(3) Combine like terms.	$24x + 18 = 6$
(4) Solve the equation.	$24x = -12$
	$x = \frac{-12}{24}$
	$x = -\frac{1}{2}$

The check is left to the student.

Answer: $x = -\frac{1}{2}$

EXERCISES

In 1–32, solve and check the equation.

1. $x + (x - 6) = 20$
2. $x - (12 - x) = 38$
3. $(15x + 7) - 12 = 4$
4. $(14 - 3c) + 7c = 94$
5. $x + (4x + 32) = 12$
6. $7x - (4x - 39) = 0$
7. $5(x + 2) = 20$
8. $3(y - 9) = 30$
9. $8(2c - 1) = 56$
10. $6(3c - 1) = -42$
11. $3y = 2(10 - y)$
12. $4(c + 1) = 32$
13. $5t - 2(t - 5) = 19$
14. $18 = -6x + 4(2x + 3)$
15. $5m - 4 = 3(m + 2)$
16. $5(x - 3) = 30 - 10x$
17. $7(x + 2) = 5(x + 4)$
18. $3(a - 5) = 2(2a + 1)$
19. $3(2b + 1) - 7 = 50$
20. $5(3c - 2) + 8 = 43$
21. $7r - (6r - 5) = 7$
22. $8y - (5y + 2) = 16$
23. $11x = 40 + (7x + 4)$
24. $10z - (3z - 11) = 17$
25. $8b - 4(b - 2) = 24$
26. $5m - 2(m - 5) = 17$
27. $9 + 2(5v + 3) = 13v$
28. $28r - 6(3r - 5) = 40$
29. $3a + (2a - 5) = 13 - 2(a + 2)$
30. $4(2r + 1) - 3(2r - 5) = 29$
31. $\frac{1}{2}(8x - 6) = 25$
32. $\frac{3}{4}(8 + 4x) - \frac{1}{3}(6x + 3) = 9$

33. The larger of two numbers is 5 more than the smaller. The smaller number plus twice the larger equals 100. Find the numbers.

34. One number is 2 less than another. If 4 times the larger is subtracted from 5 times the smaller, the result is 10. Find the numbers.

35. The larger of two numbers is 20 more than the smaller. Four times the larger is 70 more than 5 times the smaller. Find the numbers.

36. If 14 is added to a certain number and the sum is multiplied by 2, the result is equal to 8 times the number decreased by 14. Find the number.

37. The difference between two numbers is 24. Find the numbers if their sum is 88.

38. Separate 144 people into two groups such that one group will be 12 less than twice the other.

39. Separate 45 into two parts such that 5 times the smaller is 6 less than twice the greater.

40. The greater of two numbers is 1 less than 3 times the smaller. If 3 times the greater is 5 more than 8 times the smaller, find the numbers.

41. The larger of two numbers is 1 more than 3 times the smaller. The difference between 8 times the smaller and 2 times the larger is 10. Find the numbers.

42. The greater of two numbers is 1 more than twice the smaller. Three times the greater exceeds 5 times the smaller by 10. Find the numbers.

43. The second of three numbers is 2 more than the first. The third number is twice the first. The sum of the first and third exceeds the second by 2. Find the three numbers.

44. The second of three numbers is 1 less than the first. The third number is 5 less than twice the second. If the third number exceeds the first number by 12, find the three numbers.

10-4 CONSECUTIVE INTEGER PROBLEMS

Preparing to Solve Consecutive Integer Problems

As you know, an integer is any whole number or its opposite. Examples of integers are 5, -3, and 0.

Consecutive integers are integers that follow one another in order. To obtain a set of consecutive integers, start with any integer and count by ones. Each number in the set is 1 more than the previous number in the set. Each of the following is a set of consecutive integers:

1. $\{5, 6, 7, 8\}$ 2. $\{-5, -4, -3, -2\}$
3. $\{x, x + 1, x + 2, x + 3\}$ $x \in \{\text{integers}\}$

Consecutive even integers are even integers that follow one another in order. To obtain a set of consecutive even integers, start with any even integer and count by twos. Each number in the set is 2 more than the previous number in the set. Each of the following is a set of consecutive even integers:

1. $\{2, 4, 6, 8\}$ 2. $\{-12, -10, -8, -6\}$
3. $\{x, x + 2, x + 4, x + 6\}$ $x \in \{\text{even integers}\}$

Consecutive odd integers are odd integers that follow one another in order. To obtain a set of consecutive odd integers, start with any odd integer and count by twos. Each number in the set is 2 more than the previous number in the set. Each of the following is a set of consecutive odd integers:

1. $\{3, 5, 7, 9\}$ 2. $\{-5, -3, -1, 1\}$
3. $\{x, x + 2, x + 4, x + 6\}$ $x \in \{\text{odd integers}\}$

KEEP IN MIND

1. Consecutive integers differ by 1.
2. Consecutive even integers and also consecutive odd integers differ by 2.

EXERCISES

1. **a.** If $x = 3$, what numbers do x, $x + 1$, $x + 2$, $x + 3$, and $x + 4$ represent? **b.** What kind of integers are these numbers?
2. **a.** If $n = -3$, what numbers do n, $n + 1$, $n + 2$, $n + 3$, and $n + 4$ represent? **b.** What kind of integers are these numbers?
3. **a.** If $x = -1$, what numbers do x, $x + 2$, $x + 4$, $x + 6$, and $x + 8$ represent? **b.** What kind of integers are these numbers?
4. Write 4 consecutive integers beginning with each of the following integers (y is an integer):
   a. 15 **b.** 31 **c.** -10 **d.** -2 **e.** y **f.** $2y + 1$ **g.** $3y - 2$
5. Write 4 consecutive even integers beginning with each of the following integers (y is an even integer):
   a. 8 **b.** 26 **c.** -20 **d.** -4 **e.** y **f.** $2y$ **g.** $2y - 6$
6. Write 4 consecutive odd integers beginning with each of the following integers (y is an odd integer):
   a. 9 **b.** 35 **c.** -15 **d.** -3 **e.** y **f.** $2y + 1$ **g.** $2y - 1$

In 7–10, tell whether the number represented is odd or even when: **(a)** n is an odd integer; **(b)** n is an even integer.

7. $n + 1$ **8.** $n - 1$ **9.** $n + 3$ **10.** $n + 4$

11. State whether $x + y$ is odd or even when: **(a)** x and y are odd integers; **(b)** x is an odd integer and y is an even integer; **(c)** x and y are even integers.

In 12–14, replace the question mark with the word "odd" or the word "even" so that the resulting statement will be true.

12. The sum of an even number of consecutive odd integers is an _?_ integer.
13. The sum of an odd number of consecutive odd integers is an _?_ integer.
14. The sum of any number of consecutive even integers is an _?_ integer.

Solving Consecutive Integer Problems

MODEL PROBLEMS

1. Find two consecutive integers whose sum is 95.

Solution:

Let n = the first integer.
Then, $n + 1$ = the second integer.

$$\underbrace{\textit{The sum of the two integers}}_{n + (n + 1)} \; \underbrace{\textit{is}}_{=} \; \underbrace{\textit{95.}}_{95}$$

$$\begin{aligned} n + n + 1 &= 95 \\ 2n + 1 &= 95 \\ 2n + 1 - 1 &= 95 - 1 \\ 2n &= 94 \\ n &= 47, n + 1 = 48 \end{aligned}$$

Check: The sum of the consecutive integers, 47 and 48, is 95.

Answer: 47 and 48

2. Find 3 consecutive positive even integers such that 4 times the first decreased by the second is 12 more than twice the third.

Solution:

Let n = the first even integer.
Then, $n + 2$ = the second even integer.
Then, $n + 4$ = the third even integer.

$$\underbrace{\textit{4 times the first decreased by the second}}_{4n - (n + 2)} \; \underbrace{\textit{is}}_{=} \; \underbrace{\textit{12 more than twice the third.}}_{2(n + 4) + 12}$$

$$\begin{aligned} 4n - n - 2 &= 2n + 8 + 12 \\ 3n - 2 &= 2n + 20 \\ 3n - 2 + 2 &= 2n + 20 + 2 \\ 3n &= 2n + 22 \\ 3n + (-2n) &= 2n + 22 + (-2n) \\ n &= 22 \\ n + 2 &= 24, n + 4 = 26 \end{aligned}$$

Check: Show that 22, 24, and 26 satisfy the conditions in the given problem: $4(22) - 24$ is 12 more than $2(26)$.

Answer: 22, 24, 26

EXERCISES

1. Find two consecutive integers whose sum is:
   a. 61 **b.** 35 **c.** 91 **d.** 125 **e.** −17 **f.** −81
2. Find three consecutive integers whose sum is:
   a. 18 **b.** 48 **c.** 99 **d.** 0 **e.** −12 **f.** −57
3. Find four consecutive integers whose sum is 234.
4. Find two consecutive even integers whose sum is:
   a. 22 **b.** 38 **c.** 146 **d.** 206 **e.** −10 **f.** −34
5. Find three consecutive even integers whose sum is:
   a. 12 **b.** 48 **c.** 156 **d.** 258 **e.** −18 **f.** −60
6. Find four consecutive even integers whose sum is 60.
7. Find three consecutive odd integers whose sum is:
   a. 33 **b.** 45 **c.** 159 **d.** 615 **e.** −27 **f.** −105
8. Find four consecutive odd integers whose sum is 112.
9. Find three consecutive integers such that the sum of the first and the third is 40.
10. Find four consecutive integers such that the sum of the second and fourth is 132.
11. Find two consecutive odd integers such that four times the larger is 29 more than three times the smaller.
12. Find two consecutive even integers such that twice the smaller is 26 less than three times the larger.
13. Find three consecutive integers such that twice the smallest is 12 more than the largest.
14. Find three consecutive integers such that the sum of the first two integers is 24 more than the third integer.
15. Find three consecutive even integers such that the sum of the smallest and twice the second is 20 more than the third.
16. Find two consecutive integers such that 4 times the larger exceeds 3 times the smaller by 23.
17. Find four consecutive odd integers such that the sum of the first three exceeds the fourth by 18.
18. Find three positive consecutive odd integers such that the largest decreased by three times the second is 47 less than the smallest.
19. Is it possible to find 3 consecutive even integers whose sum is 40? Why?
20. Is it possible to find 3 consecutive odd integers whose sum is 59? Why?

10-5 FINDING THE VALUE OF A VARIABLE IN A FORMULA

In Section 3-6, you learned to evaluate the subject of a formula. Now that you can solve equations, you will be able to find the value, not only of the subject, but of any variable in a formula, when the values of the other variables are known.

PROCEDURE. To find the value of a variable in a formula when the values of the other variables including the subject of the formula are given:

1. Substitute the given values in the formula.
2. Solve the resulting equation.

MODEL PROBLEM

The perimeter of a rectangle is 48 cm. If the length of the rectangle is 16 cm, find its width. (Use $P = 2l + 2w$.)

Solution

$$P = 2l + 2w$$
$$48 = 2(16) + 2w$$
$$48 = 32 + 2w$$
$$16 = 2w$$
$$8 = w$$

Answer: 8 cm

Check

$$P = 2l + 2w$$
$$48 \stackrel{?}{=} 2(16) + 2(8)$$
$$48 \stackrel{?}{=} 32 + 16$$
$$48 = 48 \quad \text{(True)}$$

EXERCISES

1. If $P = a + b + c$, find c when $P = 80$, $a = 20$, and $b = 25$.

2. If $P = 4s$, find s when **(a)** $P = 20$; **(b)** $P = 32$; **(c)** $P = 6.4$.

3. If $A = lw$, find w when **(a)** $A = 100$, $l = 5$; **(b)** $A = 3.6$, $l = .9$.

4. If $A = \frac{1}{2}bh$, find h when **(a)** $A = 24$, $b = 8$; **(b)** $A = 12$, $b = 3$.

5. If $V = lwh$, find w when $V = 72$, $l = \frac{3}{4}$, and $h = 12$.

6. If $P = 2l + 2w$, find w when:
(a) $P = 20$, $l = 7$; (b) $P = 36$, $l = 9\frac{1}{2}$.

7. If $P = 2l + 2w$, find l when:
(a) $P = 28$, $w = 3$; (b) $P = 24.8$, $w = 4.7$.

8. If $P = 2a + b$, find b when $P = 80$ cm and $a = 30$ cm.

9. If $P = 2a + b$, find a when $P = 18.6$ cm and $b = 5.8$ cm.

10. If $A = \frac{1}{2}h(b + c)$, find h when $A = 30$, $b = 4$, and $c = 6$.

11. If $A = \frac{1}{2}h(b + c)$, find b when $A = 50\text{ cm}^2$, $h = 4$ cm, and $c = 11$ cm.

12. If $D = RT$, find R when (a) $D = 120$, $T = 3$; (b) $D = 40$, $T = \frac{1}{2}$.

13. If $I = prt$, find p when $I = \$135$, $r = 6\%$, and $t = 3$ yr.

14. If $F = \frac{9}{5}C + 32$, find C when (a) $F = 95°$; (b) $F = 68°$; (c) $F = 59°$; (d) $F = 32°$; (e) $F = 212°$; (f) $F = -13°$.

15. Find the length of a rectangle whose perimeter is 34.6 cm and whose width is 5.7 cm.

16. The area of a triangle is 36 cm^2. Find the measure of the altitude drawn to the base when the (a) base = 8 cm; (b) base = 12 cm.

In 17–20, the perimeter of a square is given.
a. Find the length of each side of the square.
b. Find the area of the square.

17. $P = 28$ cm 18. $P = 3$ in. 19. $P = 16.8$ cm 20. $P = 2$ ft.

10-6 PERIMETER PROBLEMS

Preparing to Solve Perimeter Problems

KEEP IN MIND

The perimeter of a geometric figure is the sum of the lengths of all of its sides.

EXERCISES

1. Represent the perimeter of each of the following figures:

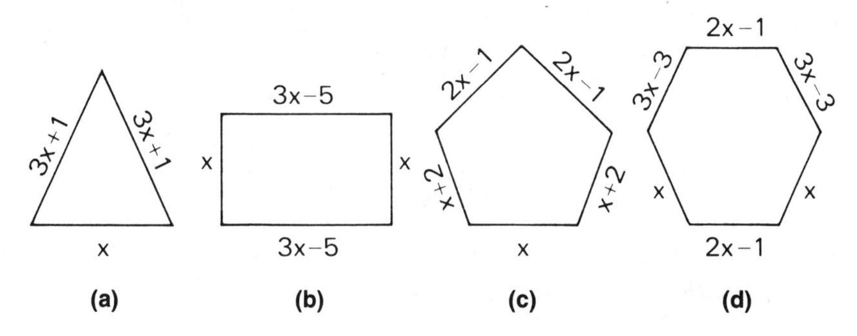

2. Represent the length and perimeter of a rectangle whose width is represented by x and whose length:
   a. is twice its width
   b. is 4 more than its width
   c. is 5 less than twice its width
   d. is 3 more than twice its width
3. The length of a side of an equilateral triangle is represented by $x + 5$. Represent the perimeter of the triangle.
4. If the length of each side of a square is represented by $2x - 1$, represent the perimeter of the square.
5. If the length of each side of an equilateral pentagon (5 sides) is represented by $2x + 3$, represent the perimeter of the pentagon.
6. The length of each side of an equilateral hexagon (6 sides) is represented by $2x - 3$. Represent the perimeter of the hexagon.
7. The perimeter of an equilateral polygon is $12x - 24$. Express the length of one side if the polygon is:
   a. a triangle **b.** a square **c.** a hexagon

Solving Perimeter Problems

In solving problems dealing with the perimeters of geometric figures, it is helpful to draw the figures.

MODEL PROBLEM

The perimeter of a rectangle is 40 feet. The length is 2 more than 5 times the width. Find the dimensions of the rectangle.

Solution

Let w = the width of the rectangle.

Then, $5w + 2$ = the length of the rectangle.

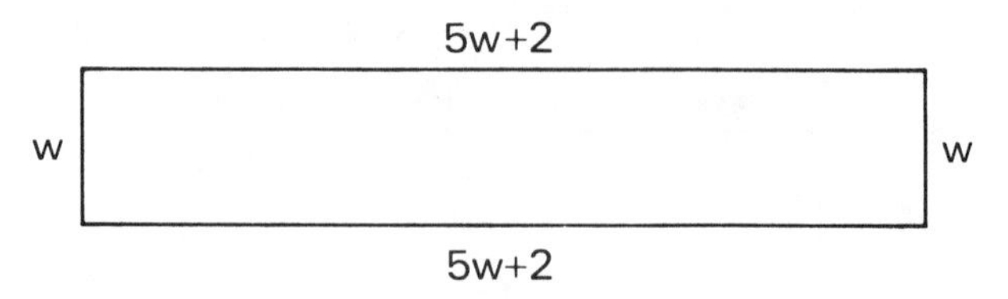

The sum of the lengths of all the sides is 40.

$$w + (5w + 2) + w + (5w + 2) = 40$$

$$\begin{aligned} w + 5w + 2 + w + 5w + 2 &= 40 \\ 12w + 4 &= 40 \\ 12w + 4 - 4 &= 40 - 4 \\ 12w &= 36 \\ w &= 3 \\ 5w + 2 &= 17 \end{aligned}$$

Check

$$3 + 17 + 3 + 17 \stackrel{?}{=} 40$$
$$40 = 40 \quad \text{(True)}$$
$$17 \stackrel{?}{=} 5(3) + 2$$
$$17 = 17 \quad \text{(True)}$$

Answer: The width is 3 feet; the length is 17 feet.

EXERCISES

1. The length of a rectangle is 3 times its width. The perimeter of the rectangle is 72 cm. Find the dimensions of the rectangle.
2. The length of a rectangle is $2\frac{1}{2}$ times its width. The perimeter of the rectangle is 84 cm. Find the dimensions of the rectangle.
3. The length of the second side of a triangle is 2 inches less than the length of the first side. The length of the third side is 12 inches more than the length of the first side. The perimeter of the triangle is 73 inches. Find the length of each side of the triangle.
4. Two sides of a triangle are equal in length. The length of the third side exceeds the length of one of the other sides by 3 centimeters. The perimeter of the triangle is 93 centimeters. Find the length of each of the shorter sides of the triangle.
5. The length of a rectangle is 5 meters more than its width. The perimeter is 66 meters. Find the dimensions of the rectangle.

6. The width of a rectangle is 3 yards less than its length. The perimeter is 130 yards. Find the length and the width of the rectangle.
7. The perimeter of a rectangular parking lot is 146 meters. Find its dimensions if the length is 7 meters less than 4 times the width.
8. The perimeter of a rectangular tennis court is 228 feet. If the length of the court exceeds twice its width by 6 feet, find its dimensions.
9. The length of the base of an isosceles triangle is 10 less than twice the length of one of its legs. If the perimeter of the triangle is 50, find the length of the base of the triangle.
10. The base of an isosceles triangle and one of its legs have lengths that are consecutive integers. The leg is longer than the base. The perimeter of the triangle is 20. Find the length of each side of the triangle.
11. The length of a rectangle is twice the width. If the length is increased by 4 inches and the width is decreased by 1 inch, a new rectangle is formed whose perimeter is 198 inches. Find the dimensions of the original rectangle.
12. The length of a rectangle exceeds its width by 4 feet. If the width is doubled and the length is diminished by 2 feet, a new rectangle is formed whose perimeter is 8 feet more than the perimeter of the original rectangle. Find the dimensions of the original rectangle.
13. A side of a square is 10 meters longer than the side of an equilateral triangle. The perimeter of the square is 3 times the perimeter of the triangle. Find the length of each side of the triangle.
14. The length of each side of a hexagon is 4 inches less than the length of a side of a square. The perimeter of the hexagon is equal to the perimeter of the square. Find the length of a side of the hexagon and the length of a side of the square.

10-7 SOLVING EQUATIONS THAT HAVE VARIABLES IN THE ANSWERS

An equation may contain more than one variable. Examples of such equations are $ax = b$, $x + c = d$, and $y - r = z$.

To solve such an equation for one of its variables means to express this particular variable in terms of the other variables. In order to plan the steps in the solution, it is helpful to use the strategy of looking at a simpler related problem, that is, compare the equation with a similar equation that contains only one variable. For example, in solving the equation $bx - c = d$ for x, compare it with $2x - 5 = 19$. The same operations are used in solving both equations.

MODEL PROBLEMS

1. Solve for x: $ax = b$ $[a \neq 0]$

Compare with $2x = 7$.	*Solution*	*Check*
$2x = 7$	$ax = b$	$ax = b$
$\frac{2x}{2} = \frac{7}{2}$	$\frac{ax}{a} = \frac{b}{a}$	$a\left(\frac{b}{a}\right) \stackrel{?}{=} b$
$x = \frac{7}{2}$	$x = \frac{b}{a}$ *Ans.*	$b = b$ (True)

2. Solve for x: $x + a = b$

Compare with $x + 5 = 9$.	*Solution*	*Check*
$x + 5 = 9$	$x + a = b$	$x + a = b$
$-5 = -5$	$-a = -a$	$b - a + a \stackrel{?}{=} b$
$x = 4$	$x = b - a$ *Ans.*	$b = b$ (True)

3. Solve for x: $2ax = 10a^2 - 3ax$

Compare with $2x = 10 - 3x$.	*Solution*
$2x = 10 - 3x$	$2ax = 10a^2 - 3ax$
$+3x = \quad + 3x$	$+3ax \quad + 3ax$
$5x = 10$	$5ax = 10a^2$
$\frac{5x}{5} = \frac{10}{5}$	$\frac{5ax}{5a} = \frac{10a^2}{5a}$
$x = 2$	$x = 2a$ *Ans.*

The check is left to the student.

EXERCISES

In 1–32, solve for x or y and check.

1. $5x = b$	**2.** $sx = 8$	**3.** $ry = s$
4. $3y = t$	**5.** $cy = 5$	**6.** $hy = m$
7. $x + 5 = r$	**8.** $x + a = 7$	**9.** $y + c = d$
10. $4 + x = k$	**11.** $d + y = 9$	**12.** $3x - q = p$
13. $x - 2 = r$	**14.** $y - a = 7$	**15.** $x - c = d$
16. $3x - e = r$	**17.** $cy - d = 4$	**18.** $ax + b = c$

19. $rx - s = 0$ **20.** $r + sy = t$ **21.** $m = 2(x + n)$
22. $4x - 5c = 3c$ **23.** $bx = 9b^2$ **24.** $cx + c^2 = 5c^2 - 3cx$
25. $bx - 5 = c$ **26.** $a = by + 6$ **27.** $ry + s = t$
28. $abx - d = 5d$ **29.** $rsx - rs^2 = 0$ **30.** $m^2x - 3m^2 = 12m^2$
31. $9x - 24a = 6a + 4x$ **32.** $8ax - 7a^2 = 19a^2 - 5ax$

10-8 TRANSFORMING FORMULAS

A formula may be expressed in more than one algebraic form. Sometimes, we must solve a formula for a variable different from the subject of the formula. This is called ***transforming*** the formula, or ***changing the subject*** of the formula. For example, the formula $D = 40t$ can be transformed into the equivalent formula $\frac{D}{40} = t$. In the formula $D = 40t$, D is expressed in terms of t; in the formula $t = \frac{D}{40}$, t is expressed in terms of D. If we know the value of D and wish to find the value of t, the computation is more convenient when we use the formula $t = \frac{D}{40}$. (See the first model problem following the Procedure.)

PROCEDURE. To transform a formula so that it is solved for a particular variable, consider the formula as an equation with several variables and solve it for the indicated variable in terms of the others.

MODEL PROBLEMS

1. a. Solve the formula $D = 40t$ for t.
b. Use the answer found in **a** to find the value of t when $D = 200$.

Solution:

a. $D = 40t$

$\frac{D}{40} = \frac{40t}{40}$

$\frac{D}{40} = t$

$t = \frac{D}{40}$ *Ans.*

b. $t = \frac{D}{40}$

$t = \frac{200}{40}$ $(D = 200)$

$t = 5$ *Ans.*

2. Solve the formula $V = \frac{1}{3}Bh$ for B.

Solution

$$V = \frac{1}{3}Bh$$
$$3V = 3\left(\frac{1}{3}Bh\right)$$
$$3V = Bh$$
$$\frac{3V}{h} = \frac{Bh}{h}$$
$$\frac{3V}{h} = B \quad \textit{Ans.}$$

3. Solve the formula $P = 2(L + W)$ for W.

Solution

$$P = 2(L + W)$$
$$P = 2L + 2W$$
$$P + (-2L) = 2L + 2W + (-2L)$$
$$P - 2L = 2W$$
$$\frac{P - 2L}{2} = \frac{2W}{2}$$
$$\frac{P - 2L}{2} = W \quad \textit{Ans.}$$

EXERCISES

In 1–19, transform the given formula by solving for the indicated letter.

1. $A = 6h$ for h **2.** $36 = bh$ for h **3.** $P = 4s$ for s
4. $D = rt$ for t **5.** $V = lwh$ for l **6.** $p = br$ for r
7. $A = BH$ for B **8.** $A = lw$ for l **9.** $V = lwh$ for h
10. $V = 4bh$ for h **11.** $i = prt$ for p **12.** $400 = BH$ for B
13. $A = \frac{1}{2}bh$ for h **14.** $V = \frac{1}{3}BH$ for H **15.** $S = \frac{1}{2}gt^2$ for g
16. $l = c - s$ for c **17.** $P = 2l + 2w$ for l **18.** $F = \frac{9}{5}C + 32$ for C
19. $2S = n(a + l)$ for a

20. If $A = BH$, express H in terms of A and B. (Solve for H.)
21. If $P = 2a + b$, express b in terms of P and a.
22. If $P = 2a + b + c$, express a in terms of the other variables.

In 23–26: **a.** Transform the given formula by solving for the variable to be evaluated. **b.** Using the result obtained in part **a**, substitute the given values to find the value of this variable.

23. If $LWH = 144$, find W when $L = 3$ and $H = 6$.

24. If $A = \frac{1}{2}bh$, find h when $A = 15$ and $b = 5$.

25. If $F = \frac{9}{5}C + 32$, find C when $F = 95$.

26. If $P = 2L + 2W$, find L when $P = 64$ and $W = 13$.

27. The formula for finding the area of a rectangle is $A = bh$. Rewrite this formula, replacing b by $4h$.

28. The formula for the area of a triangle is $A = \frac{1}{2}bh$. Rewrite this formula, replacing h by $4b$.

10-9 PROPERTIES OF INEQUALITY

The Order Property of Numbers

If two signed numbers x and y are graphed on a number line, only one of the following three situations can happen:

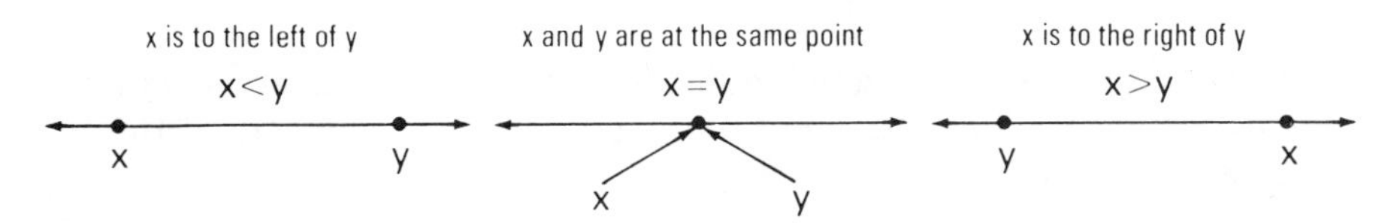

The graphs illustrate the ***order property of numbers***:

● **If x and y are two signed numbers, then one and only one of the following sentences is true:**

$$x < y \qquad\qquad x = y \qquad\qquad x > y$$

The Transitive Property of Inequality

From the graph at the right, you see that if x lies to the left of y and if y lies to the left of z, then x lies to the left of z.

The graph illustrates the ***transitive property of inequality***:

● **If x, y, and z are signed numbers, then:**

If $x < y$ and $y < z$, then $x < z$.
Similarly, if $z > y$ and $y > x$, then $z > x$.

The Addition Property of Inequality

Consider the following examples to observe the result of adding a number to both members of an inequality:

True Sentence	*Add to Both Members*	*Result*
$9 > 2$ order is $>$	$+3$ add positive	$12 > 5$ (True) same order as original
$2 < 9$ order is $<$	-3 add negative	$-1 < 6$ (True) same order as original

The examples illustrate the ***addition property of inequality***:

- **If x, y, and z are signed numbers, then:**

$$\textbf{If } x > y, \textbf{ then } x + z > y + z.$$
$$\textbf{If } x < y, \textbf{ then } x + z < y + z.$$

Since subtracting a signed number from both members of an inequality means adding its additive inverse to both members of the inequality, we can say:

- **When the same number is added to or subtracted from both members of an inequality, the order of the inequality remains unchanged.**

The Multiplication Property of Inequality

Consider the following examples to observe the result of multiplying both members of an inequality by a number:

True Sentence	*Multiply Both Members by:*	*Result*
$9 > 2$ order is $>$	$+3$ multiply by positive	$27 > 6$ (True) same order as original
$2 < 9$ order is $<$	$+3$ multiply by positive	$6 < 27$ (True) same order as original
$9 > 2$ order is $>$	-3 multiply by negative	$-27 > -6$ (False) $-27 < -6$ (True) opposite order from original
$2 < 9$ order is $<$	-3 multiply by negative	$-6 < -27$ (False) $-6 > -27$ (True) opposite order from original

The examples illustrate the ***multiplication property of inequality***:

- **If x, y, and z are signed numbers, then:**

  If $x > y$, then $xz > yz$ when z is positive ($z > 0$).
  If $x < y$, then $xz < yz$ when z is positive ($z > 0$).
  If $x > y$, then $xz < yz$ when z is negative ($z < 0$).
  If $x < y$, then $xz > yz$ when z is negative ($z < 0$).

Since dividing both members of an inequality by a nonzero signed number means to multiply by the reciprocal, or multiplicative inverse, of the number, we can say:

- **When both members of an inequality are multiplied or divided by a positive number, the order of the inequality remains unchanged; when both members are multiplied or divided by a negative number, the order of the inequality is reversed.**

EXERCISES

In 1–25, replace the question mark with the symbol > or < so that the resulting sentence will be true. All variables in Exercises 9–25 are nonzero signed numbers.

1. Since $8 > 2$, $8 + 1 \ ? \ 2 + 1$.
2. Since $-6 < 2$, $-6 + (-4) \ ? \ 2 + (-4)$.
3. Since $9 > 5$, $9 - 2 \ ? \ 5 - 2$.
4. Since $-2 > -8$, $-2 - \left(\frac{1}{4}\right) \ ? \ -8 - \left(\frac{1}{4}\right)$.
5. Since $7 > 3$, $\frac{2}{3}(7) \ ? \ \frac{2}{3}(3)$.
6. Since $-4 < 1$, $(-2)(-4) \ ? \ (-2)(1)$.
7. Since $-8 < 4$, $(-8) \div (4) \ ? \ (4) \div (4)$.
8. Since $9 > 6$, $(9) \div \left(-\frac{1}{3}\right) \ ? \ (6) \div \left(-\frac{1}{3}\right)$.
9. If $5 > x$, then $5 + 7 \ ? \ x + 7$.
10. If $y < 6$, then $y - 2 \ ? \ 6 - 2$.
11. If $20 > r$, then $4(20) \ ? \ 4(r)$.
12. If $t < 64$, then $t \div 8 \ ? \ 64 \div 8$.
13. If $x > 8$, then $-2x \ ? \ (-2)(8)$.
14. If $y < 8$, then $y \div (-4) \ ? \ 8 \div (-4)$.
15. If $x + 2 > 7$, then $x + 2 + (-2) \ ? \ 7 + (-2)$, or $x \ ? \ 5$.
16. If $y - 3 < 12$, then $y - 3 + 3 \ ? \ 12 + 3$, or $y \ ? \ 15$.
17. If $x + 5 < 14$, then $x + 5 - 5 \ ? \ 14 - 5$, or $x \ ? \ 9$.

18. If $2x > 8$, then $\frac{2x}{2}$? $\frac{8}{2}$, or x ? 4.

19. If $\frac{1}{3}y < 4$, then $3\left(\frac{1}{3}y\right)$? 3(4), or y ? 12.

20. If $-3x < 36$, then $\frac{-3x}{-3}$? $\frac{36}{-3}$, or x ? -12.

21. If $-2x > 6$, then $\left(-\frac{1}{2}\right)(-2x)$? $\left(-\frac{1}{2}\right)(6)$, or x ? -3.

22. If $x < 5$ and $5 < y$, then x ? y.

23. If $m > -7$ and $-7 > a$, then m ? a.

24. If $x < 10$ and $z > 10$, then x ? z.

25. If $a > b$ and $c < b$, then a ? c.

In 26–33, tell whether the statement is always, sometimes, or never true.

26. If $c > d$, then $c + a > d + a$.

27. If $r < s$, then $r - t < s - t$.

28. If $a > b$, then $ac > bc$.

29. If $w < d$, then $-w < -d$.

30. If $x > y$ and $z < 0$, then $\frac{x}{z} < \frac{y}{z}$.

31. If $cd > 0$, then $c > 0$ and $d > 0$.

32. If $ab < 0$, then $a < 0$ and $b < 0$.

33. If $b > c$ and $c > d$, then $b > d$.

10-10 FINDING AND GRAPHING THE SOLUTION SETS OF INEQUALITIES

Let us find the solution set of the inequality $2x > 8$, when x is a member of the set of signed numbers. To do this, we must find the set of all numbers each of which can replace x in the sentence $2x > 8$ and result in a true sentence.

$$2x > 8$$

If $x = 1$, then $2(1) > 8$, or $2 > 8$, is a false sentence.

If $x = 2\frac{1}{2}$, then $2\left(2\frac{1}{2}\right) > 8$, or $5 > 8$, is a false sentence.

If $x = 3\frac{3}{4}$, then $2\left(3\frac{3}{4}\right) > 8$, or $7\frac{1}{2} > 8$, is a false sentence.

If $x = 4$, then $2(4) > 8$, or $8 > 8$, is a false sentence.

If $x = 4.1$, then $2(4.1) > 8$, or $8.2 > 8$, is a *true* sentence.

If $x = 5$, then $2(5) > 8$, or $10 > 8$, is a *true* sentence.

Notice that if x is replaced by any number greater than 4, the resulting sentence is true. Therefore, the solution set of $2x > 8$ is the set of all signed numbers greater than 4. The solution set can be described, using set notation, as $\{x \mid x > 4\}$.

Every member of the solution set of $x > 4$ is also a member of the solution set of $2x > 8$. Therefore, we call $2x > 8$ and $x > 4$ ***equivalent inequalities***.

To find the solution set of an inequality, solve the inequality by using methods similar to those used in solving an equation. Use the properties of inequality to transform the given inequality into a simpler equivalent inequality whose solution set is evident. The following model problems show how this is done.

MODEL PROBLEMS

In 1–4, the domain of the variable is the set of signed numbers.

1. Find and graph the solution set of the inequality $x - 4 > 1$.

How to Proceed	*Solution*
(1) Write the inequality.	$x - 4 > 1$
(2) Add 4 to both members, and use the addition property of inequality.	$x - 4 + 4 > 1 + 4$ $x > 5$

Answer: The solution set is $\{x \mid x > 5\}$.

The graph of the solution set is shown below. Since 5 is not included in the graph, the circle at 5 is not filled in.

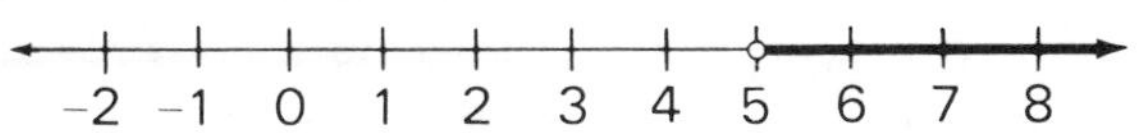

2. Find and graph the solution set of the inequality $x + 1 \leq 4$. Remember that $x + 1 \leq 4$ means $x + 1 < 4$ or $x + 1 = 4$. That is, $x + 1 \leq 4$ is equivalent to the disjunction

$$(x + 1 < 4) \vee (x + 1 = 4)$$

How to Proceed	*Solution*
(1) Write the inequality.	$x + 1 \leq 4$
(2) Add -1 to both members, and use the addition property of inequality.	$x + 1 + (-1) \leq 4 + (-1)$ $x \leq 3$

Check: If $x = 3$, then $x + 1 = 4$.
If $x < 3$, then $x + 1 < 4$.

Answer: The solution set is $\{x | x \leq 3\}$.

The graph of the solution set is shown at the right. Since 3 is included in the graph, the circle at 3 is filled in.

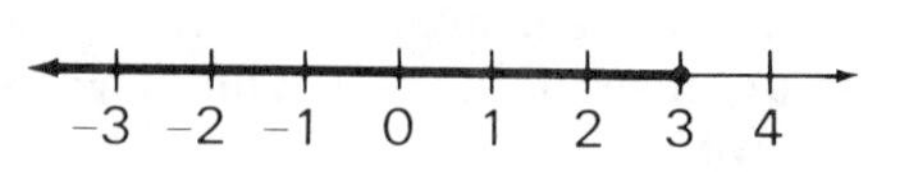

3. Find and graph the solution set of the inequality $5x + 4 \leq 11 - 2x$.

How to Proceed	*Solution*
(1) Write the inequality.	$5x + 4 \leq 11 - 2x$
(2) Add $2x$ to each member, and use the addition property of inequality.	$5x + 4 + 2x \leq 11 - 2x + 2x$ $7x + 4 \leq 11$
(3) Add -4 to both members, and use the addition property of inequality.	$7x + 4 + (-4) \leq 11 + (-4)$ $7x \leq 7$
(4) Multiply both members by $\frac{1}{7}$, and use the multiplication property of inequality.	$\frac{1}{7}(7x) \leq \frac{1}{7}(7)$ $x \leq 1$

The check is left to the student.

Answer: $\{x | x \leq 1\}$

The graph of the solution set is shown at the right. Note that 1 is included in the graph.

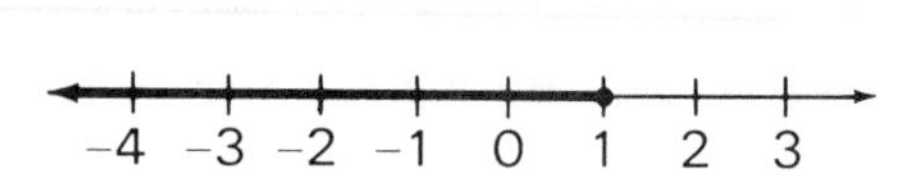

4. Find and graph the solution set of the inequality $2(2x - 8) - 8x \leq 0$.

How to Proceed	*Solution*
(1) Write the inequality.	$2(2x - 8) - 8x \leq 0$
(2) Use the distributive property.	$4x - 16 - 8x \leq 0$
(3) Combine like terms.	$-4x - 16 \leq 0$
(4) Add 16 to both members, and use the addition property of inequality.	$+ 16 = + 16$ $-4x \leq 16$

(5) Multiply both members by $-\frac{1}{4}$, and use the multiplication property of inequality. Remember to reverse the order of the inequality.

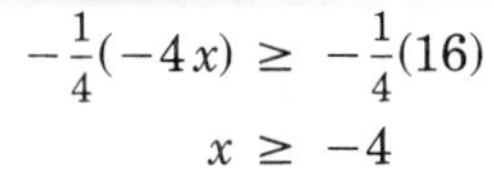

$$-\frac{1}{4}(-4x) \geq -\frac{1}{4}(16)$$
$$x \geq -4$$

The check is left to the student.

Answer: $\{x \mid x \geq -4\}$

The graph of the solution set is shown at the right. Note that -4 is included in the graph.

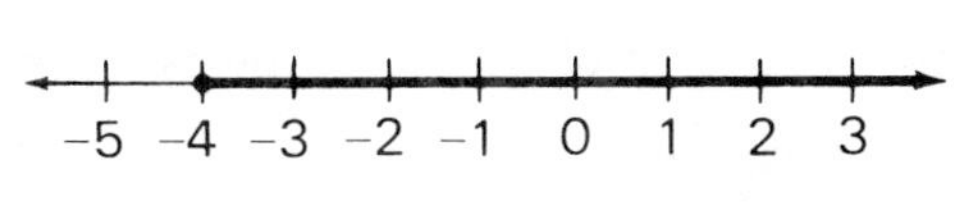

EXERCISES

In 1–51, find and graph the solution set of the inequality. Use the set of signed numbers as the domain of the variable.

1. $x - 2 > 4$ **2.** $z - 6 < 4$ **3.** $y - \frac{1}{2} > 2$

4. $x - 1.5 < 3.5$ **5.** $x + 3 > 6$ **6.** $19 < y + 17$

7. $d + \frac{1}{4} > 3$ **8.** $-3\frac{1}{2} > c + \frac{1}{2}$ **9.** $y - 4 \geq 4$

10. $25 \leq d + 22$ **11.** $3t > 6$ **12.** $2x \leq 12$

13. $15 \leq 3y$ **14.** $-10 \leq 4h$ **15.** $-6y < 24$

16. $27 > -9y$ **17.** $-10x > -20$ **18.** $12 \leq -1.2r$

19. $\frac{1}{3}x > 2$ **20.** $-\frac{2}{3}z \geq 6$ **21.** $\frac{x}{2} > 1$

22. $\frac{y}{3} \leq -1$ **23.** $\frac{1}{2} \leq \frac{z}{4}$ **24.** $-.4y \leq 4$

25. $-10 \geq 2.5z$ **26.** $2x - 1 > 5$ **27.** $3y - 6 \geq 12$

28. $5x - 1 > -31$ **29.** $-5 \leq 3y - 2$ **30.** $3x + 4 > 10$

31. $5y + 3 \geq 13$ **32.** $6c + 1 > -11$ **33.** $4d + 3 \leq 17$

34. $5x + 3x - 4 > 4$ **35.** $8y - 3y - 1 \leq 29$

36. $6x + 2 - 8x < 14$ **37.** $3x + 1 > 2x + 7$

38. $7y - 4 < 6 + 2y$ **39.** $4 - 3x \geq 16 + x$

40. $2x - 1 > 4 - \frac{1}{2}x$ **41.** $2c + 5 \geq 14 + 2\frac{1}{3}c$

42. $\frac{x}{3} - 1 \leq \frac{x}{2} + 3$ **43.** $4(x - 1) > 16$

44. $8x < 5(2x + 4)$

45. $12\left(\frac{1}{4} + \frac{x}{3}\right) > 15$

46. $8m - 2(2m + 3) \geq 0$

47. $12r - (8r - 20) > 12$

48. $3y - 6 \leq 3(7 + 2y)$

49. $5x \leq 10 + 2(3x - 4)$

50. $-3(4x - 8) > 2(3 + 2x)$

51. $4 - 5(y - 2) \leq -2(-9 + 2y)$

52. To which of the following is $y + 4 \geq 9$ equivalent?
(1) $y > 5$ (2) $y \geq 5$ (3) $y \geq 13$ (4) $y = 13$

53. To which of the following is $5x < 4x + 6$ equivalent?
(1) $x > 6$ (2) $x = 6$ (3) $x = \frac{6}{5}$ (4) $x < 6$

54. Which of the following is the smallest member of the solution set of $3x - 7 \geq 8$?
(1) 3 (2) 4 (3) 5 (4) 6

55. Which of the following is the largest member of the solution set of $4x \leq 3x + 2$?
(1) 1 (2) 2 (3) 3 (4) 4

In 56–59, write an inequality for the graph that is shown.

56.
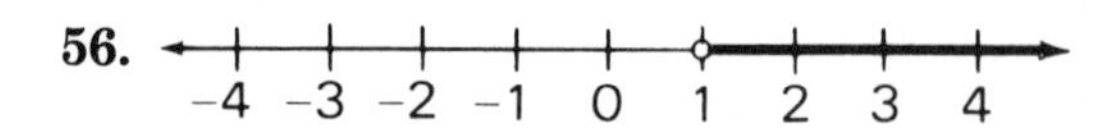

57.
-4 -3 -2 -1 0 1 2 3 4

58.
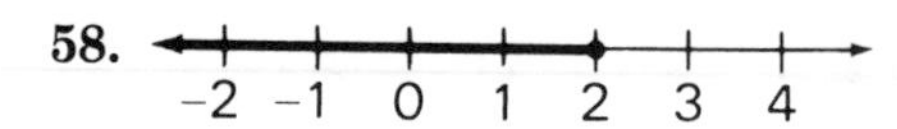

59.
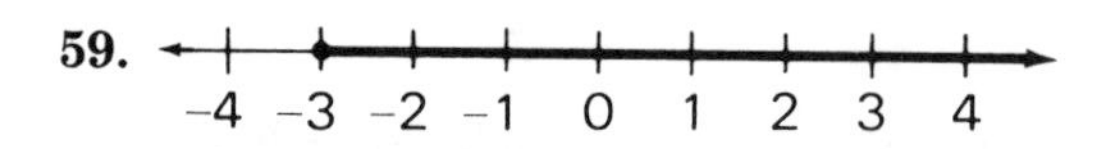

10-11 SOLVING VERBAL PROBLEMS BY USING INEQUALITIES

Preparing to Solve Problems Involving Inequalities

The following examples illustrate how to represent algebraically some sentences that involve relationships of inequality.

	Sentence	*Meaning*	*Representation*
1.	x is more than 25.	x is greater than 25.	$x > 25$
2.	x is under 25.	x is less than 25.	$x < 25$
3.	x is at least 25.	x is equal to 25, or x is greater than 25.	$x \geq 25$
4.	The minimum value of x is 25.	x is equal to 25, or x is greater than 25.	$x \geq 25$
5.	x is at most 25.	x is equal to 25, or x is less than 25.	$x \leq 25$
6.	The maximum value of x is 25.	x is equal to 25, or x is less than 25.	$x \leq 25$

EXERCISES

In 1–9, represent the sentence as an algebraic inequality.

1. x is less than or equal to 15.
2. y is greater than or equal to 4.
3. x is at most 50.
4. x is more than 50.
5. The greatest possible value of $3y$ is 30.
6. The sum of $5x$ and $2x$ is at least 70.
7. The maximum value of $4x - 6$ is 54.
8. The minimum value of $2x + 1$ is 13.
9. The product of $3x$ and $x + 1$ is less than 35.

In 10–16, write in set-builder notation the numbers that satisfy the given conditions.

10. Six less than a number is greater than 4.
11. Six less than a number is less than 4.
12. Six times a number is less than 72.
13. A number increased by 10 is greater than 50.
14. A number decreased by 15 is less than 35.
15. Twice a number, increased by 6, is less than 48.
16. Five times a number, decreased by 24, is greater than 3 times the number.

Solving Problems Involving Inequalities

MODEL PROBLEMS

1. Five times a number is less than 55. Find the greatest possible integer value for the number.

   Solution:

   Let x = the number.

   $$\underbrace{\textit{Five times a number}}_{5x}\ \underbrace{\textit{is less than}}_{<}\ \underbrace{\textit{55.}}_{55}$$

   $$x < 11$$

   Check: If a number is less than 11, 5 times the number will be less than 5(11), which is 55. The greatest integer less than 11 is 10, and $5(10) = 50$, which is less than 55.

   Answer: 10

2. The length of a rectangle is 5 cm more than its width. The perimeter of the rectangle is at least 66 cm. Find the minimum measures of the length and the width.

   Solution:

   If the perimeter is at least 66 cm, then the sum of the measures of the four sides is either equal to 66 cm or is greater than 66 cm.

   Let x = the width of the rectangle.
   Then, $x + 5$ = the length of the rectangle.

   $$\underbrace{\textit{The perimeter of the rectangle}}_{x + (x + 5) + x + (x + 5)}\ \underbrace{\textit{is at least}}_{\geq}\ \underbrace{\textit{66 cm.}}_{66}$$

   $$x + x + 5 + x + x + 5 \geq 66$$
   $$4x + 10 \geq 66$$
   $$4x \geq 56$$
   $$x \geq 14 \quad \text{(The width is at least 14 cm.)}$$
   $$x + 5 \geq 19 \quad \text{(The length is at least 19 cm.)}$$

   Answer: The length is at least 19 cm; the width is at least 14 cm.

   (The check is left to the student.)

EXERCISES

1. Six more than 4 times a whole number is less than 60. Find the maximum value of the number.
2. Six more than 2 times a certain number is less than the number increased by 20. Find the numbers that satisfy this condition.
3. Carol weighs 3 times as much as Sue. Both weights are whole numbers. The sum of their weights is less than 160 pounds. Find the greatest possible weight in pounds for each girl.
4. Mr. Burke had a sum of money in a bank. After he deposited an additional sum of $100, he had at least $550 in the bank. At least how much money did Mr. Burke have in the bank originally?
5. A club agreed to buy at least 250 tickets for a theatre party. If it agreed to buy 80 less orchestra tickets than balcony tickets, what was the least number of balcony tickets it could buy?
6. Mrs. Scott decided that she would spend no more than $120 to buy a jacket and a skirt. If the price of the jacket was $20 more than 3 times the price of the skirt, find the highest possible price of the skirt.
7. Three times a number increased by 8 is at most 40 more than the number. Find the greatest value of the number.
8. The length of a rectangle is 8 meters less than 5 times its width. If the perimeter of the rectangle is at most 104 meters, find the greatest possible width of the rectangle.
9. The length of a rectangle is 10 cm less than 3 times its width. If the perimeter of the rectangle is at most 180 cm, find the greatest possible length of the rectangle.
10. Mrs. Diaz wishes to save at least $1,500 in 12 months. If she saved $300 during the first 4 months, what is the least possible average amount that she must save in each of the remaining 8 months?
11. Two consecutive even numbers are such that their sum is greater than 98 decreased by twice the larger. Find the smallest possible values for the integers.
12. Joan needed $14 to buy some records. Her father agreed to pay her $3 an hour for gardening in addition to her $2 weekly allowance for helping around the house. What is the minimum number of hours Joan must work at gardening to earn $14 this week?
13. Fred bought 3 shirts, each at the same price, and received less than $2.00 change from a $20.00 bill. What is the minimum cost of one shirt?

14. Allison has 2 to 3 hours to spend on her homework. She has work in math, English, and social studies. She plans to spend an equal amount of time studying English and studying social studies. Allison plans to spend twice as much time studying math as the time spent in doing English.

a. What is the minimum time she can spend on English homework?

b. What is the maximum time she can spend on social studies?

c. What is the maximum time she can devote to math?

10-12 REVIEW EXERCISES

In 1–15, solve for the variable and check.

1. $x - 9 = 14$ **2.** $-18 = 3m$ **3.** $8c + 2 = 6$

4. $4 - x = -3$ **5.** $0.3d + 2 = 8$ **6.** $\frac{5}{8}b = -40$

7. $\frac{3}{5}p - 4 = 26$ **8.** $2(z - 6) = 14$ **9.** $8w = 60 - 4w$

10. $8w - 4w = 60$ **11.** $4h + 3 = 23 - h$ **12.** $5y + 3 = 2y$

13. $8z - (6z - 5) = 1$ **14.** $2(b - 4) = 4(2b + 1)$

15. $3(4x - 1) - 2 = 19$

In 16–19, solve for x in terms of a, b, and c.

16. $a + x = b + c$ **17.** $cx + a = b$

18. $bx - a = c$ **19.** $\frac{(a + c)x}{2} = b$

20. Solve for h in terms of A and b: $A = \frac{1}{2}bh$

21. If $P = 2l + 2w$, find w when $P = 17$ and $l = 5$.

22. If Joe got 36 points less than twice his test grade, he would have scored 100 points. Find Joe's test grade.

23. Five times a number is equal to 6 more than three times the number. Find the number.

In 24–31, find and graph the solution set of the inequality.

24. $\frac{1}{3}x < 1$ **25.** $-x \geq 4$

26. $6 + x > 3$ **27.** $2x - 3 \geq -5$

28. $\{x \mid -3 < x \leq 2\}$ **29.** $(x \geq 3) \wedge (x < 7)$

30. $(x \leq -2) \vee (x > 0)$ **31.** $(x - 4 \geq 1) \wedge (-2x > -18)$

In 32–35, tell if the statement is sometimes, always, or never true.

32. If $x > y$, then $a + x > a + y$.

33. If $x > y$, then $ax > ay$.

34. If $x > y$ and $y > z$, then $x > z$.

35. If $x > y$, then $-x > -y$.

In 36–38, select the numeral preceding the correct choice.

36. An inequality that is equivalent to $4x - 3 > 5$ is
(1) $x > 2$ (2) $x < 2$ (3) $x > \frac{1}{2}$ (4) $x < \frac{1}{2}$

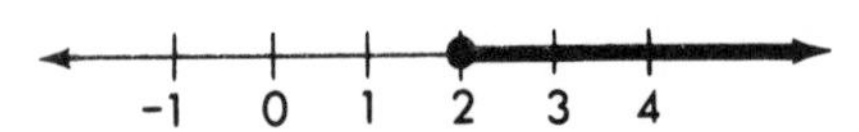

37. The solution set of which inequality is shown in the graph above?
(1) $x - 2 \geq 0$ (2) $x - 2 > 0$ (3) $x - 2 < 0$ (4) $x - 2 \leq 0$

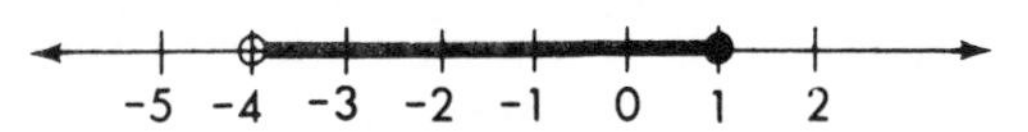

38. The above graph shows the solution set of which inequality?
(1) $-4 < x < 1$ (2) $-4 \leq x < 1$
(3) $-4 < x \leq 1$ (4) $-4 \leq x \leq 1$

39. The greater of two numbers is 8 more than three times the smaller. Their sum is at least 28. Find the smallest possible values of the numbers.

40. One of two numbers is 8 more than three times the other. Their sum is -28. Find the numbers.

41. The length of a rectangular room is 5 feet more than three times the width. The perimeter of the room is 62 feet. Find the dimensions of the room.

42. Find three consecutive integers such that three times the largest integer is twice the sum of the other two integers.

43. Find four consecutive odd integers such that the sum of the first three is one more than twice the largest.

44. Madeline spent 2 hours working on an essay for English class. On Tuesday, she spent 20 minutes longer than on Monday, and on Wednesday, she spent twice as long as she did on Tuesday. How long did she work each day if she finished the essay in the three days?

45. You have some pennies that you place in 4 piles. The first pile has 2 more pennies than the second, the second pile has 1 less than the third, and the fourth has twice as many as the second. What is the smallest number of pennies you could have?

Chapter 11

Geometry

In Chapter 1, we discussed some simple geometric figures, and we have been using these in our work. Now, we will add other geometric concepts.

11-1 POINTS, LINES, AND PLANES

Undefined Terms

We ordinarily define a word by using a simpler term. The simpler term can be defined by using one or more still simpler terms. But this process cannot go on endlessly. There comes a time when the definition must use a term whose meaning is *assumed* to be clear to all people. Because the meaning is accepted without definition, such a term is called an ***undefined term***.

In geometry, we are concerned with such ideas as *point*, *line*, and *plane*. Since we cannot give a satisfactory definition of these words using simpler defined words, we will consider them as *undefined terms*.

Although point, line, and plane are undefined words, you must have a clear understanding of what they mean. Knowing the properties and characteristics they possess helps you to do this.

The word ***point*** is often used to mean *place* or *position*. In geometry, a *point* is an idea. It has no length, no width, no thickness—only position.

A ***line*** may be considered as a set of points. The set of points that is chosen may form a curved line or it may form a straight line. Unless it is otherwise stated, the term *line* will mean *straight line.*

A ***plane*** is a special set of points suggested by a flat surface. A plane extends in all directions without end. The figure at the right represents a plane, called plane P. Actually, we have pictured only part of a plane because a plane has no boundaries.

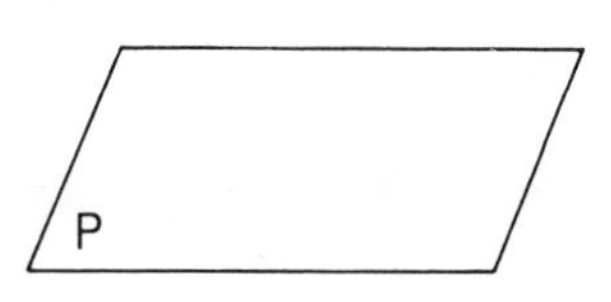

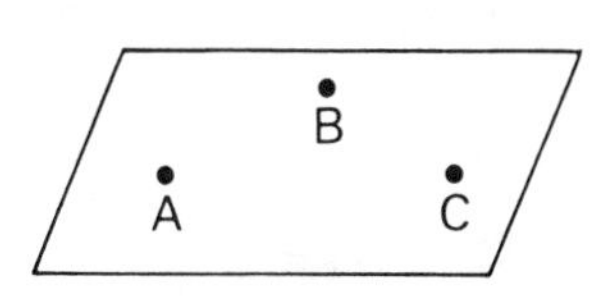

A plane may also be named by using letters that name three points in the plane, provided that the points are not on the same line. For example, the plane pictured at the left may be named plane ABC.

Using these three undefined terms, we define other geometric terms.

Line Segment

A ***line segment*** or ***segment*** is a part of a line consisting of two *endpoints* and all points on the line between these endpoints.

At the right is pictured a line segment whose endpoints are points R and S. We use these endpoints to name this segment, segment RS, which may be written as $\overline{RS}$.

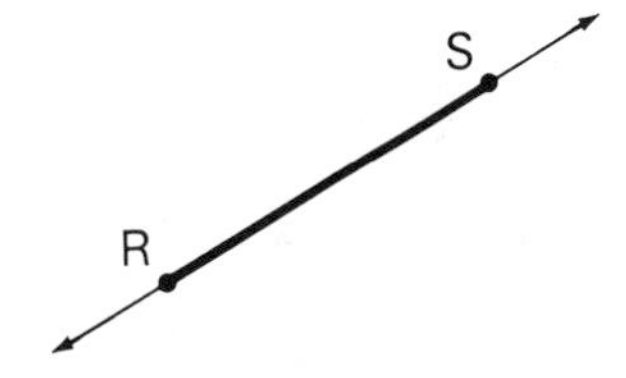

Ray

A ***ray*** is a part of a line that consists of a point on the line, called an endpoint, and all the points on one side of the endpoint. To name a ray we use two capital letters and an arrow with one arrowhead. The first letter must be the letter that names the endpoint. The second letter may be the name of any other point on the ray.

The figure at the right shows ray AB, which is represented in symbols by $\overrightarrow{AB}$.

Any point on a line is the common endpoint of two different rays, which are called ***opposite rays***. In the figure at the right, $\overrightarrow{PS}$ and $\overrightarrow{PR}$ are called opposite rays.

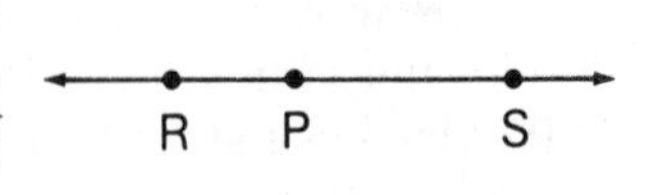

Facts About Straight Lines

A statement that is accepted as true without proof is called an ***axiom*** or a ***postulate***. If you examine the three figures pictured below, you will see that it is reasonable to accept the following three statements as postulates:

1. **In a plane, an infinite number of straight lines can be drawn through a given point.**

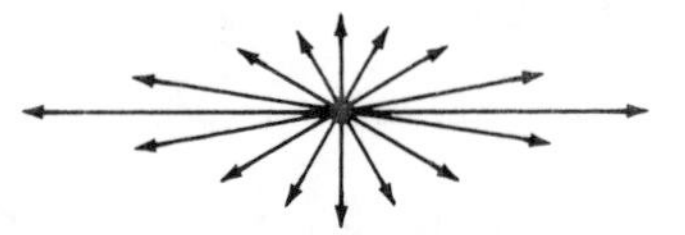

2. **One and only one straight line can be drawn that contains two given points.** (Two points determine a straight line.)

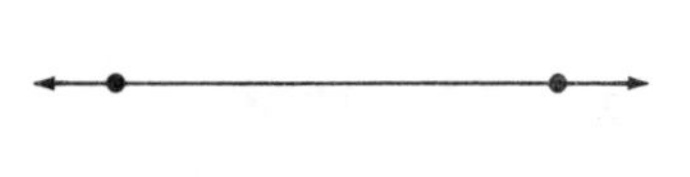

3. **Two different straight lines can intersect in only one point.**

Operations Involving Sets

The ***intersection*** of two sets of points is the set of all points that belongs to both of those sets of points. For example, the intersection of the set of points contained in line AB and the set of points contained in line CD is the set that has one element, point E. We can write $\overleftrightarrow{AB} \cap \overleftrightarrow{CD} = \{E\}$.

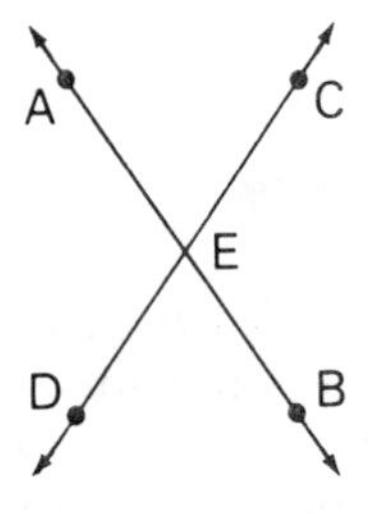

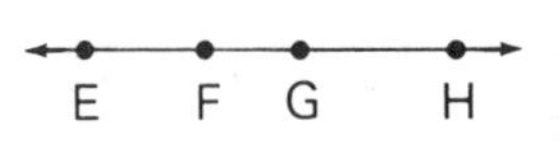

Since the intersection of segment EG and segment FH is segment FG, we can write $\overline{EG} \cap \overline{FH} = \overline{FG}$.

The ***union*** of two sets of points is the set of points that belongs to either or both of those sets of points. For example, the union of ray AB and ray AC is line BC. Hence, we can write $\overrightarrow{AB} \cup \overrightarrow{AC} = \overleftrightarrow{BC}$.

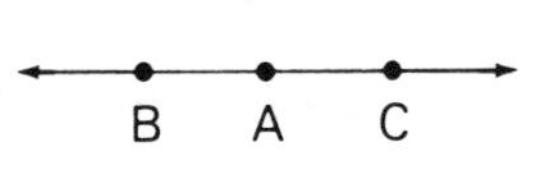

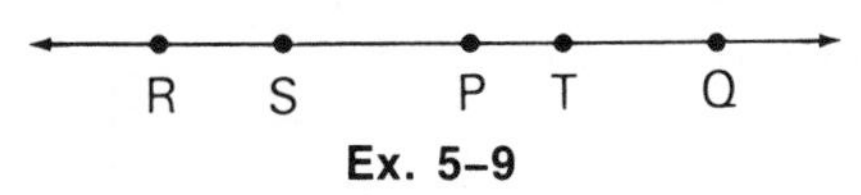

Since the union of segment EG and segment FH is segment EH, we can write $\overline{EG} \cup \overline{FH} = \overline{EH}$.

EXERCISES

In 1–4, write the meaning of the symbol.

1. $\overleftrightarrow{LM}$ **2.** $\overline{LM}$ **3.** $\overrightarrow{LM}$ **4.** LM

In 5–9, use the given figure:

R S P T Q

Ex. 5–9

5. Name two points on the same side of P.
6. Name two segments on the line.
7. Name two rays each of which has point T as an endpoint.
8. Name the opposite ray of $\overrightarrow{TQ}$.
9. Is point R in $\overrightarrow{SP}$?

10. State the number of endpoints that there are for
(**a**) a line segment (**b**) a ray (**c**) a line

11-2 ANGLES, ANGLE MEASURES, AND PERPENDICULARITY

An ***angle*** is a set of points that is the union of two rays having the same endpoint.

In the figure at the right, we can think of $\angle TOS$ as having been formed by rotating $\overrightarrow{OT}$. If $\overrightarrow{OT}$ is rotated in a counterclockwise direction about vertex O, it will assume the position $\overrightarrow{OS}$, forming $\angle TOS$.

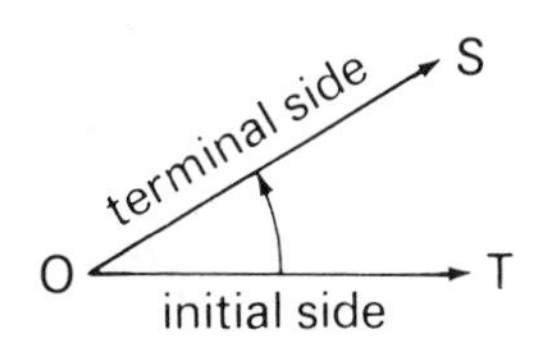

Measuring Angles

To measure an angle means to determine the number of units of measure it contains. A common standard unit of measure of an angle is a degree, written as $1°$. A degree is $\frac{1}{360}$ of a complete rotation of a ray about a point. Thus a complete rotation contains 360 degrees, written as $360°$.

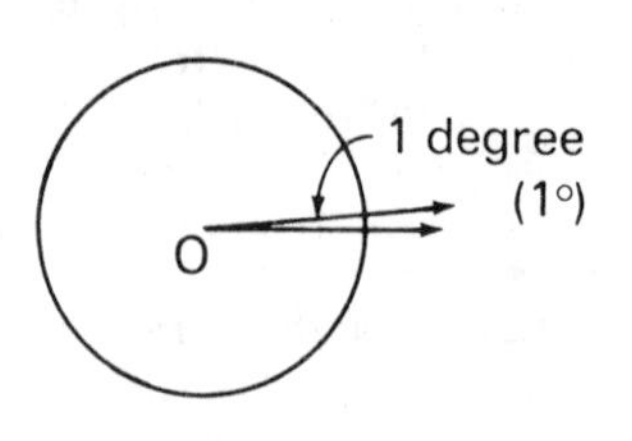

As discussed in Section 1-8, a *protractor* is an instrument used to measure angles.

The diagram shows that the measure of $\angle AOB$ is $45°$.

$$m\angle AOB = 45$$

Note: When we use the symbol for angle measure (as in $m\angle AOB$), we omit the degree symbol.

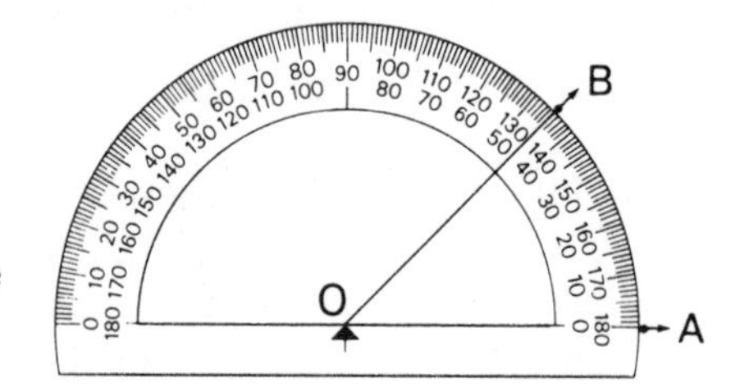

Types of Angles

Angles are classified according to their measures.

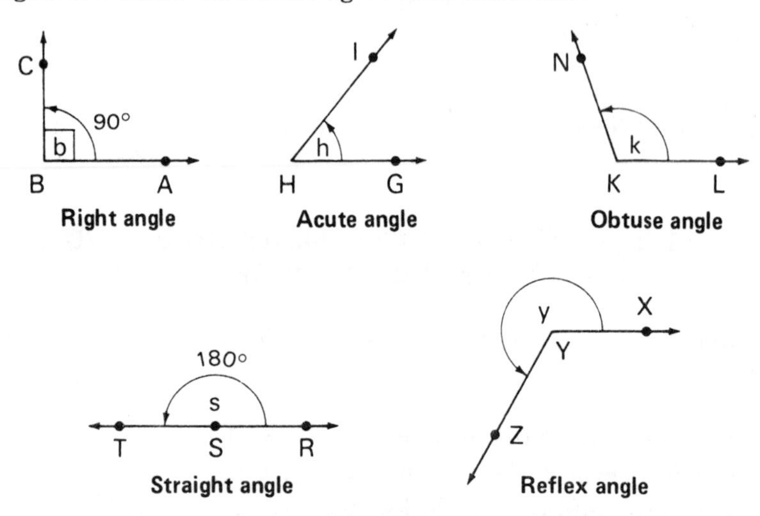

A ***right angle*** is an angle whose measure is $90°$. $\angle ABC$ pictures a right angle. Hence, we can say that $m\angle ABC = 90$, or $m\angle b = 90$. Note that the symbol ⏋ at B is used to show that $\angle ABC$ is a right angle.

An ***acute angle*** is an angle whose measure is greater than 0° and less than 90°. That is, its measure is between 0° and 90°. $\angle GHI$ is an acute angle. Hence, $0 < m\angle h < 90$.

An ***obtuse angle*** is an angle whose measure is greater than 90° and less than 180°. That is, its measure is between 90° and 180°. $\angle LKN$ is an obtuse angle. Hence, $90 < m\angle k < 180$.

A ***straight angle*** is an angle whose measure is 180°. $\angle RST$ is a straight angle. Hence, $m\angle RST = 180$, or $m\angle s = 180$.

A ***reflex angle*** is an angle whose measure is greater than 180° and less than 360°. That is, its measure is between 180° and 360°. Hence, $180 < m\angle y < 360$.

Note the following facts about angles:

1. The measure of an angle depends only upon the amount of rotation, not upon the pictured lengths of the rays forming the angle.
2. Since every right angle measures 90°, all right angles are equal in measure.
3. Since every straight angle measures 180°, all straight angles are equal in measure.

Perpendicularity

Two lines are ***perpendicular*** if and only if the two lines or parts of the lines intersect to form right angles. The symbol for perpendicular is $\perp$.

In the diagram, $\overleftrightarrow{PR}$ is perpendicular to $\overleftrightarrow{AB}$, symbolized by $\overleftrightarrow{PR} \perp \overleftrightarrow{AB}$.

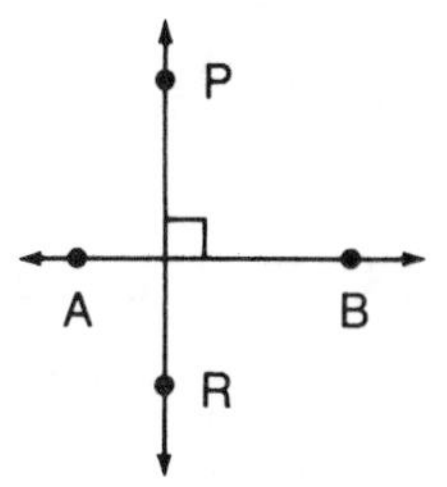

Segments of perpendicular lines that contain the point of intersection of the lines are also perpendicular. In the diagram, $\overline{PR} \perp \overleftrightarrow{AB}$. The symbol ⌉ is used to show that the lines indicated are perpendicular.

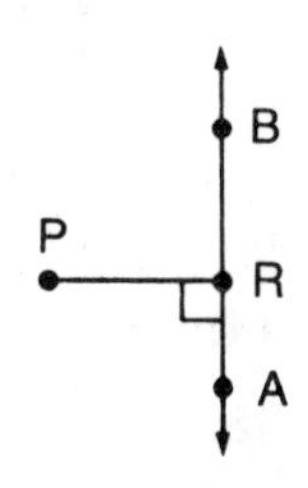

A line that is perpendicular to a line segment at the midpoint of the line segment is called the ***perpendicular bisector*** of the line segment. In the diagram, line k is the perpendicular bisector of $\overline{AB}$. Thus, line $k \perp \overline{AB}$ and $AM = MB$.

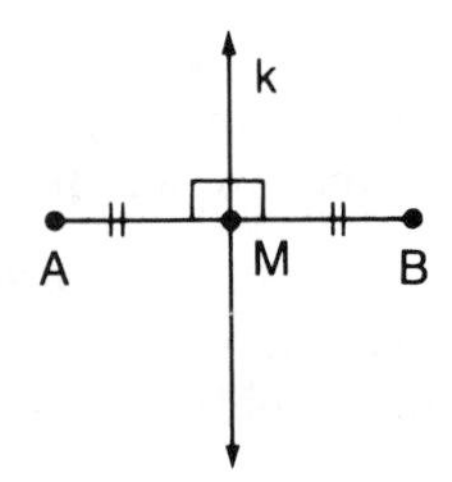

EXERCISES

In 1–5: **a.** Use a protractor to measure the angle. **b.** State what kind of angle it is.

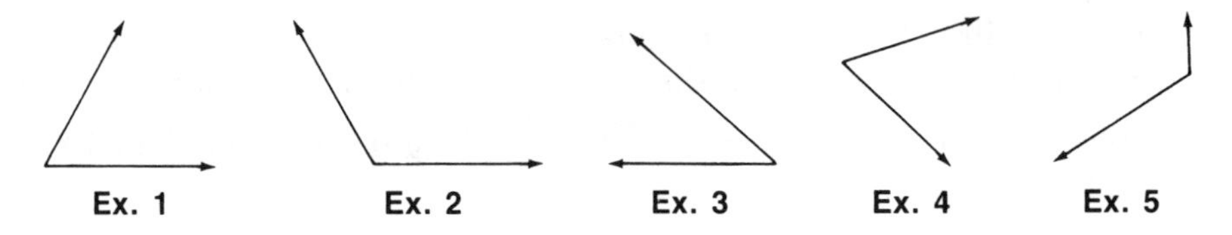

In 6–11, use a protractor to draw an angle whose measure is:

6. 30° **7.** 90° **8.** 48° **9.** 120° **10.** 180° **11.** 138°

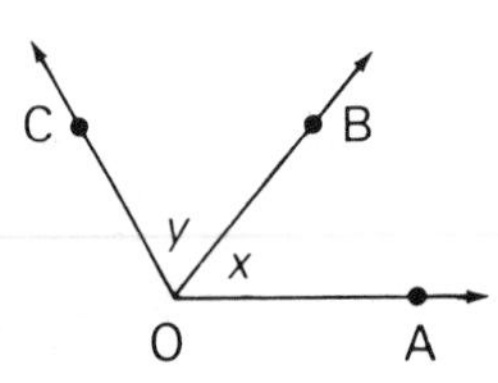

12. Using the figure at the left:
a. Name angle x, using 3 capital letters.
b. Give the shorter name for angle COB.
c. Name one acute angle.
d. Name one obtuse angle.

13. Find the number of degrees in:
a. $\frac{1}{2}$ of a complete rotation **b.** $\frac{3}{4}$ of a complete rotation
c. $\frac{1}{3}$ of a right angle **d.** $\frac{5}{6}$ of a straight angle

In 14–17, find the number of degrees in the angle formed by the hands of a clock at the given time.

14. 1 P.M. **15.** 4 P.M. **16.** 6 P.M. **17.** 5:30 P.M.

18. Name a time when the hands of a clock form an angle of 0°.

11-3 PAIRS OF ANGLES

An angle divides the points in a plane that are not points of the angle into two sets of points called regions. One region is called the ***interior of the angle***; the other is called the ***exterior of the angle***, as is illustrated in the figure at the right.

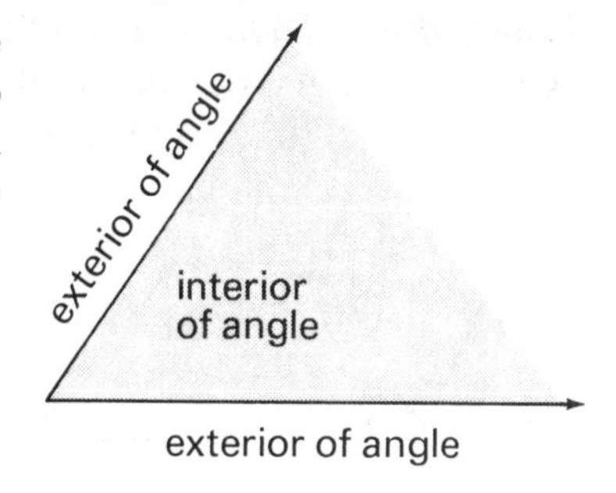

Adjacent Angles

Adjacent angles are two angles in the same plane that have a common vertex and a common side but do not have any interior points in common. In the figure at the right, $\angle ABC$ and $\angle CBD$ are adjacent angles.

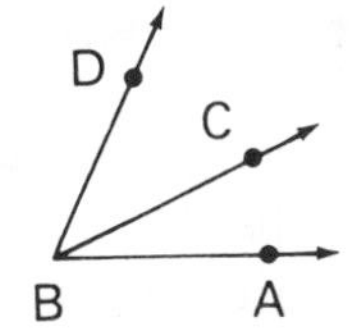

Complementary Angles

Two angles are ***complementary angles*** if and only if the sum of their measures is 90°. Each angle is the ***complement*** of the other. In the figures shown below, $\angle CAB$ and $\angle FDE$ are complementary angles because $m\angle CAB + m\angle FDE = 25 + 65 = 90$. Also, $\angle HGI$ and $\angle IGJ$ are complementary angles because $m\angle HGI + m\angle IGJ = 53 + 37 = 90$. If an angle contains 50°, its complement contains 90° − 50°, or 40°. If an angle contains x°, its complement contains $(90 - x)$°.

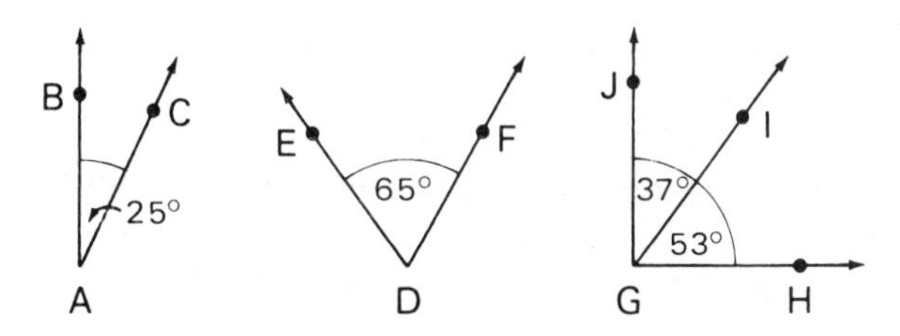

Supplementary Angles

Two angles are ***supplementary angles*** if and only if the sum of their measures is 180°. Each angle is the ***supplement*** of the other. As shown in the following figures, $\angle LKM$ and $\angle ONP$ are supplementary angles because m$\angle LKM$ + m$\angle ONP$ = 50 + 130 = 180. Also, $\angle RQS$ and $\angle SQT$ are supplementary angles because m$\angle RQS$ + m$\angle SQT$ = 115 + 65 = 180. If an angle contains 70°, its supplement contains 180° − 70°, or 110°. If an angle contains x°, its supplement contains $(180 - x)$°.

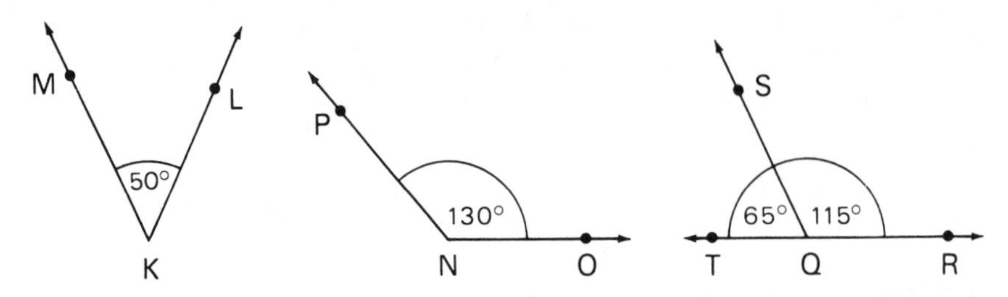

Linear Pair

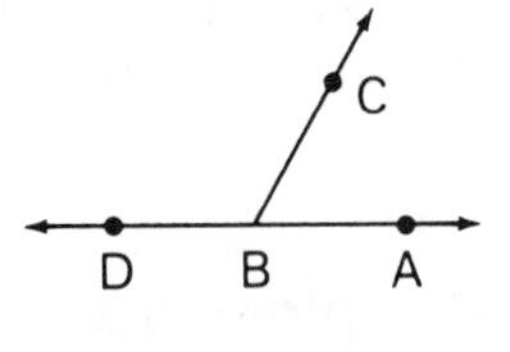

Two adjacent angles are a ***linear pair*** if and only if they have a common side and their remaining sides are opposite rays. In the figure at the right, $\angle ABC$ and $\angle CBD$ share $\overrightarrow{BC}$ as a common side. The remaining sides of these angles are $\overrightarrow{BA}$ and $\overrightarrow{BD}$, opposite rays which together form the straight line $\overleftrightarrow{AD}$. Notice that the term *linear* tells us that a *line* exists.

We also observe that a linear pair can be described as two adjacent angles that are supplementary. If m$\angle ABC$ = 56, then m$\angle CBD$ = 180 − 56 = 124. If m$\angle ABC = x$, then m$\angle CBD = 180 - x$.

Vertical Angles

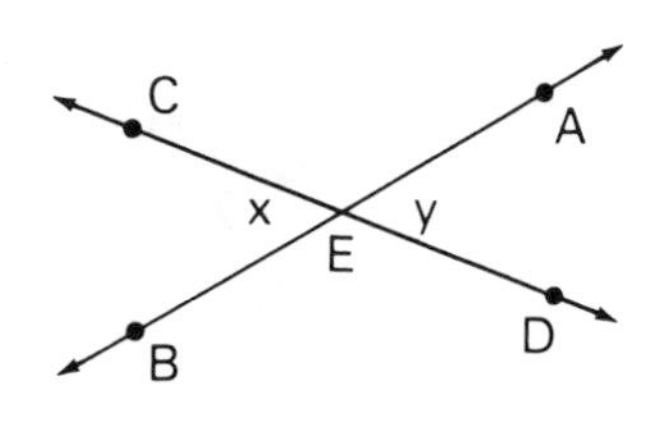

If two straight lines such as $\overleftrightarrow{AB}$ and $\overleftrightarrow{CD}$ intersect at E, $\angle x$ and $\angle y$ are opposite each other and share a common vertex at E. They are called a pair of ***vertical angles***. Two angles are *vertical angles* if and only if the sides of one angle are opposite rays of the sides of the second angle.

If two lines intersect, four angles are formed. In the diagram, $\overleftrightarrow{AB}$ and $\overleftrightarrow{CD}$ intersect at E. There are four linear pairs of angles:

$\angle AED$ and $\angle DEB$ $\qquad$ $\angle DEB$ and $\angle BEC$
$\angle BEC$ and $\angle CEA$ $\qquad$ $\angle CEA$ and $\angle AED$

The angles of each linear pair are supplementary.

If $m\angle y = 50$, then $m\angle AEC = 180 - 50 = 130$.
If $m\angle AEC = 130$, then $m\angle x = 180 - 130 = 50$.
Therefore, $m\angle x = m\angle y$.

- **When two angles have equal measure, they are congruent.**

We use the symbol $\cong$ to represent *is congruent to.* Here, we would write $\angle BEC \cong \angle AED$, read as "angle *BEC* is congruent to angle *AED*." Notice the different correct ways to indicate angles with equal measures:

1. *The angle measures are equal:* $m\angle BEC = m\angle AED$ *or* $m\angle x = m\angle y$
2. *The angles are congruent:* $\angle BEC \cong \angle AED$ *or* $\angle x \cong \angle y$

It would not be correct to say that the angles are equal, or that the angle measures are congruent.

If we were to draw and measure additional pairs of vertical angles, we would find in each case that the vertical angles would be equal in measure. No matter how many examples of a given situation we consider, we cannot *assume* that a conclusion that we draw in these examples will always be true. We must *prove* the conclusion. Statements that we prove are called ***theorems.***

Let us use algebraic expressions and properties to write an informal proof of the following statement.

- **If two lines intersect, the vertical angles formed are equal in measure, that is, they are congruent.**

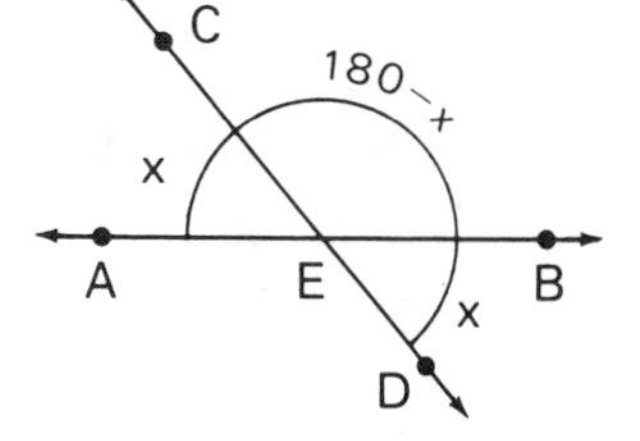

(1) If $\overleftrightarrow{AB}$ and $\overleftrightarrow{CD}$ intersect at E, then $\angle AEB$ is a straight angle whose measure is $180°$. Hence, $m\angle AEC + m\angle CEB = 180$.

(2) If $m\angle AEC = x$, then $m\angle CEB = 180 - x$.

(3) Likewise, $\angle CED$ is a straight angle whose measure is $180°$. Hence, $m\angle CEB + m\angle BED = 180$.

(4) Since $m\angle CEB = 180 - x$, then $m\angle BED = 180 - (180 - x) = 180 - 180 + x = x$.

(5) Since both $m\angle AEC = x$ and $m\angle BED = x$, then $m\angle AEC = m\angle BED$, that is, $\angle AEC \cong \angle BED$.

MODEL PROBLEMS

1. The measure of the complement of an angle is four times the measure of the angle. Find the measure of the angle.

Solution: Let x = the measure of the angle.
Then, $4x$ = the measure of the complement of the angle.

The sum of the measures of an angle and its complement is 90°.

$$\begin{aligned} x + 4x &= 90 \\ 5x &= 90 \\ x &= 18 \end{aligned}$$

Check: The measure of the first angle is 18°.
The measure of the second angle is 4(18°) or 72°.

The sum of the measures of the angles is 90°.
Hence, the angles are complementary.

Answer: The measure of the angle is 18°.

Note

The unit of measure is very important in the solution of a problem. While it is not necessary to include the unit of measure in each step of the solution, it is important that each term in an equation represent the same unit of measure and that the unit of measure be included in the answer. Since, in the statement about the sum of the measures of an angle and its complement that we used to write the equation, 90 is a measure in degrees, x and $4x$ must also represent measures in degrees.

2. Find the measure of an angle if its measure is 40° more than the measure of its supplement.

Solution: Let x = the measure of the supplement of the angle.
Then, $x + 40$ = the measure of the angle.

The sum of the measures of an angle and its supplement is 180°.

$$\begin{aligned} x + x + 40 &= 180 \\ 2x + 40 &= 180 \\ 2x &= 140 \\ x &= 70,\ x + 40 = 110 \end{aligned}$$

Answer: The measure of the angle is 110°.

3. The measures of a pair of vertical angles are represented by $5w - 20$ and $2w + 16$. **a.** Find the value of w. **b.** Find the measure of each angle.

Solution: Vertical angles are equal in measure.

$$5w - 20 = 2w + 16$$
$$3w - 20 = 16$$
$$3w = 36$$
$$w = 12$$

$$5w - 20 = 5(12) - 20 = 60 - 20 = 40$$
$$2w + 16 = 2(12) + 16 = 24 + 16 = 40$$

Check:

Since each angle has a measure of 40°, the vertical angles are equal in measure.

Answer: **a.** $w = 12$ **b.** Each angle measures 40°.

EXERCISES

Complementary Angles

In 1–10, write the measure of the complement of the angle whose measure is given.

1. 40° **2.** 25° **3.** 45° **4.** 69.5° **5.** $87\frac{1}{3}°$

6. $m°$ **7.** $d°$ **8.** $(90 - y)°$ **9.** $(x + 10)°$ **10.** $(x - 20)°$

In 11–14, $\angle A$ and $\angle B$ are complementary. Find the measure of each angle if their measures are represented by the given expressions. Solve the problem algebraically using an equation.

11. $m\angle A = x$, $m\angle B = 5x$ **12.** $m\angle B = x$, $m\angle A = x + 50$

13. $m\angle A = x$, $m\angle B = x - 40$ **14.** $m\angle B = y$, $m\angle A = 2y + 30$

In 15–19, solve the problem algebraically using an equation.

15. Two angles are complementary. One angle is twice as large as the other. Find the number of degrees in each angle.

16. The complement of an angle is 8 times as large as the angle. Find the measure of the complement.

17. The complement of an angle is one-fifth of the measure of the angle. Find the measure of the angle.

18. The complement of an angle measures 20° more than the angle. Find the number of degrees in the angle.
19. Find the number of degrees in an angle that measures 8° less than its complement.

Supplementary Angles

In 20–29, write the measure of the supplement of the angle whose measure is given.

20. 40° **21.** 69° **22.** 90° **23.** 110° **24.** $167\frac{1}{2}°$
25. $m°$ **26.** $c°$ **27.** $(2y)°$ **28.** $(180 - t)°$ **29.** $(x + 40)°$

In 30–33, $\angle A$ and $\angle B$ form a linear pair. (They are supplementary.) Find the measure of each angle if their measures are represented by the given expressions. Solve the problem algebraically using an equation.

30. $m\angle A = x$, $m\angle B = 3x$ **31.** $m\angle B = y$, $m\angle A = \frac{1}{2}y$
32. $m\angle A = w$, $m\angle B = w - 30$ **33.** $m\angle B = x$, $m\angle A = x + 80$

In 34–37, solve the problem algebraically using an equation.

34. Two angles are supplementary. The measure of one angle is twice as large as the measure of the other. Find the number of degrees in each angle.
35. The measure of the supplement of an angle is 40° more than the measure of the angle. Find the number of degrees in the supplement.
36. Find the number of degrees in the measure of an angle that is 20° less than 4 times the measure of its supplement.
37. The difference between the measures of two supplementary angles is 80°. Find the measure of the larger of the two angles.

38. The supplement of the complement of an acute angle is always
(1) an acute angle (2) a right angle
(3) an obtuse angle (4) a straight angle

Vertical Angles

In 39–43, $\overleftrightarrow{AB}$ and $\overleftrightarrow{CD}$ intersect at E. Find the measure of angle BEC when angle AED measures:

39. 30° **40.** 65° **41.** 90° **42.** 128.4° **43.** $175\frac{1}{2}°$

In 44–46, $\overleftrightarrow{MN}$ and $\overleftrightarrow{RS}$ intersect at T.

44. If $m\angle RTM = 5x$ and $m\angle NTS = 3x + 10$, find $m\angle RTM$.

45. If $m\angle MTS = 4x - 60$ and $m\angle NTR = 2x$, find $m\angle MTS$.

46. If $m\angle RTM = 7x + 16$ and $m\angle NTS = 3x + 48$, find $m\angle NTS$.

Miscellaneous

In 47–54, based on the given conditions, find the measure of each angle named.

47. *Given*: $\overleftrightarrow{EF} \perp \overrightarrow{GH}$; $m\angle EGI = 62$.

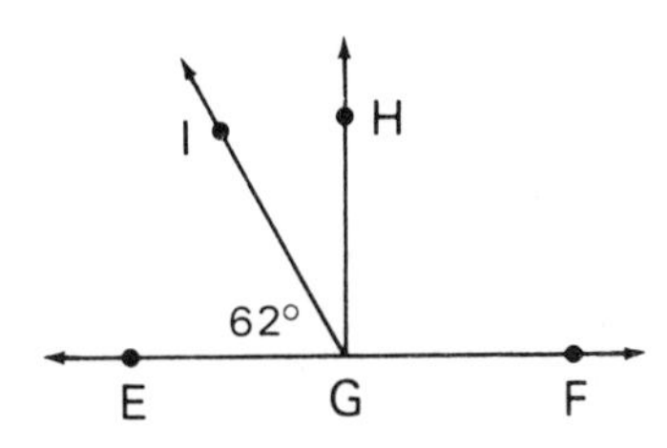

Find: $m\angle FGH$; $m\angle HGI$.

48. *Given*: $\overleftrightarrow{JK} \perp \overrightarrow{LM}$; $\overleftrightarrow{NLO}$ is a line; $m\angle NLM = 48$.

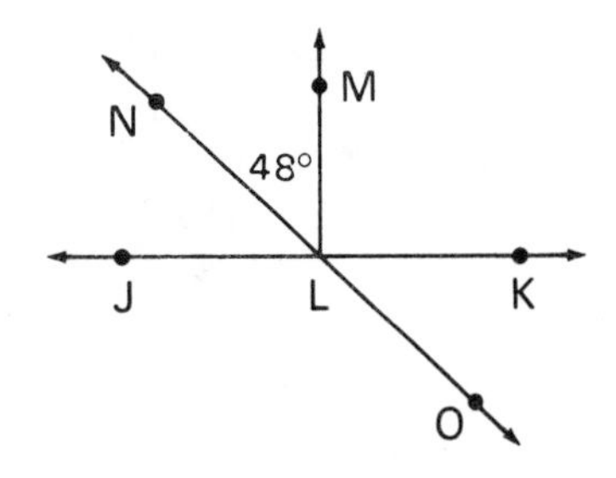

Find: $m\angle JLN$; $m\angle MLK$; $m\angle KLO$; $m\angle JLO$.

49. *Given*: $\angle GKH$ and $\angle HKI$ are a linear pair; $\overrightarrow{KH} \perp \overrightarrow{KJ}$; $m\angle IKJ = 34$.

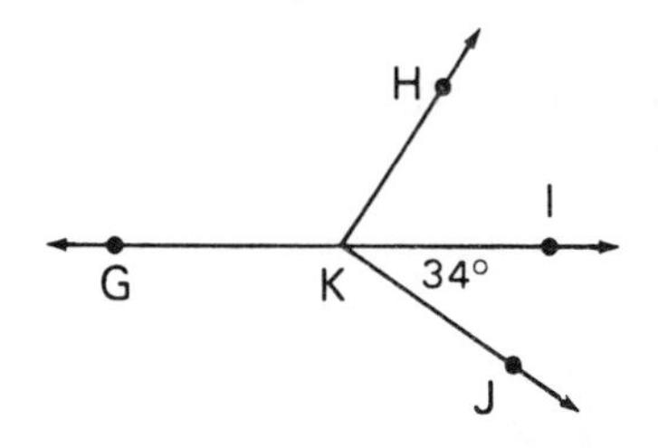

Find: $m\angle HKI$; $m\angle HKG$; $m\angle GKJ$.

50. *Given*: $\overrightarrow{MO} \perp \overrightarrow{MP}$; $\overleftrightarrow{LMN}$ is a line; $m\angle PMN = 40$.

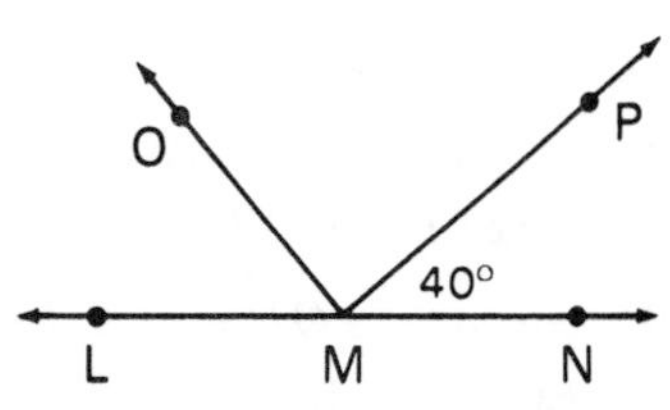

Find: $m\angle PMO$; $m\angle OML$.

51. *Given*: $\overleftrightarrow{RST} \perp \overrightarrow{SQ}$; $m\angle RSU = 89$.

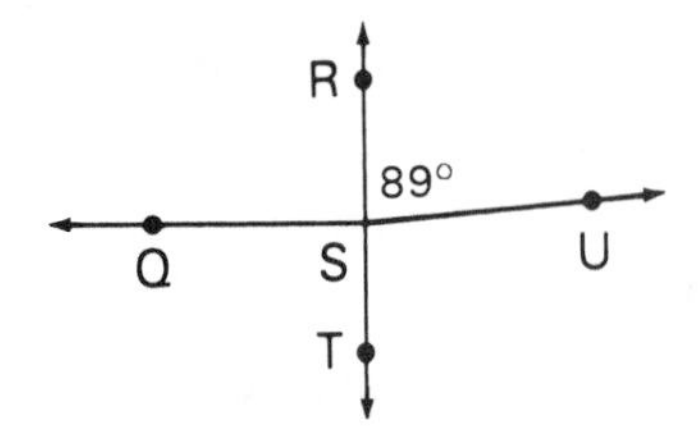

Find: $m\angle RSQ$; $m\angle QST$; $m\angle TSU$.

52. *Given*: lines $\overleftrightarrow{VWX}$ and $\overleftrightarrow{YWZ}$; $m\angle VWZ = 89$.

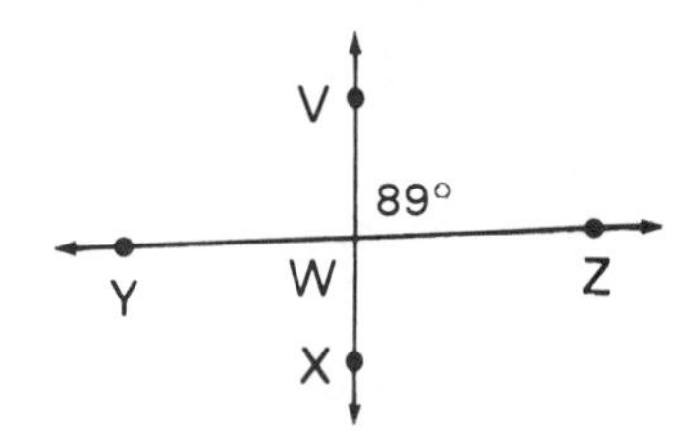

Find: $m\angle VWY$; $m\angle YWX$; $m\angle XWZ$.

53. *Given*: $\angle ABE$ and $\angle EBC$ form a linear pair; $m\angle EBC = 40$; $\angle ABD \cong \angle DBE$.

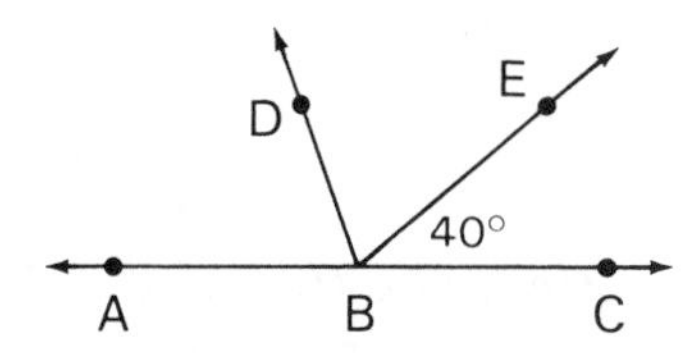

Find: $m\angle ABD$; $m\angle DBE$.

54. *Given*: $\overleftrightarrow{FI}$ intersects $\overleftrightarrow{JH}$ at K; $m\angle HKI = 40$; $\angle FKG \cong \angle FKJ$.

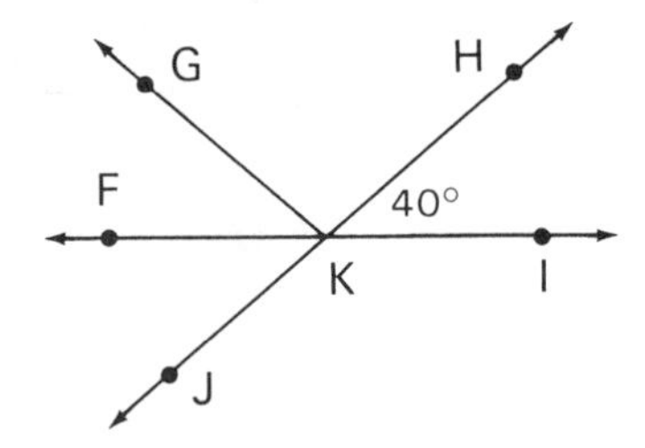

Find: $m\angle FKJ$; $m\angle FKG$; $m\angle GKH$; $m\angle JKI$.

In 55–58, sketch and label a diagram.

55. *Given*: $\overleftrightarrow{AB}$ intersects $\overleftrightarrow{CD}$ at E; $m\angle AED = 20$.
Find: $m\angle CEB$; $m\angle BED$; $m\angle CEA$.

56. *Given*: $\angle PQR$ and $\angle RQS$ are complementary; $m\angle PQR = 30$; $\overleftrightarrow{RQT}$ is a line.
Find: $m\angle RQS$; $m\angle SQT$; $m\angle PQT$.

57. *Given*: $\overleftrightarrow{LM}$ intersects $\overleftrightarrow{PQ}$ at R. The number of degrees in $\angle LRQ$ is 80° more than the number of degrees in $\angle LRP$.
Find: $m\angle LRP$; $m\angle LRQ$; $m\angle PRM$.

58. *Given*: $\overleftrightarrow{CD}$ is perpendicular to $\overleftrightarrow{AB}$ at E. Point F is in the interior of $\angle CEB$. The measure of $\angle CEF$ is 8 times the measure of $\angle FEB$.
Find: $m\angle FEB$; $m\angle CEF$; $m\angle AEF$.

11-4 ANGLES AND PARALLEL LINES

Not all lines in the same plane intersect. Two or more lines are called ***parallel lines*** if and only if the lines lie in the same plane and do not intersect.

In the figure at the right, $\overleftrightarrow{AB}$ and $\overleftrightarrow{CD}$ lie in the same plane and do not intersect. Hence, we say that **$\overleftrightarrow{AB}$ is parallel to $\overleftrightarrow{CD}$.** Using the symbol $\parallel$ for *is parallel to*, we write $\overleftrightarrow{AB} \parallel \overleftrightarrow{CD}$. When we speak of two parallel lines, we will mean two *distinct* lines. In higher courses, you will see that a line is parallel to itself.

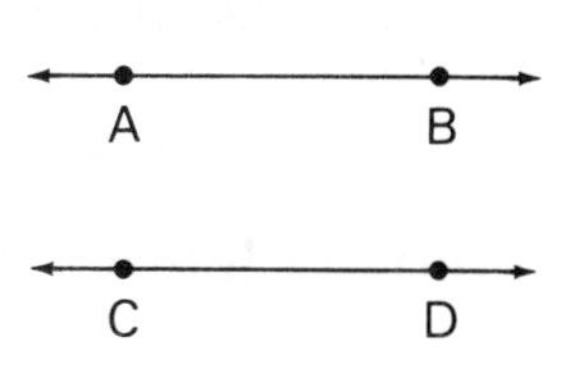

Line segments and rays are parallel if the lines that contain them are parallel.

If two lines such as $\overleftrightarrow{AB}$ and $\overleftrightarrow{CD}$ lie in the same plane, they must be either intersecting lines or parallel lines, as shown in the following figures.

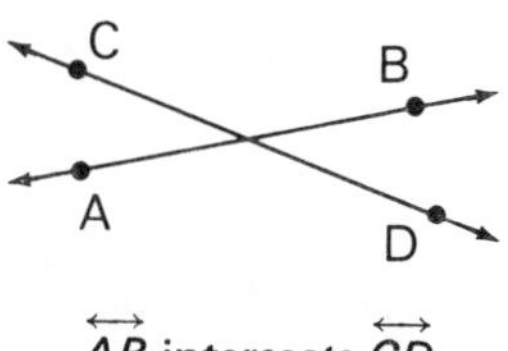

$\overleftrightarrow{AB}$ intersects $\overleftrightarrow{CD}$

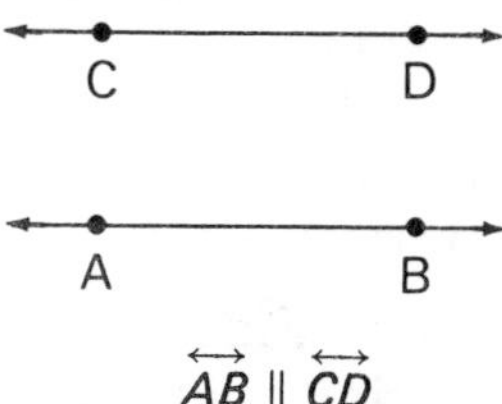

$\overleftrightarrow{AB} \parallel \overleftrightarrow{CD}$

Notice that when two lines such as $\overleftrightarrow{AB}$ and $\overleftrightarrow{CD}$ are parallel, they have no points in common. Hence, the intersection set of $\overleftrightarrow{AB}$ and $\overleftrightarrow{CD}$ is the empty set, symbolized as $\overleftrightarrow{AB} \cap \overleftrightarrow{CD} = \varnothing$.

When two lines are cut by a third line, called a ***transversal***, two sets of angles, each containing four angles, are formed.

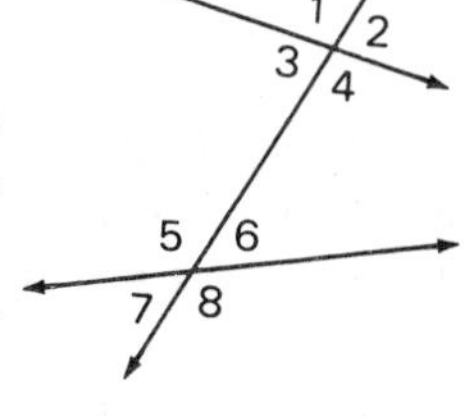

Angles 3, 4, 5, 6 are called *interior angles.*

Angles 1, 2, 7, 8 are called *exterior angles.*

Angles 4 and 5 are interior angles on opposite sides of the transversal and do not have the same vertex. They are called *alternate interior angles.* Angles 3 and 6 are another pair of alternate interior angles.

Angles 1 and 8 are exterior angles on opposite sides of the transversal and do not have the same vertex. They are called *alternate exterior angles.* Angles 2 and 7 are another pair of alternate exterior angles.

Angles 4 and 6 are *interior angles on the same side of the transversal.* Angles 3 and 5 are another pair of interior angles on the same side of the transversal.

Angles 1 and 5 are on the same side of the transversal, one interior and one exterior, and at different vertices. They are called *corresponding angles.* Other pairs of corresponding angles are 2 and 6, 3 and 7, 4 and 8.

Alternate Interior Angles and Parallel Lines

In the figure at the right, you see a transversal that intersects two parallel lines forming a pair of alternate interior angles, $\angle 3$ and $\angle 6$.

If you measure $\angle 3$ and $\angle 6$ with a protractor, you will find each angle measures $60°$. Hence, alternate interior angles 3 and 6 have equal measures, and $\angle 3 \cong \angle 6$. If you draw other pairs of parallel lines intersected by transversals, you would find again that pairs of alternate interior angles have equal measures. Yet, you would be hard pressed to prove that this is *always* true. Hence, we will accept, without proof, the following postulate:

- **If two parallel lines are cut by a transversal, then the alternate interior angles that are formed have equal measures, that is, they are congruent.**

Note that $\angle 4$ and $\angle 5$ are another pair of alternate interior angles. Hence, $\angle 4 \cong \angle 5$.

Corresponding Angles and Parallel Lines

If two lines are cut by a transversal, four pairs of corresponding angles are formed. One such pair of corresponding angles is $\angle 2$ and $\angle 6$, as shown in the figure at the right. If the original two lines are parallel, do these corresponding angles have equal measures? We are ready to prove in an informal manner that they do.

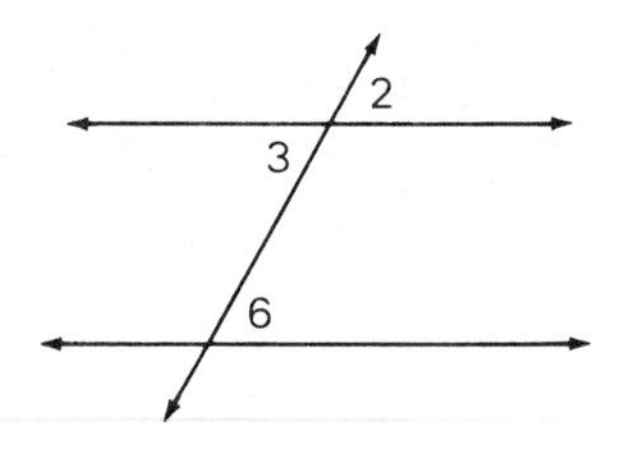

(1) Let $m\angle 2 = x$.

(2) If $m\angle 2 = x$, then $m\angle 3 = x$ (because $\angle 2$ and $\angle 3$ are vertical angles, which we have previously shown must have equal measures).
(3) If $m\angle 3 = x$, then $m\angle 6 = x$ (because $\angle 3$ and $\angle 6$ are alternate interior angles of parallel lines, and we have just accepted the postulate that they have the same measure).
(4) Therefore $m\angle 2 = m\angle 6$ (because the measure of each angle is x).

Hence, we have proved informally the following theorem:

● **If two parallel lines are cut by a transversal, then the corresponding angles formed have equal measures, that is, they are congruent.**

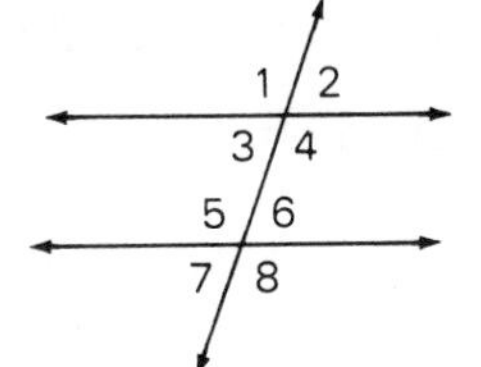

Note that this theorem is true for all pairs of corresponding angles: $\angle 1 \cong \angle 5$; $\angle 2 \cong \angle 6$; $\angle 3 \cong \angle 7$; $\angle 4 \cong \angle 8$.

Alternate Exterior Angles and Parallel Lines

If two parallel lines are cut by a transversal, we can prove informally that the alternate exterior angles formed have equal measures. One such pair of alternate exterior angles is $\angle 2$ and $\angle 7$, as shown in the figure at the right.

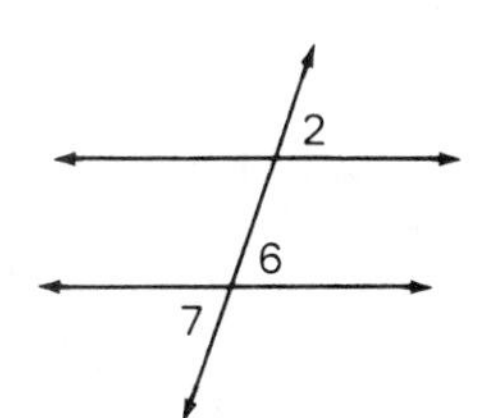

(1) Let $m\angle 2 = x$.
(2) If $m\angle 2 = x$, then $m\angle 6 = x$ (because $\angle 2$ and $\angle 6$ are corresponding angles of parallel lines, proven to have the same measure).
(3) If $m\angle 6 = x$, then $m\angle 7 = x$ (because $\angle 6$ and $\angle 7$ are vertical angles, previously proven to have the same measure).
(4) Therefore $m\angle 2 = m\angle 7$ (because the measure of each angle is x).

Hence, we have proven informally the theorem:

● **If two parallel lines are cut by a transversal, then the alternate exterior angles formed have equal measures, that is, they are congruent.**

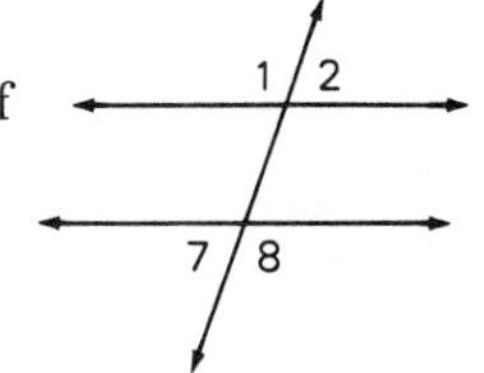

Note that this theorem is true for all pairs of alternate exterior angles: $\angle 1 \cong \angle 8$; $\angle 2 \cong \angle 7$.

Interior Angles on the Same Side of the Transversal

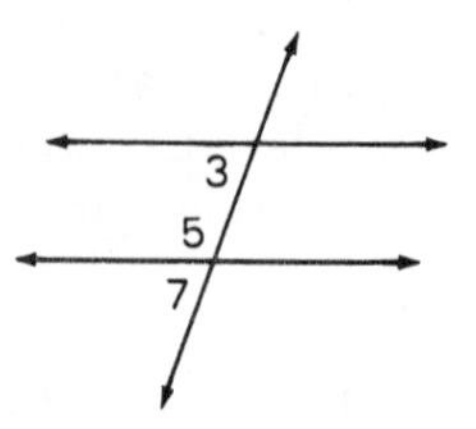

When two parallel lines are cut by a transversal, we can prove informally that the sum of the measures of the interior angles on the same side of the transversal will be 180°. One such pair of interior angles on the same side of the transversal is $\angle 3$ and $\angle 5$, as shown in the figure at the right.

(1) $m\angle 5 + m\angle 7 = 180$ ($\angle 5$ and $\angle 7$ are supplementary angles).

(2) $m\angle 7 = m\angle 3$ ($\angle 7$ and $\angle 3$ are corresponding angles).

(3) $m\angle 5 + m\angle 3 = 180$ (by substituting $m\angle 3$ for $m\angle 7$).

Hence, we have proven informally the theorem:

● **If two parallel lines are cut by a transversal, then the sum of the measures of the interior angles on the same side of the transversal is 180°.**

MODEL PROBLEM

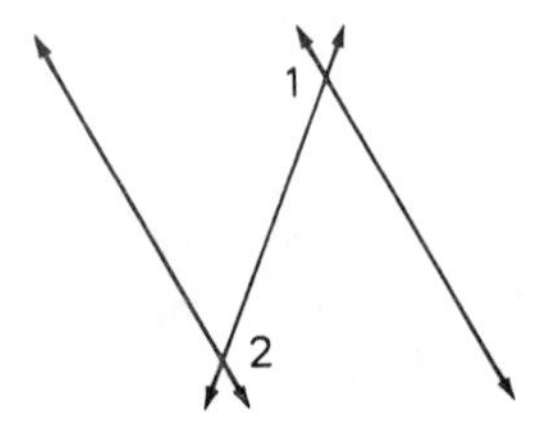

In the figure, the parallel lines are cut by a transversal. If $m\angle 1 = (5x - 10)$ and $m\angle 2 = (3x + 60)$, find the measures of $\angle 1$ and $\angle 2$.

Solution

Since the lines are parallel, the alternate interior angles, $\angle 1$ and $\angle 2$, have equal measures.

Hence: $5x - 10 = 3x + 60$
$5x = 3x + 70$
$2x = 70$
$x = 35$

Substitute: $5x - 10 = 5(35) - 10 = 175 - 10 = 165$
$3x + 60 = 3(35) + 60 = 105 + 60 = 165$

Answer: $m\angle 1 = 165$ and $m\angle 2 = 165$.

EXERCISES

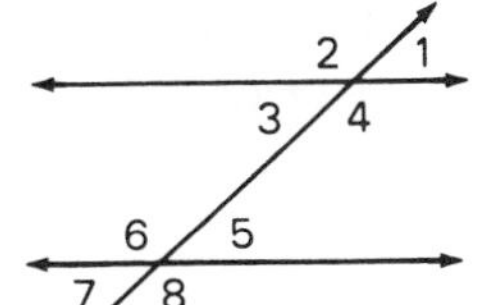

Ex. 1–4

In 1–4, the figure at the right shows two parallel lines cut by a transversal. In each exercise, find the measure of the remaining seven angles.

1. $m\angle 3 = 80$ **2.** $m\angle 6 = 150$
3. $m\angle 5 = 60$ **4.** $m\angle 1 = 75$

5. If $\overleftrightarrow{AB} \parallel \overleftrightarrow{CD}$, $m\angle 5 = 40$, and $m\angle 4 = 30$, find the measures of the remaining angles in the figure.

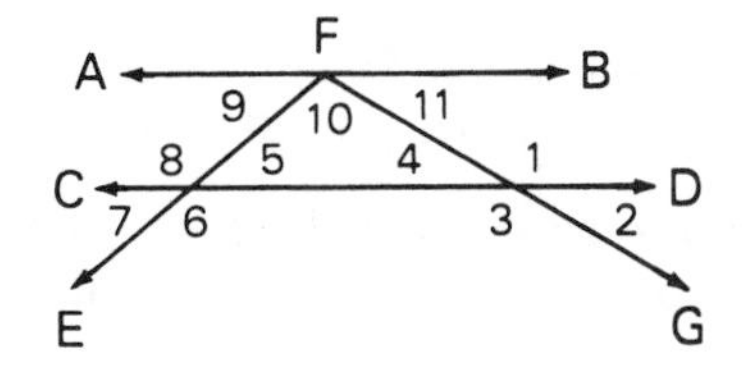

In 6–10, the figure at the right shows two parallel lines cut by a transversal. In each exercise, find the measures of all eight angles under the given conditions.

6. $m\angle 3 = 2x + 40$ and $m\angle 7 = 3x + 20$

7. $m\angle 4 = 4x - 10$ and $m\angle 6 = x + 80$

8. $m\angle 4 = 3x + 40$ and $m\angle 5 = 2x$

9. $m\angle 3 = 2x - 10$ and $m\angle 1 = x + 60$

10. $\angle 8 \cong \angle 3$

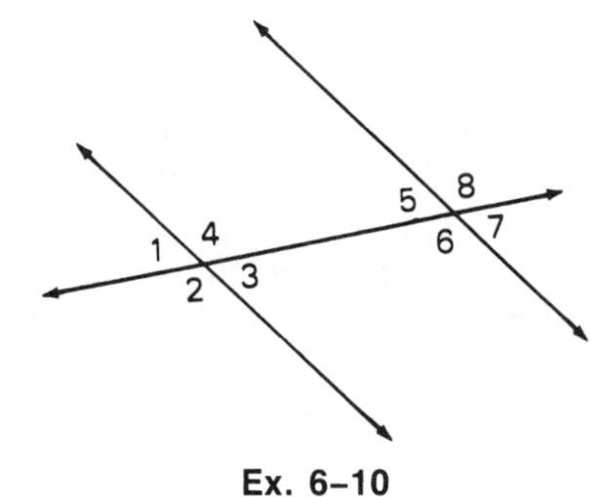

Ex. 6–10

In 11–16, tell whether the statement is always, sometimes, or never true.

11. If two distinct lines intersect, then they are parallel.
12. If two distinct lines do not intersect, then they are parallel.
13. If two angles are alternate interior angles, then they are on opposite sides of the transversal.
14. If two parallel lines are cut by a transversal, then the alternate interior angles are congruent.
15. If two parallel lines are cut by a transversal, then the alternate interior angles are complementary.
16. If two parallel lines are cut by a transversal, then the corresponding angles are supplementary.

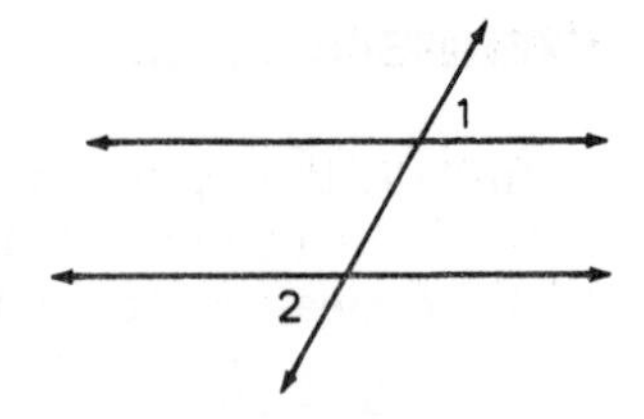

17. In the figure at the right, two parallel lines are cut by a transversal. Write an informal proof that demonstrates that $\angle 1$ and $\angle 2$ have equal measures.

11-5 GEOMETRIC FIGURES

Any set of points is a ***geometric figure***. A geometric figure may be a set of one point or a set of many points.

Plane geometric figures are figures all of whose points are in the same plane. Plane geometric figures can be pictured on a flat surface.

Curves

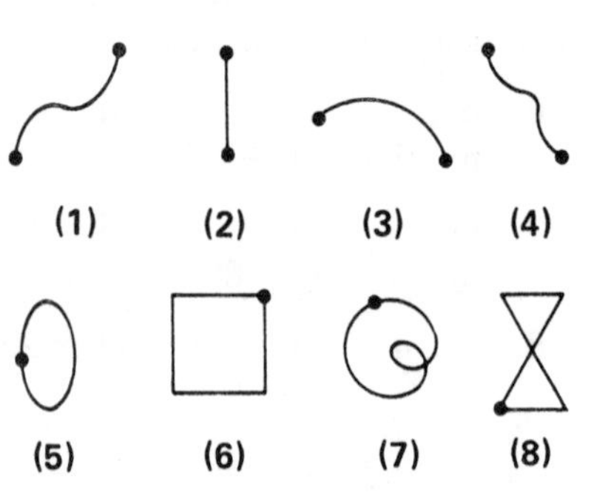

If a picture of a set of points can be drawn without removing the pencil from the paper, the figure is called a ***curve***. A curve that starts and ends at the same point is called a ***closed curve***. Among the eight curves pictured at the right only curves (5), (6), (7), and (8) are closed curves. Observe that curves (7) and (8) cross themselves; curves (5) and (6) do not cross themselves. Closed curves that do not cross themselves are called ***simple closed curves***.

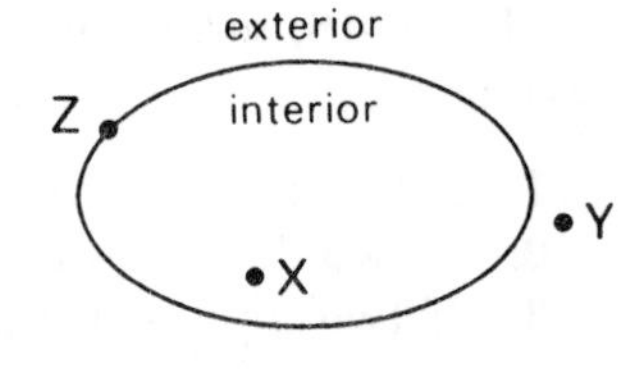

As shown in the figure at the right, a simple closed plane curve divides the plane into three sets of points:

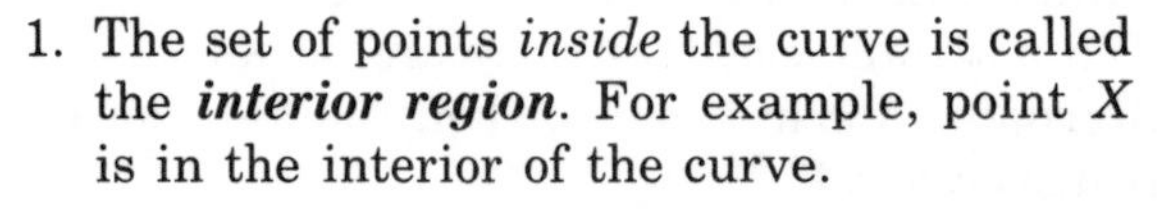

1. The set of points *inside* the curve is called the ***interior region***. For example, point X is in the interior of the curve.
2. The set of points *outside* the curve is called the ***exterior region***. For example, point Y is in the exterior of the curve.
3. The set of points that are *in* the curve is called the ***boundary*** between the interior and the exterior. For example, point Z is in the boundary of the curve.

Polygons

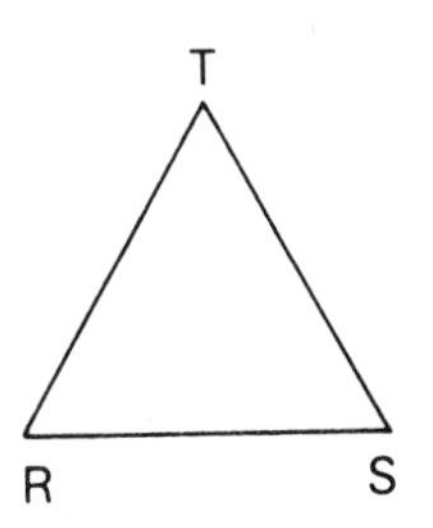

A ***polygon*** is a simple closed plane curve that consists of line segments. In a polygon, each line segment is called a ***side*** of the polygon. A common endpoint of two line segments is called a ***vertex*** of the polygon. We can name a polygon by naming each vertex with a capital letter. In the figure at the right, the polygon is named RST. Its sides are the line segments $\overline{RS}$, $\overline{ST}$, and $\overline{TR}$. Its vertices are R, S, and T.

As shown in the following figures, polygons are classified according to the number of sides.

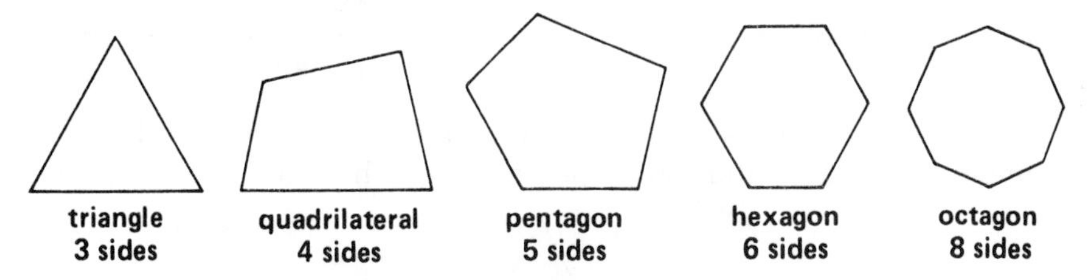

Other polygons with special names are *heptagon* (7 sides), *nonagon* (9 sides), and *decagon* (10 sides).

A ***regular polygon*** is a polygon in which all of the sides have equal measures, and all of its angles have equal measures. That is, a regular polygon is equilateral and equiangular. For example, in the figures pictured above, the triangle, the hexagon, and the octagon are regular polygons; the quadrilateral and the pentagon are not regular polygons. Of course, there are regular pentagons and regular quadrilaterals. For example, a square is a regular quadrilateral.

Previously, we said that two angles are congruent if their measures are equal. Similarly, we say that two line segments are congruent if their measures are equal. Hence, we can say that in a regular polygon all the sides are congruent and all the angles are congruent.

Notice the commonly accepted ways to indicate that two distinct line segments such as $\overline{AB}$ and $\overline{CD}$ have equal measures:

1. *The line segments are congruent:* $\overline{AB} \cong \overline{CD}$
2. *The measures of the segments are equal:* $AB = CD$

It is correct to say that $m\overline{AB} = m\overline{CD}$, but this symbolism is cumbersome. It is not correct to say that line segments $\overline{AB}$ and $\overline{CD}$ are equal.

EXERCISES

1. Which of the figures shown below are **(a)** closed curves **(b)** simple closed curves?

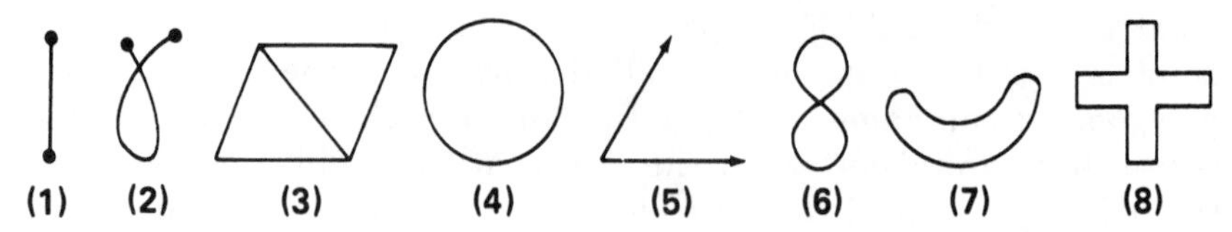

2. In the figure:

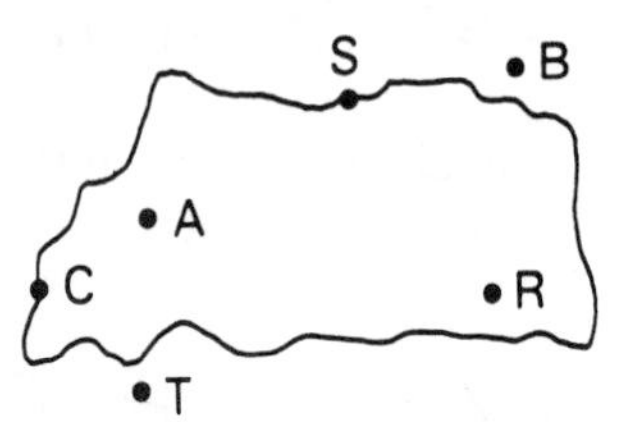

   a. Name the points in the curve.
   b. Name the points in the interior of the curve.
   c. Name the points in the exterior of the curve.

3. Which of the figures shown below represent **(a)** polygons **(b)** a triangle **(c)** a quadrilateral **(d)** a pentagon **(e)** a hexagon **(f)** an octagon?

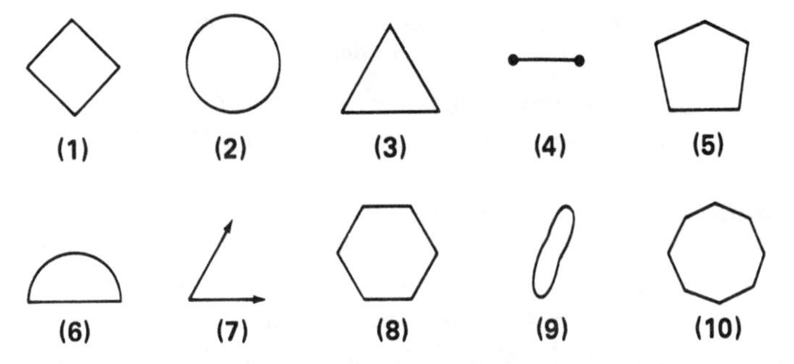

4. Tell the number of sides each of the following polygons has:
   a. hexagon **b.** quadrilateral **c.** triangle **d.** octagon **e.** pentagon
5. Each of the following is shaped like a geometric plane figure. Name the figure.
   a. a door **b.** a baseball diamond **c.** a snowflake **d.** a stop sign

11-6 THE TRIANGLE

We will begin our discussion of polygons with a study of the triangle, which is the simplest polygon in a plane.

On a practical side, there are many uses of the triangle, especially in construction work such as the building of bridges, radio towers, and airplane wings because the triangle is a ***rigid figure***. The shape of the triangle cannot be changed without changing the length of at least one of its sides.

Let us begin our discussion of the triangle by considering triangle ABC. The symbol for triangle ABC is $\triangle ABC$. In triangle ABC, the points A, B, and C are the ***vertices*** of the triangle. Line segments $\overline{AB}$, $\overline{BC}$, and $\overline{CA}$ are the ***sides*** of the triangle. $\angle A$, $\angle B$, and $\angle C$ are the ***angles*** of the triangle.

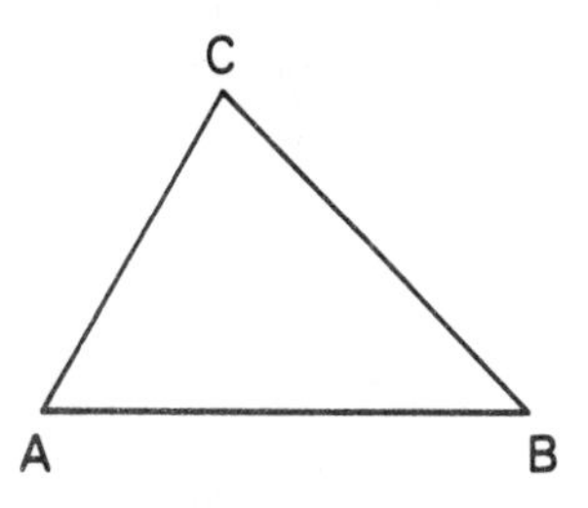

We make the following observations:

1. Side $\overline{AB}$ is included between $\angle A$ and $\angle B$.
2. Side $\overline{BC}$ is included between $\angle B$ and $\angle C$.
3. Side $\overline{CA}$ is included between $\angle C$ and $\angle A$.
4. $\angle A$ is included between sides $\overline{AB}$ and $\overline{AC}$.
5. $\angle B$ is included between sides $\overline{BA}$ and $\overline{BC}$.
6. $\angle C$ is included between sides $\overline{CA}$ and $\overline{CB}$.

Classifying Triangles According to Angles

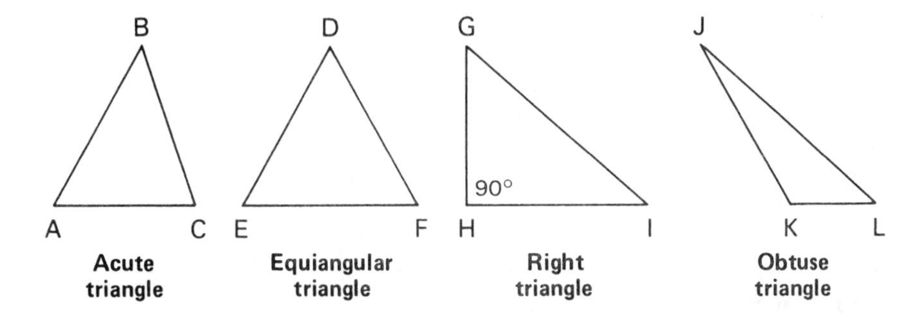

An ***acute triangle*** is a triangle that has three acute angles.

An ***equiangular triangle*** is a triangle that has three angles equal in measure.

A ***right triangle*** is a triangle that has one right angle.

An ***obtuse triangle*** is a triangle that has one obtuse angle.

In right triangle GHI above, the two sides of the triangle that form the right angle, $\overline{GH}$ and $\overline{HI}$, are called the ***legs*** of the right triangle. $\overline{GI}$, the side opposite the right angle, is called the ***hypotenuse***.

Classifying Triangles According to Sides

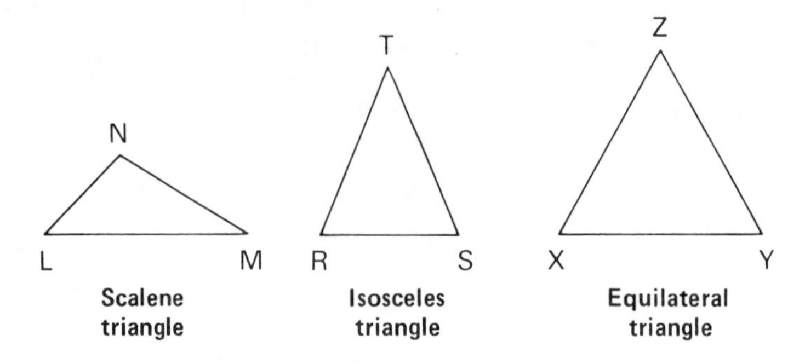

A ***scalene triangle*** is a triangle that has no sides equal in length.

An ***isosceles triangle*** is a triangle that has two sides equal in length.

An ***equilateral triangle*** is a triangle that has three sides equal in length.

The Sum of the Measures of the Angles of a Triangle

When we change the shape of a triangle, changes take place in the measures of its angles. Is there any relationship among the measures of the triangle that does not change? Let us see.

Draw several triangles of different shapes. In each triangle, measure the three angles and find the sum of the three measures. For example, in $\triangle ABC$, $m\angle A + m\angle B + m\angle C = 65 + 45 + 70 = 180$.

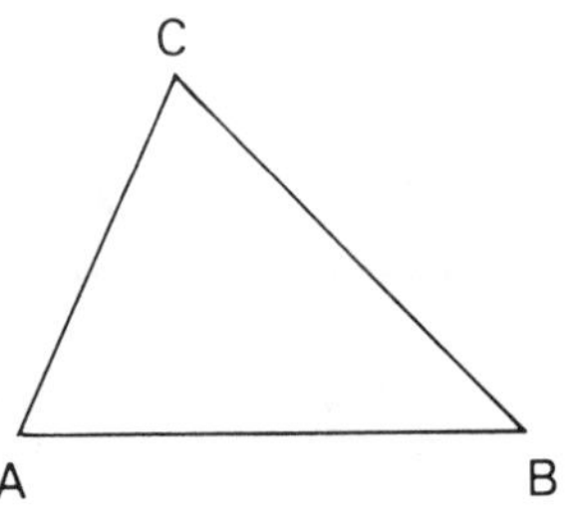

If you measured accurately, you should have found that in each triangle the sum of the measures of the three angles is 180°.

You can see that this is so by tearing off two angles of any triangle and placing them adjacent to the third angle as is shown in the figures below.

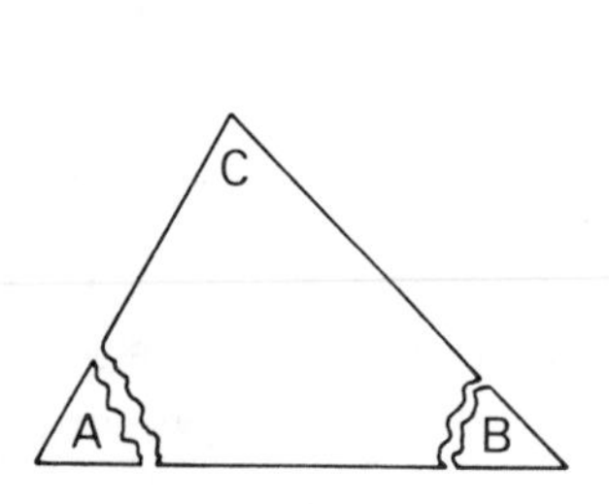

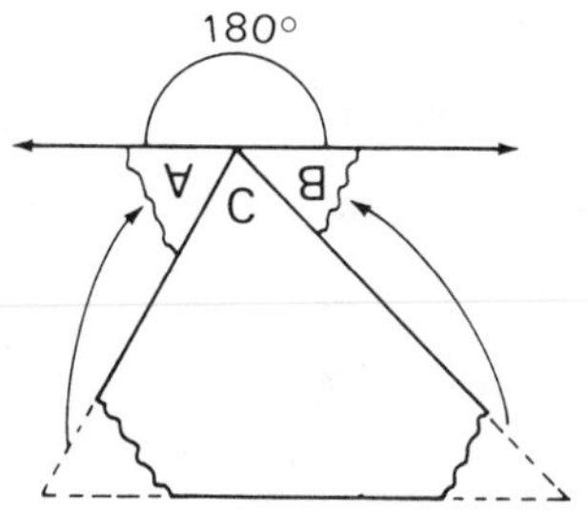

We can write an informal algebraic proof of the following statement:

- **The sum of the measures of the angles of a triangle is 180°.**

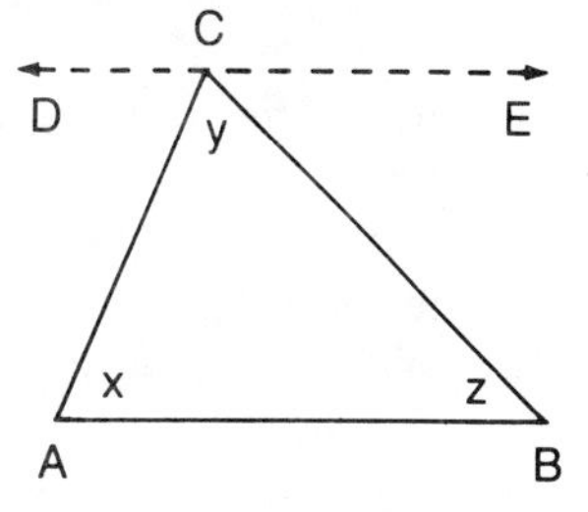

(1) In $\triangle ABC$, let m $\angle A = x$, m $\angle ACB = y$, and m $\angle B = z$.
(2) Let $\overleftrightarrow{DCE}$ be a line parallel to $\overline{AB}$.
(3) Since $\angle DCE$ is a straight angle, m $\angle DCE = 180$.
(4) m $\angle DCE$ = m $\angle DCA$ + m $\angle ACB$ + m $\angle BCE$.
(5) m $\angle DCA$ + m $\angle ACB$ + m $\angle BCE = 180$.
(6) m $\angle DCA$ = m $\angle A = x$.
(7) m $\angle BCE$ = m $\angle B = z$.
(8) Therefore, by substituting values from lines (6), (1), and (7) into the equation from line (5), we obtain $x + y + z = 180$.

MODEL PROBLEM

In triangle ABC, the measure of angle B is twice the measure of angle A, and the measure of angle C is three times the measure of angle A. Find the number of degrees in each angle of the triangle.

Solution:

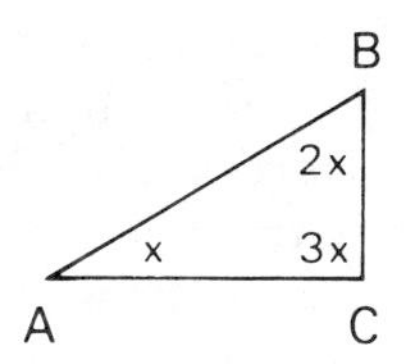

Let x = the number of degrees in angle A.
Then, $2x$ = the number of degrees in angle B.
Then, $3x$ = the number of degrees in angle C.

The sum of the measures of the angles of a triangle is 180°.

$$x + 2x + 3x = 180$$
$$6x = 180$$
$$x = 30$$
$$2x = 60$$
$$3x = 90$$

Check

$60 = 2 \times 30$
$90 = 3 \times 30$
$30 + 60 + 90 = 180$

Answer: m $\angle A = 30$, m $\angle B = 60$, m $\angle C = 90$

EXERCISES

In 1–3, state whether the three angles can be the three angles of a triangle.

1. 30°, 70°, 80° **2.** 70°, 80°, 90° **3.** 30°, 110°, 40°

In 4–7, find the measure of the third angle of the triangle if the measures of the first two angles are:

4. 60°, 40° **5.** 100°, 20° **6.** 54.5°, 82.3° **7.** $24\frac{1}{4}°$, $81\frac{3}{4}°$

8. Find the number of degrees in each angle of an equiangular triangle.

9. Can a triangle have: **(a)** two right angles? **(b)** two obtuse angles? **(c)** one right and one obtuse angle? Why?

10. What is the sum of the measures of the two acute angles of a right triangle?

11. If two angles in one triangle contain the same number of degrees as two angles in another triangle, what must be true of the third pair of angles in the two triangles? Why?

12. In a triangle, the measure of the second angle is 3 times the measure of the first angle, and the measure of the third angle is 5 times the measure of the first angle. Find the number of degrees in each angle of the triangle.

13. In a triangle, the measure of the second angle is 4 times the measure of the first angle. The measure of the third angle is equal to the sum of the measures of the first two angles. Find the number of degrees in each angle of the triangle.

14. In a triangle, the measure of the second angle is 30° more than the measure of the first angle, and the measure of the third angle is 45° more than the measure of the first angle. Find the number of degrees in each angle of the triangle.

15. In a triangle, the measure of the second angle is 5° more than twice the measure of the first angle. The measure of the third angle is 35° less than 3 times the measure of the first angle. Find the number of degrees in each angle of the triangle.

16. $\overleftrightarrow{AEFB}$ is a straight line; m $\angle AEG$ = 120, m $\angle BFG$ = 150.

a. Find m $\angle x$, m $\angle y$, and m $\angle z$.

b. What kind of triangle is triangle EFG?

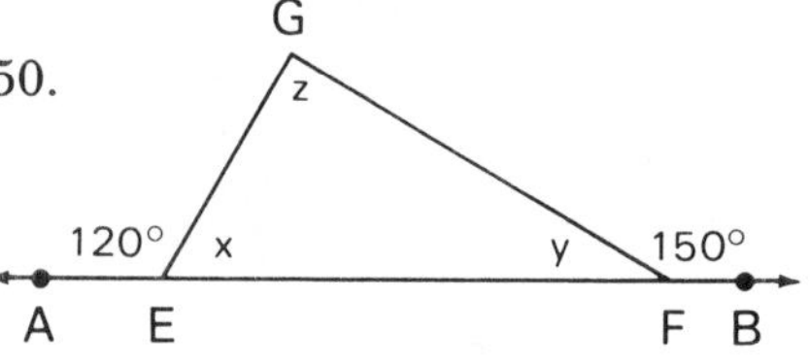

17. In $\triangle RST$, m $\angle R = x$, m $\angle S = x + 30$, m $\angle T = x - 30$.

a. Find the measures of the three angles.

b. What kind of triangle is $\triangle RST$?

18. In $\triangle KLM$, m $\angle K = 2x$, m $\angle L = x + 30$, m $\angle M = 3x - 30$.

a. Find the measures of the three angles.

b. What kind of triangle is $\triangle KLM$?

The Exterior Angle of a Triangle

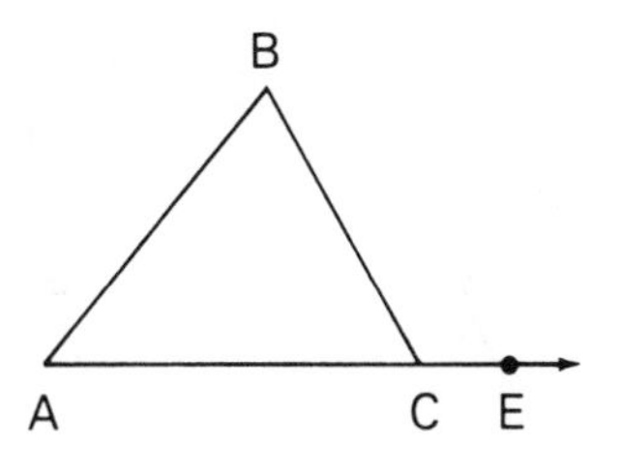

In the figure at the right, side $\overline{AC}$ is extended to form $\angle BCE$ at vertex C. Notice that $\angle BCE$ is in the *exterior* of triangle ABC. Also notice that $\angle BCE$ and $\angle BCA$ are supplementary, forming a linear pair. We call $\angle BCE$ an ***exterior angle*** of $\triangle ABC$, drawn at vertex C.

If m $\angle BCA = x$, then m $\angle BCE = 180 - x$.

● **An exterior angle of a triangle is an angle that forms a linear pair with one of the angles of the triangle.**

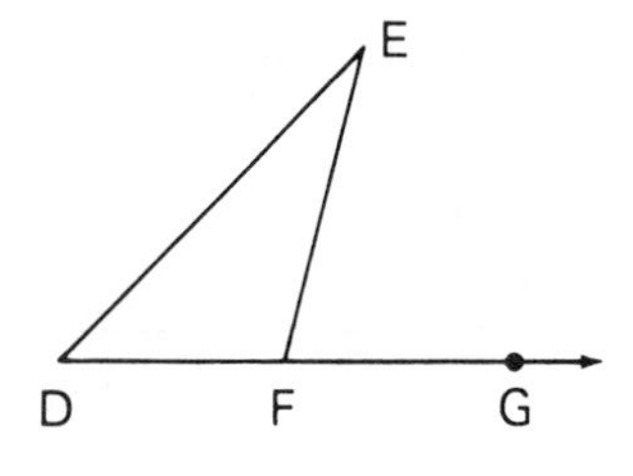

In the figure at the right, $\triangle DEF$ is shown with exterior $\angle EFG$ at vertex F. There are three angles found in the *interior* of $\triangle DEF$. You known that $\angle EFD$ is *adjacent* to exterior $\angle EFG$ because these angles form a linear pair. The two angles at the remaining vertices, $\angle D$ and $\angle E$, are called ***remote interior angles*** to the exterior $\angle EFG$. Using this triangle and exterior $\angle EFG$, observe:

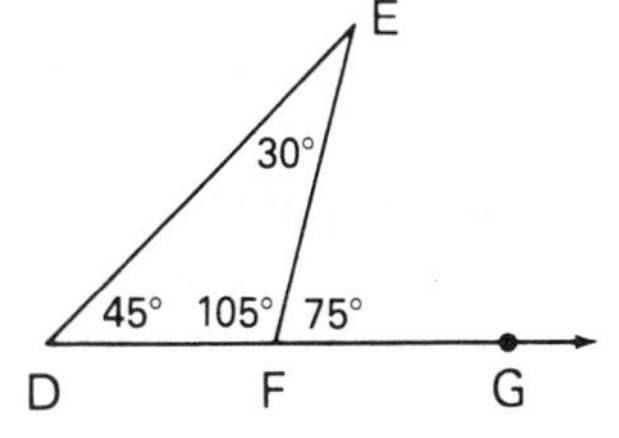

(1) If m $\angle D = 45$ and m $\angle E = 30$, then m $\angle EFD = 180 - (45 + 30) = 180 - 75 = 105$.

(2) If m $\angle EFD = 105$ and this angle forms a linear pair with exterior $\angle EFG$, then m $\angle EFG = 180 - 105 = 75$.

(3) Conclude that m $\angle D$ + m $\angle E$ = m $\angle EFG$, since $45 + 30 = 75$.

This example illustrates the truth of the following statement, which is proved informally by replacing the given values by variables.

● **The measure of an exterior angle of a triangle is equal to the sum of the measures of the two remote interior angles.**

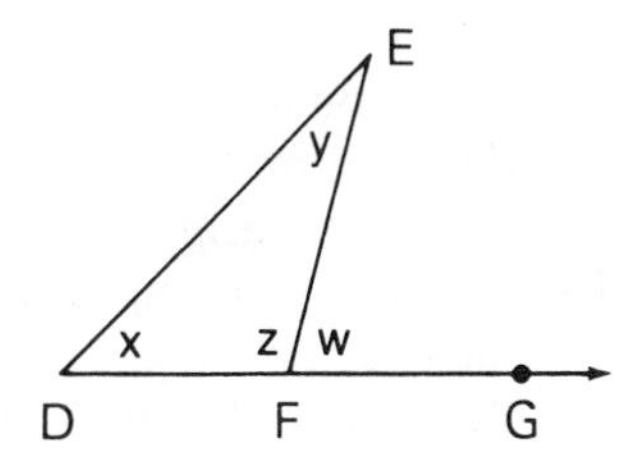

(1) Let x = m $\angle D$,
y = m $\angle E$,
z = m $\angle EFD$, and
w = m $\angle EFG$.

(2) $x + y + z = 180$ or $x + y = 180 - z$

(3) $w + z = 180$ or $w = 180 - z$

(4) Therefore, $x + y = w$.

EXERCISES

In 1–5: **a.** Name the exterior angle of the triangle that is shown in the diagram. **b.** Name the two remote interior angles to that exterior angle.

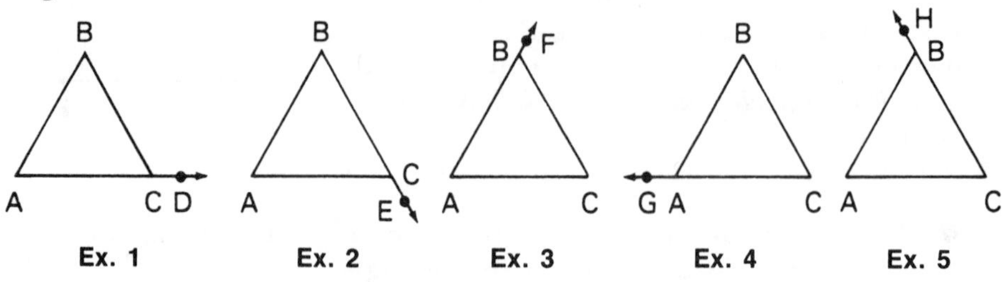

In 6–10, find the number of degrees in the value of x.

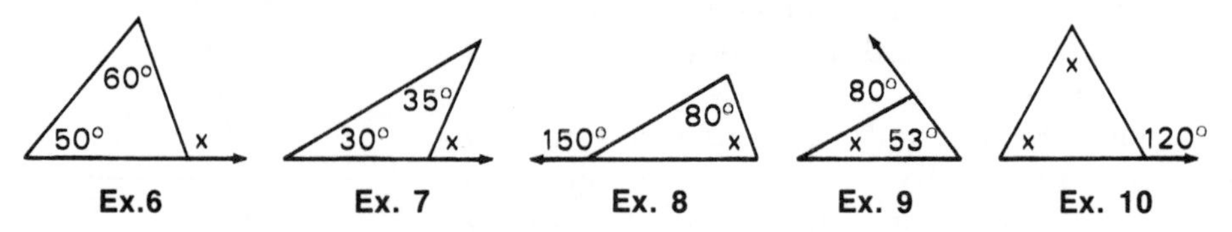

In 11–13, draw and label a diagram and find the value of x.

11. In $\triangle ABC$, m $\angle A = x$, m $\angle B = x + 10$, and the measure of an exterior angle at C is 70°.

12. In $\triangle PQR$, m $\angle P = 2x$, m $\angle Q = 3x$, and the measure of an exterior angle at R is 150°.

13. In $\triangle RST$, m $\angle R = 36$, m $\angle S = 2x$, and the measure of an exterior angle at T is $(5x)°$.

In 14 and 15, draw and label a diagram, and find the specified measure.

14. In $\triangle DEF$, m $\angle D = 80$, and the measure of an exterior angle at E is 3 times the measure of $\angle F$. Find the measure of $\angle F$.

15. In $\triangle KLM$, m $\angle K = 85$ and the measure of an exterior angle at L is 135°. Find the measure of an exterior angle at M.

16. In an equiangular triangle, what is the measure of any one of its exterior angles?

17. In a right triangle, what is the measure of the exterior angle to the right angle?

18. An exterior angle is drawn to a triangle. If this exterior angle is acute, then the triangle must be:
(1) acute (2) right (3) obtuse (4) equilateral.

19. In the figure at the right, three exterior angles are drawn to $\triangle ABC$.

a. What is the sum of the measures of the three exterior angles of $\triangle ABC$?

b. Will the answer to part **a** be true for the sum of the measures of the exterior angles of any triangle?

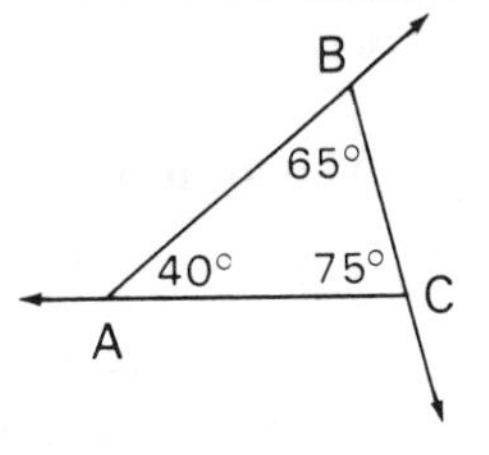

The Isosceles Triangle

In isosceles triangle ABC, the two sides that are equal in measure, $\overline{AC}$ and $\overline{BC}$, are called the ***legs***. The remaining side, $\overline{AB}$, is called the ***base***. The angle formed by the two congruent sides, $\angle C$, is called the ***vertex angle***. The two angles at the endpoints of the base, $\angle A$ and $\angle B$, are called the ***base angles***.

In isosceles triangle ABC, if you measure the base angles, $\angle A$ and $\angle B$, you will find that each angle contains 65°. Therefore, $m\angle A = m\angle B$. If you measure the base angles in any other isosceles triangle, you will find that they are again equal in measure. Thus, we will accept the truth of the following statement, which will be proved in a higher-level course as a theorem:

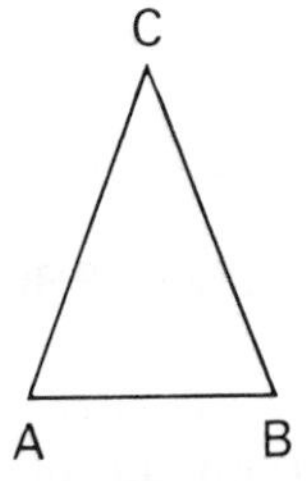

● **The base angles of an isosceles triangle are equal in measure, that is, they are congruent.**

This statement may be rephrased in a variety of ways. For example:

1. If a triangle is isosceles, then its base angles are equal in measure.
2. If two sides of a triangle are congruent, then the angles opposite these sides are congruent.

Note: The following, which is the *converse* of the previous statement, can be proved in a higher-level course:

● **If two angles of a triangle are equal in measure, then the triangle is an isosceles triangle.**

This statement may also be rephrased as follows:

If two angles of a triangle are congruent, then the sides opposite these angles are congruent.

Properties of Triangles

1. The sum of the measures of the angles of a triangle is 180°.
2. The acute angles of a right triangle are complementary.
3. If the measures of two angles of one triangle are equal, respectively, to the measures of two angles of another triangle, then the remaining angles are equal in measure.
4. If two sides of a triangle are equal in measure, the angles opposite these sides are equal in measure. (Base angles of an isosceles triangle are equal in measure.)
5. If two angles of a triangle are equal in measure, the sides opposite these angles are equal in measure and the triangle is an isosceles triangle.
6. The measure of an exterior angle of a triangle is equal to the sum of the measures of the two remote interior angles.

MODEL PROBLEM

In isosceles triangle ABC, the measure of vertex angle C is 30° more than the measure of each base angle. Find the number of degrees in each angle of the triangle.

Solution:

Let x = the number of degrees in one base angle, A.
Then, x = the number of degrees in the other base angle, B.
Then, $x + 30$ = the number of degrees in the vertex angle, C

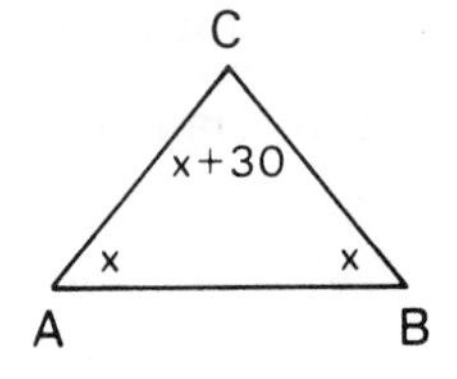

The sum of the measures of the angles of a triangle is 180°.

$$x + x + x + 30 = 180$$
$$3x + 30 = 180$$
$$3x = 150$$
$$x = 50$$
$$x + 30 = 80$$

Check

$50 + 50 + 80 = 180$

Answer: $m\angle A = 50$, $m\angle B = 50$, $m\angle C = 80$

EXERCISES

1. Name the legs, base, vertex angle, and base angles in each of the following isosceles triangles:

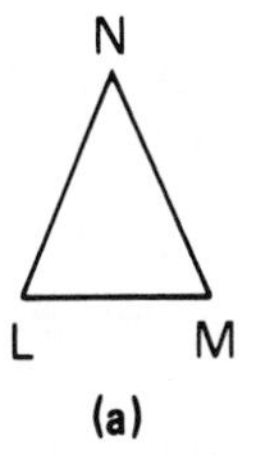

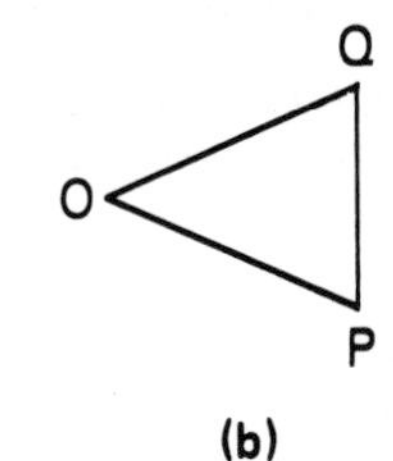

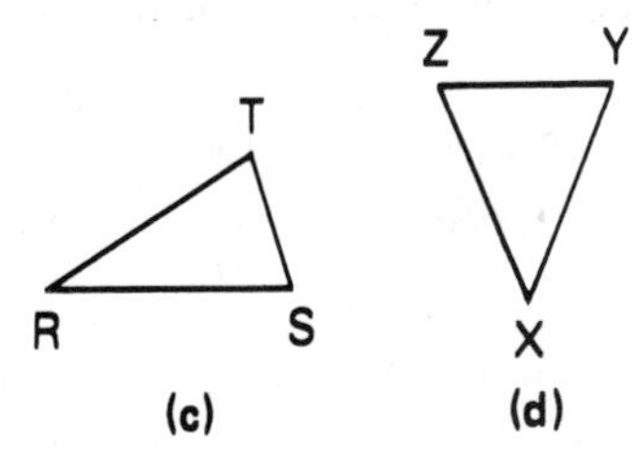

2. In $\triangle ABC$, $AC = 4$ cm, $CB = 6$ cm, and $AB = 6$ cm.
   a. What type of triangle is $\triangle ABC$?
   b. Name two angles in $\triangle ABC$ whose measures are equal.
   c. Why are they equal in measure?

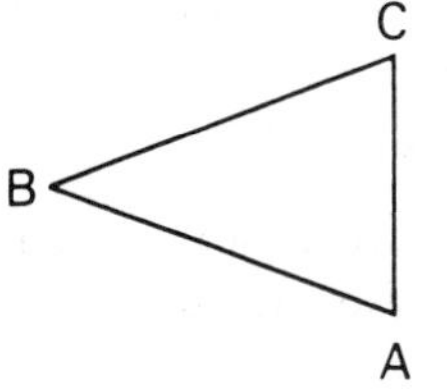

3. In $\triangle RST$, $m\angle R = 70$ and $m\angle T = 40$.
   a. Find the measure of $\angle S$.
   b. Name two sides in $\triangle RST$ that are congruent.
   c. Why are the two sides congruent?
   d. What type of triangle is $\triangle RST$?

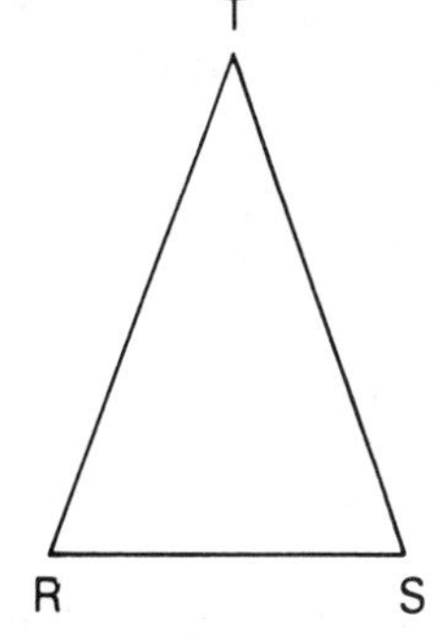

4. Draw an isosceles triangle that is **(a)** an acute triangle, **(b)** a right triangle, and **(c)** an obtuse triangle.

5. Can a base angle of an isosceles triangle be **(a)** a right angle **(b)** an obtuse angle? Why?

6. Find the measure of the vertex angle of an isosceles triangle if the measure of each base angle is:

   a. $80°$ **b.** $55°$ **c.** $42°$ **d.** $22\frac{1}{2}°$ **e.** $51.5°$

7. Find the measure of each base angle of an isosceles triangle if the vertex angle measures:

   a. $40°$ **b.** $50°$ **c.** $76°$ **d.** $100°$ **e.** $65°$

8. What is the number of degrees in each acute angle of an isosceles right triangle?

9. The measure of each base angle of an isosceles triangle is seven times the measure of the vertex angle. Find the measure of each angle of the triangle.

10. The measure of each of the congruent angles of an isosceles triangle is one-half of the measure of the vertex angle. Find the measure of each angle of the triangle.

11. The measure of the vertex angle of an isosceles triangle is 3 times as large as the measure of each base angle. Find the number of degrees in each angle of the triangle.

12. The measure of the vertex angle of an isosceles triangle is $15°$ more than the measure of each base angle. Find the number of degrees in each angle of the triangle.

13. The measure of each of the congruent angles of an isosceles triangle is $6°$ less than the measure of the vertex angle. Find the measure of each angle of the triangle.

14. The measure of each of the congruent angles of an isosceles triangle is $9°$ less than 4 times the vertex angle. Find the measure of each angle of the triangle.

15. In $\triangle ABC$, $m \angle A = x$, $m \angle B = x + 30$, and $m \angle C = 2x - 10$.

a. Find the measures of the three angles.

b. What kind of triangle is $\triangle ABC$?

16. The vertex angle of an isosceles triangle is $80°$ in measure. What is the measure of an exterior angle to one of the base angles of this triangle?

17. The measure of an exterior angle to a base angle of an isosceles triangle is $115°$. What is the measure of the vertex angle of the triangle?

18. The measure of an exterior angle to the vertex angle of an isosceles triangle is $60°$. What is the measure of one of the base angles of the triangle?

19. Given the statement: "If two sides of a triangle are equal in measure, then the angles opposite these sides are equal in measure."

a. What is the truth value of the statement?

b. Write the converse of the statement.

c. What is the truth value of the converse?

d. Using the statement and its converse, write a biconditional statement.

e. What is the truth value of this biconditional?

11-7 CONGRUENT TRIANGLES

Congruent Polygons

In modern industry, it is often necessary to make many copies of a part so that the original part and all copies will have the same size and shape. For example, a machine can stamp out many duplicates of a piece of metal, each copy having the same size and shape as the original. We say that the original and all its copies are ***congruent***.

We have already talked about congruent segments, which are segments that are equal in length. We have also discussed congruent angles, which are angles whose measures are equal. In the figures at the right that show congruent segments and congruent angles, read the symbol ′ as *prime*. We often use this symbol to show a matching.

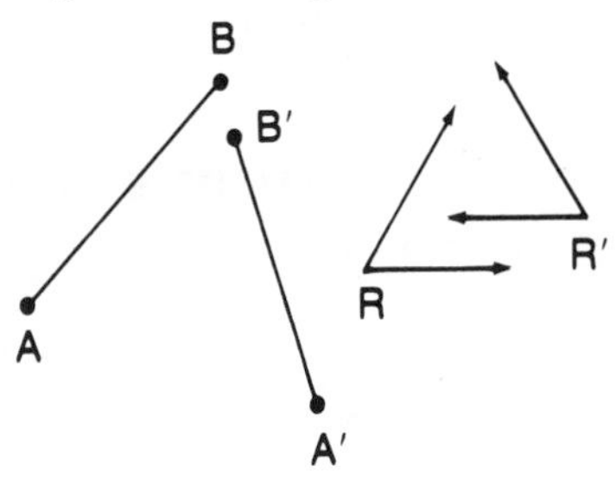

At the right, you also see two polygons that have the same size and shape. Such polygons are ***congruent polygons***: polygon $ABCD \cong$ polygon $A'B'C'D'$.

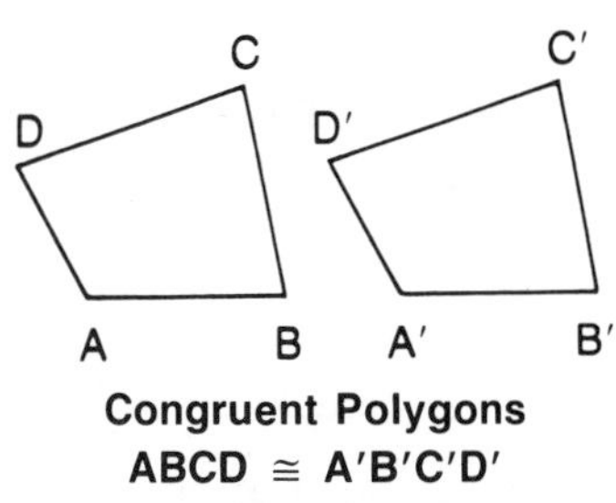

Congruent Polygons
ABCD ≅ A′B′C′D′

One way to discover whether or not two polygons have the same size and shape is to place one polygon upon the other. If the figures can be turned in such a way that the sides of one polygon *fit exactly* upon the sides of the other and the angles of one polygon *fit exactly* upon the angles of the other, we say that the polygons *coincide*. The sides that fit one upon the other are called ***corresponding sides***. For example, in the given polygons $ABCD$ and $A'B'C'D'$, the pairs of corresponding sides are:

$\overline{AB}$ and $\overline{A'B'}$ $\quad$ $\overline{BC}$ and $\overline{B'C'}$ $\quad$ $\overline{CD}$ and $\overline{C'D'}$ $\quad$ $\overline{DA}$ and $\overline{D'A'}$

The angles that fit one upon the other are called ***corresponding angles***. For example, in the given polygons $ABCD$ and $A'B'C'D'$, the pairs of corresponding angles are:

$\angle A$ and $\angle A'$ $\quad$ $\angle B$ and $\angle B'$ $\quad$ $\angle C$ and $\angle C'$ $\quad$ $\angle D$ and $\angle D'$

Thus, we can say:

● **If the sides of a first polygon are congruent to the corresponding sides of a second polygon, and if the angles of the first polygon are congruent to the corresponding angles of the second, then the two polygons are congruent.**

Conversely, we can also say:

● **If two polygons are congruent, then their corresponding sides are congruent and their corresponding angles are congruent.**

You can see that to prove two triangles congruent, you would have to prove three pairs of corresponding sides congruent and three pairs of corresponding angles congruent. Let us see whether it is possible to prove two triangles congruent by proving fewer than three pairs of sides and three pairs of angles congruent.

Congruent Triangles Involving Two Sides and the Included Angle

Let us perform the following experiment:

In $\triangle ABC$, $AB = 1$ inch, $m\angle A = 45$, and $AC = \frac{3}{4}$ inch. We say that $\angle A$ is *included* between side $\overline{AB}$ and side $\overline{AC}$ because these two segments are on the sides of the angle.

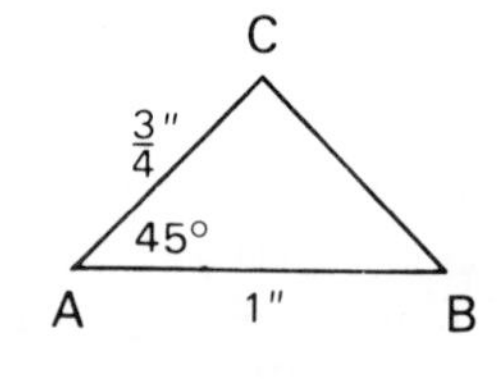

On a sheet of paper, draw $\triangle A'B'C'$ so that $A'B' = 1$ inch, $A'C' = \frac{3}{4}$ inch, and the measure of the included angle A' is 45°:

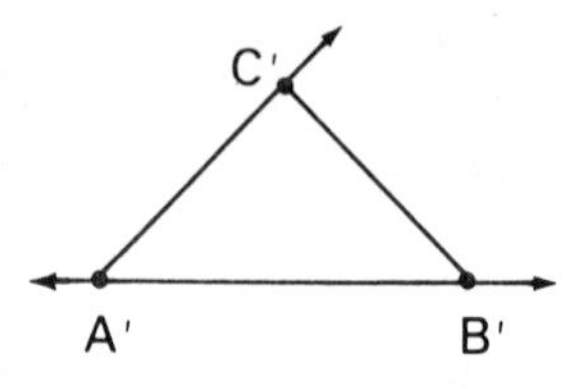

(1) Begin by drawing a working line on which to measure off 1 inch, the length of $\overline{A'B'}$.

(2) With a protractor, draw an angle of 45° whose vertex is at point A'.

(3) On the side of $\angle A'$ that was last drawn, measure off a line segment $\frac{3}{4}$ of an inch in length, beginning at point A' and ending at point C'.

(4) Then, draw side $\overline{C'B'}$ to complete the triangle.

If you measure sides $\overline{CB}$ and $\overline{C'B'}$, you will find the measures equal. Hence, $\overline{CB} \cong \overline{C'B'}$. Also, if you measure $\angle C$ and $\angle C'$, you will find their measures equal. $\angle B$ and $\angle B'$, if measured, are also found to have equal measures. Hence, $\angle C \cong \angle C'$ and $\angle B \cong \angle B'$. Also, if you cut out $\triangle A'B'C'$, you can make it coincide with $\triangle ABC$. Thus, $\triangle A'B'C'$ appears to be congruent to $\triangle ABC$.

If you repeat the same experiment several times with different sets of measurements for the two sides and the included angle, in each experiment the remaining pairs of corresponding parts of the triangles will appear to be congruent, and the triangles themselves will appear to be congruent. It seems reasonable, therefore, to accept the truth of the following statement:

- **Two triangles are congruent if two sides and the included angle of one triangle are congruent, respectively, to two sides and the included angle of the other. [s.a.s. ≅ s.a.s.]**

In $\triangle ABC$ and $\triangle A'B'C'$, if $\overline{AB} \cong \overline{A'B'}$, $\angle A \cong \angle A'$, and $\overline{AC} \cong \overline{A'C'}$, then $\triangle ABC \cong \triangle A'B'C'$. [s.a.s. ≅ s.a.s.]

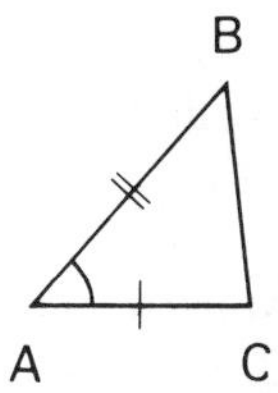

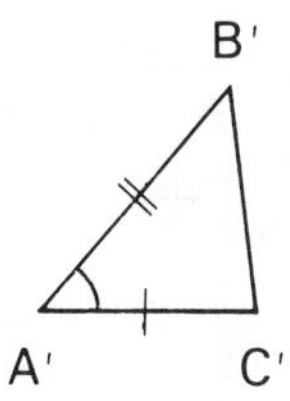

When the Angle Is Not Included Between the Sides

Now, see what happens when each of two triangles has one side 1″, a second side $\frac{3}{4}$″, and an angle that measures 45° is not included between these sides; that is, s.a.s. ≅ s.s.a. Must two such triangles always be congruent? With the use of a ruler and a protractor, you can draw two different triangles that have the given measures, as shown below.

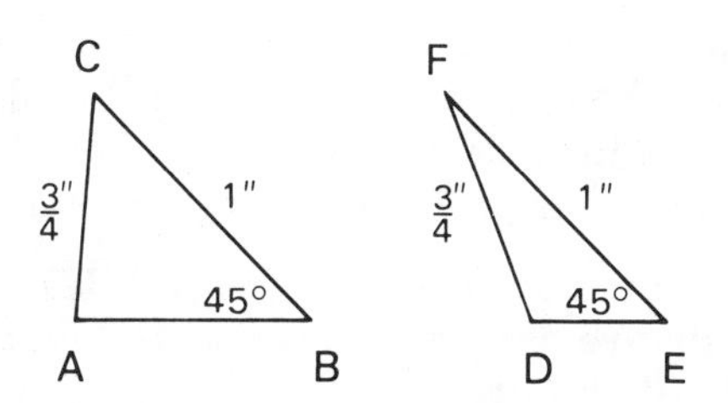

Even without cutting out triangle *ABC* and triangle *DEF*, you see that they would not coincide. Hence, when the angle is *not included* between the two sides, two triangles need not be congruent although two sides and an angle of one triangle are congruent to two sides and an angle in the other. Therefore, proving s.s.a. ≅ s.s.a. in two triangles is *not sufficient* to prove the triangles congruent.

Congruent Triangles Involving Two Angles and the Included Side

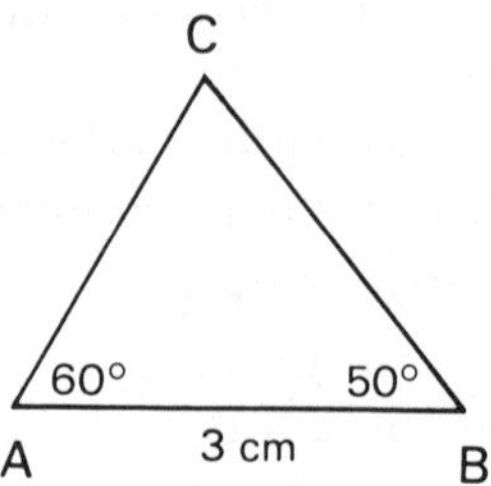

In triangle ABC, m $\angle A$ = 60, AB = 3 cm, and m $\angle B$ = 50. Side $\overline{AB}$ *is included* between $\angle A$ and $\angle B$ because side $\overline{AB}$ is drawn between vertex A and vertex B.

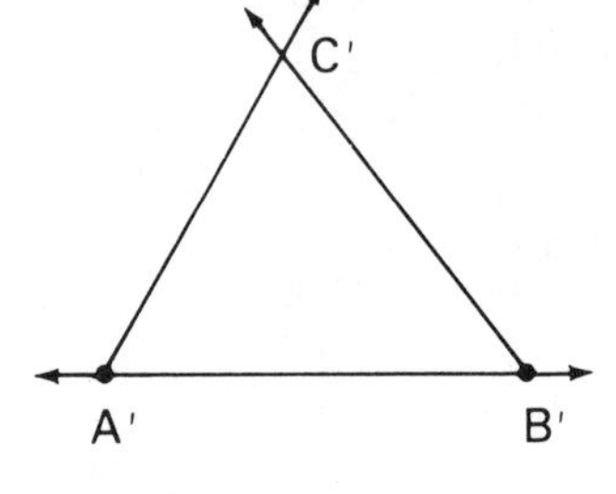

Draw $\triangle A'B'C'$ so that m $\angle A'$ = 60, m $\angle B'$ = 50, and the included side $\overline{A'B'}$ measures 3 cm.

(1) Begin by drawing a working line on which to measure off 3 cm, the length of side $\overline{A'B'}$.

(2) With a protractor, draw an angle of 60° whose vertex is at point A'.

(3) Then, draw an angle of 50° whose vertex is at point B'.

(4) To complete the triangle, draw the sides of these angles so that they intersect at point C'.

You can see that $\triangle ABC$ and $\triangle A'B'C'$ appear to have the same size and shape.

Hence, $\triangle ABC$ appears to be congruent to $\triangle A'B'C'$.

If you repeat the same experiment several times with different sets of measurements for the two angles and the included side, the triangles in each experiment will appear to be congruent. Therefore, it seems reasonable to accept the truth of the following statement:

● **Two triangles are congruent if two angles and the included side of one triangle are congruent, respectively, to two angles and the included side of the other triangle. [a.s.a. ≅ a.s.a.]**

In $\triangle RST$ and $\triangle R'S'T'$:
If $\angle R \cong \angle R'$, $\overline{RS} \cong \overline{R'S'}$, and $\angle S \cong \angle S'$, then $\triangle RST \cong \triangle R'S'T'$.
[a.s.a. ≅ a.s.a.]

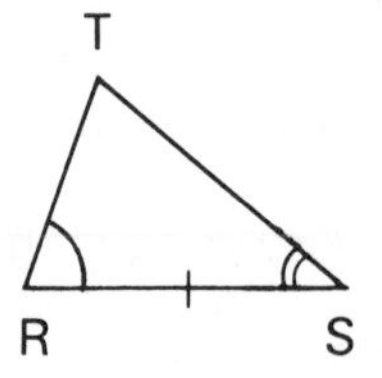

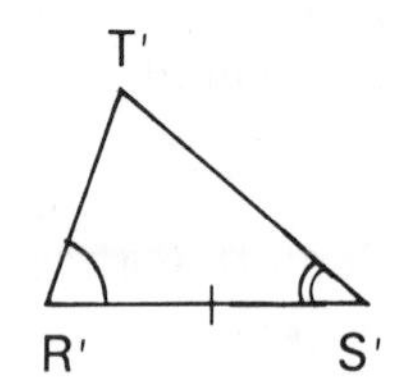

Congruent Triangles Involving Three Sides

The following statement about triangles is also true:

● **Two triangles are congruent if three sides of one triangle are congruent, respectively, to three sides of the other triangle. [s.s.s. ≅ s.s.s.]**

In $\triangle RST$ and $\triangle R'S'T'$:
If $\overline{RS} \cong \overline{R'S'}$, $\overline{ST} \cong \overline{S'T'}$, and $\overline{TR} \cong \overline{T'R'}$, then $\triangle RST \cong \triangle R'S'T'$.
[s.s.s. ≅ s.s.s.]

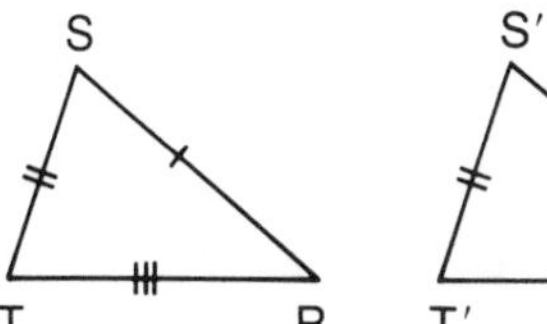

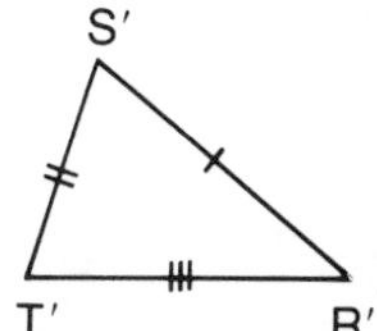

MODEL PROBLEM

In the figure, $\overline{AB}$ intersects $\overline{CD}$ at E, $\angle C$ and $\angle D$ are right angles, and $\overline{CE} \cong \overline{DE}$.

a. Prove informally that $\triangle AEC \cong \triangle BED$.

b. If $AE = 3x$ and $BE = 2x + 10$, find AE and BE.

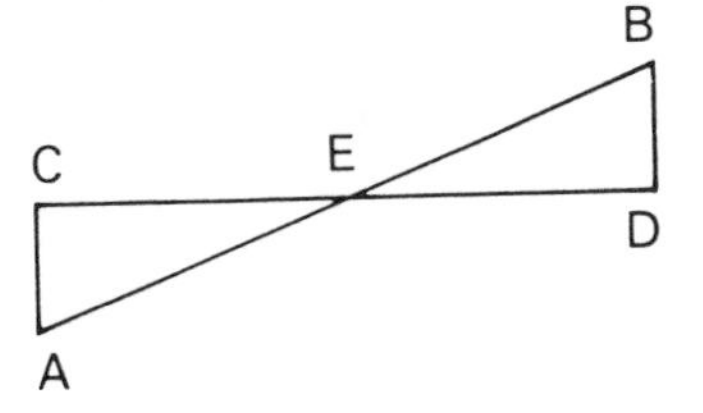

Solution

a. (1) You are told that $\overline{CE} \cong \overline{DE}$.

(2) You are told that $\angle C$ and $\angle D$ are right angles. Since all right angles are congruent, $\angle C \cong \angle D$.

(3) Since $\angle CEA$ and $\angle DEB$ are a pair of vertical angles, and vertical angles are congruent, $\angle CEA \cong \angle DEB$.

(4) $\triangle AEC \cong \triangle BED$ because two angles and the included side of one triangle are congruent to two angles and the included side of the other triangle. [a.s.a. ≅ a.s.a.]

b. Since $\triangle AEC$ and $\triangle BED$ are congruent, their corresponding sides must be congruent. Therefore, $\overline{AE} \cong \overline{BE}$.

Hence, $AE = BE$.

Thus: $3x = 2x + 10$

$x = 10$

$3x = 30$ and $2x + 10 = 2(10) + 10 = 30$

Answer: $AE = 30$ and $BE = 30$.

EXERCISES

In 1–3: Among figures **a**, **b**, **c**, **d**, and **e**, choose the figures that appear to be congruent.

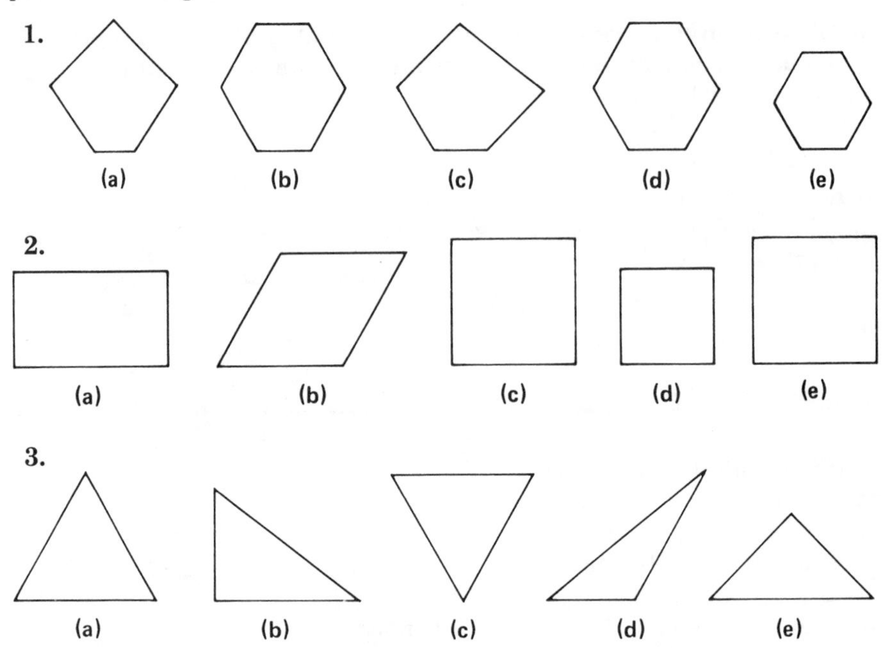

4. **a.** If several copies are made of the same photograph, will the figures in the original photograph and in the copies be congruent?
b. Why?

5. **a.** If a photograph is enlarged, what can be said of the figures in the original and the figures in the enlargement?
b. Why?

In 6–8, two triangles are to be drawn for each problem. **a.** Use a ruler and protractor to draw $\triangle ABC$ and $\triangle DEF$, starting with the measures given for each triangle. **b.** If the triangles are congruent, state the reason why they must be congruent. If the triangles are not congruent, explain why.

6. In $\triangle ABC$: $AB = 2''$, $m\angle B = 60$, and $BC = 1\frac{1}{2}''$.
In $\triangle DEF$: $DE = 1\frac{1}{2}''$, $m\angle E = 60$, and $EF = 2''$.

7. In $\triangle ABC$: $AB = 3''$, $m\angle A = 40$, and $m\angle B = 80$.
In $\triangle DEF$: $m\angle E = 80$, $EF = 3''$, and $m\angle F = 40$.

8. In $\triangle ABC$: $BC = 2.5$ cm, $m\angle C = 90$, and $m\angle B = 60$.
In $\triangle DEF$: $EF = 2.5$ cm, $m\angle E = 60$, and $\angle F$ is a right angle.

9. Sam and Rita each drew a triangle in which two sides and an angle measured, respectively, 5 cm, 8 cm, and 70°. The triangles were not congruent. Tell why this could have happened.

10. a. From the following triangles, select pairs that are congruent. Tell why they are congruent.

b. For each congruent pair, name the corresponding angles and the corresponding sides.

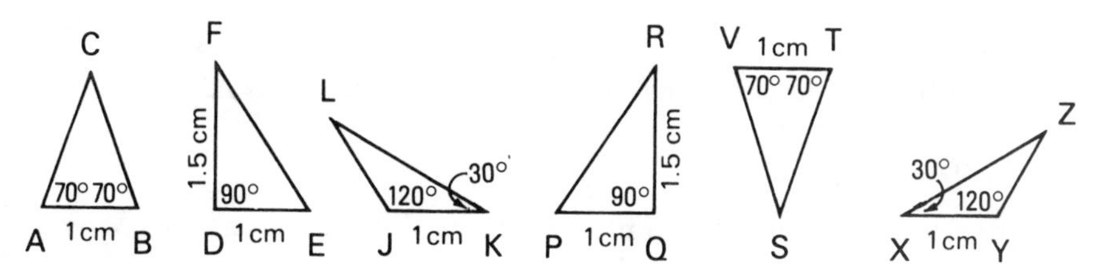

11. a. Tell why $\triangle RST \cong \triangle R'S'T'$.

b. Find $m \angle RTS$ and $m \angle R'T'S'$.

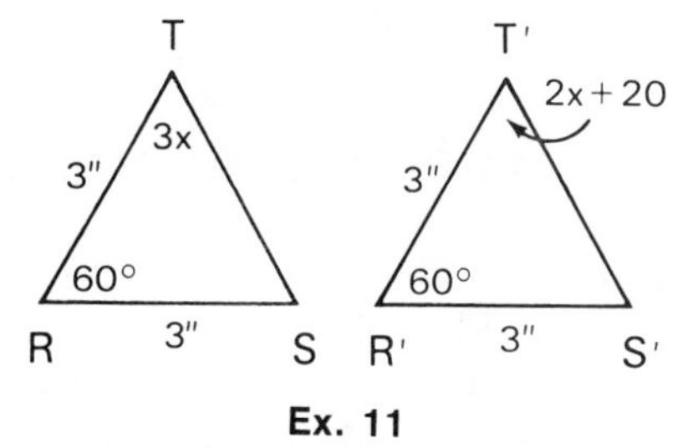

Ex. 11

12. a. Tell why $\triangle ABC \cong \triangle A'B'C'$.

b. Find BC and $B'C'$.

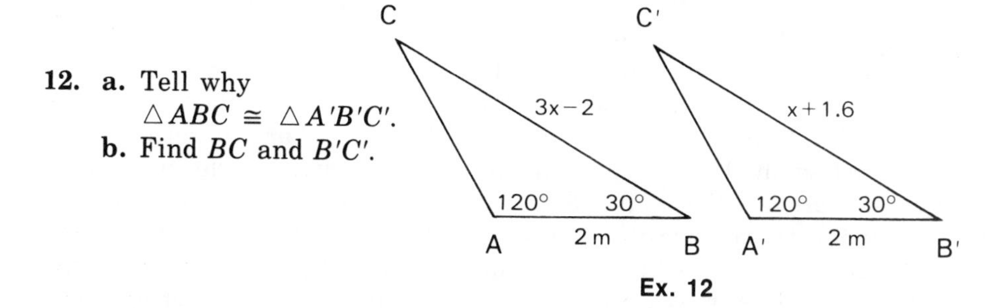

Ex. 12

13. a. $\overline{BC}$ and $\overline{AD}$ intersect at E. What is the relationship between $m \angle CED$ and $m \angle BEA$?

b. Why is $\triangle CED \cong \triangle BEA$?

c. What is the relationship between CD and BA?

d. Find BA and CD.

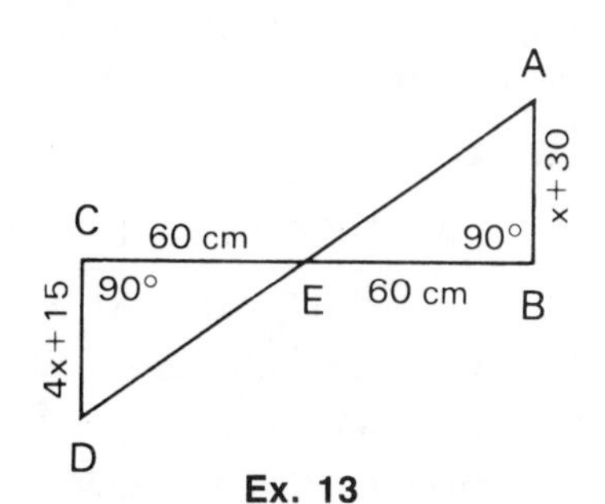

Ex. 13

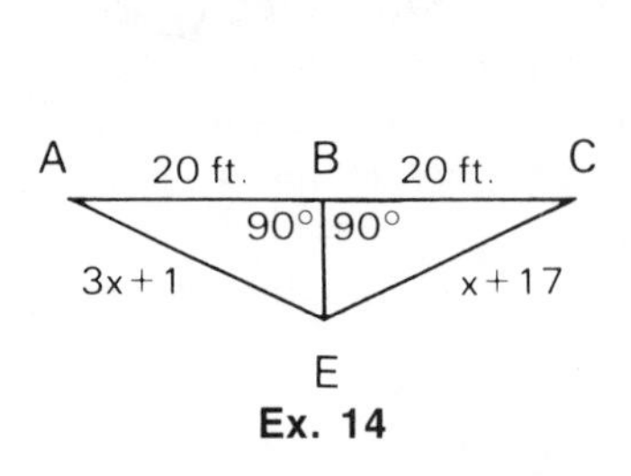

Ex. 14

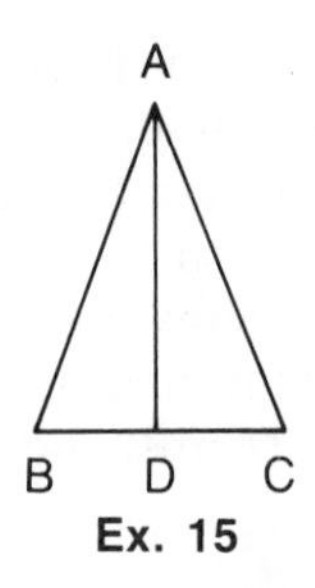

Ex. 15

14. **a.** What is the relationship that exists between $\triangle ABE$ and $\triangle CBE$? Why?
b. Find AE and CE.

15. **a.** $\triangle ABC$ is isosceles, and D is the midpoint of base $\overline{BC}$. Why is $\triangle ABD \cong \triangle ACD$?
b. If $m\angle B = 3x + 10$ and $m\angle C = 70$, find the value of x.

11-8 THE QUADRILATERAL

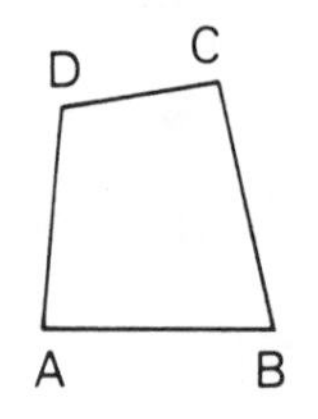

A ***quadrilateral*** is a polygon that has four sides. A point at which any two sides of the quadrilateral meet is called a ***vertex*** of the quadrilateral. At each vertex, the two sides that meet form an angle of the quadrilateral. Thus, $ABCD$ is a quadrilateral whose sides are $\overline{AB}$, $\overline{BC}$, $\overline{CD}$, and $\overline{DA}$. Its vertices are A, B, C, and D. Its angles are $\angle ABC$, $\angle BCD$, $\angle CDA$, and $\angle DAB$.

In a quadrilateral, two angles whose vertices are the endpoints of a side are called ***consecutive angles***. For example, in quadrilateral $ABCD$, $\angle A$ and $\angle B$ are consecutive angles; $\angle B$ and $\angle C$ are consecutive angles; and so on. Two angles that are not consecutive angles are called ***opposite angles***. For example, $\angle A$ and $\angle C$ are opposite angles, and $\angle B$ and $\angle D$ are opposite angles.

When we vary the shape of the quadrilateral by making some of its sides parallel, by making some of its sides equal in length, or by making its angles right angles, we get different members of the family of quadrilaterals, as shown below:

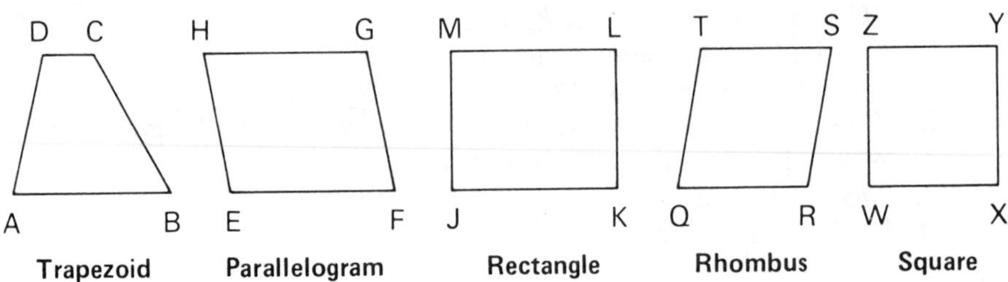

A ***trapezoid*** is a quadrilateral in which two and only two opposite sides are parallel. In trapezoid $ABCD$, $\overline{AB} \parallel \overline{CD}$. The parallel sides $\overline{AB}$ and $\overline{CD}$ are called the ***bases*** of the trapezoid.

A ***parallelogram*** is a quadrilateral in which both pairs of opposite sides are parallel. In parallelogram $EFGH$, $\overline{EF} \parallel \overline{GH}$ and $\overline{EH} \parallel \overline{FG}$. The symbol for parallelogram is ▱.

A ***rectangle*** is a parallelogram in which all four angles are right angles. Rectangle $JKLM$ is a parallelogram in which $\angle J$, $\angle K$, $\angle L$, and $\angle M$ are right angles.

A ***rhombus*** is a parallelogram in which all sides are of equal length. Rhombus $QRST$ is a parallelogram in which $QR = RS = ST = TQ$.

A ***square*** is a rectangle in which all sides are of equal length. Therefore, square $WXYZ$ is also a parallelogram in which $\angle W$, $\angle X$, $\angle Y$, and $\angle Z$ are right angles, and $WX = XY = YZ = ZW$.

Draw a large quadrilateral like the one shown at the right. Measure each of its four angles. Is the sum 360°? It should be. Do the same with several other quadrilaterals of different shapes and sizes. Is the sum of the four measures 360° in each case? It should be. Relying on what you have just verified by experimentation, it seems reasonable to make the following statement:

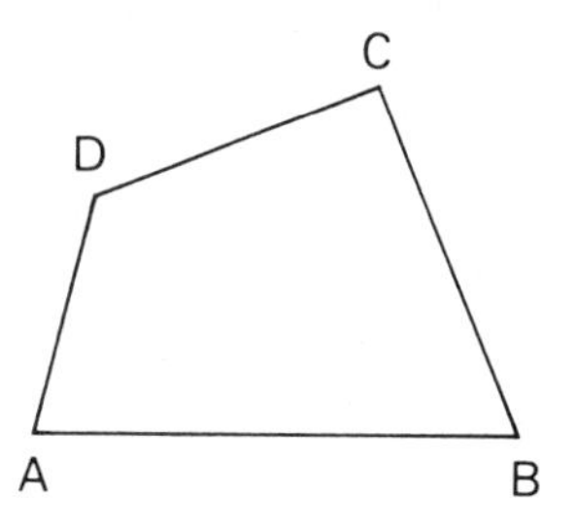

● **The sum of the measures of the angles of a quadrilateral is 360°.**

In order to prove informally that this statement is true, draw diagonal $\overline{AC}$, which is a line segment whose endpoints are the vertices of the two opposite angles, $\angle A$ and $\angle C$.

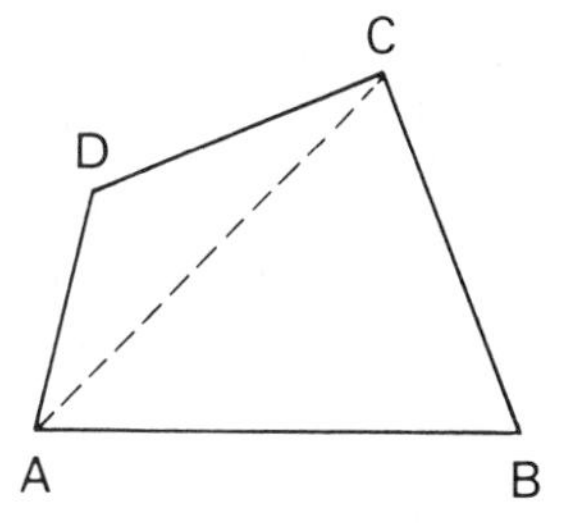

(1) Diagonal $\overline{AC}$ divides quadrilateral $ABCD$ into two triangles, $\triangle ABC$ and $\triangle ADC$.

(2) The sum of the measures of the angles of $\triangle ABC$ is 180° and the sum of the measures of the angles of $\triangle ADC$ is 180°.

(3) The sum of the measures of all the angles of $\triangle ABC$ and $\triangle ADC$ together is 360°.

(4) Hence, $m\angle A + m\angle B + m\angle C + m\angle D = 360$.

The Family of Parallelograms

Let us discuss some of the relationships that hold true in parallelograms, rectangles, rhombuses, and squares.

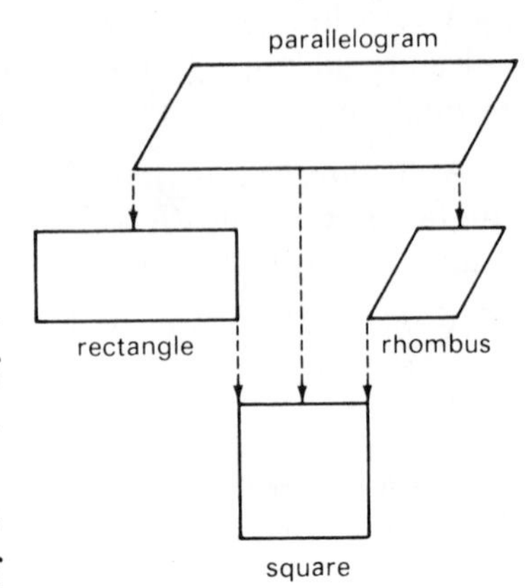

1. All rectangles, rhombuses, and squares are members of the family of parallelograms. Therefore, any property of the family of parallelograms must also be a property of rectangles, rhombuses, and squares.

2. A square is a member of the family of rectangles. Therefore, any property of the family of rectangles must also be a property of squares.

3. A square is a member of the family of rhombuses. Therefore, any property of the family of rhombuses must also be a property of squares.

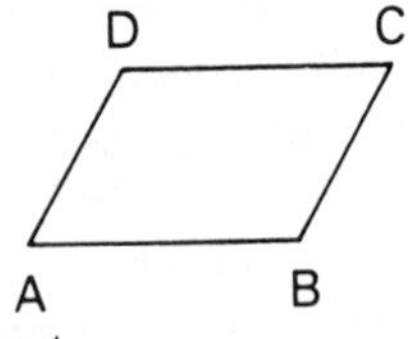

In parallelogram $ABCD$ at the right, opposite sides are parallel. Thus, $\overline{AB} \parallel \overline{DC}$ and $\overline{AD} \parallel \overline{BC}$. The following statements, which will be proved in a higher-level course, are true for any parallelogram:

1. **Opposite sides of a parallelogram are congruent.**
   Here, $\overline{AB} \cong \overline{DC}$ and $\overline{AD} \cong \overline{BC}$.

2. **Opposite angles of a parallelogram are congruent.**
   Here, $\angle A \cong \angle C$ and $\angle B \cong \angle D$.

3. **Consecutive angles of a parallelogram are supplementary.**
   Here, $m\angle A + m\angle B = 180$, $m\angle B + m\angle C = 180$, and so forth.

Since rhombuses, rectangles, and squares are members of the family of parallelograms, these statements will be true for any rhombus, any rectangle, and any square.

Informal Proofs for Statements About Angles in a Parallelogram

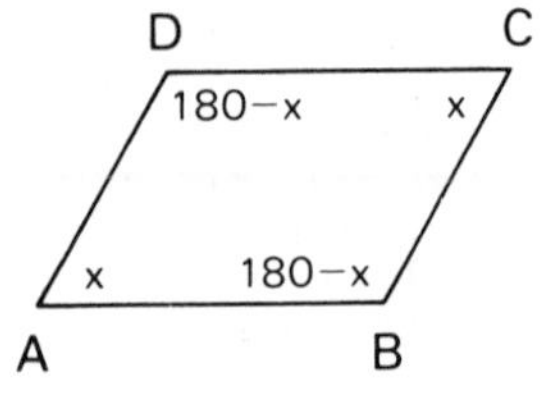

Consider parallelogram $ABCD$. The following statements can help to prove informally that *consecutive angles of a parallelogram are supplementary.*

(1) $\overline{DC} \parallel \overline{AB}$ and $\overline{AD}$ is a transversal (because opposite sides of a parallelogram are parallel and a transversal intersects these lines).
(2) If $m\angle A = x$, then $m\angle D = 180 - x$ (since $\angle A$ and $\angle D$ are interior angles on the same side of a transversal, previously shown to be supplementary).
(3) Similarly, in looking at $\overline{AD} \parallel \overline{BC}$ and the transversal $\overline{AB}$, we can say: If $m\angle A = x$, then $m\angle B = 180 - x$ for the same reason as given in step (2).
(4) Also, in looking at $\overline{AB} \parallel \overline{DC}$ and the transversal $\overline{BC}$, we say: If $m\angle B = 180 - x$, then $m\angle C = x$. Therefore, we have proved:

- **Consecutive angles of a parallelogram are supplementary.**

We can use this statement to prove informally that:

- **Opposite angles of a parallelogram are equal in measure.**

(1) In parallelogram $ABCD$, we have shown that:
$m\angle A = x$, $m\angle B = 180 - x$, $m\angle C = x$, and $m\angle D = 180 - x$.
(2) Thus, $m\angle A = m\angle C$ and $m\angle B = m\angle D$.

Remember that these statements are also true for any rhombus, any rectangle, and any square.

MODEL PROBLEM

$ABCD$ is a parallelogram where $m\angle A = 2x + 50$ and $m\angle C = 3x + 40$.

a. Find the value of x.
b. Find the measure of each angle.

Solution:

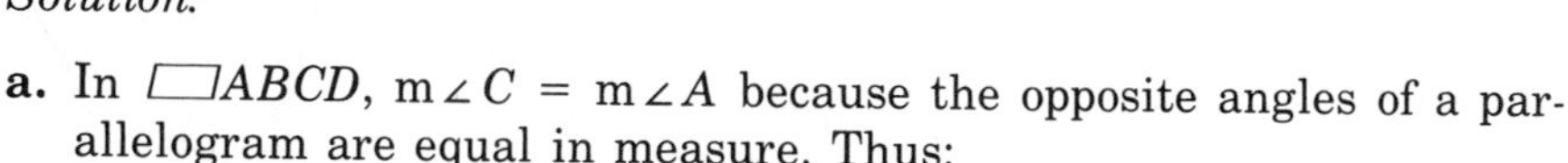

a. In ▱$ABCD$, $m\angle C = m\angle A$ because the opposite angles of a parallelogram are equal in measure. Thus:

$$3x + 40 = 2x + 50$$
$$x = 10 \quad \textit{Ans.}$$

b. By substitution:
$m\angle A = 2x + 50 = 2(10) + 50 = 70$ and
$m\angle C = 3x + 40 = 3(10) + 40 = 70$.
Since $m\angle B + m\angle A = 180$, $m\angle B + 70 = 180$, and $m\angle B = 110$.
Since $m\angle D = m\angle B$, $m\angle D = 110$.

Answer: $m\angle A = 70$, $m\angle B = 110$, $m\angle C = 70$, and $m\angle D = 110$.

EXERCISES

In 1–6:

a. Copy the given statement. Is it true or false?
b. Write the converse of the given statement. Is it true or false?
c. Write the inverse of the given statement. Is it true or false?
d. Write the contrapositive of the given statement. Is it true or false?

1. If a polygon is a trapezoid, it is a quadrilateral.
2. If a polygon is a rectangle, it is a parallelogram.
3. If a polygon is a rhombus, it is a parallelogram.
4. If a polygon is a rhombus, it is a square.
5. If a polygon is a parallelogram, it is a square.
6. If two angles are opposite angles of a parallelogram, they are congruent.

In 7–10, the angle measures are represented in each quadrilateral.
a. Find the value of x.
b. Find the measure of each angle of the quadrilateral.

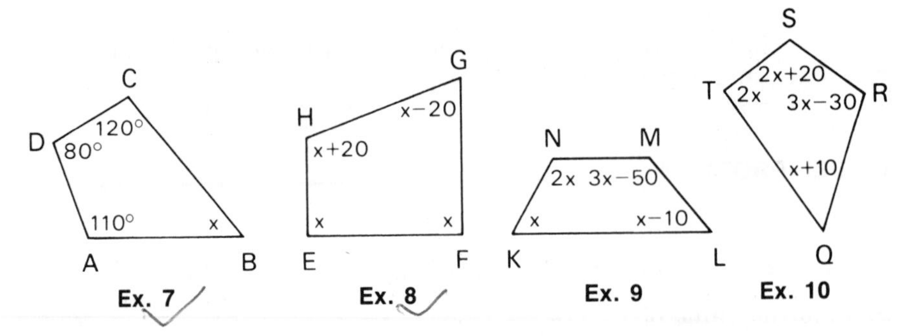

Ex. 7 Ex. 8 Ex. 9 Ex. 10

In 11 and 12, polygon $ABCD$ is a parallelogram.

11. $AB = 3x + 8$; $DC = x + 12$.
Find AB and DC.

12. $m\angle A = 5x - 40$; $m\angle C = 3x + 20$.
Find $m\angle A$, $m\angle B$, $m\angle C$, and $m\angle D$.

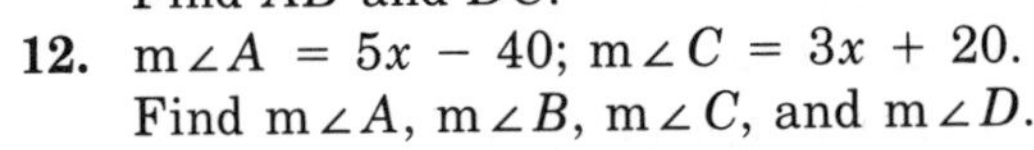

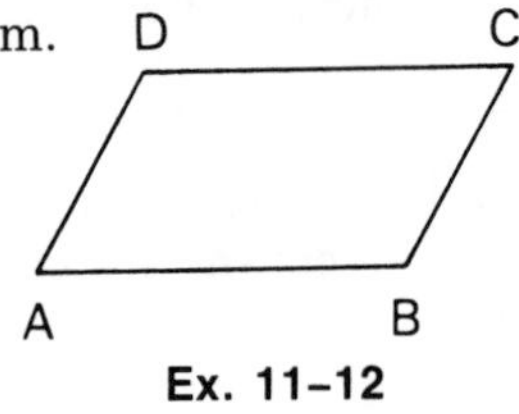

Ex. 11–12

In 13 and 14, polygon $ABCD$ is a rectangle.

13. $BC = 4x - 5$, $AD = 2x + 3$.
Find BC and AD.

14. $m\angle A = 5x - 10$. Find the value of x.

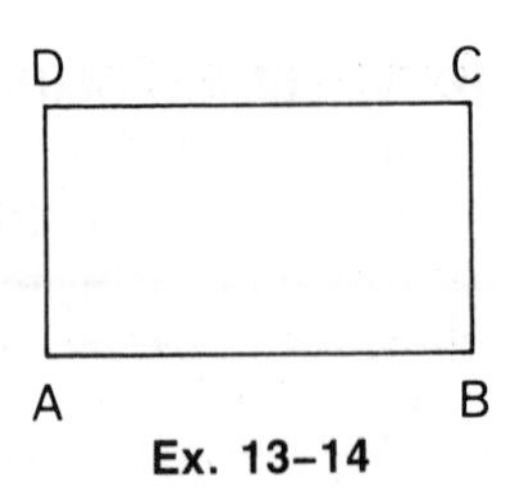

Ex. 13–14

15. $ABCD$ is a square. If $AB = 8x - 6$ and $BC = 5x + 12$, find the length of each side of the square.

16. In rhombus $KLMN$, $KL = 3x$, $LM = 2(x + 3)$. Find the length of each side of the rhombus.

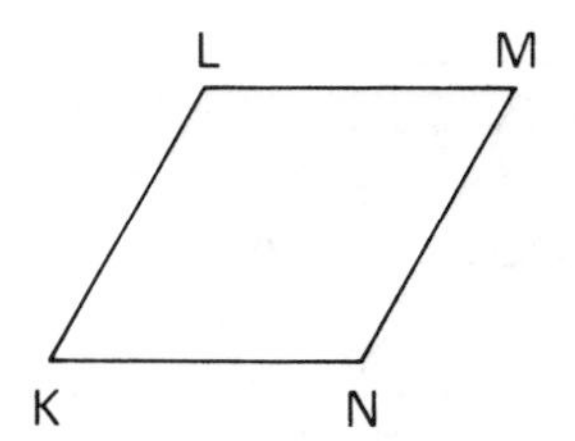

17. In the figure at the right, $\triangle MPT$ is an isosceles triangle in which $PM = TM$. $\overline{RS}$ is drawn parallel to $\overline{PT}$ forming an *isosceles trapezoid* in which $PR = TS$. (In an ***isosceles trapezoid*** exactly one pair of opposite sides is parallel, while the sides that are not parallel are congruent.)

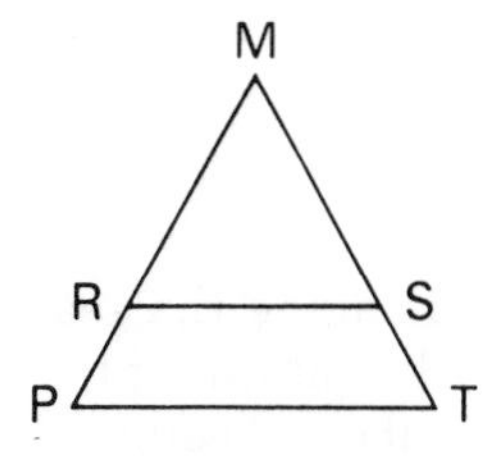

a. If $PR = x + 3$, $TS = 2x + 2$, $RS = x + 6$, and $PT = 8x + 3$, find the length of each side.

b. If $m\angle P = 60$, find the measures of all four angles in the isosceles trapezoid.

18. In the figure at the right, the angle measures of quadrilateral $ABCD$ are 70°, 65°, 85°, and 140°. Four exterior angles are drawn.

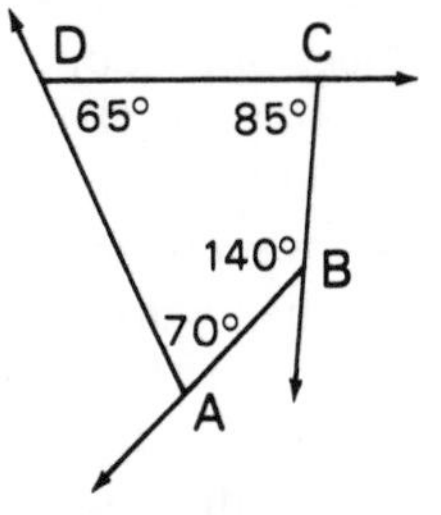

a. What is the sum of the measures of the interior angles of the quadrilateral?

b. What is the sum of the measures of the exterior angles of the quadrilateral?

c. Will the answer to part **b** be true for the sum of the measures of the exterior angles of any quadrilateral?

11-9 TRANSFORMATIONS

What Is a Transformation?

In the game of pool, fifteen object balls numbered 1 through 15 are placed in a triangular framework called a rack (see Fig. A). Let us suppose that the object balls are taken out of the rack, mixed up, and thrown back, as shown in Fig. B. Most of the object balls have changed their position. A few object balls, such as 2 and 11, remain fixed in their position. However, the new rack of fifteen object balls is merely a *change*, or a *transformation*, of the fifteen object balls found in the original rack.

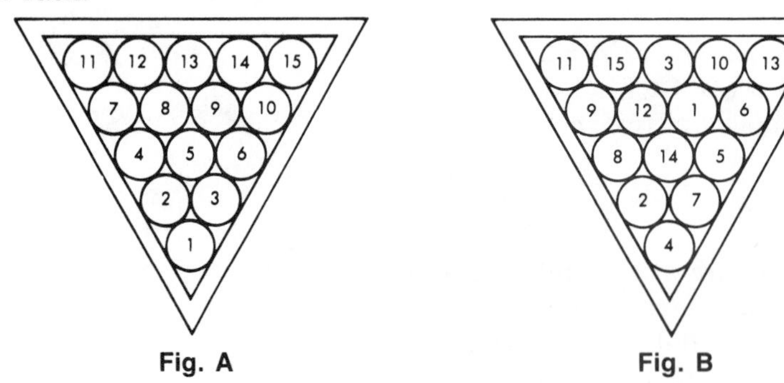

Fig. A **Fig. B**

Imagine that each object ball is like a point, and imagine that the rack containing the object balls is like a plane that contains an infinite number of points. In the same way that object balls in a rack change their position, under a *transformation of the plane*, points will move about and change their position in the plane. At times, some of the points in the plane may remain fixed. After the transformation, or change, takes place, however, the plane must once again appear full and complete, without any missing points, just as the rack of fifteen object balls appeared full and complete in Fig. B.

Compare the positions of the object balls in the two racks: 1 is replaced by 4, 2 is still 2, 3 is replaced by 7, and so on. This shows a *one-to-one correspondence* between the two sets, each of fifteen object balls. In other words, each object ball is replaced by one and only one object ball until the rack is once again complete. We will extend this idea to points in a plane.

An infinite number of transformations can take place in a plane. In this chapter we will study only a few special transformations.

Line Reflection

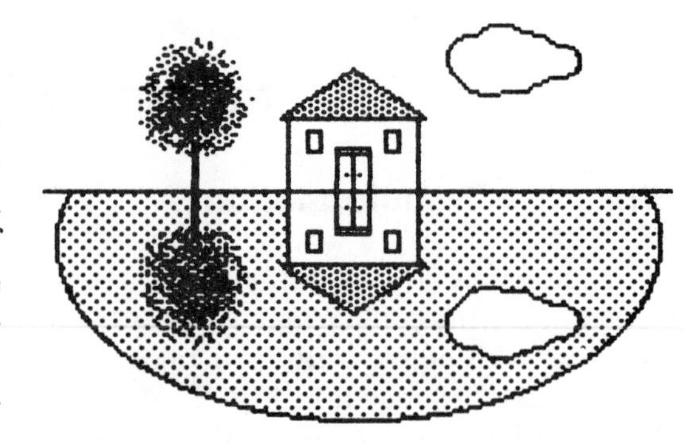

It is often possible to see the objects along the shore of a body of water reflected in the water. If a picture of such a scene is folded, the objects can be made to coincide with their images. Each point of the reflection is an image point of the corresponding point of the object. The line along which the picture is folded is the ***line of reflection***. This common experience is used in mathematics to study congruent figures.

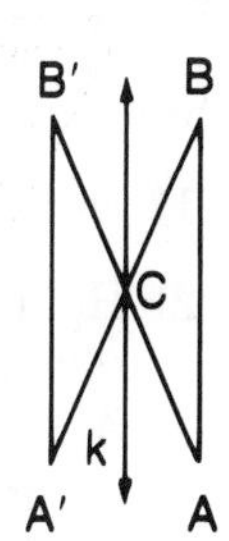

If the figure at the right were folded along line k, $\triangle ABC$ would coincide with $\triangle A'B'C$. Line k is the line of reflection, the image of A is A' (in symbols, $A \rightarrow A'$), and the image of B is B' ($B \rightarrow B'$). Point C is a fixed point because it is *on* the line of reflection. In other words, we say that C is its own image ($C \rightarrow C$). Under a reflection in line k, the image of $\triangle ABC$ is $\triangle A'B'C$ ($\triangle ABC \rightarrow \triangle A'B'C$).

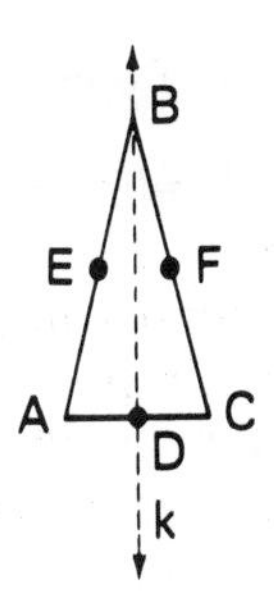

Isosceles triangle ABC is shown at the right. Imagine this triangle folded so that A falls on C. The line along which it folds is a reflection line. Every point of the triangle has as its image a point of the triangle. Points B and D are fixed points because they are on the line of reflection.

Thus, under the line reflection in k:

1. all points of the triangle are reflected so that $A \rightarrow C$, $C \rightarrow A$, $E \rightarrow F$, $F \rightarrow E$, $B \rightarrow B$, $D \rightarrow D$, and so on.
2. the sides of the triangle are reflected. That is, $\overline{AB} \rightarrow \overline{CB}$, a statement verifying that the legs of an isosceles triangle are congruent. Also, $\overline{AC} \rightarrow \overline{CA}$, showing that the base is its own image.
3. the angles of the triangle are reflected. That is, $\angle BAD \rightarrow \angle BCD$, a statement verifying that the base angles of an isosceles triangle are congruent. Also, $\angle ABC \rightarrow \angle CBA$, showing that the vertex angle is its own image.

Looking at isosceles triangle ABC and the reflection line k, we can note some properties of a line reflection:

1. Distance is preserved (unchanged).

$$\overline{AB} \rightarrow \overline{CB} \text{ and } AB = CB$$
$$\overline{AD} \rightarrow \overline{CD} \text{ and } AD = CD$$

2. Angle measure is preserved.

$$\angle BAD \rightarrow \angle BCD \text{ and } m\angle BAD = m\angle BCD$$
$$\angle BDA \rightarrow \angle BDC \text{ and } m\angle BDA = m\angle BDC$$

3. The line of reflection is the perpendicular bisector of every segment joining a point to its image.
4. A figure is always congruent to its image.

MODEL PROBLEM

On your paper, draw a segment and label the endpoints A and B. Draw any line m. Sketch the image of $\overline{AB}$ under a reflection in m.

Solution:

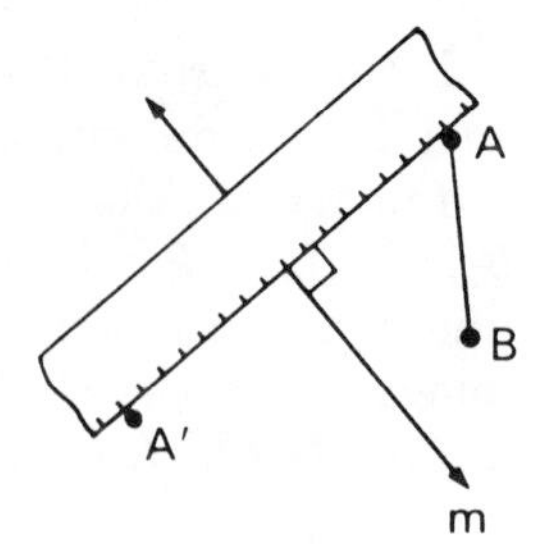

(1) Draw $\overline{AB}$ and line m.

(2) Hold a ruler perpendicular to line m and touching point A. Measure the distance from A to line m. Find a point along the ruler that is the same distance from m as A but that is on the opposite side of m. Label this point A'.

(3) Repeat step (2) for point B to locate B'.

(4) Draw $\overline{A'B'}$, the image of $\overline{AB}$.

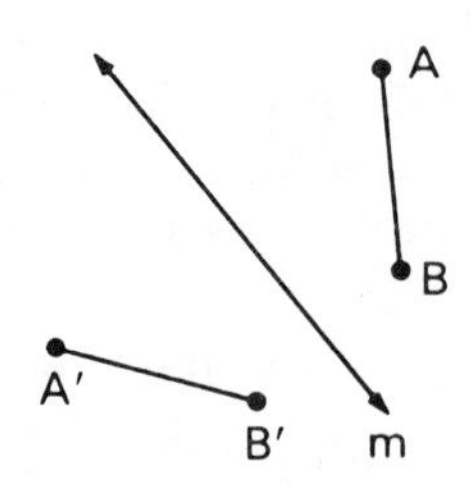

EXERCISES

In 1–4, copy the figure and line m on your paper. Using a ruler, sketch the image of the given figure under a reflection in line m.

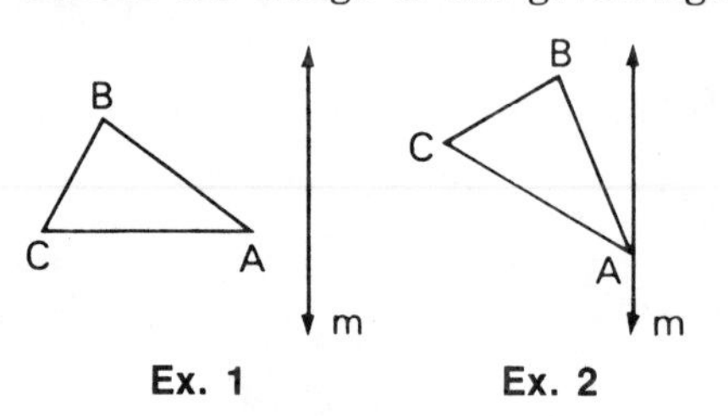

Ex. 1 Ex. 2

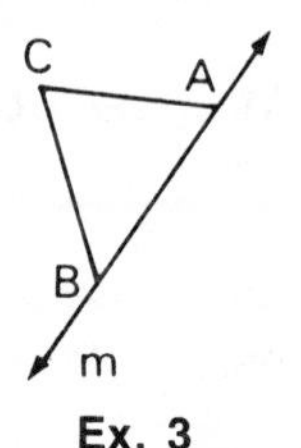

Ex. 3

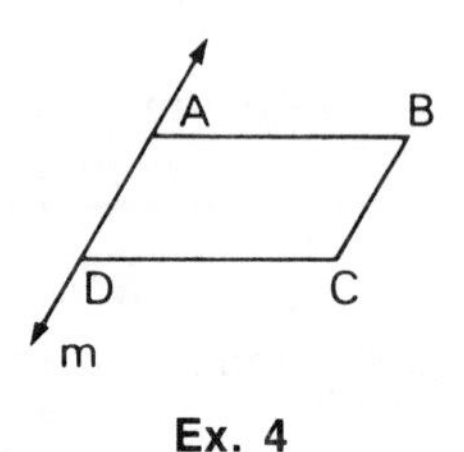

Ex. 4

5. **a.** Draw square $ABCD$.
b. Draw m, a line of reflection for which the image of A is B.
c. Draw n, a line of reflection for which the image of A is C.
d. Draw p, a line of reflection for which the image of A is D.

Line Symmetry

In nature, in art, and in industry, we find many forms that have a pleasing and attractive appearance because of a balanced arrangement of their parts. We say that such forms have symmetry.

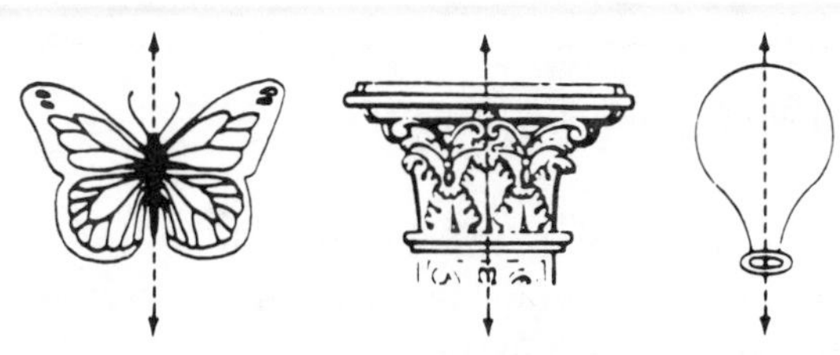

In each of the figures above, there is a line on which the figure could be folded so that the parts of the figure on opposite sides of that line will coincide. If we think of that line as a line of reflection, each point of the figure has as its image a point of the figure. This line of reflection is a line of symmetry, or ***axis of symmetry***, and the figure has line symmetry.

An isosceles triangle has line symmetry. In the diagram on page 357, the line of reflection, k, is an axis of symmetry and the isosceles triangle is symmetric with respect to the line through its vertex that is perpendicular to its base.

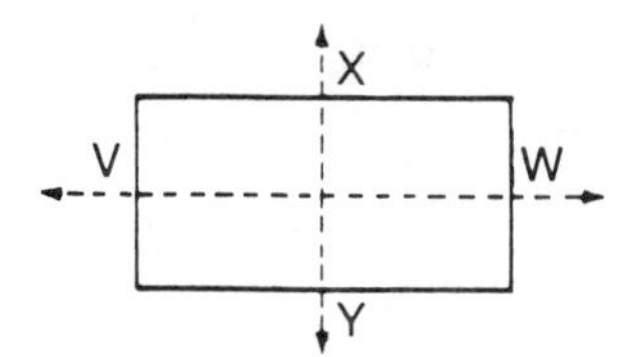

It is possible for a figure to have more than one axis of symmetry. In the rectangle at the left, the line XY is an axis of symmetry and the line VW is a second axis of symmetry.

Lines of symmetry may be found for some letters and for some words, as shown to the right.

A CODE

Not every figure has line symmetry. Consider ▱$ABCD$ and diagonal $\overline{BD}$.

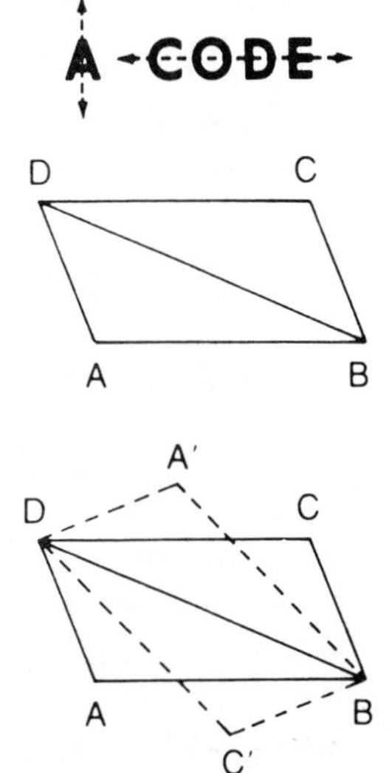

If ▱$ABCD$ is reflected in $\overline{BD}$, the image of A is A' and the image of C is C'. The points A' and C' are not points of the original parallelogram. The image of ▱$ABCD$ under a reflection in $\overline{BD}$ is ▱$A'BC'D$. Therefore, ▱$ABCD$ is not symmetric with respect to $\overline{BD}$.

We have used diagonal $\overline{BD}$ as a line of reflection, but note that it is not a line of symmetry. The parallelogram has no line of symmetry. That is, there is no line along which the parallelogram may be folded so that points of the parallelogram on one side of the line will coincide with points of the parallelogram on the other.

MODEL PROBLEM

How many lines of symmetry does the letter **H** have?

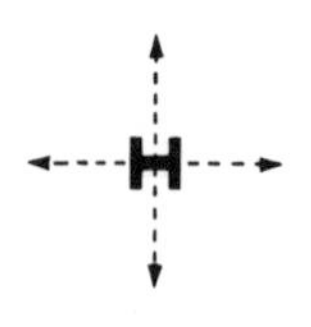

Solution:

The horizontal line through the crossbar is a line of symmetry. The vertical line midway between the vertical segments is also a line of symmetry.

Answer: The letter **H** has two lines of symmetry.

EXERCISES

1. Using the printed capital letters of the alphabet, write all letters that have line symmetry on your paper. Show the lines of symmetry.
2. Copy each of the following "words." Draw a line of symmetry or indicate that the word does not have line symmetry by writing "none."

a. MOM **b.** DAD **c.** SIS **d.** OTTO
e. BOOK **f.** RADAR **g.** un **h.** NOON
i. HIKE **j.** SWIMS **k.** OHHO **l.** CHOKED

In 3–14, for each geometric figure named: **a.** Sketch the figure. **b.** Tell the number of lines of symmetry, if any, that the figure has and sketch them on your drawing.

3. rectangle **4.** equilateral triangle **5.** parallelogram
6. isosceles triangle **7.** rhombus **8.** regular hexagon
9. trapezoid **10.** scalene triangle **11.** circle
12. regular octagon **13.** square **14.** regular pentagon

Point Reflection

Another kind of reflection is with respect to a point. In the figure at the right, $\triangle A'B'C'$ is the image of $\triangle ABC$ under a reflection in point P. If a line segment is drawn connecting any point to its image, then the point of reflection is the midpoint of that segment. In the figure:

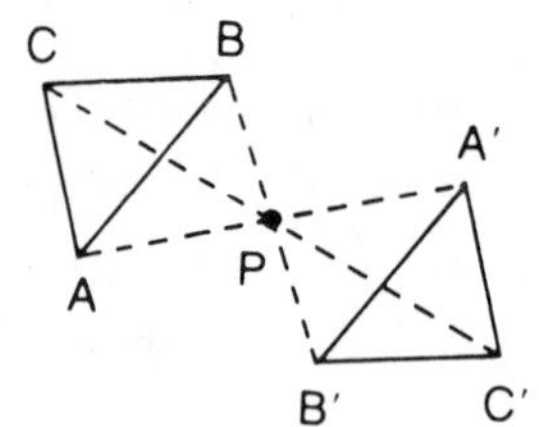

Point A' is on $\overleftrightarrow{AP}$, $AP = PA'$, and P is the midpoint of $\overline{AA'}$.

Point B' is on $\overleftrightarrow{BP}$, $BP = PB'$, and P is the midpoint of $\overline{BB'}$.

Point C' is on $\overleftrightarrow{CP}$, $CP = PC'$, and P is the midpoint of $\overline{CC'}$.

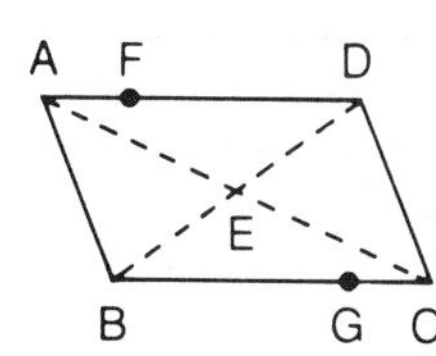

Parallelogram $ABCD$ is shown at the left. Diagonals $\overline{AC}$ and $\overline{BD}$ intersect at E. Point E is the midpoint of $\overline{AC}$ and of $\overline{BD}$. Therefore, under a reflection in point E, $A \rightarrow C$ and $C \rightarrow A$, $B \rightarrow D$ and $D \rightarrow B$. Similarly, $F \rightarrow G$ and $G \rightarrow F$, and every point of the parallelogram has its image on the parallelogram.

Looking at parallelogram $ABCD$ and the point of reflection E, we can note some properties of point reflection:

1. Distance is preserved.
   $$\overline{AB} \rightarrow \overline{CD} \text{ and } AB = CD$$
   $$\overline{AD} \rightarrow \overline{CB} \text{ and } AD = CB$$
2. Angle measure is preserved.
   $$\angle BAD \rightarrow \angle DCB \text{ and } m\angle BAD = m\angle DCB$$
   $$\angle ABC \rightarrow \angle CDA \text{ and } m\angle ABC = m\angle CDA$$
3. The point of reflection is the midpoint of every segment formed by joining a point to its image.
   $$AE = EC \text{ and } BE = ED$$
4. A figure is always congruent to its image.

MODEL PROBLEM

On your paper, draw any triangle ABC. Sketch the image of $\triangle ABC$ under a reflection in point A.

Solution:

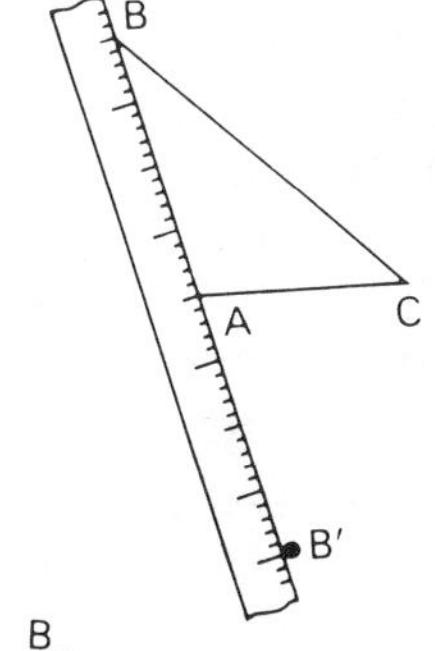

(1) Draw a triangle and label it $\triangle ABC$.

(2) Hold a ruler on $\overline{AB}$ and measure the distance from B to A. Since A is the point of reflection, the image of B is on the line $\overleftrightarrow{AB}$. Locate the point along the ruler that is the same distance from A as B but that is on the opposite side of A. Label this point B'.

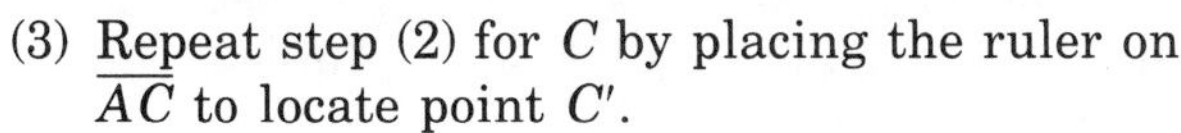

(3) Repeat step (2) for C by placing the ruler on $\overline{AC}$ to locate point C'.

(4) Draw $\triangle AB'C'$, the image of $\triangle ABC$.

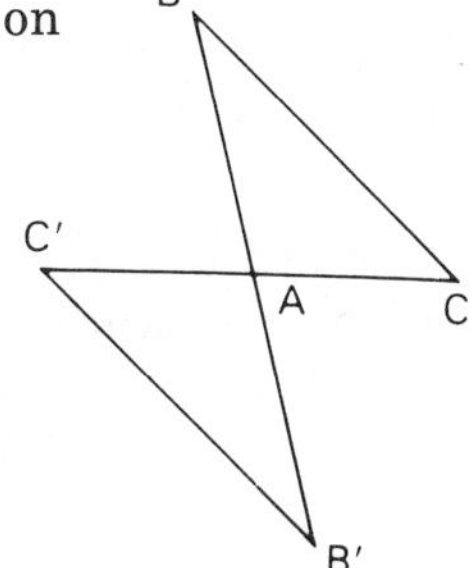

EXERCISES

1. On your paper, copy the figures in Exercises 1–4 on page 358, but omit line m. Using a ruler, sketch the image of the given figure under a reflection in point A.

2. In the figure, $\triangle ABC \cong \triangle DBE$. Find the image of each of the following under a reflection in B.
   a. A b. B c. C
   d. D e. E f. $\overline{AC}$
   g. $\overline{AB}$ h. $\overline{DE}$

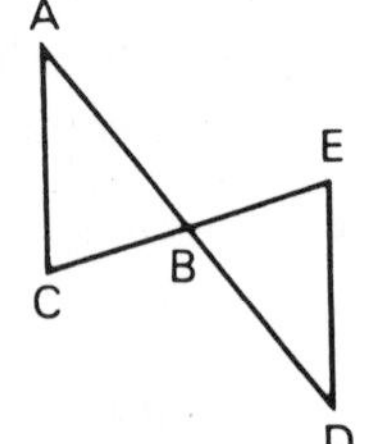

Point Symmetry

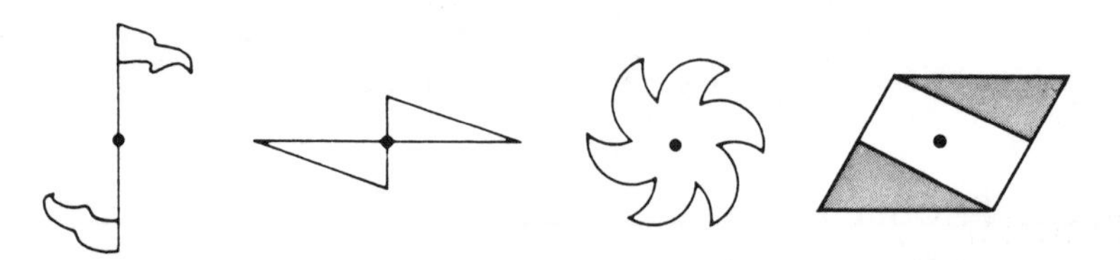

In each of the figures shown above, the design is built around a central point. For every point in the figure, there is another point at the same distance from the center so that the center is the midpoint of the segment joining the pair of points. Under a point reflection through the center, each point has as its image another point of the figure. The figure has point symmetry.

The parallelogram $ABCD$ shown on page 361 has point symmetry under a reflection in point E, the intersection of its diagonals.

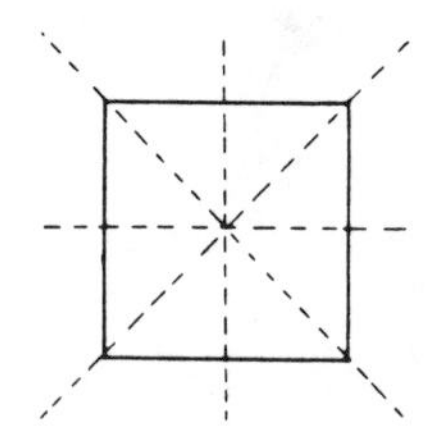

It is possible for a figure to have both line symmetry and point symmetry at the same time. In the square at the left, there are 4 lines of symmetry. Note that the point of symmetry lies at the intersection of these 4 lines.

Points of symmetry may be found for some letters and for some words, as shown.

Z MOW

EXERCISES

1. Using the printed capital letters of the alphabet, write on your paper all of the letters that have point symmetry and show the point of symmetry on your drawing.
2. For each of the words in Exercise 2 on page 360, locate a point of symmetry or indicate that the word does not have point symmetry by writing "no point symmetry."
3. For each geometric figure named in Exercises 3–14 on page 360:
   - **a.** Sketch the figure.
   - **b.** Tell if the figure has point symmetry and locate the point of symmetry if one exists.

In 4–9, tell if the figure has point symmetry.

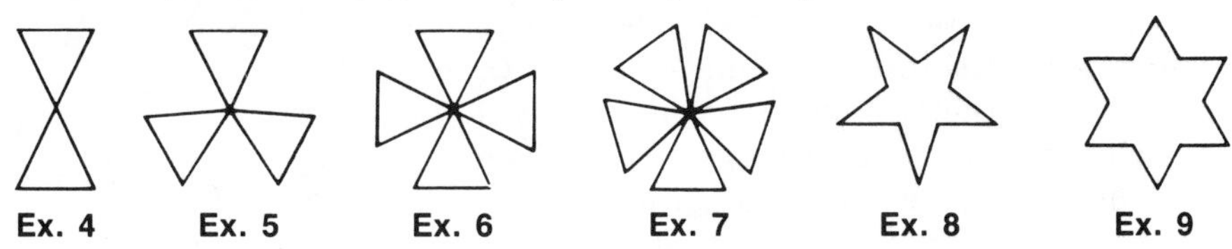

Ex. 4 **Ex. 5** **Ex. 6** **Ex. 7** **Ex. 8** **Ex. 9**

Translation

It is often useful or necessary to move objects from one place to another. If you move your desk from one place in your room to another, each leg moves the same distance in the same direction.

A translation moves every point in the plane the same distance in the same direction.

If $\triangle A'B'C'$ is the image of $\triangle ABC$ under a translation, $AA' = BB' = CC'$ and $\overline{AA'} \parallel \overline{BB'} \parallel \overline{CC'}$. The size and shape of the figure are unchanged so that $\triangle ABC \cong \triangle A'B'C'$. Thus, as with reflections, a figure is congruent to its image under a translation.

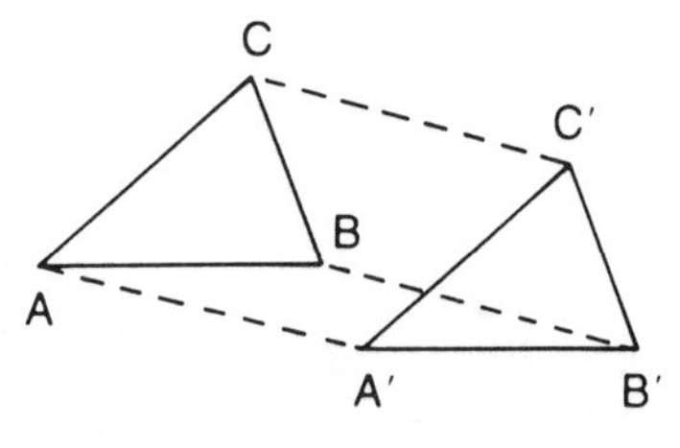

Patterns used for decorative purposes such as wallpaper or the binding on clothing often appear to have translational symmetry, but true translational symmetry is possible only if the pattern could repeat without end.

EXERCISES

In 1 and 2, the diagram consists of nine congruent rectangles.

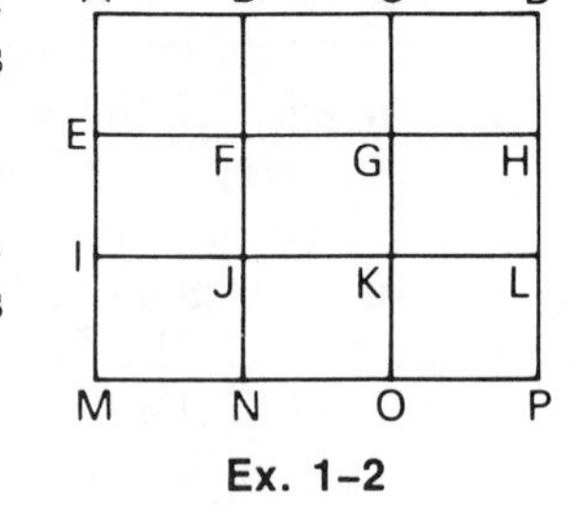

Ex. 1–2

1. Under a translation, the image of A is G. Find the image of each of the given points under the same translation.
   a. J **b.** B **c.** I **d.** F **e.** E
2. Under a translation, the image of K is J. Find the image of each of the given points under the same translation.
   a. J **b.** B **c.** O **d.** L **e.** G

3. In the diagram, $ADEH$ is a parallelogram. Points B, C, G, and F divide $\overline{AD}$ and $\overline{EH}$ into congruent segments. Under a translation, the image of A is G. Under the same translation, tell whether or not the image of the given point is a point of the diagram.
   a. G **b.** B **c.** C **d.** F

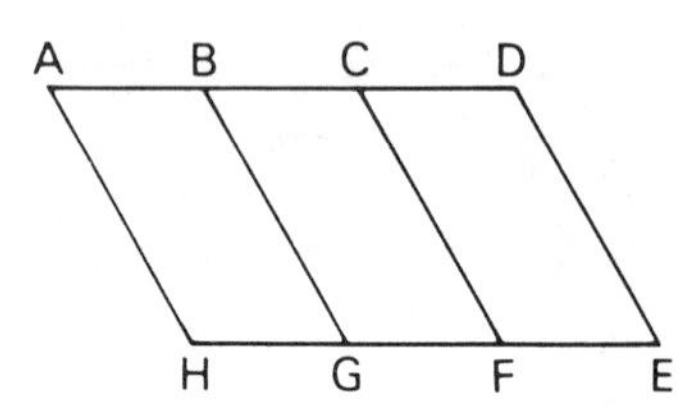

Rotations

Think of what happens to all of the points of the steering wheel of a car as the wheel is turned. Except for the fixed point in the center, every point moves through a part of a circle, or arc, so that the position of each point is changed by a rotation of the same number of degrees.

In the figure, if A is rotated to A', then B is rotated the same number of degrees to B', and $m\angle APA' = m\angle BPB'$. Since P is the center of rotation, $PA = PA'$ and $PB = PB'$.

In general, a rotation preserves distance and angle measure. Under a rotation, a figure is congruent to its image. Unless otherwise stated, a rotation is in the counterclockwise direction.

Many letters, as well as designs in the shape of wheels, stars, and polygons have rotational symmetry. Each figure shown at the right has rotational symmetry.

Any regular polygon has rotational symmetry. When regular pentagon $ABCDE$ is rotated $\frac{360°}{5}$, or $72°$, about its center, the image of every point of the figure is a point of the figure. Under this rotation, $A \rightarrow B$, $B \rightarrow C$, $C \rightarrow D$, $D \rightarrow E$, and $E \rightarrow A$.

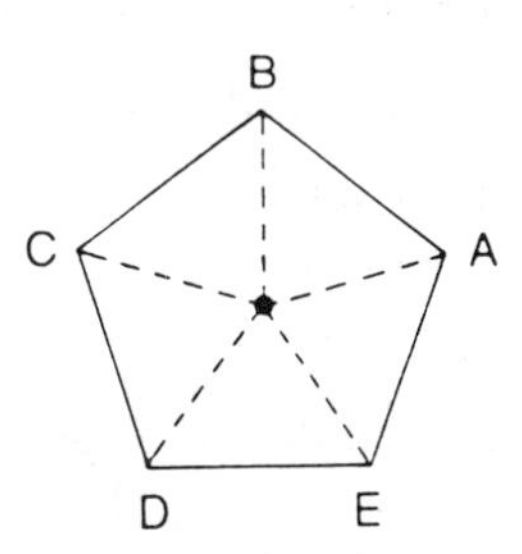

The figure would also have rotational symmetry if rotated through a multiple of $72°$ ($144°$, $216°$, or $288°$). If it were rotated through $360°$, every point would be its own image. Since this is true for every figure, we do not usually consider a $360°$ rotation as rotational symmetry.

MODEL PROBLEM

Point O is at the center of equilateral triangle ABC so that $OA = OB = OC$. Find the image of each of the following under a rotation of $120°$ about O.

a. A **b.** B **c.** C **d.** $\overline{AB}$ **e.** $\angle CAB$

Answers:

a. B **b.** C **c.** A **d.** $\overline{BC}$ **e.** $\angle ABC$

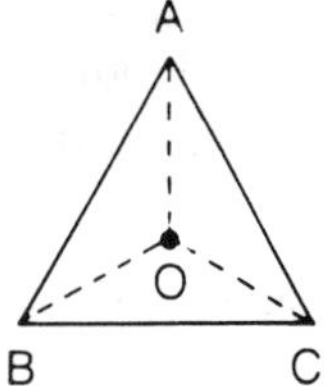

EXERCISES

Exercises Using Rotations

1. What is the image of each of the given points under a rotation of $90°$ in the counterclockwise direction about O?
a. A **b.** B **c.** C **d.** G **e.** H
f. J **g.** K

2. What is the image of each of the given points under a rotation of $90°$ in the clockwise direction about O?
a. A **b.** B **c.** C **d.** G **e.** H
f. J **g.** K

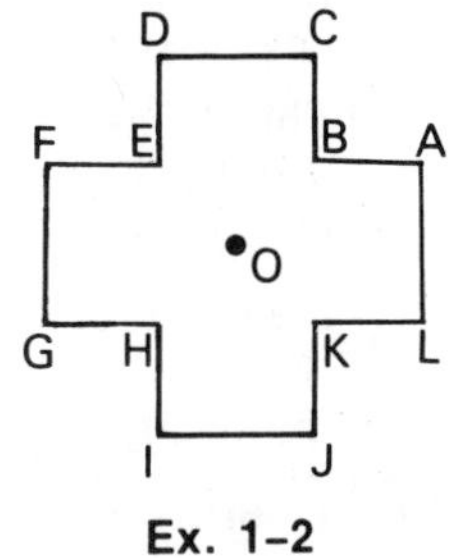

Ex. 1–2

3. For each geometric figure named in Exercises 3–14 on page 360:
a. Tell if the figure has rotational symmetry.
b. If the figure has rotational symmetry, give the measure of the smallest angle of rotation for which the symmetry exists.

General Exercises

4. Look at the handwritten word "chump" in the box at the left. Does it have any type of symmetry?

5. What is the image of each of the following under a reflection in $\overline{EG}$?
   a. A b. E c. $\overline{AO}$ d. $\angle HDO$
6. What is the image of each of the following under a reflection in O?
   a. A b. E c. $\overline{AO}$ d. $\angle HDO$
7. What is the image of each of the following under a rotation of 90° about O?
   a. A b. E c. $\overline{AO}$ d. $\angle HDO$

Ex. 5–9

8. Under a certain translation, the image of A is O. What is the image of each of the following under the same translation?
   a. H b. O c. $\overline{AH}$ d. $\angle OAH$
9. Under which transformation is the figure not symmetric?
   (1) line reflection (2) point reflection
   (3) translation (4) 90° rotation

11-10 REVIEW EXERCISES

1. Find the number of degrees in $\frac{3}{8}$ of a complete rotation.
2. If the lines $\overleftrightarrow{AB}$ and $\overleftrightarrow{CD}$ intersect at E, $m\angle AEC = x + 10$, and $m\angle DEB = 2x - 30$, find the measure of $\angle AEC$.
3. If two angles of a triangle are complementary, what is the measure of the third angle?
4. The measure of the complement of an angle is 20° less than the measure of the angle. Find the number of degrees in the angle.
5. If each base angle of an isosceles triangle measures 55°, find the measure of the vertex angle of the triangle.

In 6–8, $\overleftrightarrow{AB}$ is parallel to $\overleftrightarrow{CD}$, and these lines are cut by transversal $\overleftrightarrow{EF}$ at points G and H, respectively.

6. If $\angle AGH$ measures 73°, find $m\angle GHD$.
7. If $m\angle EGB = 70$ and $m\angle GHD = 3x - 2$, find x.
8. If $m\angle HGB = 2x + 10$ and $m\angle GHD = x + 20$, find $m\angle GHD$.

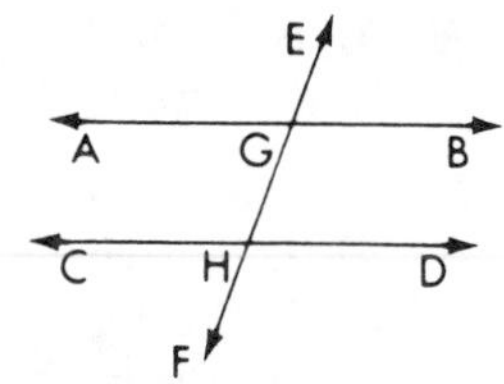

Ex. 6–8

9. In $\triangle ART$, $m\angle A = y + 10$, $m\angle R = 2y$, and $m\angle T = 2y - 30$.
a. Find the measure of each of the three angles.
b. *Choose one:* $\triangle ART$ is
(1) right (2) isosceles (3) equilateral (4) scalene

In 10–12, $\angle BCD$ is an exterior angle to triangle ABC at vertex C.

10. If $m\angle A = 57$ and $m\angle B = 62$, find $m\angle BCD$.
11. If $m\angle BCD = 118$ and $m\angle A = 54$, find $m\angle B$.
12. If $m\angle A = x + 12$, $m\angle B = x + 4$, and $m\angle BCD = 120$, find x.

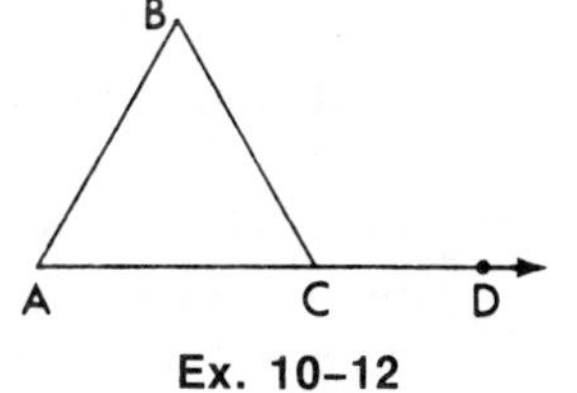

Ex. 10–12

13. If $FLAT$ is a square where $FL = 9y - 2$ and $LA = 7y + 16$, find y.
14. In quadrilateral $ABCD$, $m\angle A = 100$, $m\angle B = 70$, $m\angle C = x$, and $m\angle D = x + 50$. Find $m\angle C$.
15. Which printed capital letters of the alphabet have point symmetry but *not* line symmetry?

In 16–19, select the numeral preceding the correct answer.

16. In triangle ABC, the measure of $\angle B$ is $\frac{3}{2}$ the measure of $\angle A$ and the measure of $\angle C$ is $\frac{5}{2}$ the measure of $\angle A$. What is the measure of the smallest angle?
(1) 9° (2) 18° (3) 36° (4) 40°

17. The measure of one angle is three times that of another angle, and the sum of these measures is 120°. Which is true?
(1) One angle must be obtuse.
(2) Both angles are acute.
(3) One angle must be right.
(4) The angles are complementary.

18. The measure of the smaller of two supplementary angles is $\frac{4}{5}$ of the measure of the larger. The measure of the smaller angle is
(1) 10° (2) 20° (3) 40° (4) 80°

19. Which condition does *not* demonstrate that two triangles are congruent?
(1) a.s.a. ≅ a.s.a. (2) s.a.s. ≅ s.a.s.
(3) a.a.a. ≅ a.a.a. (4) s.s.s. ≅ s.s.s.

20. In the diagram, $\triangle ABC$ is isosceles with $AC = BC$, $m\angle ACB = 4x$, and the measure of exterior $\angle BAD = 5x$.

a. Express the measure of $\angle B$ in terms of x.

b. Find the value of x.

c. Find the measure of each of the indicated angles:

1. $m\angle ABC$ **2.** $m\angle ACB$

3. $m\angle BAD$

21. In $\triangle ABC$, D is the midpoint of $\overline{AC}$, and $\overline{BD}$ intersects $\overline{AC}$ at right angles.

a. State a reason why $\triangle ABD \cong \triangle CBD$.

b. If $AB = 3x + 1$ and $CB = 5x - 7$, find x.

c. What is the image of A under a reflection in $\overleftrightarrow{BD}$?

d. What is the line of symmetry for $\triangle ABC$?

22. In the diagram below, O is the center of the circle, $OACD$ is a rectangle, $AB = 8$ cm, and $AD = 13$ cm. Find the radius of the circle.

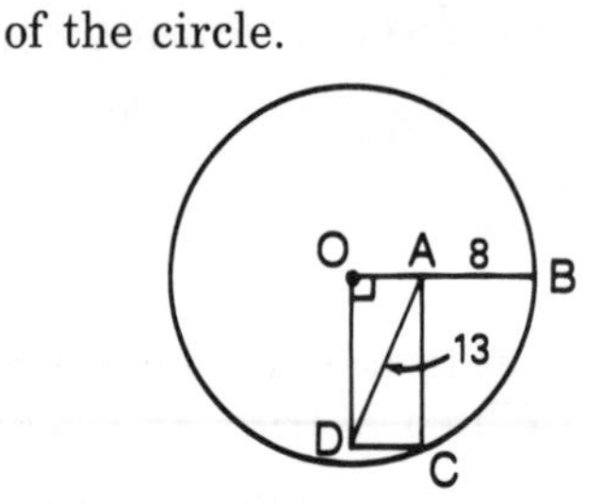

23. How many triangles are in the figure?

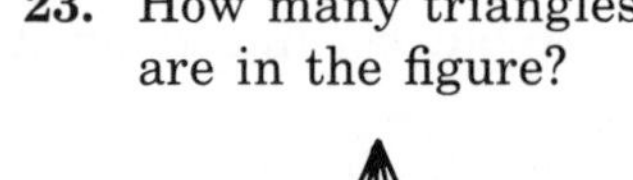

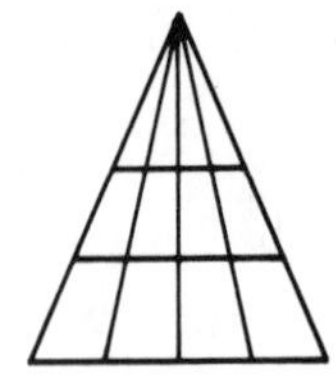

Chapter 12

Ratio and Proportion

12-1 RATIO

A ***ratio*** of one number to another nonzero number is the quotient of the first number divided by the second. Therefore, a ratio may be considered as an ordered pair of numbers, and finding the ratio of two numbers is a binary operation.

Since a ratio is the quotient of two numbers in a definite order, you must be careful to write them in the intended order. For example, the ratio of 3 to 1 is written $\frac{3}{1}$ or 3:1, whereas the ratio of 1 to 3 is written $\frac{1}{3}$ or 1:3.

In general, the ratio of a to b can be expressed as

$$\frac{a}{b} \text{ or } a \div b \text{ or } a : b$$

A ratio is used to compare two numbers by division. To find the ratio of two quantities, both quantities must be expressed in the same unit of measure before finding their quotient. For example, to compare a nickel with a penny, first convert the nickel to 5 pennies and then find the ratio, which is $\frac{5}{1}$ or 5:1. Therefore, a nickel is worth 5 times as much as a penny. The ratio has no unit of measure.

Equivalent Ratios

Since the ratio $\frac{5}{1}$ is a fraction, we can use the multiplication property of 1 to find many equivalent ratios. For example:

$$\frac{5}{1} = \frac{5}{1} \cdot \frac{2}{2} = \frac{10}{2} \qquad \frac{5}{1} = \frac{5}{1} \cdot \frac{3}{3} = \frac{15}{3} \qquad \frac{5}{1} = \frac{5}{1} \cdot \frac{x}{x} = \frac{5x}{1x} \quad (x \neq 0)$$

From the last example, you see that $5x$ and $1x$ represent two numbers whose ratio is 5:1.

In general, if a, b, and x are numbers ($b \neq 0$, $x \neq 0$), ax and bx represent two numbers whose ratio is $a:b$ because:

$$\frac{a}{b} = \frac{a}{b} \cdot 1 = \frac{a}{b} \cdot \frac{x}{x} = \frac{ax}{bx}$$

Also, since a ratio such as $\frac{24}{16}$ is a fraction, we can divide the numerator and the denominator of the fraction by the same nonzero number to find equivalent ratios. For example:

$$\frac{24}{16} = \frac{24 \div 2}{16 \div 2} = \frac{12}{8} \qquad \frac{24}{16} = \frac{24 \div 4}{16 \div 4} = \frac{6}{4} \qquad \frac{24}{16} = \frac{24 \div 8}{16 \div 8} = \frac{3}{2}$$

A ratio is expressed in ***simplest form*** when both terms of the ratio are whole numbers and when there is no whole number other than 1 that is a factor of both of these terms. Therefore, to express the ratio $\frac{24}{16}$ in simplest form, divide both terms by 8, the greatest common factor of 24 and 16, thus obtaining the ratio $\frac{3}{2}$ (as shown above).

Continued Ratio

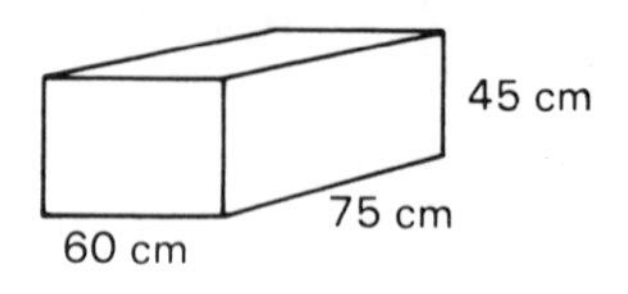

In a rectangular solid, the length is 75 cm, the width is 60 cm, and the height is 45 cm. The ratio of the length to the width is 75:60, and the ratio of the width to the height is 60:45. We can write these two ratios in an abbreviated form as the continued ratio 75:60:45. Thus, the ratio of the measures of the length, width, and height of the rectangular solid is 75:60:45, or, in simplest form, 5:4:3.

In general, the ratio of the numbers a, b, and c (b, $c \neq 0$) is $\boldsymbol{a:b:c}$.

MODEL PROBLEMS

1. An oil tank with a capacity of 200 gallons has 50 gallons of oil in it. **a.** Find the ratio of the number of gallons of oil in the tank to the capacity of the tank. **b.** What part of the tank is full?

Solution:

a. ratio $= \dfrac{\text{number of gallons of oil in the tank}}{\text{capacity of the tank}} = \dfrac{50}{200} = \dfrac{1}{4}$ *Ans.*

b. *Answer:* $\frac{1}{4}$ of the tank is full.

2. Compute the ratio of 6.4 ounces to 1 pound.

Solution:

First, express both quantities in the same unit of measure.

$$1 \text{ pound} = 16 \text{ ounces}$$

$$\text{ratio} = \frac{6.4 \text{ ounces}}{1 \text{ pound}} = \frac{6.4 \text{ ounces}}{16 \text{ ounces}} = \frac{6.4}{16} = \frac{64}{160} = \frac{2}{5}$$

Answer: The ratio is 2:5.

3. Express the ratio $1\frac{3}{4}$ to $1\frac{1}{2}$ in simplest form.

Solution: ratio $= 1\frac{3}{4} : 1\frac{1}{2} = 1\frac{3}{4} \div 1\frac{1}{2} = \frac{7}{4} \div \frac{3}{2} = \frac{7}{4} \times \frac{2}{3} = \frac{14}{12} = \frac{7}{6}$ *Ans.*

EXERCISES

In 1–5, express each ratio **(a)** as a fraction and **(b)** using a colon.

1. 36 to 12 **2.** 48 to 24 **3.** 40 to 25 **4.** 2 to 3 **5.** 5 to 4

6. Express each ratio in simplest form.

a. $\frac{8}{32}$ **b.** $\frac{40}{5}$ **c.** $\frac{12}{28}$ **d.** $\frac{36}{27}$ **e.** $\frac{36}{24}$

f. 20:10 **g.** 15:45 **h.** 18:18 **i.** 48:20 **j.** 21:35

k. $3x:2x$ **l.** $1y:4y$ **m.** $3c:5c$ **n.** $7x:7y$ **o.** $12s:4s$

7. The larger number is how many times the smaller number?
a. 10, 5 **b.** 18, 6 **c.** 12, 8 **d.** 25, 10 **e.** 15, 25

8. If the ratio of two numbers is 10:1, how many times the smaller number is the larger number?

9. If the ratio of two numbers is 8:1, the smaller number is what fractional part of the larger number?

10. In each part, tell whether the ratio is equal to $\frac{3}{2}$.

a. $\frac{30}{20}$ **b.** $\frac{9}{4}$ **c.** $\frac{8}{12}$ **d.** 9:6 **e.** $\frac{45}{30}$ **f.** 18:6

11. In each part, name the ratios that are equal.

a. $\frac{2}{3}, \frac{6}{9}, \frac{10}{30}, \frac{28}{36}, \frac{50}{75}$ **b.** 10:8, 20:16, 15:13, 4:5, 50:40

12. In each part, find three pairs of numbers such that each pair of numbers has the given ratio.

a. $\frac{1}{2}$ **b.** $\frac{1}{5}$ **c.** 3:1 **d.** 4:1 **e.** $\frac{3}{4}$ **f.** 2:3

13. Using a colon, express in simplest form the ratio of all pairs of:
a. equal numbers (not zero)
b. nonzero numbers whose difference is 0.

14. Express each ratio in simplest form.

a. $\frac{3}{4}$ to $\frac{1}{4}$ **b.** $1\frac{1}{8}$ to $\frac{3}{8}$ **c.** 1.2 to 2.4 **d.** .75 to .25 **e.** 6:.25

15. Express each ratio in simplest form.
a. 80 m to 16 m **b.** 75 g to 100 g **c.** 36 cm to 72 cm
d. 54 g to 90 g **e.** 75 cm to 35 mm **f.** 32 cg to 80 mg
g. 36 ml to 3 l **h.** 150 m to 5 km **i.** 500 g to 2 kg

16. Express each ratio in simplest form.

a. $1\frac{1}{2}$ hr. to $\frac{1}{2}$ hr. **b.** 3 in. to $\frac{1}{2}$ in. **c.** 1 ft. to 1 in.

d. 1 yd. to 1 ft. **e.** $\frac{1}{3}$ yd. to 6 in. **f.** 12 oz. to 3 lb.

g. 1 hr. to 15 min. **h.** \$6 to 50 cents **i.** 2 mi. to 880 yd.

17. A baseball team played 162 games and won 90.
a. What is the ratio of the number of games won to the number of games played?
b. For every 9 games played, how many games were won?

18. A student did 6 out of 10 problems correctly.
a. What is the ratio of the number right to the number wrong?
b. For every two answers that were wrong, how many answers were right?

19. A cake recipe calls for $1\frac{1}{2}$ cups of milk to $1\frac{3}{4}$ cups of flour. What is the ratio of the number of cups of milk to the number of cups of flour in this recipe?

20. The perimeter of a rectangle is 30 ft. and the width is 5 ft. Find the ratio of the length of the rectangle to its width.

21. In a freshman class, there are b boys and g girls. Express the ratio of the number of boys to the total number of pupils.

22. The length of a rectangle is represented by $3x$ and its width by $2x$. Find the ratio of the width of the rectangle to its perimeter.

23. Represent in terms of x two numbers whose ratio is:
a. 3 to 4 **b.** 5 to 3 **c.** 1 to 4 **d.** 1:2 **e.** 3:5

24. Represent in terms of x three numbers that have the continued ratio: **a.** 1 to 2 to 3 **b.** 3 to 4 to 5 **c.** 1:3:4 **d.** 2:3:5

12-2 USING A RATIO TO EXPRESS A RATE

You have learned how to use a ratio to compare two quantities that are measured in the same unit. It is also possible to compare two quantities of different types. If a plane flies 1,920 kilometers in 3 hours, the ratio of the distance traveled to the time that the plane was in flight is $\frac{1{,}920 \text{ kilometers}}{3 \text{ hours}} = \frac{640}{1} = 640$ kilometers per hour. We say that the plane was flying at the *rate* of 640 kilometers per hour. When the numbers in a ratio are expressed in simplest form to state a rate, we say that the rate is expressed in ***lowest terms***. The rate should always be written with its unit of measure.

MODEL PROBLEMS

1. Clyde Champion scored 175 points in 7 basketball games. Express, in lowest terms, the ratio of the number of points Clyde scored to the number of games Clyde played.

Solution: ratio $= \frac{175}{7} = \frac{25}{1} = 25$ points per game *Ans.*

2. There are 5 grams of salt in 100 cm^3 of a solution of salt and water. Express, in lowest terms, the ratio of the number of grams of salt to the number of cm^3 in the solution.

Solution: ratio $= \frac{5}{100} = \frac{.05}{1} = .05$ g per cm^3 *Ans.*

Note that $\frac{5}{100}$ could have been reduced to $\frac{1}{20}$. Therefore, the answer could also be written as $\frac{1}{20}$ g per cm^3.

EXERCISES

In 1–6, express the ratio in lowest terms.

1. the ratio of 36 apples to 18 people
2. the ratio of 48 patients to 6 nurses
3. the ratio of $1.50 to 3 liters
4. the ratio of 96 cents to 16 grams
5. the ratio of 6.75 ounces to $2.25
6. the ratio of 62 miles to 100 kilometers

7. If there are 240 tennis balls in 80 cans, how many tennis balls are there in each can?
8. If an 11-ounce can of shaving cream costs 88 cents, what is the cost of each ounce of shaving cream in the can?
9. If, in traveling 31 miles, you travel 50 kilometers, how many miles are there in each kilometer?
10. In a supermarket, the regular size of Cleanright cleanser contains 14 ounces and costs 49 cents. The giant size of Cleanright cleanser, which contains 20 ounces, costs 66 cents.
    a. Find, correct to the nearest tenth of a cent, the cost per ounce for the regular can.
    b. Find, correct to the nearest tenth of a cent, the cost per ounce for the giant can.
    c. Which is the better buy?
11. Sue types 1,800 words in 30 minutes. Rita types 1,000 words in 20 minutes. Which girl is the faster typist?
12. Ronald runs 300 meters in 40 seconds. Carlos runs 200 meters in 30 seconds. Which boy is the faster runner for short races?

12-3 SOLVING VERBAL PROBLEMS INVOLVING RATIOS

1. The perimeter of a triangle is 60 cm. If the sides are in the ratio 3:4:5, find the length of each side of the triangle.

Solution: Let $3x$ = length of the first side.
Let $4x$ = length of the second side.
Let $5x$ = length of the third side.

The perimeter of the triangle is 60 cm.

$$3x + 4x + 5x = 60$$
$$12x = 60$$
$$x = 5$$
$$3x = 15$$
$$4x = 20$$
$$5x = 25$$

Check

$15:20:25 = 3:4:5$

$15 + 20 + 25 = 60$

Answer: The lengths of the sides are 15 cm, 20 cm, and 25 cm.

2. Two numbers have the ratio 2:3. The larger is 30 more than $\frac{1}{2}$ of the smaller. Find the numbers.

Solution: Let $2x$ = the smaller number.
Let $3x$ = the larger number.

The larger number is 30 more than $\frac{1}{2}$ of the smaller number.

$$\begin{aligned} 3x &= \frac{1}{2}(2x) + 30 \\ 3x &= x + 30 \\ 3x - x &= x + 30 - x \\ 2x &= 30 \\ x &= 15 \\ 2x &= 30 \\ 3x &= 45 \end{aligned}$$

Check

The ratio of 30 to 45 is 30:45 or 2:3. The larger number, 45, is 30 more than 15, which is $\frac{1}{2}$ of the smaller number.

Answer: The numbers are 30 and 45.

EXERCISES

1. Two numbers are in the ratio 4:3. Their sum is 70. Find the numbers.
2. Find two numbers whose sum is 160 and that have the ratio 5:3.
3. Two numbers have the ratio 7:5. Their difference is 12. Find the numbers.
4. Find two numbers whose ratio is 4:1 and whose difference is 36.
5. A piece of wire 32 centimeters in length is divided into two parts that are in the ratio 3:5. Find the length of each part.
6. The sides of a triangle are in the ratio of 6:6:5. The perimeter of the triangle is 34 cm. Find the length of each side of the triangle.
7. The ratio of the number of boys in a school to the number of girls is 11 to 10. If there are 525 pupils in the school, how many of them are boys?
8. The perimeter of a triangle is 48 cm. The lengths of the sides are in the ratio 3:4:5. Find the length of each side.
9. The perimeter of a rectangle is 360 centimeters. If the ratio of its length to its width is 11:4, find the dimensions of the rectangle.
10. The ratio of the measures of two complementary angles is 2:3. Find the measure of each angle.
11. The ratio of the measures of two supplementary angles is 4:5. Find the measure of each angle.

12. In the figure, $\overleftrightarrow{AB}$ and $\overleftrightarrow{CD}$ are two parallel lines cut by a transversal. If the ratio of m∠3 to m∠6 is 1:2, find the measures of all eight angles.

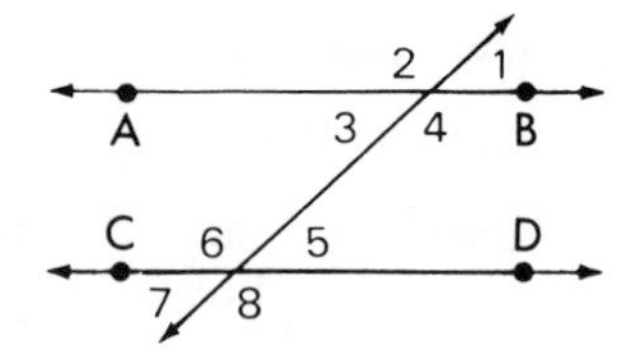

13. In △*DEF*, the ratio of the measures of the three angles is 2:2:5.
    a. Find the measures of the three angles.
    b. What kind of triangle is △*DEF*?
14. In isosceles △*ABC*, the ratio of the measures of vertex ∠*A* and base ∠*B* is 7:4. Find the measure of each angle of the triangle.
15. In isosceles △*RST*, the ratio of the measures of vertex angle *T* and the exterior angle at *R* is 1:5. Find the measure of each interior angle of the triangle.
16. The ratio of Carl's money to Donald's money is 7:3. If Carl gives Donald $20, the two then have equal amounts. Find the original amount that each one had.
17. Two numbers are in the ratio 3:7. The larger exceeds the smaller by 12. Find the numbers.
18. Two numbers are in the ratio 3:5. If 9 is added to their sum, the result is 41. Find the numbers.
19. In a basketball foul-shooting contest, the points made by Sam and Wilbur were in the ratio 7:9. Wilbur made 6 more points than Sam. Find the number of points made by each.
20. A chemist wishes to make $12\frac{1}{2}$ liters of an acid solution by using water and acid in the ratio 3:2. How many liters of each should he use?
21. In a triangle, two sides have the same length. The ratio of each of these sides to the third side is 5:3. If the perimeter of the triangle is 65 in., find the length of each side of the triangle.

12-4 PROPORTION

A ***proportion*** is an equation that states that two ratios are equal. Since the ratio $\frac{4}{20}$ is equal to the ratio $\frac{1}{5}$, we may write the proportion $\frac{4}{20} = \frac{1}{5}$ or 4:20 = 1:5. Both of these proportions are read *4 is to 20 as 1 is to 5*.

The proportion $\frac{a}{b} = \frac{c}{d}$ $(b \neq 0,\ d \neq 0)$, or $a:b = c:d$, is read *a is to b as c is to d*. There are four terms in this proportion, namely, a, b, c, and d. The first and fourth terms, a and d, are called the ***extremes*** of the proportion. The second and third terms, b and c, are called the ***means***.

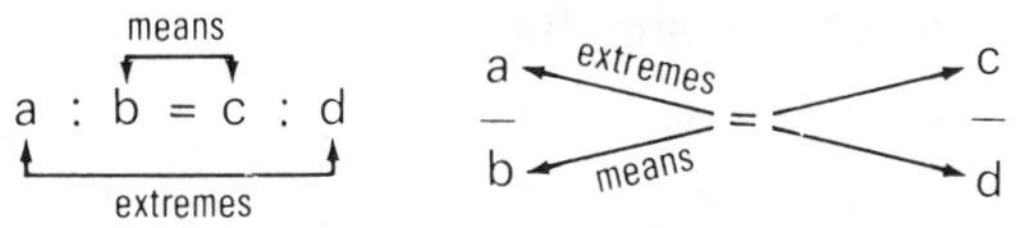

means

a : b = c : d

extremes

$\frac{a}{b}$ extremes = means $\frac{c}{d}$

In the proportion $4:20 = 1:5$, the product of the means, $20(1)$, is equal to the product of the extremes, $4(5)$.

In the proportion $\frac{5}{15} = \frac{10}{30}$, the product of the means, $15(10)$, is equal to the product of the extremes, $5(30)$.

In any proportion $\frac{a}{b} = \frac{c}{d}$, we can show that the product of the means is equal to the product of the extremes, $ad = bc$.

Since $\frac{a}{b} = \frac{c}{d}$ is an equation, we can multiply both members by bd, the least common denominator of the fractions in the equation.

$$\frac{a}{b} = \frac{c}{d}$$

$$bd\left(\frac{a}{b}\right) = bd\left(\frac{c}{d}\right)$$

$$bd\left(a \cdot \frac{1}{b}\right) = bd\left(c \cdot \frac{1}{d}\right)$$

$$\left(b \cdot \frac{1}{b}\right)(a \cdot d) = \left(d \cdot \frac{1}{d}\right)(bc) \qquad \text{(Commutative and associative properties)}$$

$$1 \cdot (ad) = 1 \cdot (bc)$$

$$ad = bc$$

Therefore, we have shown that the following statement is always true:

● **In a proportion, the product of the means is equal to the product of the extremes.**

MODEL PROBLEMS

1. Tell whether $\frac{4}{16} = \frac{5}{20}$ is a proportion.

Solution:

Method 1

Reduce each ratio to simplest form.

$$\frac{4}{16} = \frac{1}{4} \text{ and } \frac{5}{20} = \frac{1}{4}$$

Therefore, $\frac{4}{16}$ and $\frac{5}{20}$ are equal ratios and $\frac{4}{16} = \frac{5}{20}$ is a proportion.

Method 2

In the equation $\frac{4}{16} = \frac{5}{20}$, the product of the second and third terms is 16(5), or 80. The product of the first and fourth terms, 4(20), is also 80. Therefore, $\frac{4}{16} = \frac{5}{20}$ is a proportion.

Answer: Yes

2. Solve for q in the proportion $25:q = 5:2$.

Solution

If $25:q = 5:2$ is a proportion, then $5q = 25(2)$ (*the product of the means is equal to the product of the extremes*).

$$\begin{aligned} 5q &= 25(2) \\ 5q &= 50 \\ q &= 10 \end{aligned}$$

Answer: $q = 10$

Check

$$\begin{aligned} 25:q &= 5:2 \\ 25:10 &\stackrel{?}{=} 5:2 \\ 5:2 &= 5:2 \quad \text{(True)} \end{aligned}$$

3. Solve for x: $\dfrac{12}{x-2} = \dfrac{32}{x+8}$

Solution

$$\begin{aligned} \frac{12}{x-2} &= \frac{32}{x+8} \\ 32(x-2) &= 12(x+8) \\ 32x - 64 &= 12x + 96 \\ 32x - 12x &= 96 + 64 \\ 20x &= 160 \\ x &= 8 \end{aligned}$$

In a proportion, the product of the means is equal to the product of the extremes.

Answer: $x = 8$

Check

$$\begin{aligned} \frac{12}{x-2} &= \frac{32}{x+8} \\ \frac{12}{8-2} &\stackrel{?}{=} \frac{32}{8+8} \\ \frac{12}{6} &\stackrel{?}{=} \frac{32}{16} \\ 2 &= 2 \quad \text{(True)} \end{aligned}$$

4. The denominator of a fraction exceeds the numerator by 7. If 3 is subtracted from the numerator of the fraction and the denominator is unchanged, the value of the resulting fraction becomes $\frac{1}{3}$. Find the original fraction.

Solution:

Let x = the numerator of the original fraction.
Then, $x + 7$ = the denominator of the original fraction.

And, $\frac{x}{x+7}$ = the original fraction.

And, $\frac{x-3}{x+7}$ = the new fraction.

The value of the new fraction is $\frac{1}{3}$.

$$\begin{aligned} \frac{x-3}{x+7} &= \frac{1}{3} \\ 1(x+7) &= 3(x-3) \\ x+7 &= 3x-9 \\ 7+9 &= 3x-x \\ 16 &= 2x \\ 8 &= x \\ 15 &= x+7 \end{aligned}$$

Check: The original fraction was $\frac{8}{15}$.

The new fraction is $\frac{8-3}{15} = \frac{5}{15} = \frac{1}{3}$.

Answer: The original fraction was $\frac{8}{15}$.

EXERCISES

In 1–6, state whether or not the given ratios may form a proportion.

1. $\frac{3}{4}, \frac{30}{40}$ **2.** $\frac{2}{3}, \frac{10}{5}$ **3.** $\frac{4}{5}, \frac{16}{25}$ **4.** $\frac{2}{5}, \frac{5}{2}$ **5.** $\frac{14}{18}, \frac{28}{36}$ **6.** $\frac{36}{30}, \frac{18}{15}$

In 7–14, find the missing term in each proportion.

7. $\frac{1}{2} = \frac{?}{8}$ **8.** $\frac{3}{5} = \frac{18}{?}$ **9.** $1:4 = 6:?$ **10.** $4:6 = ?:42$

11. $\frac{4}{?} = \frac{12}{60}$ **12.** $\frac{?}{9} = \frac{35}{63}$ **13.** $?:60 = 6:10$ **14.** $16:? = 12:9$

In 15–23, solve the equation.

15. $\frac{x}{60} = \frac{3}{20}$

16. $\frac{5}{4} = \frac{x}{12}$

17. $\frac{30}{4x} = \frac{10}{24}$

18. $\frac{5}{15} = \frac{x}{x + 8}$

19. $\frac{x}{12 - x} = \frac{10}{30}$

20. $\frac{16}{8} = \frac{21 - x}{x}$

21. $12:15 = x:45$

22. $\frac{5}{x + 2} = \frac{4}{x}$

23. $\frac{3x + 3}{3} = \frac{7x - 1}{5}$

In 24–26, solve for x in terms of the other variables.

24. $a:b = c:x$

25. $2r:s = x:3s$

26. $2x:m = 4r:s$

In 27–34, use a proportion to solve the problem.

27. The numerator of a fraction is 8 less than the denominator of the fraction. The value of the fraction is $\frac{3}{5}$. Find the fraction.

28. The denominator of a fraction exceeds twice the numerator of the fraction by 10. The value of the fraction is $\frac{5}{12}$. Find the fraction.

29. The denominator of a fraction is 30 more than the numerator of the fraction. If 10 is added to the numerator of the fraction and the denominator is unchanged, the value of the resulting fraction becomes $\frac{3}{5}$. Find the original fraction.

30. The numerator of a certain fraction is three times the denominator. If the numerator is decreased by 1 and the denominator is increased by 2, the value of the resulting fraction is $\frac{5}{2}$. Find the original fraction.

31. What number must be added to both the numerator and denominator of the fraction $\frac{7}{19}$ to make the resulting fraction equal to $\frac{3}{4}$?

32. The numerator of a fraction exceeds the denominator by 3. If 3 is added to the numerator and 3 is subtracted from the denominator, the resulting fraction is equal to $\frac{5}{2}$. Find the original fraction.

33. The numerator of a fraction is 7 less than the denominator. If 3 is added to the numerator and 9 is subtracted from the denominator, the new fraction is equal to $\frac{3}{2}$. Find the original fraction.

34. Slim Johnson was usually the best free-throw shooter on his basketball team. Early in the season, however, he had made only 9 out of 20 shots. By the end of the season, he had made all the additional shots he had taken, thereby ending with a season record of 75%. How many additional shots had he taken?

12-5 DIRECT VARIATION

If the length of a side of a square, s, is 1 in., then the perimeter of the square, p, is 4 in. Also, if s is 2 in., p is 8 in.; if s is 3 in., p is 12 in. These pairs of values are shown in the table at the right.

s	1	2	3
p	4	8	12

From the table, observe that as s varies, p also varies. Comparing the values of p to the values of s, you see that all three sets of values result in the same ratio.

$$\frac{4}{1} \qquad \frac{8}{2} = \frac{4}{1} \qquad \frac{12}{3} = \frac{4}{1}$$

Since the ratio $\frac{p}{s}$ is always the same, $\frac{4}{1}$, we say that the ratio $\frac{p}{s}$ is a *constant*. The proportion $\frac{p}{s} = \frac{4}{1}$ can be written as $p = 4s$, which is the formula for the perimeter of a square. Such a relation is called a ***direct variation***.

When two variables always change so that the ratio of the value of one variable to the corresponding value of the other is constant, we say that one variable ***varies directly*** as the other. We also say that one variable is ***directly proportional*** to the other. The constant ratio is called the ***constant of variation***. In the preceding example, the constant of variation is 4.

Note that when two quantities vary directly, the first quantity increases as the second quantity increases by the same factor. That is, if s is multiplied by a number, then p is multiplied by the same number.

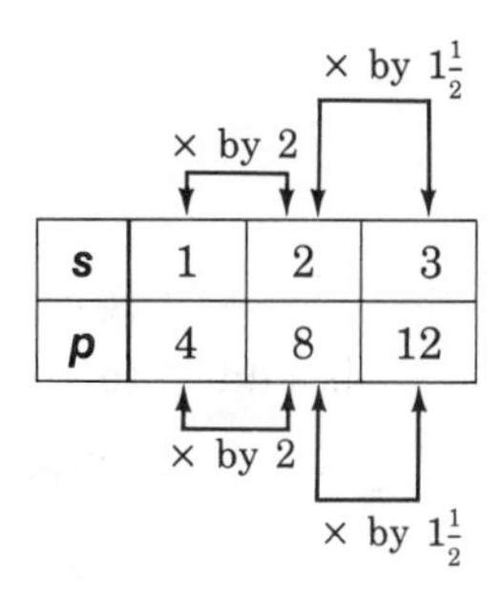

Or, the first quantity decreases as the second quantity decreases by the same factor. That is, if p is divided by a number, then s is divided by the same number.

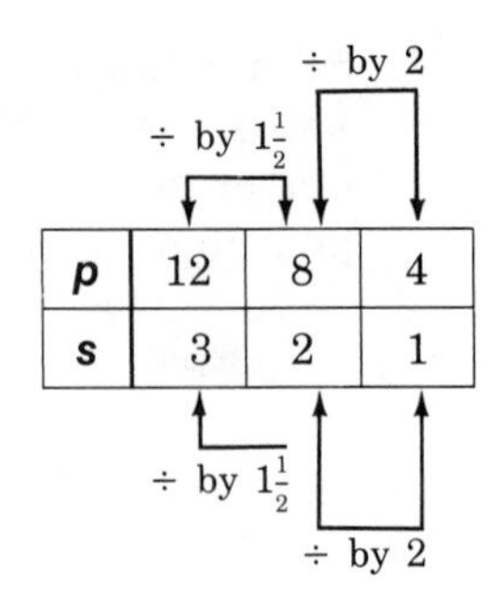

MODEL PROBLEMS

1. If x varies directly as y, and $x = 1.2$ when $y = 7.2$, find the constant of variation.

Solution: Constant of variation $= \dfrac{x}{y}$

$$= \frac{1.2}{7.2}$$

$$= \frac{1}{6}$$

Answer: The constant of variation is $\frac{1}{6}$.

2. The table gives pairs of values for the variables x and y.

x	1	2	3	10	?
y	8	16	24	?	1,600

a. Show that one variable varies directly as the other.
b. Express the relationship between the variables as a formula.
c. Find the missing values of the table.

Solution
a. Compare the x-values to the y-values.

$$\frac{1}{8} \quad \frac{2}{16} = \frac{1}{8} \quad \frac{3}{24} = \frac{1}{8}$$

Since all the given pairs of values result in a constant ratio, x and y vary directly.

b. Write the proportion that tells the constant of variation.

$$\frac{x}{y} = \frac{1}{8}$$

$$y = 8x \quad \textit{Ans.}$$

c. The missing values must fit into the constant ratio.

$$\frac{10}{y} = \frac{1}{8} \qquad\qquad \frac{x}{1{,}600} = \frac{1}{8}$$

$$y = 80 \quad \textit{Ans.} \qquad\qquad 1{,}600 = 8x$$

$$200 = x \quad \textit{Ans.}$$

3. There are about 90 calories in 20 grams of a cheese. Reggie ate 70 grams of this cheese. About how many calories were there in the cheese she ate if the number of calories varies directly as the weight of the cheese?

Solution

Let x = the number of calories in 70 grams of cheese.

$$\frac{x}{70} = \frac{90}{20} \begin{array}{l} \leftarrow \text{number of calories} \\ \leftarrow \text{number of grams of cheese} \end{array}$$

$20x = 90(70)$ In a proportion, the product of the means is equal to the product of the extremes.

$20x = 6{,}300$

$x = 315$

Check: $\frac{315}{70} \stackrel{?}{=} \frac{90}{20}$

$\frac{9}{2} = \frac{9}{2}$ (True)

Answer: There were about 315 calories in 70 grams of the cheese.

EXERCISES

In 1–9, find the constant of variation if the first variable varies directly as the second.

1. $x = 12$, $y = 3$ **2.** $d = 120$, $t = 3$ **3.** $y = 2$, $z = 18$

4. $P = 12.8$, $s = 3.2$ **5.** $t = 12$, $n = 8$ **6.** $i = 51$, $t = 6$

7. $s = 88$, $t = 110$ **8.** $A = 212$, $P = 200$ **9.** $r = 87$, $s = 58$

In 10–15, tell whether one variable varies directly as the other. If it does, express the relation between the variables by means of a formula.

10.

p	3	6	9
s	1	2	3

11.

n	3	4	5
c	6	8	10

12.

x	4	5	6
y	6	8	10

13.

t	1	2	3
d	20	40	60

14.

x	2	3	4
y	−6	−9	−12

15.

x	1	2	3
y	1	4	9

In 16–18, one variable varies directly as the other. Find the missing numbers and write the formula that relates the variables.

16.

h	1	2	?
A	5	?	25

17.

h	4	8	?
S	6	?	15

18.

L	2	8	?
W	1	?	7

In 19–22, state whether the relation between the variables is a direct variation. Give a reason for your answer.

19. $R + T = 80$ **20.** $15T = D$ **21.** $\frac{e}{i} = 20$ **22.** $bh = 36$

23. $C = 7N$ is a formula for the cost of any number of articles that sell for $7 each.
- **a.** How do C and N vary?
- **b.** How will the cost of 9 articles compare with the cost of 3 articles?
- **c.** If N is doubled, what change takes place in C?

24. $A = 12L$ is a formula for the area of any rectangle whose width is 12.
- **a.** Describe how A and L vary.
- **b.** How will the area of a rectangle whose length is 8 in. compare with the area of a rectangle whose length is 4 in.?
- **c.** If L is tripled, what change takes place in A?

25. d varies directly as t. If $d = 520$ when $t = 13$, find d when $t = 9$.

26. Y varies directly as x. If $Y = 35$ when $x = -5$, find Y when $x = -20$.

27. A varies directly as h. $A = 48$ when $h = 4$. Find h when $A = 36$.

28. N varies directly as d. $N = 10$ when $d = 8$. Find N when $d = 12$.

In 29–46, the quantities vary directly. Solve algebraically.

29. If 3 pounds of apples cost $.89, what is the cost of 15 pounds of apples at the same rate?

30. If 4 tickets to a show cost $17.60, what is the cost of 7 such tickets?

31. If 2 pounds of chopped meat sell for $3.50, how much chopped meat can be bought for $8.75?

32. Willis scores an average of 7 foul shots out of every 10 attempts. At the same rate, how many shots would he score in 200 attempts?

33. There are about 60 calories in 30 grams of canned salmon. About how many calories are there in a 210-gram can?

34. There are 81 calories in a slice of bread that weighs 30 grams. How many calories are there in a package of this bread that weighs 600 grams?

35. There are about 17 calories in three medium shelled peanuts. Joan ate 30 such peanuts. How many calories were there in the peanuts she ate?

36. A train traveled 90 miles in $1\frac{1}{2}$ hours. At the same rate, how long will it take the train to travel 330 miles?

37. The weight of 20 meters of copper wire is .9 kilogram. Find the weight of 170 meters of the same wire.

38. A recipe calls for $1\frac{1}{2}$ cups of sugar for a 3-pound cake. How many cups of sugar should be used for a 5-pound cake?

39. If a 7.5-pound breast of veal sells for $11.25, how much should Mr. Daniels pay for a 5.5-pound breast of veal?

40. A house that is assessed for $12,000 pays $960 in realty taxes. What should be the realty tax on a house that is assessed for $16,500?

41. The scale on a map is: 5 cm represents 3.5 km. How far apart are two towns if the distance between these two towns on the map is 8 cm?

42. David received $8.75 in dividends on 25 shares of a stock. How much should Marie receive in dividends on 60 shares of the same stock?

43. A picture $3\frac{1}{4}$ inches long and $2\frac{1}{8}$ inches wide is to be enlarged so that its length will become $6\frac{1}{2}$ inches. What will be the width of the enlarged picture?

44. In a certain concrete mixture, the ratio of cement to sand is 1:4. How many bags of cement would be used with 100 bags of sand?

45. If a man can buy p kilograms of candy for d dollars, represent the cost of n kilograms of this candy.

46. If a family consumes q liters of milk in d days, represent the amount of milk consumed in h days.

12-6 PERCENT AND PERCENTAGE PROBLEMS

Problems dealing with discounts, commissions, and taxes frequently involve percents. For example, to find the amount of tax when $60 is taxed at the rate of 8%, three quantities are involved:

1. the sum of money being taxed, the ***base***, $60
2. the rate of tax, the ***rate***, 8% or .08
3. the amount of tax, the ***percentage***, to be found

Recall that a percent is the ratio of a number to 100. For example, 8%, which means $\frac{8}{100}$, is the ratio of 8 to 100. An 8% tax means that a tax of $8 must be paid on a base of $100. We can use this fact in the following manner to find how much an 8% tax on $60 is.

Let t = amount of tax.

Then, $\frac{\text{amount of the tax}}{\text{sum to be taxed}} = \frac{8}{100}$ This is a proportion because it consists of two equal ratios.

$$\frac{t}{60} = \frac{8}{100}$$

$$100t = 480$$ In a proportion, the product of the means is equal to the product of the extremes.

$$t = 4.80$$

Answer: The tax is $4.80.

It is sometimes helpful to think of rate, expressed as a percent, as the ratio of the percentage to the base.

$$\frac{\textbf{percentage}}{\textbf{base}} = \textbf{rate}$$

or $$\textbf{percentage} = \textbf{rate} \times \textbf{base}$$

MODEL PROBLEMS

1. Represent $\frac{3}{5}$ as a percent.

Solution: Let x = the number of percent.

$$\frac{3}{5} = \frac{x}{100}$$

$$5x = 300$$

$$x = 60$$

Check

Does $\frac{3}{5} = \frac{60}{100}$? Yes.

Answer: 60%

2. If 25% of a number is 80, find the number.

Solution: Let n = the number, or base.

$$\frac{80}{n} = \frac{25}{100}$$

$$25n = 8{,}000$$

$$n = 320$$

Check: 25% of 320 is 80.

Answer: The number is 320.

3. Of the 560 seniors in Village High School, 476 attended the senior prom. What percent of the senior class attended the prom?

Solution

Let $\frac{x}{100}$ = the percent of the senior class that attended the dance.

$$\frac{476}{560} = \frac{x}{100}$$

$$560x = 47{,}600$$

$$x = 85$$

$$\frac{x}{100} = \frac{85}{100} = 85\%$$

Check: 85% of 560 is 476.

Answer: 85% of the seniors attended.

EXERCISES

In 1–9, find the indicated percentage.

1. 2% of 36
2. 6% of 150
3. 15% of 48
4. 2.5% of 400
5. 60% of 56
6. 100% of 7.5
7. $12\frac{1}{2}\%$ of 128
8. $33\frac{1}{3}\%$ of 72
9. 150% of 18

In 10–17, find the number.

10. 20 is 10% of what number?
11. 64 is 80% of what number?
12. 8% of what number is 16?
13. 72 is 100% of what number?

14. 125% of what number is 45?

15. $37\frac{1}{2}\%$ of what number is 60?

16. $66\frac{2}{3}\%$ of what number is 54?

17. 3% of what number is 1.86?

In 18–25, find the percent.

18. 6 is what percent of 12?

19. 9 is what percent of 30?

20. What percent of 10 is 6?

21. What percent of 35 is 28?

22. 5 is what percent of 15?

23. 22 is what percent of 22?

24. 18 is what percent of 12?

25. 2 is what percent of 400?

26. A newspaper has 80 pages. If 20 of the 80 pages are devoted to advertising, what percent of the newspaper is advertising?

27. A test was passed by 90% of a class. If 27 students passed the test, how many students were in the class?

28. Marie bought a dress that was marked $24. The sales tax is 8%. **a.** Find the sales tax. **b.** Find the total amount Marie had to pay.

29. There were 120 planes on an airfield. If 75% of the planes took off for a flight, how many planes took off?

30. One year, the Ace Manufacturing Company made a profit of $480,000. This represented 6% of the volume of business for the year. What was the volume of business for the year?

31. The price of a new motorcycle is $5,430. Mr. Klein made a down payment of 15% of the price of the motorcycle when he bought it. How much was his down payment?

32. How much silver is in 75 kilograms of an alloy that is 8% silver?

33. In a factory, 54,650 parts were made. When these were tested, 4% were found to be defective. How many parts were good?

34. A baseball team won 8 games, which was 50% of the total number of games it played. How many games did the team play?

35. Helen bought a coat at a "20% off" sale and saved $24. What was the marked price of the coat?

36. A businessman is required to collect an 8% sales tax. One day, he collected $280 in taxes. Find the total amount of sales he made that day.

37. A merchant sold a stereo speaker for $150, which was 25% above its cost to him. Find the cost of the stereo speaker to the dealer.

38. Bill bought a wooden chess set at a sale. The original price was $120; the sale price was $90. By what percent was the original price reduced?

39. If the sales tax on $150 is $7.50, what is the percent of the sales tax?

40. Mr. Taylor took a 2% discount on a bill. He paid the balance with a check for $76.44. What was the original amount of the bill?

41. After the price of a pound of meat was increased 10%, the new price was $1.98. What was the price of a pound of meat before the increase?

42. After Mrs. Sims lost 15% of her investment, she had $2,550 left. How much did she invest originally?

43. When a salesman sold a vacuum cleaner for $110, he received a commission of $8.80. What was the rate of commission?

44. Alicia bought a bond for $4,800. At the end of a year, the value of the bond had decreased $960. By what percent had the bond decreased in value?

45. Restful Motel's basic room charge increased from $200 per week to $275 per week. Find the percent of increase in this charge.

46. At a sale, a camera was reduced $8. This represented 10% of the original price. On the last day of the sale, the camera was sold for 75% of the original price. What was the final selling price of the camera?

12-7 SIMILAR POLYGONS

We have already talked about congruent polygons and have seen their use when we want to have copies that are the same size and the same shape as an original.

Now, let us discuss geometric figures that have the same shape but do not have the same size. Figures are related in this way when, for example, a photographer enlarges a small picture or an architect makes a small scale drawing of the floor plan of a room.

Polygons that have the same shape but not the same size are called ***similar polygons***. The symbol for *similar* or *is similar to* is $\sim$.

Consider the similar polygons shown below. To study why they have the same shape, we start by pairing vertices: A with A', B with B', C with C', and D with D'.

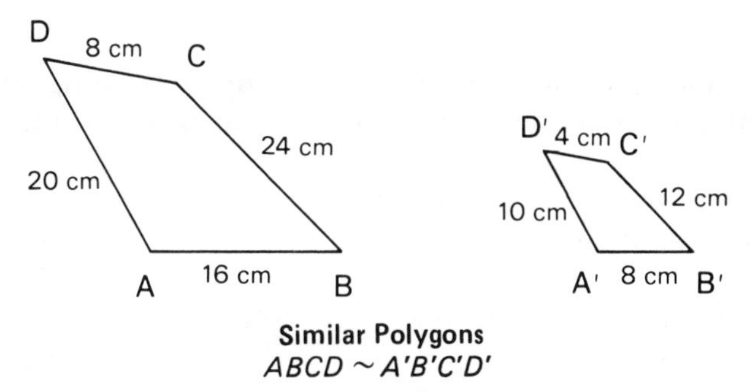

Similar Polygons
$ABCD \sim A'B'C'D'$

Thus, we have established *corresponding parts.* Four pairs of angles correspond. If you measure these pairs of corresponding angles, you will find the corresponding angles equal in measure, or congruent.

$$\angle A \cong \angle A', \angle B \cong \angle B', \angle C \cong \angle C', \text{ and } \angle D \cong \angle D'.$$

Now, examining the ratios of the lengths of the corresponding sides, we find:

For sides $\overline{AB}$ and $\overline{A'B'}$, the ratio of their lengths $= \frac{AB}{A'B'} = \frac{16}{8} = \frac{2}{1}$.

For sides $\overline{BC}$ and $\overline{B'C'}$, the ratio of their lengths $= \frac{BC}{B'C'} = \frac{24}{12} = \frac{2}{1}$.

For the remaining sides, what is $\frac{CD}{C'D'}$? What is $\frac{DA}{D'A'}$?

You should notice that, although corresponding sides are not congruent, the *ratios* of the measures of all pairs of corresponding sides are equal.

$$\frac{AB}{A'B'} = \frac{BC}{B'C'} = \frac{CD}{C'D'} = \frac{DA}{D'A'}$$

Therefore, we say:

- **Two polygons are similar if the measures of their corresponding angles are equal and the ratio of the measures of any pair of corresponding sides is equal to the ratio of the measures of all the other pairs of corresponding sides.**

Another way of stating this is:

- **Two polygons are similar if their corresponding angles are congruent and corresponding sides are in proportion.**

Conversely, we may also say:

- **If two polygons are similar, then corresponding angles are congruent and corresponding sides are in proportion.**

MODEL PROBLEM

If rectangle $ABCD$ is similar to rectangle $A'B'C'D'$, find the length of $\overline{A'D'}$.

Solution:

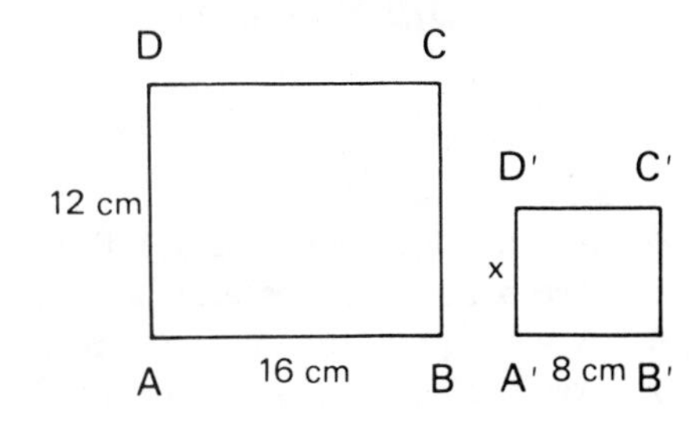

Let x represent the length of $\overline{A'D'}$.

The ratio of $A'D'$ to $AD = \frac{x}{12}$.

The ratio of $A'B'$ to $AB = \frac{8}{16}$.

Since the figures are similar,

$$\frac{x}{12} = \frac{8}{16}$$
$$16(x) = 8(12)$$
$$16x = 96$$
$$x = 6$$

Check

$$\frac{6}{12} \stackrel{?}{=} \frac{8}{16}$$
$$\frac{1}{2} = \frac{1}{2} \quad \text{(True)}$$

Answer: $A'D' = 6$ cm

EXERCISES

1. Is square $ABCD$ similar to rectangle $EFGH$? Why?

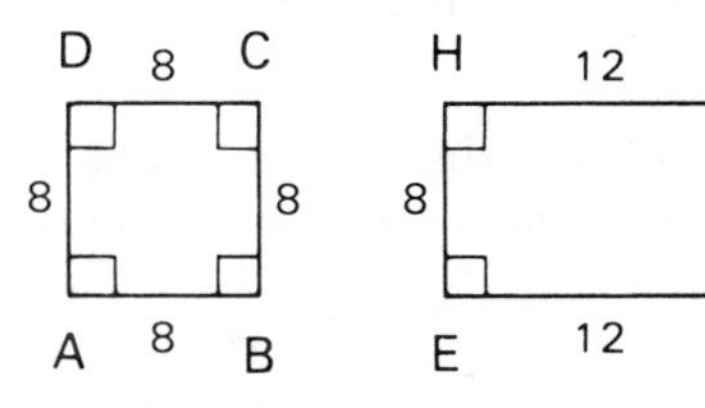

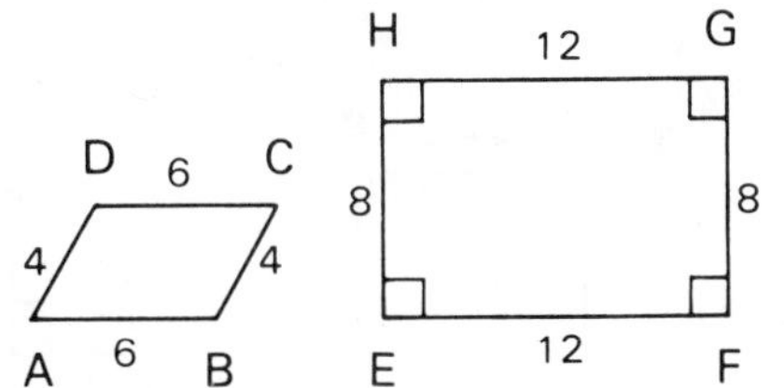

2. Is parallelogram $ABCD$ similar to rectangle $EFGH$? Why?

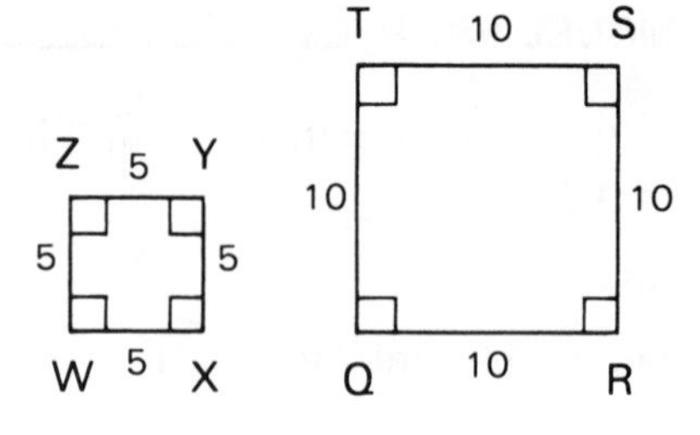

3. Is square $WXYZ$ similar to square $QRST$? Why?

4. Are all squares similar? Why?
5. Are all rectangles similar? Why?
6. Are all parallelograms similar? Why?
7. Are all rhombuses similar? Why?

8. Using the dimensions shown on the similar rectangles, find the number of centimeters in the length of the longer side $\overline{ZY}$.

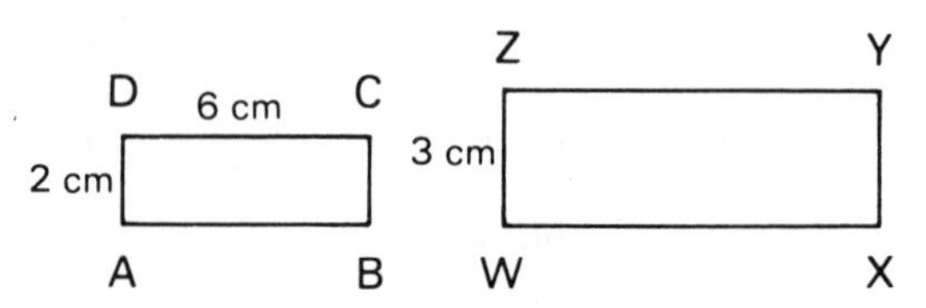

9. A picture 12 centimeters long and 9 centimeters wide is to be enlarged so that its length will be 16 centimeters. How wide will the enlarged picture be?

10. Jenny wishes to enlarge a rectangular photograph that is $4\frac{1}{2}$ inches long and $2\frac{3}{4}$ inches wide so that it will be 9 inches long. How wide will the enlargement be?

In 11–14, the two polygons are similar. Pairs of corresponding vertices are noted by using primes. Find the length of every side in each polygon.

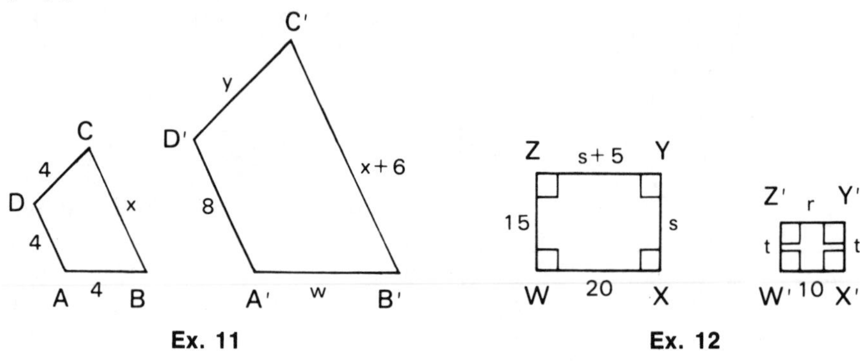

Ex. 11

Ex. 12

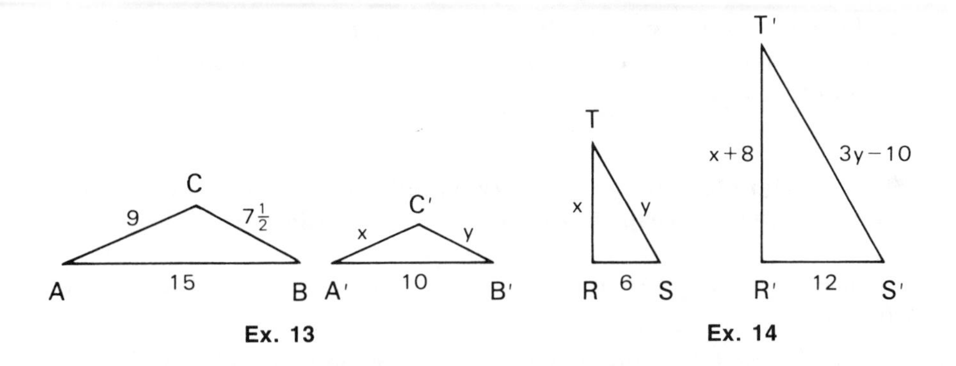

Ex. 13 **Ex. 14**

12-8 SIMILAR TRIANGLES; DILATION

Since similar triangles are often used to solve problems in measurement, let us see whether we can discover a short method of showing that two triangles are similar.

In triangle ABC, the figure at the right, $m \angle A = 27$ and $m \angle B = 63$.

Let us draw another triangle in which the measures of two angles are also 27° and 63°.

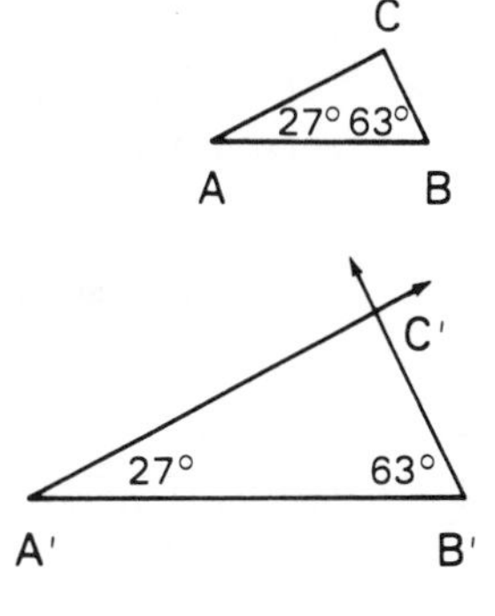

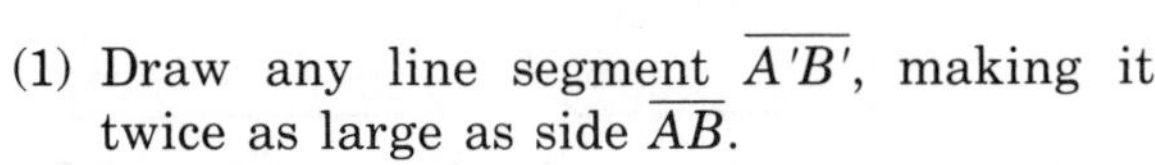

(1) Draw any line segment $\overline{A'B'}$, making it twice as large as side $\overline{AB}$.
(2) Using a protractor, measure an angle of 27° at A'.
(3) Measure an angle of 63° at B'.
(4) Let C' name the point of intersection of the two rays just drawn.

Observe that triangle $A'B'C'$ appears to be similar to triangle ABC. If you wish to be more certain of this, measure $\angle C$ and $\angle C'$. How do their measures compare? They should be equal in measure. Recall that the sum of the measures of the angles of a triangle is always 180°. Hence, if two angles in one triangle are equal in measure to two angles in another triangle, the third pair of angles must also be equal in measure.

How do the ratios of the measures of the corresponding sides of the two triangles compare? They should be equal to 2:1. Therefore, in $\triangle ABC$ and $\triangle A'B'C'$, the corresponding angles are equal in measure and the measures of corresponding sides are in proportion. Now, it seems reasonable to accept the truth of the following statement:

● **Two triangles are similar if two angles of one triangle are equal in measure to two corresponding angles of the other triangle. [a.a. ≅ a.a.]**

or

● **Two triangles are similar if two angles of one triangle are congruent to two corresponding angles of the other triangle.**

Since two right triangles have a pair of congruent right angles, we can say:

● **Two right triangles are similar if an acute angle in one triangle is equal in measure to an acute angle in the other triangle.**

MODEL PROBLEMS

1. $\triangle ABC$ and $\triangle DEF$ are similar.

a. Name three pairs of corresponding angles.

b. Name three pairs of corresponding sides.

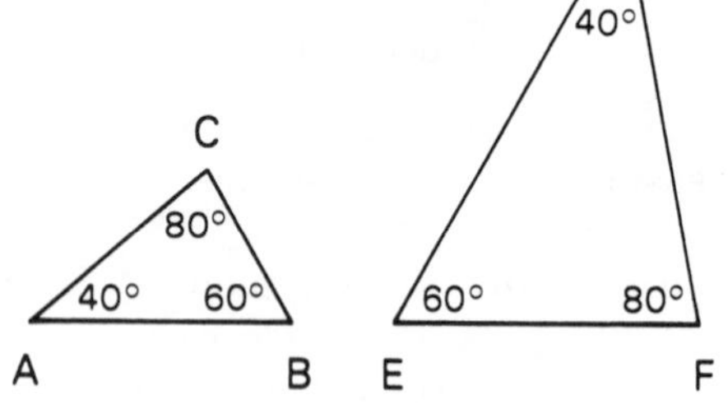

Solution

a. The corresponding angles are the pairs of angles whose measures are equal, that is, the angles that are congruent.

Since $m\angle A = 40$ and $m\angle D = 40$, $\angle D$ corresponds to $\angle A$.
Since $m\angle B = 60$ and $m\angle E = 60$, $\angle E$ corresponds to $\angle B$.
Since $m\angle C = 80$ and $m\angle F = 80$, $\angle F$ corresponds to $\angle C$.

b. The sides that join the vertices of two pairs of congruent angles are corresponding sides.

Since $\angle A \cong \angle D$ and $\angle B \cong \angle E$, $\overline{AB}$ corresponds to $\overline{DE}$.
Since $\angle B \cong \angle E$ and $\angle C \cong \angle F$, $\overline{BC}$ corresponds to $\overline{EF}$.
Since $\angle C \cong \angle F$ and $\angle A \cong \angle D$, $\overline{CA}$ corresponds to $\overline{FD}$.

Note that when we are not using prime notation, it is especially important to be sure that corresponding vertices are kept in proper order.

2. The measures of the sides of a triangle are 2, 8, and 7 inches. If the longest side of a similar triangle measures 4 inches, find the length of the shortest side of that triangle.

Solution

If $\triangle RST \sim \triangle DEF$, their corresponding sides are in proportion. The longest sides $\overline{RS}$ and $\overline{DE}$ are corresponding sides; the shortest sides $\overline{RT}$ and $\overline{DF}$ are corresponding sides.

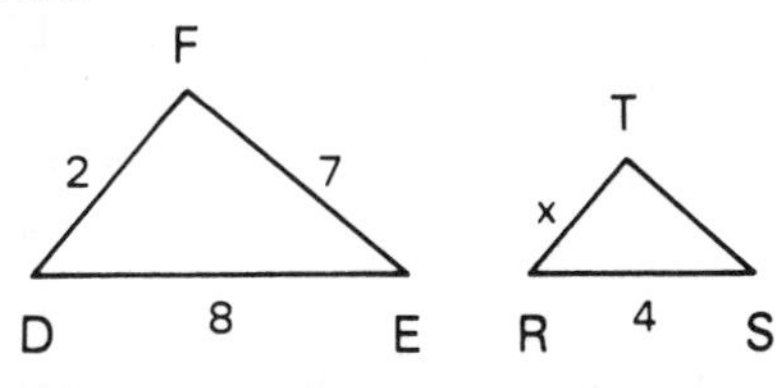

$$\frac{RT}{DF} = \frac{RS}{DE} \quad [RS = 4, DE = 8, DF = 2]$$

$$\frac{x}{2} = \frac{4}{8} \quad \text{Let } x = \text{length } RT.$$

$$8(x) = 2(4)$$

$$8x = 8$$

$$x = 1$$

Answer: The length of the shortest side of $\triangle RST$ is 1 inch.

3. At the same time that a vertical flagpole casts a shadow 15 feet long, a vertical pole that is 6 feet high casts a shadow that is 5 feet long. Find the height of the flagpole.

Solution

(1) Since the poles represented by $\overline{BC}$ and $\overline{EF}$ are vertical, m $\angle ABC = 90$ and m $\angle DEF = 90$.

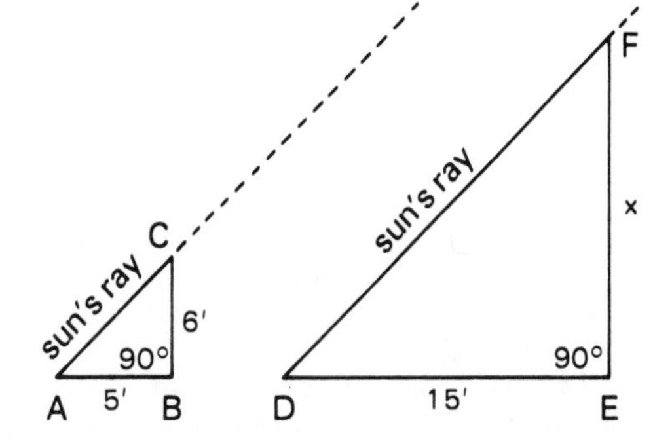

(2) We will say that the sun's rays are parallel and that the ground acts as a transversal. Hence, the angles that the sun's rays make with the ground, $\angle A$ and $\angle D$, have equal measures because they are corresponding angles of parallel lines.

(3) Hence, $\triangle DEF \sim \triangle ABC$ because two angles in one triangle are equal in measure to two corresponding angles in the other triangle.

(4) The corresponding sides of $\triangle DEF$ and $\triangle ABC$ are in proportion. Therefore:

$$\frac{EF}{BC} = \frac{DE}{AB}$$

$$\frac{x}{6} = \frac{15}{5} \quad \text{Let } x = \text{length } EF.$$

$$5(x) = 6(15)$$

$$5x = 90$$

$$x = 18$$

Answer: The height of the flagpole is 18 feet.

Dilation

In Section 11–9, we discussed four ways in which a geometric figure can be transformed, namely, under a line reflection, a point reflection, a translation, and a rotation. In each of these, the original figure was congruent to its image, since distance and angle measure were unchanged. A fifth type of transformation, called a ***dilation***, keeps angle measure unchanged, but changes distance. Thus, under a dilation, the original figure is similar to, but not congruent to, its image. A common example of a dilation occurs in projecting photographic images.

As light passes through a frame of movie film or a photographic slide, it projects a larger, or dilated, image on a screen. In the diagram, $\overline{AB}$ represents the film and $\overline{A'B'}$ represents the image on the screen.

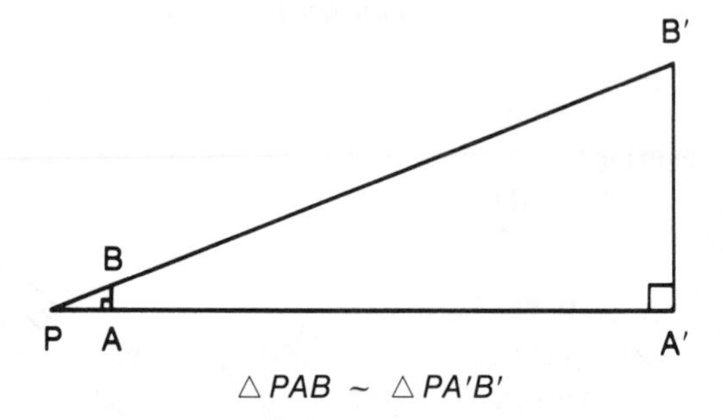

$\triangle PAB \sim \triangle PA'B'$

Two pairs of angles are congruent ($\angle P \cong \angle P$ and $\angle PAB \cong \angle PA'B'$). Therefore, the two triangles are similar, and their corresponding sides are in proportion. The proportion $\frac{A'B'}{AB} = \frac{PA'}{PA}$ states that the ratio of the height of the image to the height of the film is equal to the ratio of the distance of the image from the source of light to the distance of the film from the source of light. This ratio is called the ***constant of dilation***.

If A is 10 cm from P and A' is 1 m (100 cm) from P, then the constant of dilation, $\frac{PA'}{PA}$, is $\frac{100}{10}$, or 10. Each figure in the film is magnified 10 times when projected on the screen.

MODEL PROBLEM

In the diagram, $\angle ADE$ and $\angle ABC$ are right angles, $\triangle ADE \sim \triangle ABC$, $AB = 24$ cm, $BC = 18$ cm, and $AD = 16$ cm.

a. Find DE.

b. If B is the image of D under a dilation with center at A, what is the constant of dilation?

c. Find the area of $\triangle ABC$.

d. Find the area of $\triangle ADE$.

e. Find the area of trapezoid $BCED$.

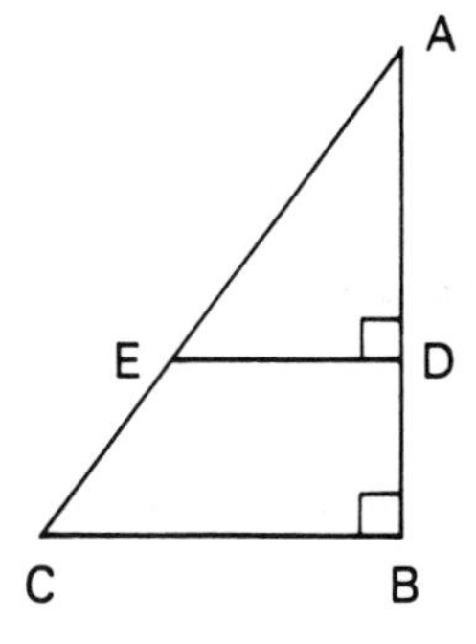

Solution

a. Since $\triangle ADE \sim \triangle ABC$, $\frac{DE}{BC} = \frac{AD}{AB}$.

Let $x = DE$.

$$\frac{x}{18} = \frac{16}{24}$$

$$24x = 16(18)$$

$$24x = 288$$

$$x = 12 \text{ cm} \quad \textit{Ans.}$$

b. Constant of dilation $= \frac{AB}{AD}$

$$= \frac{24}{16} = \frac{3}{2}$$

Note that:
The image of E is C and the image of $\overline{DE}$ is $\overline{BC}$.

$$BC = \text{constant of dilation} \times DE$$

$$18 = \frac{3}{2} \times 12$$

c. Area of

$$\triangle ABC = \frac{1}{2} \text{ base} \cdot \text{height}$$

$$= \frac{1}{2} \cdot BC \cdot AB$$

$$= \frac{1}{2}(18)(24)$$

$$= \frac{1}{2}(432) = 216$$

Area of $\triangle ABC = 216 \text{ cm}^2$ *Ans.*

d. Area of

$$\triangle ADE = \frac{1}{2} \text{ base} \cdot \text{height}$$

$$= \frac{1}{2} \cdot DE \cdot AD$$

$$= \frac{1}{2}(12)(16)$$

$$= \frac{1}{2}(192) = 96$$

Area of $\triangle ADE = 96 \text{ cm}^2$ *Ans.*

e. Area of trapezoid $BCED$ = area of $\triangle ABC$ − area of $\triangle ADE$

$$\text{Area of trapezoid } BCED = 216 - 96$$
$$= 120$$

Or, use the formula for the area of a trapezoid.

$$\text{The height } BD = AB - AD = 24 - 16 = 8$$

$$\text{Area of a trapezoid} = \frac{1}{2}h(b_1 + b_2)$$

$$\text{Area of } BCED = \frac{1}{2}(BD)(BC + DE)$$
$$= \frac{1}{2}(8)(18 + 12)$$
$$= \frac{1}{2}(8)(30)$$
$$= 120$$

Answer: Area of trapezoid $BCED$ = 120 cm^2

EXERCISES

In 1–4, the two triangles are similar. **a.** Name three pairs of corresponding angles. **b.** Name three pairs of corresponding sides.

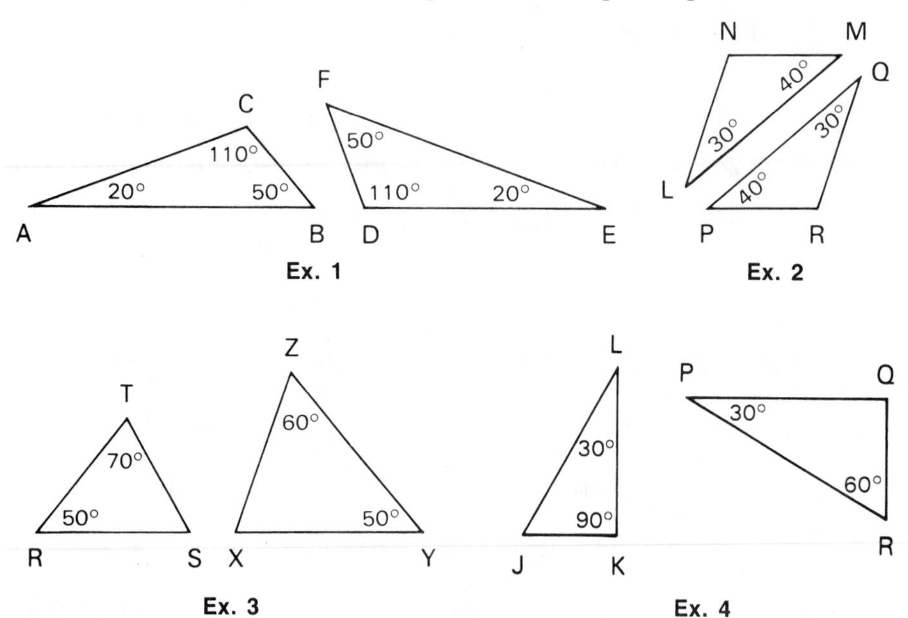

In 5 and 6, use a ruler and a protractor to draw the triangles.

5. **a.** Draw triangle ABC in which $AB = 2$ in., $m\angle A = 50$, and $m\angle B = 70$.

b. Draw triangle $A'B'C'$ similar to triangle ABC so that $\angle A'$ corresponds to $\angle A$, $\angle B'$ corresponds to $\angle B$, and $A'B' = 3$ inches.

6. **a.** Draw right triangle RST in which angle S is a right angle, $m\angle R = 70$, and $RS = 3$ inches.

b. Draw $\triangle R'S'T' \sim \triangle RST$ so that $\overline{R'S'}$ corresponds to $\overline{RS}$ and $R'S' = 1\frac{1}{2}$ inches.

7. In triangle RST, $m\angle R = 90$ and $m\angle S = 40$.
In triangle XYZ, $m\angle Y = 40$ and $m\angle Z = 50$.
a. Is triangle RST similar to triangle XYZ ? **b.** Why?

In 8 and 9, select the triangles that are similar and tell why they are similar.

8.

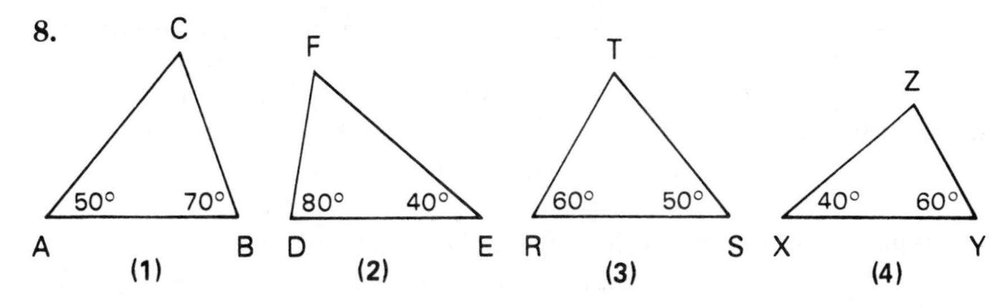

9.

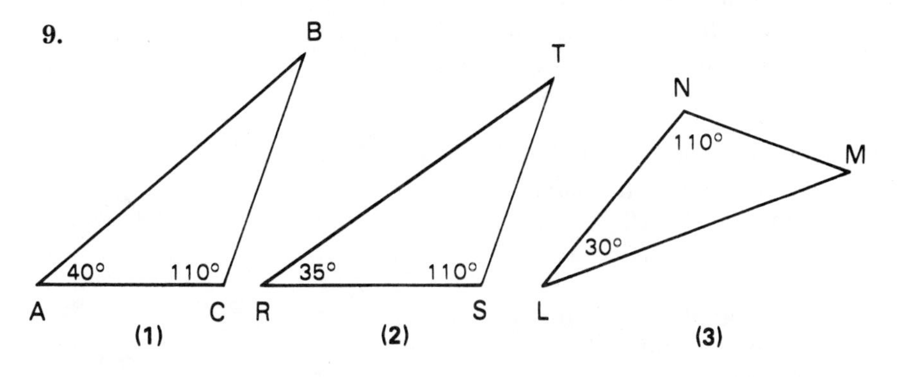

10. $\triangle ABC \sim \triangle RST$. $\angle A$ corresponds to $\angle R$, $\angle B$ to $\angle S$, and $\angle C$ to $\angle T$. If $AB = 9$ cm, $AC = 6$ cm, and $RS = 3$ cm, find RT.

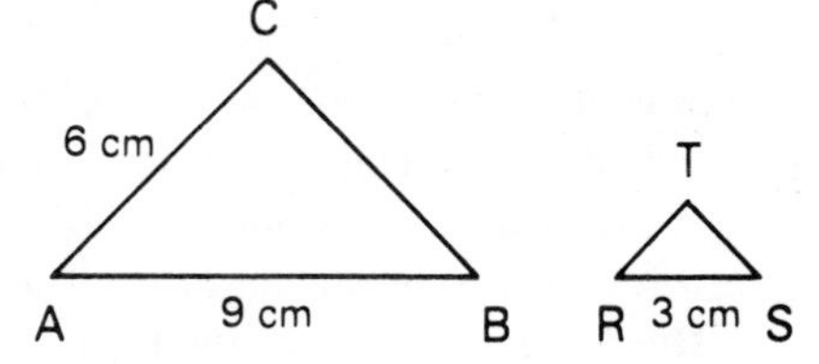

11. **a.** Is triangle ABC similar to triangle DEF?
b. State the reason for the answer given in part **a.**
c. Find BC.

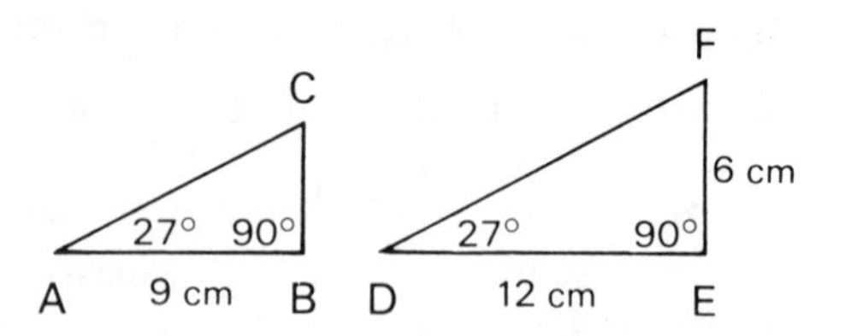

12. In the figure, $m \angle A = m \angle D$ and $m \angle C = m \angle F$. If $AB = 12$, $AC = 16$, $BC = 20$, and $DE = 6$, find DF and EF.

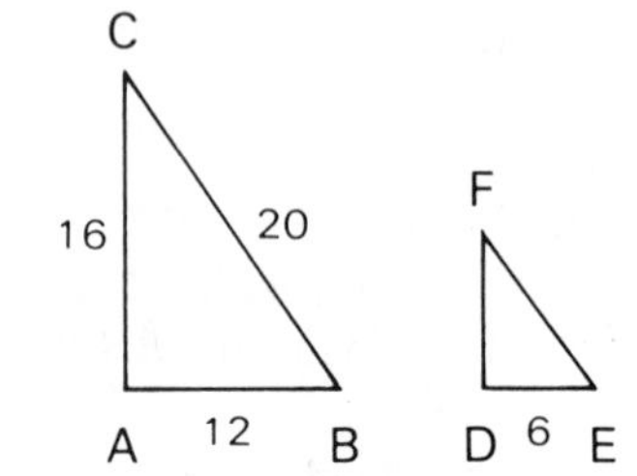

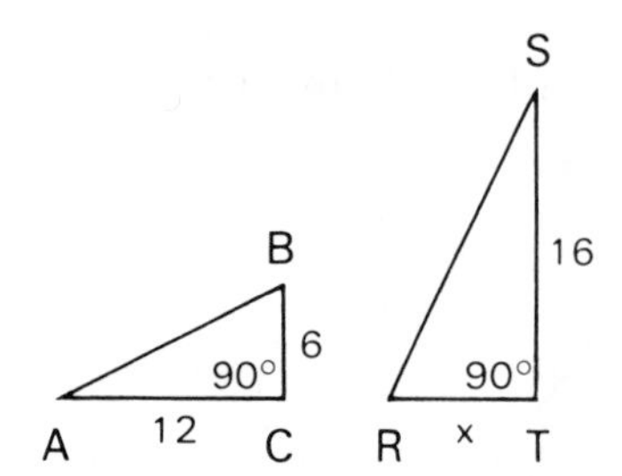

13. In right triangles ABC and RST, $m \angle A = m \angle S$.
a. Why is triangle ABC similar to triangle SRT?
b. Find the value of x.

14. The lengths of the sides of a triangle are 24, 16, and 12. If the shortest side of a similar triangle is 6, what is the length of the longest side of this triangle?

15. The lengths of the sides of a triangle are 36, 30, and 18. If the longest side of a similar triangle is 9, what is the length of the shortest side of this triangle?

16. To find the height of a tree, a hiker measured its shadow and found it to be 12 feet. At the same time, a vertical rod 8 feet high cast a shadow of 6 feet.
a. Is triangle ABC similar to triangle DEF?
b. Why?
c. Find the height of the tree.

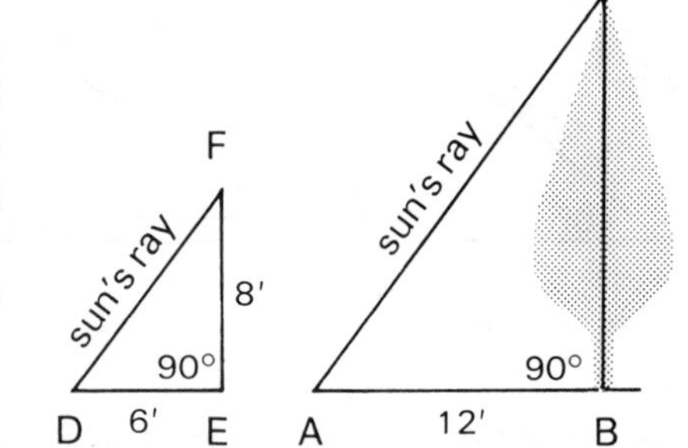

17. A certain tree casts a shadow 6 m long. At the same time, a nearby boy 2 m tall casts a shadow 4 m long. Find the height of the tree.

18. A building casts a shadow 18 feet long. At the same time, a woman 5 feet tall casts a shadow 3 feet long. Find the number of feet in the height of the building.

19. $\overline{AB}$ represents the width of a river. $\angle B$ and $\angle D$ are right angles. $\overline{AE}$ and $\overline{BD}$ are line segments. $BC = 16$ m, $CD = 8$ m, and $DE = 4$ m.

a. Is triangle ABC similar to triangle EDC? Why?

b. Find AB, the width of the river.

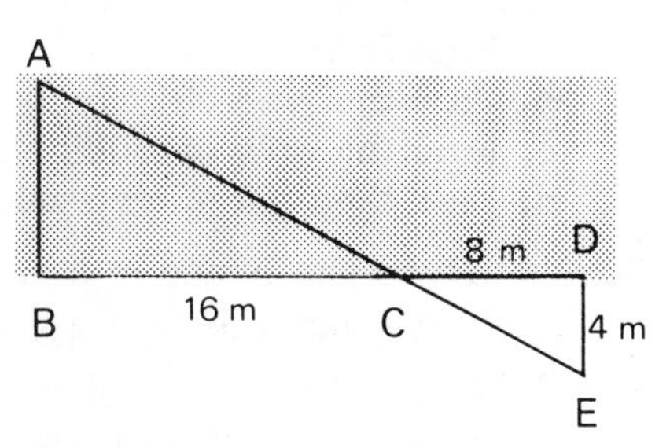

20. Using the figure in Exercise 19, find AB, the width of the river, if $BC = 240$ ft., $CD = 80$ ft., and $DE = 25$ ft.

21. In the figure, $\overline{AD}$ and $\overline{CB}$ are line segments, and $\overline{AB} \parallel \overline{DC}$.

a. Why is $\triangle ABE \sim \triangle DCE$?

b. Find AB, the distance across the pond, if:

(1) $CE = 40$ m, $EB = 120$ m, and $CD = 50$ m

(2) $AE = 75$ m, $ED = 30$ m, and $CD = 36$ m

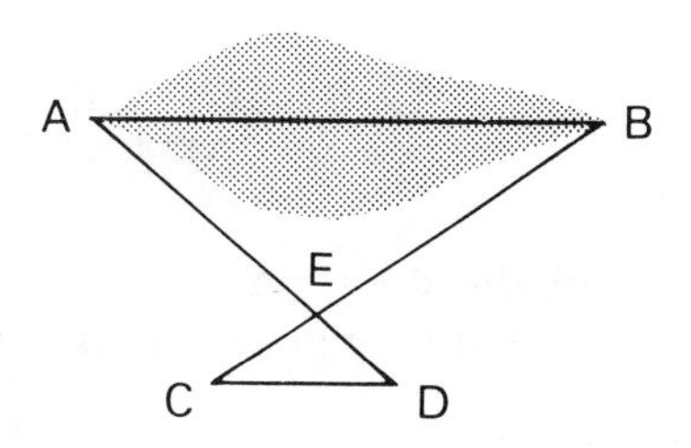

In 22–24, under a dilation with center at O, the image of A is A' and the image of B is B'.

22. Find OA' and OB' when $OA = 12$, $OB = 18$, and the constant of dilation is:

a. 4 **b.** 1.5 **c.** $\frac{4}{3}$

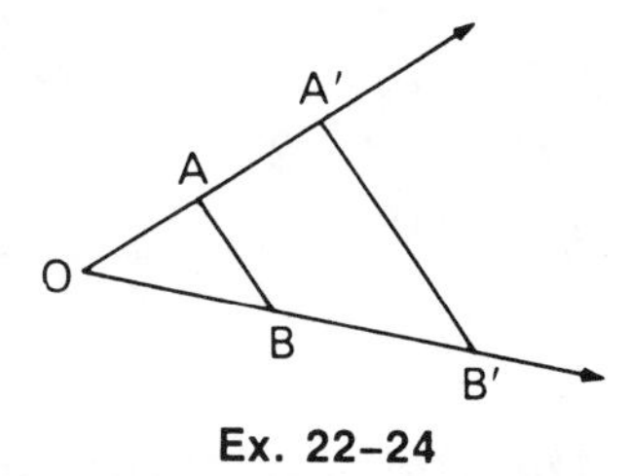

Ex. 22–24

23. Find the constant of dilation when $OA = 16$, $OB = 20$, and:
a. $OA' = 48$ **b.** $OB' = 24$ **c.** $OA' = 40$

24. Find AB if $A'B' = 20$ and the constant of dilation is $\frac{5}{4}$.

In 25–30, use the figure at the right.

25. Prove informally that $\triangle ABC \sim \triangle AB'C'$.

26. $BC = 3$ in., $AC = 4$ in., and $AC' = 8$ in. Find $B'C'$.

27. $AC = 5$ m, $BC = 2$ m, and $B'C' = 8$ m. Find AC'.

28. $BC = 10$ cm and $AC = 8$ cm. Find the ratio of $B'C'$ to AC'.

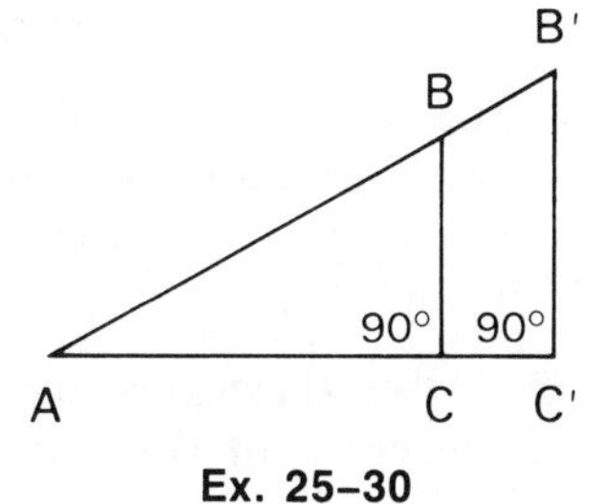

Ex. 25–30

29. $AC' = 48$ m, $B'C' = 36$ m, and $AC = 32$ m.
   a. Find BC.
   b. Find the area of triangle $AC'B'$.
   c. Find the area of triangle ACB.
   d. Find the area of trapezoid $BCC'B'$.

30. Let B' be the image of B under a dilation with center at A. Let $AC = 12$, $BC = 9$, and $AB = 15$. The constant of dilation is $\frac{4}{3}$.
   a. What is the image of C under this dilation?
   b. Find $B'C'$, AC', and AB', using the constant of dilation.

12-9 RATIOS OF PERIMETERS AND OF AREAS OF SIMILAR POLYGONS

In the diagram, $\triangle ABC \sim \triangle A'B'C'$. Therefore, the corresponding angles are congruent and the corresponding sides are in proportion.

$$\frac{AB}{A'B'} = \frac{40}{20} = \frac{2}{1}$$

$$\frac{AC}{A'C'} = \frac{30}{15} = \frac{2}{1}$$

$$\frac{BC}{B'C'} = \frac{36}{18} = \frac{2}{1}$$

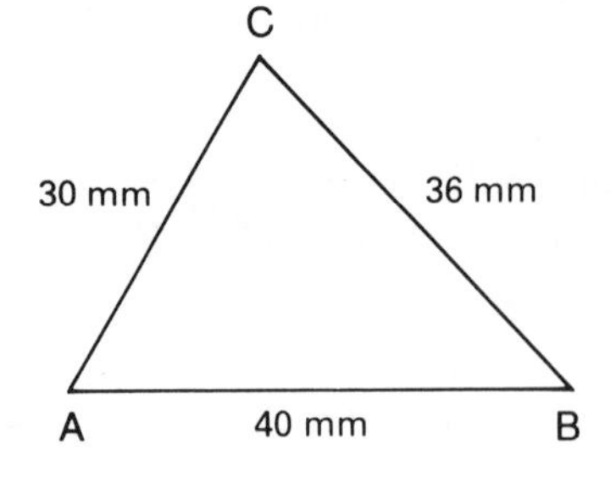

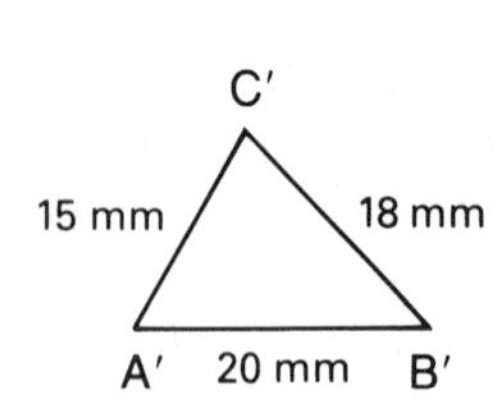

The Ratio of Perimeters

The perimeter of $\triangle ABC = 40 + 30 + 36 = 106$ mm.
The perimeter of $\triangle A'B'C' = 20 + 15 + 18 = 53$ mm.

$$\frac{\text{perimeter of } \triangle ABC}{\text{perimeter of } \triangle A'B'C'} = \frac{106}{53} = \frac{2}{1}$$

The ratio of the perimeters of these similar triangles is the same as the ratio of the measures of the corresponding sides. If you were to look at other numerical examples, you would reach the same conclusion.

Thus, we say:

● **If two triangles are similar, the ratio of the perimeters is equal to the ratio of the measures of the corresponding sides.**

This statement is also true for similar polygons of more than three sides. We say:

● **If two polygons are similar, the ratio of the perimeters is equal to the ratio of the measures of the corresponding sides.**

MODEL PROBLEMS

1. Right $\triangle ABC$ is similar to right $\triangle A'B'C'$. As shown in the diagram, the measures in inches are $AB = 13$, $AC = 5$, $BC = 12$, and $B'C' = 8$.
   a. Find the ratio of the measures of the corresponding sides.
   b. Find $A'B'$ and $A'C'$.
   c. Find the ratio of perimeters.

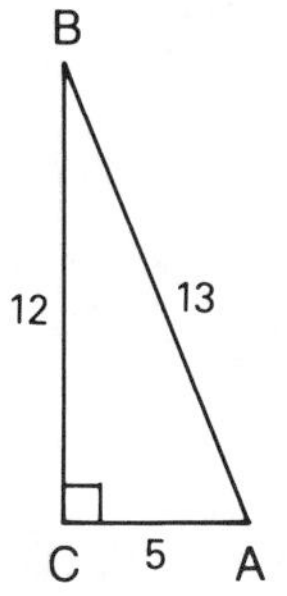

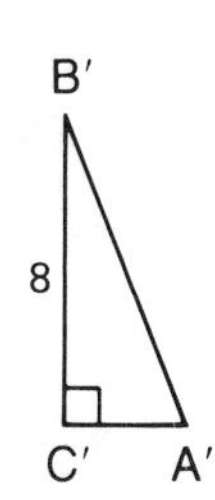

Solution

a. Since corresponding sides of similar triangles are in proportion, the ratio of the measures of each pair of sides of these triangles is the same as the ratio of the two given sides, $\frac{12}{8}$ or $\frac{3}{2}$.

Answer: The ratio of the measures of the corresponding sides is 3:2.

b. The ratio of the measures of each pair of corresponding sides of these triangles is 3:2.

$$\frac{AB}{A'B'} = \frac{3}{2} \qquad\qquad \frac{AC}{A'C'} = \frac{3}{2}$$

$$\frac{13}{A'B'} = \frac{3}{2} \qquad\qquad \frac{5}{A'C'} = \frac{3}{2}$$

$$3A'B' = 26 \qquad\qquad 3A'C' = 10$$

$$A'B' = \frac{26}{3} \text{ in. } \textit{Ans.} \qquad\qquad A'C' = \frac{10}{3} \text{ in. } \textit{Ans.}$$

c. In similar triangles, the ratio of the perimeters is equal to the ratio of the measures of the corresponding sides.

The ratio of perimeters is $\frac{3}{2}$.

Or, the ratio of the perimeters could be found by computing the perimeter of each triangle and finding the ratio.

$$\frac{\text{perimeter of } \triangle ABC}{\text{perimeter of } \triangle A'B'C'} = \frac{12 + 13 + 5}{8 + \frac{26}{3} + \frac{10}{3}} = \frac{30}{20} = \frac{3}{2}$$

Answer: The ratio of the perimeters is 3:2.

2. Each side of a square is tripled. What is the ratio of the perimeter of the new square to that of the original?

Solution:

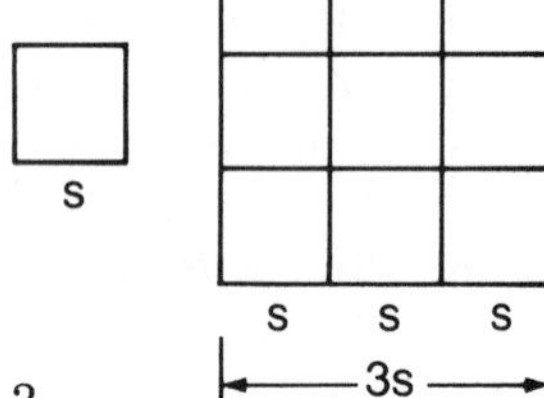

The two squares are similar because the corresponding angles are congruent and the ratios of the measures of the corresponding sides are equal.

$$\frac{\text{perimeter of new square}}{\text{perimeter of old square}} = \frac{\text{new side}}{\text{old side}} = \frac{3s}{s} = \frac{3}{1} \quad \textit{Ans.}$$

3. Two triangles are similar. The sides of the smaller triangle have lengths of 4, 6, and 8 cm. The perimeter of the larger triangle is 27 cm. Find the length of the shortest side of the larger triangle.

Solution

(1) In the smaller triangle, the length of the shortest side is 4 cm. Find the perimeter p:

$$p = 4 + 6 + 8 = 18$$

4 6 8 x

$p = 18$ $p' = 27$

(2) In the larger triangle, the perimeter p' is 27 cm. Let x = the length of the shortest side.

(3) Since the triangles are similar, use the following proportion:

$$\frac{\text{perimeter of smaller } \triangle}{\text{perimeter of larger } \triangle} = \frac{\text{length of shortest side in smaller } \triangle}{\text{length of shortest side in larger } \triangle}$$

$$\frac{18}{27} = \frac{4}{x}$$

(4) Solve for x:

$$18x = 27(4)$$
$$18x = 108$$
$$x = 6$$

Answer: The length of the shortest side of the larger triangle is 6 cm.

EXERCISES

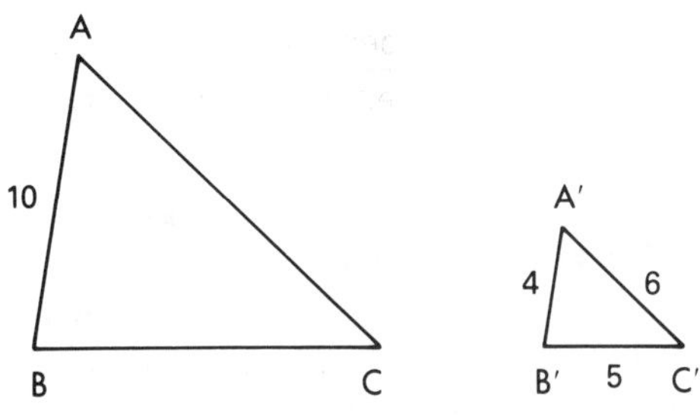

1. $\triangle ABC \sim \triangle A'B'C'$.
   The measures in inches are $AB = 10$, $A'B' = 4$, $B'C' = 5$, and $C'A' = 6$.
   a. Find the ratio of the measures of the corresponding sides.
   b. Find BC and CA.
   c. Find the ratio of the measures of the perimeters.

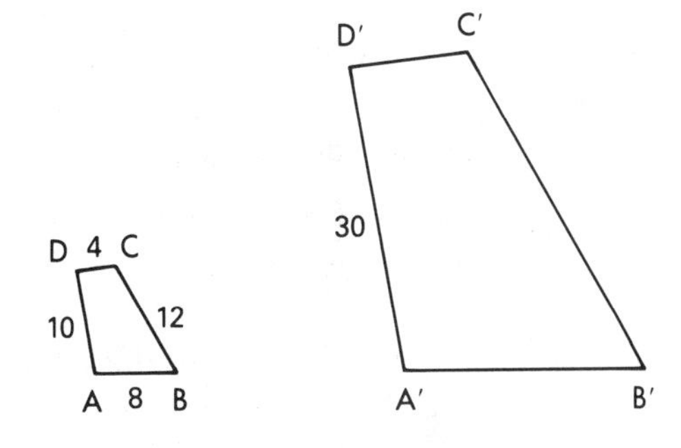

2. $ABCD \sim A'B'C'D'$.
   The measures in inches are $AB = 8$, $BC = 12$, $CD = 4$, $DA = 10$, and $D'A' = 30$.
   a. Find the ratio of the measures of the corresponding sides.
   b. Find $A'B'$, $B'C'$, and $C'D'$.
   c. Find the ratio of the measures of the perimeters.
3. Each side of a square has length 9 cm. If the length of each side is doubled, what is the ratio of the perimeter of the larger square thus formed to the perimeter of the smaller square?
   (1) 36:1 (2) 18:1 (3) 9:1 (4) 2:1
4. The sides of a triangle have lengths 6, 8, and 10. What is the length of the shortest side of a similar triangle whose perimeter is 12?
   (1) 6 (2) 8 (3) 3 (4) 4
5. Two triangles are similar. The sides of the smaller triangle have lengths of 6, 7, and 12. The perimeter of the larger triangle is 75. The length of the longest side of the larger triangle is:
   (1) 18 (2) 2 (3) 36 (4) 4
6. The sides of a triangle measure 7, 9, and 11. Find the perimeter of a similar triangle in which the shortest side has a length of 21.
7. The sides of a triangle measure 5, 12, and 13. Find the perimeter of a similar triangle in which the longest side has a length of 39.
8. The lengths of the sides of a triangle are 8, 10, and 12. If the length of the shortest side of a similar triangle is 6, find the length of its longest side.
9. The lengths of corresponding sides of two similar polygons are in the ratio 3:1. If the perimeter of the larger polygon exceeds the perimeter of the smaller polygon by 48 mm, find the perimeter of the smaller polygon.

10. The lengths of corresponding sides of two similar polygons are in the ratio 5:8. If the perimeter of the larger polygon is 10 inches less than twice the perimeter of the smaller polygon, find the perimeter of each polygon.

The Ratio of Areas

From page 402, let us reconsider similar triangles ABC and $A'B'C'$, in which the ratio of the corresponding sides is 2:1.

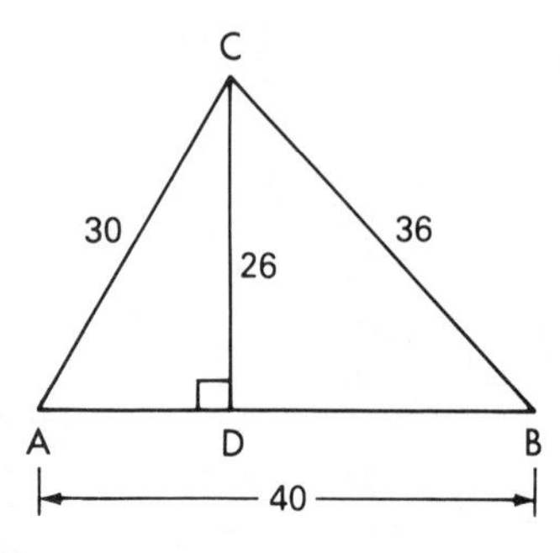

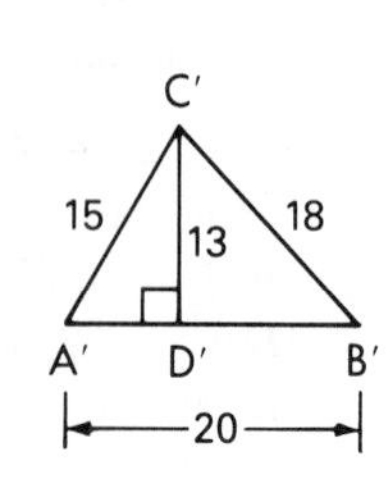

In these similar triangles, let us include corresponding altitudes $\overline{CD}$ and $\overline{C'D'}$. Note that the ratio of the corresponding altitudes $\left(\frac{CD}{C'D'} = \frac{26}{13} = \frac{2}{1}\right)$ is the same as the ratio of the corresponding sides, 2:1.

$$\text{The area of a triangle} = \frac{1}{2}\text{ base} \times \text{altitude}$$

$$\text{The area of } \triangle ABC = \frac{1}{2}(40)(26) = 520$$

$$\text{The area of } \triangle A'B'C' = \frac{1}{2}(20)(13) = 130$$

$$\frac{\text{area of } \triangle ABC}{\text{area of } \triangle A'B'C'} = \frac{520}{130} = \frac{4}{1} = \left(\frac{2}{1}\right)^2$$

The ratio of the areas of these similar triangles is the *square* of the ratio of the measures of the corresponding sides. If you were to look at other numerical examples, you would verify this conclusion.

Thus, we say:

- **If two triangles are similar, the ratio of the areas is the *square* of the ratio of the measures of the corresponding sides.**

Since this statement is also true for similar polygons of more than three sides, we say:

- **If two polygons are similar, the ratio of the areas is the *square* of the ratio of the measures of the corresponding sides.**

MODEL PROBLEMS

1. Polygon $WXYZ \sim$ polygon $W'X'Y'Z'$.
If $WX = 3$ cm and $W'X' = 12$ cm, find:
a. the ratio of the measures of the areas.
b. the ratio of the measures of the perimeters.

Solution:

a. $$\frac{\text{area of polygon } WXYZ}{\text{area of polygon } W'X'Y'Z'} = \left(\frac{WX}{W'X'}\right)^2 = \left(\frac{3}{12}\right)^2 = \frac{9}{144} = \frac{1}{16} \quad \textit{Ans.}$$

b. $$\frac{\text{perimeter of polygon } WXYZ}{\text{perimeter of polygon } W'X'Y'Z'} = \frac{WX}{W'X'} = \frac{3}{12} = \frac{1}{4} \quad \textit{Ans.}$$

2. The lengths of a pair of corresponding sides of two similar polygons are 4 inches and 6 inches. If the area of the first polygon is 20 square inches, find the area of the second polygon.

Solution:

Let x = the measure of the area of the second polygon.

Since the polygons are similar:

$$\frac{\text{area of 1st poly.}}{\text{area of 2nd poly.}} = \left(\frac{\text{length of side of 1st poly.}}{\text{length of side of 2nd poly.}}\right)^2$$

$$\frac{20}{x} = \left(\frac{4}{6}\right)^2$$

$$\frac{20}{x} = \frac{16}{36}$$

$$16x = 720$$

$$x = 45$$

Answer: The area of the second polygon is 45 square inches.

EXERCISES

In 1–6, the ratio of the measures of the sides of two similar polygons is given. **a.** What is the ratio of the perimeters? **b.** What is the ratio of the areas?

1. $\frac{3}{1}$ **2.** $\frac{4}{1}$ **3.** $\frac{1}{2}$ **4.** $\frac{1}{4}$ **5.** $\frac{6}{5}$ **6.** $\frac{10}{3}$

7. The lengths of a pair of corresponding sides of two similar polygons are 3 cm and 11 cm. If the area of the first polygon is 18 cm^2, find the area of the second polygon.
8. The ratio of the perimeters of two similar polygons is 5:9. What is the ratio of the areas?
9. The ratio of the perimeters of two similar polygons is 2:3. If the area of the smaller polygon is 6 square feet, what is the area of the larger polygon?
10. The length and width of a rectangle are each doubled, forming a new rectangle.
    a. What is the ratio of the perimeter of the new rectangle to the perimeter of the original rectangle?
    b. What is the ratio of the area of the new rectangle to the area of the original rectangle?
11. All regular polygons with the same number of sides are similar. The measure of the side of a regular pentagon is 12 cm and the measure of the side of a larger regular pentagon is 15 cm.
    a. What is the ratio of the measures of the sides?
    b. What is the ratio of the areas?
12. The length of a large rectangle is twice the length of a smaller rectangle. The width of the larger rectangle is 3 times the width of the smaller.
    a. What is the ratio of the length of the larger rectangle to the length of the smaller rectangle?
    b. What is the ratio of the width of the larger rectangle to the width of the smaller one?
    c. Are the rectangles similar? Why?
    d. What is the ratio of the area of the larger rectangle to the area of the smaller one?

12-10 REVIEW EXERCISES

In 1–4, express the ratio in simplest form.

1. 30:35 **2.** $8w$ to $12w$ **3.** $\frac{3}{8}$ to $\frac{5}{8}$ **4.** 75 mm to 15 cm

5. If 10 paper clips weigh 11 grams, what is the weight in grams of 150 paper clips?
6. Thelma can type 150 words in 3 minutes. At this rate, how many words can she type in 10 minutes?
7. If $1\frac{1}{2}$ pounds of beef cost \$2.40, how many pounds of beef can be purchased for \$8.00?

In 8–10, solve for x and check.

8. $\frac{8}{2x} = \frac{12}{9}$ **9.** $\frac{x}{x + 5} = \frac{1}{2}$ **10.** $\frac{4}{x} = \frac{6}{x + 3}$

11. The ratio of two numbers is 1:4 and the sum of these numbers is 40. Find the numbers.

In 12–14, select the numeral preceding the choice that makes the statement true.

12. In a class of 9 boys and 12 girls, the ratio of the number of girls to the number of students in the class is
(1) 3:4 (2) 4:3 (3) 4:7 (4) 7:4

13. The perimeter of a triangle is 45 cm, and the lengths of its sides are in the ratio 2:3:4. The length of the longest side is
(1) 5 cm (2) 10 cm (3) 20 cm (4) 30 cm

14. If $a:x = b:c$, then x equals
(1) $\frac{ac}{b}$ (2) $\frac{bc}{a}$ (3) $ac - b$ (4) $bc - a$

15. Seven percent of what number is 21?

16. What percent of 36 is 45?

17. The sales tax collected on each sale varies directly as the amount of the sale. What is the constant of variation if the sales tax of \$.63 is collected on a sale of \$9.00?

18. In the diagram, $\triangle DEF \sim \triangle GHJ$. The measures in inches are $DE = 12$, $DF = 10$, $EF = 15$, and $GH = 9$.

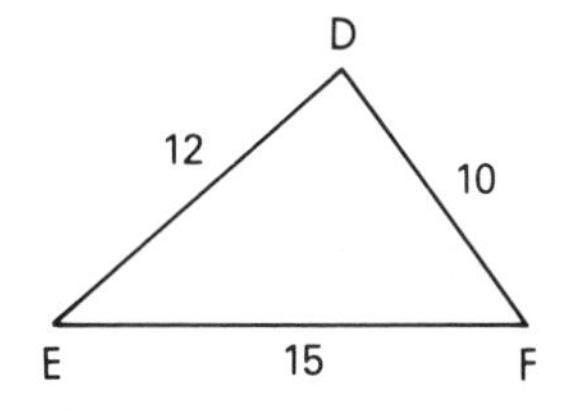

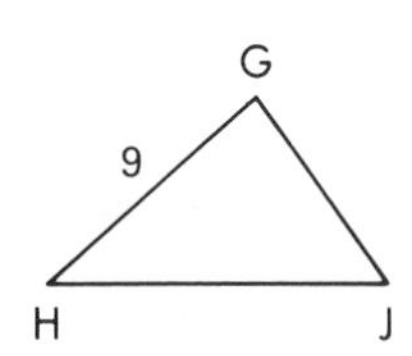

a. Find HJ and GJ.
b. What is the ratio of the measures of the corresponding sides?
c. What is the ratio of the perimeters?
d. What is the ratio of the areas?

19. On a stormy February day, 28% of the students enrolled at Southside High School were absent. How many students are enrolled at Southside High School if 476 students were absent?

20. After a 5″-by-7″ photograph is enlarged, its shorter side measures 15 inches. Find the length in inches of its longer side.

21. In the diagram, the two triangles are similar.
a. Name 3 pairs of corresponding angles.
b. Name 3 pairs of corresponding sides.
c. If $AB = 4$, $AC = 6$, and $ST = 9$, find RT.

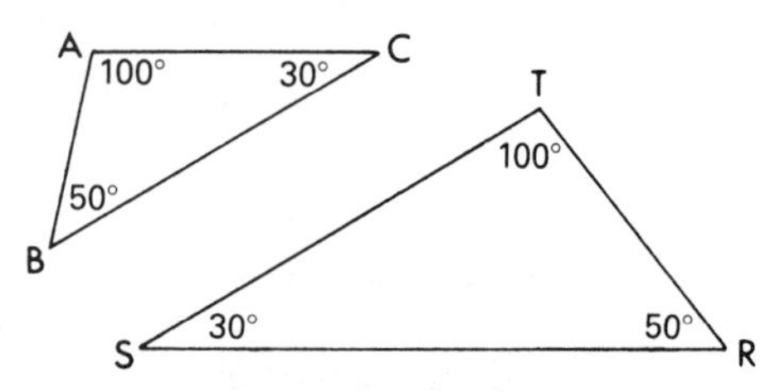

22. The sides of a triangle measure 18, 20, and 24. If the shortest side of a similar triangle measures 12, find the length of its longest side.

23. $\triangle ABC \sim \triangle A'B'C'$. Find:
a. y **b.** z **c.** AC
d. $A'C'$ **e.** AB **f.** $A'B'$

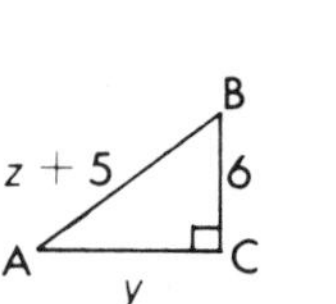

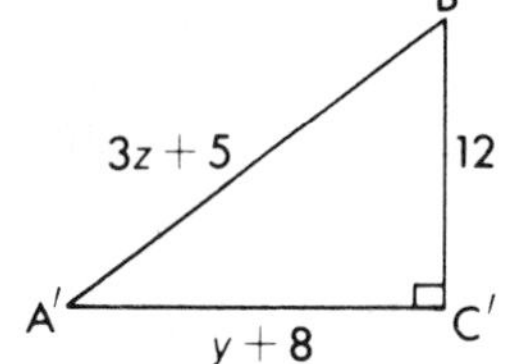

24. A student who is 5 feet tall casts an 8-foot shadow. At the same time, a tree casts a 40-foot shadow. How many feet tall is the tree?

25. In $\triangle ABC$, $m\angle A = 70$ and $m\angle B = 45$. In $\triangle DEF$, $m\angle D = 65$ and $m\angle E = 70$.
a. Is $\triangle ABC$ similar to $\triangle EFD$? Why?
b. If $AB = 5$, $DE = 12$, and $EF = 20$, find the length of $\overline{AC}$.

26. In the diagram, $\angle Y$ and $\angle ATR$ are right angles. The measures in inches are $YT = 4$, $AT = 3$, and $TR = 6$.
a. State a reason why $\triangle CRY \sim \triangle ART$.
b. Find CY.
c. Find the area of:
1. $\triangle ART$ **2.** $\triangle CRY$ **3.** trapezoid $CATY$

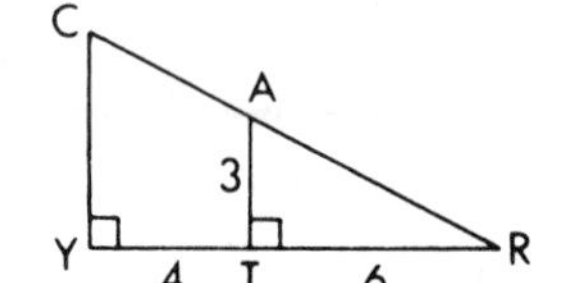

27. If 4 carpenters can build 4 tables in 4 days, how long will it take 1 carpenter to build 1 table?

28. How many girls would have to leave a room in which there are 99 girls and 1 boy in order that 98% of the remaining persons are girls?

Chapter 13

Special Products and Factoring

13-1 UNDERSTANDING THE MEANING OF FACTORING

When two numbers are multiplied, the result is called their ***product***. The numbers that are multiplied are called ***factors*** of the product. Since $3(5) = 15$, the numbers 3 and 5 are factors of 15.

Factors of a product can be found by using division. If the remainder is 0, then the divisor and the quotient are factors of the dividend. For example, $35 \div 5 = 7$. Thus, $35 = 5(7)$, and 5 and 7 are factors of 35.

Factoring a number is the process of finding those numbers whose product is a given number. Usually, when we factor, we are finding the factors of an integer and we find only those factors that are integers. We call this *factoring over the set of integers.*

Every positive integer that is the product of two positive integers is also the product of the opposites of those integers.

$$+21 = +3(+7) \qquad +21 = -3(-7)$$

Every negative integer that is the product of a positive integer and a negative integer is also the product of the opposites of those integers.

$$-21 = +3(-7) \qquad -21 = -3(+7)$$

Usually, when we factor a positive integer, we write only the positive integral factors.

Recall that a *prime number* is an integer greater than 1 that has no positive integral factors other than itself and 1. The first prime numbers are 2, 3, 5, 7, 11, 13, 17, ... Integers greater than 1 that are not prime are called ***composites***.

In general, a positive integer can be expressed as the product of prime factors. Although the factors may be written in any order, there is one and only one combination of prime factors whose product is a given composite. Note that a prime factor may occur in the product more than once.

$21 = 3 \cdot 7$

$20 = 2 \cdot 2 \cdot 5$ or $2^2 \cdot 5$

To express a positive integer, for example 280, as the product of primes, start with any pair of positive integers whose product is the given number. Then, factor these factors. Continue to factor the factors until all are primes. Finally, rearrange these factors in numerical order.

$280 = 28 \cdot 10$

$280 = 2 \cdot 14 \cdot 2 \cdot 5$

$280 = 2 \cdot 2 \cdot 7 \cdot 2 \cdot 5$

$280 = 2 \cdot 2 \cdot 2 \cdot 5 \cdot 7$

or

$280 = 2^3 \cdot 5 \cdot 7$

Expressing each of two integers as the product of prime factors makes it possible to discover the greatest integer that is a factor of both of them. We call this factor their ***greatest common factor***.

Let us find the greatest common factor of 180 and 54.

$$180 = 2 \cdot 2 \cdot 3 \cdot 3 \cdot 5 \text{ or } 2^2 \cdot 3^2 \cdot 5$$

$$54 = 2 \cdot 3 \cdot 3 \cdot 3 \quad \text{or } 2 \cdot 3^3$$

$$\text{Greatest common factor} = 2 \cdot 3 \cdot 3 \quad \text{or } 2 \cdot 3^2 \quad \text{or } 18$$

You see that the greatest number of times that 2 appears as a factor in both 180 and 54 is once; the greatest number of times that 3 appears as a factor in both 180 and 54 is twice. Therefore, the greatest common factor of 180 and 54 is $2 \cdot 3 \cdot 3$, or $2 \cdot 3^2$, or 18.

The *greatest common factor* of two or more monomials is the product of the greatest common factor of their numerical coefficients and the highest power of every variable that is a factor of each monomial.

For example, let us find the greatest common factor of $24a^3b^2$ and $18a^2b$.

$$24a^3b^2 = 2 \cdot 2 \cdot 2 \cdot 3 \cdot a \cdot a \cdot a \cdot b \cdot b$$

$$18a^2b = 2 \cdot 3 \cdot 3 \cdot a \cdot a \cdot b$$

$$\text{Greatest common factor} = 2 \cdot 3 \cdot a \cdot a \cdot b = 6a^2b$$

The greatest common factor of $24a^3b^2$ and $18a^2b$ is $6a^2b$.

When we are expressing an algebraic factor, such as $6a^2b$, we will agree that:

- **Numerical coefficients need not be factored.** (6 need not be written as $2 \cdot 3$.)

- **Powers of variables need not be represented as the product of several equal factors.** (a^2b need not be written as $a \cdot a \cdot b$.)

MODEL PROBLEMS

1. Express 700 as a product of prime factors.

Solution:

$$700 = 2 \cdot 350$$
$$700 = 2 \cdot 2 \cdot 175$$
$$700 = 2 \cdot 2 \cdot 5 \cdot 35$$
$$700 = 2 \cdot 2 \cdot 5 \cdot 5 \cdot 7 \text{ or } 2^2 \cdot 5^2 \cdot 7 \quad \textit{Ans.}$$

2. Find the greatest common factor of the monomials $60r^2s^4$ and $36rs^2t$.

Solution:

$$60r^2s^4 = 2 \cdot 2 \cdot 3 \cdot 5 \cdot r \cdot r \cdot s \cdot s \cdot s \cdot s \text{ or } 2^2 \cdot 3 \cdot 5 \cdot r^2 \cdot s^4$$
$$36rs^2t = 2 \cdot 2 \cdot 3 \cdot 3 \cdot r \cdot s \cdot s \cdot t \text{ or } 2^2 \cdot 3^2 \cdot r \cdot s^2 \cdot t$$

The greatest common factor is $2 \cdot 2 \cdot 3 \cdot r \cdot s \cdot s$ or $2^2 \cdot 3 \cdot r \cdot s^2$ or $12rs^2$. *Ans.*

3. Write all the positive integral factors of 98.

Solution:

$$98 = 1 \cdot 98$$
$$98 = 2 \cdot 49$$
$$98 = 7 \cdot 14$$

Answer: 1, 2, 7, 14, 49, and 98.

EXERCISES

In 1–10, tell whether or not the integer is a prime.

1. 5 **2.** 8 **3.** 13 **4.** 18 **5.** 73
6. 36 **7.** 41 **8.** 49 **9.** 57 **10.** 1

In 11–20, express the integer as a product of prime numbers.

11. 35 **12.** 18 **13.** 144 **14.** 77 **15.** 128
16. 400 **17.** 202 **18.** 129 **19.** 590 **20.** 316

In 21–26, write all the positive integral factors of the number.

21. 26 **22.** 50 **23.** 36 **24.** 88 **25.** 100 **26.** 242

27. The product of two integers is 144. Find the second factor if the first factor is: **a.** 2 **b.** 8 **c.** -18 **d.** 36 **e.** -48

28. The product of two monomials is $36x^3y^4$. Find the second factor if the first factor is:
a. $3x^2y^3$ **b.** $6x^3y^2$ **c.** $12xy^2$ **d.** $-9x^3y$ **e.** $18x^3y^2$

In 29–36, find the greatest common factor of the given integers.

29. 10; 15 **30.** 12; -28 **31.** 14; 35 **32.** 18; -24; 36
33. 75; 50 **34.** 72; 108 **35.** -144; 200 **36.** 96; 156; 175

In 37–48, find the greatest common factor of the given monomials.

37. $4x$; $4y$ **38.** 6; $12a$ **39.** $4r$; $6r^2$
40. $8xy$, $6xz$ **41.** $10x^2$; $15xy^2$ **42.** $7c^3d^3$; $-14c^2d$
43. $36xy^2z$; $-27xy^2z^2$ **44.** $50m^3n^2$; $75m^3n$ **45.** $24ab^2c^3$; $18ac^2$
46. $14a^2b$; $13ab$ **47.** $36xyz$; $25xyz$ **48.** $2ab^2c$; $3x^2yz$

13-2 FACTORING POLYNOMIALS WHOSE TERMS HAVE A COMMON MONOMIAL FACTOR

To ***factor a polynomial*** over a designated set of numbers means to express that polynomial as a product of polynomials whose coefficients are members of that set. Let us factor $2x + 2y$ over the integers. The distributive property tells us that $2(x + y) = 2x + 2y$. Therefore, when $2x + 2y$ is expressed as a product of factors, the result is $2(x + y)$. Notice that the monomial 2 is a factor of each term of the polynomial $2x + 2y$. Therefore, 2 is called a ***common monomial factor*** of the polynomial $2x + 2y$.

To factor a polynomial, look first for the ***greatest common monomial factor***, that is, the greatest monomial that is a factor of each term of the polynomial. For example:

1. Factor $4rs + 8st$. There are many common factors such as 2, 4, $2s$, and $4s$. The greatest common monomial factor is $4s$. Hence, divide $4rs + 8st$ by $4s$ to obtain the quotient $(r + 2t)$, which is the second factor. Therefore, $4rs + 8st = 4s(r + 2t)$.

2. Factor $3x + 4y$. Notice that 1 is the only common factor. The second factor is $3x + 4y$. We say that $3x + 4y$ is a ***prime polynomial***. A polynomial with integers as coefficients is a prime polynomial if its only factors are 1 and the polynomial itself.

PROCEDURE. To factor a polynomial whose terms have a common monomial factor:

1. Find the greatest monomial that is a factor of each term of the polynomial.
2. Divide the polynomial by the monomial factor. The quotient is the other factor.
3. Express the polynomial as the indicated product of the two factors.

We can check by multiplying the factors to obtain the original polynomial.

MODEL PROBLEM

Write in factored form: $6c^3d - 12c^2d^2 + 3cd$

Solution

(1) $3cd$ is the greatest common factor of $6c^3d$, $12c^2d^2$, and $3cd$.

(2) To find the other factor, divide $6c^3d - 12c^2d^2 + 3cd$ by $3cd$.

$(6c^3d - 12c^2d^2 + 3cd) \div 3cd = 2c^2 - 4cd + 1$

(3) $6c^3d - 12c^2d^2 + 3cd = 3cd(2c^2 - 4cd + 1)$ *Ans.*

EXERCISES

In 1–51, write the expression in factored form.

1. $2a + 2b$
2. $5c + 5d$
3. $8m + 8n$
4. $3x - 3y$
5. $7l - 7n$
6. $6R - 6r$
7. $bx + by$
8. $sr - st$
9. $xc - xd$
10. $4x + 8y$
11. $3m - 6n$
12. $12t - 6r$
13. $15c - 10d$
14. $12x - 18y$
15. $18c - 27d$
16. $8x + 16$
17. $6x - 18$
18. $8x - 12$
19. $7y - 7$
20. $8 - 4y$
21. $6 - 18c$
22. $y^2 - 3y$
23. $2x^2 + 5x$
24. $3x^2 - 6x$
25. $32x + x^2$
26. $rs^2 - 2r$
27. $ax - 5ab$
28. $3y^4 + 3y^2$
29. $10x - 15x^3$
30. $2x - 4x^3$
31. $p + prt$
32. $s - sr$
33. $hb + hc$
34. $\pi r^2 + \pi R^2$
35. $\pi r^2 + \pi rl$
36. $\pi r^2 + 2\pi rh$
37. $4x^2 + 4y^2$
38. $3a^2 - 9$
39. $5x^2 + 5$

40. $3ab^2 - 6a^2b$

41. $10xy - 15x^2y^2$

42. $21r^3s^2 - 14r^2s$

43. $2x^2 + 8x + 4$

44. $3x^2 - 6x - 30$

45. $ay - 4aw - 12a$

46. $c^3 - c^2 + 2c$

47. $2ma + 4mb + 2mc$

48. $9ab^2 - 6ab - 3a$

49. $15x^3y^3z^3 - 5xyz$

50. $8a^4b^2c^3 + 12a^2b^2c^2$

51. $28m^4n^3 - 70m^2n^4$

52. The perimeter of a rectangle is represented by $2L + 2W$. Express the perimeter as a product of two factors.

In 53–56, the expression represents the area of a rectangle. Write this expression as the product of two factors.

53. $5x + 5y$ **54.** $18x + 6$ **55.** $x^2 + 2x$ **56.** $4x^3 + 6x^2$

13-3 SQUARING A MONOMIAL

To square a monomial means to use that monomial as a factor two times. For example:

$(3x)^2 = (3x)(3x) = (3)(3)(x)(x) = (3)^2(x)^2$ or $9x^2$

$(5y^2)^2 = (5y^2)(5y^2) = (5)(5)(y^2)(y^2) = (5)^2(y^2)^2$ or $25y^4$

$(-6b^4)^2 = (-6b^4)(-6b^4) = (-6)(-6)(b^4)(b^4) = (-6)^2(b^4)^2$ or $36b^8$

$(4c^2d^3)^2 = (4c^2d^3)(4c^2d^3) = (4)(4)(c^2)(c^2)(d^3)(d^3) = (4)^2(c^2)^2(d^3)^2$ or $16c^4d^6$

When a monomial is a square, its numerical coefficient is a square and the exponent of each variable is an even number. This is the case with each of the previous results.

MODEL PROBLEMS

In each of the following, square the monomial mentally.

		Think	*Write*
1.	$(4a^3)^2$	$= (4)^2 \cdot (a^3)^2$	$= 16a^6$
2.	$\left(\frac{2}{5}ab\right)^2$	$= \left(\frac{2}{5}\right)^2 \cdot (a)^2(b)^2$	$= \frac{4}{25}a^2b^2$
3.	$(-7xy^2)^2$	$= (-7)^2 \cdot (x)^2 \cdot (y^2)^2$	$= 49x^2y^4$
4.	$(.3y^2)^2$	$= (.3)^2(y^2)^2$	$= .09y^4$

EXERCISES

In 1–20, square the monomial mentally.

1. $(a^2)^2$ **2.** $(b^3)^2$ **3.** $(-d^5)^2$ **4.** $(rs)^2$

5. $(m^2n^2)^2$ **6.** $(-x^3y^2)^2$ **7.** $(3x^2)^2$ **8.** $(-5y^4)^2$

9. $(9ab)^2$ **10.** $(10x^2y^2)^2$ **11.** $(-12cd^3)^2$ **12.** $\left(\frac{3}{4}a\right)^2$

13. $\left(\frac{5}{7}xy\right)^2$ **14.** $\left(-\frac{7}{8}a^2b^2\right)^2$ **15.** $\left(\frac{x}{6}\right)^2$ **16.** $\left(-\frac{4x^2}{5}\right)^2$

17. $(.8x)^2$ **18.** $(.5y^2)^2$ **19.** $(.1xy)^2$ **20.** $(-.6a^2b)^2$

21. Represent the area of a square whose side is represented by:

a. $4x$ **b.** $10y$ **c.** $\frac{2}{3}x$ **d.** $1.5x$ **e.** $3x^2$ **f.** $4x^2y^2$

13-4 MULTIPLYING THE SUM AND DIFFERENCE OF TWO TERMS

When two binomials are multiplied, four terms result. Usually, two of these terms are similar terms and can be combined so that the result is a trinomial. However, if the sum of the similar terms is 0, the result is a binomial.

Study the following examples:

$$\begin{aligned}(a + 4)(a - 4) &= a(a - 4) + 4(a - 4)\\ &= a^2 - 4a + 4a - 16\\ &= a^2 - 16\end{aligned}$$

$$\begin{aligned}(3x^2 + 5y)(3x^2 - 5y) &= 3x^2(3x^2 - 5y) + 5y(3x^2 - 5y)\\ &= 9x^4 - 15x^2y + 15x^2y - 25y^2\\ &= 9x^4 - 25y^2\end{aligned}$$

These examples illustrate the following procedure, which will enable us to find the products mentally:

PROCEDURE. To multiply the sum of two terms by the difference of the same two terms:

1. Square the first term.
2. From this result, subtract the square of the second term.

KEEP IN MIND

$$(a + b)(a - b) = a^2 - b^2$$

MODEL PROBLEMS

In 1 and 2, find the product mentally.

		Think	*Write*
1.	$(y + 7)(y - 7)$	$= (y)^2 - (7)^2$	$= y^2 - 49$
2.	$(3a + 4b)(3a - 4b)$	$= (3a)^2 - (4b)^2$	$= 9a^2 - 16b^2$

EXERCISES

In 1–22, find the product mentally.

1. $(x + 8)(x - 8)$ **2.** $(y + 10)(y - 10)$ **3.** $(m - 4)(m + 4)$

4. $(n - 9)(n + 9)$ **5.** $(10 + a)(10 - a)$ **6.** $(12 - b)(12 + b)$

7. $(c + d)(c - d)$ **8.** $(r - s)(r + s)$

9. $(3x + 1)(3x - 1)$ **10.** $(5c + 4)(5c - 4)$

11. $(8x + 3y)(8x - 3y)$ **12.** $(5r - 7s)(5r + 7s)$

13. $(x^2 + 8)(x^2 - 8)$ **14.** $(3 - 5y^2)(3 + 5y^2)$ **15.** $\left(a + \frac{1}{2}\right)\left(a - \frac{1}{2}\right)$

16. $(r + .5)(r - .5)$ **17.** $(.3 + m)(.3 - m)$ **18.** $(ab + 8)(ab - 8)$

19. $(r^3 - 2s^4)(r^3 + 2s^4)$ **20.** $(a + 5)(a - 5)(a^2 + 25)$

21. $(x - 3)(x + 3)(x^2 + 9)$ **22.** $(a + b)(a - b)(a^2 + b^2)$

In 23–26, express the area of the rectangle whose length L and width W are given.

23. $L = x + 7$, $W = x - 7$ **24.** $L = 2x + 3$, $W = 2x - 3$

25. $L = c + d$, $W = c - d$ **26.** $L = 2a + 3b$, $W = 2a - 3b$

13-5 FACTORING THE DIFFERENCE OF TWO SQUARES

An expression of the form $a^2 - b^2$ is called a ***difference of two squares***. Factoring an expression that is the difference of two squares is the reverse of multiplying the sum of two terms by the difference of the same two terms. Since the product $(a + b)(a - b)$ is $a^2 - b^2$, the factors of $a^2 - b^2$ are $(a + b)$ and $(a - b)$. Therefore:

$$a^2 - b^2 = (a + b)(a - b)$$

Remember that for a monomial to be a square, its numerical coefficient must be a square and the exponent of each of its variables must be an even number.

PROCEDURE. To factor a binomial that is a difference of two squares:

Express each of its terms as the square of a monomial; then apply the rule $a^2 - b^2 = (a + b)(a - b)$.

MODEL PROBLEMS

In 1–3, factor the polynomials mentally.

		Think	*Write*
1.	$r^2 - 9$	$= (r)^2 - (3)^2$	$= (r + 3)(r - 3)$
2.	$25x^2 - \frac{1}{49}y^2$	$= (5x)^2 - \left(\frac{1}{7}y\right)^2$	$= \left(5x + \frac{1}{7}y\right)\left(5x - \frac{1}{7}y\right)$
3.	$.04 - c^6d^4$	$= (.2)^2 - (c^3d^2)^2$	$= (.2 + c^3d^2)(.2 - c^3d^2)$

4. Express $x^2 - 100$ as the product of two binomials.

Solution: Since $x^2 - 100$ is a difference of two squares, the factors of $x^2 - 100$ are $(x + 10)$ and $(x - 10)$.

Answer: $(x + 10)(x - 10)$

EXERCISES

In 1–9: If possible, express the binomial as the difference of the squares of monomials as in the first step of the model problems; if not possible, tell why.

1. $y^2 - 64$ **2.** $4r^2 - b^2$ **3.** $r^2 + s^2$

4. $t^2 - 7$ **5.** $9n^2 - 16m^2$ **6.** $c^2 - .09d^2$

7. $p^2 - \frac{9}{25}q^2$ **8.** $16a^4 - 25b^6$ **9.** $-9 + m^2$

In 10–45, factor the binomial.

10. $a^2 - 4$ **11.** $b^2 - 25$ **12.** $c^2 - 100$

13. $r^2 - 16$ **14.** $s^2 - 49$ **15.** $t^2 - 81$

16. $9 - x^2$ **17.** $144 - c^2$ **18.** $121 - m^2$
19. $16a^2 - b^2$ **20.** $25m^2 - n^2$ **21.** $d^2 - 4c^2$
22. $r^4 - 9$ **23.** $x^4 - 64$ **24.** $25 - s^4$
25. $100x^2 - 81y^2$ **26.** $64e^2 - 9f^2$ **27.** $r^2s^2 - 144$
28. $w^2 - \frac{1}{64}$ **29.** $s^2 - \frac{1}{100}$ **30.** $\frac{1}{81} - t^2$
31. $49x^2 - \frac{1}{9}$ **32.** $\frac{4}{25} - \frac{49d^2}{81}$ **33.** $\frac{1}{9}r^2 - \frac{64s^2}{121}$
34. $x^2 - .64$ **35.** $y^2 - 1.44$ **36.** $.04 - 49r^2$
37. $.16m^2 - 9$ **38.** $81n^2 - .01$ **39.** $.81x^2 - y^2$
40. $64a^2b^2 - c^2d^2$ **41.** $25r^2s^2 - 9t^2u^2$ **42.** $81m^2n^2 - 49x^2y^2$
43. $49m^4 - 64n^4$ **44.** $25x^6 - 121y^{10}$ **45.** $x^4y^8 - 144a^6b^{10}$

In 46–50, the given polynomial represents the area of a rectangle. Express the area as the product of two binomials.

46. $x^2 - 4$ **47.** $y^2 - 9$ **48.** $t^2 - 49$ **49.** $t^2 - 64$ **50.** $4x^2 - y^2$

In 51–53, express the area of the shaded region as **(a)** the difference of the areas shown and **(b)** the product of two binomials.

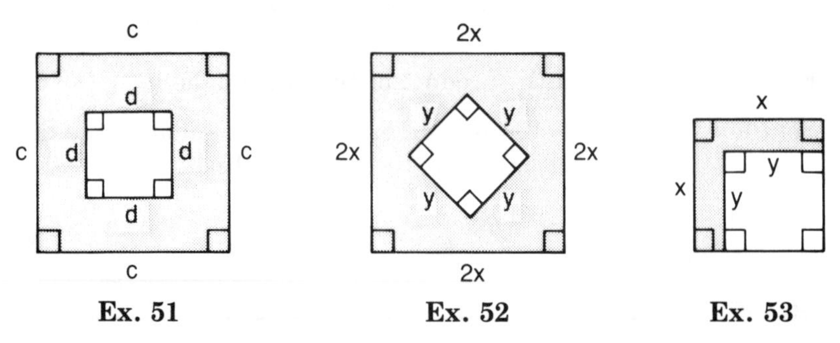

Ex. 51 **Ex. 52** **Ex. 53**

In 54 and 55, express the area of the shaded region as the product of two binomials.

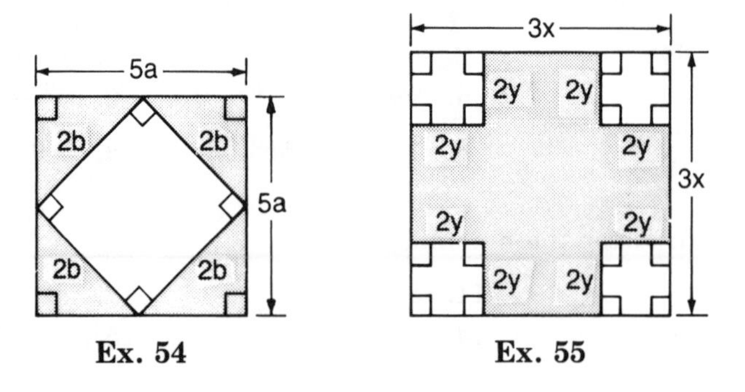

Ex. 54 **Ex. 55**

13-6 FINDING THE PRODUCT OF TWO BINOMIALS

We have already discussed how to use the distributive property to multiply two binomials of the form $ax + b$ and $cx + d$. Now, you will see how to find such a product mentally.

Study carefully the multiplication below that makes use of the distributive property.

$$\begin{aligned}(2x - 3)(4x + 5) &= 2x(4x + 5) - 3(4x + 5)\\ &= 2x(4x) + 2x(5) - 3(4x) - 3(5)\\ &= 8x^2 + 10x - 12x - 15\\ &= 8x^2 - 2x - 15\end{aligned}$$

Note the following:

1. The first term of the trinomial is the product of the first terms of the binomials.

$$2x(4x) = 8x^2$$

2. The middle term of the trinomial is the sum of two terms, the product of the outer terms and the product of the inner terms of the binomials.

$$\overset{-12x}{(2x - \overbrace{3)(4x} + 5)}$$
$$\underbrace{(2x \quad\quad\quad 5)}_{+10x} \qquad \text{Think: } (-12x) + (+10x) = -2x$$

3. The last term of the trinomial is the product of the last terms of the binomials.

$$-3(+5) = -15$$

PROCEDURE. To find the product of two binomials of the form $ax + b$ and $cx + d$:

1. Multiply the first terms of the binomials.
2. Multiply the first term of each binomial by the last term of the other binomial (the outer terms and the inner terms), and add these products.
3. Multiply the last terms of the binomials.
4. Combine the results obtained in steps 1, 2, and 3.

MODEL PROBLEMS

1. Multiply: $(x - 5)(x - 7)$

Solution:

$-5x$

$(x - 5)(x - 7)$

$-7x$

Think:

1. $(x)(x) = x^2$
2. $(-5x) + (-7x) = -12x$
3. $(-5)(-7) = +35$

Write: $(x - 5)(x - 7) = x^2 - 12x + 35$ *Ans.*

2. Multiply: $(3y - 8)(4y + 3)$

Solution:

$-32y$

$(3y - 8)(4y + 3)$

$+9y$

Think:

1. $(3y)(4y) = 12y^2$
2. $(-32y) + (+9y) = -23y$
3. $(-8)(+3) = -24$

Write: $(3y - 8)(4y + 3) = 12y^2 - 23y - 24$ *Ans.*

EXERCISES

In 1–36, perform the indicated operation mentally.

1. $(x + 5)(x + 3)$ **2.** $(y + 9)(y + 2)$ **3.** $(6 + d)(3 + d)$
4. $(x - 10)(x - 5)$ **5.** $(y - 1)(y - 9)$ **6.** $(8 - c)(3 - c)$
7. $(x + 7)(x - 2)$ **8.** $(y + 11)(y - 4)$ **9.** $(m - 15)(m + 2)$
10. $(n - 20)(n + 3)$ **11.** $(5 - t)(9 + t)$ **12.** $(2x + 1)(x + 1)$
13. $(3x + 2)(x + 5)$ **14.** $(c - 5)(3c - 1)$ **15.** $(m - 6)(3m + 2)$
16. $(y + 8)^2$ **17.** $(Z - 4)^2$ **18.** $(y + 5)^2$
19. $(1 - t)^2$ **20.** $(2x + 1)^2$ **21.** $(3x - 2)^2$
22. $(7x + 3)(2x - 1)$ **23.** $(2y + 3)(3y + 2)$ **24.** $(5Z - 3)(2Z - 5)$
25. $(2y + 3)(2y + 3)$ **26.** $(3x + 4)^2$ **27.** $(2x - 5)^2$
28. $(3t - 2)(4t + 7)$ **29.** $(5y - 4)(5y - 4)$ **30.** $(2t + 3)(5t + 1)$
31. $(2c - 3d)(5c - 2d)$ **32.** $(4a - 3b)(3a + b)$
33. $(5a + 7b)(5a - 7b)$ **34.** $(5a + 7b)(5a + 7b)$
35. $(5a + 7b)(7a + 5b)$ **36.** $(5a + 7b)(7a - 5b)$

37. Represent the area of a rectangle whose length and width are:
a. $(x + 5)$ and $(x + 4)$ **b.** $(2x + 3)$ and $(x - 1)$

38. Represent the area of a square each of whose sides is:
a. $(x + 6)$ **b.** $(x - 2)$ **c.** $(2x + 1)$ **d.** $(3x - 2)$

13-7 FACTORING TRINOMIALS OF THE FORM $ax^2 + bx + c$

You have learned that $(x + 3)(x + 5) = x^2 + 8x + 15$. Therefore, the factors of $x^2 + 8x + 15$ are $(x + 3)$ and $(x + 5)$. Factoring a trinomial of the form $ax^2 + bx + c$ is the reverse of multiplying binomials of the form $(dx + e)$ and $(fx + g)$. When we factor a trinomial of this form, we list the possible pairs of factors using combinations of factors of the first and last terms and test them, one by one, until we find the correct middle term.

For example, let us factor $x^2 + 7x + 10$.

1. The product of the first terms of the binomials must be x^2. Therefore, for each first term, we use x. We write:

$$x^2 + 7x + 10 = (x \quad)(x \quad)$$

2. Since the product of the last terms of the binomials must be $+10$, these last terms must be either both positive or both negative.

   The pairs of integers whose product is $+10$ are:

   $(+1)(+10)$ $\qquad$ $(+5)(+2)$ $\qquad$ $(-1)(-10)$ $\qquad$ $(-2)(-5)$

3. From the products obtained in steps 1 and 2, we see that the possible pairs of factors are:

$$(x + 10)(x + 1) \qquad (x - 10)(x - 1)$$
$$(x + 5)(x + 2) \qquad (x - 5)(x - 2)$$

4. Now, we test each pair of factors. For example,

   $(x + 10)(x + 1)$ is not correct because the middle term, $(+10x) + (+1x)$, is $+11x$, not $+7x$.

   +10x
   (x + 10)(x + 1)
   +1x

   $(x + 5)(x + 2)$ is correct because the middle term, $(+5x) + (+2x)$, is $+7x$.

   +5x
   (x + 5)(x + 2)
   +2x

   None of the remaining pairs of factors is correct because each would have a middle term that is negative.

5. $x^2 + 7x + 10 = (x + 5)(x + 2)$ *Ans.*

Observe that, in this trinomial, the first and last terms are both positive: x^2 and $+10$. Since the middle term of the trinomial is *positive*, the last terms of both binomial factors must be *positive* ($+5$ and $+2$).

PROCEDURE. To factor a trinomial of the form $ax^2 + bx + c$, find two binomials that have the following characteristics:

1. The product of the first terms of the binomials must be equal to the first term in the trinomial (ax^2).
2. The product of the last terms of the binomials must be equal to the last term of the trinomial (c).
3. When the first term of each binomial is multiplied by the second term of the other and the sum of these products is found, this result must equal the middle term of the trinomial (bx).

MODEL PROBLEMS

1. Factor: $y^2 - 8y + 12$

Solution:

1. The product of the first terms of the binomials must be y^2. Therefore, for each first term, we use y. We write:

$$y^2 - 8y + 12 = (y \quad)(y \quad)$$

2. Since the product of the last terms of the binomials must be $+12$, these last terms must be either both positive or both negative. The pairs of integers whose product is $+12$ are:

$(+1)(+12)$ $(+2)(+6)$ $(+3)(+4)$
$(-1)(-12)$ $(-2)(-6)$ $(-3)(-4)$

3. The possible factors are:

$(y + 1)(y + 12)$ $(y + 2)(y + 6)$ $(y + 3)(y + 4)$
$(y - 1)(y - 12)$ $(y - 2)(y - 6)$ $(y - 3)(y - 4)$

4. When we find the middle term in each of the trinomial products, we find that only the factors $(y - 6)(y - 2)$ yield a middle term of $-8y$.

$-6y$
$(y - 6)(y - 2)$
$-2y$

5. $y^2 - 8y + 12 = (y - 6)(y - 2)$ *Ans.*

When the first and last terms are both positive (y^2 and $+12$) and the middle term of the trinomial is *negative*, the last terms of both binomial factors must be *negative* (-6 and -2).

2. Factor: $c^2 + 5c - 6$

Solution:

1. The product of the first terms of the binomials must be c^2. Therefore, for each first term, we use c. We write:

$$c^2 + 5c - 6 = (c \quad)(c \quad)$$

2. Since the product of the last terms of the binomials must be -6, one of these last terms must be positive, the other negative. The pairs of integers whose product is -6 are $(+1)$ and (-6); (-1) and $(+6)$; $(+3)$ and (-2); (-3) and $(+2)$.

3. The possible factors are: $(c + 1)(c - 6)$ $(c + 3)(c - 2)$
$(c - 1)(c + 6)$ $(c - 3)(c + 2)$

4. When we find the middle term of each of the trinomial products, we find that only the factors $(c - 1)(c + 6)$ yield a middle term of $+5c$.

$-1c$
$(c - 1)(c + 6)$
$+6c$

5. $c^2 + 5c - 6 = (c - 1)(c + 6)$ *Ans.*

3. Factor: $2x^2 - 7x - 15$

Solution:

1. Since the product of the first terms of the binomials must be $2x^2$, we use as one of these terms $2x$, and as the other, x. We write:

$$2x^2 - 7x - 15 = (2x \quad)(x \quad)$$

2. Since the product of the last terms of the binomials must be -15, one of these last terms must be positive, the other negative. The pairs of integers whose product is -15 are $(+1)$ and (-15); (-1) and $(+15)$; $(+3)$ and (-5); (-3) and $(+5)$.

   Notice that these four pairs of integers will form eight pairs of binomial factors since the way in which the integers are combined with the first terms will produce different pairs of factors: $(2x + 1)(x - 15)$ is not the same product as $(2x - 15)(x + 1)$.

3. The possible pairs of factors are:

$(2x + 1)(x - 15)$ $(2x + 3)(x - 5)$ $(2x - 1)(x + 15)$ $(2x - 3)(x + 5)$
$(2x + 15)(x - 1)$ $(2x + 5)(x - 3)$ $(2x - 15)(x + 1)$ $(2x - 5)(x + 3)$

4. When we find the middle term of each of the trinomial products, we find that only the factors $(2x + 3)(x - 5)$ yield a middle term of $-7x$.

$$\overset{+3x}{(2x + 3)(x - 5)}$$
$$-10x$$

5. $2x^2 - 7x - 15 = (2x + 3)(x - 5)$ *Ans.*

KEEP IN MIND

In factoring a trinomial of the form $ax^2 + bx + c$, when a is a positive integer ($a > 0$):

1. If the last term is positive, the last terms of the binomial factors must be either both positive or both negative.
2. If the last term is negative, one of the last terms in the binomial factors must be positive, the other negative.

EXERCISES

In 1–45, factor.

1. $a^2 + 3a + 2$ **2.** $c^2 + 6c + 5$ **3.** $x^2 + 8x + 7$
4. $r^2 + 12r + 11$ **5.** $m^2 + 5m + 4$ **6.** $y^2 + 12y + 35$
7. $x^2 + 11x + 24$ **8.** $a^2 + 11a + 18$ **9.** $16 + 17c + c^2$
10. $x^2 + 2x + 1$ **11.** $z^2 + 10z + 25$ **12.** $a^2 - 8a + 7$
13. $a^2 - 6a + 5$ **14.** $x^2 - 5x + 6$ **15.** $x^2 - 11x + 10$
16. $y^2 - 6y + 8$ **17.** $15 - 8y + y^2$ **18.** $x^2 - 10x + 24$
19. $c^2 - 14c + 40$ **20.** $x^2 - 16x + 48$ **21.** $x^2 - 14x + 49$
22. $x^2 - x - 2$ **23.** $x^2 - 6x - 7$ **24.** $y^2 + 4y - 5$
25. $z^2 - 12z - 13$ **26.** $c^2 - 2c - 15$ **27.** $c^2 + 2c - 35$
28. $x^2 - 7x - 18$ **29.** $z^2 + 9z - 36$ **30.** $x^2 - 13x - 48$
31. $x^2 - 16x + 64$ **32.** $2x^2 + 5x + 2$ **33.** $2x^2 + 7x + 6$
34. $3x^2 + 10x + 8$ **35.** $16x^2 + 8x + 1$ **36.** $2x^2 + x - 3$
37. $3x^2 + 2x - 5$ **38.** $2x^2 + x - 6$ **39.** $4x^2 - 12x + 5$
40. $10a^2 - 9a + 2$ **41.** $18y^2 - 23y - 6$ **42.** $x^2 + 3xy + 2y^2$
43. $r^2 - 3rs - 10s^2$ **44.** $3a^2 - 7ab + 2b^2$ **45.** $4x^2 - 5xy - 6y^2$

In 46–48, express the polynomial as the product of two binomial factors.

46. $x^2 + 9x + 18$ **47.** $x^2 - 9x + 14$ **48.** $y^2 - 5y - 24$

In 49–51, the trinomial represents the area of a rectangle. Find the binomials that could be expressions for the dimensions of the rectangle.

49. $x^2 + 8x + 7$ **50.** $x^2 + 9x + 18$ **51.** $3x^2 + 14x + 15$

In 52–54, the trinomial represents the area of a square. Find the binomial that could be an expression for the measure of each side of the square.

52. $x^2 + 10x + 25$ **53.** $81x^2 + 18x + 1$ **54.** $4x^2 + 12x + 9$

13-8 FACTORING COMPLETELY

Some polynomials, such as $x^2 + 4$ and $x^2 + x + 1$, cannot be factored into other polynomials with integral coefficients. We say that these polynomials are *prime over the set of integers.*

To factor a polynomial completely means to find the *prime factors* of the polynomial over a designated set of numbers. In this text, whenever we factor a polynomial, we will continue the process of factoring until all factors other than monomial factors are prime factors over the set of integers.

PROCEDURE. To factor a polynomial completely, use the following steps:

1. Look for the greatest common factor. If there is one, factor the given polynomial. Then examine each factor.
2. If one of these factors is a binomial, see if it is a difference of two squares. If it is, factor it as such.
3. If one of these factors is a trinomial, see if it can be factored. If it can, find its binomial factors.
4. Write the answer as the product of all the factors. Make certain that in the answer all factors other than monomial factors are prime factors.

MODEL PROBLEMS

1. Factor: $by^2 - 4b$

How to Proceed	*Solution*
1. Find the greatest common factor.	$by^2 - 4b = b(y^2 - 4)$
2. Factor the difference of 2 squares.	$by^2 - 4b = b(y + 2)(y - 2)$ *Ans.*

2. Factor: $3x^2 - 6x - 24$

How to Proceed	*Solution*
1. Find the greatest common factor.	$3x^2 - 6x - 24 = 3(x^2 - 2x - 8)$
2. Factor the trinomial.	$3x^2 - 6x - 24 = 3(x - 4)(x + 2)$ *Ans.*

3. Factor: $x^4 - 16$

How to Proceed	*Solution*
1. Factor $x^4 - 16$ as the difference of 2 squares.	$x^4 - 16 = (x^2 + 4)(x^2 - 4)$
2. Factor $x^2 - 4$ as the difference of 2 squares.	$x^4 - 16 = (x^2 + 4)(x + 2)(x - 2)$ *Ans.*

EXERCISES

In 1–45, factor completely.

1. $2a^2 - 2b^2$
2. $6x^2 - 6y^2$
3. $4x^2 - 4$
4. $ax^2 - ay^2$
5. $cm^2 - cn^2$
6. $st^2 - s$
7. $2x^2 - 18$
8. $2x^2 - 32$
9. $3x^2 - 27y^2$
10. $18m^2 - 8$
11. $12a^2 - 27b^2$
12. $63c^2 - 7$
13. $x^3 - 4x$
14. $y^3 - 25y$
15. $z^3 - z$
16. $4a^3 - ab^2$
17. $4c^3 - 49c$
18. $9db^2 - d$
19. $4a^2 - 36$
20. $x^4 - 1$
21. $y^4 - 81$
22. $\pi R^2 - \pi r^2$
23. $\pi c^2 - \pi d^2$
24. $100x^2 - 36y^2$
25. $ax^2 + 3ax + 2a$
26. $3x^2 + 6x + 3$
27. $4r^2 - 4r - 48$
28. $x^3 + 7x^2 + 10x$
29. $4x^2 - 6x - 4$
30. $a^2y + 10ay + 25y$
31. $d^3 - 8d^2 + 16d$
32. $2ax^2 - 2ax - 12a$
33. $abx^2 - ab$
34. $z^6 - z^2$
35. $16x^2 - x^2y^4$
36. $x^4 + x^2 - 2$
37. $a^4 - 10a^2 + 9$
38. $y^4 - 13y^2 + 36$
39. $2x^2 + 12x + 8$
40. $5x^4 + 10x^2 + 5$
41. $2a^2b + 7ab + 3b$
42. $16x^2 - 16x + 4$
43. $25x^2 + 100xy + 100y^2$
44. $18m^2 + 24m + 8$
45. $12a^2 - 5ab - 2b^2$

13-9 REVIEW EXERCISES

1. Express 250 as a product of prime numbers.
2. What is the greatest common factor of $8ax$ and $4ay$?
3. Find the greatest common factor of $16a^3bc^2$ and $24a^2bc^4$.

In 4–6, square the monomial

4. $(3g^3)^2$
5. $(-4x^4)^2$
6. $(0.2c^2y)^2$

In 7–12, find the product.

7. $(x - 5)(x + 9)$
8. $(y - 8)(y - 6)$
9. $(ab + 4)(ab - 4)$
10. $(3d + 1)(d - 2)$
11. $(2w + 1)^2$
12. $(2x + 3c)(x + 4c)$

In 13–27, factor completely.

13. $6x + 27b$
14. $3y^2 + 10y$
15. $m^2 - 81$
16. $x^2 - 16h^2$
17. $x^2 - 4x - 5$
18. $y^2 - 9y + 14$
19. $64b^2 - 9$
20. $121 - k^2$
21. $x^2 - 8x + 16$
22. $a^2 - 7a - 30$
23. $x^2 - 16x + 60$
24. $16y^2 - 16$
25. $x^2 + 6bx - 16b^2$
26. $2x^2 - 9xy - 5y^2$
27. $3x^3 - 6x^2 - 24x$

28. Express the product $(k + 15)(k - 15)$ as a binomial.

29. Express as a binomial: $4ez^2(4e - z)$

30. If the length and width of a rectangle are represented by $(2x - 3)$ and $(3x - 2)$, respectively, express the area of the rectangle as a trinomial.

31. Find the trinomial that represents the area of a square if the measure of a side is $8m + 1$.

32. If $9x^2 + 30x + 25$ represents the square of a number, find the binomial that represents the number.

33. Factor completely: $60a^2 + 37a - 6$

34. A group of people wanted to form committees having the same number of persons on each committee. They found that there was one extra person when they tried to make 2, 3, 4, 5, or 6 committees. However, they were able to make more than 6 and fewer than 12 equal committees. What is the smallest possible number of persons in the group? What was the size of the committees that were formed?

Chapter 14

Fractions, and First-Degree Equations and Inequalities Involving Fractions

14-1 THE MEANING OF AN ALGEBRAIC FRACTION

A *fraction* is a symbol that indicates the quotient of any number divided by any nonzero number. For example, the arithmetic fraction $\frac{3}{4}$ indicates the quotient of 3 and 4.

An ***algebraic fraction*** is a quotient of polynomials. An algebraic fraction is sometimes called a ***rational expression***.

Examples of algebraic fractions are: $\frac{2}{5}, \frac{x}{2}, \frac{2}{x}, \frac{a}{b}, \frac{4c}{3d}, \frac{x+5}{x-2}, \frac{x^2+4x+3}{x+1}$

The fraction $\frac{a}{b}$ means that the number represented by a, the numerator, is to be divided by the number represented by b, the denominator. Since division by zero is not possible, the value of the denominator, b, may not be zero. An algebraic fraction is defined or has meaning only for those values of the variables for which the denominator is not zero.

MODEL PROBLEM

Find the value of x for which $\frac{12}{x-9}$ has no meaning.

Solution: $\frac{12}{x-9}$ is not defined when the denominator $x - 9$ is equal to 0.

Let $x - 9 = 0$. Then $x = 9$. *Answer:* 9

EXERCISES

In 1–10, find the value of the variable for which the fraction is not defined.

1. $\frac{2}{x}$
2. $\frac{-5}{6x}$
3. $\frac{12}{y^2}$
4. $\frac{1}{x - 5}$
5. $\frac{x}{x - 8}$
6. $\frac{7}{2 - x}$
7. $\frac{y + 5}{y + 2}$
8. $\frac{10}{2x - 1}$
9. $\frac{2y + 3}{4y + 2}$
10. $\frac{1}{x^2 - 4}$

In 11–15, represent the answer to the problem as a fraction.

11. Represent the cost of 1 piece of candy if 5 pieces cost c cents.
12. Represent the cost of 1 meter of lumber if p meters cost 98 cents.
13. If a piece of lumber $10x + 20$ centimeters in length is cut into y pieces of equal length, represent the length of each of the pieces.
14. What fractional part of an hour is m minutes?
15. If the perimeter of a square is represented by $4x + 2y$, represent the length of each side of the square.

14-2 REDUCING FRACTIONS TO LOWEST TERMS

A fraction is said to be ***reduced to lowest terms*** when its numerator and denominator have no common factor other than 1 or -1.

Each of the fractions $\frac{5}{10}$ and $\frac{a}{2a}$ can be expressed in lowest terms as $\frac{1}{2}$. Use the multiplication property of 1 to show that $\frac{5}{10}$ names the same number as $\frac{1}{2}$ and that $\frac{a}{2a}$ also names the same number as $\frac{1}{2}$. Remember that any nonzero number divided by itself equals 1.

$$\frac{5}{10} = \frac{5 \div 5}{10 \div 5} = \frac{1}{2} \quad \text{and} \quad \frac{a}{2a} = \frac{1a \div a}{2a \div a} = \frac{1}{2}$$

These examples illustrate the ***division property of a fraction:*** If the numerator and denominator of a fraction are divided by the same nonzero number, the resulting fraction is equal to the original fraction.

In general, for any numbers x, y, and a, where $y \neq 0$ and $a \neq 0$:

$$\frac{ax}{ay} = \frac{ax \div a}{ay \div a} = \frac{x}{y}$$

Note the following examples of the division property of fractions:

$$\frac{4x}{5x} = \frac{4x \div x}{5x \div x} = \frac{4}{5} \qquad \frac{cy}{dy} = \frac{cy \div y}{dy \div y} = \frac{c}{d}$$

$$\frac{3(x + 5)}{18} = \frac{3(x + 5) \div 3}{18 \div 3} = \frac{x + 5}{6}$$

When reducing a fraction, the division of the numerator and the denominator by a common factor may be indicated by a ***cancellation***. For example:

$$\frac{3(x + 5)}{18} = \frac{\overset{1}{\cancel{3}}(x + 5)}{\underset{6}{\cancel{18}}} = \frac{x + 5}{6}$$

$$\frac{a^2 - 9}{3a - 9} = \frac{\overset{1}{\cancel{(a - 3)}}(a + 3)}{3\underset{1}{\cancel{(a - 3)}}} = \frac{a + 3}{3}$$

PROCEDURE. To reduce a fraction to its lowest terms:

Method 1

1. Factor both its numerator and its denominator.
2. Examine the factors and determine the greatest common factor of the numerator and the denominator.
3. Express the given fraction as the product of two fractions, one of which has as its numerator and its denominator the greatest common factor determined in step 2.
4. Use the multiplication property of 1.

Method 2

1. Factor both its numerator and its denominator.
2. Divide both the numerator and the denominator by their greatest common factor.

MODEL PROBLEMS

1. Reduce $\frac{15x^2}{35x^4}$ to lowest terms.

Solution

Method 1

$$\frac{15x^2}{35x^4} = \frac{3}{7x^2} \cdot \frac{5x^2}{5x^2}$$

$$= \frac{3}{7x^2} \cdot 1$$

$$= \frac{3}{7x^2} \quad \textit{Ans.}$$

Method 2

$$\frac{15x^2}{35x^4} = \frac{3 \cdot 5x^2}{7x^2 \cdot 5x^2}$$

$$= \frac{3 \cdot \overset{1}{\cancel{5x^2}}}{7x^2 \cdot \underset{1}{\cancel{5x^2}}} = \frac{3}{7x^2} \quad \textit{Ans.}$$

2. Express $\frac{2x^2 - 6x}{10x}$ as an equivalent fraction in lowest terms.

Solution

Method 1

$$\frac{2x^2 - 6x}{10x} = \frac{2x(x - 3)}{2x \cdot 5}$$

$$= \frac{(x - 3)}{5} \cdot \frac{2x}{2x}$$

$$= \frac{(x - 3)}{5} \cdot 1$$

$$= \frac{x - 3}{5} \quad \textit{Ans.}$$

Method 2

$$\frac{2x^2 - 6x}{10x} = \frac{2x(x - 3)}{10x}$$

$$= \frac{\overset{1}{\cancel{2x}}(x - 3)}{\underset{5}{\cancel{10x}}}$$

$$= \frac{x - 3}{5} \quad \textit{Ans.}$$

3. Reduce: $\frac{x^2 - 16}{x^2 - 5x + 4}$

Solution

$$\frac{x^2 - 16}{x^2 - 5x + 4} = \frac{(x + 4)(x - 4)}{(x - 1)(x - 4)}$$

$$= \frac{(x + 4)\overset{1}{\cancel{(x - 4)}}}{(x - 1)\underset{1}{\cancel{(x - 4)}}}$$

$$= \frac{x + 4}{x - 1} \quad \textit{Ans.}$$

4. Reduce: $\frac{2 - x}{4x - 8}$

Solution

$$\frac{2 - x}{4x - 8} = \frac{-x + 2}{4x - 8}$$

$$= \frac{-1(x - 2)}{4(x - 2)}$$

$$= \frac{-1\overset{1}{\cancel{(x - 2)}}}{4\underset{1}{\cancel{(x - 2)}}} = -\frac{1}{4} \quad \textit{Ans.}$$

EXERCISES

In 1–54, reduce the fraction to lowest terms.

1. $\frac{4}{12}$
2. $\frac{27}{36}$
3. $\frac{24c}{36d}$
4. $\frac{9r}{10r}$
5. $\frac{ab}{cb}$
6. $\frac{3ay^2}{6by^2}$
7. $\frac{5xy}{9xy}$
8. $\frac{2abc}{4abc}$
9. $\frac{15x^2}{5x}$
10. $\frac{5x^2}{25x^4}$
11. $\frac{27a}{36a^2}$
12. $\frac{8xy^2}{24x^2y}$
13. $\frac{+12a^2b}{-8ac}$
14. $\frac{-20x^2y^2}{-90xy^2}$
15. $\frac{-32a^3b^3}{+48a^3b^3}$
16. $\frac{+5xy}{+45x^2y^2}$
17. $\frac{3x + 6}{4}$
18. $\frac{8y - 12}{6}$
19. $\frac{5x - 35}{5x}$
20. $\frac{8m^2 + 40m}{8m}$
21. $\frac{2ax + 2bx}{6x^2}$
22. $\frac{5a^2 - 10a}{5a^2}$
23. $\frac{12ab - 3b^2}{3ab}$
24. $\frac{6x^2y + 9xy^2}{12xy}$
25. $\frac{18b^2 + 30b}{9b^3}$
26. $\frac{4x}{4x + 8}$
27. $\frac{7d}{7d + 14}$
28. $\frac{5y}{5y + 5x}$
29. $\frac{2a^2}{6a^2 - 2ab}$
30. $\frac{14}{7r - 21s}$
31. $\frac{12a + 12b}{3a + 3b}$
32. $\frac{x^2 - 9}{3x + 9}$
33. $\frac{x^2 - 1}{5x - 5}$
34. $\frac{1 - x}{x - 1}$
35. $\frac{3 - b}{b^2 - 9}$
36. $\frac{2s - 2r}{s^2 - r^2}$
37. $\frac{16 - a^2}{2a - 8}$
38. $\frac{x^2 - y^2}{3y - 3x}$
39. $\frac{2b(3 - b)}{b^2 - 9}$
40. $\frac{r^2 - r - 6}{3r - 9}$
41. $\frac{x^2 + 7x + 12}{x^2 - 16}$
42. $\frac{x^2 + x - 2}{x^2 + 4x + 4}$
43. $\frac{3y - 3}{y^2 - 2y + 1}$
44. $\frac{x^2 - 3x}{x^2 - 4x + 3}$
45. $\frac{x^2 - 25}{x^2 - 2x - 15}$

46. $\dfrac{a^2 - a - 6}{a^2 - 9}$

47. $\dfrac{a^2 - 6a}{a^2 - 7a + 6}$

48. $\dfrac{2x^2 - 50}{x^2 + 8x + 15}$

49. $\dfrac{r^2 - 4r - 5}{r^2 - 2r - 15}$

50. $\dfrac{48 + 8x - x^2}{x^2 + x - 12}$

51. $\dfrac{2x^2 - 7x + 3}{(x - 3)^2}$

52. $\dfrac{3x^2 - 15x + 18}{x^2 - x - 6}$

53. $\dfrac{x^2 - 7xy + 12y^2}{x^2 + xy - 20y^2}$

54. $\dfrac{18c^2 - 32d^2}{6c^2 - cd - 12d^2}$

14-3 MULTIPLYING FRACTIONS

The product of two fractions is a fraction with the following properties:

1. The numerator of the product is the product of the numerators of the given fractions.
2. The denominator of the product is the product of the denominators of the given fractions.

In general, for any numbers a, b, x, and y, when $b \neq 0$ and $y \neq 0$:

$$\frac{a}{b} \cdot \frac{x}{y} = \frac{ax}{by}$$

You can find the product of $\frac{7}{27}$ and $\frac{9}{4}$ in lowest terms by using either one of the following two methods:

Method 1

$$\frac{7}{27} \cdot \frac{9}{4} = \frac{7 \cdot 9}{27 \cdot 4} = \frac{63}{108} = \frac{7 \cdot \overset{1}{\cancel{9}}}{12 \cdot \underset{1}{\cancel{9}}} = \frac{7}{12}$$

Method 2

$$\frac{7}{27} \cdot \frac{9}{4} = \frac{7 \cdot \overset{1}{\cancel{9}}}{\underset{3}{\cancel{27}} \cdot 4} = \frac{7}{12}$$

Notice that Method 2 requires less computation than Method 1 since the reduced form of the product was obtained by dividing the numerator and the denominator by a common factor *before* the product was found. This method may be called the ***cancellation method***.

When you multiply algebraic fractions, the product has the same properties as when you multiply arithmetic fractions.

Thus, to multiply $\frac{5x^2}{7y}$ by $\frac{14y^2}{15x^3}$, you may use either one of the following two methods:

Method 1

$$\frac{5x^2}{7y} \cdot \frac{14y^2}{15x^3} = \frac{5x^2 \cdot 14y^2}{7y \cdot 15x^3} = \frac{70x^2y^2}{105x^3y} = \frac{2y}{3x} \cdot \frac{35x^2y}{35x^2y} = \frac{2y}{3x} \cdot 1 = \frac{2y}{3x}$$

Method 2 (the cancellation method)

$$\frac{5x^2}{7y} \cdot \frac{14y^2}{15x^3} = \frac{\overset{1}{\cancel{5x^2}}}{\underset{1}{\cancel{7y}}} \cdot \frac{\overset{2y}{\cancel{14y^2}}}{\underset{3x}{\cancel{15x^3}}} = \frac{2y}{3x}$$

In problems in which the numerator, the denominator, or both are polynomials that are not monomials, it is helpful to factor the numerator and the denominator before applying the cancellation method. For example:

$$\frac{3x + 15}{4y} \cdot \frac{2}{3} = \frac{\overset{1}{\cancel{3}}(x + 5)}{\underset{2}{\cancel{4}}y} \cdot \frac{\overset{1}{\cancel{2}}}{\underset{1}{\cancel{3}}} = \frac{x + 5}{2y}$$

$$\frac{x^2 - 4}{2x} \cdot \frac{2}{x^2 - 3x + 2} = \frac{\overset{1}{\cancel{(x - 2)}}(x + 2)}{\underset{1}{\cancel{2}}x} \cdot \frac{\overset{1}{\cancel{2}}}{(x - 1)\underset{1}{\cancel{(x - 2)}}} = \frac{x + 2}{x(x - 1)}$$

MODEL PROBLEMS

1. Multiply and express the product in reduced form: $\frac{5a^3}{9bx} \cdot \frac{6bx}{a^2}$

How to Proceed	*Solution*
(1) Divide the numerators and the denominators by the common factors $3bx$ and a^2.	$\frac{5a^3}{9bx} \cdot \frac{6bx}{a^2} = \frac{\overset{5a}{\cancel{5a^3}}}{\underset{3}{\cancel{9bx}}} \cdot \frac{\overset{2}{\cancel{6bx}}}{\underset{1}{\cancel{a^2}}}$
(2) Multiply the remaining numerators and then multiply the remaining denominators.	$= \frac{10a}{3}$ *Ans.*

2. Multiply and simplify the product: $\dfrac{x^2 - 5x + 6}{3x} \cdot \dfrac{2}{4x - 12}$

Solution: $\dfrac{x^2 - 5x + 6}{3x} \cdot \dfrac{2}{4x - 12}$

$$= \frac{\overset{1}{\cancel{(x - 3)}}(x - 2)}{3x} \cdot \frac{\overset{1}{\cancel{2}}}{\underset{2}{\cancel{4}}\underset{1}{\cancel{(x - 3)}}} = \frac{x - 2}{6x} \quad \textit{Ans.}$$

EXERCISES

In 1–38, find the product in lowest terms.

1. $\frac{8}{12} \cdot \frac{30}{36}$

2. $36 \cdot \frac{5}{9}$

3. $\frac{1}{2} \cdot 20x$

4. $\dfrac{5}{d} \cdot d^2$

5. $\dfrac{x^2}{36} \cdot 20$

6. $mn \cdot \dfrac{8}{m^2n^2}$

7. $\dfrac{24x}{35y} \cdot \dfrac{14y}{8x}$

8. $\dfrac{12x}{5y} \cdot \dfrac{15y^2}{36x^2}$

9. $\dfrac{m^2}{8} \cdot \dfrac{32}{3m}$

10. $\dfrac{6r^2}{5s^2} \cdot \dfrac{10rs}{6r^3}$

11. $\dfrac{30m^2}{18n} \cdot \dfrac{6n}{5m}$

12. $\dfrac{24a^3b^2}{7c^3} \cdot \dfrac{21c^2}{12ab}$

13. $\dfrac{7}{8} \cdot \dfrac{2x + 4}{21}$

14. $\dfrac{3a + 9}{15a} \cdot \dfrac{a^3}{18}$

15. $\dfrac{5x - 5y}{x^2y} \cdot \dfrac{xy^2}{25}$

16. $\dfrac{12a - 4}{b} \cdot \dfrac{b^3}{12}$

17. $\dfrac{ab - a}{b^2} \cdot \dfrac{b^3 - b^2}{a}$

18. $\dfrac{x^2 - 1}{x^2} \cdot \dfrac{3x^2 - 3x}{15}$

19. $\dfrac{2r}{r - 1} \cdot \dfrac{r - 1}{10}$

20. $\dfrac{7s}{s + 2} \cdot \dfrac{2s + 4}{21}$

21. $\dfrac{8x}{2x + 6} \cdot \dfrac{x + 3}{x^2}$

22. $\dfrac{1}{x^2 - 1} \cdot \dfrac{2x + 2}{6}$

23. $\dfrac{a^2 - 9}{3} \cdot \dfrac{12}{2a - 6}$

24. $\dfrac{x^2 - x - 2}{3} \cdot \dfrac{21}{x^2 - 4}$

25. $\dfrac{a(a - b)^2}{4b} \cdot \dfrac{4b}{a(a^2 - b^2)}$

26. $\frac{(a-2)^2}{4b} \cdot \frac{16b^3}{4-a^2}$

27. $\frac{a^2-7a-8}{2a+2} \cdot \frac{5}{a-8}$

28. $\frac{x^2+6x+5}{9y^2} \cdot \frac{3y}{x+1}$

29. $\frac{y^2-2y-3}{2c^3} \cdot \frac{4c^2}{2y+2}$

30. $\frac{4a-6}{4a+8} \cdot \frac{6a+12}{5a-15}$

31. $\frac{x^2-25}{4x^2-9} \cdot \frac{2x+3}{x-5}$

32. $\frac{4x+8}{6x+18} \cdot \frac{5x+15}{x^2-4}$

33. $\frac{y^2-81}{(y+9)^2} \cdot \frac{10y+90}{5y-45}$

34. $\frac{8x}{2x^2-8} \cdot \frac{8x+16}{32x^2}$

35. $\frac{2-x}{2x} \cdot \frac{3x}{3x-6}$

36. $\frac{d^2-25}{4-d^2} \cdot \frac{5d^2-20}{d+5}$

37. $\frac{b^2+81}{b^2-81} \cdot \frac{81-b^2}{81+b^2}$

38. $\frac{a^2+12a+36}{a^2-36} \cdot \frac{36-a^2}{36+a^2}$

39. For what value(s) of a is $\frac{a^2-25}{5} \cdot \frac{10}{2a-10}$ undefined?

14-4 DIVIDING FRACTIONS

We know that the operation of division may be defined by means of the multiplicative inverse, the reciprocal. A quotient can be expressed as the product of the dividend and the reciprocal of the divisor. Thus,

$$8 \div 5 = \frac{8}{1} \cdot \frac{1}{5} = \frac{8 \cdot 1}{1 \cdot 5} = \frac{8}{5} \quad \text{and} \quad \frac{8}{7} \div \frac{5}{3} = \frac{8}{7} \cdot \frac{3}{5} = \frac{8 \cdot 3}{7 \cdot 5} = \frac{24}{35}$$

In general, for any numbers a, b, c, and d, when $b \neq 0$, $c \neq 0$, and $d \neq 0$:

$$\frac{a}{b} \div \frac{c}{d} = \frac{a}{b} \cdot \frac{d}{c} = \frac{ad}{bc}$$

PROCEDURE. To divide by an algebraic fraction, multiply the dividend by the reciprocal of the divisor.

MODEL PROBLEMS

1. Divide: $\frac{16c^3}{21d^2} \div \frac{24c^4}{14d^3}$

How to Proceed | *Solution*

Multiply the dividend by the reciprocal of the divisor.

$$\frac{16c^3}{21d^2} \div \frac{24c^4}{14d^3} = \frac{\overset{2}{\cancel{16c^3}}}{\underset{3}{\cancel{21d^2}}} \cdot \frac{\overset{2d}{\cancel{14d^3}}}{\underset{3c}{\cancel{24c^4}}} = \frac{4d}{9c} \quad \textit{Ans.}$$

2. Divide: $\frac{8x + 24}{x^2 - 25} \div \frac{4x}{x^2 + 8x + 15}$

How to Proceed | *Solution*

1. Multiply the dividend by the reciprocal of the divisor.

$$\frac{8x + 24}{x^2 - 25} \div \frac{4x}{x^2 + 8x + 15}$$

$$= \frac{8x + 24}{x^2 - 25} \cdot \frac{x^2 + 8x + 15}{4x}$$

2. Factor the numerators and denominators. Divide by the common factors.

$$= \frac{\overset{2}{\cancel{8}}(x + 3)}{\underset{1}{\cancel{(x + 5)}}(x - 5)} \cdot \frac{\overset{1}{\cancel{(x + 5)}}(x + 3)}{\underset{1}{\cancel{4}}x}$$

3. Multiply the remaining numerators and then the remaining denominators.

$$= \frac{2(x + 3)^2}{x(x - 5)} \quad \textit{Ans.}$$

EXERCISES

In 1–25, divide and express the quotient in lowest terms.

1. $\frac{7}{10} \div \frac{21}{5}$ **2.** $\frac{12}{35} \div \frac{4}{7}$ **3.** $8 \div \frac{1}{2}$ **4.** $\frac{x}{9} \div \frac{x}{3}$

5. $\frac{3x}{5y} \div \frac{21x}{2y}$ **6.** $\frac{7ab^2}{10cd} \div \frac{14b^3}{5c^2d^2}$ **7.** $\frac{xy^2}{x^2y} \div \frac{x}{y^3}$ **8.** $\frac{6a^2b^2}{8c} \div 3ab$

9. $\frac{4x + 4}{9} \div \frac{3}{8x}$ **10.** $\frac{3y^2 + 9y}{18} \div \frac{5y^2}{27}$ **11.** $\frac{a^3 - a}{b} \div \frac{a^3}{4b^3}$

12. $\frac{x^2 - 1}{5} \div \frac{x - 1}{10}$ **13.** $\frac{x^2 - 5x + 4}{2x} \div \frac{2x - 2}{8x^2}$

14. $\dfrac{4a^2 - 9}{10} \div \dfrac{10a + 15}{25}$

15. $\dfrac{b^2 - b - 6}{2b} \div \dfrac{b^2 - 4}{b^2}$

16. $\dfrac{a^2 - ab}{4a} \div (a^2 - b^2)$

17. $\dfrac{12y - 6}{8} \div (2y^2 - 3y + 1)$

18. $\dfrac{(x - 2)^2}{4x^2 - 16} \div \dfrac{21x}{3x + 6}$

19. $\dfrac{x^2 - 2xy - 8y^2}{x^2 - 16y^2} \div \dfrac{5x + 10y}{3x + 12y}$

20. $\dfrac{x^2 - 4x + 4}{3x - 6} \div (2 - x)$

21. $(9 - y^2) \div \dfrac{y^2 + 8y + 15}{2y + 10}$

22. $\dfrac{x - 1}{x + 1} \cdot \dfrac{2x + 2}{x + 2} \div \dfrac{4x - 4}{x + 2}$

23. $\dfrac{x + y}{x^2 + y^2} \cdot \dfrac{x}{x - y} \div \dfrac{(x + y)^2}{x^4 - y^4}$

24. $\dfrac{2a + 6}{a^2 - 9} \div \dfrac{3 + a}{3 - a} \cdot \dfrac{a + 3}{4}$

25. $\dfrac{(a + b)^2}{a^2 - b^2} \div \dfrac{a + b}{b^2 - a^2} \cdot \dfrac{a - b}{(a - b)^2}$

26. For what value(s) of a is $\dfrac{a^2 - 2a + 1}{a^2} \div \dfrac{a^2 - 1}{a}$ undefined?

14-5 ADDING OR SUBTRACTING FRACTIONS THAT HAVE THE SAME DENOMINATOR

We know that the sum (or difference) of two arithmetic fractions that have the same denominator is a fraction whose numerator is the sum (or difference) of the numerators and whose denominator is the common denominator of the given fractions. We use the same rule to add algebraic fractions that have the same nonzero denominator. Thus:

Arithmetic fractions	*Algebraic fractions*
$\dfrac{5}{7} + \dfrac{1}{7} = \dfrac{5 + 1}{7} = \dfrac{6}{7}$	$\dfrac{a}{x} + \dfrac{b}{x} = \dfrac{a + b}{x}$
$\dfrac{5}{7} - \dfrac{1}{7} = \dfrac{5 - 1}{7} = \dfrac{4}{7}$	$\dfrac{a}{x} - \dfrac{b}{x} = \dfrac{a - b}{x}$

PROCEDURE. To add (or subtract) fractions that have the same denominator:

1. Write a fraction whose numerator is the sum (or difference) of the numerators and whose denominator is the common denominator of the given fractions.
2. Reduce the resulting fraction to lowest terms.

MODEL PROBLEMS

Add or subtract as indicated. Reduce answers to lowest terms.

1. $\frac{5}{4x} + \frac{9}{4x} - \frac{8}{4x}$

Solution

$$\frac{5}{4x} + \frac{9}{4x} - \frac{8}{4x}$$

$$= \frac{5 + 9 - 8}{4x}$$

$$= \frac{6}{4x}$$

$$= \frac{3}{2x} \quad \textit{Ans.}$$

2. $\frac{4x + 7}{6x} - \frac{2x - 4}{6x}$

Solution

$$\frac{4x + 7}{6x} - \frac{2x - 4}{6x}$$

$$= \frac{(4x + 7) - (2x - 4)}{6x}$$

$$= \frac{4x + 7 - 2x + 4}{6x}$$

$$= \frac{2x + 11}{6x} \quad \textit{Ans.}$$

Note: In Model Problem 2, since the fraction bar is a symbol of grouping, we place numerators that have more than one term in parentheses. In this way, we can see all the signs that need to be changed for the subtraction.

EXERCISES

In 1–38, add or subtract (combine) the fractions as indicated. Reduce answers to lowest terms.

1. $\frac{1}{8} + \frac{4}{8}$ **2.** $\frac{9}{15} - \frac{6}{15}$ **3.** $\frac{2}{x} + \frac{3}{x}$ **4.** $\frac{11}{4c} + \frac{5}{4c} - \frac{6}{4c}$

5. $\frac{3x}{4} + \frac{2x}{4}$ **6.** $\frac{12y}{5} - \frac{4y}{5}$ **7.** $\frac{2c}{5} - \frac{3d}{5}$

8. $\frac{x}{2} - \frac{y}{2} + \frac{z}{2}$ **9.** $\frac{x}{a} + \frac{y}{a}$ **10.** $\frac{5r}{t} - \frac{2s}{t}$

11. $\frac{9}{8x} + \frac{6}{8x}$ **12.** $\frac{8}{9y} + \frac{4}{9y} - \frac{3}{9y}$ **13.** $\frac{6a}{4x} + \frac{5a}{4x}$

14. $\frac{11b}{3y} - \frac{4b}{3y}$ **15.** $\frac{19c}{12d} + \frac{9c}{12d}$ **16.** $\frac{6}{10c} + \frac{9}{10c} - \frac{3}{10c}$

17. $\frac{2x + 1}{2} + \frac{3x + 6}{2}$ **18.** $\frac{4x + 12}{16x} + \frac{8x + 4}{16x}$

19. $\dfrac{5x - 4}{3} - \dfrac{2x + 1}{3}$

20. $\dfrac{12a - 15}{12a} - \dfrac{9a - 6}{12a}$

21. $\dfrac{5}{x + 2} + \dfrac{3}{x + 2}$

22. $\dfrac{2}{a - b} - \dfrac{1}{a - b}$

23. $\dfrac{r}{y - 2} + \dfrac{x}{y - 2}$

24. $\dfrac{x}{x + 1} + \dfrac{1}{x + 1}$

25. $\dfrac{2x}{x + 3} + \dfrac{6}{x + 3}$

26. $\dfrac{y}{y^2 - 4} - \dfrac{2}{y^2 - 4}$

27. $\dfrac{4x + 1}{3x + 2} + \dfrac{6x - 3}{3x + 2}$

28. $\dfrac{3c - 7}{2c - 3} + \dfrac{c + 9}{2c - 3}$

29. $\dfrac{6y - 4}{4y + 3} + \dfrac{7 - 2y}{4y + 3}$

30. $\dfrac{9d + 6}{2d + 1} - \dfrac{7d + 5}{2d + 1}$

31. $\dfrac{8x - 4}{2x + 6} - \dfrac{4x - 6}{2x + 6}$

32. $\dfrac{6x - 5}{x^2 - 1} - \dfrac{5x - 6}{x^2 - 1}$

33. $\dfrac{a^2 + 3ab}{a + b} + \dfrac{b^2 - ab}{a + b}$

34. $\dfrac{x^2 - 2xy}{x - 2y} - \dfrac{xy - 2y^2}{x - 2y}$

35. $\dfrac{8x - 8}{6x - 5} - \dfrac{2x - 7}{6x - 5} + \dfrac{6x - 9}{6x - 5}$

36. $\dfrac{a + 4b}{a^2 - b^2} + \dfrac{4a - 7b}{a^2 - b^2} - \dfrac{3a - b}{a^2 - b^2}$

37. $\dfrac{r^2 + 4r}{r^2 - r - 6} + \dfrac{8 - r^2}{r^2 - r - 6}$

38. $\dfrac{4m^2 + 7m}{2m^2 + 5m + 2} - \dfrac{1 + 7m}{2m^2 + 5m + 2}$

In 39–42, copy and complete the table, showing the results of adding, subtracting, multiplying, and dividing the expressions that represent *A* and *B*.

	A	**B**	**A + B**	**A − B**	**A · B**	**A ÷ B or $\frac{A}{B}$**
39.	$\frac{12}{y}$	$\frac{3}{y}$				
40.	$\frac{3x}{8}$	$\frac{x}{8}$				
41.	$\frac{r}{t}$	$\frac{p}{t}$				
42.	$\frac{7k}{2x}$	$\frac{5k}{2x}$				

14-6 ADDING OR SUBTRACTING FRACTIONS THAT HAVE DIFFERENT DENOMINATORS

Each of the fractions $\frac{2}{8}$, $\frac{3}{12}$, and $\frac{a}{4a}$ is equivalent to the fraction $\frac{1}{4}$ since each one names the fraction $\frac{1}{4}$. Let us use the multiplication property of 1 to show that this statement is true (remember that $\frac{a}{a} = 1$ when $a \neq 0$):

$$\frac{1}{4} = \frac{1}{4} \cdot 1 = \frac{1}{4} \cdot \frac{2}{2} = \frac{2}{8} \qquad \frac{1}{4} = \frac{1}{4} \cdot 1 = \frac{1}{4} \cdot \frac{3}{3} = \frac{3}{12} \qquad \frac{1}{4} = \frac{1}{4} \cdot 1 = \frac{1}{4} \cdot \frac{a}{a} = \frac{a}{4a}$$

These examples illustrate the ***multiplication property of a fraction***: If the numerator and the denominator of a fraction are multiplied by the same nonzero number, the resulting fraction is equivalent to the original fraction.

To add $\frac{5}{4}$ and $\frac{7}{6}$, first transform them to equivalent fractions that have a common denominator. Any integer that has both 4 and 6 as factors can become a common denominator. To simplify the work, use the ***lowest common denominator*** (L.C.D.). For 4 and 6, the L.C.D. is 12.

To find the integer by which to multiply the numerator and the denominator of $\frac{5}{4}$ to transform it into an equivalent fraction whose denominator is the L.C.D. 12, divide 12 by the denominator 4. The result is 3. (Or, you can think "By what must I multiply 4 to obtain 12?") Then, $\frac{5}{4} = \frac{5 \cdot 3}{4 \cdot 3} = \frac{15}{12}$.

To find the integer by which to multiply the numerator and the denominator of $\frac{7}{6}$ to transform it into an equivalent fraction whose denominator is the L.C.D. 12, divide 12 by 6. The result is 2. Then, $\frac{7}{6} = \frac{7 \cdot 2}{6 \cdot 2} = \frac{14}{12}$.

Now, add $\frac{15}{12}$ and $\frac{14}{12}$, and obtain $\frac{15 + 14}{12}$, or $\frac{29}{12}$, as the result.

The entire solution may be written as follows:

$$\frac{5}{4} + \frac{7}{6} = \frac{5 \cdot 3}{4 \cdot 3} + \frac{7 \cdot 2}{6 \cdot 2} = \frac{15}{12} + \frac{14}{12} = \frac{15 + 14}{12} = \frac{29}{12} \quad \textit{Ans.}$$

Algebraic fractions are added in the same manner as arithmetic fractions.

PROCEDURE. To add (or subtract) fractions that have different denominators:

1. Factor each denominator in order to find the lowest common denominator, L.C.D.
2. Transform each fraction to an equivalent fraction by multiplying its numerator and denominator by the quotient that is obtained when the L.C.D. is divided by the denominator of the fraction. (That is, in each fraction, multiply the numerator and denominator by the number or expression needed to change that denominator to the L.C.D.)
3. Write a fraction whose numerator is the sum (or difference) of the numerators of the new fractions and whose denominator is the L.C.D.
4. Reduce the resulting fraction to lowest terms.

MODEL PROBLEMS

1. Add: $\frac{5}{a^2b} + \frac{2}{ab^2}$

Solution

$a^2b = a^2 \cdot b$; $ab^2 = a \cdot b^2$

L.C.D. $= a^2 \cdot b^2 = a^2b^2$

$$\frac{5}{a^2b} + \frac{2}{ab^2}$$

$$= \frac{5(b)}{a^2b(b)} + \frac{2(a)}{ab^2(a)}$$

$$= \frac{5b}{a^2b^2} + \frac{2a}{a^2b^2}$$

$$= \frac{5b + 2a}{a^2b^2} \quad \textit{Ans.}$$

2. Subtract: $\frac{2x + 5}{3} - \frac{x - 2}{4}$

Solution

L.C.D. $= 3 \cdot 4 = 12$

$$\frac{2x + 5}{3} - \frac{x - 2}{4}$$

$$= \frac{4(2x + 5)}{4(3)} - \frac{3(x - 2)}{3(4)}$$

$$= \frac{8x + 20}{12} - \frac{3x - 6}{12}$$

$$= \frac{(8x + 20) - (3x - 6)}{12}$$

$$= \frac{8x + 20 - 3x + 6}{12} = \frac{5x + 26}{12} \quad \textit{Ans.}$$

3. Express as a fraction in simplest form:

$$y + 1 - \frac{1}{y - 1}$$

Solution

L.C.D. $= y - 1$

$$y + 1 - \frac{1}{y - 1}$$

$$= \frac{y + 1}{1} - \frac{1}{y - 1}$$

$$= \frac{(y + 1)(y - 1)}{1(y - 1)} - \frac{1}{y - 1}$$

$$= \frac{y^2 - 1}{y - 1} - \frac{1}{y - 1}$$

$$= \frac{y^2 - 1 - 1}{y - 1}$$

$$= \frac{y^2 - 2}{y - 1} \quad \textit{Ans.}$$

4. Subtract: $\frac{5x}{x^2 - 4} - \frac{3}{x - 2}$

Solution

$x^2 - 4 = (x - 2)(x + 2)$
$x - 2 = 1 \cdot (x - 2)$
L.C.D. $= (x - 2)(x + 2)$

$$\frac{5x}{x^2 - 4} - \frac{3}{x - 2}$$

$$= \frac{5x}{(x - 2)(x + 2)} - \frac{3}{(x - 2)}$$

$$= \frac{5x}{(x - 2)(x + 2)} - \frac{3(x + 2)}{(x - 2)(x + 2)}$$

$$= \frac{5x - (3x + 6)}{(x - 2)(x + 2)}$$

$$= \frac{5x - 3x - 6}{(x - 2)(x + 2)}$$

$$= \frac{2x - 6}{(x - 2)(x + 2)} \text{ or } \frac{2x - 6}{x^2 - 4} \quad \textit{Ans.}$$

EXERCISES

In 1–10, find the lowest common denominator for two fractions whose denominators are:

1. 2; 3 **2.** 6; 5 **3.** x; $4x$ **4.** $12r$; 8 **5.** xy; yz
6. $12x^2$; $15y^2$ **7.** $5x$; $15(x + y)$ **8.** $4(a + b)$; $12a$
9. $4(y + z)$; $12(y + z)$ **10.** $(x^2 - 9)$; $(x + 3)$

In 11–46, add or subtract (combine) the fractions as indicated. Reduce answers to lowest terms.

11. $\frac{5}{3} + \frac{3}{2}$ **12.** $\frac{9}{5} - \frac{2}{3}$ **13.** $\frac{7}{4} + \frac{10}{3}$ **14.** $\frac{4}{10} - \frac{7}{100}$

15. $\frac{5}{6} + \frac{1}{12}$ **16.** $\frac{7}{8} - \frac{1}{4}$ **17.** $\frac{1}{3} - \frac{1}{6}$ **18.** $\frac{5}{4} + \frac{3}{2} - \frac{1}{3}$

19. $\frac{x}{3} + \frac{x}{2}$ **20.** $\frac{d}{3} - \frac{d}{5}$ **21.** $\frac{5x}{6} - \frac{2x}{3}$ **22.** $\frac{y}{6} + \frac{y}{5} - \frac{y}{2}$

23. $\frac{ab}{5} + \frac{ab}{4}$

24. $\frac{8x}{5} - \frac{3x}{4} + \frac{7x}{10}$

25. $\frac{5a}{6} - \frac{3a}{4}$

26. $\frac{a}{7} + \frac{b}{14}$

27. $\frac{9}{4x} + \frac{3}{2x}$

28. $\frac{1}{2x} - \frac{1}{x} + \frac{3}{8x}$

29. $\frac{9a}{8b} - \frac{3a}{4b}$

30. $\frac{1}{a} + \frac{1}{b}$

31. $\frac{2}{a^2} - \frac{5}{b}$

32. $\frac{1}{xy} + \frac{1}{yz}$

33. $\frac{5}{rs} + \frac{9}{st}$

34. $\frac{x}{3ab} - \frac{y}{2bc}$

35. $\frac{9}{ab} + \frac{2}{bc} - \frac{3}{ac}$

36. $\frac{1}{x^2} + \frac{3}{xy} - \frac{5}{y^2}$

37. $\frac{a - 3}{3} + \frac{a + 1}{6}$

38. $\frac{x + 7}{3} - \frac{2x - 3}{5}$

39. $\frac{3y - 4}{5} - \frac{y - 2}{4}$

40. $\frac{a - b}{4} - \frac{a + b}{6}$

41. $\frac{x + 5}{2x} + \frac{2x - 1}{4x}$

42. $\frac{d + 6}{d} + \frac{d - 3}{4d}$

43. $\frac{b - 3}{5b} - \frac{b + 2}{10b}$

44. $\frac{3b + 1}{5b} - \frac{4b - 3}{4b}$

45. $\frac{y - 4}{4y^2} + \frac{3y - 5}{3y}$

46. $\frac{3c - 7}{2c} - \frac{3c - 3}{6c^2}$

In 47–49, represent the perimeter of each polygon in simplest form.

47. A triangle whose sides are represented by $\frac{x}{2}$, $\frac{3x}{5}$, and $\frac{7x}{10}$.

48. A rectangle whose length is represented by $\frac{x + 3}{4}$, and whose width is represented by $\frac{x - 4}{3}$.

49. An isosceles triangle whose equal legs are each represented by $\frac{2x - 3}{7}$ and whose base is represented by $\frac{6x - 18}{21}$.

In 50 and 51, find a representation in simplest form for the indicated length.

50. If the perimeter of a triangle is represented by $\frac{17x}{24}$, and two of the sides are represented by $\frac{3x}{8}$ and $\frac{2x - 5}{12}$ respectively, find a representation for the third side.

51. If the perimeter of a rectangle is represented by $\frac{14x}{15}$, and each length is represented by $\frac{x + 2}{3}$, find a representation for each width.

In 52–70, write the expression as a fraction in lowest terms.

52. $5\frac{2}{3}$ **53.** $9\frac{3}{4}$ **54.** $5 + \frac{1}{x}$ **55.** $9 - \frac{7}{s}$

56. $m + \frac{1}{m}$ **57.** $d - \frac{7}{5d}$ **58.** $\frac{a}{b} + c$ **59.** $3 + \frac{5}{x + 1}$

60. $6 - \frac{4}{x - y}$ **61.** $7 + \frac{2a}{b + c}$ **62.** $t + \frac{1}{t + 1}$ **63.** $s - \frac{1}{s - 1}$

64. $5 - \frac{2x}{x + y}$ **65.** $\frac{4}{y - 2} + 4$ **66.** $8 + \frac{c + 2}{c - 3}$ **67.** $7 - \frac{x + y}{x - y}$

68. $a + 1 + \frac{1}{a + 1}$ **69.** $x - 5 - \frac{x}{x + 3}$ **70.** $\frac{2x - 1}{x + 2} + 2x - 3$

In 71–97, combine the fractions as indicated. Reduce answers to lowest terms.

71. $\frac{5}{x - 3} + \frac{7}{2x - 6}$

72. $\frac{9}{y + 1} - \frac{3}{4y + 4}$

73. $\frac{2}{3a - 1} + \frac{7}{15a - 5}$

74. $\frac{10}{3x - 6} + \frac{3}{2x - 4}$

75. $\frac{11x}{8x - 8} - \frac{3x}{4x - 4}$

76. $\frac{3}{2x - 3y} + \frac{5}{3y - 2x}$

77. $\frac{2a}{4a - 8b} + \frac{3b}{3a - 6b}$

78. $\frac{3x - 2}{2x + 2} + \frac{4x - 1}{3x + 3}$

79. $\frac{5x + 2}{6x - 3} - \frac{3x - 5}{8x - 4}$

80. $\frac{1}{x - 5} + \frac{1}{x + 5}$

81. $\frac{9}{y + 4} - \frac{6}{y - 4}$

82. $\frac{7}{a + 3} + \frac{4}{2 - a}$

83. $\frac{7}{x - 2} + \frac{3}{x}$

84. $\frac{9}{c + 8} - \frac{2}{c}$

85. $\dfrac{2a + b}{a - b} + \dfrac{a}{b}$

86. $\dfrac{5}{y^2 - 9} + \dfrac{3}{y - 3}$

87. $\dfrac{6}{y^2 - 16} - \dfrac{5}{y + 4}$

88. $\dfrac{9}{a^2 - b^2} + \dfrac{3}{b - a}$

89. $\dfrac{3y}{y^2 - 4} - \dfrac{4}{2y - 4}$

90. $\dfrac{x}{x^2 - 36} - \dfrac{4}{3x + 18}$

91. $\dfrac{9}{a^2 - ab} + \dfrac{3}{ab - b^2}$

92. $\dfrac{1}{y - 3} + \dfrac{2}{y + 4} + \dfrac{2}{3}$

93. $\dfrac{1}{(x + 2)^3} - \dfrac{1}{(x + 2)^2} + \dfrac{1}{x + 2}$

94. $\dfrac{7a}{(a - 1)(a + 3)} + \dfrac{2a - 5}{(a + 3)(a + 2)}$

95. $\dfrac{5}{r^2 - 4} - \dfrac{3}{r^2 + 3r - 10}$

96. $\dfrac{x + 2y}{3x + 12y} - \dfrac{6x - y}{x^2 + 3xy - 4y^2}$

97. $\dfrac{2a + 7}{a^2 - 2a - 15} - \dfrac{3a - 4}{a^2 - 7a + 10}$

14-7 SOLVING EQUATIONS CONTAINING FRACTIONAL COEFFICIENTS

Examples of equations that contain fractional coefficients are:

$$\frac{1}{2}x = 10 \quad or \quad \frac{x}{2} = 10 \qquad \frac{1}{3}x + 60 = \frac{5}{6}x \quad or \quad \frac{x}{3} + 60 = \frac{5x}{6}$$

Each of these equations can be solved by transforming it into an equivalent equation that does not contain fractional coefficients. This can be done by multiplying both members of the equation by a common denominator for all the fractions present in the equation. We usually multiply by the lowest common denominator, the L.C.D.

PROCEDURE. To solve an equation that contains fractional coefficients:

1. Find the L.C.D.
2. Multiply both members of the equation by the L.C.D.
3. Solve the resulting equation using the usual methods.

MODEL PROBLEMS

1. Solve and check: $\frac{x}{3} + \frac{x}{5} = 8$

How to Proceed	*Solution*
(1) Write the equation.	$\frac{x}{3} + \frac{x}{5} = 8$
(2) Find the L.C.D.	L.C.D. $= 3 \cdot 5 = 15$
(3) Multiply both members of the equation by the L.C.D.	$15\left(\frac{x}{3} + \frac{x}{5}\right) = 15(8)$
(4) Use the distributive property.	$15\left(\frac{x}{3}\right) + 15\left(\frac{x}{5}\right) = 15(8)$
(5) Simplify.	$5x + 3x = 120$
(6) Solve for x.	$8x = 120$
	$x = 15$ *Ans.*

Check

$$\frac{x}{3} + \frac{x}{5} = 8$$

$$\frac{15}{3} + \frac{15}{5} \stackrel{?}{=} 8$$

$$5 + 3 \stackrel{?}{=} 8$$

$$8 = 8$$

(True)

2. Solve: $\frac{3x}{4} = 20 + \frac{x}{4}$

Solution

$$\frac{3x}{4} = 20 + \frac{x}{4}$$

L.C.D. $= 4$

$$4\left(\frac{3x}{4}\right) = 4\left(20 + \frac{x}{4}\right)$$

$$4\left(\frac{3x}{4}\right) = 4(20) + 4\left(\frac{x}{4}\right)$$

$$3x = 80 + x$$

$$2x = 80$$

$$x = 40 \quad Ans.$$

3. Solve: $\frac{2x + 7}{6} - \frac{2x - 9}{10} = 3$

Solution

$$\frac{2x + 7}{6} - \frac{2x - 9}{10} = 3$$

L.C.D. $= 30$

$$30\left(\frac{2x + 7}{6} - \frac{2x - 9}{10}\right) = 30(3)$$

$$30\left(\frac{2x + 7}{6}\right) - 30\left(\frac{2x - 9}{10}\right) = 30(3)$$

$$5(2x + 7) - 3(2x - 9) = 90$$

$$10x + 35 - 6x + 27 = 90$$

$$4x + 62 = 90$$

$$4x = 28$$

$$x = 7 \quad Ans.$$

In Problems 2 and 3, the check is left to the student.

4. A father divided his shares of stock in AAA Company among his four daughters. The first daughter got one-half the number of shares, the second daughter got one-quarter of the number of shares, the third daughter got one-fifth of the number of shares, and the fourth daughter got 7 shares of stock. How many shares of this stock did the father have originally?

Solution

Let x = the total number of the father's shares.

Then, $\frac{1}{2}x$ = the number of the 1st daughter's shares.

And, $\frac{1}{4}x$ = the number of the 2nd daughter's shares.

And, $\frac{1}{5}x$ = the number of the 3rd daughter's shares.

The sum of all the daughters' shares is the father's total.

1st + 2nd + 3rd + 4th = father's total

$$\frac{1}{2}x + \frac{1}{4}x + \frac{1}{5}x + 7 = x$$

$$20\left(\frac{1}{2}x + \frac{1}{4}x + \frac{1}{5}x + 7\right) = 20(x)$$

$$10x + 5x + 4x + 140 = 20x$$

$$19x + 140 = 20x$$

$$140 = x \quad \text{number of father's shares}$$

$$\frac{1}{2}x = \frac{140}{2} = 70 \quad \text{1st daughter}$$

$$\frac{1}{4}x = \frac{140}{4} = 35 \quad \text{2nd daughter}$$

$$\frac{1}{5}x = \frac{140}{5} = 28 \quad \text{3rd daughter}$$

Check: $70 + 35 + 28 + 7 = 140$

Answer: The father had 140 shares of stock.

5. In a child's coin bank, there is a collection of nickels, dimes, and quarters that amounts to \$3.20. There are 3 times as many quarters as nickels, and 5 more dimes than nickels. How many coins of each kind are there?

Solution

Let x = the number of nickels.
Then, $3x$ = the number of quarters.
And, $x + 5$ = the number of dimes.

Then, $.05x$ = the value of the nickels.
And, $.25(3x)$ = the value of the quarters.
And, $.10(x + 5)$ = the value of the dimes.

The total value of the coins is $3.20.

$$.05x + .25(3x) + .10(x + 5) = 3.20$$

$$100[.05x + .25(3x) + .10(x + 5)] = 100[3.20]$$

$$5x + 25(3x) + 10(x + 5) = 320$$

$$5x + 75x + 10x + 50 = 320$$

$$90x + 50 = 320$$

$$90x = 270$$

$x = 3$ number of nickels

$3x = 3(3) = 9$ number of quarters

$x + 5 = 3 + 5 = 8$ number of dimes

Check: The value of 3 nickels = .05(3) = .15
The value of 9 quarters = .25(9) = 2.25
The value of 8 dimes = .10(8) = .80
$3.20

Answer: There are 3 nickels, 9 quarters, and 8 dimes.

EXERCISES

In 1–50, solve and check.

1. $\frac{x}{7} = 3$ **2.** $\frac{1}{6}t = 18$ **3.** $\frac{3x}{5} = 15$

4. $\frac{5}{7}n = 35$ **5.** $\frac{x + 8}{4} = 6$ **6.** $\frac{m - 2}{9} = 3$

7. $.8 = 8y$ **8.** $2z = .08$ **9.** $6t = 18.6$

10. $.4r = 16$ **11.** $1.3t = .39$ **12.** $.25a = 6$

13. $\frac{2r + 6}{5} = -4$ **14.** $\frac{5y - 30}{7} = 0$ **15.** $\frac{5x}{2} = \frac{15}{4}$

16. $\frac{m - 5}{35} = \frac{5}{7}$ **17.** $\frac{2x + 1}{3} = \frac{6x - 9}{5}$ **18.** $\frac{3y + 1}{4} = \frac{44 - y}{5}$

19. $\frac{x}{5} + \frac{x}{3} = \frac{8}{15}$ **20.** $10 = \frac{x}{3} + \frac{x}{7}$ **21.** $\frac{r}{3} - \frac{r}{6} = 2$

22. $\frac{3t}{5} - \frac{t}{5} = 3$ **23.** $1 = \frac{7r}{8} - \frac{3r}{8}$

24. $\frac{3t}{4} - 6 = \frac{t}{12}$ **25.** $\frac{a}{2} + \frac{a}{3} + \frac{a}{4} = 26$

26. $\frac{7y}{12} - \frac{1}{4} = 2y - \frac{5}{3}$ **27.** $\frac{y + 2}{4} - \frac{y - 3}{3} = \frac{1}{2}$

28. $\frac{t - 3}{6} - \frac{t - 25}{5} = 4$ **29.** $\frac{3m + 1}{4} = 2 - \frac{3 - 2m}{6}$

30. $.7x - .4 = 1$ **31.** $.03y - 1.2 = 8.7$

32. $.4x + .08 = 4.24$ **33.** $.5x - .3x = 8$

34. $2c + .5c = 50$ **35.** $.08y - .9 = .02y$

36. $1.7x = 30 + .2x$ **37.** $1.5y - 1.69 = .2y$

38. $.08c = 1.5 + .07c$

39. $.8m + 2.6 = .2m + 9.8$ **40.** $.05x - .25 = .02x + .44$

41. $.13x - 1.4 = .08x + 7.6$ **42.** $.06y + 40 - .03y = 70$

43. $.02(x + 5) = 8$ **44.** $.05(x - 8) = .07x$

45. $.4(x - 9) = .3(x + 4)$ **46.** $.06(x - 5) = .04(x + 8)$

47. $.04x + .03(2{,}000 - x) = 75$ **48.** $.02x + .04(1{,}500 - x) = 48$

49. $.05x + 10 = .06(x + 50)$ **50.** $.08x = .03(x + 200) - 4$

51. The sum of one-half of a number and one-third of that number is 25. Find the number.

52. The difference between one-fifth of a positive number and one-tenth of that number is 10. Find the number.

53. If one-half of a number is increased by 20, the result is 35. Find the number.

54. If two-thirds of a number is decreased by 30, the result is 10. Find the number.

55. If the sum of two consecutive integers is divided by 3, the quotient is 9. Find the integers.

56. If the sum of two consecutive odd integers is divided by 4, the quotient is 10. Find the integers.

57. In an isosceles triangle, each of the congruent sides is two-thirds of the base. The perimeter of the triangle is 42. Find the length of each side of the triangle.

58. The larger of two numbers is 12 less than 5 times the smaller. If the smaller number is equal to one-third of the larger number, find the numbers.

59. The larger of two numbers exceeds the smaller by 14. If the smaller number is equal to three-fifths of the larger, find the numbers.

60. Separate 90 into two parts such that one part is one-half of the other part.

61. Separate 150 into two parts such that one part is two-thirds of the other part.

1	2	3	4

62. Four vegetable plots of unequal lengths and of equal widths are arranged as shown. The length of the third plot is one-fourth the length of the second plot. The length of the fourth plot is one-half the length of the second plot. The length of the first plot is 10 feet more than the length of the fourth plot. If the total length of the four plots is 100 feet, find the length of each plot.

63. Sam is now one-sixth as old as his father. In 4 years, Sam will be one-fourth as old as his father will be then. Find the ages of Sam and his father now.

64. Robert is one-half as old as his father. Twelve years ago, he was one-third as old as his father was then. Find their present ages.

65. A coach finds that, of the students who try out for track, 65% qualify for the team and 90% of those who qualify remain on the team throughout the season. What is the smallest number of students who must try out for track in order to have 30 on the team at the end of the season?

66. Sally spent half of her money on a present for her mother, and she spent half of what remained on a treat for herself. If Sally had \$2.00 left, how much money did she have originally?

67. A bus that runs once daily between the villages of Alpaca and Down makes only two stops between, at Billow and at Comfort. Today, the bus left Alpaca with some passengers. At Billow, one-half of the passengers got off, and 6 new ones got on. At Comfort, again one-half of the passengers got off, and, this time, 5 new ones got on. At Down, the last 13 passengers on the bus got off. How many passengers were aboard when the bus left Alpaca?

68. Bob planted some lettuce seedlings in his garden. A few days later, after one-third of these seedlings had been eaten by a rabbit, Bob planted 15 new lettuce seedlings. A week later, again one-third of the seedlings had been eaten, leaving 22 seedlings unharmed. How many lettuce seedlings had Bob planted originally?

69. May has 3 times as many dimes as nickels. In all, she has $1.40. How many coins of each type does she have?
70. Mr. Jantzen bought some cans of soup at 39¢ per can, and some packages of frozen vegetables at 59¢ per package. He bought twice as many packages of vegetables as cans of soup. If the total bill was $9.42, how many cans of soup did he buy?
71. Roger has $2.30 in dimes and nickels. There are 5 more dimes than nickels. Find the number of each kind of coin that he has.
72. Bess has $2.80 in quarters and dimes. The number of dimes is 7 less than the number of quarters. Find the number of each kind of coin that she has.
73. A movie theater sold student tickets for $3.25 and full-price tickets for $5. On Saturday, the theater sold 16 more full-price tickets than student tickets. If the total sales on Saturday were $740, how many of each kind of ticket were sold?
74. Is it possible to have $4.50 in dimes and quarters, and have twice as many quarters as dimes? Explain.
75. Is it possible to have $6.00 in nickels, dimes, and quarters, and have the same number of each kind of coin? Explain.
76. Mr. Symms invested a sum of money in 7% bonds. He invested $400 more than this sum in 8% bonds. If the total annual interest from these two investments is $257, how much did he invest at each rate?
77. Mr. Charles borrowed a sum of money at a 10% interest rate. He borrowed a second sum, which was $1,500 less than the first sum, at an 11% interest rate. If the total annual interest he has to pay on these two loans is $202.50, how much did he borrow at each rate?

14-8 SOLVING INEQUALITIES CONTAINING FRACTIONAL COEFFICIENTS

PROCEDURE. To solve an inequality that contains fractional coefficients:

1. Find the L.C.D., a positive number.
2. Multiply both members of the inequality by the L.C.D.
3. Solve the resulting inequality using the usual methods.

MODEL PROBLEMS

1. Solve, and graph the solution set on a number line:

$$\frac{x}{3} - \frac{x}{6} > 2$$

Solution

$$\frac{x}{3} - \frac{x}{6} > 2$$

$$6\left(\frac{x}{3} - \frac{x}{6}\right) > 6(2)$$

$$6\left(\frac{x}{3}\right) - 6\left(\frac{x}{6}\right) > 12$$

$$2x - x > 12$$

$$x > 12 \quad Ans.$$

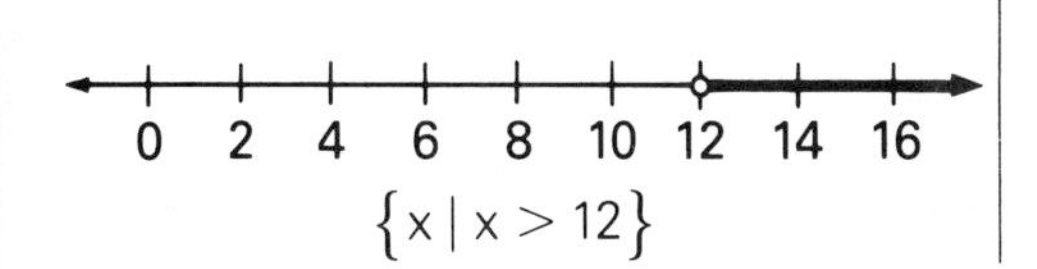

2. Solve, and graph the solution set on a number line:

$$\frac{3y}{2} + \frac{8 - 4y}{7} \le 3$$

Solution

$$\frac{3y}{2} + \frac{8 - 4y}{7} \le 3$$

$$14\left(\frac{3y}{2} + \frac{8 - 4y}{7}\right) \le 14(3)$$

$$14\left(\frac{3y}{2}\right) + 14\left(\frac{8 - 4y}{7}\right) \le 42$$

$$21y + 16 - 8y \le 42$$

$$13y \le 26$$

$$y \le 2 \quad Ans.$$

−1 0 1 2 3

$\{y \mid y \le 2\}$

3. The smaller of two integers is one-third of the larger, and their sum is greater than 100. Find the smallest possible integers.

Solution

Let x = the larger integer.

Then, $\frac{1}{3}x$ = the smaller integer.

Their sum is greater than 100.

$$x + \frac{1}{3}x \quad > \quad 100$$

$$3\left(x + \frac{1}{3}x\right) > 3(100)$$

$$3x + x > 300$$

$$4x > 300$$

$$x > 75$$

$$\frac{1}{3}x > 25$$

The smallest integer greater than 25 is 26. This is one-third of 78.

Check: 78 and 26 $26 \stackrel{?}{=} \frac{1}{3}(78)$ | $78 + 26 \stackrel{?}{>} 100$

$26 = 26$ (True) | $104 > 100$ (True)

Answer: The smallest possible integers are 26 and 78.

KEEP IN MIND

Words	Symbols
a is greater than b	$a > b$
a is less than b	$a < b$
a is at least b a is no less than b	$a \geq b$
a is at most b a is no greater than b	$a \leq b$

EXERCISES

In 1–20, solve the inequality, and graph the solution set on a number line.

1. $\frac{1}{4}x - \frac{1}{5}x > \frac{9}{20}$
2. $y - \frac{2}{3}y < 5$
3. $\frac{5}{6}c > \frac{1}{3}c + 3$
4. $\frac{x}{4} - \frac{x}{8} \leq \frac{5}{8}$
5. $\frac{y}{6} \geq \frac{y}{12} + 1$
6. $\frac{y}{9} - \frac{y}{4} > \frac{5}{36}$
7. $\frac{t}{10} \leq 4 + \frac{t}{5}$
8. $1 + \frac{2x}{3} \geq \frac{x}{2}$
9. $2.5x - 1.6x > 4$
10. $2y + 3 \geq .2y$
11. $\frac{3x - 1}{7} > 5$
12. $\frac{5y - 30}{7} \leq 0$
13. $2d + \frac{1}{4} < \frac{7d}{12} + \frac{5}{3}$
14. $\frac{4c}{3} - \frac{7}{9} \geq \frac{c}{2} + \frac{7}{6}$
15. $\frac{2m}{3} \geq \frac{7 - m}{4} + 1$
16. $\frac{3x - 30}{6} < \frac{x}{3} - 2$
17. $\frac{6x - 3}{2} > \frac{37}{10} + \frac{x + 2}{5}$
18. $\frac{2y - 3}{3} + \frac{y + 1}{2} < 10$

19. $\frac{2r - 3}{5} - \frac{r - 3}{3} \leq 2$

20. $\frac{3t - 4}{3} \geq \frac{2t + 4}{6} + \frac{5t - 1}{9}$

In 21–35, solve the problem.

21. If one-third of an integer is increased by 7, the result is at most 13. Find the greatest possible integer.
22. If two-fifths of an integer is decreased by 11, the result is at least 4. Find the smallest possible integer.
23. The sum of one-fifth of an integer and one-tenth of that integer is less than 40. Find the greatest possible integer.
24. The difference between three-fourths of a positive integer and one-half of that integer is greater than 28. Find the smallest possible integer.
25. The smaller of two integers is two-fifths of the larger, and their sum is less than 40. Find the greatest possible integers.
26. The smaller of two positive integers is five-sixths of the larger, and their difference is greater than 3. Find the smallest possible integers.

27. Talk and Tell Answering Service offers customers the following two monthly options.

    Option 1: Measured Service
    base rate = \$15
    each call = 10¢
    Option 2: Unmeasured Service
    unlimited calls = \$20

    Find the least number of calls for which unmeasured service is cheaper than measured service.

28. Nobody's Home Answering Service offers customers the following two monthly options.

    Option 1: Measured Service
    base rate = \$17
    first 20 calls = no charge
    each additional call = 10¢
    Option 2: Unmeasured Service
    unlimited calls = \$20

    Find the least number of calls for which unmeasured service is cheaper than measured service.

29. Mary bought some cans of vegetables at 89¢ per can, and some cans of soup at 99¢ per can. If she bought twice as many cans of vegetables as cans of soup, and paid at least \$10, what is the least number of cans of vegetables she could have bought?

30. A coin bank contains nickels, dimes, and quarters. The number of dimes is 7 more than the number of nickels, and the number of quarters is twice the number of dimes. If the total value of the coins is no greater than \$7.20, what is the greatest possible number of nickels in the bank?

31. Rhoda is two-thirds as old as her sister Alice. Five years from now, the sum of their ages will be less than 60. What is the largest possible integral value for each sister's present age?

32. Bill is $1\frac{1}{4}$ times as old as his cousin Mary. Four years ago, the difference between their ages was greater than 3. What is the smallest possible integral value for each cousin's present age?

33. Mr. Drew invested a sum of money at $7\frac{1}{2}\%$. He invested a second sum, which was \$200 less than the first, at 7%. If the total annual interest from these two investments is at least \$160, what is the smallest amount he could have invested at $7\frac{1}{2}\%$?

34. Paul spent one-half of his pocket money for a book, and then spent one-half of what remained for a record. If he had less than \$3 left over, what was the greatest amount of money he could have had originally?

35. An express train started from the depot with some passengers. At the first stop, one-third of the passengers got off, and 3 new passengers got on. At the second stop, again one-third of the passengers got off and, this time, 6 new ones got on. At the next and last stop, the remaining passengers, fewer than 45, left the train. What was the greatest possible number of passengers on the train when it started?

14-9 SOLVING FRACTIONAL EQUATIONS

An equation is called a ***fractional equation*** when a variable appears in the *denominator* of one, or more than one, of its terms. For example, $\frac{1}{3} + \frac{1}{x} = \frac{1}{2}$ and $\frac{2}{3d} + \frac{1}{3} = \frac{11}{6d} - \frac{1}{4}$ are called fractional equations. To solve such an equation, we can clear the equation of fractions by multiplying both of its members by the lowest common denominator (L.C.D.) for the denominators of the fractions present in the equation.

As with all algebraic fractions, a fractional equation has meaning only for those values of the variable that do not lead to a denominator of 0.

MODEL PROBLEMS

1. Solve and check: $\frac{1}{3} + \frac{1}{x} = \frac{1}{2}$

Solution: Multiply both members of the equation by the L.C.D., $6x$.

$$\frac{1}{3} + \frac{1}{x} = \frac{1}{2}$$

$$6x\left(\frac{1}{3} + \frac{1}{x}\right) = 6x\left(\frac{1}{2}\right)$$

$$6x\left(\frac{1}{3}\right) + 6x\left(\frac{1}{x}\right) = 6x\left(\frac{1}{2}\right)$$

$$2x + 6 = 3x$$

$$6 = x$$

Answer: $x = 6$

Check

$$\frac{1}{3} + \frac{1}{x} = \frac{1}{2}$$

$$\frac{1}{3} + \frac{1}{6} \stackrel{?}{=} \frac{1}{2}$$

$$\frac{2}{6} + \frac{1}{6} \stackrel{?}{=} \frac{1}{2}$$

$$\frac{3}{6} \stackrel{?}{=} \frac{1}{2}$$

$$\frac{1}{2} = \frac{1}{2} \quad \text{(True)}$$

2. Solve and check: $\frac{5x + 10}{x + 2} = 7$

Solution: Multiply both members of the equation by the L.C.D., $x + 2$.

$$\cancel{(x + 2)}\left(\frac{5x + 10}{\cancel{x + 2}}\right) = (x + 2)(7)$$

$$5x + 10 = 7x + 14$$

$$10 = 2x + 14$$

$$-4 = 2x$$

$$-2 = x$$

Check

$$\frac{5x + 10}{x + 2} = 7$$

$$\frac{5(-2) + 10}{-2 + 2} \stackrel{?}{=} 7$$

$$\frac{-10 + 10}{-2 + 2} \stackrel{?}{=} 7$$

$$\frac{0}{0} = 7 \quad \text{(False)}$$

Since the only possible value of x is a value for which the equation has no meaning because it leads to a denominator of 0, there is no solution for this equation.

Answer: The solution set is $\varnothing$.

KEEP IN MIND

When both members of an equation are multiplied by a variable expression that may represent zero, the resulting equation may not be equivalent to the given equation. Each solution, therefore, must be checked in the given equation.

EXERCISES

In 1–18, solve and check.

1. $\frac{10}{x} = 5$ **2.** $\frac{15}{y} = 3$ **3.** $\frac{6}{x} = 12$

4. $\frac{8}{b} = -2$ **5.** $\frac{3}{2x} = \frac{1}{2}$ **6.** $\frac{15}{4x} = \frac{1}{8}$

7. $\frac{7}{3y} = -\frac{1}{3}$ **8.** $\frac{4}{5y} = -\frac{1}{10}$ **9.** $\frac{10}{x} + \frac{8}{x} = 9$

10. $\frac{15}{y} - \frac{3}{y} = 4$ **11.** $\frac{7}{c} + \frac{1}{c} = 16$ **12.** $\frac{9}{2x} = \frac{7}{2x} + 2$

13. $\frac{30}{x} = 7 + \frac{18}{2x}$ **14.** $\frac{y - 2}{2y} = \frac{3}{8}$ **15.** $\frac{5}{c} + 6 = \frac{17}{c}$

16. $\frac{y + 9}{2y} + 3 = \frac{15}{y}$ **17.** $\frac{5 + x}{2x} - 1 = \frac{x + 1}{x}$

18. $\frac{2 + x}{6x} = \frac{3}{5x} + \frac{1}{30}$

In 19–21, explain why each fractional equation has no solution.

19. $\frac{6x}{x} = 3$ **20.** $\frac{4a + 4}{a + 1} = 5$ **21.** $\frac{2}{x} = 4 + \frac{2}{x}$

In 22–32, solve and check.

22. $\frac{6}{3x - 1} = \frac{3}{4}$ **23.** $\frac{2}{3x - 4} = \frac{1}{4}$ **24.** $\frac{5x}{x + 1} = 4$

25. $\frac{3}{5 - 3a} = \frac{1}{2}$ **26.** $\frac{4z}{7 + 5z} = \frac{1}{3}$ **27.** $\frac{1 - r}{1 + r} = \frac{2}{3}$

28. $\frac{3}{y} = \frac{2}{5 - y}$ **29.** $\frac{5}{a} = \frac{7}{a - 4}$ **30.** $\frac{2}{m} = \frac{5}{3m - 1}$

31. $\frac{y}{y + 1} - \frac{1}{y} = 1$ **32.** $\frac{x}{x^2 - 9} = \frac{1}{x + 3}$

33. If 24 is divided by a number, the result is 6. Find the number.
34. If 10 is divided by a number, the result is 30. Find the number.
35. The sum of 20 divided by a number, and 7 divided by the same number, is 9. Find the number.
36. If 3 times a number is increased by one-third of that number, the result is 280. Find the number.
37. When the reciprocal of a number is decreased by 2, the result is 5. Find the number.
38. The numerator of a fraction is 8 less than the denominator of the fraction. The value of the fraction is $\frac{3}{5}$. Find the fraction.
39. If one-half of a number is 8 more than one-third of the number, find the number.

14-10 EQUATIONS AND FORMULAS INVOLVING SEVERAL VARIABLES

To solve an equation involving several variables for one of those variables, express this variable in terms of the other variables. In order to determine what steps should be used in the solution, use a simpler related equation in which all of the variables except the one for which you are solving are replaced by constants. Solve the given equation using the same procedure that you used for the simpler equation. For example, to solve $\frac{3x}{2a} + b = 3b$, follow the same procedure that you would use for solving $\frac{3x}{2} + 1 = 3$.

$$\frac{3x}{2} + 1 = 3 \qquad\qquad \frac{3x}{2a} + b = 3b$$

$$2\left(\frac{3x}{2} + 1\right) = 2(3) \qquad\qquad 2a\left(\frac{3x}{2a} + b\right) = 2a(3b)$$

$$2\left(\frac{3x}{2}\right) + 2(1) = 2(3) \qquad\qquad 2a\left(\frac{3x}{2a}\right) + 2a(b) = 2a(3b)$$

$$3x + 2 = 6 \qquad\qquad 3x + 2ab = 6ab$$

$$-2 = -2 \qquad\qquad -2ab = -2ab$$

$$3x = 4 \qquad\qquad 3x = 4ab$$

$$x = \frac{4}{3} \quad \textit{Ans.} \qquad\qquad x = \frac{4ab}{3} \quad \textit{Ans.}$$

EXERCISES

In 1–12, solve for x and check.

1. $\frac{x}{5} = t$
2. $\frac{x}{c} = d$
3. $\frac{x}{3a} = b$
4. $\frac{x}{3} = \frac{b}{4}$
5. $\frac{x}{a} - \frac{b}{3} = 0$
6. $\frac{x}{3} + b = 4b$
7. $\frac{r}{x} = t$
8. $\frac{t}{x} - k = 0$
9. $\frac{x - 4b}{5} = 8b$
10. $\frac{a + b}{x} = c$
11. $\frac{mx}{r} + d = 2d$
12. $\frac{x}{3a} + \frac{x}{5a} = 8$

In 13–18, solve the formula for the indicated variable.

13. $C = \frac{360}{n}$ for n
14. $R = \frac{E}{I}$ for I
15. $v = \frac{s}{t}$ for t
16. $F = \frac{mv^2}{gr}$ for m
17. $V = \frac{L}{RA}$ for R
18. $F = 32 + \frac{9}{5}C$ for C

19. If $S = \frac{1}{2}at^2$, express a in terms of S and t.
20. If $A = p + prt$, express t in terms of A, r, and p.

In 21 and 22, find the value of the variable indicated when the other variables have the stated values.

21. If $\frac{D}{R} = T$, find the value of R when $D = 120$ and $T = 4$.
22. If $A = \frac{1}{2}h(b + c)$, find the value of h when $A = 48$, $b = 12$, and $c = 4$.

14-11 REVIEW EXERCISES

1. What fractional part of one centimeter is x millimeters?
2. For what value of y is the fraction $\frac{y - 1}{y - 4}$ undefined?

In 3–6, reduce the fraction to lowest terms.

3. $\frac{8bg}{12bg}$
4. $\frac{14d}{7d^2}$
5. $\frac{5x^2 - 60}{5}$
6. $\frac{8y^2 - 12y}{8y}$

In 7–18, perform the indicated operation, and express the answer in lowest terms.

7. $\frac{3x^2}{4} \cdot \frac{8}{9x}$ **8.** $\frac{2x - 2}{3y} \cdot \frac{3xy}{2x}$ **9.** $6c^2 \div \frac{c}{2}$

10. $\frac{3a}{7b} \div \frac{18a}{35}$ **11.** $\frac{5m}{6} - \frac{m}{6}$ **12.** $\frac{9}{k} - \frac{3}{k} + \frac{4}{k}$

13. $\frac{ax}{3} + \frac{ax}{4}$ **14.** $\frac{2}{3y} + \frac{2}{9y}$ **15.** $\frac{5}{xy} - \frac{2}{yz}$

16. $\frac{w + 3}{w} + \frac{w - 2}{3w}$ **17.** $\frac{4x + 5}{3x} + \frac{2x + 1}{3x}$ **18.** $\frac{2}{5x} + \frac{x - 3}{5x^2} - \frac{x - 4}{25x^2}$

19. If the sides of a triangle are represented by $\frac{b}{2}$, $\frac{5b}{6}$, and $\frac{2b}{3}$, express the perimeter of the triangle in simplest form.

20. If $a = 2$, $b = 3$, and $c = 4$, what is the sum of $\frac{b}{a}$ and $\frac{a}{c}$?

In 21–26, solve and check.

21. $\frac{k}{20} = \frac{3}{4}$ **22.** $\frac{x - 3}{10} = \frac{4}{5}$ **23.** $\frac{y}{2} - \frac{y}{6} = 4$

24. $\frac{6}{m} = \frac{20}{m} - 2$ **25.** $\frac{2t}{5} - \frac{t - 2}{10} = 2$ **26.** $\frac{3}{x} + 4 = \frac{3}{x}$

In 27–29, solve for r in terms of the other variables.

27. $\frac{S}{h} = 2\pi r$ **28.** $\frac{a}{r} - n = 0$ **29.** $\frac{c}{2r} = \pi$

In 30–35, perform the indicated operation and express the answer in lowest terms.

30. $\frac{x^2 - 5x}{x^2} \cdot \frac{x}{2x - 10}$ **31.** $\frac{2a}{a + b} + \frac{2b}{a + b}$

32. $\frac{y + 7}{5} - \frac{y + 3}{4}$ **33.** $\frac{x^2 - 25}{12} \div \frac{x^2 - 10x + 25}{4}$

34. $\frac{c - 3}{12} + \frac{c + 3}{8}$ **35.** $\frac{3a - 9a^2}{a} \div (1 - 9a^2)$

In 36–40, solve the problem.

36. The numerator of a fraction is 4 less than the denominator. If the value of the fraction is $\frac{3}{4}$, find the fraction.

37. Mr. Vroman deposited a sum of money in the bank. After a few years, he found that the interest equaled $\frac{1}{4}$ of his original deposit and he had a total, deposit plus interest, of $2,400 in the bank. What was the original deposit?

38. One-third of the result obtained by adding 5 to a certain number is equal to one-half of the result obtained when 5 is subtracted from the number. Find the number.

39. When Victoria went shopping, she spent one-half of her money in the first store, one-half of what she had left in the second store, and two-thirds of what remained in the third store. When she returned home without spending any more money, she had $2.10. How much money did she spend?

40. Of the total number of points scored by the winning team in a basketball game, one-fifth were scored in the first quarter, one-sixth were scored in the second quarter, one-third were scored in the third quarter, and 27 were scored in the fourth quarter. How many points did the winning team score?

Chapter 15

Probability

15-1 EMPIRICAL PROBABILITY

A decision is sometimes reached by the toss of a coin. "Heads, we'll go to the movies; tails, we'll go bowling." When we toss a coin, we don't know whether the coin will land with heads facing upward or with tails up. However, we believe that heads and tails have an *equal chance* of happening whenever we toss a fair coin. We can describe this situation by saying that the ***probability*** of heads is $\frac{1}{2}$ and the ***probability*** of tails is $\frac{1}{2}$, symbolized as:

$$P(\text{heads}) = \frac{1}{2} \quad \text{or} \quad P(H) = \frac{1}{2} \qquad P(\text{tails}) = \frac{1}{2} \quad \text{or} \quad P(T) = \frac{1}{2}$$

Before we define probability, let us consider a few more situations.

1. Suppose a girl tosses a coin and it lands heads up. If she were to toss the coin a second time, will the coin now land tails up? Is your answer "I don't know"? Good! We cannot say that the coin must now be tails because *we cannot predict the next result with certainty*.

2. Suppose we take a card made of stiff cardboard, such as an index card, and fold it down the center. When we toss the card and let it fall, there will be only three possible results. The card may land on its side, or it may land on its edge, or it may form a tent when it lands. Can we say $P(\text{edge}) = \frac{1}{3}$, $P(\text{side}) = \frac{1}{3}$, and $P(\text{tent}) = \frac{1}{3}$?

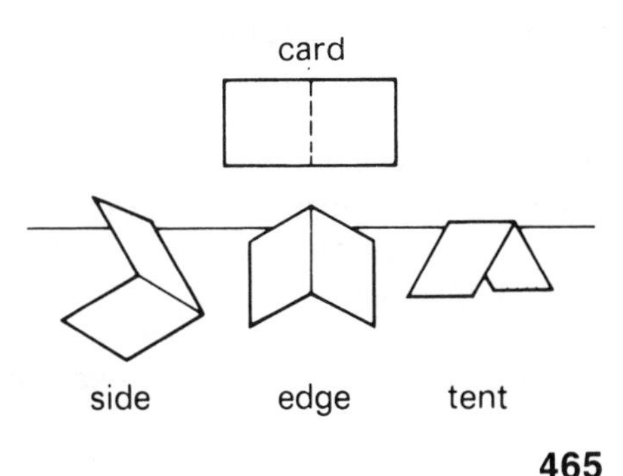

Again, is your answer "I don't know"? *We cannot assign a number as a probability until we have some evidence to support our claim.* In fact, if we were to gather evidence by tossing this card, we would find that the probabilities are *not* $\frac{1}{3}$, $\frac{1}{3}$, and $\frac{1}{3}$.

An Empirical Study

Let us go back to the problem of tossing a coin. While we cannot predict the result of one toss of a coin, we can still say that the probability of heads is $\frac{1}{2}$, based on observations made in an empirical study. In an ***empirical study***, we perform an experiment many times, keep records of the results, and analyze these results. For example, ten students decided to take turns tossing a coin. Each student completed 20 tosses and the number of heads was recorded as shown.

	Number of Heads	Number of Tosses
Albert	8	20
Peter	13	20
Thomas	12	20
Maria	10	20
Elizabeth	6	20
Joanna	12	20
Kathy	11	20
Jeanne	7	20
Debbie	13	20
James	9	20

If we look at the results and try to think of the probability of heads as a fraction comparing the number of heads to the total number of tosses, only Maria with 10 heads out of 20 tosses had results where the probability was $\frac{10}{20}$, or $\frac{1}{2}$. This fraction is called the ***relative frequency***. Elizabeth had the lowest relative frequency of heads, $\frac{6}{20}$. Peter and Debbie tied for the highest relative frequency with $\frac{13}{20}$.

This table does *not* mean that Maria had correct results while the other students were incorrect; the coins simply fell that way. The students decided to put their results together, by expanding the chart, to see what happened with 200 tosses of the coin. As shown in columns 3 and 4 of the next chart, the ***cumulative*** results are found by adding the results up to that point. For example, in the second row, by adding the

8 heads that Albert tossed and the 13 heads that Peter tossed, we find that the cumulative number of heads is 21, the total number of heads tossed by Albert and Peter together. By adding the 20 tosses that Albert made and the 20 tosses that Peter made, we find the cumulative number of tosses is 40.

	(Col. 1) Number of Heads	(Col. 2) Number of Tosses	(Col. 3) Cumulative Number of Heads	(Col. 4) Cumulative Number of Tosses	(Col. 5) Cumulative Relative Frequency
Albert	8	20	8	20	8/20 = .400
Peter	13	20	21	40	21/40 = .525
Thomas	12	20	33	60	33/60 = .550
Maria	10	20	43	80	43/80 = .538
Elizabeth	6	20	49	100	49/100 = .490
Joanna	12	20	61	120	61/120 = .508
Kathy	11	20	72	140	72/140 = .514
Jeanne	7	20	79	160	79/160 = .494
Debbie	13	20	92	180	92/180 = .511
James	9	20	101	200	101/200 = .505

In column 5, the ***cumulative relative frequency*** is found by dividing the total number of heads at or above a row by the total number of tosses at or above that row. Notice that the cumulative relative frequency is shown as a fraction and then, for easy comparison, as a decimal. The decimal is given to the nearest thousandth.

While the relative frequency for individual students varied greatly from $\frac{6}{20}$ to $\frac{13}{20}$, the cumulative relative frequency is $\frac{101}{200}$, a number very close to $\frac{1}{2}$.

A ***graph*** of the results of columns 4 and 5 will tell us even more. In the graph shown on the next page, the horizontal axis is labeled "Number of Tosses" to show the cumulative results of column 4; the vertical axis is labeled "Cumulative Relative Frequency of Heads" to show the results of column 5.

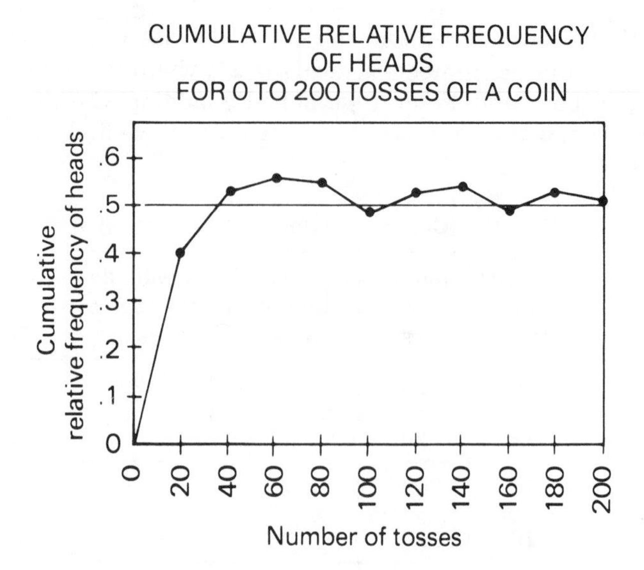

In the graph, we have plotted the points that represent the data in columns 4 and 5 of the previous chart, and we have connected these points to form a line graph. Notice how the line moves up and down around the relative frequency of .5, or $\frac{1}{2}$. The graph shows that the more times the coin is tossed, the closer the relative frequency comes to $\frac{1}{2}$. The line seems to *level out* at a relative frequency of $\frac{1}{2}$.

What would probably have happened if the students had tossed the coin 400 times? Or 1,000 times? Or 10,000 times? The line would appear to become a straight horizontal line, very close to the horizontal line at a cumulative relative frequency of .5. Hence, we can say that the cumulative relative frequency ***converges*** to the number $\frac{1}{2}$ and the coin will land heads up about $\frac{1}{2}$ of the time.

When a line graph levels out, or converges to a number, we say that the cumulative relative frequency has ***stabilized***. Even though the cumulative relative frequency of $\frac{101}{200}$ is not exactly $\frac{1}{2}$, we sense that the line will approach the number $\frac{1}{2}$. When we use this carefully collected evidence about tossing a fair coin to guess that the probability of heads is $\frac{1}{2}$, we have arrived at this conclusion ***empirically***, that is, by experimentation and observation.

● **Empirical probability may be defined as the most accurate scientific estimate, based on a large number of trials, of the cumulative relative frequency of an event happening.**

Experiments in Probability

A single attempt at doing something, such as tossing a coin only once, is called a ***trial***. We perform ***experiments*** in probability by repeating the same trial many times. Experiments are aimed at finding the probabilities to be assigned to different events occurring, such as heads or tails on a coin. The objects used in an experiment may be classified into one of two categories:

1. ***Fair and unbiased objects*** have not been weighted or made unbalanced. An object is fair when the different results have an *equal chance* of happening. Objects such as coins, cards, and spinners will always be treated in this book as fair objects, unless otherwise noted.

2. ***Biased objects*** are those that have been tampered with or are weighted to give one result a *better chance* of happening than another. The folded index card, described earlier in this chapter, is a biased object because the probability of each of three results is not $\frac{1}{3}$. The card is weighted so that it will fall on its side more often than it will fall on its edge.

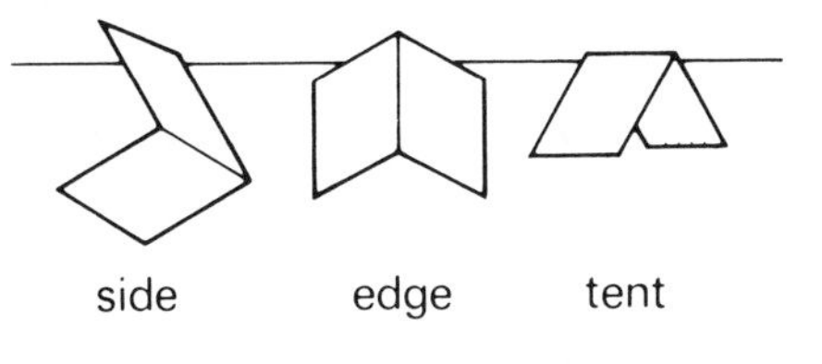

Uses of Probability

Mathematicians first studied probability by looking at situations involving games. Today, probability is used in a wide variety of fields. In medicine, it helps us to know the chances of catching an infection, of controlling an epidemic, and the rate of effectiveness of a drug in curing a disease. In industry, probability tells us how long a manufactured product should last. We can predict when more tellers are needed at bank windows, when and where traffic jams are likely to occur, and the expected weather for the next few days. In biology, the study of genes inherited from one's parents and grandparents is a direct application of probability. While this list is almost endless, all of these applications demand a strong knowledge of higher mathematics. Like the early mathematicians, you will begin your formal study of probability by looking at games and rather simple applications.

MODEL PROBLEMS

You have seen how to determine empirical probability by the charts and graphs previously shown in this chapter. Sometimes, it is possible to guess the probability that should be assigned to the result described before you start an experiment.

In 1–5, use common sense to guess the probability that might be assigned to the result described. (The solutions are given without comment here. You will learn how to determine these probabilities in the next section.)

1. A ***die*** is a six-sided solid object. Each side (or face) is a square. The sides are numbered 1, 2, 3, 4, 5, and 6. The plural of die is *dice*.

   In rolling a fair die, find the probability of getting a 4, or $P(4)$.

   Solution: $P(4) = \frac{1}{6}$

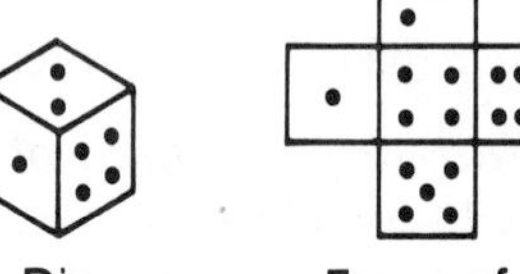

2. A ***standard deck of cards*** contains 52 cards. There are 4 suits called hearts, diamonds, spades, and clubs. Each suit contains 13 cards: 2, 3, 4, 5, 6, 7, 8, 9, 10, jack, queen, king, and ace. The diamonds and hearts are red; the spades and clubs are black.

   In selecting a card from the deck without looking, find the probability of drawing:

   a. the seven of diamonds **b.** a seven **c.** a diamond

   Solution: **a.** $P(\text{seven of diamonds}) = \frac{1}{52}$

   b. $P(\text{seven}) = \frac{4}{52}$, or $\frac{1}{13}$

   c. $P(\text{diamond}) = \frac{13}{52}$, or $\frac{1}{4}$

3. There are 10 ***digits*** in our numeral system: 0, 1, 2, 3, 4, 5, 6, 7, 8, and 9. In selecting a digit without looking, what is the probability it will be: **a.** the 8? **b.** an odd digit?

   Solution: **a.** $P(8) = \frac{1}{10}$ **b.** $P(\text{odd}) = \frac{5}{10}$, or $\frac{1}{2}$

4. An ***urn***, or a jar, contains 8 marbles: 3 are white and the remaining 5 are blue. In selecting a marble without looking, what is the probability it will be blue? (All marbles are the same size.)

   Solution: $P(\text{blue}) = \frac{5}{8}$

5. The English alphabet contains 26 letters. There are 5 ***vowels*** (A, E, I, O, and U) and the remaining 21 letters are ***consonants***. If a person turns 26 tiles from a word game face down and each tile represents a different letter of the alphabet, what is the probability of turning over: **a.** the A? **b.** a vowel? **c.** a consonant?

Solution: **a.** $P(\text{A}) = \frac{1}{26}$ **b.** $P(\text{vowel}) = \frac{5}{26}$ **c.** $P(\text{consonant}) = \frac{21}{26}$

EXERCISES

1. The figure shows a disk, with an arrow that can be spun so that it has an equal chance of landing on one of four regions on the disk. The regions are equal in size and are numbered 1, 2, 3, and 4.

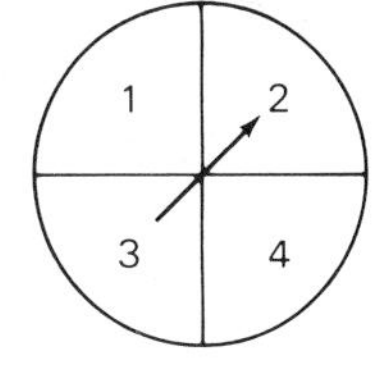

An experiment was conducted by five people to find the probability that the arrow will land on the 2. Each person spun the arrow 100 times. When the arrow landed on a line, it did not count and the arrow was spun again.

a. Before doing the experiment, what probability would you assign to the arrow landing on the 2? (In symbols, $P(2) = \underline{\ ?\ }$)

b. Copy and complete the chart to find the *cumulative* results of this experiment. In the last column, record the cumulative relative frequencies as fractions and as decimals to the *nearest thousandth.*

	Number of Times 2 Appeared	Number of Spins	Cumulative Number of Times 2 Appeared	Cumulative Number of Spins	Cumulative Relative Frequency
Barbara	29	100	29	100	$\frac{29}{100} = .290$
Tom	31	100	60	200	
Ann	19	100			
Eddie	23	100			
Cathy	24	100			

c. Did the experiment provide evidence that the probability assigned in part **a** was correct?

In 2–7, a result is described for a *fair, unbiased object.* What probability should be assigned to the result described? These questions should be answered *without* conducting an experiment; take a guess.

2. A six-sided die is rolled; the sides are numbered 1, 2, 3, 4, 5, and 6. In rolling the die, find $P(5)$.

3. In drawing a card from a standard deck without looking, find $P(\text{any heart})$.

4. Each of ten pieces of paper contains a different number from {0, 1, 2, 3, 4, 5, 6, 7, 8, 9}. The pieces of paper are folded and placed into a paper bag. In selecting a piece of paper without looking, find $P(7)$.

5. An urn contains 5 marbles, all the same size. Two marbles are black and the other three are white. In selecting a marble without looking, find $P(\text{black})$.

6. Using the lettered tiles from a word game, a boy places 26 tiles face down on a table, one tile for each letter of the alphabet. After mixing up the tiles, he takes one. What is the probability that the tile contains one of the letters in the word MATH?

7. A ***tetrahedron*** is a four-sided object. Each side, or face, is an equilateral triangle. The numerals 1, 2, 3, and 4 are used to number the different faces. A trial consists of rolling the tetrahedron and reading the number that is face down. Find $P(4)$.

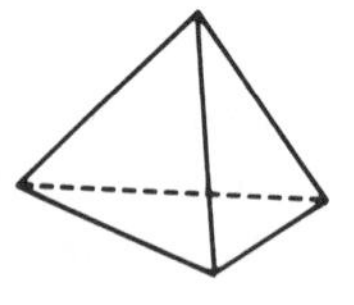

Tetrahedron

Faces of a tetrahedron

8. By yourself, or with some classmates, conduct any of the experiments described in Exercises 2–7 to verify that you have assigned the correct probability to the event or result described. A good experiment should contain at least 100 trials.

In 9–13, a *biased object* is described. A probability can be assigned to a result described only by conducting an experiment to determine the cumulative relative frequency of the event. While you may wish to guess at the probability of the event before starting the experiment, conduct at least 100 trials to determine the best probability to be assigned.

9. An index card is folded in half and tossed. As described earlier in this chapter, the card may land in one of three positions: on its side, on its edge, or in the form of a tent. In tossing a folded card, find P(tent), the probability that the card will form a tent when it lands.

10. A paper cup is tossed. It can land in one of three positions: on its top, on its bottom, or on its side. In tossing the cup, find P(top), the probability of landing on its top.

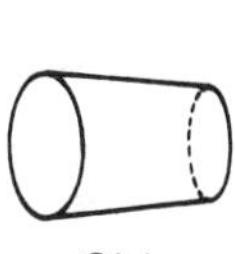

11. A nickel and a quarter are glued or taped together so that the two faces seen are the head of the quarter and the tail of the nickel. This is a very crude model of a weighted coin. In tossing the coin, find P(head).

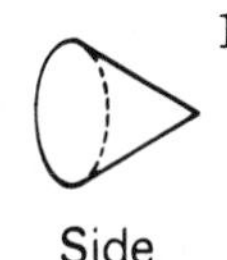

12. A paper cup in the shape of a cone is tossed. It can land in one of two positions: on its base or on its side. In tossing this cup, find P(side), the probability of landing on its side.

13. A thumbtack is tossed. It may land either with the pin up or with the pin down. In tossing a thumbtack, find P(pin is up).

15-2 THEORETICAL PROBABILITY

An empirical approach to probability is necessary whenever we deal with biased objects. However, common sense tells us that there is a simple way to define the probability of an event when we deal with fair, unbiased objects. For example:

Alma is playing a game in which each player must roll a die. To win, Alma must roll a number *greater than 4*. What is the probability that Alma will win on her next turn?

Common sense tells us:

(1) The die has an equal chance of falling in any one of *six* ways: **1**, **2**, **3**, **4**, **5**, and **6**.

(2) There are *two* ways for Alma to win: rolling a **5** or a **6**.

(3) Therefore: $P(\text{Alma wins}) = \dfrac{\text{number of winning results}}{\text{number of possible results}} = \dfrac{2}{6} = \dfrac{1}{3}$

Terms and Definitions

Using the details of the previous example, let us examine the correct terminology to be used.

An ***outcome*** is a result of some activity or experiment. In rolling a die, 1 is an outcome, 2 is an outcome, 3 is an outcome, and so on. There are six outcomes here.

A ***sample space*** is a set of all possible outcomes for the activity. In rolling a die, there are six possible outcomes in the sample space: 1, 2, 3, 4, 5, and 6. We also say the ***outcome set*** is {1, 2, 3, 4, 5, 6}.

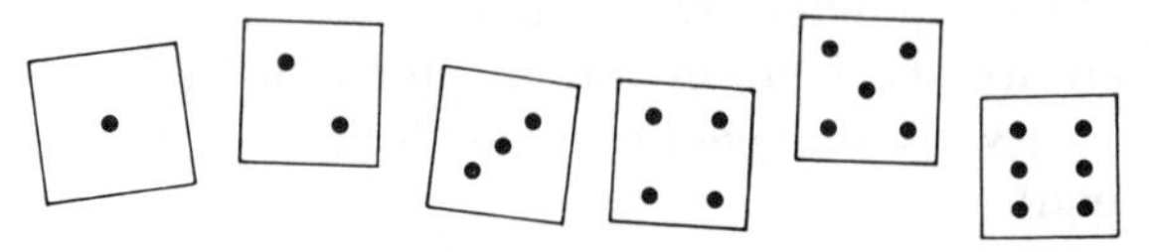

An ***event*** is a subset of the sample space. We use the term *event* in two ways. In ordinary conversation, it means a situation or happening. In the technical language of probability, it is the subset of the sample space that lists all of the outcomes for a given situation.

When we roll a die, we may define many different situations. Each of these is called an event.

1. For Alma, the event of rolling a number *greater than 4* contains only two outcomes: 5 and 6.

2. For Lee, a different event might be rolling a number *less than 5*. This event contains four outcomes: 1, 2, 3, and 4.

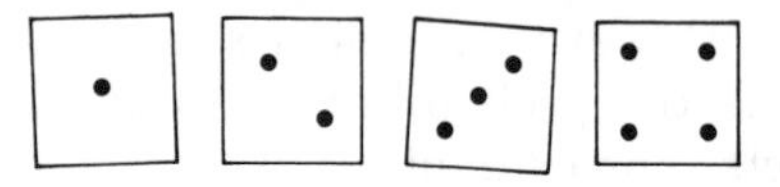

3. For Sandy, the event of rolling *a 2* contains only one outcome: 2. When there is only one outcome, we call this a ***singleton event***.

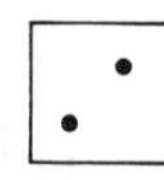

We can now define ***theoretical probability*** for fair, unbiased objects:

● **The theoretical probability of an event is the number of ways that the event can occur, divided by the total number of possibilities in the sample space.**

In symbolic form, we write:

$$P(E) = \frac{n(E)}{n(S)} \text{ where } \begin{cases} P(E) \text{ represents the probability of an event } E; \\ n(E) \text{ represents the number of ways that event } E \text{ can occur;} \\ n(S) \text{ represents the total number of possibilities, or the total number of possible outcomes in the sample space } S. \end{cases}$$

For Alma's problem, there are 2 ways to roll a number greater than 4, and there are 6 possible ways that the die may fall. We say:

E = the set of numbers greater than 4 = {5, 6}. Then, $n(E)$ **= 2.**

S = the set of 6 possible outcomes = {1, 2, 3, 4, 5, 6}. So, $n(S)$ **= 6.**

Therefore:

$$P(E) = \frac{n(E)}{n(S)} = \frac{\text{number of ways to roll a number greater than 4}}{\text{total number of outcomes for the die}} = \frac{2}{6}, \text{ or } \frac{1}{3}$$

Similarly, for Lee, we can say that the probability of rolling a number *less than 5* is found by $P(E) = \frac{n(E)}{n(S)} = \frac{4}{6}$, or $\frac{2}{3}$. Likewise, for Sandy, the probability of rolling *a 2* is $P(E) = \frac{n(E)}{n(S)} = \frac{1}{6}$.

Notice that the theoretical probability of an event is the *ratio* of the number of ways the event can occur to the total number of possibilities in the sample space.

Uniform Probability

A sample space is said to have ***uniform probability***, or to contain ***equally likely outcomes***, when each of the possible outcomes has an *equal chance* of occurring. In rolling a die, there are six possible outcomes in the sample space; each is equally likely to occur. So,

$P(1) = \frac{1}{6}$; $P(2) = \frac{1}{6}$; $P(3) = \frac{1}{6}$; $P(4) = \frac{1}{6}$; $P(5) = \frac{1}{6}$; $P(6) = \frac{1}{6}$; and, we say that the die has uniform probability.

If a die is weighted to make it biased, then one or more sides will have a probability greater than $\frac{1}{6}$, while one or more sides will have a probability less than $\frac{1}{6}$. A weighted die does not have uniform probability. Remember that the rule for theoretical probability does not apply to weighted objects.

Random Selection

When we select an object without looking, we are making a ***random selection***. Random selections are made when drawing a marble from a bag, when taking a card from a deck, or when picking a name out of a hat. In the same way, we may use the word *random* to describe outcomes when tossing a coin or rolling a die; the outcomes happen without any special selection on our part.

Procedure for finding the simple probability of an event:

1. Count the total number of outcomes in the sample space: $n(S)$
2. Count all the possible outcomes of the event E: $n(E)$
3. Substitute these values in the formula for the probability of event E: $P(E) = \frac{n(E)}{n(S)}$

MODEL PROBLEMS

1. A standard deck of 52 cards is shuffled. Lillian draws a single card from the deck at random. What is the probability that the card is a jack?

Solution

S = sample space of all possible outcomes. There are 52 cards in the deck. Hence, $n(S) = 52$.

J = event of selecting a jack. There are 4 jacks in the deck: jack of hearts, jack of diamonds, jack of spades, and jack of clubs. So, $n(J) = 4$.

$$P(J) = \frac{n(J)}{n(S)} = \frac{\text{number of possible jacks}}{\text{total number of possible cards}} = \frac{4}{52}, \text{ or } \frac{1}{13} \quad \textit{Ans.}$$

2. A spinner contains eight regions, numbered 1 through 8. The arrow has an equally likely chance of landing on any of the eight regions. If the arrow lands on a line, it is not counted and the arrow is spun again.

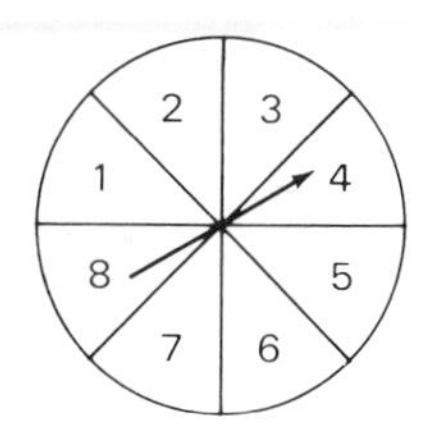

Solution

a. How many possible outcomes are in the sample space S?

a. $n(S) = 8$

b. What is the probability that the arrow lands on the 4? Simply, what is $P(4)$?

b. Since there is only 1 way to land on the 4 out of 8 possible numbers, $P(4) = \frac{1}{8}$.

c. List the set of possible outcomes for event O in which the arrow lands on an odd number.

c. Event $O = \{1, 3, 5, 7\}$.

d. Find the probability that the arrow lands on an odd number.

d. Since event $O = \{1, 3, 5, 7\}$, then $n(O) = 4$.

$$P(O) = \frac{n(O)}{n(S)} = \frac{4}{8}, \text{ or } \frac{1}{2}$$

EXERCISES

1. A fair coin is tossed.
a. List the sample space.
b. What is $P(\text{head})$, the probability that a head will appear?
c. What is $P(\text{tail})$?

2. A fair die is tossed. For each part of this question, (1) list the outcomes for the event and (2) state the probability of the event.
a. The number 3 appears.
b. An even number appears.
c. A number less than 3 appears.
d. An odd number appears.
e. A number greater than 3 appears.
f. A number greater than or equal to 3 appears.

3. A spinner is divided into 5 equal regions, numbered 1 through 5. An arrow is spun and lands in one of the regions. For each part of this question, (1) list the outcomes for the event; (2) state the probability of the event.

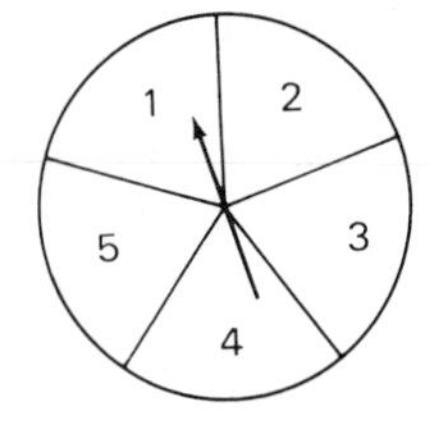

a. The number 3 appears.
b. An even number appears.
c. A number less than 3 appears. **d.** An odd number appears.
e. A number greater than 3 appears.
f. A number greater than or equal to 3 appears.

4. A standard deck of 52 cards is shuffled and one card is drawn. What is the probability that the card is:
a. the queen of hearts? **b.** a queen? **c.** a heart?
d. a red card? **e.** the seven of clubs? **f.** a club?
g. an ace? **h.** a red seven? **i.** a black ten?
j. a picture card (king, queen, jack)?

5. A person does not know the answer to a test question and takes a guess. Find the probability that the answer is correct if the question is: **a.** a multiple-choice question with 4 choices **b.** a true-false question **c.** a question where the choices given are "sometimes, always, or never."

6. A marble is drawn at random from a bag. Find the probability that the marble is green if the bag contains marbles whose colors are:
a. 3 blue, 2 green **b.** 4 blue, 1 green
c. 5 red, 2 green, 3 blue **d.** 6 blue, 4 green
e. 3 green, 9 blue **f.** 5 red, 2 green, 93 blue

7. The digits of the number 1776 are written on disks and placed into a jar. What is the probability that the digit 7 will be chosen on a single random draw?

8. A class contains 16 boys and 14 girls. The teacher calls students at random to the chalkboard. What is the probability that the first person called is: **a.** a boy? **b.** a girl?

9. There are 840 tickets sold in a raffle. Jay bought 5 tickets and Lynn bought 4 tickets. What is the probability that: **a.** Jay has the winning ticket? **b.** Lynn has the winning ticket?

10. A letter is chosen at random from a given word. Find the probability that the letter is a vowel if the word is:
a. APPLE **b.** BANANA **c.** GEOMETRY **d.** MATHEMATICS

11. In the figure on the next page, there are eight polygons shown: a square; a rectangle; a parallelogram (which is not a rectangle); a right triangle; an isosceles triangle (not containing a right angle); a trapezoid (not containing a right angle); an equilateral triangle; a regular hexagon.

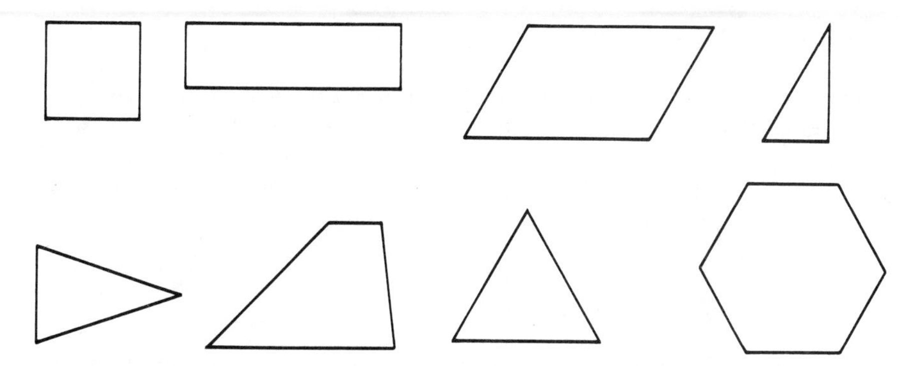

One of the figures is selected at random. What is the probability that the polygon:

a. contains a right angle?
b. is a quadrilateral?
c. is a triangle?
d. has at least one acute angle?
e. has all sides congruent?
f. has at least two sides congruent?
g. has less than five sides?
h. has an odd number of sides?
i. has four or more sides?
j. has at least two obtuse angles?

12. Explain why each of the following statements is *incorrect*.

a. Since there are 50 states, the probability that someone born in the United States will be born in New Jersey is $\frac{1}{50}$.

b. Since there are 12 months in a year, the probability of being born in September is $\frac{1}{12}$.

c. A pin containing a small head is tossed, as shown at the right. Since the pin can fall either point up or point down, the probability of falling point down is $\frac{1}{2}$.

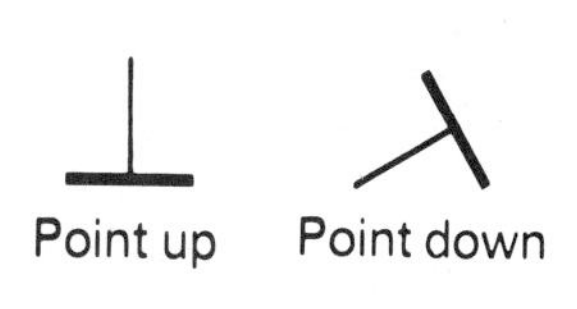

d. Since there are 7 days in a week, the probability that a person attends religious services on a Wednesday is $\frac{1}{7}$.

15-3 EVALUATING SIMPLE PROBABILITIES

We have called an event that can occur in only one way a *singleton*. For example, in rolling a single die, there is only one way to roll the number 3.

However, when rolling a single die, there can be more than one way for an event to occur. For example:

1. The event of rolling an even number on a die = {2, 4, 6}.
2. The event of rolling a number less than 6 on a die = {1, 2, 3, 4, 5}.

The Impossible Case

On a single roll of a die, what is the probability that the number 7 will appear? We call this an ***impossibility*** because there are no ways in which this event can occur. In this example, event E = rolling a 7. $E = \{\ \}$ and $n(E) = 0$. The sample space S for rolling a die contains six possible outcomes, and $n(S) = 6$. Therefore:

$$P(E) = \frac{n(E)}{n(S)} = \frac{\text{number of ways to roll a 7}}{\text{total number of outcomes for the die}} = \frac{0}{6} = 0$$

In general, for any sample space S containing k possible outcomes, we say $n(S) = k$. For any impossible event E in which there are no ways for the event E to occur, we say $n(E) = 0$. Thus, the probability of an impossible event is $P(E) = \frac{n(E)}{n(S)} = \frac{0}{k} = 0$, and we say:

- **The probability of an impossible event is zero.**

There are many other examples of impossibilities where the probability must equal zero. For example, the probability of selecting the letter E from the word PROBABILITY is zero.

Also, selecting a coin worth 9¢ from a bank containing a nickel, a dime, and a quarter is an impossible event.

The Certain Case

On a single roll of a die, what is the probability that a number less than 7 will appear? We call this a ***certainty*** because every one of the possible outcomes in the sample space is also an outcome for this event. In this example, event E = rolling a number less than 7, and $n(E) = 6$. The sample space S for rolling a die contains six possible outcomes, and $n(S) = 6$. Therefore:

$$P(E) = \frac{n(E)}{n(S)} = \frac{\text{number of ways to roll a number less than 7}}{\text{total number of outcomes on the die}} = \frac{6}{6} = 1$$

Note that to describe an event that is certain, we may use the sample space itself, that is, in this case $E = S$.

In general, for any sample space S containing k possible outcomes, we say $n(S) = k$. When the event E is certain, then every possible outcome for the sample space is also an outcome for event E, or $n(E) = k$.

Thus, the probability of a certainty is given as $P(E) = \dfrac{n(E)}{n(S)} = \dfrac{k}{k} = 1$, and we say:

- **The probability of an event that is certain to occur is 1.**

There are many other examples of certainties where the probability must equal 1. Such examples include the probability of selecting a consonant from the letters JFK or selecting a red sweater from a drawer containing only red sweaters.

The Probability of Any Event

Let us consider a series of events for rolling a six-sided die:

Event A, a number less than 1, has no outcomes.

$$A = \{\ \} \qquad P(A) = \frac{0}{6} = 0$$

Event B, a number less than 2, has one outcome:

$$B = \{1\} \qquad P(B) = \frac{1}{6}$$

Event C, a number less than 3, has two outcomes:

$$C = \{1, 2\} \qquad P(C) = \frac{2}{6}$$

Event D, a number less than 4, has three outcomes:

$$D = \{1, 2, 3\} \qquad P(D) = \frac{3}{6}$$

Event F, a number less than 5, has four outcomes:

$$F = \{1, 2, 3, 4\} \qquad P(F) = \frac{4}{6}$$

Event G, a number less than 6, has five outcomes:

$$G = \{1, 2, 3, 4, 5\} \qquad P(G) = \frac{5}{6}$$

Event H, a number less than 7, has six outcomes:

$$H = \{1, 2, 3, 4, 5, 6\} \qquad P(H) = \frac{6}{6} = 1$$

The smallest probability here is 0. The largest probability is 1. Each of the other events has a probability that falls between 0 and 1. This example illustrates the following:

- **The probability of any event E must be equal to or greater than zero, and less than or equal to one. Or, simply:**

$$\mathbf{0 \leq P(E) \leq 1}$$

MODEL PROBLEMS

1. A bank contains a nickel, a dime, and a quarter. A person selects one of the coins. What is the probability that the coin is worth:
a. exactly 10¢? **b.** exactly 3¢? **c.** more than 3¢?

Solution

a. There is only one coin worth exactly 10¢: the dime. There are three coins in the bank. Thus:

$$P(\text{coin is worth } 10¢) = \frac{n(E)}{n(S)} = \frac{1}{3} \quad \textit{Ans.}$$

b. There are no coins worth exactly 3¢. This is an impossible event.
Thus: $P(\text{coin is worth } 3¢) = \frac{n(E)}{n(S)} = \frac{0}{3} = 0$ *Ans.*

c. All three coins are worth more than 3¢. This is a certain event.
Thus: $P(\text{coin is worth more than } 3¢) = \frac{n(E)}{n(S)} = \frac{3}{3} = 1$ *Ans.*

2. In the Sullivan family, there are 2 more girls than boys. At random, Mrs. Sullivan asks one of her children to go to the store. If she is equally likely to have asked any one of her children, and the probability that she asked a girl is $\frac{2}{3}$, how many boys and how many girls are there in the Sullivan family?

Solution:

Let x = the number of boys.
Then, $x + 2$ = the number of girls.
And, $2x + 2$ = the number of children.

$$P(\text{girl}) = \frac{\text{number of girls}}{\text{number of children}}$$

$$\frac{2}{3} = \frac{x + 2}{2x + 2}$$

$$2(2x + 2) = 3(x + 2)$$

$$4x + 4 = 3x + 6$$

$$x = 2$$

$$x + 2 = 4$$

$$2x + 2 = 6$$

Answer: There are 2 boys and 4 girls.

3. An arrow is spun once and lands on one of three equally likely regions, numbered 1, 2, and 3.
a. List the sample space for this experiment.
b. List all 8 possible events for one spin of the arrow.

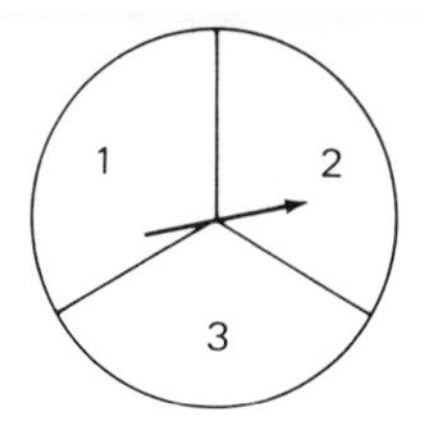

Solution

a. The sample space $S = \{1, 2, 3\}$.

b. Since events are subsets of the sample space S, the 8 possible events are the 8 subsets of S:

$\{\ \}$ = the empty set, for *impossible* events.

$\{1, 2, 3\}$ = the set itself, for events with *certainty*.

$\{1\}, \{2\}, \{3\}$ = the *singleton* events.

$\{1, 2\}, \{1, 3\}, \{2, 3\}$ = events with *two possible* outcomes. An event such as getting an odd number has two possible outcomes: 1 and 3.

Subscripts in Sample Spaces

A sample space may sometimes contain two or more objects that are exactly alike. To distinguish one object from another, we make use of a device called a *subscript*. A ***subscript*** is a number, usually written in smaller size, that always appears to the lower right of a term. For example:

A box contains six jellybeans: 2 red, 3 green, and 1 yellow. Using R, G, and Y to represent the colors red, green, and yellow, respectively, we can list this sample space in full, using subscripts:

$$\{R_1, R_2, G_1, G_2, G_3, Y_1\}$$

Since there is only one yellow jellybean, we could have listed the last element as Y instead of Y_1.

MODEL PROBLEM

A letter is chosen at random from the word REED.
a. List the sample space using subscripts.
b. Find the probability of choosing an E.

Solution:
a. The sample space is $\{\text{R}, \text{E}_1, \text{E}_2, \text{D}\}$.
b. Since there are 2 ways to choose the letter E from the 4 letters:

$P(\text{E}) = \frac{2}{4}$, or $\frac{1}{2}$ *Ans.*

EXERCISES

1. A fair coin is tossed and its sample space $S = \{H, T\}$.
   a. List all 4 possible events for the toss of a coin.
   b. Find the probability of each event named in part **a**.

2. A spinner is divided into 7 equal regions, numbered 1 through 7. An arrow is spun to fall into one of the regions. For each part of this question:
   (1) List the elements of the event, shown as a subset of $\{1, 2, 3, 4, 5, 6, 7\}$.
   (2) Find the probability that the arrow lands on the number described.

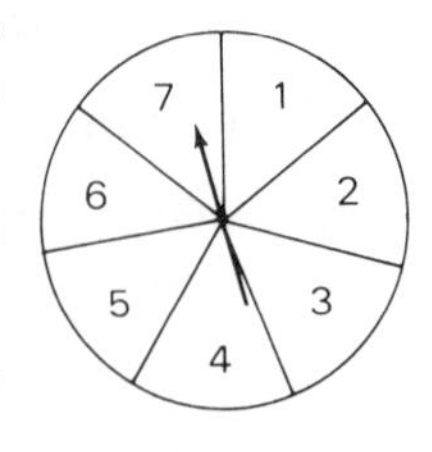

   a. the number 5
   b. an even number
   c. a number less than 5
   d. an odd number
   e. a number greater than 5
   f. a number greater than 1
   g. a number greater than 7
   h. a number less than 8

3. A marble is drawn at random from a bag. Find the probability that the marble is black if the bag contains marbles whose colors are:
   a. 5 black, 2 green
   b. 2 black, 1 green
   c. 3 black, 4 green, 1 red
   d. 9 black
   e. 3 green, 4 red
   f. 3 black

4. Ted has 2 quarters, 3 dimes, and 1 nickel in his pocket. He pulls out a coin at random. Find the probability that the coin is worth:
   a. exactly 5¢
   b. exactly 10¢
   c. exactly 25¢
   d. exactly 50¢
   e. less than 25¢
   f. less than 50¢
   g. more than 25¢
   h. more than 1¢
   i. less than 1¢

5. A single fair die is rolled. Find the probability for each event.
   a. The number 8 appears.
   b. A whole number appears.
   c. The number is less than 5.
   d. The number is less than 1.
   e. The number is less than 10.
   f. The number is negative.

6. A standard deck of 52 cards is shuffled and you pick a card at random. Find the probability that the card is:
   a. a jack
   b. a club
   c. a star
   d. a red club
   e. a card from the deck
   f. a black club
   g. the jack of stars
   h. a 17
   i. a red 17

7. A class contains 15 girls and 10 boys. The teacher calls on a student at random to answer a question. Express, in decimal form, the probability that the student called upon is: **a.** a girl **b.** a boy **c.** a pupil in the class **d.** the teacher of the class

8. The last digit of a telephone number can be any of the following: 0, 1, 2, 3, 4, 5, 6, 7, 8, or 9. Express, as a percent, the probability that the last digit is: **a.** 7 **b.** odd **c.** more than 5 **d.** a whole number **e.** the letter R

9. A girl is holding 5 cards in her hand. They are the 3 of hearts, 3 of diamonds, 3 of clubs, 4 of diamonds, 7 of clubs. A player to her left takes one of these cards at random. (If the player takes the 3 of hearts, we describe the probability of this event as $\frac{1}{5}$.) Find the probability that the card selected from the 5 cards in the girl's hand is:

a. a 3 **b.** a diamond **c.** a 4
d. a black 4 **e.** a club **f.** 4 of hearts
g. a 5 **h.** 7 of clubs **i.** a red card
j. a number card **k.** a spade **l.** a number less than 8

10. The measures of three interior angles of a triangle are given as 40°, 60°, and 80°. The measures of the exterior angles of the triangle are 140°, 120°, and 100°, respectively. One of the six angles is chosen at random. Find the probability that the angle is:

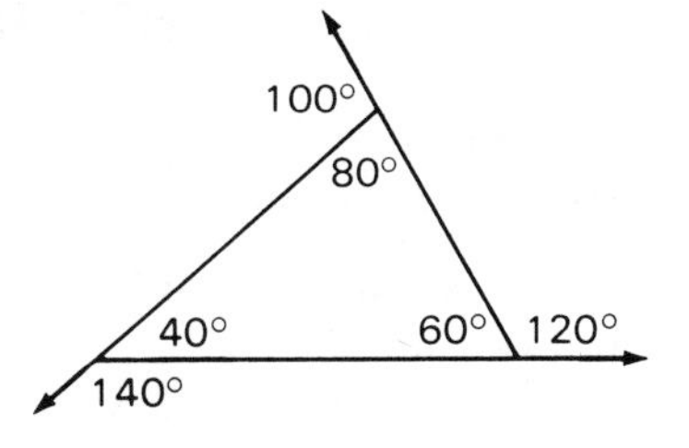

a. an interior angle **b.** a straight angle
c. a right angle **d.** an acute angle
e. a 60° angle **f.** an acute exterior angle
g. an angle whose measure is less than 180°

11. A sack contains 20 marbles. The probability of drawing a green marble is $\frac{2}{5}$. How many green marbles are in the sack?

12. There are 3 more boys than girls in the chess club. A member of the club is to be chosen at random to play in a tournament. Each member is equally likely to be chosen. If the probability that a girl is chosen is $\frac{3}{7}$, how many boys and how many girls are there?

13. A box of candy contains caramels and nut clusters. There are 6 more caramels than nut clusters. If a piece of candy is to be chosen at random, the probability that it will be a caramel is $\frac{3}{5}$. How many caramels and how many nut clusters are in the box?

14. List three situations where the probability of an event is 0.

15. List three situations where the probability of an event is 1.
16. Explain why the following sentence is *incorrect:* "I did so well on the test that the probability of my passing is greater than 1."
17. Explain why the following sentence is *incorrect:* "The probability that we'll go swimming in Maine next December is less than 0."
18. Explain why this sentence is *correct:* "The probability that Tuesday comes after Monday is 100%."
19. A letter is chosen at random from a given word. For each part of this question: (1) List the elements of the event, using subscripts if needed. (2) Find the probability of the event.
    a. selecting the letter E from the word EVENT
    b. selecting the letter S from the word MISSISSIPPI
    c. selecting a vowel from the word TRIANGLE
    d. selecting a vowel from the word RECEIVE
    e. selecting a consonant from the word SPRY

15-4 THE PROBABILITY OF *A AND B*

The connective ***and*** has been used in logic. As we will see, the connective *and* is sometimes used to describe events in probability.

For example, a fair die is rolled. What is the probability of obtaining an even number?

Call this event A. Since there are three ways to obtain an even number:

$$P(A) = \frac{n(A)}{n(S)} = \frac{3}{6}$$

When a fair die is rolled, what is the probability of obtaining a number less than 3?

Call this event B. Since there are two ways to obtain a number less than 3:

$$P(B) = \frac{n(B)}{n(S)} = \frac{2}{6}$$

Now, what is the probability of obtaining a number on the die that is even **and** less than 3? We may think of this as the **event *A and B*.**

In logic, you learned that a sentence ***p and q***, written $p \wedge q$, is true only when p is true *and* q is true.	In probability, an outcome is in event ***A and B*** only when the outcome is in event A *and* the outcome is also in event B.

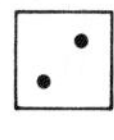

The only outcome in the event *A and B* is 2, because 2 is even and 2 is less than 3. No other outcome on the die is true for both events. Since $n(A \text{ and } B) = 1$ and there are 6 outcomes on the die, or $n(S) = 6$:

$$P(A \text{ and } B) = \frac{n(A \text{ and } B)}{n(S)} = \frac{1}{6}$$

Consider another example in which a fair die is rolled.

Event C = {die numbers that are odd}

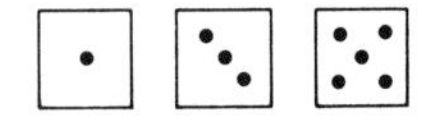

$$P(C) = \frac{n(C)}{n(S)} = \frac{3}{6}$$

Event D = {4}

$$P(D) = \frac{n(D)}{n(S)} = \frac{1}{6}$$

Then, event ***C and D*** = {die numbers that are odd *and* 4}

Since there are *no* outcomes common to both event C and event D, we can say that there are *no* outcomes in the event *C and D*, that is, $n(C \text{ and } D) = 0$. Therefore, $P(C \text{ and } D) = \dfrac{n(C \text{ and } D)}{n(S)} = \dfrac{0}{6} = 0$.

Using Sets to Look at *P(A and B)*

You have seen that an outcome is in the event *A and B* only when the outcome is in event *A and* in event *B*. Using event *A* and event *B* as previously described, the following set diagram illustrates the event *A and B*.

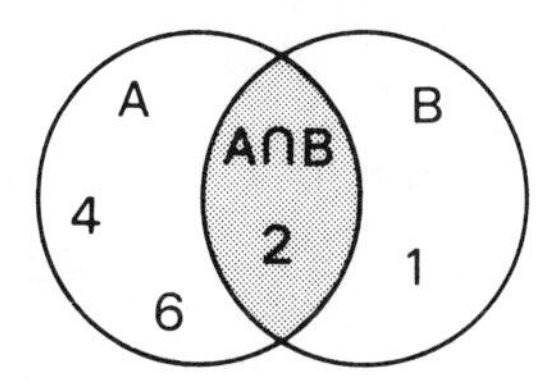

Event *A and B* is the set of numbers that are even *and* less than 3. Since event A is {2, 4, 6} and event B is {1, 2}, you can see that the event *A and B* is {2}, the ***intersection*** of the two sets, or $A \cap B$. By counting the number of outcomes in this intersection, we say:

$$P(A \text{ and } B) = P(A \cap B) = \frac{n(A \cap B)}{n(S)} = \frac{1}{6}$$

Event C *and* D is the set of numbers that are odd and 4. Here, the intersection of event C and event D is empty, or $C \cap D = \{\ \}$. Thus:

$$P(C \text{ and } D) = P(C \cap D) = \frac{n(C \cap D)}{n(S)} = \frac{0}{6} = 0$$

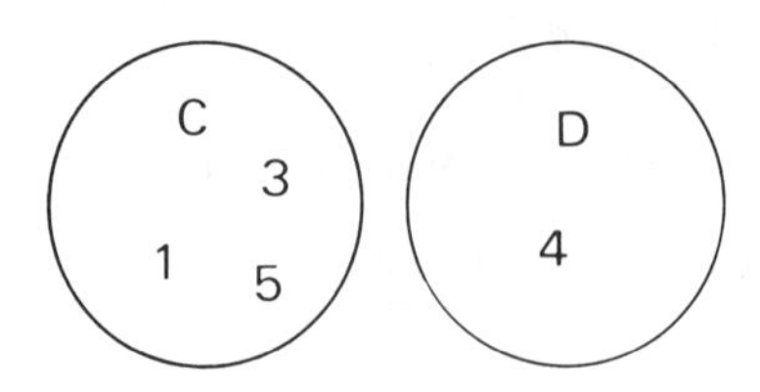

Observe that *there is no simple rule or formula that works for all problems* whereby you can use the values of $P(A)$ and $P(B)$ to find $P(A \text{ and } B)$. You must simply observe the intersection of the two sets and count the number of elements in that intersection. Or, you can count the number of outcomes that are common to both events.

KEEP IN MIND

The event ***A and B*** consists of outcomes that are in event A *and* in event B. Event A *and* B may be regarded as the *intersection* of sets, namely $\boldsymbol{A \cap B}$.

MODEL PROBLEM

A fair die is rolled once. Find the probability of obtaining a number that is greater than 3 and less than 6.

Solution

Event A = {numbers greater than 3} = {4, 5, 6}.

Event B = {numbers less than 6} = {1, 2, 3, 4, 5}.

Event A *and* B = {outcomes common to both events} = {4, 5}, or $(A \cap B) = \{4, 5\}$.

Therefore:

$$P(A \text{ and } B) = \frac{n(A \text{ and } B)}{n(S)} = \frac{2}{6} \quad \text{or} \quad \frac{n(A \cap B)}{n(S)} = \frac{2}{6} \quad \textit{Ans.}$$

EXERCISES

1. A fair die is rolled once. The sides are numbered 1, 2, 3, 4, 5, and 6. Find the probability that the number rolled is:

a. greater than 2 and odd
b. less than 4 and even
c. greater than 2 and less than 4
d. less than 2 and even
e. less than 6 and odd
f. less than 4 and greater than 3

2. From a standard deck of cards, one card is drawn. Find the probability that the card will be:

a. the king of hearts
b. a red king
c. a club king
d. a black jack
e. a diamond ten
f. a red club
g. the two of spades
h. a black two
i. a red picture card

3. A set of polygons consists of an equilateral triangle, a square, a rhombus, and a rectangle as shown. One of the polygons is selected at random. Find the probability that the polygon contains:

a. all sides congruent and all angles congruent
b. all sides congruent and all right angles
c. all sides congruent and two angles not congruent
d. at least 2 congruent sides and at least 2 congruent angles
e. at least 3 congruent sides and at least 2 congruent angles

4. At a St. Patrick's Day party, some of the boys and girls take turns singing songs. Of the 5 boys: Patrick and Terence are teenagers; younger boys include Brendan, Drew, and Kevin. Of the 7 girls: Heather and Claudia are teenagers; younger girls include Maureen, Elizabeth, Gwen, Caitlin, and Kelly. Find the probability that the first song is sung by:

a. a girl
b. a boy
c. a teenager
d. someone under 13 years old
e. a boy under 13
f. a girl whose initial is C
g. a teenage girl
h. a girl under 13
i. a boy whose initial is C
j. a teenage boy

15-5 THE PROBABILITY OF *A OR B*

The connective **or** has been used in logic. As you will see, the connective *or* is sometimes used to describe events in probability.

For example, a fair die is rolled. What is the probability of obtaining an even number?

Calling this event A, you have seen:

$$P(A) = \frac{n(A)}{n(S)} = \frac{3}{6}$$

When a die is rolled, what is the probability of obtaining a number less than 2?

Call this event C. Since there is only one possible outcome for this event (the number 1): $P(C) = \frac{n(C)}{n(S)} = \frac{1}{6}$

Now, what is the probability of obtaining a number on the die that is even **or** less than 2? We may think of this as the **event *A or C*.**

In logic, you learned that a sentence ***p or q***, written $p \vee q$, is true when p is true, or when q is true, or when both p and q are true.

In probability, an outcome is in event ***A or C*** when the outcome is in event A, or the outcome is in event C, or when the outcome is in both event A and event C.

There are four outcomes in event A *or* C: 1, 2, 4, and 6. Each of these numbers is even or it is less than 2. Since $n(A \text{ or } C) = 4$, and there are 6 outcomes on the die, or $n(S) = 6$:

$$P(A \text{ or } C) = \frac{n(A \text{ or } C)}{n(S)} = \frac{4}{6}$$

Observe that $P(A) = \frac{3}{6}$, $P(C) = \frac{1}{6}$, and $P(A \text{ or } C) = \frac{4}{6}$. In this case, it appears that $P(A) + P(C) = P(A \text{ or } C)$. Will this simple addition rule be true for all problems?

Consider another example in which a fair die is rolled.

Event A = {even numbers}

$$P(A) = \frac{n(A)}{n(S)} = \frac{3}{6}$$

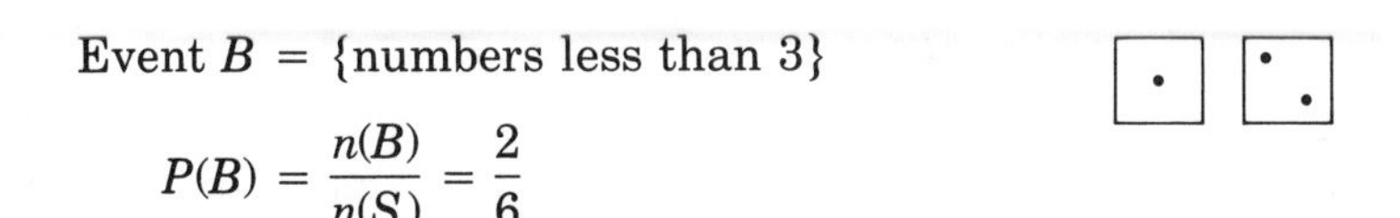

Event B = {numbers less than 3}

$$P(B) = \frac{n(B)}{n(S)} = \frac{2}{6}$$

Then, event ***A or B*** = {numbers that are even or less than 3}.

There are still only four outcomes in this new event: 1, 2, 4, and 6. Each of these numbers is even or it is less than 3. Therefore:

$$P(A \text{ or } B) = \frac{n(A \text{ or } B)}{n(S)} = \frac{4}{6}$$

Observe that $P(A) = \frac{3}{6}$, $P(B) = \frac{2}{6}$, and $P(A \text{ or } B) = \frac{4}{6}$. In this case, the simple rule of addition does not work. $P(A) + P(B) \neq P(A \text{ or } B)$. What made this example different from $P(A \text{ or } C)$, shown previously?

A Rule for the Probability of *A or B*

Probability is based upon counting the outcomes in the event. For the event *A or B*, we observe that the outcome 2 is found in event *A* *and* in event *B*. Therefore, we may describe the outcome 2 as the event *A and B*.

We realize that the simple addition rule does not work for the event *A or B* because we have counted the outcome 2 twice: first in event *A*, then again in event *B*. Since we counted this outcome twice, we must take it away, or subtract it, *once*.

Hence, the rule becomes: $n(A) + n(B) - n(A \text{ and } B) = n(A \text{ or } B)$

For this example: $3 + 2 - 1 = 4$

Dividing each term by $n(S)$, we get an equivalent equation:

$$\frac{n(A)}{n(S)} + \frac{n(B)}{n(S)} - \frac{n(A \text{ and } B)}{n(S)} = \frac{n(A \text{ or } B)}{n(S)}$$

For this example:

$$\frac{3}{6} + \frac{2}{6} - \frac{1}{6} = \frac{4}{6}$$

Since $P(A \text{ or } B) = \frac{n(A \text{ or } B)}{n(S)}$, we can write a general rule:

$$\mathbf{P(A \text{ or } B) = P(A) + P(B) - P(A \text{ and } B)}$$

Using Sets to Look at *P*(*A or B*)

A set diagram can help you to understand the event *A or B* just discussed. Event *A or B* = {numbers that are even *or* less than 3}.

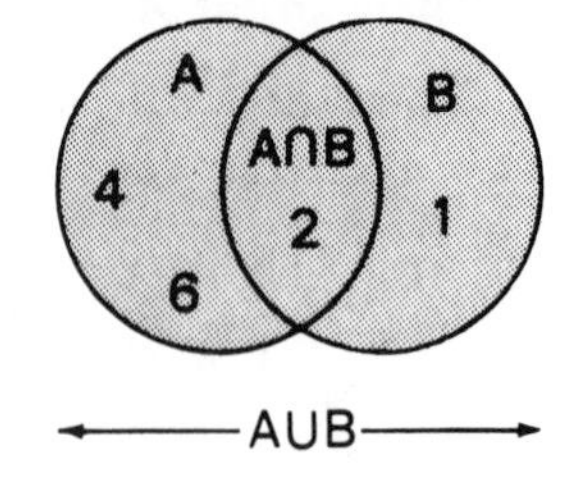

Since event *A or B* = {1, 2, 4, 6}, recognize that *A or B* is the ***union*** of the two sets, or $A \cup B$, as shown by the shaded region.

Recall that the event *A and B* = {2}, which is the intersection of the two sets, or $A \cap B$. You can again see how the counting procedure works:

1. Count the number of elements in one set.
2. Add the number of elements from the second set.
3. Subtract the number of elements in their intersection since they were counted twice.
4. This result is the number of elements in the union of the two sets, or $n(A \cup B)$.

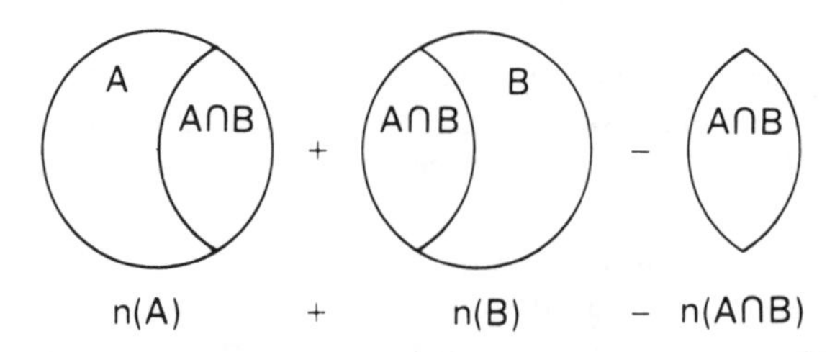

In set terminology, the rule for probability becomes:

$$P(A \text{ or } B) = P(A \cup B) = P(A) + P(B) - P(A \cap B)$$

Disjoint Sets

Event *A or C* = {numbers that are even or less than 2}

Here, event *A* = {2, 4, 6} and event *C* = {1}, shown below by disjoint sets. Since *A* and *C* are disjoint, there are no elements in their intersection, or $(A \cap C) = \{\ \}$.

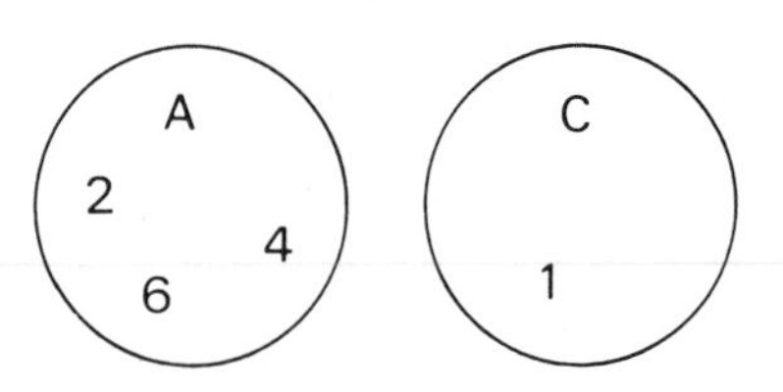

Recall that the probability of the empty set is zero. Thus:

$$P(A \cup C) = P(A) + P(C) - P(A \cap C)$$
$$P(A \cup C) = P(A) + P(C) - 0 = P(A) + P(C)$$

This tells us that the simple addition rule first observed will always be true for events that do not intersect, or that have no outcomes in common.

For disjoint events, $P(A \text{ or } C) = P(A \cup C) = P(A) + P(C)$.

KEEP IN MIND

The event ***A or B*** consists of all outcomes that are in event A, or in event B, or in both event A and event B. Event A *or* B may be regarded as the *union* of sets, namely $\boldsymbol{A \cup B}$.

MODEL PROBLEMS

1. A standard deck of 52 cards is shuffled. One card is drawn at random. Find the probability that the card is:
a. a king or an ace **b.** red or an ace

Solution

a. There are 4 kings in the deck, so $P(\text{king}) = \frac{4}{52}$.

There are 4 aces in the deck, so $P(\text{ace}) = \frac{4}{52}$.

Kings and aces are disjoint events, having no outcomes in common.

So, $P(\text{king or ace}) = P(\text{king}) + P(\text{ace}) = \frac{4}{52} + \frac{4}{52} = \frac{8}{52}$. *Ans.*

b. There are 26 red cards in the deck, so $P(\text{red}) = \frac{26}{52}$.

There are 4 aces in the deck, so $P(\text{ace}) = \frac{4}{52}$.

Two cards in the deck are red aces, so $P(\text{red and ace}) = \frac{2}{52}$.

Then: $P(\text{red or ace}) = P(\text{red}) + P(\text{ace}) - P(\text{red and ace})$

$$= \frac{26}{52} + \frac{4}{52} - \frac{2}{52} = \frac{28}{52}, \text{ or } \frac{7}{13} \quad \textit{Ans.}$$

or

By a counting procedure, there are 26 red cards and 2 more aces not already counted (the ace of spades; the ace of clubs). Therefore, there are 26 + 2, or 28, cards in this event.

So, $P(\text{red or ace}) = \frac{28}{52}$, or $\frac{7}{13}$. *Ans.*

2. There are two events, A and B. Given that $P(A) = .3$, $P(B) = .5$, and $P(A \cap B) = .1$, find $P(A \cup B)$.

Solution: $P(A \cup B) = P(A) + P(B) - P(A \cap B)$
$= .3 + .5 - .1 = .8 - .1 = .7$ *Ans.*

EXERCISES

1. A spinner consists of 5 regions as shown, equally likely to occur when an arrow is spun. For a single spin of the arrow, find the probability of the given event.

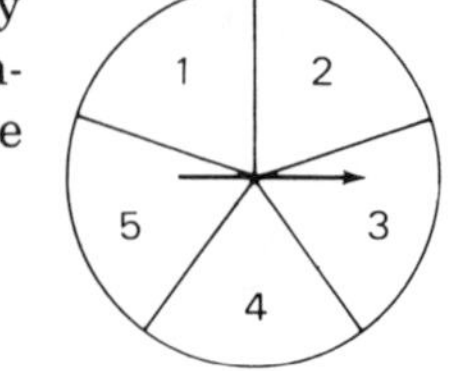

a. 4 **b.** 3 or 4
c. an odd number **d.** an odd number or 2
e. less than 4 **f.** 4 or less
g. 2 or 3 or 4 **h.** an odd number or 3

2. A fair die is rolled once. The sides are numbered 1, 2, 3, 4, 5, and 6. Find the probability of the event described.
a. 4 **b.** 3 or 4 **c.** an odd number
d. an odd number or 2 **e.** less than 4 **f.** 4 or less
g. 2 or 3 or 4 **h.** an odd number or 3
i. less than 2 or more than 5 **j.** less than 5 or more than 2

3. From a standard deck of cards, one card is drawn. Tell what the probability is that the card will be:
a. a queen or an ace **b.** a queen or a 7 **c.** a heart or a spade
d. a queen or a spade **e.** a queen or a red card
f. jack or queen or king **g.** a 7 or a diamond
h. a club or a red card **i.** an ace or a picture card

4. A bank contains 2 quarters, 6 dimes, 3 nickels, and 5 pennies. A coin is drawn at random. Find the probability that the coin is:
a. a quarter **b.** a quarter or a dime **c.** a dime or a nickel
d. worth 10 cents **e.** worth more than 10 cents
f. worth 10 cents or less **g.** worth 1 cent or more
h. worth more than 1 cent **i.** a quarter, nickel, or penny

In 5–10, choose the correct numeral preceding the word or expression that best completes the statement or answers the question.

5. If a single card is drawn from a standard deck, what is the probability that it is a 4 or a 9?

(1) $\frac{2}{52}$ (2) $\frac{8}{52}$ (3) $\frac{13}{52}$ (4) $\frac{26}{52}$

6. If a single card is drawn from a standard deck, what is the probability that it is a 4 or a diamond?
(1) $\frac{8}{52}$ (2) $\frac{16}{52}$ (3) $\frac{17}{52}$ (4) $\frac{26}{52}$

7. If $P(A) = .2$, $P(B) = .5$, and $P(A \cap B) = .1$, then $P(A \cup B) =$
(1) .6 (2) .7 (3) .8 (4) .9

8. If $P(A) = \frac{1}{3}$, $P(B) = \frac{1}{2}$, and $P(A \text{ and } B) = \frac{1}{6}$, then $P(A \text{ or } B) =$
(1) $\frac{2}{5}$ (2) $\frac{2}{3}$ (3) $\frac{5}{6}$ (4) 1

9. If $P(A) = \frac{1}{4}$, $P(B) = \frac{1}{2}$, and $P(A \cap B) = \frac{1}{8}$, then $P(A \cup B) =$
(1) $\frac{1}{8}$ (2) $\frac{5}{8}$ (3) $\frac{3}{4}$ (4) $\frac{7}{8}$

10. If $P(A) = .3$, $P(B) = .35$, and $(A \cap B) = \varnothing$, then $P(A \text{ or } B) =$
(1) .05 (2) .38 (3) .65 (4) 0

15-6 THE PROBABILITY OF *NOT A*; PROBABILITY AS A SUM

The Probability of *Not A*

In rolling a fair die, let A represent rolling the number 4. We know that $P(A) = P(4) = \frac{1}{6}$ since there is only one outcome for this event.

Since it is *certain* that we roll a 4 or do not roll a 4, we can say:

$$P(4) + P(\text{not getting } 4) = 1$$

Hence, by subtracting, $P(\text{not getting } 4) = 1 - P(4)$

$$= 1 - \frac{1}{6} = \frac{5}{6}$$

The event ***not A*** is seen as the ***complement*** of set A, namely $\overline{A}$. Therefore, we may also say:

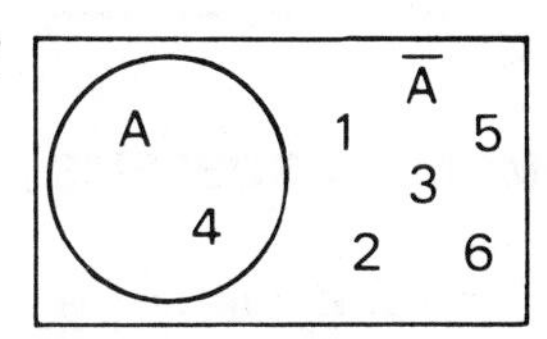

$$P(\textit{not } A) = P(\overline{A}) = 1 - P(A) = 1 - \frac{1}{6} = \frac{5}{6}$$

A Rule for the Probability of *Not A*

In general, if $P(A)$ is the probability that some given result will occur, and $P(\textit{not } A)$ is the probability that result will not occur, then:

$$\mathbf{P(A) + P(\textit{not } A) = 1} \quad \textit{or} \quad \mathbf{P(A) = 1 - P(\textit{not } A)} \quad \textit{or} \quad \mathbf{P(\textit{not } A) = 1 - P(A)}$$

Probability as a Sum

When sets are disjoint, you have seen that the probability of a union can be found by the rule, $P(A \cup B) = P(A) + P(B)$. Since the possible outcomes that are singletons represent disjoint sets, we can say:

- **The probability of any event is equal to the sum of the probabilities of the singleton outcomes in the event.**

For example, when you draw a card from a standard deck, there are 52 singleton outcomes, each with a probability of $\frac{1}{52}$. Since all singleton events are disjoint, we can say:

$$P(\text{king}) = P(\text{king of hearts}) + P(\text{king of diamonds}) + P(\text{king of spades}) + P(\text{king of clubs})$$

$$P(\text{king}) = \frac{1}{52} + \frac{1}{52} + \frac{1}{52} + \frac{1}{52}$$

$$P(\text{king}) = \frac{4}{52}, \text{ or } \frac{1}{13}$$

We also say:

- **The sum of the probabilities of all possible singleton outcomes for any sample space must always equal 1.**

For example, in tossing a coin, $P(S) = P(\text{head}) + P(\text{tail}) = \frac{1}{2} + \frac{1}{2} = 1$. Also, in rolling a die, $P(S) = P(1) + P(2) + P(3) + P(4) + P(5) + P(6)$

$$= \frac{1}{6} + \frac{1}{6} + \frac{1}{6} + \frac{1}{6} + \frac{1}{6} + \frac{1}{6} = 1.$$

KEEP IN MIND

Event ***not A*** consists of outcomes that are not true for event A. Event *not A* may be regarded as the *complement of set A*, namely $\overline{A}$.

MODEL PROBLEMS

1. A fair die is tossed. Find the probability of not rolling a number less than 5.

Solution: Rolling a number less than 5 can be done in four ways: 1, 2, 3, and 4. Hence, $P(\text{less than } 5) = \frac{4}{6}$, and we can say:

$P(\text{not less than } 5) = 1 - P(\text{less than } 5) = 1 - \frac{4}{6} = \frac{2}{6}$, or $\frac{1}{3}$ *Ans.*

2. A letter is drawn at random from the letters in the word ERROR.

a. Find the probability of drawing each of the letters used in the word.

b. Demonstrate that the sum of these probabilities is 1.

Solution:

a. $P(E) = \frac{1}{5}$; $P(R) = \frac{3}{5}$; $P(O) = \frac{1}{5}$

b. $P(E) + P(R) + P(O) = \frac{1}{5} + \frac{3}{5} + \frac{1}{5} = \frac{5}{5} = 1$

EXERCISES

1. A fair die is rolled once. Find the probability that the number is:

a. 3 **b.** not 3 **c.** even **d.** not even **e.** less than 3 **f.** not less than 3 **g.** odd or even **h.** not odd or even

2. The weather bureau predicted a 30% chance of rain. Express in fractional form: **(a)** the probability that it will rain; **(b)** the probability that it will not rain.

3. From a standard deck of cards, one card is drawn. Find the probability that the card will be:

a. a club **b.** not a club **c.** a picture card **d.** not a picture card **e.** not an 8 **f.** not a red 6 **g.** not the queen of spades

4. A bank contains 3 quarters, 4 dimes, and 5 nickels. A coin is drawn at random.

a. Find the probability of drawing: (1) a quarter; (2) a dime; (3) a nickel.

b. Demonstrate that the sum of the three probabilities given as answers in part **a** is 1.

5. One letter is selected at random from the letters in the word PICNICKING.

a. Find the probability of drawing each of the different letters used in the word.

b. Demonstrate that the sum of these probabilities is 1.

6. If the probability of an event happening is $\frac{1}{7}$, what is the probability of that event not happening?

7. If the probability of an event happening is .093, what is the probability of that event not happening?

8. A square dart board whose side measures 30 inches has at its center a shaded square region whose side measures 10 inches. If darts directed at the board are equally likely to land anywhere on the board, what is the probability that a dart does not land in the shaded region?

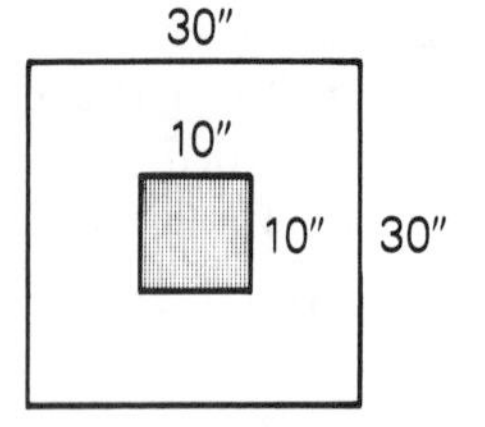

General Exercises

9. A jar contains 7 marbles, all the same size: 3 are red and 4 are green. If a marble is chosen at random, find the probability that it is:
a. red **b.** green **c.** not red **d.** red or green **e.** red and green

10. A box contains three times as many black marbles as green, all the same size. If a marble is drawn at random, find the probability that it is:
a. black **b.** green **c.** not black **d.** black or green **e.** not green

11. The mail contained 2 letters, 3 bills, and 5 ads. Mr. Jacobsen picked up the first piece of mail without looking at it. Express, in decimal form, the probability that this piece of mail is:
a. a letter **b.** a bill **c.** an ad
d. a letter or an ad **e.** a bill or an ad **f.** not a bill
g. not an ad **h.** a bill and an ad

12. A letter is chosen at random from the word PROBABILITY. Find the probability that the letter chosen is:
a. A **b.** B **c.** C
d. A or B **e.** A or I **f.** a vowel
g. not a vowel **h.** A or B or L **i.** A or not A

13. A single card is drawn at random from a well-shuffled deck of 52 cards. Find the probability that the card is:
a. a six **b.** a club **c.** six of clubs
d. a six or a club **e.** not a club **f.** not a six
g. six or seven **h.** not the six of clubs
i. a six and a seven **j.** a black six
k. a six or a black card

14. A telephone dial contains the ten digits: 0, 1, 2, 3, 4, 5, 6, 7, 8, 9. Mabel is dialing a friend. Find the probability that the last digit in the telephone number is:
a. 6 **b.** 6 or more **c.** less than 6 **d.** 6 or odd
e. 6 or less **f.** not 6 **g.** 6 and odd **h.** not more than
i. less than 2 and more than 6 **j.** less than 2 or more than 6
k. less than 6 and more than 2 **l.** less than 6 or more than 2

15-7 THE COUNTING PRINCIPLE AND SAMPLE SPACES

So far, we have looked at simple problems involving a *single* activity, such as the roll of one die or choosing one card. More realistic problems occur when there are *two or more* activities, such as rolling two dice or holding a hand of five cards. Before studying the probability of such events, let us study an easy way to count the number of elements in a sample space when two or more activities are involved. For example:

> A store offers 5 flavors of ice cream: vanilla, chocolate, strawberry, peach, and raspberry. A sundae can be made with either a hot fudge topping or a marshmallow topping. If a sundae consists of one flavor of ice cream and one topping, how many different sundaes are possible?

Let us use initials to represent the 5 flavors of ice cream (V, C, S, P, R) and the 2 toppings (F, M). We can show the number of elements in the sample space in three ways:

(1) The ***tree diagram*** at the right first branches out to show 5 flavors of ice cream. For each of these flavors the tree continues to branch out to show the 2 toppings. At the far right, there are 10 paths or branches to follow, each with one flavor of ice cream and one topping. These 10 branches show that the sample space consists of 10 possible sundaes.

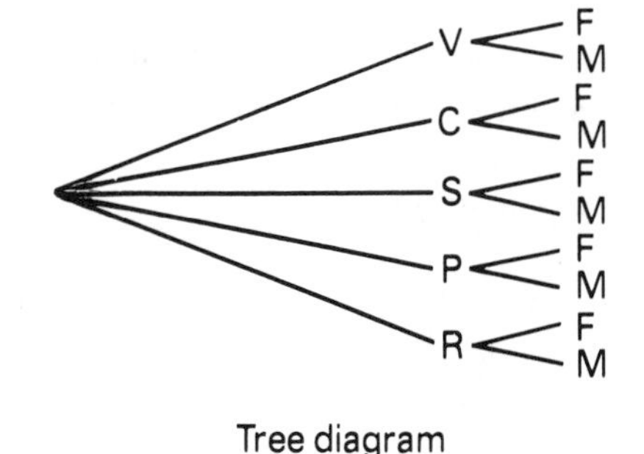

Tree diagram

(2) It is usual to order a sundae by telling the clerk the flavor of ice cream and the type of topping. This suggests a ***listing of ordered pairs***. The first component of the ordered pair is the ice-cream flavor, and the second component is the type of topping. The set of pairs (ice cream, topping) is shown above.

$$\left\{ \begin{matrix} (V, F), & (C, F), & (S, F), & (P, F), & (R, F), \\ (V, M), & (C, M), & (S, M), & (P, M), & (R, M) \end{matrix} \right\}$$

List of ordered pairs

These 10 ordered pairs show that the sample space consists of 10 possible sundaes.

(3) Instead of listing ordered pairs, we may construct a ***graph of the ordered pairs***. At the right, the 5 flavors of ice cream appear on a horizontal scale or line, and the 2 toppings appear on a vertical line.

F	•	•	•	(•)	•
M	•	•	•	•	•
	V	C	S	P	R

Graph of ordered pairs

Each point in the graph represents an ordered pair. For example, the point circled shows the ordered pair (P, F), or (peach ice cream, fudge topping). This graph of 10 points, or 10 ordered pairs, shows that the sample space consists of 10 possible sundaes.

Whether using a tree diagram, a list of ordered pairs, or a graph of ordered pairs, recognize that the sample space consists of 10 sundaes. The number of elements in the sample space can be found by multiplication:

$$\underbrace{\text{number of ice-cream flavors}}_{5} \times \underbrace{\text{number of toppings}}_{2} = \underbrace{\text{number of possible sundaes}}_{10}$$

Suppose the store offered 30 flavors of ice cream and 7 possible toppings. To find the number of elements in the sample space, we may say $30 \times 7 = 210$ possible sundaes. This simple multiplication procedure is known as the ***counting principle***, because it helps us to count the number of elements in a sample space.

● **The Counting Principle: If one activity can occur in any of m ways and, following this, a second activity can occur in any of n ways, then both activities can occur in the order given in $m \cdot n$ ways.**

We can extend this rule to include three or more activities by extending the multiplication process. We can also display three or more activities by extending the branches on a tree diagram, or by listing ordered elements such as ordered triples and ordered quadruples.

For example, a coin is tossed three times in succession.

On the first toss,
the coin may fall in either of 2 ways: a head or a tail.
On the second toss, the coin may also fall in either of 2 ways.
On the third toss, the coin may still fall in either of 2 ways.

By the counting principle, the sample space contains $2 \cdot 2 \cdot 2$, or 8, possible outcomes.

By letting ***H*** represent a head and ***T*** represent a tail, we can illustrate the sample space by a tree diagram or by a set of ordered triples, both shown on the next page. Observe that each entry in the list of triples corresponds to a branch in the tree diagram, in the order given.

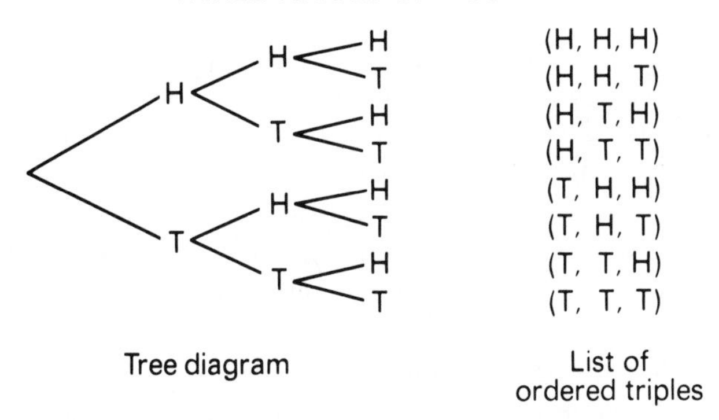

Notice that we did *not* attempt to draw a graph of this sample space since we would need a horizontal scale, a vertical scale, and a third scale making the graph three-dimensional. Although such a graph can be drawn, it is too difficult for you at this time. We can conclude:

1. *Tree diagrams*, or *lists of ordered elements*, are effective ways to indicate any compound event of two or more activities.
2. *Graphs should be limited to ordered pairs*, or only to those events consisting of exactly two activities.

MODEL PROBLEMS

1. The school cafeteria offers 4 types of salads, 3 types of beverages, and 5 types of desserts. If a lunch consists of 1 salad, 1 beverage, and 1 dessert, how many possible lunches can be chosen?

Solution: By the counting principle, we multiply the number of possibilities for each choice:

$$4 \cdot 3 \cdot 5 = 12 \cdot 5 = 60$$

This sample space consists of 60 possible lunches. *Ans.*

2. There are 12 staircases going from the first floor to the second in our school. Roger goes up one staircase and then goes down a different staircase. How many possible ways can this event occur?

Solution: Roger can choose any of 12 staircases going up. Since he goes down a different staircase, he now has 11 choices left for going down. By applying the counting principle, we see that there are $12 \cdot 11 = 132$ ways. *Ans.*

EXERCISES

1. Tell how many possible outfits consisting of one shirt and one pair of pants Terry can choose if he owns:
   a. 5 shirts, 2 pairs of pants **b.** 10 shirts, 4 pairs of pants
   c. 6 shirts, an equal number of pairs of pants
2. There are 10 doors into the school and 8 staircases from the first floor to the second. How many possible ways are there for a student to go from outside the school to a classroom on the second floor?
3. A tennis club has 15 members, 8 women and 7 men. How many different teams may be formed consisting of 1 woman and 1 man on each team?
4. A dinner menu lists 2 soups, 7 meats, and 3 desserts. How many different meals consisting of 1 soup, 1 meat, and 1 dessert are possible?
5. There are 3 ways to go from town A to town B. There are 4 ways to go from town B to town C. How many different ways are there to go from town A to town C, passing through town B?
6. The school cafeteria offers the menu shown.

Main Course	Dessert	Drink
Pizza	Ice cream	Milk
Frankfurter	Cookies	Juice
Ham sandwich	Jello	
Tuna sandwich	Apple pie	
Jelly sandwich		

   a. How many meals consisting of one main course, one dessert, and one drink can be selected from this menu?
   b. Joe hates ham and jelly. How many meals (again, one main course, one dessert, and one drink) can Joe select, not having ham and not having jelly?
   c. JoAnn is at the end of the lunch line. The pizza, frankfurters, ice cream, and cookies have been sold out. How many menus can JoAnn select?
7. A quarter and a penny are tossed simultaneously. Each coin may fall heads or tails. The tree diagram shows the sample space involved. **a.** List the sample space as a set of ordered pairs. **b.** Use the counting principle to demonstrate that there are four outcomes in the sample space. **c.** In how many outcomes do the coins both fall heads up? **d.** In how many outcomes do the coins land showing one head and one tail?

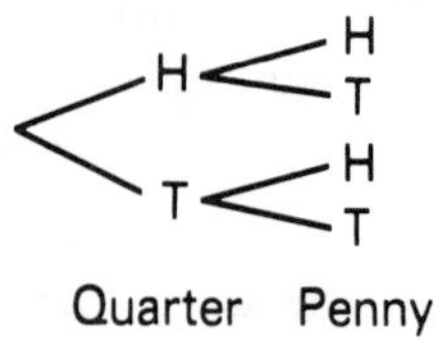

H — H, T
T — H, T
Quarter Penny

8. A teacher gives a quiz consisting of two questions. Each question has as its answer either true or false. **a.** Using T and F, draw a tree diagram to show all possible ways the questions can be answered. **b.** List this sample space as a set of ordered pairs. **c.** State the relationship that exists between this sample space and the p and q columns in a table of truth values for two statements developed in the study of logic.

9. A quiz consists of three true-false questions. **a.** How many possible ways are there to answer the questions on this test? **b.** Draw a tree diagram to show the sample space. **c.** List the ordered triples to show this sample space.

10. Elizabeth has a number of possible routes to school. She can take either A Street or B Street, and then turn onto Avenue X, Avenue Y, or Avenue Z. **a.** Draw a tree diagram to show all the possible routes. **b.** List the ordered pairs to show this sample space.

11. A test consists of multiple-choice questions. Each question has 4 choices. Tell how many possible ways there are to answer the questions on the test if the test consists of:
    a. 1 question **b.** 3 questions **c.** n questions

12. Options on a bicycle include 2 types of handlebars, 2 types of seats, and a choice of 15 colors. The bike may also be ordered in ten-speeds, in three-speeds, or standard. How many possible versions of a bicycle can customers choose from, if they select the type of handlebars, seat, color, and speed?

13. Two dice are rolled simultaneously. Each die may land with one of six numbers face up. **a.** Use the counting principle to determine the number of outcomes in this sample space. **b.** Display the sample space by constructing a graph of the set of ordered pairs.

14. A state issues license plates consisting of letters and numbers. There are 26 letters and the letters may be repeated in a plate; there are 10 digits and the digits may be repeated. Tell how many possible license plates the state may issue when a license consists of:
    a. two letters, followed by three numbers
    b. three numbers, followed by three letters
    c. four numbers, followed by two letters
    (*Note:* The license 1-ID is actually 0001-ID.)

15. An ice-cream company offers 31 different flavors. Hilda orders a double-scoop cone. In how many different ways can the clerk put the ice cream on the cone if: **a.** Hilda wanted two different flavors? **b.** Hilda wanted the same flavor on both scoops? **c.** Hilda could not make up her mind and told the clerk, "Anything at all"?

15-8 PROBABILITIES AND THE COUNTING PRINCIPLE; PREDICTING OUTCOMES

We know that the probability of rolling 1 on the single toss of a die is $\frac{1}{6}$, or $P(1) = \frac{1}{6}$. What is the probability of rolling a pair of ones when two dice are tossed?

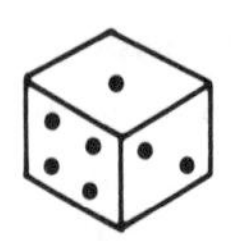
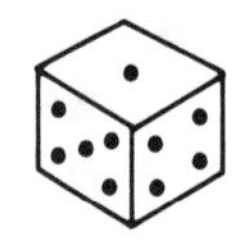

When we roll two dice, the number obtained on one die is completely, absolutely, without question, *independent* of the result obtained on the second die.

When the result of one activity in no way influences the result of a second activity, these activities are ***independent events***. In cases where two activities are independent, we may extend the counting principle to find the probability that both independent events occur at the same time. For example:

$$P(1 \text{ on first die}) = \frac{1}{6}$$

$$P(1 \text{ on second die}) = \frac{1}{6}$$

$$P(1 \text{ on both dice}) = P(1 \text{ on first}) \cdot P(1 \text{ on second}) = \frac{1}{6} \cdot \frac{1}{6} = \frac{1}{36}$$

When an event consists of two or more independent events or activities, it is called a ***compound event***. Compound events are illustrated when we are rolling 2 dice, or tossing 3 coins, or spinning an arrow 3 times in succession.

We can extend the counting principle to help us find the probability of any compound event consisting of two or more independent events.

- **The Counting Principle for Probability:**

When E and F are independent events, and when the probability of event E is m ($0 \leq m \leq 1$) and the probability of event F is n ($0 \leq n \leq 1$), then the probability of the compound event in which E and F occur jointly is the product $m \cdot n$.

Note 1: The product $m \cdot n$ is within the range of values for a probability, namely ($0 \leq m \cdot n \leq 1$).

Note 2: Not all events are independent. Hence, this simple product rule cannot be used to find the probability of every compound event.

MODEL PROBLEM

Mr. Gillen may take any of three buses to get to the same train station. The buses are marked A or B or C. He may then take the 6th Avenue train or the 8th Avenue train to get to work. The buses and trains arrive at random and are equally likely to arrive. What is the probability that Mr. Gillen takes the B bus and the 6th Avenue train to get to work?

Solution

$P(B \text{ bus}) = \frac{1}{3}$ and $P(\text{6th Ave. train}) = \frac{1}{2}$. Since the train taken is independent of the bus taken:

$$P(B \text{ bus and 6th Ave. train}) = P(B \text{ bus}) \cdot P(\text{6th Ave. train})$$

$$= \frac{1}{3} \cdot \frac{1}{2} = \frac{1}{6} \quad \textit{Ans.}$$

Predicting Outcomes

It is often useful to use past experience to predict what to expect in the future. For example, insurance companies establish their rates based on expectancies. The branch of biology known as genetics has established certain probability ratios that are useful to plant and animal breeders. In daily life, we decide how much we will buy of a particular item based on how much we have used in the past.

MODEL PROBLEM

The owner of a garden nursery knows that the color red will occur in a certain plant once out of about 15 seedlings. If the nursery wants to have about 200 red seedlings of this plant to sell, how many seedlings of this variety should be planted?

Solution: Set up a proportion based on the established ratio:

$$\frac{\text{number of red}}{\text{number planted}}$$

Let x = the number of seedlings to plant.

$$\frac{1}{15} = \frac{200}{x}$$

$$x = 3{,}000$$

Answer: To have about 200 red seedlings, the nursery must plant 3,000 seedlings.

EXERCISES

(All the events described in Exercises 1–16 are independent events.)

1. A fair coin and a six-sided die are tossed simultaneously. What is the probability of obtaining:
   a. a head on the coin?
   b. a 4 on the die?
   c. a head on the coin and a 4 on the die jointly?
2. A fair coin and a six-sided die are tossed simultaneously. What is the probability of obtaining jointly:
   a. a head and a 3?
   b. a head and an even number?
   c. a tail and a number less than 5?
   d. a tail and a number more than 4?
3. Two fair coins are tossed. What is the probability that both land heads up?
4. When I enter school, 3 out of 4 times I use the main door. When I leave school, I use the main door only 1 out of 3 times. On any given day, what is the probability that I both enter and leave by the main door?
5. In our school cafeteria the menu rotates so that $P(\text{hamburger}) = \frac{1}{4}$, $P(\text{apple pie}) = \frac{2}{3}$, and $P(\text{soup}) = \frac{4}{5}$. On any given day, what is the probability that the cafeteria offers hamburger, apple pie, and soup on the same menu?
6. A quiz consists of true-false questions only. Harry has not studied, and he guesses every answer. Find the probability that he will guess correctly to get a perfect paper if the test consists of:
   a. 1 question **b.** 4 questions **c.** n questions
7. The probability of the Tigers beating the Cougars is $\frac{2}{3}$. The probability of the Tigers beating the Mustangs is $\frac{1}{4}$. If the Tigers play one game with the Cougars and one game with the Mustangs, find the probability of the Tigers:
   a. winning both games **b.** losing both games
8. Three fair coins are tossed.
   a. Find $P(H, H, H)$. **b.** Find $P(T, T, T)$.
9. A fair spinner contains equal regions, numbered 1 through 8. If the arrow is spun twice, find the probability that:
   a. it lands on 7 both times **b.** it does not land on 7 either time

10. As shown in the diagram, 6th Avenue runs north and south. The lights are not timed to accommodate traffic traveling along 6th Avenue; they are independent of one another. At each of the intersections shown, $P(\text{red light}) = .7$ and $P(\text{green light}) = .3$ for cars traveling along 6th Avenue.

Find the probability that a car traveling north on 6th Avenue will be faced with the given conditions at the two traffic lights shown.

a. both lights are red
b. the first is red and the second is green
c. both lights are green
d. the first is green and the second is red
e. one of the two lights is red
f. both lights are the same color

11. A manufacturer of radios knows that the probability of a defect in any of his products is $\frac{1}{400}$. If 10,000 radios are manufactured in January, how many are likely to be defective?

12. Past records from the weather bureau indicate that it has rained 2 out of every 7 days in August on Cape Cod. If Joan goes to Cape Cod for two weeks in August, how many days will it probably rain if the records hold true?

13. In order to advertise a new store, chances on gift certificates for the store are to be given away. If the store owner wants the probability of winning to be $\frac{1}{50}$, how many of the 1,000 chances that are given away should be marked as winners?

14. Dr. Velez knows that she recommends a flu shot for 20% of her patients, and that an additional 15% elect to take a flu shot. If she expects to work with 500 patients during this flu season, how many flu shots will she probably administer?

15. About how many times can you expect to turn up a head in 100 consecutive tosses of a coin?

16. About how many times can you expect to roll a 5 in 100 consecutive rolls of a single standard die?

15-9 PROBABILITIES WITH TWO OR MORE ACTIVITIES

The counting principle, involving simple probabilities, works for a limited number of situations, those in which the events are independent. We will now examine a procedure that can be used to find the probability of any compound event.

For example, a family moves in next door. We have heard that they have three children, but we do not know how many are boys and how many are girls. The counting principle tells us that the sample space consists of $2 \cdot 2 \cdot 2$, or 8, possibilities for the sex of the three children in this family.

By letting G represent a girl and B represent a boy, we can illustrate the sample space by a tree diagram, or by a set of ordered triples, both shown below.

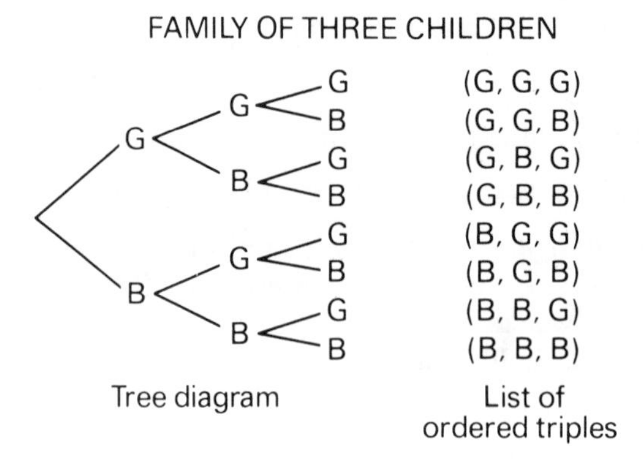

The Probability of Different Events

To find the probability of any event, you must know two values:

1. the number of ways in which event E can occur, or $n(E)$;
2. the total number of possible outcomes in sample space S, or $n(S)$.

You have just seen that the sample space for a family of three children contains 8 ordered triples, or $n(S) = 8$. Let us consider the probabilities of some possible events for this sample space.

What is the probability that the family contains two girls and one boy?

By examining the 8 outcomes in the sample space, we see that this event can happen in 3 possible ways:

$$(G, G, B) \qquad (G, B, G) \qquad (B, G, G)$$

Therefore: $n(E) = 3$, and $P(2 \text{ girls and } 1 \text{ boy}) = \dfrac{n(E)}{n(S)} = \dfrac{3}{8}$

What is the probability that the family contains at least one boy?

When we say *at least one*, there may be 1 boy or 2 boys or 3 boys. By examining the sample space of 8 outcomes, we see that there are 7 possible ways for this event to happen: (G, G, B); (G, B, G); (G, B, B); (B, G, G); (B, G, B); (B, B, G); (B, B, B).

Therefore: $n(E) = 7$, and $P(\text{at least one boy}) = \dfrac{n(E)}{n(S)} = \dfrac{7}{8}$

Note: Since *at least one boy* has the same meaning as *not all girls*, it is also correct to use the following alternate approach to the problem:

$P(\text{at least one boy}) = P(\text{not all girls}) = 1 - P(\text{all girls}) = 1 - \frac{1}{8} = \frac{7}{8}$

Procedure for Finding Probabilities of Compound Events:

1. List the sample space by constructing a tree diagram or a set of ordered elements.
2. Count the number of elements in the sample space, $n(S)$.
3. For the event E being described, count the number of elements from the sample space that are elements of the event, $n(E)$.
4. Substitute these numbers in the rule for the probability of an event E, namely $P(E) = \dfrac{n(E)}{n(S)}$.

MODEL PROBLEMS

1. A fair coin is tossed two times in succession.

a. List the sample space by using: (1) a tree diagram (2) a set of ordered pairs (3) a graph of ordered pairs

b. Find the probability of each event: (1) Event A = the coin is heads both times (2) Event B = 1 head and 1 tail are tossed

Solution

a. (1) Tree diagram

H — H, T
T — H, T

(2) Set of ordered pairs

$\{(H, H), (H, T), (T, H), (T, T)\}$

(3) Graph of ordered pairs

	T	H
H	•	•
T	•	•

b. (1) Event A consists of one pair, heads on both tosses: (H, H)

Since $n(A) = 1$, and $n(S) = 4$: $P(A) = \dfrac{n(A)}{n(S)} = \dfrac{1}{4}$ *Ans.*

(2) Event B consists of two pairs with 1 head and 1 tail:
$(H, T)\ (T, H)$

Since $n(B) = 2$, and $n(S) = 4$: $P(B) = \dfrac{n(B)}{n(S)} = \dfrac{2}{4}$ *Ans.*

2. In an experiment, the first step is to pick one number from the set $\{1, 2, 3\}$. The second step of the experiment is to pick one number from the set $\{3, 5\}$.

a. Draw a tree diagram or list the sample space of all possible pairs that are outcomes.

b. Determine the probability that:
(1) both numbers are the same
(2) the sum of the numbers is even
(3) the first number is larger than the second

Solution

a.

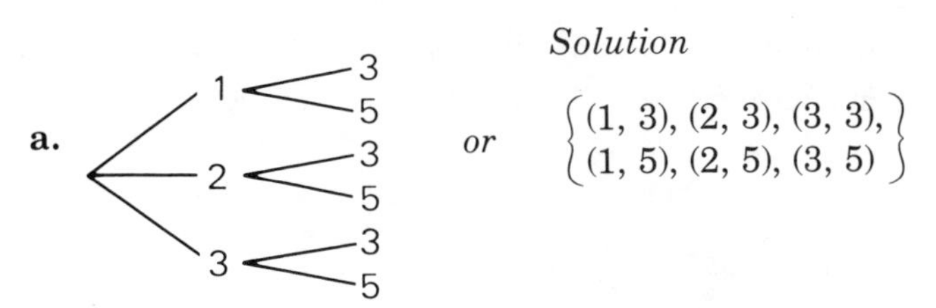

$\left\{\begin{matrix}(1, 3), (2, 3), (3, 3),\\ (1, 5), (2, 5), (3, 5)\end{matrix}\right\}$

b. (1) Event A consists of only one pair with both numbers the same: (3, 3)

Since $n(A) = 1$, and $n(S) = 6$: $P(A) = \dfrac{n(A)}{n(S)} = \dfrac{1}{6}$ *Ans.*

(2) Event B consists of four pairs with an even sum:
(1, 3) (1, 5) (3, 3) (3, 5)

Since $n(B) = 4$, and $n(S) = 6$: $P(B) = \dfrac{n(B)}{n(S)} = \dfrac{4}{6}$ *Ans.*

(3) Event C is the empty set since there are no pairs in which the first number is larger than the second. Since $n(C) = 0$:

$P(C) = \dfrac{n(C)}{n(S)} = \dfrac{0}{6} = 0$ *Ans.*

3. Two standard dice are rolled. Find the probability that the sum of the numbers on the dice is 8.

Solution

The sample space consists of $6 \cdot 6$, or 36, outcomes, as shown by a graph of ordered pairs.

For this event E, there are 5 ordered pairs in which the sum of the numbers on the dice is 8. These 5 pairs, encircled on the graph, are: (2, 6), (3, 5), (4, 4), (5, 3), (6, 2)

Since $n(S) = 36$, and $n(E) = 5$:

$$P(E) = \frac{n(E)}{n(S)} = \frac{5}{36} \quad Ans.$$

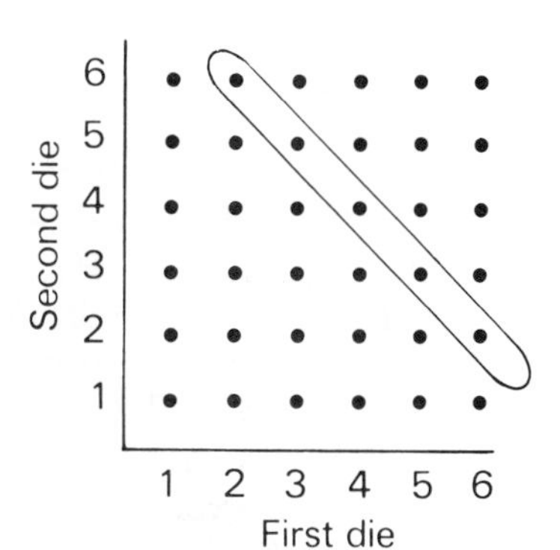

EXERCISES

1. Two fair coins are tossed simultaneously. **a.** Draw a tree diagram or list the sample space of all possible pairs of outcomes. **b.** Find *P*(both coins are tails). **c.** Find *P*(no tails). **d.** Find *P*(at least one coin is a head).

2. In a family of two children, determine the following probabilities:
a. both are boys
b. both are girls
c. there is one boy and one girl
d. both are of the same sex
e. there is at least one girl
f. the younger is a boy
g. the older is a girl and the younger is a boy

3. In a family of three children, determine the probability that:
a. all are boys
b. all are girls
c. all are of the same sex
d. exactly two are boys
e. the youngest is a girl
f. there is at least one boy
g. the oldest and the youngest are both girls

4. Draw a tree diagram or list the ordered elements of the sample space to indicate the possible sexes of four children in a family. Then reanswer the questions in **3(a–g)** for the sample space of *four* children.

5. Three fair coins are tossed simultaneously.
a. Indicate the sample space as a tree diagram or as a set of ordered triples.
b. Find *P*(all are tails).
c. Find *P*(there are exactly 2 tails).
d. Find *P*(there are at least 2 tails).

6. In a certain game, darts are thrown at two boards so that *each board* will contain exactly *one dart.* If any darts miss or land on a line, they are not counted and the person tries again.

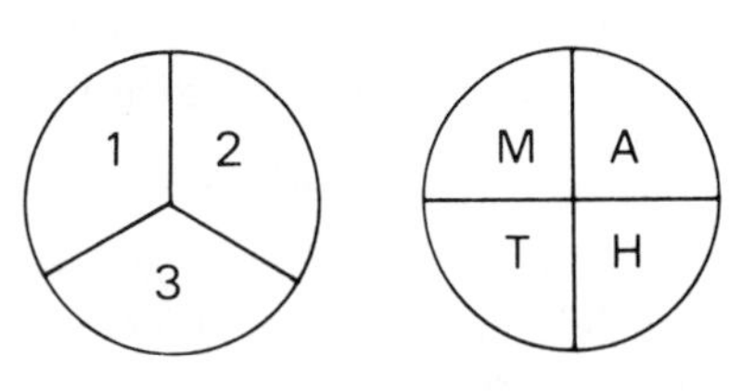

The first board contains three equally likely regions numbered {1, 2, 3} and the second board contains four equally likely regions lettered {M, A, T, H}.

a. Draw a tree diagram or list the sample space of all possible pairs that are outcomes for placing one dart on each board.
b. Find $P(2, M)$, the probability of obtaining the result $(2, M)$.
c. Find $P(3, T)$. **d.** Find $P(\text{odd number}, H)$.
e. Find $P(\text{even number, vowel})$. **f.** Find $P(5, \text{vowel})$.
g. Find $P(\text{odd number, consonant})$.

7. Two standard dice are rolled. To describe the sample space: **a.** Draw a tree diagram. **b.** List the ordered pairs.

8. Two standard dice are rolled. When the numbers on the dice are added, the smallest possible sum is 2, from the pair (1, 1). The largest possible sum is 12, from the pair (6, 6). Find the probability of rolling two dice to get a sum of:
a. 2 **b.** 3 **c.** 4 **d.** 5 **e.** 6 **f.** 7 **g.** 8 **h.** 9 **i.** 10 **j.** 11 **k.** 12

9. What is the sum of all the probabilities obtained in Exercise 8?

10. One standard die is rolled twice. Verify that the sample space is the same as when two standard dice are rolled once.

11. One standard die is rolled twice. Find the probability of getting, on two rolls of the die:
a. the pair (3, 3) **b.** the pair (5, 2) **c.** the pair (7, 1)
d. a pair of identical even numbers
e. a pair of identical odd numbers
f. a pair whose sum is even **g.** a pair whose sum is odd
h. a sum less than 5 **i.** a sum less than 15

In 12–14, select the choice that best answers the question.

12. When a coin and a standard die are tossed simultaneously, the number of outcomes in the sample space is:
(1) 8 (2) 2 (3) 12 (4) 36

13. A spinner shows three regions, numbered {1, 2, 3}, all equally likely to occur. When the arrow is spun twice, the number of pairs in the outcome set is:
(1) 6 (2) 2 (3) 3 (4) 9

14. Two coins and a standard die are tossed simultaneously. The number of outcomes in the sample space is:
(1) 10 (2) 24 (3) 3 (4) 8

15-10 PERMUTATIONS

Mrs. Hendrix, a teacher, has announced that she will call upon three students of her class to give oral reports today. The students are Al, Betty, and Chris. How many possible ways are there for Mrs. Hendrix to choose the order in which these students will give their reports?

Let us use a tree diagram to picture the possible orders. From the diagram, we see that there are 6 possible arrangements. For example, Al, Betty, Chris is one possible arrangement; Al, Chris, Betty is another possible arrangement. Each of these arrangements is called a ***permutation***. A permutation is an arrangement of objects in some specific order. By letting A represent Al, B represent Betty, and C represent Chris, we may show the 6 possible permutations as a set of ordered triples:

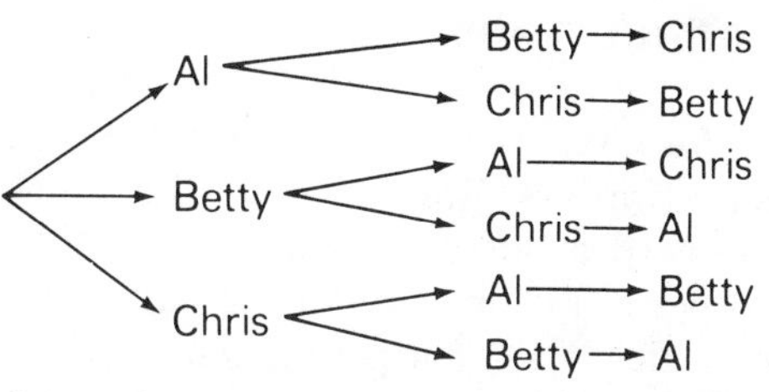

$$\{(A, B, C), (A, C, B), (B, A, C), (B, C, A), (C, A, B), (C, B, A)\}$$

Let us see, from another point of view, why there are 6 possible orders. We know that any one of 3 students can be called to give the first report. Once the first report is given, the teacher may call on any one of the 2 remaining students. After the second report is given, the teacher must call the 1 remaining student. Using the counting principle, there are $3 \cdot 2 \cdot 1$, or 6 possible orders.

Consider another situation. A chef is preparing a recipe with 10 ingredients. The chef puts all of one ingredient into a bowl, followed by all of another ingredient, and so on. How many possible orders are there for placing the 10 ingredients into a bowl, using the stated procedure? By using the counting principle, we have:

$10 \cdot 9 \cdot 8 \cdot 7 \cdot 6 \cdot 5 \cdot 4 \cdot 3 \cdot 2 \cdot 1$, or 3,628,800 possible ways

Factorials

If there are more than 3 million possible ways of placing 10 ingredients into a bowl, can you imagine in how many ways 300 people who want to buy tickets for a football game can be arranged in a straight line? Using the counting principle, we would have the 300 factors $300 \cdot 299 \cdot 298 \cdot 297 \ldots 3 \cdot 2 \cdot 1$. To deal with such an example, we make use of a ***factorial symbol*** !. We represent the product of these 300 numbers by the symbol 300!, read *three hundred factorial* or *factorial 300*.

In general, for any natural number n, we define n factorial or factorial n as follows:

$$n! = n(n - 1)(n - 2)(n - 3) \ldots 3 \cdot 2 \cdot 1$$

Note that 1! is the natural number 1.

Permutations

We have said that permutations are arrangements of objects in different orders. For example, the number of different ways for 4 people to board a bus can be shown as 4!, or $4 \cdot 3 \cdot 2 \cdot 1$, or 24. There are 24 permutations, that is, 24 different arrangements of these 4 people, in which all the 4 people get on the bus.

We also represent this number of permutations by the symbol ${}_4P_4$, read as *the permutation of four objects taken four at a time.* In ${}_4P_4$:

The letter P represents the word *permutation.*

The small $_4$ written to the lower left of P tells us there are 4 objects available to be used in an arrangement, as in 4 waiting for a bus.

The small $_4$ written to the lower right of P tells us how many of these objects are to be used in each arrangement, as in all 4 getting on the bus.

Thus, ${}_4P_4 = 4! = 4 \cdot 3 \cdot 2 \cdot 1 = 24$.

Similarly, ${}_5P_5 = 5! = 5 \cdot 4 \cdot 3 \cdot 2 \cdot 1 = 120$.

In the next section, you will study examples where not all the objects are used in the arrangement. For now, let us make the following observation:

● **For any natural number n, the number of permutations of n objects taken n at a time can be represented as:**

$$ {}_nP_n = n! = n(n - 1)(n - 2) \ldots 3 \cdot 2 \cdot 1$$

MODEL PROBLEMS

1. Compute the value of each expression.

a. $6!$ **b.** $\dfrac{7!}{3!}$ **c.** ${}_2P_2$

Solution:

a. $6! = 6 \cdot 5 \cdot 4 \cdot 3 \cdot 2 \cdot 1 = 720$

b. $\dfrac{7!}{3!} = \dfrac{7 \cdot 6 \cdot 5 \cdot 4 \cdot \cancel{3 \cdot 2 \cdot 1}}{\cancel{3 \cdot 2 \cdot 1}} = 840$

c. ${}_2P_2 = 2! = 2 \cdot 1 = 2$

2. Let any arrangement of letters be called a word even if it has no meaning. Consider the letters in {N, O, W}:

   a. How many three-letter words can be formed if each letter is used only once in the word?

   b. List the words.

   Solution: **a.** Because we are arranging letters in different orders, this is a permutation. Thus, ${}_3P_3 = 3! = 3 \cdot 2 \cdot 1 = 6$ possible words.

   b. NOW; NWO; ONW; OWN; WNO; WON.

3. Paul wishes to call Virginia, but he has forgotten her unlisted telephone number. He knows that the exchange is 555, and he knows that the last four digits are 1, 4, 7, and 9, but he cannot remember their order. What is the maximum number of telephone calls that Paul may have to make in order to dial the correct number?

   Solution: The telephone number is 555-__ __ __ __. Since the last four digits will be an arrangement of 1, 4, 7, and 9, this is a permutation of four numbers, taken four at a time.

   Thus, ${}_4P_4 = 4! = 4 \cdot 3 \cdot 2 \cdot 1 = 24$ possible orders.

   Answer: The maximum number of telephone calls that Paul may have to make is 24.

EXERCISES

1. Compute the value of each expression.

   a. $2!$ **b.** $4!$ **c.** $6!$ **d.** $7!$

   e. $3! + 2!$ **f.** $(3 + 2)!$ **g.** ${}_3P_3$ **h.** ${}_8P_8$

   i. ${}_5P_5$ **j.** $\frac{8!}{5!}$ **k.** $\frac{15!}{15!}$ **l.** $({}_3P_3) \cdot ({}_4P_4)$

2. Using the letters E, M, I, T: **a.** How many words of four letters can be found if each letter is used only once in the word? **b.** List these words.
3. In how many different ways can 5 students be arranged in a row?
4. How many possible three-letter arrangements of the letters X, Y, and Z can be made if each letter is used only once in each arrangement?
5. How many different 4-digit numbers can be made using the digits 2, 4, 6, and 8 if each digit appears only once in each number?

6. In a game of cards, Gary held exactly one club, one diamond, one heart, and one spade. In how many different ways can Gary arrange these four cards in his hand?
7. There are 9 players on a baseball team. The manager must establish a batting order for the players at each game. The pitcher will bat last. How many different batting orders are possible for the 8 remaining players on the team?

In 8–10, numerical answers may be left in factorial form.

8. In how many different ways may 60 people line up to buy tickets at a theater?
9. We learn the alphabet in an order, starting with A, B, C, and going down to Z. How many possible orders are there for saying the letters of the English alphabet?
10. In how many different ways can the librarian put 35 different novels on a shelf, with one book following another?

15-11 MORE ABOUT PERMUTATIONS

At times, we deal with situations involving permutations in which we are given n objects, but we use fewer than n objects in each arrangement. For example:

Mr. Brown has announced that he will call students from the first row to explain homework problems at the board. The students in the first row are George, Helene, Jay, Karla, and Lou. If there are only two homework problems, and each problem is to be explained by a different student, in how many ways may Mr. Brown select students to go to the board?

We know that the first problem can be assigned to any of 5 students. Once this problem is explained, the second problem can be assigned to any of the 4 remaining students. By the counting principle, there are $5 \cdot 4$, or 20, possible selections.

There are 8 basketball players on a team. In how many ways can 3 of them be seated on a bench?

Seat	*First*	*Second*	*Third*
Players to choose from	8	7	6

By the counting principle, the number of possible seating arrangements is $8 \cdot 7 \cdot 6 = 336$. Observe that the first factor 8 is the number of players on the team; once that player is chosen, the next factor is 7, and so on. Each factor is one less than the previous factor since each time a player is seated there is one less possible player for the next seat.

The number of factors, 3, is the number of players in the seating arrangement or the number of seats available.

Using the language of permutations, we say that the number of permutations of 8 different things taken 3 at a time is 336. In symbols:

$$_8P_3 = 8 \cdot 7 \cdot 6 = 336$$

The Symbols for Permutations

In general, if we have a set of n different objects, and we make arrangements of r objects from this set, we represent the number of arrangements by the symbol $_nP_r$. Notice that r, the number of factors being used, must be less than or equal to n, the total number of objects in the set. Thus:

● **For numbers n and r where $r \le n$, the number of permutations of n things, taken r at a time, is found by the formula:**

$$_nP_r = \underbrace{n(n-1)(n-2)\ldots}_{r \text{ factors}}$$

MODEL PROBLEMS

1. Evaluate $_6P_2$.

Solution: This is a permutation of 6 things, taken 2 at a time. We begin with 6 and write 2 factors.

$$_6P_2 = \underbrace{6 \cdot 5}_{2 \text{ factors}} = 30 \quad \textit{Ans.}$$

2. There are 12 horses in a race. Winning horses are those crossing the finish line in first, second, and third place, commonly called win, place, and show. How many possible winning orders are there for a race with 12 horses?

Solution: This is a permutation of 12 things, taken 3 at a time, since there are 3 winning positions in a race. Thus:

$$_{12}P_3 = \underbrace{12 \cdot 11 \cdot 10}_{3 \text{ factors}} = 1{,}320 \text{ possible orders} \quad \textit{Ans.}$$

3. How many 3-letter words can be formed from the letters L, O, G, I, C if each letter is used only once in a word?

Solution: Forming 3-letter words from a set of 5 letters is a permutation of 5 things, taken 3 at a time. Thus:

$$_5P_3 = \underbrace{5 \cdot 4 \cdot 3}_{\text{3 factors}} = 60 \text{ words} \quad \textit{Ans.}$$

4. A lottery ticket contains a four-digit number. How many possible four-digit numbers are there when:
   a. a digit may appear only once in the number?
   b. digits may appear more than once in the number?

Solution

a. If a digit appears only once in a four-digit number, this is a permutation of 10 digits, taken 4 at a time. Hence:

$$_{10}P_4 = 10 \cdot 9 \cdot 8 \cdot 7 = 5{,}040 \quad \textit{Ans.}$$

b. If a digit may appear more than once, we can choose any of 10 digits for the first position, then any of 10 digits for the second position, and so forth. By the counting principle:

$$10 \cdot 10 \cdot 10 \cdot 10 = 10{,}000 \text{ possible four-digit numbers} \quad \textit{Ans.}$$

EXERCISES

1. Evaluate each expression.
   a. $_6P_3$ **b.** $_{10}P_2$ **c.** $_{25}P_2$ **d.** $_4P_3$ **e.** $_{20}P_2$ **f.** $_{11}P_4$
   g. $_{22}P_3$ **h.** $_{10}P_4$ **i.** $_7P_6$ **j.** $_{101}P_3$ **k.** $_8P_5$ **l.** $_6P_6$

2. How many 3-letter words can be formed from the given letters, if each letter is used only once in a word?
   a. LION **b.** TIGER **c.** MONKEY **d.** LEOPARD **e.** MAN

3. There are 30 students in a class. Every day, the teacher calls on different students to write homework problems on the board, with each problem done by only one student. In how many ways can the teacher call students to the board if the homework consists of:
   a. only 1 problem? **b.** 2 problems? **c.** 3 problems?

4. Tell how many possible winning orders there are for a horse race where 3 horses finish in winning positions and there are:
   a. 7 horses **b.** 9 horses **c.** 11 horses **d.** n horses

5. How many different ways are there to label the 3 vertices of the scalene triangle that is shown, using no letter more than once, when:
   a. we use the letters R, S, T?
   b. we use all the letters of the English alphabet?

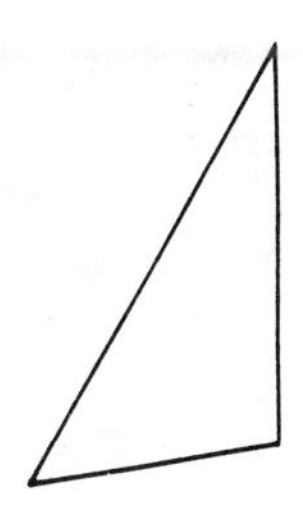

6. A class has 31 students. They elect 4 people to office, namely the President, Vice-President, Secretary, and Treasurer. In how many possible ways can 4 people be elected from this class? (Answer may be left as a series of factors.)

7. How many possible ways are there to write 2 initials, using the letters of the English alphabet, if: **a.** an initial may appear only once in each pair? **b.** the same initial may be used twice?

15-12 PROBABILITY WITHOUT REPLACEMENT; PROBABILITY WITH REPLACEMENT

Without Replacement

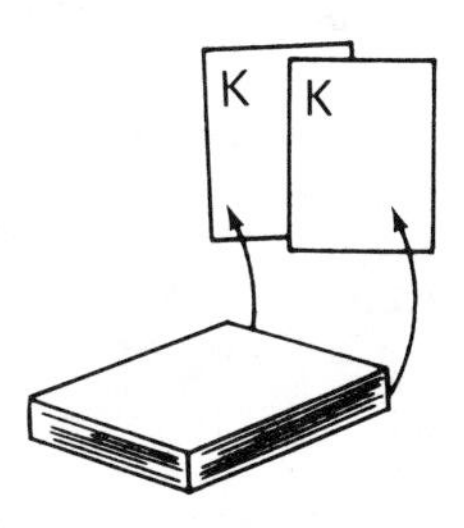

Two cards are drawn at random from an ordinary pack of 52 cards. In this situation, you should understand that a single card is drawn from a deck of 52 cards, and then a second card is drawn from the remaining 51 cards in the deck. What is the probability that both cards drawn are kings?

On the first draw, there are 4 kings in a deck of 52 cards. Hence, $P(\text{first king}) = \frac{4}{52}$. If a second card is drawn without replacing the first king selected, there are now only 3 kings in the deck of 51 cards remaining. Hence, $P(\text{second king}) = \frac{3}{51}$.

By the counting principle:

$$\begin{aligned} P(\text{both kings}) &= P(\text{first king}) \cdot P(\text{second king}) \\ &= \frac{4}{52} \cdot \frac{3}{51} \\ &= \frac{1}{13} \cdot \frac{1}{17} \\ &= \frac{1}{221} \end{aligned}$$

This is called a problem *without replacement* because the first king drawn was not placed back into the deck. Typical problems without replacement include spending coins from your pocket, eating jelly beans from a jar, and choosing students to give reports.

With Replacement

A card is drawn at random from an ordinary deck, placed back into the deck, and a second card is then drawn. In this situation, you should understand that the deck contains 52 cards each time that a card is drawn. What is the probability that each time the card drawn is a king?

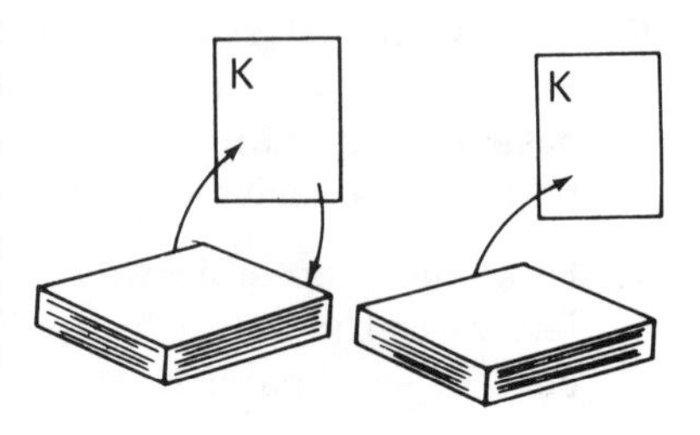

On the first draw, there are 4 kings in a deck of 52 cards. Hence, $P(\text{first king}) = \frac{4}{52}$. If the first king drawn is now placed back into the deck, then on the second draw, there are again 4 kings in a deck of 52 cards. Hence, $P(\text{second king}) = \frac{4}{52}$.

By the counting principle:

$$\begin{aligned} P(\text{both kings}) &= P(\text{first king}) \cdot P(\text{second king}) \\ &= \frac{4}{52} \cdot \frac{4}{52} \\ &= \frac{1}{13} \cdot \frac{1}{13} \\ &= \frac{1}{169} \end{aligned}$$

This is called a problem *with replacement* because the first card drawn was placed back into the deck. Since the card drawn is replaced, the number of cards in the deck remains constant.

Rolling two dice is similar to drawing two cards with replacement because the number of faces on each of the dice remains constant, as did the number of cards in the deck. Typical problems with replacement include rolling dice, tossing coins (each coin always has two sides), and spinning arrows.

KEEP IN MIND

1. If the problem does not specifically mention *with replacement* or *without replacement*, ask yourself: "Is this problem with or without replacement?"
2. For many compound events, the probability can be determined most easily by using the counting principle.
3. Every probability problem can always be solved by: counting the number of elements in the sample space, $n(S)$; counting the number in the event, $n(E)$; and substituting in the probability formula,

$$P(E) = \frac{n(E)}{n(S)}$$

MODEL PROBLEMS

1. A fair die is thrown three times. What is the probability that a 5 comes up each time?

Solution

This is a problem with replacement. On the first toss, there is only one way to obtain a 5 from the six possible outcomes of 1, 2, 3, 4, 5, and 6. Thus: $P(5 \text{ on } \textit{first} \text{ toss}) = \frac{1}{6}$

On the second toss, there is again one way to obtain a 5. So:

$$P(5 \text{ on second toss}) = \frac{1}{6}$$

On the third toss, there is again one way to obtain a 5. So:

$$P(5 \text{ on third toss}) = \frac{1}{6}$$

By the counting principle:

$$P(\text{rolling 5 each time}) = P(5 \text{ on first}) \cdot P(5 \text{ on second}) \cdot P(5 \text{ on third})$$

$$= \frac{1}{6} \cdot \frac{1}{6} \cdot \frac{1}{6}$$

$$= \frac{1}{216} \quad \textit{Ans.}$$

2. If two cards are drawn from an ordinary deck without replacement, what is the probability that the cards form a pair?

Solution

On the first draw, any card at all may be chosen. So:

$$P(\text{any card}) = \frac{52}{52}$$

There are now 51 cards left in the deck. Of these 51, there are 3 that match the first card taken, to form a pair. So:

$$P(\text{second card forms a pair}) = \frac{3}{51}$$

Then: $P(\text{pair}) = P(\text{any card}) \cdot P(\text{second card forms a pair})$

$$= \frac{52}{52} \cdot \frac{3}{51} = \frac{1}{1} \cdot \frac{1}{17} = \frac{1}{17} \quad \textit{Ans.}$$

3. An urn contains 4 white marbles and 2 blue marbles, all the same size. A marble is drawn at random and not replaced. A second marble is then drawn from the urn. Find the probability that:

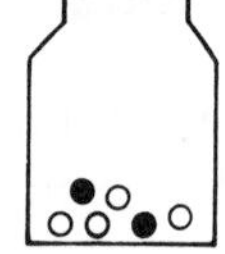

a. both marbles are white
b. both marbles are blue
c. both marbles are the same color

Solution

a. On the first draw, $P(\text{white}) = \frac{4}{6}$. Since the white marble drawn is not replaced, there are now 5 marbles left in the urn of which 3 are white.

So, on the second draw: $P(\text{white}) = \frac{3}{5}$

Then: $P(\text{both white}) = \frac{4}{6} \cdot \frac{3}{5} = \frac{12}{30}$, or $\frac{2}{5}$ *Ans.*

b. When we start with a full urn of 6 marbles, on the first draw:

$$P(\text{blue}) = \frac{2}{6}$$

Since the blue marble drawn is not replaced, there are now 5 marbles left in the urn of which only 1 is blue.

So, on the second draw: $P(\text{blue}) = \frac{1}{5}$

Then: $P(\text{both blue}) = \frac{2}{6} \cdot \frac{1}{5} = \frac{2}{30}$, or $\frac{1}{15}$ *Ans.*

c. If both marbles are the same color, then both are white or both are blue. These are disjoint events. So, $P(A \text{ or } B) = P(A) + P(B)$. Therefore:

$$\begin{aligned} P(\text{both white or both blue}) &= P(\text{both white}) + P(\text{both blue}) \\ &= \frac{4}{6} \cdot \frac{3}{5} + \frac{2}{6} \cdot \frac{1}{5} \\ &= \frac{12}{30} + \frac{2}{30} \\ &= \frac{14}{30}, \text{ or } \frac{7}{15} \quad \textit{Ans.} \end{aligned}$$

4. Fred has 2 quarters and 1 nickel in his pocket. The pocket has a hole in it and a coin drops out. Fred picks up the coin and puts it back into his pocket. A few minutes later, a coin drops out of his pocket again.

a. Draw a tree diagram or list the sample space for all possible pairs that are outcomes to describe the coins that fell.

b. What is the probability that the same coin fell out of his pocket both times?

c. What is the probability that the two coins that fell have a total value of 30 cents?

d. What is the probability that a quarter fell out at least once?

Solution

a. Because there are 2 quarters, use subscripts. The three coins are $\{Q_1, Q_2, N\}$ where Q represents a quarter and N represents a nickel. This is a problem with replacement.

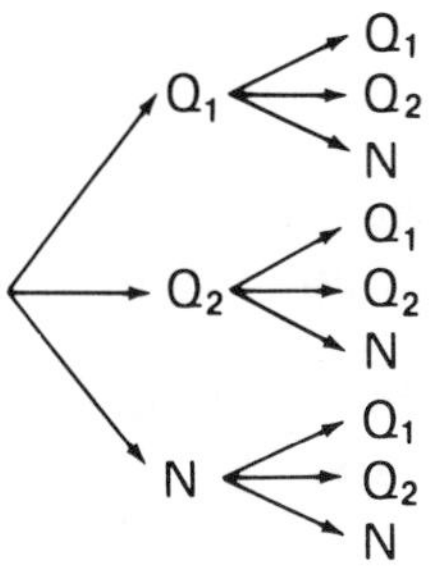

Q1 → Q1, Q2, N
Q2 → Q1, Q2, N
N → Q1, Q2, N

or

$$\left\{\begin{matrix} (Q_1, Q_1), (Q_1, Q_2), (Q_1, N), \\ (Q_2, Q_1), (Q_2, Q_2), (Q_2, N), \\ (N, Q_1), (N, Q_2), (N, N) \end{matrix}\right\}$$

b. Of the 9 outcomes, 3 name the same coin both times:

(Q_1, Q_1); (Q_2, Q_2); (N, N). So: $P(\text{same coin}) = \frac{3}{9}$, or $\frac{1}{3}$ *Ans.*

c. Of the 9 outcomes, 4 consist of a quarter and a nickel, which total 30 cents: (Q_1, N); (Q_2, N); (N, Q_1); (N, Q_2). So:

$$P(\text{coins total 30 cents}) = \frac{4}{9} \quad Ans.$$

d. Of the 9 outcomes, 8 contain one or more quarters, the equivalent of at least one quarter. The only outcome not counted is (N, N). So: $P(\text{at least one quarter}) = \frac{8}{9}$ *Ans.*

EXERCISES

1. An urn contains 2 red and 5 yellow marbles. If one marble is drawn at random, what is the probability that it is:
a. red? **b.** yellow?

2. An urn contains 2 red and 5 yellow marbles. A marble is drawn at random and then replaced. A second draw is made at random. Find the probability that:
a. both marbles are red
b. both marbles are yellow
c. both marbles are the same color
d. the marbles are different in color

3. An urn contains 2 red and 5 yellow marbles. A marble is drawn at random. Without replacement, a second draw is made at random. Find the probability that:
a. both marbles are red
b. both marbles are yellow
c. both marbles are the same color
d. the marbles are different in color

4. In an experiment, the arrow is spun twice on a wheel containing four equally likely regions, numbered 1 through 4.

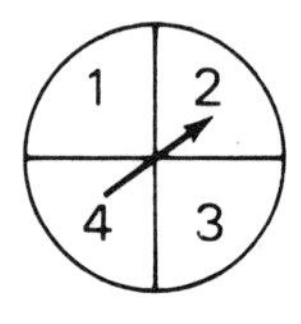

a. Indicate the sample space by drawing a tree diagram or writing a set of ordered pairs.
b. Find the probability of spinning the digits 2 and 3 in that order.
c. Find the probability that the digits are the same.
d. What is the probability that the first digit is larger than the second?

5. Sal has a bag of hard candies: 3 are lemon and 2 are grape. He ate 2 of the candies while waiting for a bus, selecting them at random one after another.

a. Using subscripts, draw a tree diagram or list the sample space of all possible outcomes showing which candies were eaten.

b. Find the probability that:

(1) both candies were lemon
(2) neither candy was lemon
(3) the candies were the same flavor
(4) at least one candy was lemon

6. Carol has 5 children, 3 girls and 2 boys. One of her children came late for lunch. Later that day, one of her children came late for supper.

a. Indicate the sample space by a tree diagram or list of ordered pairs showing which children were late.

b. If each child was equally likely to be late, find the probability that:

(1) both children who came late were girls
(2) both children who came late were boys
(3) the same child came late both times
(4) at least one of the children who came late was a boy

7. Several players start playing a game with a full deck of 52 cards. Each player draws two cards at random, one at a time. Find the probability that:

a. Flo draws two jacks.
b. Frances draws two hearts
c. Jerry draws two red cards
d. Mary draws two picture cards
e. Carrie draws a 5 and a 10 in that order
f. Bill draws a heart and a club in that order
g. Ann does not draw a pair
h. Stephen draws two black kings

8. Saverio had 4 coins, a half dollar, a quarter, a dime, and a nickel. He chose one of the coins and put it into a bank. He chose another coin later on and also put that into the bank.

a. Indicate the sample space of coins saved.

b. If each coin was equally likely to have been saved, find the probability that the coins saved:

(1) were worth a total of 35¢
(2) added to an even amount
(3) included the half dollar
(4) were worth a total of less than 30¢

9. Farmer Brown must wake up before sunrise to start his chores. Dressing in the dark, he reaches into a drawer and pulls out 2 loose socks. There are 8 white socks and 6 red socks in the drawer.
a. Find the probability that both socks are:
(1) white (2) red (3) the same color
b. Find the maximum number of socks Farmer Brown must pull out of the drawer to guarantee that he will get a matching pair.

10. Tillie had 3 quarters and 4 dimes in her purse. She took out a coin at random, but it slipped from her hand and fell back into her purse. She reached in and picked out a coin at random. Find the probability that:
a. both coins were quarters **b.** both coins were dimes
c. the coins were a dime and a quarter, in any order
d. the same coin was picked both times
e. the coins picked totaled less than 40¢
f. the coins picked totaled exactly 30¢
g. at least one of the coins picked was a quarter

11. An urn contained 9 orange disks and 3 blue disks. A girl chose one at random and, without replacing it, chose another. Letting o represent orange and b represent blue:
a. Find the probability of each of the following outcomes:
(1) (o, o) (2) (o, b) (3) (b, o) (4) (b, b)
b. Now, find the probability that, for the disks chosen:
(1) neither was orange (2) only one was blue
(3) at least one was orange (4) they were the same color
(5) at most one was orange (6) they were the same disk

15-13 REVIEW EXERCISES

1. If 3 fair coins are tossed simultaneously, what is the total number of outcomes in the sample space?

2. There are 5 staircases going from the first to the second floor in school. Find the total number of ways that Andrea can go from the first floor to the second and then return to the first.

3. The probability that Glennon will get a hit the next time at bat is 35%. What is the probability that Glennon will not get a hit?

4. In right triangle ABC, one of the three angles is chosen at random. Find the probability that the angle is:
a. acute **b.** right **c.** obtuse **d.** reflex

5. An urn contains 2 red, 3 white, and 4 blue marbles. If a marble is drawn at random, find the probability that the marble is:
a. red **b.** not red **c.** red or blue **d.** green

6. A single card is drawn from a standard deck of 52 playing cards. Find the probability that the card is:
   a. red **b.** a four **c.** a red four **d.** a heart
   e. a four or a heart **f.** red or a four **g.** not a club
7. How many different two-digit numbers can be formed using the digits of 1492 if each digit appears only once in a number?
8. If 2 fair six-sided dice are tossed, and their sum is noted, find:
   a. *P*(sum is 8) **b.** *P*(sum is greater than 9)
   c. *P*(sum is odd) **d.** *P*(sum is 12 or less)
9. Assume that *P*(male) = *P*(female). In a family of 3 children, what is the probability that all 3 children are of the same sex?
10. Compute the value of each expression.
   a. 5! **b.** $\frac{10!}{8!}$ **c.** ${}_4P_4$ **d.** ${}_{30}P_2$ **e.** ${}_7P_3$

In 11–13, select the numeral preceding the correct answer.

11. How many different arrangements of 4 letters can be made from the letters in HELP if each letter is used only once in each arrangement?
   (1) 1 (2) 256 (3) 24 (4) 4
12. If $P(A) = 0.4$, $P(B) = 0.3$, and $P(A \cap B) = 0.2$, then $P(A \cup B) =$
   (1) 0.9 (2) 0.7 (3) 0.5 (4) 0.12
13. If $P(E) = \frac{n}{5}$ for event E, which can*not* be a value for n?
   (1) 1 (2) 0 (3) 5 (4) 15

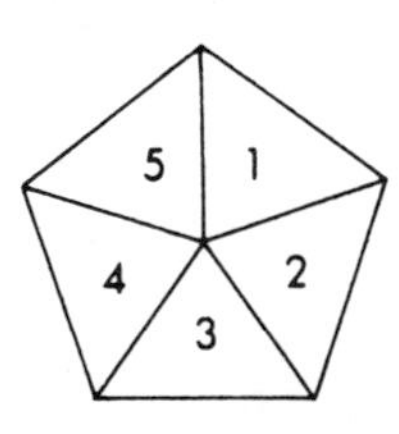

14. Two darts were thrown at a pentagon-shaped dartboard until they each held in one of 5 equally likely numerical regions.
   a. Draw a tree diagram or write a sample space to show the numbered regions for the two darts.
   b. Find the probability that both darts landed in the same region.
   c. Find the probability that both darts landed in odd-numbered regions.
   d. Find the probability that at least one dart landed in the region marked 5.
   e. If the score is the sum of the numbers of the regions hit, find the probability of getting an odd-numbered score.

15. A bank contains a quarter, a dime, and a nickel. It is shaken until one coin falls out, and it is shaken again until a second coin falls out. All coins are equally likely to fall out.

a. Draw a tree diagram or list the sample space to show the pairs of the first and second coins that fell out.

b. Find the probability that the *sum* of the values of the coins is:

1. exactly 30¢ 2. exactly 20¢ 3. 30¢ or less
4. less than 40¢ 5. equal to an odd number of cents

c. Find the probability that the first coin is a quarter.

d. Find the probability that neither coin is a dime.

16. About how many times can you expect to turn up a head in 50 consecutive tosses of a fair coin?

17. An urn contains red and white marbles. The number of white marbles is 5 more than twice the number of red marbles. If the probability of drawing a red marble at random is $\frac{2}{7}$, how many red marbles are in the urn?

18. From a class of girls and boys, the probability that one student chosen at random will be a girl is $\frac{1}{3}$. If 4 boys leave the class, the probability that a student chosen at random will be a girl is $\frac{2}{5}$. How many boys and girls are there in the class before the 4 boys leave?

Chapter 16

Statistics

16-1 THE COLLECTION OF DATA

In our daily lives, we often deal with problems that involve many related pieces of numerical information called ***data***. For example, in the daily newspaper we can find data dealing with sports, with business, with politics, even with the weather.

Statistics is the study of numerical data. The typical steps in a statistical study are:

Step 1. The collection of data
Step 2. The organization of this data into tables, charts, and graphs
Step 3. The drawing of conclusions from an analysis of this data

When these three steps, which describe and summarize a set of data, are included in a statistical study, the study is often called ***descriptive statistics***. You will study these steps in this first course. In some cases, a fourth step is added in which the analyzed data is used to predict trends and future events. You will not study this type of statistics in this course.

Data can be collected in a number of ways, including:

1. a written ***questionnaire*** or list of questions in which a person can check one of several categories as an answer or fill in some written response;
2. an ***interview***, either in person or by telephone, in which answers are given verbally and responses are recorded by the person asking the questions;
3. a ***log*** or a diary, such as a hospital chart or an hourly recording of the outdoor temperature, in which a person records information on a regular basis.

The Census

Starting in 1790, and every ten years thereafter, the United States has conducted a ***census*** to count the number of people in the country and to determine the geographic regions in which they live. To collect this data, a questionnaire is mailed to every household in the country. For a portion of the citizens, more detailed questionnaires are mailed and, in some cases, workers from the Census Bureau visit homes to conduct interviews. Important decisions are made on the basis of the data collected. For example, geographic regions with larger populations receive a greater share of the billions of dollars distributed in federal and state funds, and these regions are also entitled to more seats in the House of Representatives. Since there are about 250,000,000 people living in the United States, the census is a major statistical study.

Sampling

Not all statistical studies are as large as the census. However, every statistical study demands that data be collected carefully and correctly if the study is to be useful. The following are some problems where a statistical study may be helpful:

1. A doctor wishes to know which medicine will be most effective in curing a disease.
2. A manufacturer wants to know the expected life span of a flashlight battery that his company makes.
3. A company advertising on television wishes to know the most frequently watched TV shows so that their ads will be seen by the greatest number of people.

Unlike the census, where every person is counted, these examples of statistical studies demand that only a ***sample***, or a portion of the items to be counted, be actually considered. To find effective medicines, tests are usually conducted with a sample, or portion, of patients having the same disease. Some patients receive one medicine and other patients receive different medicines.

The manufacturer of flashlight batteries cannot test the life span of every battery made because he would soon have a warehouse filled with dead batteries. He tests only a sample of the batteries to determine their average life span.

An advertiser cannot contact every person owning a TV set to see which shows are being watched. The advertiser will study TV ratings released by a firm that conducts polls based upon a small sample of TV viewers.

Techniques of Sampling

We must be careful when choosing samples:

1. The sample must be fair, to reflect the entire population being studied.

To know what an apple pie tastes like, it is not necessary to eat the entire pie. A sample, such as a piece of apple pie, would be a fair way of knowing how the pie tastes. However, eating only the crust or only the apples would be an unfair sample; these samples would not truly tell us what the entire pie tastes like.

2. The sample must contain a reasonable number of items being tested or counted.

If a medicine is generally effective, it must work for many people. The sample tested cannot include only one or two patients. Similarly, the manufacturer of flashlight batteries cannot make claims based on 5 or 10 batteries tested. A better sample might include 100 batteries.

3. Patterns of sampling or random selection should be employed in a study.

The manufacturer of flashlight batteries might set up a pattern to test every 1,000th battery to come off the assembly line. He may also select the batteries to be tested at random.

These techniques will help to make the sample, or the small group, *representative* of the entire group of items being studied. From the study of the small group, reasonable conclusions can be drawn about the entire group.

MODEL PROBLEM

To determine which television shows are the most popular in a large city, a poll is conducted by selecting people at random at a street corner and interviewing them. Outside of which location would we find the most fair sample?

(1) a ball park (2) a concert hall (3) a supermarket

Solution

People outside a ball park may be going to a game or purchasing tickets for a game in the future; this sample might be biased in favor of sports shows. Similarly, those outside a concert hall may favor musical or cultural shows. The best sample or cross section of people for the three choices given would probably be found outside a supermarket.

Answer: (3)

EXERCISES

In 1–8, a sample of students is to be selected and the height of each student is to be measured to determine the average height of a student in high school. **a.** Tell if the sample is fair or unfair. **b.** If the sample is unfair, explain why.

1. the basketball team
2. the senior class
3. all 14-year-old students
4. all girls
5. every tenth person selected from an alphabetical list of all students
6. every fifth person selected from an alphabetical list of all boys
7. the first three students who report to the nurse on Monday
8. the first three students who enter each homeroom on Tuesday

In 9–14, the Student Organization wishes to interview a sample of students to determine the general interests of the student body. Two questions will be asked: "Do you want more pep rallies for sports events? Do you want more dances?"

Tell whether the Student Organization would find a fair sample at the given location.

9. the gym, after a game
10. the library
11. the lunchroom
12. the cheerleaders' meeting
13. the next meeting of the Junior Prom committee
14. a homeroom section chosen at random

15. A statistical study is useful when reliable data is collected. At times, people may exaggerate or lie when answering a question. Of the six questions that follow, find the *three* questions that will most probably produce the largest number of *unreliable* answers.
a. What is your height?
b. What is your weight?
c. What is your age?
d. In which state do you live?
e. What is your income?
f. How many people are in your family?

16. List the three steps necessary to conduct a statistical study.

16-2 THE ORGANIZATION OF DATA INTO TABLES

Data is often collected in an unorganized and random manner. For example, as a teacher marked a set of 32 test papers, the grades or scores earned by the students were:

90, 85, 74, 86, 65, 62, 100, 95, 77, 82, 50, 83, 77, 93, 73, 72, 98,

66, 45, 100, 50, 89, 78, 70, 75, 95, 80, 78, 83, 81, 72, 75.

How many test scores are 60 or less? Are most of the scores around 70 or around 80? These types of questions are difficult to answer, using ungrouped data. To answer such questions, we organize or *group* the data into a table.

The accompanying table contains six ***intervals*** of equal size: 41 to 50; 51 to 60; 61 to 70; 71 to 80; 81 to 90; 91 to 100. Each interval has a ***length*** of 10, found by subtracting the starting point of any interval from the starting point of the next higher interval. For example, 91 − 81 = 10; 81 − 71 = 10; and so on.

Interval	Tallies
91–100	~~||||~~ |
81–90	~~||||~~ |||
71–80	~~||||~~ ~~||||~~ |
61–70	||||
51–60	
41–50	|||

For each test score, a ***tally*** mark, |, is placed in the interval containing that score. For example, since the first two test grades are 90 and 85, we place two tallies, | |, in the interval 81–90. Since the third test score is 74, we place one tally, |, in the interval 71–80. We follow this process until all test scores are grouped into their proper intervals. To simplify counting, every fifth tally is written as a mark passing through four other tallies, as in ~~||||~~.

Once the data has been organized, we can convert the tally marks to counting numbers. Each counting number tells us the ***frequency*** or the number of scores that will fall into each of the intervals. When no scores fall into one of the established intervals, as in 51–60, the frequency for this interval is zero. The sum of all the frequencies is called the ***total frequency***. Here, 6 + 8 + 11 + 4 + 0 + 3 = 32. Thus, the total frequency is 32. It is always wise to check the total frequency to see that a score was not overlooked in tallying. This table, containing a series of intervals and the corresponding frequency for each interval, is an example of ***grouped data***.

Interval	Frequency (*Number of Scores*)
91–100	6
81–90	8
71–80	11
61–70	4
51–60	0
41–50	3

From the table, we now see that exactly 3 students scored 60 or less. Also, the largest number of test scores fell into the interval 71–80.

Rules for Grouping Data

When unorganized data is grouped into intervals, we must follow certain rules in setting up the intervals:

1. The intervals must cover the complete range of values. The ***range*** is the difference between the highest and lowest values.
2. The intervals must be equal in size.
3. The number of intervals should be between 5 and 15. The use of too many intervals or too few intervals does not make for effective grouping of data. We usually use a large number of intervals, such as 15 intervals, only when we have a large set of data, such as hundreds of scores.
4. Every score to be tallied, from the highest to the lowest, must fall into one and only one interval. Thus, the intervals should not overlap each other. When an interval ends with a counting number, the following interval will begin with the next counting number.

These rules tell us that there are many ways to set up tables, all of them correct, for the same set of data. For example, here is another correct way to group the 32 unorganized test scores given at the beginning of this section. Note that the length of the interval is 8.

Interval	Tallies	Frequency (*Number*)
93–100	~~||||~~ |	6
85–92	||||	4
77–84	~~||||~~ ||||	9
69–76	~~||||~~ ||	7
61–68	|||	3
53–60		0
45–52	|||	3

MODEL PROBLEM

The following data consists of weights (in kilograms) of a group of 30 students:

70, 43, 48, 72, 53, 81, 76, 54, 58, 64, 51, 53, 75, 62, 84, 67, 72, 80, 88, 65, 60, 43, 53, 42, 57, 61, 55, 75, 82, 71.

a. Copy and complete the table to group this data.

b. Based on the grouped data, which interval contains the greatest number of students?

c. How many students weigh less than 70 kilograms?

Interval	Tallies	Frequency (*Number*)
80–89		
70–79		
60–69		
50–59		
40–49		

Solution: **a.**

Interval	Tallies	Frequency (*Number*)
80–89	𝍸	5
70–79	𝍸 II	7
60–69	𝍸 I	6
50–59	𝍸 III	8
40–49	IIII	4

b. The interval 50–59 contains the greatest number of students, 8.

c. The three lowest intervals, namely 40–49, 50–59, and 60–69, show weights less than 70 kilograms. Add the frequencies in these three intervals: 4 + 8 + 6 = 18.

Answer: There are 18 students, each weighing less than 70 kg.

EXERCISES

1. a. Copy and complete the table to group the data, giving heights (in centimeters) of 36 students:

162, 173, 178, 181,
155, 162, 168, 147,
180, 171, 168, 183,
157, 158, 180, 164,
160, 171, 183, 174,
166, 175, 169, 180,
149, 170, 150, 158,
162, 175, 171, 163,
158, 163, 164, 177.

Interval	Tallies	Frequency (*Number*)
180–189		
170–179		
160–169		
150–159		
140–149		

b. Use the grouped data to answer the following questions:
1. How many students are less than 160 centimeters in height?
2. How many students are 160 centimeters or more in height?
3. Which interval contains the greatest number of students?
4. Which interval contains the least number of students?

2. **a.** Copy and complete the table to group the data that gives the life span, in hours, of 50 flashlight batteries:

73, 81, 92, 80, 108,
76, 84, 102, 58, 72,
82, 100, 70, 72, 95,
105, 75, 84, 101, 62,
63, 104, 97, 85, 106,
72, 57, 85, 82, 90, 54,
75, 80, 52, 87, 91, 85,
103, 78, 79, 91, 70, 88,
73, 67, 101, 96, 84, 53, 86.

Interval	Tallies	Frequency (*Number*)
50–59		
60–69		
70–79		
80–89		
90–99		
100–109		

b. Use the grouped data to answer the following questions:
1. How many flashlight batteries lasted for 80 or more hours?
2. How many flashlight batteries lasted less than 80 hours?
3. Which interval contains the greatest number of batteries?
4. Which interval contains the least number of batteries?

3. The following data shows test scores for 30 students:

90, 83, 87, 71, 62, 46, 67, 72, 75, 100, 93, 81, 74, 75, 82,
83, 83, 84, 92, 58, 95, 98, 81, 88, 72, 59, 95, 50, 73, 93.

a. Copy and complete the table, using these intervals of length 10.

Interval	Frequency
91–100	
81–90	
71–80	
61–70	
51–60	
41–50	

b. Copy and complete the table, using these intervals of length 12.

Interval	Frequency
89–100	
77–88	
65–76	
53–64	
41–52	

c. For the grouped data in part **a**, which interval contains the greatest number of students?
d. For the grouped data in part **b**, which interval contains the greatest number of students?
e. Do the answers for parts **c** and **d** indicate the same general region of test scores, such as "scores in the eighties"? Explain your answer.

4. The following data consists of the hours spent each week watching television, as reported by a group of 38 teenagers:

13, 20, 17, 36, 25, 21, 9, 32, 20, 17, 12, 19, 5, 8, 11, 28, 25, 18,

19, 22, 4, 6, 0, 10, 16, 3, 27, 31, 15, 18, 20, 17, 3, 6, 19, 25, 4, 7.

a. Construct a table to group this data, using intervals of 0–4; 5–9; 10–14; 15–19; 20–24; 25–29; 30–34; 35–39.

b. Construct a table to group this data, using intervals of 0–7; 8–15; 16–23; 24–31; 32–39.

5. For the ungrouped data from Exercise 4, tell why each of the following sets of intervals is *not correct* for grouping the data.

a.

Interval
20–39
0–19

b.

Interval
30–36
20–29
11–19
0–10

c.

Interval
30–40
20–30
10–20
0–10

d.

Interval
31–40
21–30
11–20
1–10

16-3 USING GRAPHS TO PRESENT ORGANIZED DATA

You have seen that grouped data, or data organized into tables, is more useful than a set of unorganized data. Going one step further, a ***graph*** or picture of organized data can often present the collected information in a way that is easier to understand than a table of numbers. You can find a variety of graphs in newspapers and magazines. Among these graphs are the bar graph, the picture graph, the line graph, and the circle graph.

Using Graphs to Compare Different Items

One of the uses of statistics is to compare *different* items, such as the sales records for different salespeople in a company, the heights of different buildings, the prices of different stocks, and so forth. The two most commonly used graphs to compare different items are the *bar graph* and the *picture graph*.

The Bar Graph

In a ***bar graph***, the length of a bar is used to represent a numerical fact, that is, some piece of data. The length of the bar depends upon the size of the number that is to be represented. A ***scale***, beginning with zero and consisting of equally spaced intervals, must accompany the bar graph so that the approximate size of a number may be read. The bars are drawn in the same direction, either all horizontally or all vertically.

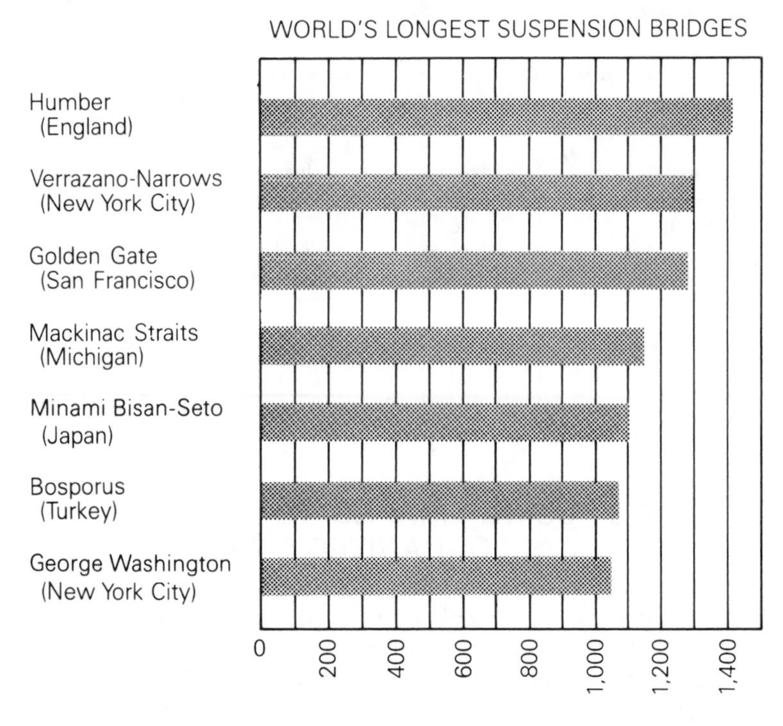

The actual lengths of the main spans of the world's seven longest suspension bridges are not stated in the bar graph above. However, the scale at the bottom of the graph allows us to compare the lengths of these seven different bridges. We can observe many numerical facts, such as:

1. There are only three suspension bridges in the world whose main spans are longer than 1,200 meters.
2. The span closest to 1,300 meters in length is on the Verrazano Narrows Bridge.
3. The longest span in the world, over 1,400 meters in length, is on the Humber Bridge.
4. Each of the world's seven longest suspension bridges has a main span that is longer than 1,000 meters.

The Picture Graph

The ***picture graph*** is a graph in which a *symbol*, or picture, is used to represent a definite number. The symbol, its meaning, and the quantity that it represents must be stated.

1985: MOST POPULATED COUNTRIES

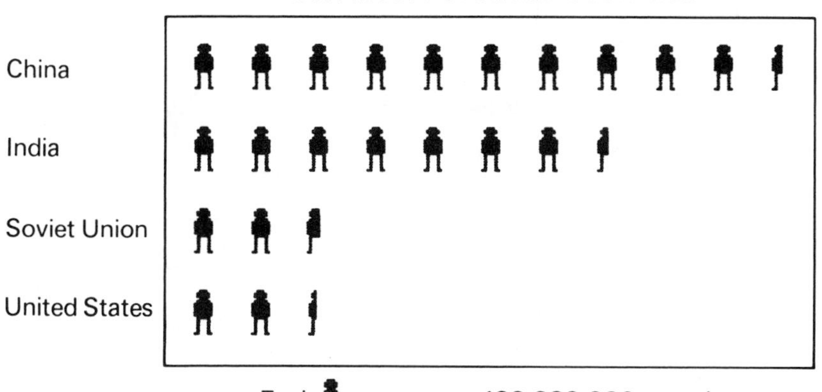

Each [symbol] represents 100,000,000 people

In this picture graph, each symbol [symbol] represents one hundred million people. While the actual populations of these countries are not stated, we can observe many numerical facts, such as:

1. The $10\frac{1}{2}$ symbols, or pictures, shown for China tell us that its population is approximately $10\frac{1}{2} \times 100,000,000$, which is 1,050,000,000 people.

2. India, with the second largest population, numbers approximately $7\frac{1}{2} \times 100{,}000{,}000$, or 750,000,000 people.

3. Both the Soviet Union and the United States have a population over 200,000,000, with the Soviet Union closer to 300,000,000.

Using a Line Graph to Compare Changes in the Same Item

To show how a particular item, such as temperature, price of a stock, or population, changes, we most often use the ***line graph***, sometimes called a ***broken line graph***. In making a line graph, a point is used to represent each numerical fact. Line segments are drawn connecting the consecutive points on the graph. A rising line segment shows that the item being studied is increasing. A falling line segment shows that the item being studied is decreasing.

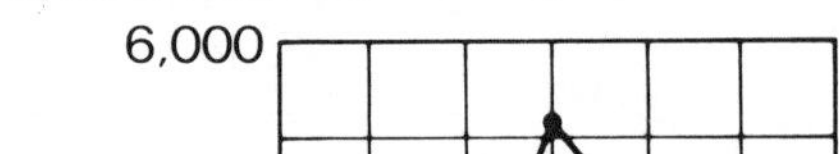

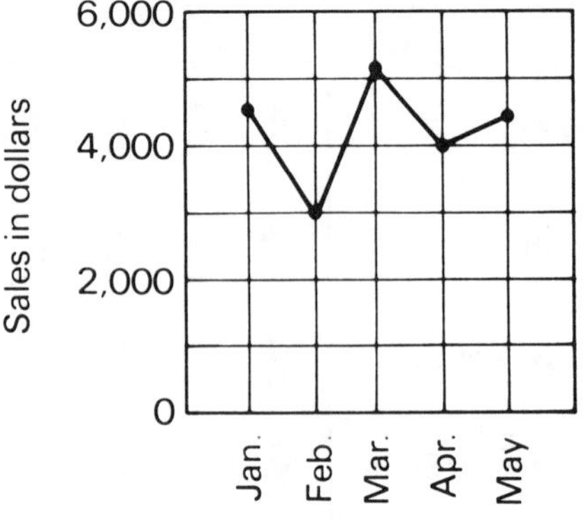

The graph at the right shows the sales of Mr. Greenwald, a salesman, during the first five months of last year.

From the graph, we can observe many numerical facts, such as:

1. His highest sales occurred in March.
2. His lowest sales occurred in February.
3. The two months closest in sales were January and May.
4. The month closest to $4,000 in sales was April.
5. The ratio of his sales in February to his sales in April is approximately $3,000 to $4,000, or 3 to 4.
6. The greatest decrease occurred from January to February. His sales dropped by approximately $1,500 during that period.
7. The smallest increase occurred from April to May. His sales went up by less than $500 during that period.

Note that it is *not* appropriate to use a line graph to compare *different* items, such as the lengths of the seven suspension bridges previously studied.

EXERCISES

1. Of the highest buildings in the U.S.A., three are in Chicago: the Sears Tower, the Standard Oil Building, and the John Hancock Building. Three others are in New York City: the Empire State Building, the World Trade Center, and the Chrysler Building. Use the graph to answer the following questions:

HIGHEST BUILDINGS IN THE U.S.A.

Height in feet
1,500
1,200
900
600
300
0
Standard Oil
John Hancock
Sears Tower
Empire State
World Trade
Chrysler

   a. Find the height of each building, to the nearest 50 feet.
   b. Name the tallest building.
   c. What is the approximate difference in height between the World Trade Center and the Empire State Building?
   d. What building is closest in height to the Standard Oil Building?
   e. What is the approximate difference between the height of the John Hancock Building and the height of the Sears Tower?
   f. Which buildings, if any, are less than 1,000 feet tall?

2. Use the graph below to answer the following questions:
   a. How many telephones were there in each of the towns, to the nearest 5,000?
   b. Which two towns had about the same number of telephones?
   c. Parr City had how many times as many telephones as Tyne?
   d. What was the ratio of the number of telephones in Sampler to the number of telephones in Tyne?
   e. How did the number of telephones in Calcot compare to the total number of telephones in the other three towns?

TELEPHONES IN VARIOUS TOWNS

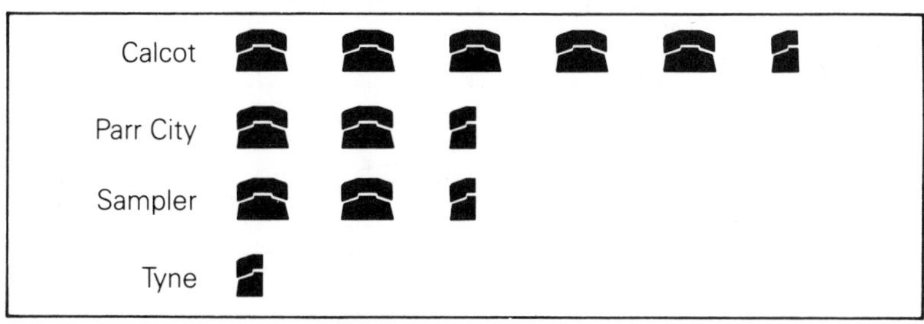

Each symbol, ☎, represents 10,000 telephones

3. The bar graph shows the number of acres of different vegetables that a farmer planted during one season.
   a. How many acres of carrots were planted?
   b. Which vegetable had the smallest number of acres planted?
   c. What was the total number of acres planted?
   d. What percent of the total number of acres planted contained carrots?
   e. What is the ratio of the number of acres of beets planted to the number of acres of spinach planted?

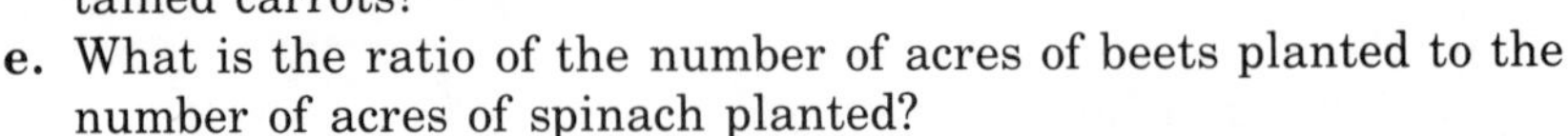

4. In a bar graph, 1 centimeter represents 30 kilometers. Find the length of the bar needed to represent the given distance.
   a. 60 kilometers
   b. 300 kilometers
   c. 15 kilometers
   d. 75 kilometers

5. In a bar graph, $\frac{1}{4}$ inch represents 100 people. Find the length of the bar needed to represent the given number of people.
   a. 200 people
   b. 400 people
   c. 2,000 people
   d. 500 people
   e. 1,500 people
   f. 50 people

6. The following graph shows the price per share of a certain stock over a period of 11 days.

a. Approximately how much was the price per share on the tenth day?
b. On what day was the price the lowest?
c. Between what two days did the price decrease most sharply?
d. Between what two days did the price remain constant?
e. For what two days was the price closest to $.40 per share?
f. What was the range of the price over the 11-day period?

7. In the scale of the following line graph, a break appears between 0 and 60 to allow for a closer look at the increases in life expectancies for U.S. males, over the period from 1940 to 1990, inclusive. If this same scale were used without a break, the graph would extend beyond the length of this page in the book. Answer the following questions related to the line graph.

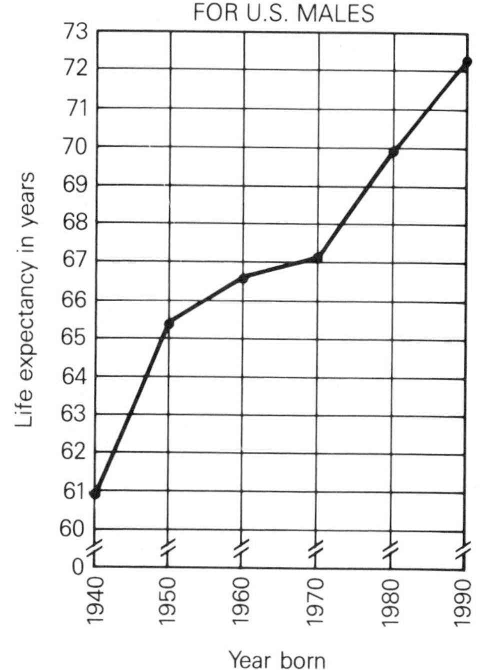

a. In what decade (ten-year period) did the life expectancy increase the most?
b. In what decade did the life expectancy increase the least?
c. What was the life expectancy for U.S. males born in 1940, to the nearest year?
d. What was the life expectancy for U.S. males born in 1950, to the nearest year?
e. By about how much does the life expectancy increase between 1980 and 1990?
(1) 0 years (2) 2 years
(3) 4 years (4) 6 years
f. Which of the following ratios best compares the life expectancy of U.S. males in 1950 to that in 1940?
(1) 3 to 1 (2) 5 to 1 (3) 61 to 65 (4) 65 to 61
g. It is true that females have a greater life expectancy than males. Which statement could be true for U.S. females?
(1) In 1960, their life expectancy was 66 years.
(2) In 1980, their life expectancy was greater than 69.9 years.
(3) In 1970, their life expectancy was less than 67 years.

8. The heights of five dams are: Boulder, 730 ft.; Kensico, 310 ft.; Shasta, 600 ft.; Grand Coulee, 450 ft.; and Gatun, 120 ft.
   a. Make a bar graph to represent this data.
   b. Tell why it is not appropriate to display this data with a line graph.

9. The unemployment rate in the United States is given in five-year intervals for the years 1960–1985, as shown in the table at the right.
   a. Construct a bar graph to represent this data.
   b. Tell why it is appropriate to display this data with a line graph.
   c. Construct a line graph.

U.S. Unemployment Rate *(per 100 workers)*	
1960	5.4
1965	4.4
1970	4.8
1975	8.3
1980	7.0
1985	7.2

16-4 THE HISTOGRAM

In Section 2 of this chapter, you learned how to organize data by grouping the data into intervals of equal length.

The table at the right consists of six intervals, each having a length of ten. This grouped data shows the distribution of test scores for 32 students in a class.

Test Scores *(Intervals)*	**Frequency** *(Number of Scores)*
91–100	6
81–90	8
71–80	11
61–70	4
51–60	0
41–50	3

The data can be displayed graphically by means of a *histogram*. A ***histogram*** is simply a vertical bar graph in which the bars are placed next to each other. We bring these bars together to show that as one interval ends, the next interval begins. Essentially, the bars of the histogram show *changes in the same item*, such as the distribution of scores for a single test.

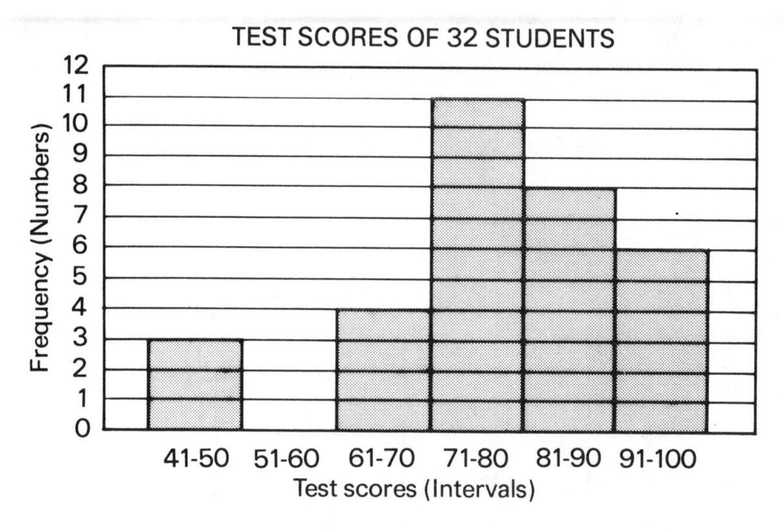

In the above histogram, note that the intervals are placed on the horizontal axis in the order of increasing values of the scores. The first bar shows that 3 students had test scores in the interval 41–50. Since no students scored in the interval 51–60, there is no bar at this location. Then, 4 students scored between 61 and 70; 11 students had test scores between 71 and 80; 8 had scores between 81 and 90; and finally, 6 students scored between 91 and 100. With the exception of an interval having a frequency of zero, as in 51–60, there are no gaps between the bars drawn in a histogram.

Since the histogram displays the frequency, or number, of scores in each interval, we sometimes call this graph a ***frequency histogram***.

MODEL PROBLEMS

1. The table shown represents the number of miles per gallon (mpg) of gasoline obtained by 40 drivers of compact cars in a large city.

Construct a frequency histogram based on the data.

Interval	Number (*Frequency*)
16–19	5
20–23	11
24–27	8
28–31	5
32–35	7
36–39	3
40–43	1

Solution

Step 1. Draw and label a vertical scale to show frequencies. The scale starts at 0 and increases to include the highest frequency in any one interval (here, it is 11).

Step 2. Draw and label intervals of equal length on a horizontal scale. Give a title to the horizontal scale, telling what the numbers represent.

Step 3. Draw the bars vertically, leaving no gaps between the intervals.

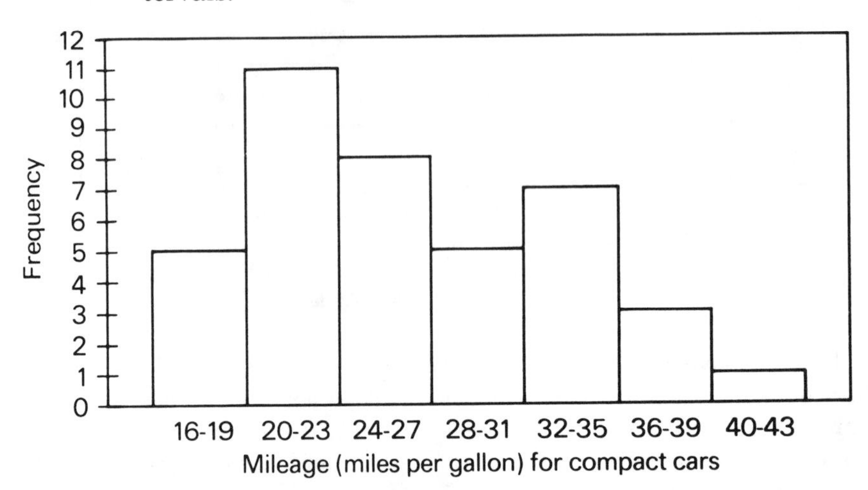

2. Use the histogram constructed in Model Problem 1 to answer the following questions:

a. In what interval is the greatest frequency found?

Answer: 20–23

b. What is the number (or frequency) of cars reporting mileage between 28 and 31 miles per gallon?

Answer: 5

c. In what interval are the fewest cars reported?

Answer: 40–43

d. How many of the cars reported mileages greater than 31 miles per gallon?

Solution: Add the frequencies for the three highest intervals. With 7 in (32–35), 3 in (36–39), and 1 in (40–43), we obtain $7 + 3 + 1 = 11$. *Ans.*

e. What percent of the cars reported mileages from 24 to 27 mpg?

Solution: The interval 24–27 contains a frequency of 8. The total frequency for this survey is 40. Then, $\frac{8}{40} = \frac{1}{5} = 20\%$. *Ans.*

EXERCISES

In 1–3, construct a frequency histogram for the grouped data.

1.

Interval	Number (*Frequency*)
91–100	5
81–90	9
71–80	7
61–70	2
51–60	4

2.

Interval	Frequency
30–34	5
25–29	10
20–24	10
15–19	12
10–14	0
5–9	2

3.

Interval	Number
1–3	24
4–6	30
7–9	28
10–12	41
13–15	19
16–18	8

4. For the table of grouped scores given in Exercise 3, answer the following questions:

a. What is the total frequency, or the total number of pieces of data, in the table?

b. What interval contains the greatest frequency?

c. The number of scores reported for the interval 4–6 is what percent of the total number of scores?

d. How many scores were reported from 10 through 18?

5. Towering Ted McGurn is the star of the school's basketball team. The number of points scored by Ted in his last 20 games are:

36, 32, 28, 30, 33, 36, 24, 33, 29, 30,
30, 25, 34, 36, 34, 31, 36, 29, 30, 34.

a. Copy and complete the table to find the number (or frequency) in each interval.

b. Construct a frequency histogram based on the data found in part **a**.

Interval	Tallies	Number (*Frequency*)
35–37		
32–34		
29–31		
26–28		
23–25		

c. Answer the following questions for this set of data:
   1. Which interval contains the greatest frequency?
   2. In how many games did Ted score 32 or more points?
   3. In what percent of these 20 games did Ted score fewer than 26 points?

6. Thirty students on the track team were timed in the 200-meter dash. Each student's time was recorded to the nearest tenth of a second. The times were:

   29.3, 31.2, 28.5, 37.6, 30.9, 26.0, 32.4, 31.8, 36.6, 35.0,

   38.0, 37.0, 22.8, 35.2, 35.8, 37.7, 38.1, 34.0, 34.1, 28.8,

   29.6, 26.9, 36.9, 39.6, 29.9, 30.0, 36.0, 36.1, 38.2, 37.8.

   a. Copy and complete the table to find the number (or frequency) in each interval.
   b. Construct a frequency histogram for the given data.
   c. From the data, determine the number of students who ran the 200-meter dash in under 29 seconds.

Interval	Tallies	Number *(Frequency)*
37.0–40.9		
33.0–36.9		
29.0–32.9		
25.0–28.9		
21.0–24.9		

16-5 THE MEAN, THE MEDIAN, AND THE MODE

In a statistical study, after we have collected the data, organized it, and presented it graphically, we then analyze the data and summarize our findings. To do this, we often look for a representative, or typical, score.

Averages in Arithmetic

In your previous study of arithmetic, you have solved problems in which you found the average of two or more numbers.

For example, find the average of 17, 25, and 30.

First, add these three numbers: $17 + 25 + 30 = 72$

Then, divide this sum by 3 since there are 3 numbers: $72 \div 3 = 24$

The average of the three numbers is 24. We can state the procedure in general terms.

PROCEDURE. To find the average of N numbers, add the numbers and divide the sum by N.

Averages in Statistics

The word **average** has many different meanings. For example, there is an *average* of test scores, a batting *average*, the *average* television viewer, an *average* intelligence, the *average* size of a family, and so on. These averages are *not* necessarily found by the same rule or procedure. Because of this confusion, we speak of the ***measures of central tendency***. These measures are numbers that usually fall somewhere in the center of a set of organized data.

We will discuss three measures of central tendency called the *mean*, the *median*, and the *mode*.

The Mean

In statistics, the arithmetic average you previously studied is called the ***mean*** of a set of numbers. It is also called the *arithmetic mean* or the *numerical average*. It is found in the same way as the arithmetic average is found. Hence, we can say the following:

PROCEDURE. To find the mean of a set of N numbers, add the numbers and divide the sum by N.

For example, if Ralph's grades on five tests in science during this marking period are 93, 80, 86, 72, and 94, he can find the mean (or mean average) of his test grades as follows:

Step 1. Add the five pieces of data: $93 + 80 + 86 + 72 + 94 = 425$

Step 2. Divide this sum by 5, the number of tests: $425 \div 5 = 85$
Hence, the mean is 85.

Let us consider another example.

In a car wash, there are seven employees whose ages are 17, 19, 20, 17, 46, 17, and 18. What is the mean (or mean average) of the ages of these employees?

Here, add the seven ages to get a sum of 154. Then, $154 \div 7 = 22$. While the mean age of 22 is the correct answer, this measure does *not* truly represent the data. Notice that only one person is older than 22 while six of these people are under 22. For this reason, we will look at another measure of central tendency that will eliminate the extreme case (the employee at age 46) that is *distorting* the data.

The Median

The ***median*** is the middle score for a set of data arranged in numerical order. For example, the median of the ages 17, 19, 20, 17, 46, 17, and 18 can be found in the following manner:

Step 1. Arrange the ages in numerical order: 17, 17, 17, 18, 19, 20, 46

Step 2. Find the middle number: 17, 17, 17, 18, 19, 20, 46

The median is 18 because there are three ages below 18, and three ages above 18.

Notice that the median 18 is a better indication of the typical age of the employees than the mean of 22 because there are so many younger people working at the car wash.

Let us find the median for the following scores: 6, 6, 21, 11, 8, 14.

Step 1. Arrange the scores in numerical order: 6, 6, 8, 11, 14, 21

Step 2. Find the middle number. In this case, there are *two* middle numbers: 6, 6, 8, 11, 14, 21

Step 3. Find the mean (arithmetic average) of the two middle scores: $\frac{8 + 11}{2} = \frac{19}{2} = 9\frac{1}{2}$

Hence, the median is $9\frac{1}{2}$.

Notice that, in both cases, the number of values of the data that are greater than the median is equal to the number that are smaller than the median.

PROCEDURE. To find the median of N numbers:

1. Arrange the numbers in numerical order.
2. If N is odd, the median is the middle number.
3. If N is even, the median is the mean (arithmetic average) of the two middle numbers.

The Mode

The ***mode*** is the score that appears most often in a given set of data. It is usually best to arrange the data in numerical order before finding the mode.

Let us consider some examples of finding the mode:

1. The ages of employees in a car wash are 17, 17, 17, 18, 19, 20, 46. The mode, which is the number appearing most often, is 17.
2. The hours spent by six students in reading a book are 6, 6, 8, 11, 14, 21. The mode, or number appearing most frequently, is 6. Notice that, in this case, the mode is not a useful measure of central tendency. Better indications can be given by the mean or the median.
3. The number of photographs printed from Renee's last six rolls of film are 8, 8, 9, 11, 11, and 12. Since 8 appears twice and 11 appears twice, we say that there are two modes: 8 and 11. We do not take the average of these two numbers since the mode tells us where most of the scores appear; we simply report both numbers. When *two modes* appear within a set of data, we say that the data is ***bimodal***.
4. The number of people living in houses on Meryl's street are 2, 2, 3, 3, 4, 5, 5, 6, 8. This data has *three* modes: 2, 3, and 5.
5. Ralph's test scores in science are 72, 80, 86, 93, and 94. Here, every number appears the same number of times, once. Since *no* number appears more often than others, we define such data as having ***no mode***.

PROCEDURE. To find the mode for a set of data:

1. Select the score that appears most often in the data.
2. If two or more numbers appear the same number of times, and this is more than any other numbers, then each of these numbers is a mode.
3. If every number in the set of data appears the same number of times, there is no mode.

KEEP IN MIND

Three measures of central tendency are:

1. the *mean*, or *mean average*, found by adding N pieces of data and then dividing the sum by N;
2. the *median*, or *middle* score, found only when data is arranged in numerical order;
3. the *mode*, or the score appearing *most often*.

MODEL PROBLEMS

1. The weights of five players on the basketball team are 168 lb., 174 lb., 181 lb., 195 lb., and 182 lb. Find the average weight of a player on this team.

Solution

The word *average*, by itself, indicates the *mean*. Therefore:

(1) Add the five weights: $168 + 174 + 181 + 195 + 182 = 900$

(2) Divide the sum by 5, the number of players: $900 \div 5 = 180$

Answer: 180 lb.

2. Renaldo has marks of 75, 82, and 90 in three mathematics tests. What mark must he obtain on the next test to have an average of exactly 85 for the four math tests?

Solution

The word *average* by itself indicates the *mean* average.
Let x = Renaldo's mark on the fourth test.

The sum of the four test marks divided by 4 is 85.

$$\frac{75 + 82 + 90 + x}{4} = 85$$

$$\frac{247 + x}{4} = 85$$

$$247 + x = 340$$

$$x = 93$$

Check

$$\frac{75 + 82 + 90 + 93}{4} \stackrel{?}{=} 85$$

$$\frac{340}{4} \stackrel{?}{=} 85$$

$$85 = 85$$

(True)

Answer: Renaldo must obtain a mark of 93 on his fourth math test.

3. Find the median for each distribution.
a. 3, 2, 5, 5, 1 **b.** 9, 8, 8, 7, 4, 3, 3, 2, 0, 0

Solution

a. Arrange the data in numerical order: 1, 2, 3, 5, 5.
The median is the middle score. 1, 2, 3, 5, 5

Answer: median = 3

b. Since there is an even number of scores, there are two middle scores. Find the mean (average) of the two middle scores.

9, 8, 8, 7, 4, 3, 3, 2, 0, 0

$$\frac{4 + 3}{2} = \frac{7}{2} = 3\frac{1}{2}$$

Answer: median $= 3\frac{1}{2}$ or 3.5.

4. Find the mode for each distribution.
a. 2, 9, 3, 7, 3 **b.** 3, 4, 5, 4, 3, 7, 2 **c.** 1, 2, 3, 4, 5, 6, 7

Solution

a. Arrange the data in numerical order: 2, 3, 3, 7, 9.
The mode, or most frequent score, is 3. *Answer:* mode = 3

b. Arrange the data in numerical order: 2, 3, 3, 4, 4, 5, 7.
Both 3 and 4 appear twice. There are two modes.
Answer: modes = 3 and 4

c. Every number occurs the same number of times in the data set 1, 2, 3, 4, 5, 6, 7. *Answer:* no mode

EXERCISES

1. Sid received grades of 92, 84, and 70 on three tests. Find his test average.
2. Sarah received the mark 80 on two of her tests and 90 on each of three other tests. Find her test average.
3. Louise received the mark of x on two of her tests and y on each of three other tests. Represent her average for all the tests in terms of x and y.
4. Andy has grades of 84, 65, and 76 on three social studies tests. What grade must be obtain on the next test to have an average of exactly 80 for the four tests?
5. Rosemary has grades of 90, 90, 92, and 78 on four English tests. What grade must she obtain on the next test so that her average for the five tests will be 90?

6. The first three test scores are shown for each of four students. A fourth test will be given and averages taken for all four tests. Each student hopes to maintain an average of 85. Find the score needed by each student on the fourth test to have an 85 average, or explain why such an average is not possible.
   a. Pat: 78, 80, 100
   b. Bernice: 79, 80, 81
   c. Helen: 90, 92, 95
   d. Al: 65, 80, 80
7. The smallest of three consecutive even integers is 32. Find the mean average of the three integers.
8. The average of the weights of Sue, Pam, and Nancy is 55 kilograms. How much does Agnes weigh if the mean weight of the four girls is 60 kilograms?
9. The average of three consecutive even integers is 20. Find the integers.
10. The mean average of three numbers is 31. The second is one more than twice the first. The third is four less than three times the first. Find the numbers.
11. Find the mean for each set of data.
   a. 7, 3, 9, 21, 10
   b. 7, 3, 9, 21
   c. 7, 3, 9, 21, 0
   d. 16, 16, 17, 19, 84
   e. $2\frac{1}{2}, 2\frac{1}{2}, 3\frac{1}{2}, 4, 7\frac{1}{2}$
   f. 2, .2, 2.2, .02
12. If the heights of a group of students are 180 cm, 180 cm, 173 cm, 170 cm, and 167 cm, what is the mean height of these students?
13. Find the median for each set of data.
   a. 3, 4, 7, 8, 12
   b. 2, 9, 10, 10, 12
   c. 3.2, 4, 4.1, 5, 5
   d. 3, 4, 7, 8, 12, 13
   e. 2, 9, 10, 10
   f. 3.2, 4, 4.1, 5
14. Find the median for each set of data after placing the data into numerical order.
   a. 1, 2, 5, 3, 4
   b. 2, 9, 2, 9, 7
   c. 3, 8, 12, 7, 1, 0, 4
   d. 80, 83, 97, 79, 25
   e. 3.2, 8.7, 1.4
   f. 2, .2, 2.2, .02, 2.02
   g. 21, 24, 23, 22, 20, 24, 23, 21, 22, 23
   h. 5, 7, 9, 3, 8, 7, 5, 6
15. What is the median age of a family whose members are 42, 38, 14, 13, 10, and 8 years old?
16. What is the median age of a class where 14 students are 14 years old and 16 students are 15 years old?
17. In a charity collection, ten people gave amounts of $1, $2, $1, $1, $3, $1, $2, $1, $1, and $1.50. What was the median donation?
18. The test results of an examination were 62, 67, 67, 70, 90, 93, and 98. What is the median test score?
19. What is the median for the digits 1, 2, 3, . . . , 9?

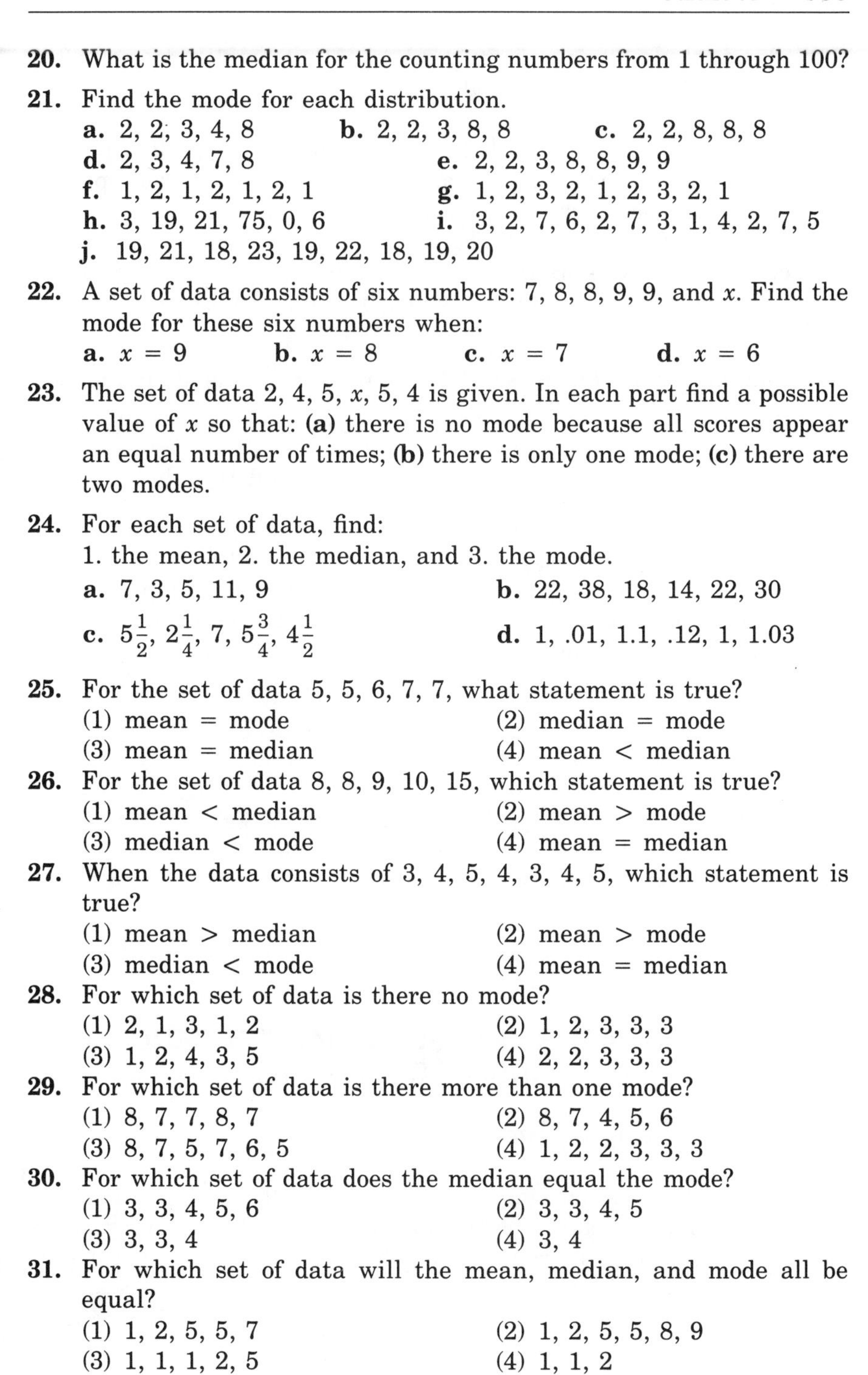

20. What is the median for the counting numbers from 1 through 100?

21. Find the mode for each distribution.

a. 2, 2, 3, 4, 8 **b.** 2, 2, 3, 8, 8 **c.** 2, 2, 8, 8, 8
d. 2, 3, 4, 7, 8 **e.** 2, 2, 3, 8, 8, 9, 9
f. 1, 2, 1, 2, 1, 2, 1 **g.** 1, 2, 3, 2, 1, 2, 3, 2, 1
h. 3, 19, 21, 75, 0, 6 **i.** 3, 2, 7, 6, 2, 7, 3, 1, 4, 2, 7, 5
j. 19, 21, 18, 23, 19, 22, 18, 19, 20

22. A set of data consists of six numbers: 7, 8, 8, 9, 9, and x. Find the mode for these six numbers when:

a. $x = 9$ **b.** $x = 8$ **c.** $x = 7$ **d.** $x = 6$

23. The set of data 2, 4, 5, x, 5, 4 is given. In each part find a possible value of x so that: **(a)** there is no mode because all scores appear an equal number of times; **(b)** there is only one mode; **(c)** there are two modes.

24. For each set of data, find:
1. the mean, 2. the median, and 3. the mode.

a. 7, 3, 5, 11, 9 **b.** 22, 38, 18, 14, 22, 30
c. $5\frac{1}{2}$, $2\frac{1}{4}$, 7, $5\frac{3}{4}$, $4\frac{1}{2}$ **d.** 1, .01, 1.1, .12, 1, 1.03

25. For the set of data 5, 5, 6, 7, 7, what statement is true?
(1) mean = mode (2) median = mode
(3) mean = median (4) mean $<$ median

26. For the set of data 8, 8, 9, 10, 15, which statement is true?
(1) mean $<$ median (2) mean $>$ mode
(3) median $<$ mode (4) mean = median

27. When the data consists of 3, 4, 5, 4, 3, 4, 5, which statement is true?
(1) mean $>$ median (2) mean $>$ mode
(3) median $<$ mode (4) mean = median

28. For which set of data is there no mode?
(1) 2, 1, 3, 1, 2 (2) 1, 2, 3, 3, 3
(3) 1, 2, 4, 3, 5 (4) 2, 2, 3, 3, 3

29. For which set of data is there more than one mode?
(1) 8, 7, 7, 8, 7 (2) 8, 7, 4, 5, 6
(3) 8, 7, 5, 7, 6, 5 (4) 1, 2, 2, 3, 3, 3

30. For which set of data does the median equal the mode?
(1) 3, 3, 4, 5, 6 (2) 3, 3, 4, 5
(3) 3, 3, 4 (4) 3, 4

31. For which set of data will the mean, median, and mode all be equal?
(1) 1, 2, 5, 5, 7 (2) 1, 2, 5, 5, 8, 9
(3) 1, 1, 1, 2, 5 (4) 1, 1, 2

32. The weekly salaries of six employees in a small firm are \$340, \$345, \$345, \$350, \$350, and \$520.

a. For these six salaries, find:
1. the mean 2. the median 3. the mode

b. If negotiations for new salaries are in session and you represent management, which measure of central tendency would you use as the average salary? Tell why.

c. If negotiations are in session and you represent the labor union, which measure of central tendency would you use as an average salary? Tell why.

33. In a certain school, bus service is provided for students living more than $1\frac{1}{2}$ miles from school. The distances from school to home for ten students are 0, $\frac{1}{2}$, $\frac{1}{2}$, 1, 1, 1, 1, $1\frac{1}{2}$, $3\frac{1}{2}$, and 10 miles.

a. For this data, find: 1. the mean 2. the median 3. the mode

b. How many of these ten students are entitled to bus service?

c. Explain why the mean is not a good measure of central tendency to describe the average distance between home and school for these students.

34. Last month, a carpenter used 12 boxes of nails each of which contained nails of only one size. The sizes marked on the boxes were $\frac{3''}{4}$, $\frac{3''}{4}$, $\frac{3''}{4}$, $\frac{3''}{4}$, $\frac{3''}{4}$, $\frac{3''}{4}$, $\frac{3''}{4}$, $\frac{3''}{4}$, 1″, 1″, 2″, and 2″.

a. For this data, find: 1. the mean 2. the median 3. the mode

b. Describe the average size nail used by the carpenter, using at least one of these measures of central tendency. Explain your answer.

16-6 MEASURES OF CENTRAL TENDENCY AND GROUPED DATA

Intervals of Length One

In a statistical study, when the range is small, we use intervals of length one to group data. For example, a class of 25 students reported on the number of books read during the first half of the school year. The data was given as:

5, 3, 5, 3, 1, 8, 2, 4, 2, 6, 3, 8, 8,
5, 3, 4, 5, 8, 5, 3, 3, 5, 6, 2, 3

Interval	Frequency (*Number*)
8	4
7	0
6	2
5	6
4	2
3	7
2	3
1	1

$N = 25$

This data was organized into a table as shown. Since 25 students were included in this study, we indicate the total frequency by writing $N = 25$ at the bottom of the Frequency column.

Let us now find the mode, median, and mean for this grouped data.

Mode

Since the greatest frequency, 7, appears in interval 3, the mode for this data is 3.

In general:

- **For a set of grouped data, the mode is the value of the interval that contains the greatest frequency.**

Median

You have learned that the median for a set of data in numerical order is the middle score.

$\underbrace{1, 2, 2, 2, 3, 3, 3, 3, 3, 3, 3, 4,}\ 4,\ \underbrace{5, 5, 5, 5, 5, 5, 6, 6, 8, 8, 8, 8}$

Hence, for these 25 scores, the median, or middle score, is 4. That is, there are 12 scores larger than or equal to the median 4, and 12 scores smaller than or equal to it.

When the data is grouped in the table that follows, a simple counting procedure can be used to find the median. Since the total frequency is 25, the median will be the 13th score. There are 12 scores above the median, and 12 scores below it.

Interval	Frequency
8	4
7	0
6	2
5	6
4	2
3	7
2	3
1	1

Counting the frequencies from the uppermost interval and moving down: $4 + 0 + 2 + 6 = 12$

Since these are the 12 scores above the median, the median must lie in the next lower interval, namely the interval that contains the score 4.

Counting the frequencies from the lowermost interval and moving up: $1 + 3 + 7 = 11$

The next higher interval contains two scores, both the 12th number and the median or middle number. Again, this is the interval that contains the score 4.

In general:

● **For a set of grouped data, the median is the value of the interval that contains the middle score.**

Mean

By adding the four 8's in the ungrouped data, we see that 4 students reading 8 books each gives a total of 32 books: $8 + 8 + 8 + 8 = 32$. Notice that we can arrive at this same number by using the grouped intervals in the table; multiply the interval 8 by the frequency 4, thus, $(4)(8) = 32$. Apply this multiplication shortcut to each line of the table, obtaining the third column of the following table:

Interval	Frequency	(Interval) · (Frequency)
8	4	$8 \cdot 4 = 32$
7	0	$7 \cdot 0 = 0$
6	2	$6 \cdot 2 = 12$
5	6	$5 \cdot 6 = 30$
4	2	$4 \cdot 2 = 8$
3	7	$3 \cdot 7 = 21$
2	3	$2 \cdot 3 = 6$
1	1	$1 \cdot 1 = 1$
	$N = 25$	Total $= 110$

To find the total number of books read, add the eight products obtained: $32 + 0 + 12 + 30 + 8 + 21 + 6 + 1 = 110$

The **total** (110) represents **the sum of all 25 items of data**. We can check this by adding the 25 scores in the unorganized data.

Finally, to find the mean, divide the total number, 110, by the number of scores, 25. Thus, $110 \div 25 = 4.4$, which is the mean.

PROCEDURE. To find the mean for N scores in a table of grouped data when the length of the interval is one:

1. For each of the intervals, multiply the interval value by its corresponding frequency.
2. Find the sum of these products.
3. Divide this sum by the total frequency N.

Intervals Other Than Length One

While there are definite mathematical procedures to find the mean, median, and mode for grouped data with intervals other than length one, you will not study them at this time. Instead, you will simply identify the intervals that contain some of these measures of central tendency. Here is an example:

A small industrial plant surveyed 50 workers to find out the number of miles each person commuted to work. The commuting distances were reported to the nearest mile as:

0, 0, 1, 1, 2, 2, 2, 3, 3, 4, 4, 4, 5, 5, 6, 6, 6, 7, 7, 7, 9,
10, 10, 10, 10, 10, 10, 10, 10, 12, 12, 14, 15, 17, 17,
18, 22, 23, 25, 28, 30, 32, 32, 33, 34, 34, 36, 37, 37, 52

This data was organized into a table with intervals of length ten, as follows:

Interval (*Commuting Distance*)	**Frequency** (*Number of Workers*)
50–59	1
40–49	0
30–39	9
20–29	4
10–19	15
0–9	21

$N = 50$

Modal Interval

In the table, the interval 0–9 contains the greatest frequency, 21. We say that the interval 0–9 is the ***group mode***, or ***modal interval***, because this group of numbers has the greatest frequency.

Notice that the modal interval is *not* the same as the mode. The modal interval is a group of numbers; the mode is generally a single number. For this particular problem, the original data (before being placed into the table) shows that the number appearing most often is 10. Hence, the mode is 10. The modal interval, which is 0–9, tells us that, of the six intervals in the table, the most frequently occurring commuting distance is 0 to 9 miles. The modal interval for this grouping is 0–9.

Both the mode and the modal interval depend upon the concept of greatest frequency. For the mode, we look for a single number that has the greatest frequency. For the modal interval, we look for the interval that has the greatest frequency.

Interval Containing the Median

To find the interval containing the median, follow the same procedure described earlier in this section. For 50 scores, the median, or middle number, will be at a point where 25 scores are above the median and 25 scores are below the median.

Interval	Frequency
50–59	1
40–49	0
30–39	9
20–29	4
10–19	15
0–9	21

$N = 50$

Counting the frequencies from the uppermost interval and moving downward, we add $1 + 0 + 9 + 4 = 14$. Since there are 15 scores in the next lower interval, and $14 + 15 = 29$, we see that 25 scores will be reached somewhere in that interval, 10–19.

Counting from the bottom interval and moving up, we have 21 scores in the first interval. Since there are 15 scores in the next higher interval, and $21 + 15 = 36$, we see that 25 scores will be reached somewhere in that interval, 10–19. Notice that this is the same result that we previously obtained when we moved downward. The interval containing the median for this grouping is 10–19.

In this course, we will not deal with problems in which the median is not found in any interval.

As a final note, there is no simple procedure to identify the interval that contains the mean when data is grouped using intervals other than length one. This problem will also be studied in higher-level courses.

MODEL PROBLEM

In the table, data is given to indicate the heights (in inches) of 17 basketball players. For this data find:

a. the mode **b.** the median **c.** the mean

Height *(Inches)*	**Frequency** *(Number)*
77	2
76	0
75	5
74	3
73	4
72	2
71	1

Solution

a. The greatest frequency, 5, occurs for the interval where heights are 75 inches. The mode, or height appearing most often, is 75.

Answer: mode = 75

Height	**Frequency**
77	2
76	0
75	5
74	3
73	4
72	2
71	1

b. For 17 players, the median is the middle number (the 9th number) so that there are 8 numbers greater than or equal to the median, and 8 numbers less than or equal to the median.

Counting the frequencies going down, $2 + 0 + 5 = 7$ and $2 + 0 + 5 + 3 = 10$. Thus, the middle number must lie in the interval 74.

Counting the frequencies going up, $1 + 2 + 4 = 7$ and $1 + 2 + 4 + 3 = 10$. Thus, the middle number must lie in the interval 74.

Answer: median = 74

c. (1) Multiply each height given as an interval by the frequency for that interval.

(2) Find the total of these products. Here, it is 1,258.

(3) Divide this total of 1,258 by the total frequency (here, $N = 17$) to obtain the mean of 74.

$77 \cdot 2 = 154$ $76 \cdot 0 = 0$ $75 \cdot 5 = 375$ $74 \cdot 3 = 222$ $73 \cdot 4 = 292$ $72 \cdot 2 = 144$ $71 \cdot 1 = 71$ Total = 1,258	$17\overline{)1258}$ = 74 119 68 68

Answer: mean = 74

EXERCISES

In 1–3, data is grouped in intervals of length one. Find:

a. the total frequency
b. the mean
c. the median
d. the mode

1.

Interval	Frequency
10	1
9	2
8	3
7	2
6	4
5	3

2.

Interval	Frequency
15	3
16	2
17	4
18	1
19	5
20	6

3.

Interval	Frequency
25	4
24	0
23	3
22	2
21	4
20	5
19	2

4. On a test consisting of 20 questions, 15 students received the following scores:

17, 14, 16, 18, 17, 19, 15, 15,
16, 13, 17, 12, 18, 16, 17.

a. On your answer paper, copy and complete the table shown at the right.

b. Find the median score.
c. Find the mode.
d. Find the mean.

Grade	Frequency *(Number)*
20	
19	
18	
17	
16	
15	
14	
13	
12	

Interval (*Minutes*)	Number of People (*Frequency*)
6	12
5	20
4	36
3	20
2	12

5. A questionnaire was distributed to 100 people. The table at the left shows the time taken in minutes to complete the questionnaire.

a. For this data, find:
1. the mean
2. the median
3. the mode

b. How are these three measures related for this data?

6. A store owner kept a tally of the sizes of suits purchased in her store, as shown in the table at the right.

a. For this set of data find:
1. the total frequency
2. the mean
3. the median
4. the mode

b. Which measure of central tendency should the store owner use to describe the average suit sold?

Size of Suit (*Interval*)	Number Sold (*Frequency*)
48	1
46	1
44	3
42	5
40	3
38	8
36	2
34	2

In 7–9, data is grouped in intervals other than length one. Find:
a. the total frequency **b.** the interval which contains the median
c. the modal interval

7.

Interval	Frequency
55–64	3
45–54	8
35–44	7
25–34	6
15–24	2

8.

Interval	Frequency
4–9	12
10–15	13
16–21	9
22–27	12
28–33	15
34–39	10

9.

Interval	Frequency
126–150	4
101–125	6
76–100	6
51–75	3
26–50	7
1–25	2

10. Test scores for a class of 20 students are given as:

93, 84, 97, 98, 100, 78, 86,

100, 85, 92, 72, 55, 91, 90,

75, 94, 83, 60, 81, 95.

Test Scores	Frequency (*Number*)
91–100	
81–90	
71–80	
61–70	
51–60	

a. On your answer paper, copy and complete the table shown at the right.
b. Find the modal interval.
c. Find the interval that contains the median.

11. The following data consists of weights, in pounds, of 35 adults:

176, 154, 161, 125, 138, 142,

108, 115, 187, 158, 168, 162,

135, 120, 134, 190, 195, 117,

142, 133, 138, 151, 150, 168,

172, 115, 148, 112, 123, 137,

186, 171, 166, 166, 179.

Interval	Frequency
180–199	
160–179	
140–159	
120–139	
100–119	

a. Copy and complete the table shown above.
b. Construct a frequency histogram based on the grouped data.
c. In what interval is the median for this grouped data?
d. What is the modal interval?

16-7 CUMULATIVE FREQUENCY HISTOGRAMS AND PERCENTILES

In a school, a final examination was given to all students taking biology. The total number of students taking this examination was 240. The test grades of these students were grouped into a table as shown at the right. At the same time, a histogram of the results was constructed, shown below.

Interval (*Test Scores*)	Frequency (*Number*)
91–100	45
81–90	60
71–80	75
61–70	40
51–60	20

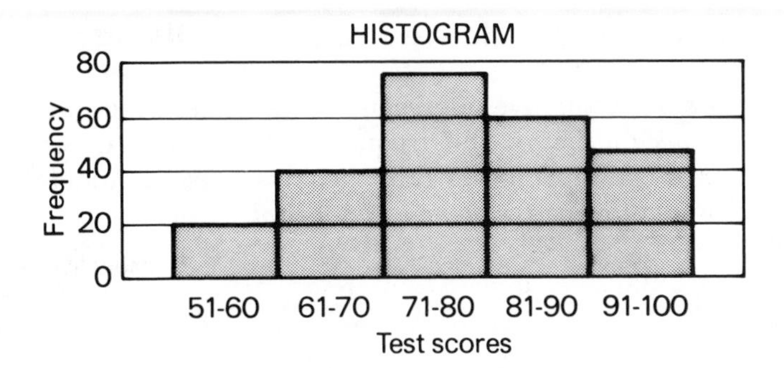

From the table and the histogram, we can see that 20 students scored in the interval 51–60, 40 students scored in the interval 61–70, and so forth. We can use this data to construct a new type of histogram that will answer the question:

"How many students scored ***below*** *a certain grade?"*

By answering the following questions, we will gather some information before constructing the new histogram.

1. How many students scored 60 or less on the test?
   From the bottom interval 51–60, we know the answer is 20 students. *Answer:* 20

2. How many students scored 70 or less on the test?
   By adding the frequencies in the two lowest intervals, 51–60 and 61–70, we see that 20 + 40, or 60, students scored 70 or less on the test. *Answer:* 60

3. How many students scored 80 or less on the test?
   By adding the frequencies in the three lowest intervals, 51–60, 61–70, and 71–80, we see that 20 + 40 + 75, or 135, students scored 80 or less on the test. *Answer:* 135

4. How many students scored 90 or less on the test?
   Here, we add the frequencies in the four lowest intervals. Thus, 20 + 40 + 75 + 60, or 195, students scored 90 or less. *Answer:* 195

5. How many students scored 100 or less on the test?
   By adding the five lowest frequencies, 20 + 40 + 75 + 60 + 45, we see that 240 students scored 100 or less. This makes sense because 240 students took the test and all of them scored 100 or less. *Answer:* 240

Constructing a Cumulative Frequency Histogram

The answers to the five questions we have just asked were found by adding or *accumulating* the frequencies from the intervals in the grouped data. The histogram that displays these accumulated figures is called a ***cumulative frequency histogram***.

Interval (*Test Scores*)	Frequency (*Number*)	Cumulative Frequency
91–100	45	240
81–90	60	195
71–80	75	135
61–70	40	60
51–60	20	20

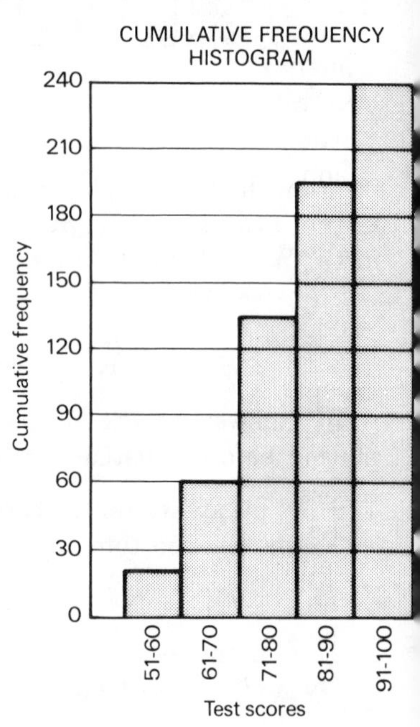

To find the cumulative frequency for each interval, add the frequency from that interval to the frequencies for the intervals with lower scores.

To draw a cumulative frequency histogram, use these cumulative frequencies to determine the heights of the bars.

Percentiles, Quartiles, and the Median

Notice that the frequency scale for this cumulative frequency histogram goes from 0 to 240 (the total frequency for this data).

It is also possible to place a different scale to the left of the cumulative frequency histogram, namely one involving *percents*. Since 240 students represent 100% of the population, we write 100% to correspond to 240 students. Since no students, or zero students, represent 0% of the population, we write 0% to correspond with 0 students.

If we divide the percent scale into four equal parts, we can label these divisions 25%, 50%, and 75%. The graph relates each cumulative frequency to a percent of the population. For example, 120 students (half of the population) correspond to 50% of the population.

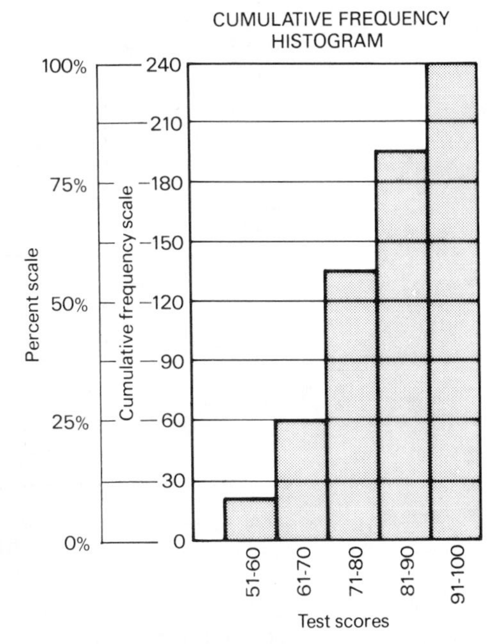

Let us examine the percent scale in relation to the question:

"What percent of the students scored 70 or below on the test?"

The height of each bar represents both the percent and the number of students who had scores in or below the range of numbers with which the bar is labeled. The bar for the interval 61–70 shows that 25% or 60 students out of a total population of 240 had scores of 70 or less: $\left(\frac{60}{240} = \frac{1}{4} = 25\%\right)$.

Since 25%, or a quarter, of the scores were 70 or below, we say that the score of 70 is the ***lower quartile*** or ***first quartile*** for these scores.

From the histogram, we can see that 50%, or one half, of the students had a score that was at or below a number in the interval 71–80. Thus, the second quartile is a score between 71 and 80. The ***second quartile*** is the same as the median.

The histogram also shows that 75% of the students had a score that was at or below a number in the interval 81–90. The ***third quartile*** is that number at or below which 75%, or three quarters, of the scores fall. For this set of data, the third quartile is some number between 81 and 90. The third quartile is also called the ***upper quartile***. If 75% of the students had a score at or below this number, then 25% of the students had a score at or above this number.

A ***percentile*** is a score or a measure that tells us what percent of the total population or the total frequency scored at or below that measure.

From the cumulative frequency histogram, we can also conveniently read the approximate percentiles for the scores that are the end values of the intervals.

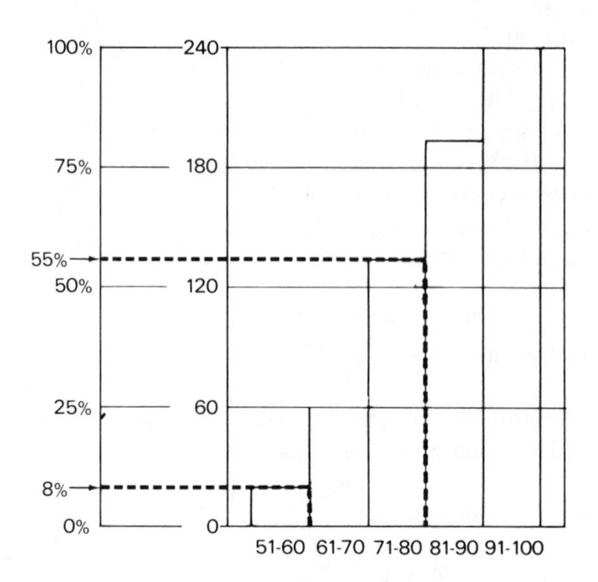

For example, to find the percentile for a score of 60, the right-end score of the first interval, draw a vertical line segment beginning at 60 on the score scale and ending at the height of the first interval. From this point, draw a horizontal line to the percent scale. We estimate this value on the percent scale to be about 8% (since the horizontal line crosses the percent scale at about one-third the distance between 0% and 25%). This tells us that approximately 8% of the students scored 60 or below 60. Hence, *60 is the score for the 8th percentile.*

Similarly, we can read from the graph that 80 on the score scale corresponds approximately to 55% on the percent scale. This tells us that approximately 55% of the students scored 80 or less than 80. Hence, *80 is the score for the 55th percentile.*

Notice that:

1. **The upper quartile is the 75th percentile.**
2. **The median is the 50th percentile.**
3. **The lower quartile is the 25th percentile.**

MODEL PROBLEM

For the given table of values, in what interval is the lower quartile found?

Interval	Frequency
41–50	4
31–40	3
21–30	6
11–20	7
1–10	4

Solution

The total frequency = $4 + 3 + 6 + 7 + 4 = 24$.

The lower quartile is the score below which 25% of the total frequency falls. Since 25% of $24 = \frac{1}{4} \times 24 = 6$, we have to find the interval (starting from the interval with the lowest scores) in which the 6th score falls.

In the lowest interval, 1–10, there are only 4 scores. Hence, the 6th score is found in the next interval, 11–20.

Answer: The lower quartile is in the interval 11–20.

EXERCISES

1. In the table, data is given for the heights in inches of 20 basketball players.

   a. Copy and complete the table.

   b. Draw a cumulative frequency histogram.

   c. Find the height that is the lower quartile.

   d. Find the height that is the 90th percentile.

Height	Frequency	Cumulative Frequency
77	2	
76	1	
75	7	
74	5	
73	2	
72	2	
71	1	

In 2–4, data is grouped into tables. For each set of data:

a. Construct a cumulative frequency histogram.
b. Find the interval in which the lower quartile lies.
c. Find the interval in which the median lies.
d. Find the interval in which the upper quartile lies.

2.

Interval	Frequency
41–50	8
31–40	5
21–30	2
11–20	5
1–10	4

3.

Interval	Frequency
25–29	3
20–24	1
15–19	3
10–14	9
5–9	4

4.

Interval	Frequency
1–4	6
5–8	3
9–12	7
13–16	2
17–20	2

5. For the data shown at the right:

a. Construct a histogram.

b. Construct a cumulative frequency histogram.

c. In what interval is the median found?

d. Find the lower quartile.

Interval	Frequency
21–25	5
16–20	4
11–15	6
6–10	3
1–5	2

6. For the cumulative frequency histogram shown at the right:

a. In what interval is the median found?
b. What is the value of the 25th percentile?
c. What is the value of the 75th percentile?
d. Approximately what percent of scores are 50 or below?
e. What percent of scores are 125 or below?

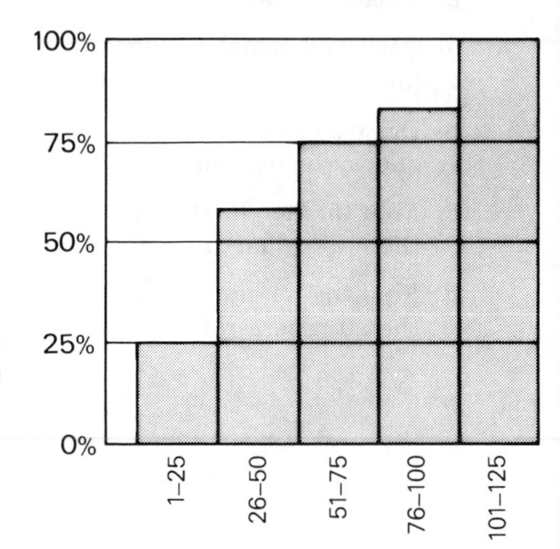

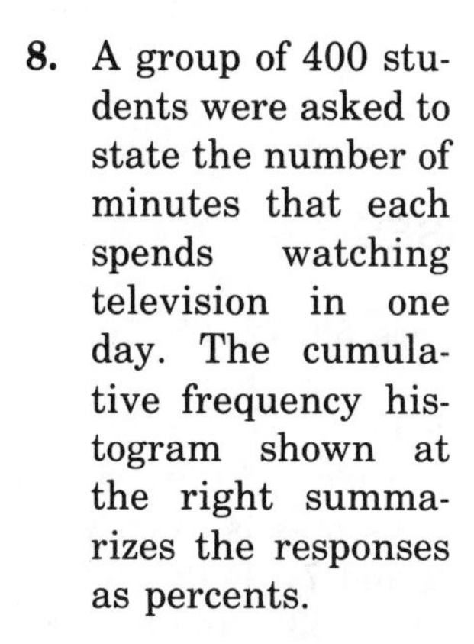

Interval	Frequency
33–37	4
28–32	3
23–27	7
18–22	12
13–17	8
8–12	5
3–7	1

7. For the data given in the table at the left:
   a. Construct a cumulative frequency histogram.
   b. In what interval is the median found?
   c. In what interval is the upper quartile found?
   d. Approximately what percent of scores are 17 or less?
   e. In what interval is the 25th percentile found?
   f. What is the value of the 90th percentile?

8. A group of 400 students were asked to state the number of minutes that each spends watching television in one day. The cumulative frequency histogram shown at the right summarizes the responses as percents.

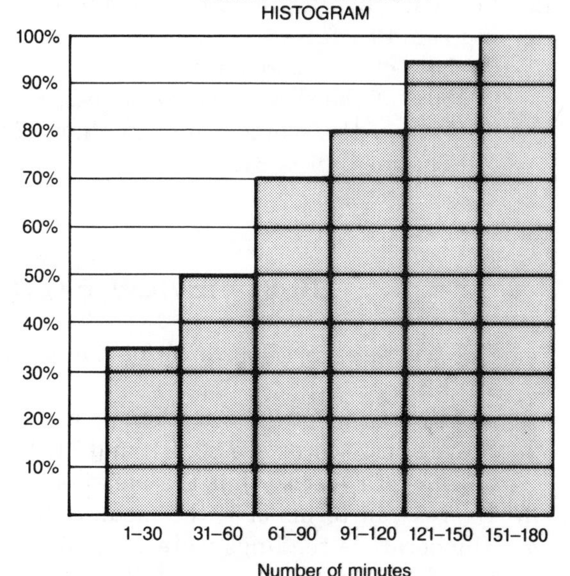

   a. What is the median number of minutes of television that these students watch each day?
   b. What percent of the students questioned watch television for 90 minutes or less each day?
   c. How many of the students watch television for 90 minutes or less each day?
   d. In what interval does the upper quartile lie?
   e. In what interval does the lower quartile lie?
   f. If one of these students is picked at random, what is the probability that he or she watches 30 minutes or less of television each day?

9. Cecilia's average for four years is 86. Her average is the upper quartile for her class of 250 students. About how many students in her class have an average that is less than or equal to Cecilia's?
10. The lower quartile for a group of data was 40. This was determined by recording the heights, in inches, of 680 children. How many of these children measured more than 40 inches?
11. On a standarized test, Sally scored at the 80th percentile. This means that
    (1) Sally answered 80 questions correctly.
    (2) Sally answered 80% of the questions correctly.
    (3) Of the students who took the test, about 80% had the same score as Sally.
    (4) Of the students who took the test, about 80% had a score that was less than or equal to Sally's score.
12. For a set of data, the 50th percentile was 87. Which of the following could be false?
    (1) 50% of the scores are 87.
    (2) 50% of the scores are 87 or less.
    (3) Half of the scores are more than 87.
    (4) The median is 87.

16-8 REVIEW EXERCISES

1. The weights, in kilograms, for five adults are 53, 72, 68, 70, and 72. Find:
   a. the mean **b.** the median **c.** the mode
2. Steve's test scores are 82, 94, and 91. What grade must Steve earn on a fourth test so that the mean of his four scores is exactly 90?
3. Express, in terms of y, the mean of $3y - 2$ and $7y + 18$.
4. Temperature readings were 72°, 75°, 79°, 83°, 83°, and 88°. For this data, find:
   a. the mean **b.** the median **c.** the mode

In 5–7, select the numeral preceding the correct answer.

5. Which set of data has more than one mode?
   (1) 3, 4, 3, 4, 3, 5 (2) 1, 3, 5, 7, 1, 2, 4
   (3) 9, 3, 2, 8, 3, 3 (4) 9, 3, 2, 3, 8, 2, 7
6. For which set of data are the mean, median, and mode all equal?
   (1) 2, 2, 5 (2) 2, 5, 5 (3) 1, 3, 3, 5 (4) 1, 1, 5, 5
7. For the set of data 3, 2, 7, 1, 2, which statement is true?
   (1) median = mean (2) median = mode
   (3) median > mean (4) median > mode

8. Paul worked the following numbers of hours each week over a 20-week period:

15, 3, 7, 6, 2, 14, 9, 25, 8, 12,
8, 8, 15, 0, 8, 12, 28, 10, 14, 10

Interval	Frequency
24–29	
18–23	
12–17	
6–11	
0–5	

a. Copy and complete the table.

b. Draw a frequency histogram of the data.

c. In what interval does the median lie?

d. Which interval contains the lower quartile?

9. Scores of 24 papers are given.

a. Find the mean score.
b. Find the median score.
c. Find the mode.
d. Copy and complete the table.
e. Draw a cumulative frequency histogram.
f. Find the 75th percentile.
g. What is the probability that a paper chosen at random has a score of 80?

Scores	Frequency	Cumulative Frequency
60	1	
70	9	
80	8	
90	1	
100	5	

10. The ages of 21 high school students are shown in the table. When the ages of these students are combined with the ages of 20 additional students, the median age remains unchanged. What is the smallest possible number of students under 16 in the second group?

Age	Frequency
18	1
17	4
16	2
15	7
14	2
13	5

Chapter 17

The Coordinate Plane

17-1 ORDERED NUMBER PAIRS AND POINTS IN A PLANE

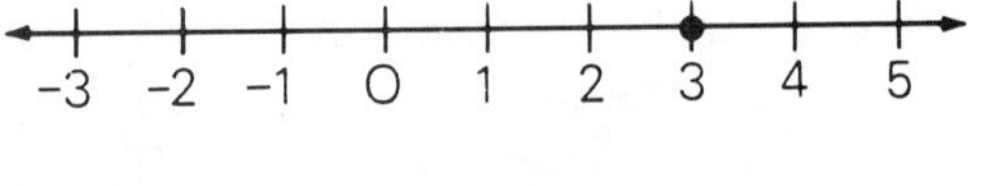

To graph a number, we use a point on a number line. To graph a pair of numbers, we use a point in a plane. We start with two signed number lines called ***coordinate axes***, which are perpendicular to each other. The horizontal line is called the ***x-axis***. The vertical line is called the ***y-axis***. In a ***coordinate plane***, or ***Cartesian plane***, the point O at which the two axes intersect is called the ***origin***. We label each axis with the same scale, letting the origin be zero on each scale. Positive numbers are to the right of zero on the x-axis, and above zero on the y-axis.

The Cartesian plane is named after the seventeenth-century mathematician René Descartes, who demonstrated how algebraic relations between variables can be described by points and lines or curves in a plane.

The x-axis and the y-axis separate the plane into four regions called ***quadrants***. These quadrants are numbered I, II, III, and IV in a counterclockwise order as shown in the drawing above. The points on the axes are not in any quadrant.

Each point of the plane is the graph of an ordered pair of numbers. The first number of the ordered pair is called the ***x-coordinate***, or the ***abscissa***. The second number is the ***y-coordinate***, or ***ordinate***. The two numbers, the abscissa and the ordinate, are called the ***coordinates*** of the point.

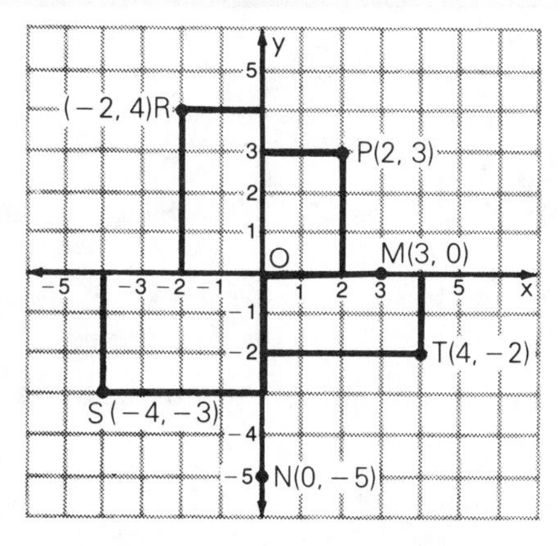

At the right, the point P in quadrant I is called the ***graph of the ordered pair*** (2, 3). In general, the coordinates of a point may be represented as (x, y).

To find the coordinates of a point, draw a vertical line from the point to the x-axis and a horizontal line to the y-axis. The number assigned to the point at which the vertical line intersects the x-axis is the x-coordinate of the point. The number assigned to the point at which the horizontal line intersects the y-axis is the y-coordinate of the point.

In the graph above, the coordinates of point R in quadrant II are $(-2, 4)$; that is, point R is the graph of the ordered pair $(-2, 4)$. Similarly, point S in quadrant III is the graph of $(-4, -3)$, and point T in quadrant IV is the graph of $(4, -2)$. A point on the x-axis has a y-coordinate of 0; point M has coordinates $(3, 0)$. A point on the y-axis has an x-coordinate of 0; point N has coordinates $(0, -5)$.

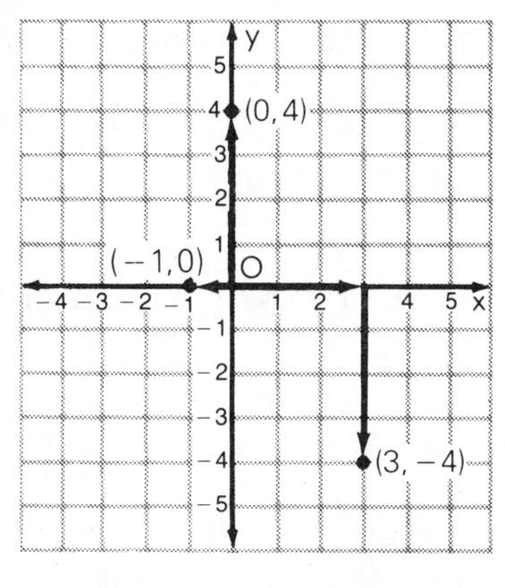

When we graph a point described by an ordered pair, we are ***plotting the point***. To plot the point $(3, -4)$, we begin at the origin, move three units to the right along the x-axis, and then move -4 units (4 units down) parallel to the y-axis. The point at which we end is the graph of $(3, -4)$. To plot the point $(-1, 0)$, we start at the origin, move -1 unit (1 unit to the left) along the x-axis and 0 units parallel to the y-axis. The graph of $(-1, 0)$ is on the x-axis. To plot the point $(0, 4)$, we start at the origin, move 0 units along the x-axis and 4 units up the y-axis. The graph of $(0, 4)$ is on the y-axis.

EXERCISES

1. Write as ordered number pairs the coordinates of points A, B, C, D, E, F, G, H, and O in the graph.

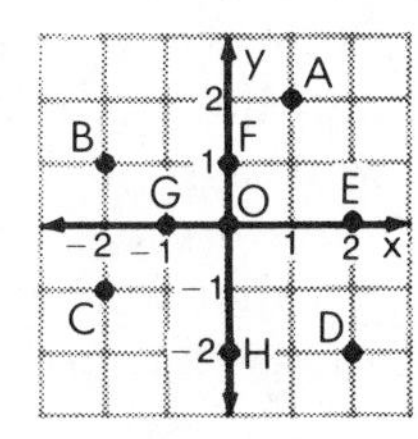

In 2–21, draw a pair of coordinate axes on a sheet of graph paper and graph the point associated with the ordered number pair. Label each point with its coordinates.

2. (5, 4) **3.** (−3, 2) **4.** (2, −6) **5.** (−4, −5)
6. (1, 6) **7.** (−8, 5) **8.** (4, −4) **9.** (−2, −7)
10. (−1.5, −2.5) **11.** (5, 0) **12.** (−3, 0) **13.** (8, 0)
14. (−10, 0) **15.** (0, 4) **16.** (0, −6) **17.** (0, 1)
18. (0, −4) **19.** (0, 0) **20.** $(|2|, |4|)$ **21.** $(|-5|, |3|)$

In 22–26, name the quadrant in which the graph of the point described appears.

22. (5, 7) **23.** (−3, −2) **24.** (−7, 4) **25.** (1, −3) **26.** $(|-2|, |-3|)$

27. Graph several points on the x-axis. What is the value of the ordinate for every point in the set of points on the x-axis?
28. Graph several points on the y-axis. What is the value of the abscissa for every point in the set of points on the y-axis?
29. What are the coordinates of the origin in the coordinate plane?
30. Name the quadrant in which the graph of point $P(x, y)$ lies when:
a. $x > 0$ and $y > 0$ **b.** $x > 0$ and $y < 0$
c. $x < 0$ and $y > 0$ **d.** $x < 0$ and $y < 0$
31. Name the quadrant in which the graph of the point $P(|x|, |y|)$ lies when:
a. $x > 0$ and $y > 0$ **b.** $x > 0$ and $y < 0$
c. $x < 0$ and $y > 0$ **d.** $x < 0$ and $y < 0$

17-2 GRAPHING POLYGONS AND FINDING AREAS

A polygon can be represented in the coordinate plane by plotting its vertices and then drawing the line segments connecting the vertices in order.

The graph at the right shows the quadrilateral $ABCD$. The vertices are $A(3, 2)$, $B(-3, 2)$, $C(-3, -2)$, and $D(3, -2)$. From the graph, note the following:

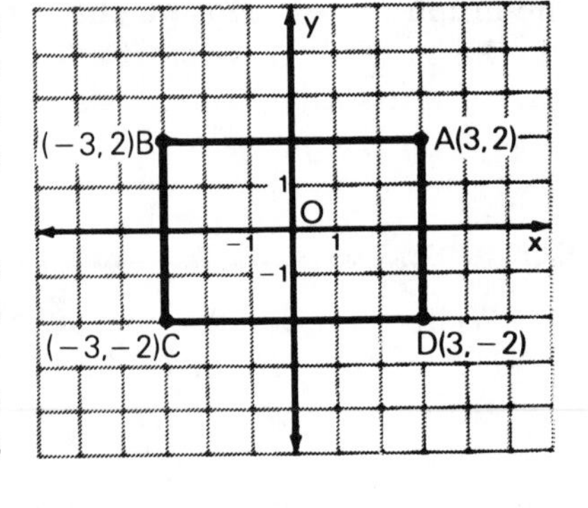

1. Points A and B, which have the same ordinate, are on a line that is parallel to the x-axis.
2. Points C and D, which have the same ordinate, are on a line that is parallel to the x-axis.
3. Lines that are parallel to the x-axis are parallel to each other.

4. Lines that are parallel to the x-axis are perpendicular to the y-axis.
5. Points B and C, which have the same abscissa, are on a line that is parallel to the y-axis.
6. Points A and D, which have the same abscissa, are on a line that is parallel to the y-axis.
7. Lines that are parallel to the y-axis are parallel to each other.
8. Lines that are parallel to the y-axis are perpendicular to the x-axis.

Now, we know that quadrilateral $ABCD$ is a rectangle, because it is a parallelogram with right angles.

From the graph, we can find the dimensions of this rectangle.

To find the length of the rectangle, we can count the number of units from A to B or from C to D. $AB = CD = 6$. Because points on the same horizontal line have the same y-coordinate, we can also find these measures by subtracting their x-coordinates.

$$AB = CD = 3 - (-3) = 3 + 3 = 6$$

To find the width of the rectangle, we can count the number of units from B to C or from D to A. $BC = DA = 4$. Because points on the same vertical line have the same x-coordinate, we can find these measures by subtracting their y-coordinates.

$$BC = DA = 2 - (-2) = 2 + 2 = 4$$

MODEL PROBLEM

Graph the following points: $A(4, 1)$, $B(1, 5)$, $C(-2, 1)$. Draw $\triangle ABC$ and find its area.

Solution:

The graph at the right shows $\triangle ABC$.

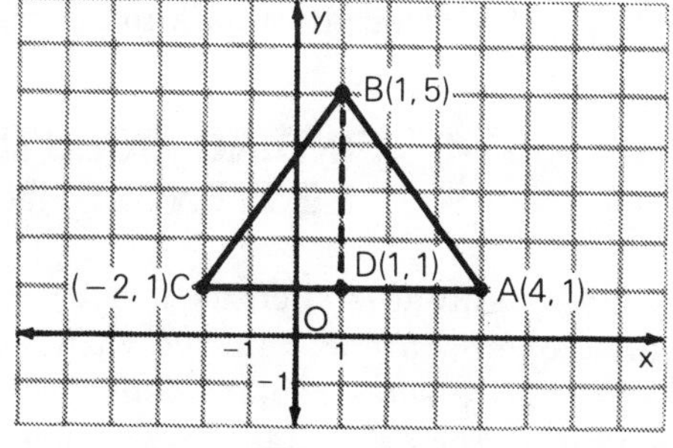

To find the area of the triangle, we need to know the lengths of the base and the altitude drawn to that base.

The base of $\triangle ABC$ is $\overline{AC}$.

$AC = 4 - (-2) = 4 + 2 = 6$

The altitude to the base is the line segment, $\overline{BD}$, drawn from B perpendicular to $\overline{AC}$. $BD = 5 - 1 = 4$

$$\text{Area} = \frac{1}{2}(AC)(BD)$$

$$= \frac{1}{2}(6)(4) = 12$$

Answer: The area of $\triangle ABC$ is 12 square units.

EXERCISES

In 1–10: **a.** Graph the points and connect them with straight lines in order, forming a polygon. **b.** Tell what kind of polygon is drawn. **c.** Find the area of the polygon.

1. $A(1, 1)$, $B(8, 1)$, $C(1, 5)$
2. $P(0, 0)$, $Q(5, 0)$, $R(5, 4)$, $S(0, 4)$
3. $C(8, -1)$, $A(9, 3)$, $L(4, 3)$, $F(3, -1)$
4. $H(-4, 0)$, $O(0, 0)$, $M(0, 4)$, $E(-4, 4)$
5. $H(5, -3)$, $E(5, 3)$, $N(-2, 0)$
6. $F(5, 1)$, $A(5, 5)$, $R(0, 5)$, $M(-2, 1)$
7. $B(-2, -2)$, $A(2, -2)$, $R(2, 2)$, $N(-2, 2)$
8. $P(-3, 0)$, $O(0, 0)$, $N(2, 2)$, $D(-1, 2)$
9. $R(-4, 2)$, $A(0, 2)$, $M(0, 7)$
10. $M(-1, -1)$, $I(3, -1)$, $L(3, 3)$, $K(-1, 3)$

11. Graph the points $A(1, 1)$, $B(5, 1)$, $C(5, 4)$. What must be the coordinates of point D if $ABCD$ is a rectangle?
12. Graph the points $P(-2, -4)$ and $Q(2, -4)$. What are the coordinates of R and S if $PQRS$ is a square? (Two answers are possible.)
13. **a.** Graph the points $S(3, 0)$, $T(0, 4)$, $A(-3, 0)$, and $R(0, -4)$ and draw the rhombus $STAR$.
b. Find the area of $STAR$ by adding the areas of the triangles into which the axes separate the rhombus.
14. **a.** Graph the points $P(2, 0)$, $L(1, 1)$, $A(-1, 1)$, $N(-2, 0)$, $E(-1, -1)$, and $T(1, -1)$ and draw the hexagon $PLANET$.
b. Find the area of $PLANET$.
Hint: Use the x-axis to separate the hexagon into two parts.

17-3 FINDING SOLUTION SETS OF OPEN SENTENCES IN TWO VARIABLES

Open sentences such as $y = 3x$, $y - 2x = 4$, and $y > 2x + 4$ are called open sentences in two variables.

There are some replacements for x and y that make $y = 3x$ a true sentence. For example, if x is replaced by 1 and y by 3, the resulting sentence $3 = 3(1)$ is a true sentence. Therefore, the pair of numbers $x = 1$, $y = 3$ is said to ***satisfy*** the open sentence $y = 3x$. Such a pair of numbers is called a ***solution*** of $y = 3x$. We can write the solution $x = 1$, $y = 3$ as an ordered pair $(1, 3)$ if we agree that the first number in the pair always represents a value of the variable x, and the second number always represents a value of the variable y.

When the replacement set for x and for y is $\{1, 2, 3, 4, 5, 6, 7, 8, 9\}$, the solutions of $y = 3x$ are the ordered pairs $(1, 3)$, $(2, 6)$, and $(3, 9)$.

This set of ordered pairs $\{(1, 3), (2, 6), (3, 9)\}$ is the ***solution set*** of the sentence $y = 3x$. The solution set of a sentence involving two variables is the set of all the ordered pairs that are solutions of the sentence. If there are no ordered pairs that are solutions of the sentence, the solution set is the empty set.

If the replacement set is the set of signed numbers, there are infinitely many members in the solution set of $y = 3x$. Since it is impossible to list all the members of the solution set, we can describe the solution set as $\{(x, y) | y = 3x\}$, which is read "the set of all ordered pairs (x, y) such that $y = 3x$."

MODEL PROBLEMS

1. Determine if the given ordered pairs are solutions of the sentence $y - 2x = 4$.

a. $(4, 0)$ **b.** $(-1, 2)$ **c.** $(0, 4)$

Solution:

a.
$$y - 2x = 4$$
$$0 - 2(4) \stackrel{?}{=} 4$$
$$0 - 8 \stackrel{?}{=} 4$$
(False)

b.
$$y - 2x = 4$$
$$2 - 2(-1) \stackrel{?}{=} 4$$
$$2 + 2 \stackrel{?}{=} 4$$
(True)

c.
$$y - 2x = 4$$
$$4 - 2(0) \stackrel{?}{=} 4$$
$$4 - 0 \stackrel{?}{=} 4$$
(True)

Answer: **a.** not a solution **b.** a solution **c.** a solution

2. Find five members of the solution set of the sentence $3x + y = 7$.

How to Proceed / *Solution*

(1) Transform the equation into an equivalent equation with y alone as one member.

$$\begin{aligned} 3x + y &= 7 \\ -3x &= -3x \\ \hline y &= -3x + 7 \end{aligned}$$

(2) Since no replacement set is given, any signed numbers can be used.

Choose any five values for x. For each selected value of x, determine y.

x	−3x + 7	y
−2	$-3(-2) + 7$	13
0	$-3(0) + 7$	7
$\frac{1}{3}$	$-3\left(\frac{1}{3}\right) + 7$	6
3	$-3(3) + 7$	−2
5	$-3(5) + 7$	−8

Answer: $(-2, 13), (0, 7), (\frac{1}{3}, 6), (3, -2), (5, -8)$

Note that many other solutions are possible.

3. Find the solution set of the sentence $y - 2x > 4$ when the replacement set for x is $R = \{1, 2, 3, 4, 5\}$ and the replacement set for y is $S = \{6, 7, 8, 9, 10\}$.

How to Proceed — *Solution*

(1) Transform the sentence into an equivalent sentence that has y alone as one member.

$$y - 2x > 4$$
$$y - 2x + 2x > 4 + 2x$$
$$y > 2x + 4$$

(2) Replace x by each member of R, the replacement set for x. Then, compute the corresponding y-values.

(3) If any y-values computed in Step (2) are members of S, the replacement set for y, then each ordered pair consisting of an x-value and its corresponding y-value is a solution of the sentence.

x	2x + 4	y > 2x + 4	y
1	2(1) + 4	$y > 6$	7, 8, 9, 10
2	2(2) + 4	$y > 8$	9, 10
3	2(3) + 4	$y > 10$	no values in S
4	2(4) + 4	$y > 12$	no values in S
5	2(5) + 4	$y > 14$	no values in S

Answer: The solution set is $\{(1, 7), (1, 8), (1, 9), (1, 10), (2, 9), (2, 10)\}$.

EXERCISES

In 1–5, find the missing member in each ordered pair if the second member of the pair is twice the first member.

1. (3, ?) **2.** (0, ?) **3.** (−2, ?) **4.** (?, 11) **5.** (?, −8)

In 6–10, find the missing member in each ordered pair if the first member of the pair is 4 more than the second member.

6. (?, 5) **7.** $\left(?, \frac{1}{2}\right)$ **8.** (?, 0) **9.** $\left(9\frac{1}{4}, ?\right)$ **10.** (−8, ?)

In 11–26, state whether or not the given ordered pair of numbers is a solution of the sentence.

11. $y = 5x$; (3, 15)
12. $y = 4x$; (16, 4)
13. $y = 3x + 1$; (7, 22)
14. $3x - 2y = 0$, (3, 2)
15. $y > 4x$; (2, 10)
16. $y < 2x + 3$; (0, 2)
17. $3y > 2x + 1$; (4, 3)
18. $2x + 3y \leq 9$; (0, 3)

19. $x + y = 8$; $(4, 5)$

20. $4x + 3y = 2$; $\left(\frac{1}{4}, \frac{1}{3}\right)$

21. $3x = y + 4$; $(-7, -1)$

22. $x - 2y = 15$; $(1, -7)$

23. $y > 6x$; $(-1, -2)$

24. $3x < 4y$; $(5, 2)$

25. $y \geq 3 - 2x$; $(-1, 6)$

26. $5x - 2y \leq 19$; $(3, -2)$

27. Which of the ordered pairs of numbers $(8, -2)$, $(2, -6)$, $(7, 13)$, $(3, 9)$ is a member of the solution set of $x + y = 6$?

28. Which of the ordered number pairs $(5, 3)$, $(7, 2)$, $(1, -1)$, $\left(-2, -\frac{1}{4}\right)$ is a member of the solution set of $y > 4x$?

29. Which of the ordered number pairs $(1, 8)$, $(5, 2)$, $(3, -1)$, $(0, -4)$ is not a member of the solution set of $y < 2x + 1$?

In 30–33, find the solution set of the sentence.

30. $x + y = 4$ when the replacement set for x is $\{5, 7\}$ and the replacement set for y is {natural numbers}.

31. $y = 3x - 1$ when the replacement set for x is $\{-3, -1, 2\}$ and the replacement set for y is {signed numbers}.

32. $y < 2x - 1$ when the replacement set for x is $\{5, 6\}$ and the replacement set for y is $\{8, 9, 10, 11\}$.

33. $x + y \geq 12$ when the replacement set for x is $\{-7, 10, 12\}$ and the replacement set for y is $\{-2, 2, 6, 10\}$.

In 34–39, use set notation to describe the solution set when the replacement set for x and for y is the set of signed numbers.

34. $y = 6x$

35. $y = x + 9$

36. $3x + y = 11$

37. $y > 10x$

38. $y < 3x - 1$

39. $y - x \geq 4$

17-4 GRAPHING A LINEAR EQUATION IN TWO VARIABLES BY MEANS OF ITS SOLUTIONS

If the replacement set for both x and y is {signed numbers}, we can find an infinite number of ordered pairs that are solutions of the sentence $x + y = 6$. Some of these solutions are shown in the table that follows.

x	7	6	5	4	3	$2\frac{1}{2}$	2	1	$\frac{1}{2}$	0	-1
y	-1	0	1	2	3	$3\frac{1}{2}$	4	5	$5\frac{1}{2}$	6	7

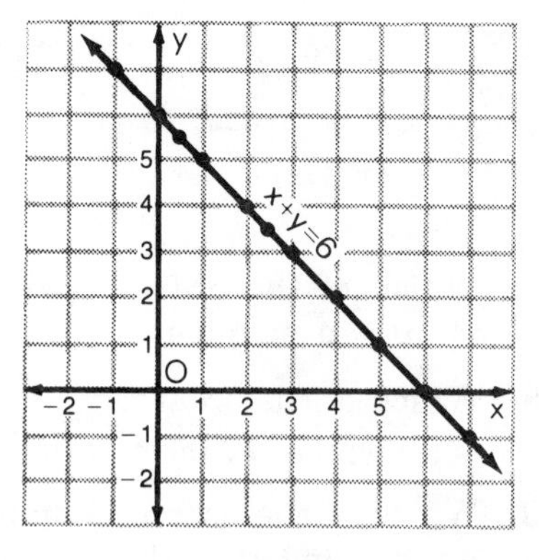

Let us plot the points associated with the ordered number pairs that are shown in the table. Notice that these points seem to lie on a straight line. In fact, if {signed numbers} is the replacement set for both x and y, then the following is true:

The graphs of all ordered pairs (x, y) that are solutions of $x + y = 6$ lie on this same line; the graphs of all ordered pairs that are not solutions of $x + y = 6$ do not lie on this line.

This line, which is the set of all those points and only those points whose coordinates satisfy the equation $x + y = 6$, is called the graph of $x + y = 6$; that is, this line is the graph of $\{(x, y) | x + y = 6\}$.

A first-degree equation in two variables may be written in the form $Ax + By + C = 0$, where A, B, and C are signed numbers, with A and B not both zero. For example, $x + y = 6$ can be written in this form, with $A = 1$, $B = 1$, and $C = -6$. It can be proved that the graph of such an equation is a straight line. We, therefore, call such an equation a ***linear equation***.

When we graph a linear equation, we can determine the line by plotting two points whose coordinates satisfy that equation. However, we always plot a third point as a check on the first two. If the third point lies on the line determined by the first two points, we have probably made no error.

PROCEDURE. To graph a linear equation by means of its solutions:

1. Transform the equation into an equivalent equation that is solved for y in terms of x.
2. Find three solutions of the equation by choosing values for x and computing corresponding values for y.
3. In the coordinate plane, graph the ordered pairs of numbers found in Step 2.
4. Draw the line that passes through the points graphed in Step 3.

Note: When we graph a linear equation involving the variables x and y, the replacement set of both variables is {signed numbers} unless otherwise indicated.

KEEP IN MIND

1. Every ordered pair of numbers that satisfies an equation represents the coordinates of a point on the graph of the equation.
2. Every point on the graph of an equation has as its coordinates an ordered pair of numbers that satisfies the equation.

MODEL PROBLEMS

1. Does the point $(2, -3)$ lie on the graph of $x - 2y = -4$?

Solution:

If the point $(2, -3)$ is to lie on the graph of $x - 2y = -4$, it must be a solution of $x - 2y = -4$.

$$\begin{aligned} x - 2y &= -4 \\ (2) - 2(-3) &\stackrel{?}{=} -4 \\ 2 + 6 &\stackrel{?}{=} -4 \\ 8 &\neq -4 \end{aligned}$$

Since $8 = -4$ is not true, the point $(2, -3)$ does not lie on the line $x - 2y = -4$. *Answer:* No

2. What must be the value of d if $(d, 4)$ lies on the line $3x + y = 10$?

Solution: The coordinate $(d, 4)$ must satisfy $3x + y = 10$.

Hence: $3d + 4 = 10$

$3d = 6$

$d = 2$ *Ans.*

3. a. Write the following verbal sentence as an equation: The sum of twice the abscissa of a point and the ordinate of that point is 4.

b. Graph the equation written in part **a**.

a. *Solution:* Let x = the abscissa of the point.

Let y = the ordinate of the point.

Then, $2x + y = 4$ *Ans.*

b.

How to Proceed	*Solution*
(1) Transform the equation into an equivalent equation that has y alone as one member.	$2x + y = 4$ $y = -2x + 4$

(2) Determine three solutions of the equation by choosing values for x and computing the corresponding values for y.

x	−2x + 4	y
0	$-2(0) + 4$	4
1	$-2(1) + 4$	2
2	$-2(2) + 4$	0

(3) Plot the points that are associated with the three solutions.

(4) Draw a line through the points that were plotted. Label the line with its equation.

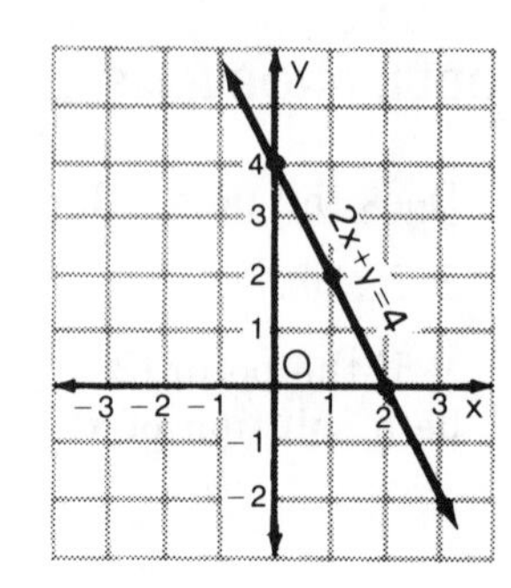

EXERCISES

1. In each part, state whether the pair of values for x and y satisfies the equation $2x - y = 6$.
a. $x = 4, y = 2$ **b.** $x = 0, y = 6$ **c.** $x = 4, y = -2$

In 2–5, state whether the point whose coordinates are given is on the graph of the given equation.

2. $x + y = 7$; (4, 3)
3. $2y + x = 7$; (1, 3)
4. $3x - 2y = 8$; (2, 1)
5. $2y = 3x - 5$; (−1, −4)

In 6–9, find the number that can replace k so that the resulting ordered number pair will be on the graph of the given equation.

6. $x + 2y = 5$; $(k, 2)$
7. $3x + 2y = 22$; $(k, 5)$
8. $x + 3y = 10$; $(13, k)$
9. $x - y = 0$; (k, k)

In 10–13, find a value that can replace k so that the graph of the resulting equation will pass through the point whose coordinates are given.

10. $x + y = k$; (2, 5)
11. $x - y = k$; (5, −3)
12. $5y - 2x = k$; (−2, −1)
13. $x + y = k$; (0, 0)

In 14–19, solve the equation for y in terms of x.

14. $3x + y = -1$ **15.** $4x - y = 6$ **16.** $2y = 6x$

17. $12x = \frac{3}{2}y$ **18.** $4x + 2y = 8$ **19.** $6x - 3y = 5$

In 20–22, find the missing values of the variable needed to complete the table. Plot the points described by the pairs of values in the completed table; then, draw a line through the points.

20. $y = 4x$

x	y
0	?
1	?
2	?

21. $y = 3x + 1$

x	y
−1	?
0	?
1	?

22. $x + 2y = 3$

x	y
−1	?
2	?
5	?

In 23–43, graph the equation.

23. $y = 2x$ **24.** $y = 5x$ **25.** $y = -3x$

26. $y = -x$ **27.** $x = 2y - 3$ **28.** $x = \frac{1}{2}y + 1$

29. $y = x + 3$ **30.** $y = 2x - 1$ **31.** $y = 3x + 1$

32. $y = -2x + 4$ **33.** $x + y = 8$ **34.** $x - y = 5$

35. $y - x = 0$ **36.** $3x + y = 12$ **37.** $x - 2y = 0$

38. $y - 3x = -5$ **39.** $2x - y = 6$ **40.** $3x - y = -6$

41. $x + 3y = 12$ **42.** $2x + 3y = 6$ **43.** $3x - 2y = -4$

In 44 and 45, graph the indicated set of points.

44. $\{(x, y) | y = 3x\}$ **45.** $\{(x, y) | x - y = 6\}$

In 46–50: **a.** Write the verbal sentence as an equation. **b.** Graph the equation.

46. The ordinate of a point is twice the abscissa.
47. The ordinate of a point is 2 more than the abscissa.
48. The sum of the ordinate and abscissa of a point is 6.
49. The difference of the ordinate and abscissa of a point is 1.
50. Twice the ordinate of a point decreased by 3 times the abscissa is 6.

17-5 GRAPHING LINES PARALLEL TO THE X-AXIS OR Y-AXIS

Lines Parallel to the X-Axis

An equation such as $y = 2$ can be graphed in the coordinate plane. Any pair of values whose y-coordinate is 2, no matter what the x-coordinate is, makes the equation $y = 2$ true. Therefore, $(-3, 2)$, $(-2, 2)$, $(-1, 2)$, $(0, 2)$, $(1, 2)$, or any pair $(a, 2)$ for all values of a are points on the graph of $y = 2$. Note that the graph of $y = 2$ is a horizontal line parallel to the x-axis and two units above it.

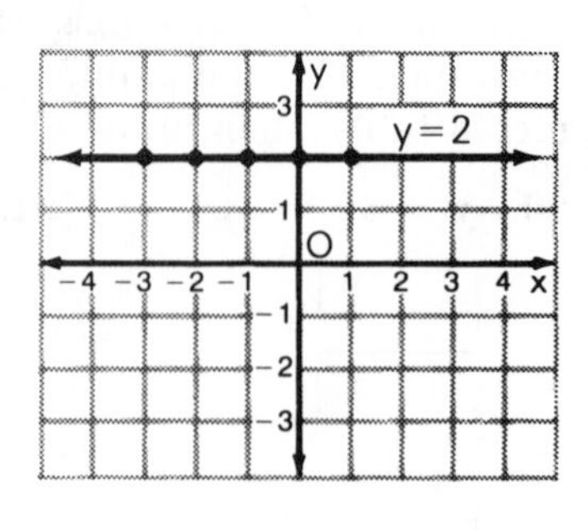

The ***y*-intercept** of a line is the y-coordinate of the point at which the line intersects the y-axis. From the graph, we see that the y-intercept of the line $y = 2$ is 2. In general:

- **An equation of the line parallel to the x-axis with y-intercept b is $y = b$.**

Lines Parallel to the Y-Axis

An equation such as $x = -3$ can be graphed in the coordinate plane. Any pair of values whose x-coordinate is -3, no matter what the y-coordinate is, makes the equation $x = -3$ true. Therefore, $(-3, -2)$, $(-3, -1)$, $(-3, 0)$, $(-3, 1)$, $(-3, 2)$, or any pair $(-3, b)$ for all values of b are points on the graph of $x = -3$. Note that the graph of $x = -3$ is a vertical line parallel to the y-axis and three units to the left of it.

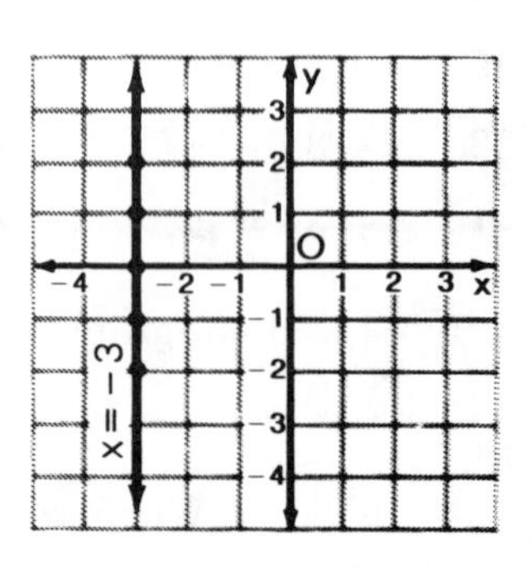

The ***x*-intercept** of a line is the x-coordinate of the point at which the line intersects the x-axis. From the graph, we see that the x-intercept of the line $x = -3$ is -3. In general:

- **An equation of the line parallel to the y-axis with x-intercept a is $x = a$.**

EXERCISES

In 1–10, draw the graph of the equation.

1. $x = 6$ **2.** $x = \frac{2}{3}$ **3.** $x = 0$ **4.** $x = -3$ **5.** $x = -5$

6. $y = 4$ **7.** $y = 2\frac{1}{4}$ **8.** $y = 0$ **9.** $y = -4$ **10.** $y = -7$

In 11–14, graph the indicated set of points.

11. $\{(x, y)|x = 4\}$ **12.** $\{(x, y)|x = -1.5\}$

13. $\{(x, y)|y = -5\}$ **14.** $\{(x, y)|y = 2\frac{1}{2}\}$

15. Write an equation of the line that is parallel to the x-axis and whose y-intercept is:
a. 1 **b.** 5 **c.** -4 **d.** -8 **e.** -2.5

16. Write an equation of the line that is parallel to the y-axis and whose x-intercept is:
a. 3 **b.** 10 **c.** $4\frac{1}{2}$ **d.** -6 **e.** -10

17. Which statement is true about the graph of the equation $y = 6$?
(1) It is parallel to the y-axis. (2) It is parallel to the x-axis.
(3) It goes through the origin. (4) It has an x-intercept.

18. Which statement is true about the graph of the equation $x = 5$?
(1) It goes through the origin. (2) It is parallel to the x-axis.
(3) It is parallel to the y-axis. (4) It has a y-intercept.

19. Which statement is true about the equation $y = x$?
(1) It is parallel to the x-axis. (2) It is parallel to the y-axis.
(3) It goes through $(2, -2)$. (4) It goes through the origin.

17-6 THE SLOPE OF A LINE

Meaning of the Slope of a Line

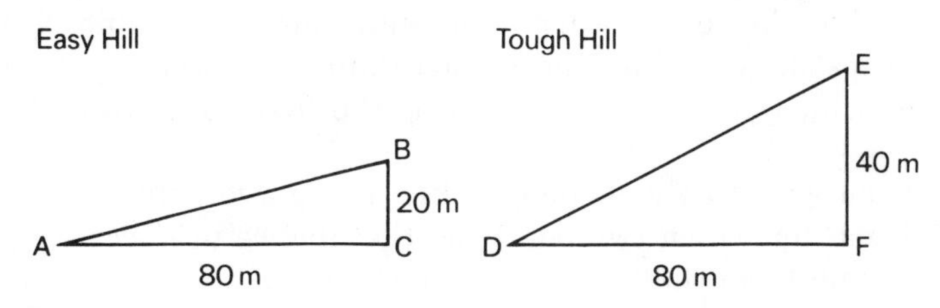

It is more difficult to hike up Tough Hill than it is to hike up Easy Hill. Tough Hill rises 40 m vertically over a horizontal distance of

80 m, whereas Easy Hill rises only 20 m vertically over the same horizontal distance of 80 m. Therefore, Tough Hill is steeper than Easy Hill. To compare the steepness of roads $\overline{AB}$ and $\overline{DE}$, roads that lead up the two hills, we compare their ***slopes***.

The slope of road $\overline{AB}$ is the ratio of the change in vertical distance, CB, to the change in horizontal distance, AC:

$$\text{slope of road } \overline{AB} = \frac{\text{change in vertical distance, } CB}{\text{change in horizontal distance, } AC} = \frac{20 \text{ m}}{80 \text{ m}} = \frac{1}{4}$$

Also:

$$\text{slope of road } \overline{DE} = \frac{\text{change in vertical distance, } FE}{\text{change in horizontal distance, } DF} = \frac{40 \text{ m}}{80 \text{ m}} = \frac{1}{2}$$

Finding the Slope of a Line

To find the slope of the line determined by the two points (2, 3) and (5, 8), write the ratio of the difference in y-values to the difference in x-values, as follows:

$$\text{slope} = \frac{\text{difference in } y\text{-values}}{\text{difference in } x\text{-values}}$$

$$= \frac{8 - 3}{5 - 2} = \frac{5}{3}$$

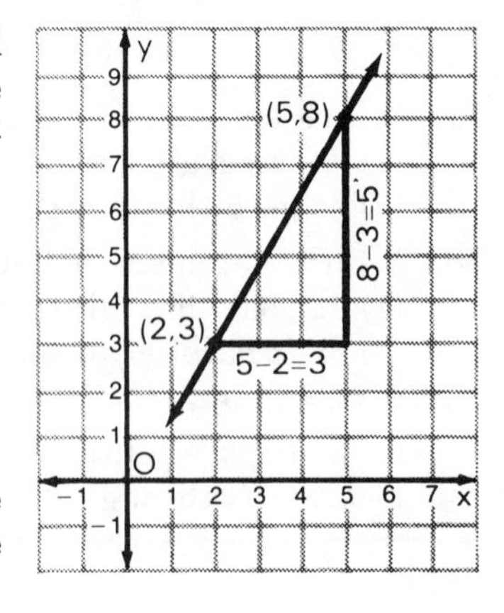

Suppose we change the order of the points in performing the computation. We would then have:

$$\text{slope} = \frac{\text{difference in } y\text{-values}}{\text{difference in } x\text{-values}} = \frac{3 - 8}{2 - 5} = \frac{-5}{-3} = \frac{5}{3}$$

Observe that the result of both computations is the same. When we compute the slope of a line that is determined by two points, it does not matter which point is considered as the first point and which the second.

Also, when we find the slope of a line using two points on the line, it does not matter which two points on the line we use because all segments of a line have the same slope as the line.

PROCEDURE. To find the slope of a line:
1. Select any two points on the line.
2. Find the vertical change, the change in y-values, in going from the point on the left to the point on the right.
3. Find the horizontal change, the change in x-values, in going from the point on the left to the point on the right.
4. Write the ratio of the vertical change to the horizontal change.

For example, in the figure:

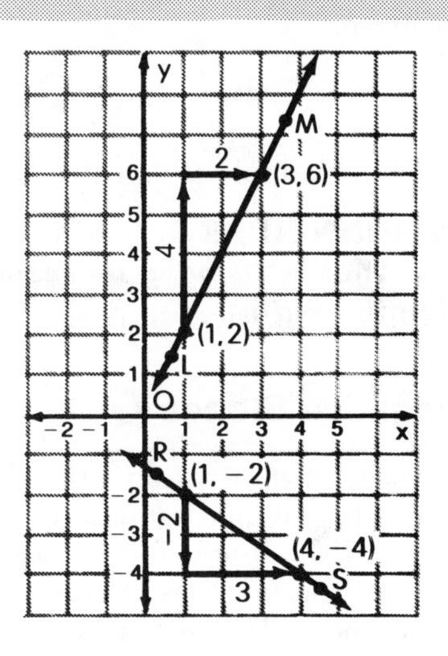

$$\text{slope of } \overleftrightarrow{LM} = \frac{\text{vertical change}}{\text{horizontal change}}$$

$$= \frac{4}{2} = \frac{2}{1}, \text{ or } 2$$

$$\text{slope of } \overleftrightarrow{RS} = \frac{\text{vertical change}}{\text{horizontal change}}$$

$$= \frac{-2}{3} = -\frac{2}{3}$$

In general, the slope, m, of a line that passes through any two points $P_1(x_1, y_1)$ and $P_2(x_2, y_2)$ when $x_1 \neq x_2$ is the ratio of the difference of the y-values of these points to the difference of the corresponding x-values. Thus:

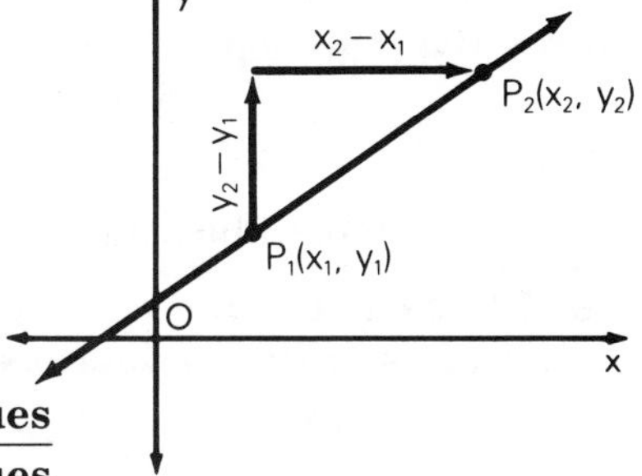

$$\textbf{slope of a line} = \frac{\textbf{difference in } y\textbf{-values}}{\textbf{difference in } x\textbf{-values}}$$

$$\textbf{slope of } \overleftrightarrow{P_1P_2} = m = \frac{y_2 - y_1}{x_2 - x_1}$$

The difference in x-values, $x_2 - x_1$, can be represented by Δx, read *delta x*. Similarly, the difference in y-values, $y_2 - y_1$, can be represented by Δy, read *delta y*. Therefore, we write:

$$\textbf{slope of a line} = m = \frac{\Delta y}{\Delta x}$$

Positive Slopes

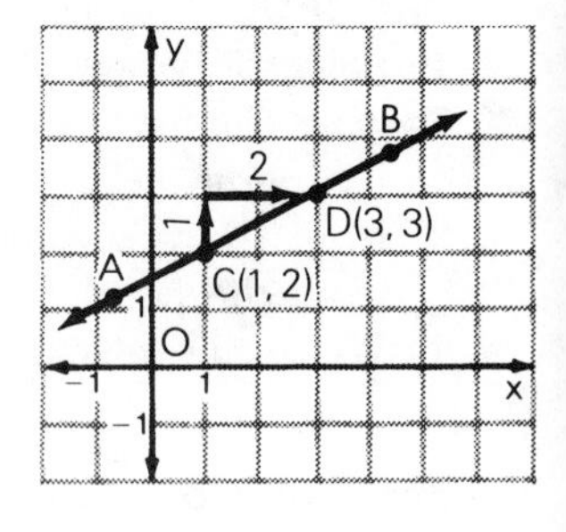

Examining $\overleftrightarrow{AB}$ from left to right and observing the path of a point from C to D, for example, we see that the line is rising. As the x-values increase, the y-values also increase. Between point C and point D, the change in y is 1, and the change in x is 2. Since both Δy and Δx are positive, the slope of $\overleftrightarrow{AB}$ must be positive.

$$\text{slope} = m = \frac{\Delta y}{\Delta x} = \frac{1}{2}$$

This example illustrates:

● **PRINCIPLE 1. As a point moves from left to right along a line that is *rising*, *y* increases as *x* increases and the slope of the line is *positive.***

Negative Slopes

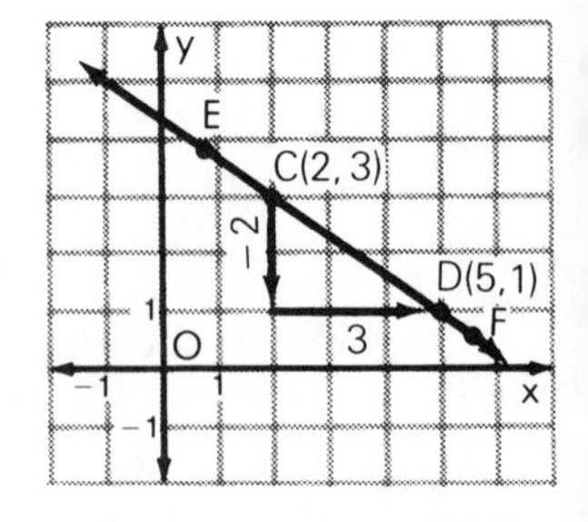

Now, examining $\overleftrightarrow{EF}$ from left to right and observing the path of a point from C to D, we see that the line is falling. As the x-values increase, the y-values decrease. Between point C and point D, the change in y is -2, and the change in x is 3. Since Δy is negative and Δx is positive, the slope of $\overleftrightarrow{EF}$ must be negative.

$$\text{slope} = m = \frac{\Delta y}{\Delta x} = \frac{-2}{3} = -\frac{2}{3}$$

This example illustrates:

● **PRINCIPLE 2. As a point moves from left to right along a line that is *falling*, *y* decreases as *x* increases and the slope of the line is *negative.***

Zero Slope

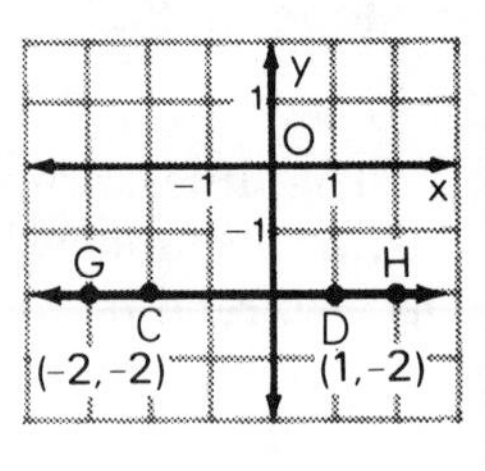

$\overleftrightarrow{GH}$ is parallel to the x-axis. Consider a point moving along $\overleftrightarrow{GH}$ from left to right, for example, from C to D: As the x-values increase, the y-values are unchanged. Between point C and point D, the change in y is 0, and the change in x is 3. Since Δy is 0 and Δx is 3, the slope of $\overleftrightarrow{GH}$ must be 0.

$$\text{slope} = m = \frac{\Delta y}{\Delta x} = \frac{0}{3} = 0$$

This example illustrates:

● **PRINCIPLE 3. If a line is parallel to the x-axis, its slope is 0.**

Note: The slope of the x-axis itself is also 0.

No Slope

$\overleftrightarrow{LM}$ is parallel to the y-axis. Consider a point moving upward along $\overleftrightarrow{LM}$, for example, from C to D: The x-values are unchanged, but the y-values increase. Between point C and point D, the change in y is 3, and the change in x is 0. Since the slope of $\overleftrightarrow{LM} = \frac{\Delta y}{\Delta x} = \frac{3}{0}$, and a number cannot be divided by 0, line $\overleftrightarrow{LM}$ has no defined slope.

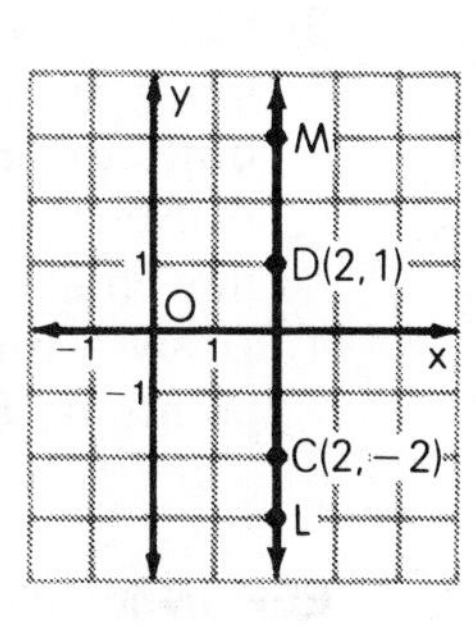

This example illustrates:

● **PRINCIPLE 4. If a line is parallel to the y-axis, it has no defined slope.**

Note: The y-axis itself has no defined slope.

MODEL PROBLEMS

1. Find the slope of the line that is determined by the points $(-2, 4)$ and $(4, 2)$.

Solution: Plot the points $(-2, 4)$ and $(4, 2)$. Let the point $(-2, 4)$ be $P_1(x_1, y_1)$, and let the point $(4, 2)$ be $P_2(x_2, y_2)$. Then, $x_1 = -2$, $y_1 = 4$ and $x_2 = 4$, $y_2 = 2$.

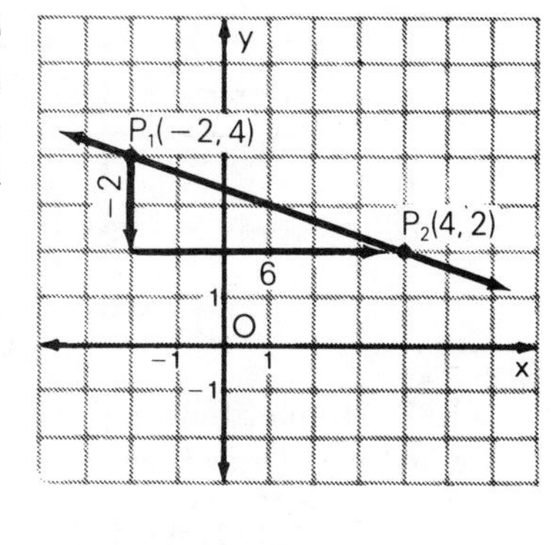

$$\text{slope of } \overleftrightarrow{P_1P_2} = \frac{\Delta y}{\Delta x} = \frac{y_2 - y_1}{x_2 - x_1}$$

$$= \frac{(2) - (4)}{(4) - (-2)}$$

$$= \frac{2 - 4}{4 + 2} = \frac{-2}{6} = -\frac{1}{3} \quad \textit{Ans.}$$

2. Through the point $(2, -1)$, draw the line whose slope is $\frac{3}{2}$.

How to Proceed

(1) Graph the point $A(2, -1)$.

(2) Since slope $= \frac{\Delta y}{\Delta x} = \frac{3}{2}$, when y changes 3, then x changes 2.

(3) Start at point $A(2, -1)$ and move 3 units upward and 2 units to the right to locate point B. Start at B and repeat these movements to locate point C.

(4) Draw a line that passes through points A, B, and C.

Solution

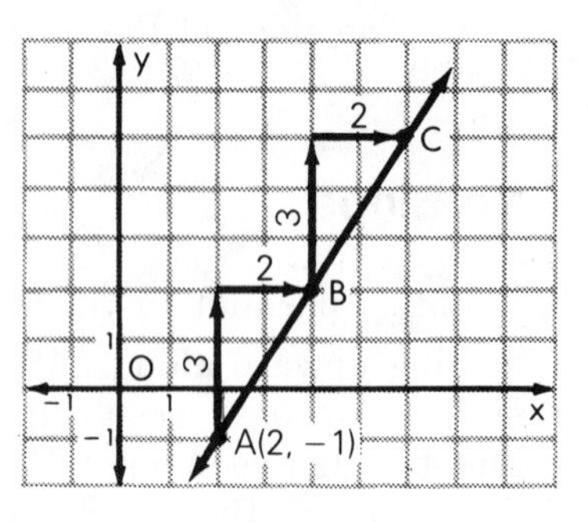

KEEP IN MIND

A fundamental property of a straight line is that its slope is constant. Therefore, any two points on a line may be used to compute the slope of the line.

EXERCISES

In 1–6: **a.** Tell whether the line has a positive slope, a negative slope, a slope of zero, or no slope. **b.** Find the slope of the line. If the line has no slope, indicate that fact.

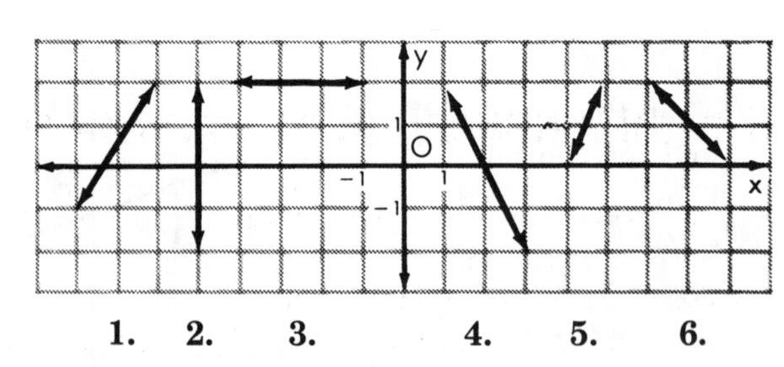

In 7–15, plot both points, draw the line that they determine, and find the slope of this line.

7. $(0, 0)$ and $(4, 4)$

8. $(0, 0)$ and $(4, 8)$

9. $(0, 0)$ and $(3, -6)$

10. $(1, 5)$ and $(3, 9)$

11. (7, 3) and (1, −1)
12. (−2, 4) and (0, 2)
13. (5, −2) and (7, −8)
14. (4, 2) and (8, 2)
15. (−1, 3) and (2, 3)

In 16–24, draw a line with the given slope, m, through the given point.

16. (0, 0); $m = 2$
17. (1, 3); $m = 3$
18. (2, −5); $m = 4$
19. (−4, 5); $m = \frac{2}{3}$
20. (3, 1); $m = 0$
21. (−3, −4); $m = -2$
22. (1, −5); $m = -1$
23. (2, 4); $m = -\frac{3}{2}$
24. (−2, 3); $m = -\frac{1}{3}$

25. The points $A(2, 4)$, $B(8, 4)$, and $C(5, 1)$ are the vertices of triangle ABC. Find the slope of each side of triangle ABC.

26. The points $A(3, -2)$, $B(9, -2)$, $C(7, 4)$, and $D(1, 4)$ are the vertices of a quadrilateral.
a. Graph the points and draw quadrilateral $ABCD$.
b. What type of quadrilateral does $ABCD$ appear to be?
c. Compute the slope of $\overline{BC}$ and the slope of $\overline{AD}$.
d. What is true of the slope of $\overline{BC}$ and the slope of $\overline{AD}$?
e. If two segments such as $\overline{AD}$ and $\overline{BC}$, or two lines such as $\overleftrightarrow{AD}$ and $\overleftrightarrow{BC}$, are parallel, what appears to be true of their slopes?
f. Since $\overline{AB}$ and $\overline{CD}$ are parallel, what might be true of their slopes?
g. Compute the slope of $\overline{AB}$ and the slope of $\overline{DC}$.
h. Is the slope of $\overline{AB}$ equal to the slope of $\overline{DC}$?

17-7 GRAPHING DIRECT VARIATION

Recall that when the ratio of two variables is a constant, the two variables represent quantities that are *directly proportional* or that *vary directly*. For example, if 3 cups of lemonade, y, can be made by using 1 cup of frozen concentrate, x, and y is directly proportional to x, then the relationship can be expressed as:

$$\frac{y}{x} = 3 \quad \text{or} \quad y = 3x$$

The *constant of variation* is 3.

The equation $\frac{y}{x} = 3$, or $y = 3x$, can be represented by a line in the coordinate plane.

x	3x	y
0	3(0)	0
1	3(1)	3
2	3(2)	6
3	3(3)	9

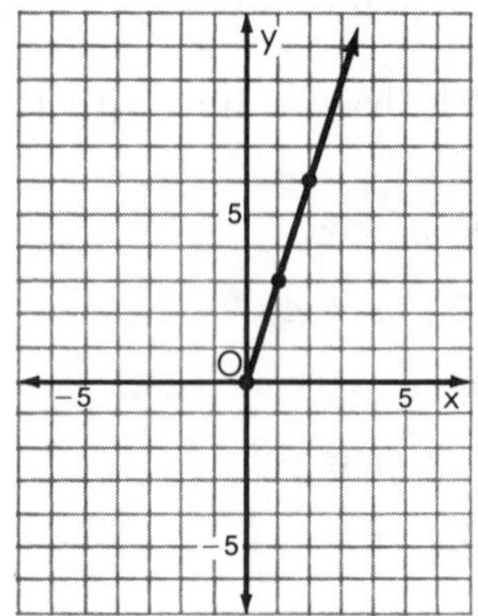

In this example, the replacement set for x and for y is the set of positive numbers and zero. Thus, the graph does not include points in the third quadrant because the negative numbers are not members of the replacement set for x and y.

If 0 cups of frozen concentrate are used, 0 cups of lemonade can be made. Thus, the ordered pair (0, 0) is a member of the solution set of $y = 3x$.

Using any two points from the table above, for example, (0, 0) and (1, 3), we can write:

$$\frac{\Delta y}{\Delta x} = \frac{3 - 0}{1 - 0} = \frac{3}{1} = 3$$

Thus, we see that the slope of the line is also the constant of variation.

Note that the units of measure for the lemonade and for the frozen concentrate are the same, cups, and there is no unit of measure associated with the ratio. In this example, no matter what unit of measure is used to express the amounts of lemonade and frozen concentrate, as long as the same unit is used for both, the ratio is always $\frac{3}{1}$.

There are many applications of direct variation in business and science. It is important to recognize how the choice of unit can affect the constant of variation. For example, if a machine is used to pack boxes of cereal in cartons, the rate at which the machine works can be expressed in cartons per minute or in cartons per second. If the machine fills a carton every 6 seconds, it will fill 10 cartons in one minute. The rate can be expressed as:

$$\frac{1 \text{ carton}}{6 \text{ seconds}} = \frac{1}{6} \text{ cartons/sec.} \quad \text{or} \quad \frac{10 \text{ cartons}}{1 \text{ minute}} = 10 \text{ cartons/min.}$$

Each of these rates can be represented by a graph, where the rate, or constant of variation, is the slope of the line.

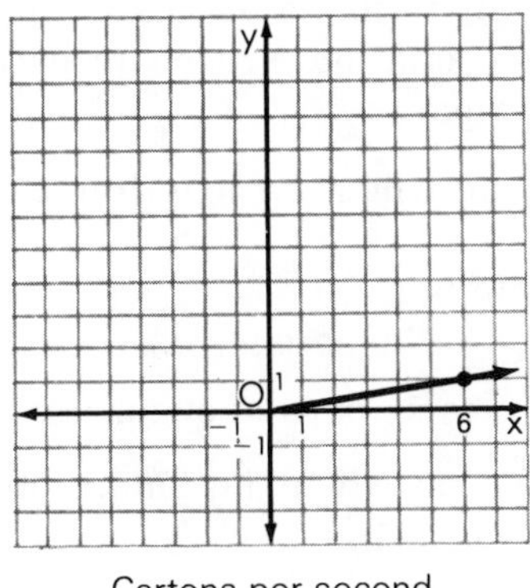

Cartons per second

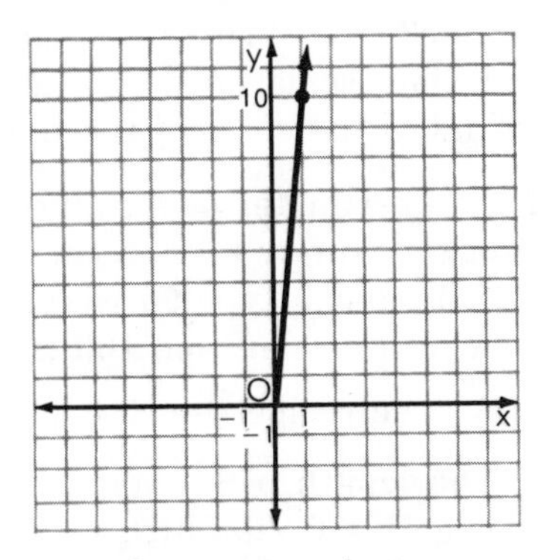

Cartons per minute

The legend of the graph must be clearly stated if the graph is to be meaningful.

MODEL PROBLEM

The amount of flour needed to make a white sauce varies directly as the amount of milk used. To make a white sauce, a chef used 2 cups of flour and 8 cups of milk. Write an equation and draw the graph of the relationship.

Solution:

Let x = the number of cups of milk.

Let y = the number of cups of flour.

$$\frac{y}{x} = \frac{2}{8}$$

$$8y = 2x$$

$$y = \frac{2}{8}x$$

$$y = \frac{1}{4}x$$

x	$\frac{1}{4}x$	y
2	$\frac{1}{4}(2)$	$\frac{1}{2}$
4	$\frac{1}{4}(4)$	1
6	$\frac{1}{4}(6)$	$\frac{3}{2}$

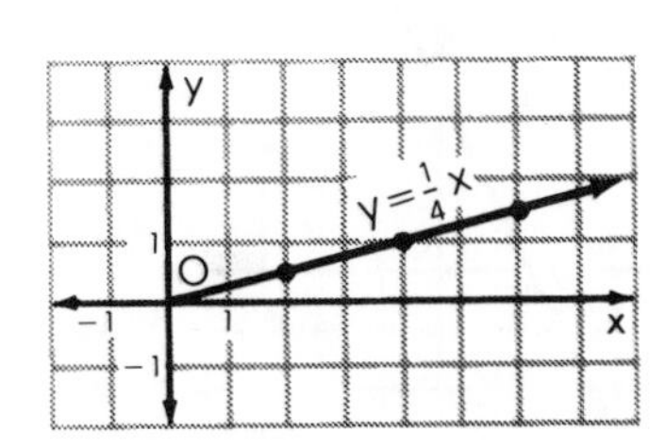

EXERCISES

In 1–10, y varies directly as x. **a.** What is the constant of variation? **b.** Write an equation for y in terms of x. **c.** Using an appropriate scale, draw the graph of the equation written in part **b.** **d.** What is the slope of the line drawn in part **c**?

1. The perimeter of a square (y) is 12 cm when the length of a side of the square (x) is 3 cm.
2. Jeanne can type 90 words (y) in 2 minutes (x).
3. A printer can type 160 characters (y) in 10 seconds (x).
4. A cake recipe uses 2 cups of flour (y) to $1\frac{1}{2}$ cups of sugar (x).
5. The length of a photograph (y) is 12 cm when the length of the negative from which it is developed (x) is 1.2 cm.
6. There are 20 slices (y) in 12 ounces of bread (x).
7. Three pounds of meat (y) will serve 15 people (x).
8. Twelve slices of cheese (y) weigh 8 ounces (x).
9. Willie averages 3 hits (y) for every 12 times at bat (x).
10. There are about 20 calories (y) in 3 crackers (x).
11. If a car travels at a constant rate of speed, the distance that it travels varies directly as time. If a car travels 75 miles in 2.5 hours, it will travel 110 feet in 2.5 seconds.
    a. Find the constant of variation in miles per hour.
    b. Find the constant of variation in feet per second.

17-8 THE SLOPE AND Y-INTERCEPT OF A LINE

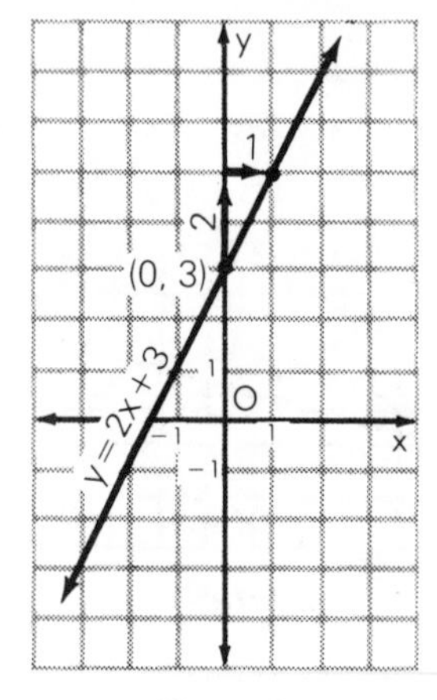

Figure 1

$$\frac{\Delta y}{\Delta x} = \frac{2}{1} = 2$$

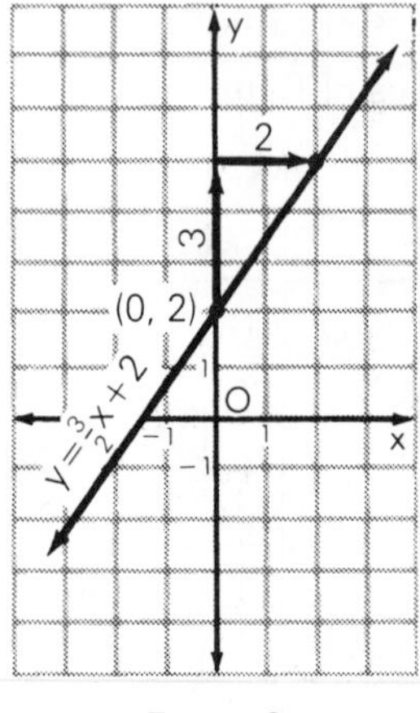

Figure 2

$$\frac{\Delta y}{\Delta x} = \frac{3}{2}$$

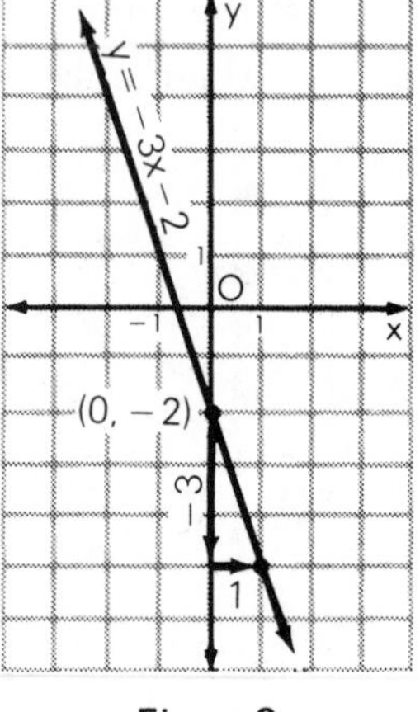

Figure 3

$$\frac{\Delta y}{\Delta x} = \frac{-3}{1} = -3$$

Each of the preceding figures shows the line that is the graph of the indicated equation. We can see that the slope of each line is the coefficient of the x-term in the equation and that the y-intercept of each line is the *constant* that follows the x-term in the equation.

	Equation	Slope $\left(\frac{\Delta y}{\Delta x}\right)$	y-intercept
In Fig. 1:	$y = 2x + 3$	2	3
In Fig. 2:	$y = \frac{3}{2}x + 2$	$\frac{3}{2}$	2
In Fig. 3:	$y = -3x - 2$	-3	-2

These examples illustrate the following general principle:

● **If a linear equation is expressed in the form $y = mx + b$, then the slope of the line is m, the coefficient of x; the y-intercept is b, the constant term.**

The following general principle is also true:

● **The equation of the line whose slope is m and whose y-intercept is b can be written in the form $y = mx + b$.**

Since the slope of a line describes the *slant* of a line, it is reasonable to believe that the following statements (which can be proved) are true:

1. **If two lines are parallel, their slopes are equal.**
2. **If the slopes of two lines are equal, the lines are parallel.**

These statements, a conditional and its converse, can be written as a biconditional:

● **Two lines are parallel if and only if their slopes are equal.**

MODEL PROBLEMS

1. Find the slope and y-intercept of the line that is the graph of the equation $4x + 2y = 10$.

How to Proceed	*Solution*
(1) Transform the equation into an equivalent equation of the form $y = mx + b$ by solving for y in terms of x.	$4x + 2y = 10$ $2y = -4x + 10$ $y = -2x + 5$

(2) m (the coefficient of x) is the slope. slope $= -2$

(3) b (the constant term) is the y-intercept. y-intercept $= 5$

Answer: slope $= -2$, y-intercept $= 5$

2. Write an equation of a line whose slope is $\frac{1}{2}$ and whose y-intercept is -4.

How to Proceed	*Solution*
(1) Write the equation $y = mx + b$.	$y = mx + b$
(2) Replace m by the numerical value of the slope; replace b by the numerical value of the y-intercept.	$m = \text{slope} = \frac{1}{2}$ $b = y\text{-intercept} = -4$ $y = \frac{1}{2}x + (-4)$ or $y = \frac{1}{2}x - 4$ *Ans.*

EXERCISES

In 1–15, find the slope and y-intercept of the line that is the graph of the equation.

1. $y = 3x + 1$ **2.** $y = x - 3$ **3.** $y = 2x$

4. $y = x$ **5.** $y = \frac{1}{2}x + 5$ **6.** $y = -2x + 3$

7. $y = -3x$ **8.** $y = -2$ **9.** $y = -\frac{2}{3}x + 4$

10. $y - 3x = 7$ **11.** $2x + y = 5$ **12.** $3y = 6x + 9$

13. $2y = 5x - 4$ **14.** $\frac{1}{2}x + \frac{3}{4} = \frac{1}{3}y$ **15.** $4x - 3y = 0$

In 16–23, write an equation of the line whose slope and y-intercept are respectively:

16. 2 and 7 **17.** -1 and -3 **18.** 0 and -5 **19.** -3 and 0

20. $\frac{2}{3}$ and 1 **21.** $\frac{1}{2}$ and 0 **22.** $-\frac{1}{3}$ and 2 **23.** $-\frac{3}{2}$ and 0

24. Write equations for three lines so that the slope of each line is 2.

25. Write equations for three lines so that the y-intercept of each line is -4.

26. What do the graphs of the equations $y = 4x$, $y = 4x + 2$, and $y = 4x - 2$ all have in common?
27. How are the graphs of $y = mx + b$ affected when m is always replaced by the same number and b is replaced by different numbers?
28. What do the lines that are the graphs of the equations $y = 2x + 1$, $y = 3x + 1$, and $y = -4x + 1$ all have in common?
29. How are the graphs of $y = mx + b$ affected when b is always replaced by the same number and m is replaced by different numbers?
30. If two lines are parallel, how are their slopes related?
31. What will be true of two lines whose slopes are equal?

In 32–35, state whether or not the lines are parallel.

32. $y = 3x + 2$, $y = 3x - 5$
33. $y = -2x - 6$, $y = 2x + 6$
34. $y = 4x - 8$, $y - 4x = 3$
35. $y = 2x$, $2y - 4x = 9$

36. Which of the following statements is true of the graph of the equation $y = -3x$?
(1) It is parallel to the x-axis. (2) It is parallel to the y-axis.
(3) Its slope is -3. (4) It does not have a y-intercept.
37. Which of the following statements is true of the graph of the equation $y = 8$?
(1) It is parallel to the x-axis. (2) It is parallel to the y-axis.
(3) It has no slope. (4) It goes through the origin.

17-9 GRAPHING A LINEAR EQUATION IN TWO VARIABLES BY THE SLOPE-INTERCEPT METHOD

The slope and y-intercept of a line can be used to draw the graph of a linear equation.

MODEL PROBLEM

Draw the graph of $2x + 3y = 9$ using the slope-intercept method.

How to Proceed	*Solution*
(1) Transform the equation into the form $y = mx + b$.	$2x + 3y = 9$ $3y = -2x + 9$ $y = \frac{-2}{3}x + 3$

(2) Find the slope of the line (the coefficient of x). slope $= \frac{-2}{3}$

(3) Find the y-intercept of the line (the constant). y-intercept $= 3$

(4) On the y-axis, graph a point A whose ordinate is the y-intercept.

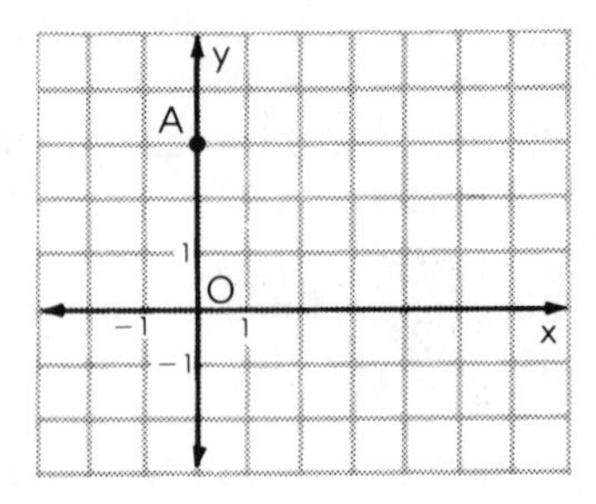

(5) Use the slope to find two more points on the line. Since slope $= \frac{\Delta y}{\Delta x} = \frac{-2}{3}$, when y changes -2, x changes 3. Start at point A and move 2 units down and 3 units to the right to locate point B. Start at point B and repeat this procedure to locate point C.

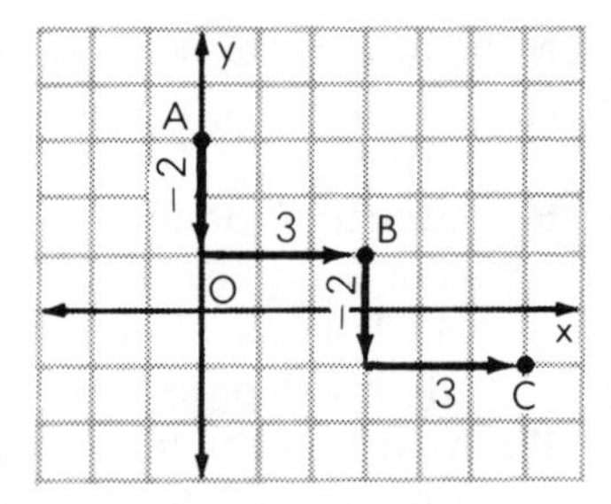

(6) Draw the line that passes through the three points.
This line is the graph of $\{(x, y)|2x + 3y = 9\}$.

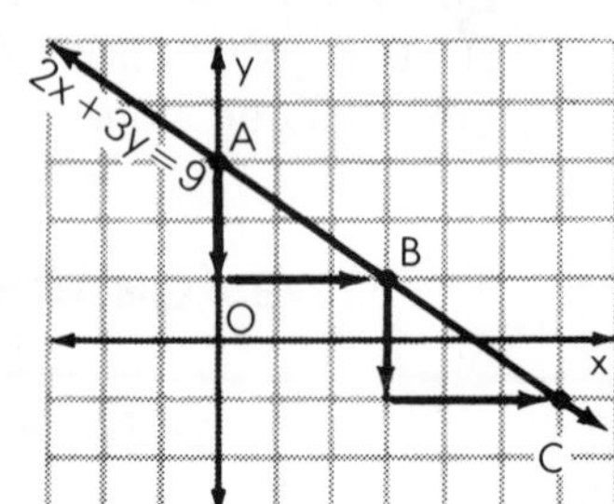

EXERCISES

In 1–24, graph each equation using the slope-intercept method.

1. $y = 2x + 3$ **2.** $y = 2x - 5$ **3.** $y = 2x$
4. $y = x - 2$ **5.** $y = 2x - 2$ **6.** $y = 3x - 2$
7. $y = 3x$ **8.** $y = 5x$ **9.** $y = -2x$
10. $y = \frac{2}{3}x + 2$ **11.** $y = \frac{1}{2}x - 1$ **12.** $y = \frac{3}{2}x$

13. $y = \frac{1}{3}x$
14. $y = -\frac{4}{3}x + 5$
15. $y = -\frac{3}{4}x$
16. $y - 2x = 8$
17. $3x + y = 4$
18. $2y = 4x + 6$
19. $3y = 4x + 9$
20. $4x - y = 3$
21. $3x + 4y = 12$
22. $2x = 3y + 6$
23. $4x + 3y = 0$
24. $2x - 3y - 6 = 0$

17-10 WRITING AN EQUATION FOR A LINE

PROCEDURE. To write an equation for a line, determine its slope and y-intercept. Then, use the slope-intercept formula: $y = mx + b$.

MODEL PROBLEMS

1. Write an equation of the line whose slope is 4, and that passes through the point (3, 5).

How to Proceed	*Solution*
(1) In the equation of a line $y = mx + b$, replace m by the given slope, 4.	$y = mx + b$ $y = 4x + b$
(2) Since the given point (3, 5) is on the line, its coordinates satisfy the equation $y = 4x + b$. Replace x by 3 and y by 5. Solve the resulting equation to find the value of b, the y-intercept.	$(5) = 4(3) + b$ $5 = 12 + b$ $-7 = b$
(3) In $y = 4x + b$, replace b by -7.	$y = 4x - 7$

Answer: $y = 4x - 7$

2. Write an equation of the line that passes through the points (2, 5) and (4, 11).

How to Proceed	*Solution*
(1) Find the slope of the line that passes through the two given points, (2, 5) and (4, 11).	Let P_1 be (2, 5). $[x_1 = 2, y_1 = 5]$ Let P_2 be (4, 11). $[x_2 = 4, y_2 = 11]$ $m = \frac{y_2 - y_1}{x_2 - x_1}$ $m = \frac{11 - 5}{4 - 2} = \frac{6}{2} = 3$

(2) In $y = mx + b$, replace m by the slope, 3.

$$y = mx + b$$
$$y = 3x + b$$

(3) Select one point that is on the line, for example, (2, 5). Its coordinates must satisfy the equation $y = 3x + b$. Replace x by 2 and y by 5.
Solve the resulting equation to find the value of b, the y-intercept.

$$(5) = 3(2) + b$$
$$5 = 6 + b$$
$$-1 = b$$

(4) In $y = 3x + b$, replace b by -1.

$$y = 3x - 1$$

(5) Check whether the coordinates of the second point (4, 11) satisfy the equation $y = 3x - 1$.

$$11 \stackrel{?}{=} 3(4) - 1$$
$$11 = 11 \quad \text{(True)}$$

Answer: $y = 3x - 1$

EXERCISES

In 1–6, write an equation of the line that has the given slope, m, and that passes through the given point.

1. $m = 2$; (1, 4) **2.** $m = 2$; $(-3, 4)$ **3.** $m = -3$; $(-2, -1)$

4. $m = \frac{1}{2}$; (4, 2) **5.** $m = \frac{-3}{4}$; (0, 0) **6.** $m = \frac{-5}{3}$; $(-3, 0)$

In 7–12, write an equation of the line that passes through the given points.

7. (1, 4); (3, 8) **8.** (3, 1); (9, 7) **9.** (1, 2); (10, 14)

10. $(0, -1)$; (6, 8) **11.** $(-2, -5)$; $(-1, -2)$ **12.** (0, 0); $(-3, 5)$

13. Write an equation of the line that is:
a. parallel to the line $y = 2x - 4$, and whose y-intercept is 7.
b. parallel to the line $y - 3x = 6$, and whose y-intercept is -2.
c. parallel to the line $2x + 3y = 12$, and that passes through the origin.

14. Write an equation of the line that is:

a. parallel to the line $y = 4x + 1$, and that passes through the point (2, 3).

b. parallel to the line $2y - 6x = 9$, and that passes through the point (−2, 1).

c. parallel to the line $y = 4x + 3$, and that has the same y-intercept as the line $y = 5x - 3$.

d. parallel to the line $y = -\frac{1}{2}x$, and that has the same y-intercept as the line $2y = 7x + 6$.

15. A triangle is determined by the three points $A(3, 5)$, $B(6, 4)$, and $C(1, -1)$. Write the equation of each line:

a. $\overleftrightarrow{AB}$ **b.** $\overleftrightarrow{BC}$ **c.** $\overleftrightarrow{CA}$

17-11 GRAPHING A LINEAR INEQUALITY IN TWO VARIABLES

When a line is graphed in the coordinate plane, the line is a ***plane divider*** because it separates the plane into two regions called ***half-planes***. One of these regions is a half-plane above the line; the other is a half-plane below the line.

Let us consider, for example, the horizontal line $y = 3$ as a plane divider. The line $y = 3$ and the two half-planes that it forms determine three sets of points:

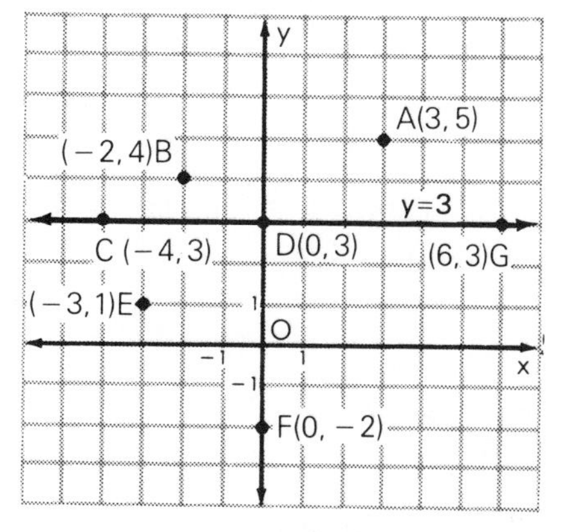

1. The half-plane above the line $y = 3$ is the set of all points whose y-coordinate is greater than 3, that is, $y > 3$. For example, at point A, $y = 5$; at point B, $y = 4$.

2. The line $y = 3$ is the set of all points whose y-coordinate is equal to 3. For example, $y = 3$ at each of the points (−4, 3), (0, 3), (6, 3).

3. The half-plane below the line $y = 3$ is the set of all points whose y-coordinate is less than 3, that is, $y < 3$. For example, at point E, $y = 1$; at point F, $y = -2$.

Together, the three sets of points form the entire plane.

To **graph an inequality** in the coordinate plane, use the following procedure:

1. On the plane, represent the plane divider, for example, $y = 3$, by a *dashed line* to show that this divider does not belong to the graph of the half-plane.

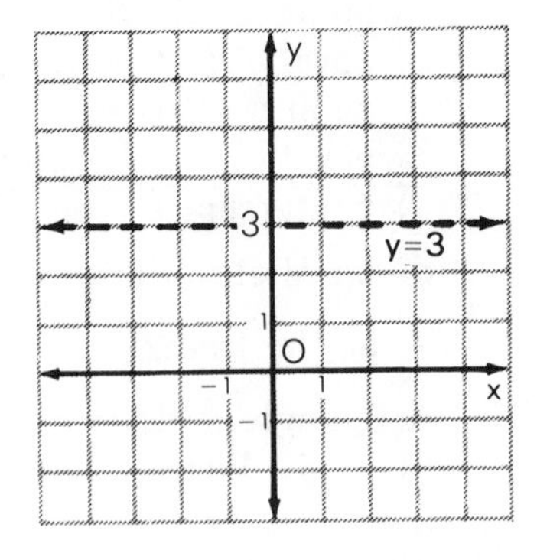

2. *Shade the region* of the half-plane whose points satisfy the inequality.

   To graph $y > 3$, shade the region *above* the plane divider.

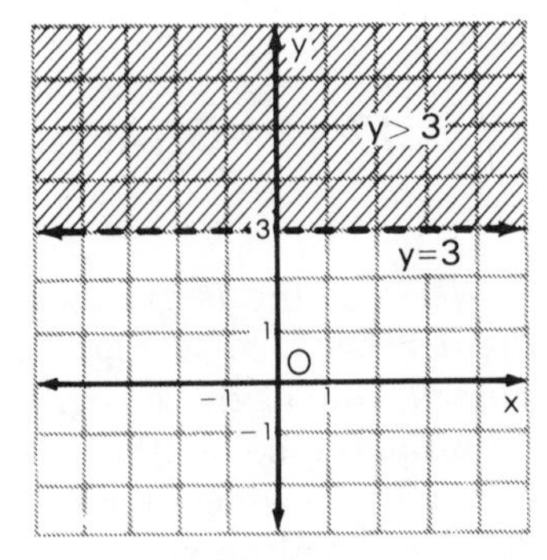

Graph of $y > 3$

To graph $y < 3$, shade the region *below* the plane divider.

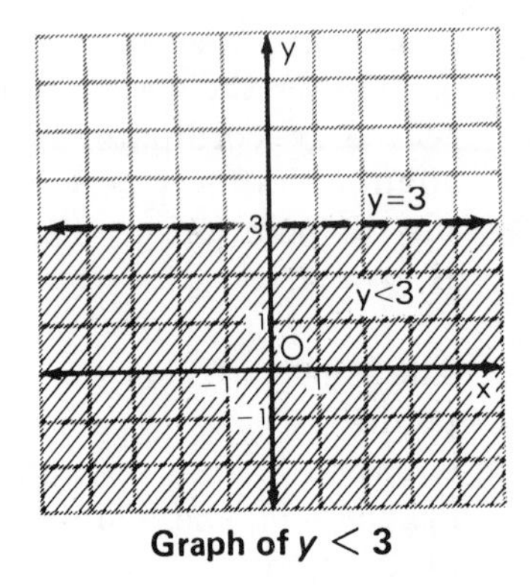

Graph of $y < 3$

Consider another example, where the plane divider is not a horizontal line.

To graph the inequality $y > 2x$ or $y < 2x$, use a dashed line to indicate that the line $y = 2x$ is not a part of the graph. This dashed line acts as a boundary line for the half-plane being graphed.

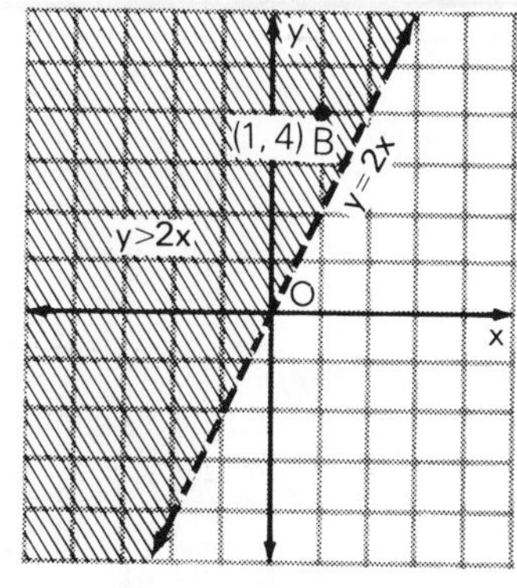

Graph of $y > 2x$

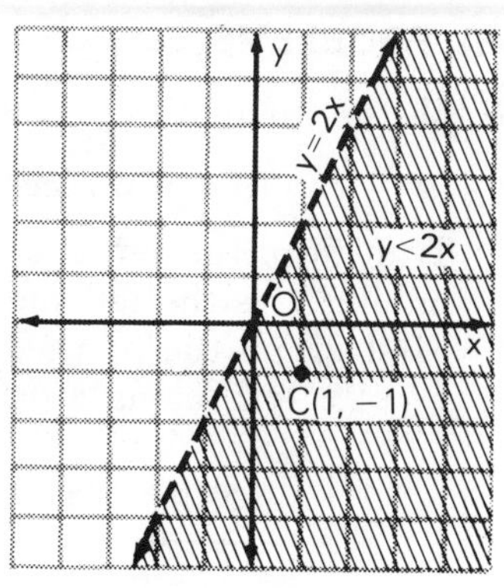

Graph of $y < 2x$

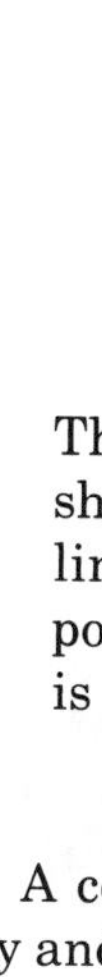

The graph of $y > 2x$ is the shaded half-plane above the line $y = 2x$. It is the set of all points in which the y-coordinate is *greater than* twice the x-coordinate.

The graph of $y < 2x$ is the shaded half-plane below the line $y = 2x$. It is the set of all points in which the y-coordinate is *less than* twice the x-coordinate.

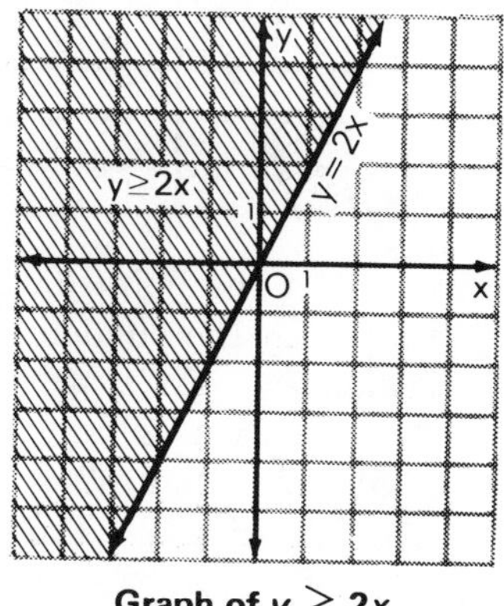

Graph of $y \geq 2x$

A condition involving both inequality and equality, such as $y \geq 2x$, means $y > 2x$ **or** $y = 2x$. The graph is the *union* of the two disjoint sets of points. To indicate that $y = 2x$ is part of the graph of $y \geq 2x$, draw the graph of $y = 2x$ as a solid line. The region above $y = 2x$ is shaded to indicate that $y > 2x$ is part of the graph of $y \geq 2x$.

In general:

- **When the equation of a line is written in the form $y = mx + b$, the half-plane above the line is the graph of $y > mx + b$ and the half-plane below the line is the graph of $y < mx + b$.**

To check whether the correct half-plane has been chosen as the graph of a linear inequality, select any point in that half-plane. If the selected point satisfies the inequality, every point in that half-plane satisfies the inequality. On the other hand, if the point chosen does not satisfy the inequality, then the other half-plane is the graph of the inequality.

MODEL PROBLEMS

1. Graph the inequality $y - 2x \geq 2$.

How to Proceed

(1) Transform the sentence into one having y as the left member.

Solution

$$y - 2x \geq 2$$
$$y \geq 2x + 2$$

(2) First graph the plane divider, $y = 2x + 2$, using a dashed line.

x	2x + 2	y
−1	−2 + 2	0
0	0 + 2	2
1	2 + 2	4

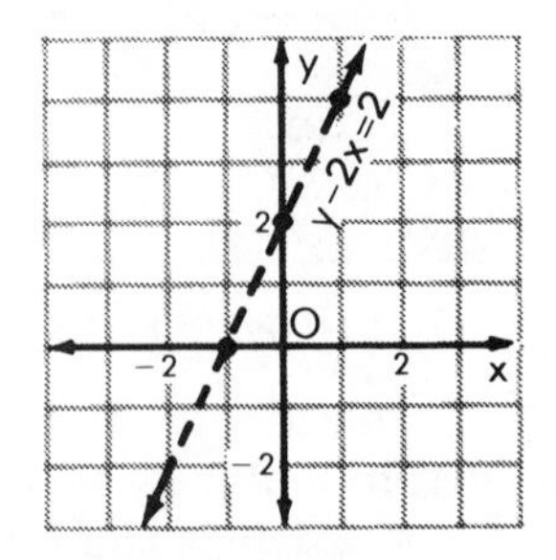

(3) Shade the half-plane above the line. This region and the line are the required graph: the half-plane is the graph of $y - 2x > 2$, and the line is the graph of $y - 2x = 2$. Note that the line is now drawn solid to show that it is part of the graph.

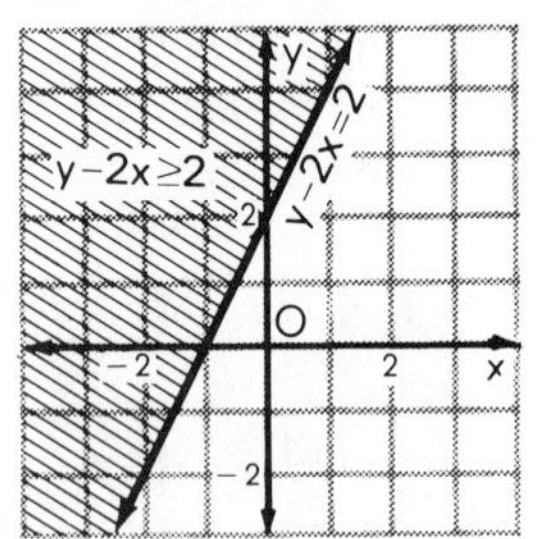

(4) Check the solution. Choose any point in the half-plane selected as the solution to see whether it satisfies the original inequality $y - 2x \geq 2$.

Select the point (0, 5) that is in the shaded region.

$$y - 2x \geq 2$$
$$(5) - 2(0) \geq 2 \quad (?)$$
$$5 \geq 2 \quad \text{(True)}$$

The above graph is the graph of $\{(x, y) | y - 2x \geq 2\}$.

2. Graph each of the following sentences in the coordinate plane:
a. $x > 1$ **b.** $x \leq 1$ **c.** $y \geq 1$ **d.** $y < 1$

Solution

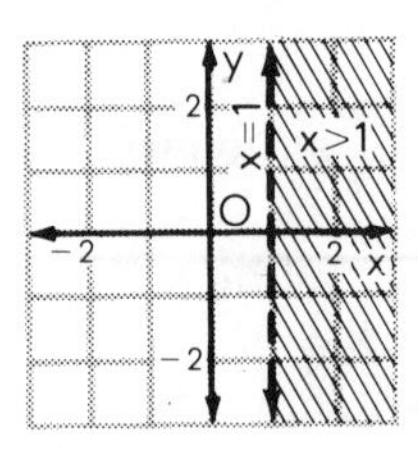

a. $x > 1$

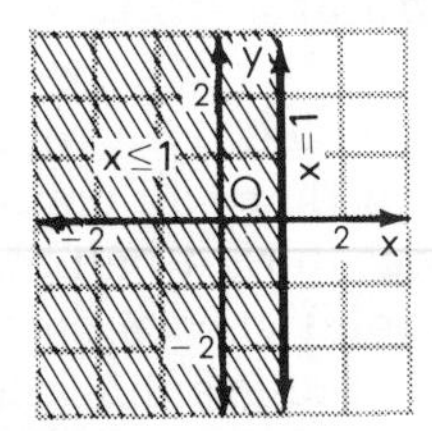

b. $x \leq 1$

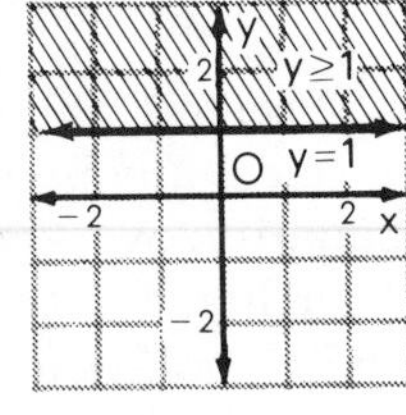

c. $y \geq 1$

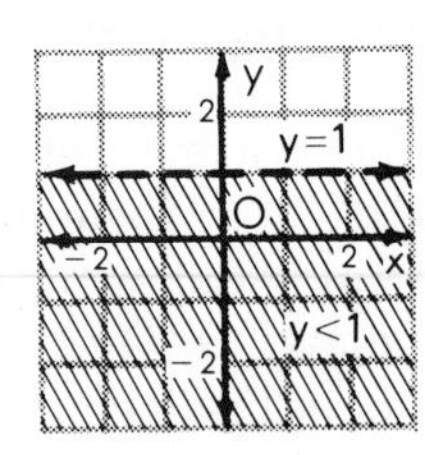

d. $y < 1$

EXERCISES

In 1–6, transform the sentence into one whose left member is y.

1. $y - 2x > 0$ **2.** $5x > 2y$ **3.** $y - x \geq 3$
4. $2x + y \leq 0$ **5.** $3x - y \geq 4$ **6.** $4y - 3x \leq 12$

In 7–27, graph the sentence in the coordinate plane.

7. $x > 4$ **8.** $x \leq -2$ **9.** $y > 5$
10. $y \leq -3$ **11.** $x \geq 6$ **12.** $y \leq 0$
13. $y > 4x$ **14.** $y \leq 3x$ **15.** $y < x - 2$
16. $y \geq \frac{1}{2}x + 3$ **17.** $x + y < 4$ **18.** $x + y \geq 4$
19. $x + y \leq -3$ **20.** $y - x \geq 5$ **21.** $x - y \leq -1$
22. $x - 2y \leq 4$ **23.** $2x + y - 4 \leq 0$ **24.** $y - x + 6 > 0$
25. $2y - 6x > 0$ **26.** $3x + 4y \leq 0$ **27.** $2x - 3y \geq 6$

In 28–30, graph the indicated set.

28. $\{(x, y) | y \leq 3x\}$ **29.** $\{(x, y) | x + y \geq 4\}$ **30.** $\{(x, y) | x - 2y \leq 4\}$

In 31–33: **a.** Write the verbal sentence as an open sentence. **b.** Graph the open sentence in the coordinate plane.

31. The ordinate of a point is equal to or greater than 3 more than the abscissa.
32. The sum of the abscissa and ordinate of a point is less than or equal to 5.
33. The ordinate of a point decreased by three times the abscissa is greater than or equal to 2.

17-12 COORDINATES AND TRANSFORMATIONS

Reflection in the *Y*-Axis

In the figure, $\triangle ABC$ is reflected in the y-axis. Its image under the reflection is $\triangle A'B'C'$. From the figure, we see that:

$A(1, 2) \rightarrow A'(-1, 2)$
$B(3, 4) \rightarrow B'(-3, 4)$
$C(1, 5) \rightarrow C'(-1, 5)$

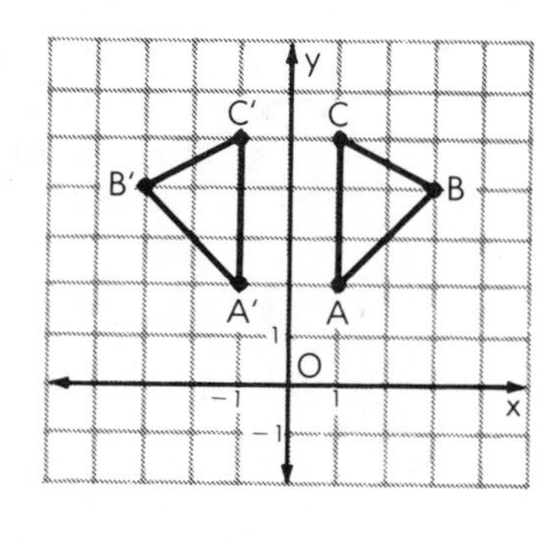

For each point and its image under a reflection in the y-axis, the y-coordinate of the image is the same as the y-coordinate of the point; the x-coordinate of the image is the opposite of the x-coordinate of the point.

From these examples, we form a general rule:

- **Under a reflection in the y-axis, the image of $P(x, y)$ is $P'(-x, y)$.**

Reflection in the X-Axis

In the figure, $\triangle ABC$ is reflected in the x-axis. Its image under the reflection is $\triangle A''B''C''$. From the figure, we see that:

$$A(1, 2) \rightarrow A''(1, -2)$$
$$B(3, 4) \rightarrow B''(3, -4)$$
$$C(1, 5) \rightarrow C''(1, -5)$$

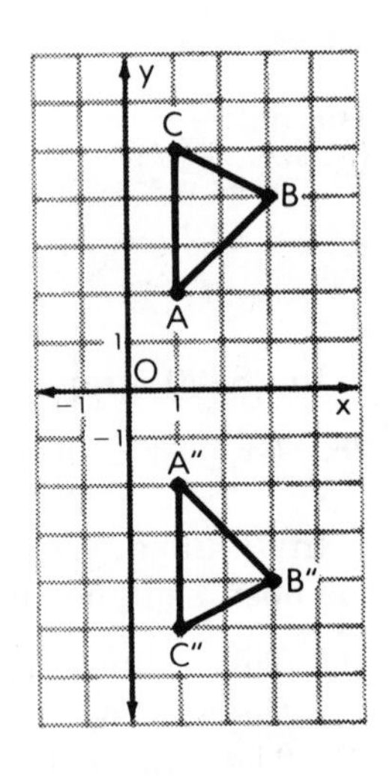

For each point and its image under a reflection in the x-axis, the x-coordinate of the image is the same as the x-coordinate of the point; the y-coordinate of the image is the opposite of the y-coordinate of the point.

From these examples, we form a general rule:

- **Under a reflection in the x-axis, the image of $P(x, y)$ is $P'(x, -y)$.**

Translation

In the figure, $\triangle ABC$ is translated by moving every point to the right 4 units and down 5 units. From the figure, we see that:

$$A(1, 2) \rightarrow A'''(5, -3)$$
$$B(3, 4) \rightarrow B'''(7, -1)$$
$$C(1, 5) \rightarrow C'''(5, 0)$$

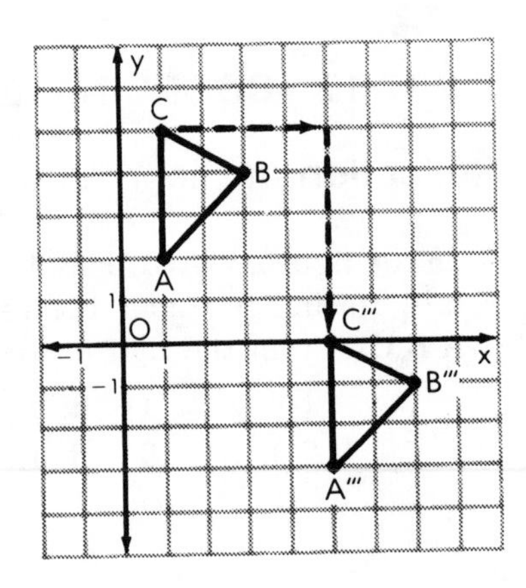

For each point and its image under a translation that moves every point 4 units to the right (+4) and 5 units down (−5), the x-coordinate of the image is 4 more than the x-coordinate of the point ($x \rightarrow x + 4$); the y-coordinate of the image is 5 less than the y-coordinate of the point ($y \rightarrow y - 5$).

From this example, we form a general rule:

- **Under a translation of a units in the horizontal direction and b units in the vertical direction, the image of $P(x, y)$ is $P'(x + a, y + b)$.**

MODEL PROBLEMS

1. On graph paper: **a.** Plot $A(3, -1)$. **b.** Plot A', the image of A under a reflection in the y-axis, and write its coordinates. **c.** Plot A'', the image of A under a reflection in the x-axis, and write its coordinates. **d.** Plot A''', the image of A under a translation of 2 units to the left and 3 units up, and write its coordinates.

Solution:

a.–b.

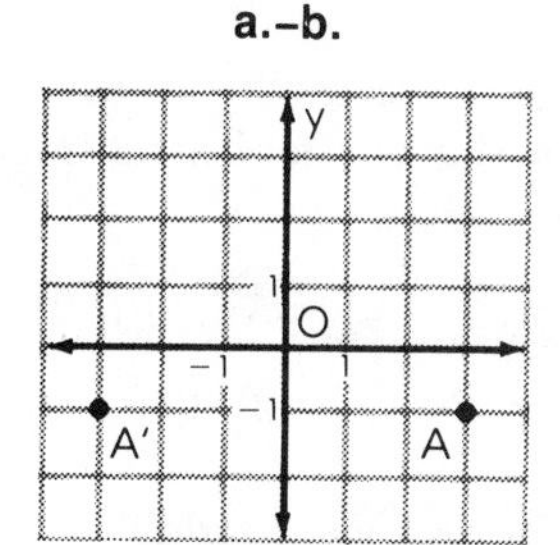

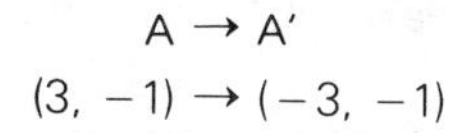

c.

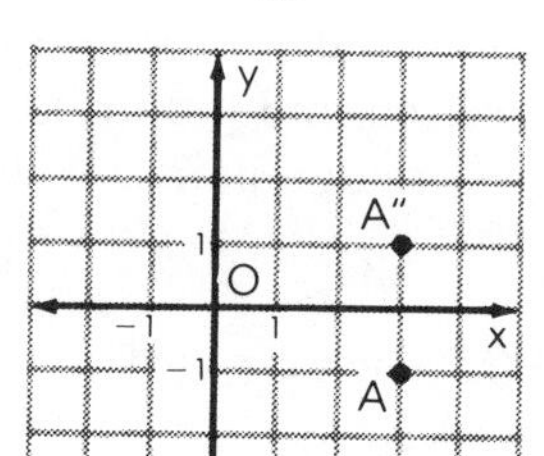

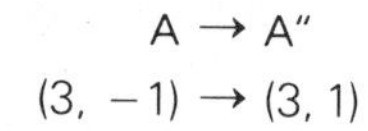

d.

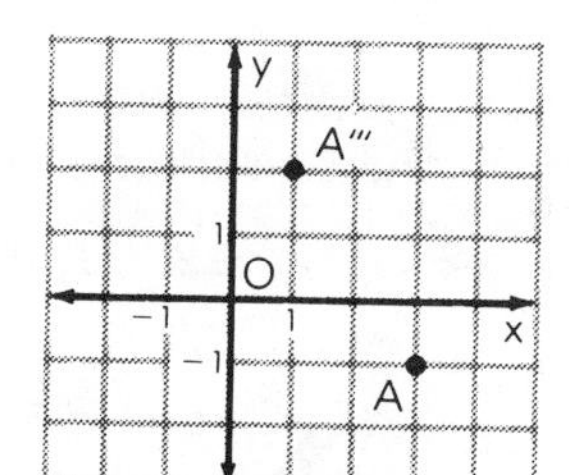

A → A'''
(3, −1) → (1, 2)

2. The graph of rectangle $ABCD$ is shown.

a. Write an equation of each line of symmetry of rectangle $ABCD$.

b. Write the coordinates of the point of symmetry of rectangle $ABCD$.

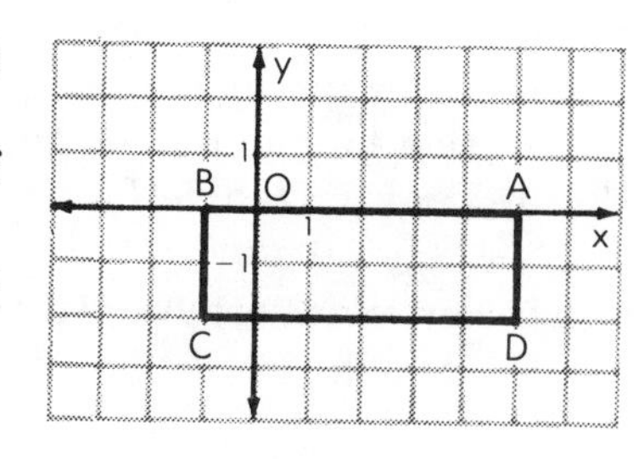

Solution:

a. Equations of the lines of symmetry are $x = 2$ and $y = -1$.

b. The point of symmetry is $(2, -1)$.

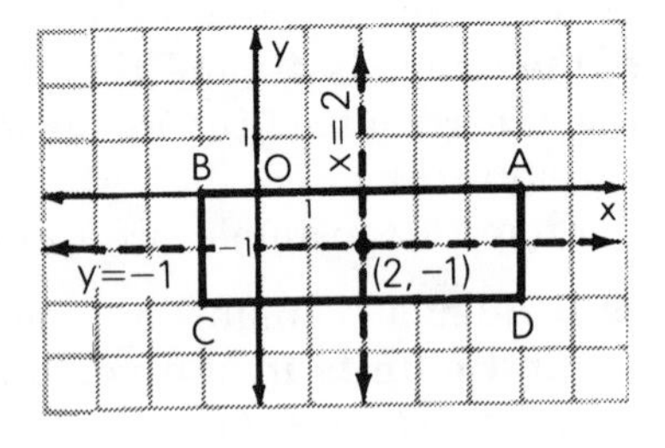

EXERCISES

Use graph paper for these exercises.

In 1–5, graph each point and its image under a reflection in the y-axis. Write the coordinates of the image point.

1. (3, 5) **2.** (1, 4) **3.** (2, −3) **4.** (−2, 3) **5.** (−1, 0)

In 6–10, graph each point and its image under a reflection in the x-axis. Write the coordinates of the image point.

6. (2, 5) **7.** (1, 3) **8.** (−2, 3) **9.** (2, −4) **10.** (0, 2)

In 11–15, graph each point and its image under a translation of +2 units in the horizontal direction and +3 units in the vertical direction. Write the coordinates of the image point.

11. (2, 5) **12.** (1, 3) **13.** (−2, 3) **14.** (2, −4) **15.** (0, 2)

16. a. Draw rectangle $PQRS$ whose vertices are $P(-5, -2)$, $Q(5, -2)$, $R(5, 2)$, and $S(-5, 2)$.

b. What two lines are lines of symmetry for the rectangle?

17. a. Draw rectangle $ABCD$ whose vertices are $A(2, 0)$, $B(2, 5)$, $C(4, 5)$, and $D(4, 0)$.

b. Draw the lines that are axes of symmetry for the rectangle, and write equations of the axes of symmetry.

18. a. Draw $\triangle ABC$ whose vertices are $A(1, 1)$, $B(5, 1)$, and $C(3, 8)$.

b. Draw the line of symmetry for the triangle, and write an equation for the line of symmetry.

19. a. Draw rectangle $LMNO$ whose vertices are $L(0, -4)$, $M(-3, -4)$, $N(-3, 0)$, and $O(0, 0)$.

b. Write equations for two lines of symmetry of the rectangle.

c. Write the coordinates of the point of symmetry of the rectangle.

20. **a.** Draw quadrilateral $RSTU$ whose vertices are $R(1, 1)$, $S(6, 1)$, $T(7, 3)$, and $U(2, 3)$.

b. Does $RSTU$ have line symmetry? If so, write an equation of any line of symmetry.

c. Does $RSTU$ have point symmetry? If so, write the coordinates of the point of symmetry.

17-13 GRAPHS INVOLVING ABSOLUTE VALUE

To draw the graph of the equation $y = |x|$, we can choose values of x and, then, find the corresponding values of y.

Consider the possible choices for x, and the resulting y values:

(1) Choose $x = 0$.

Since the absolute value of 0 is 0, y will be 0.

(2) Choose x as any positive number.

Since the absolute value of any positive number is that positive number, y will have the same value as x. For example, if $x = 5$, then $y = |5| = 5$.

(3) Choose x as any negative number.

Since the absolute value of any negative number is positive, y will have the opposite value of x. For example, if $x = -3$, then $y = |-3| = 3$.

Thus, we conclude that x can be 0, positive, or negative, but y will only be 0 or positive.

Here is a table of values and the corresponding graph:

x	$\|x\|$	y
-5	$\|-5\|$	5
-3	$\|-3\|$	3
-1	$\|-1\|$	1
0	$\|0\|$	0
1	$\|1\|$	1
3	$\|3\|$	3
5	$\|5\|$	5

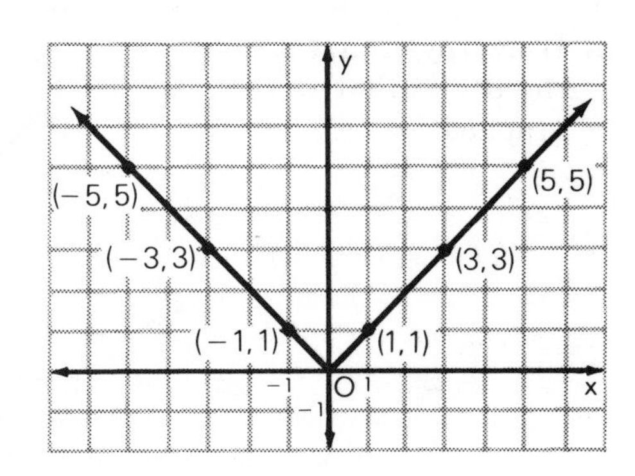

Notice that for positive values of x, the graph of $y = |x|$ is the same as the graph of $y = x$. For negative values of x, the graph of $y = |x|$ is the same as the graph of $y = -x$.

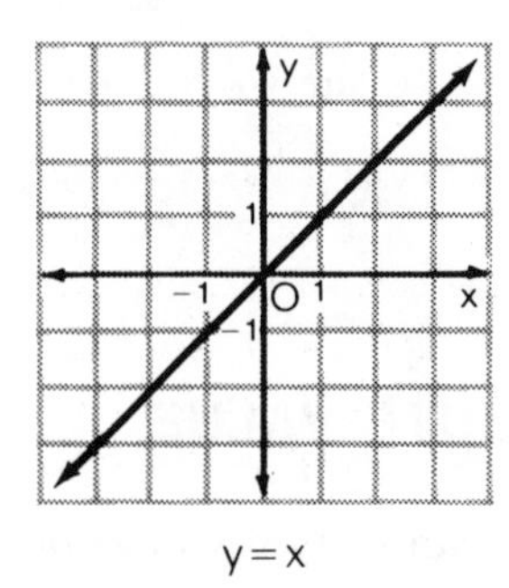

y=x

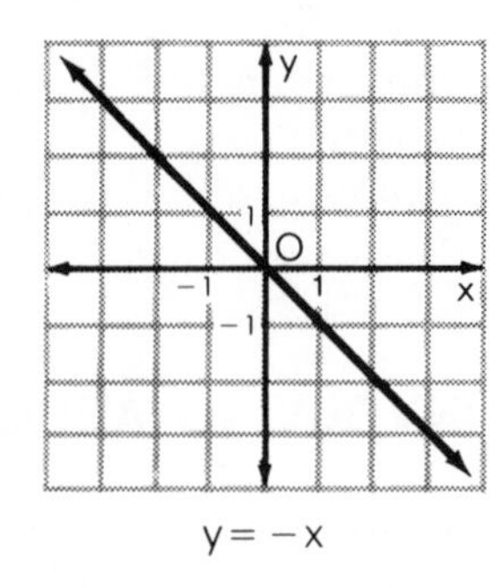

y= −x

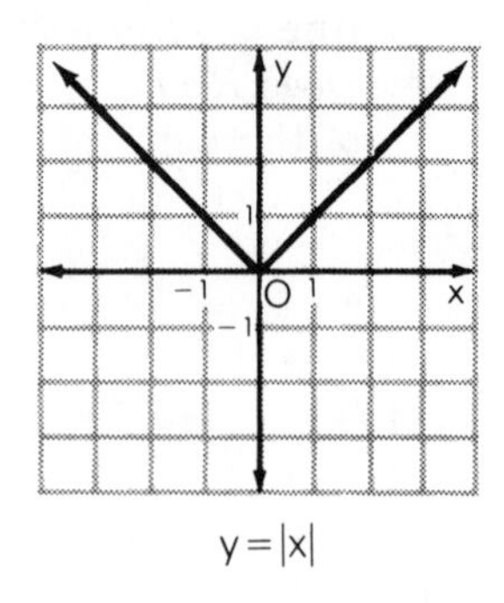

y=|x|

MODEL PROBLEMS

1. Draw the graph of $y = |x| + 2$.

How to Proceed | *Solution*

(1) Make a table of values.

x	|x| + 2	y
−4	|−4| + 2	6
−2	|−2| + 2	4
−1	|−1| + 2	3
0	|0| + 2	2
1	|1| + 2	3
3	|3| + 2	5
5	|5| + 2	7

(2) Plot the points.

(3) Draw rays connecting the points that were graphed.

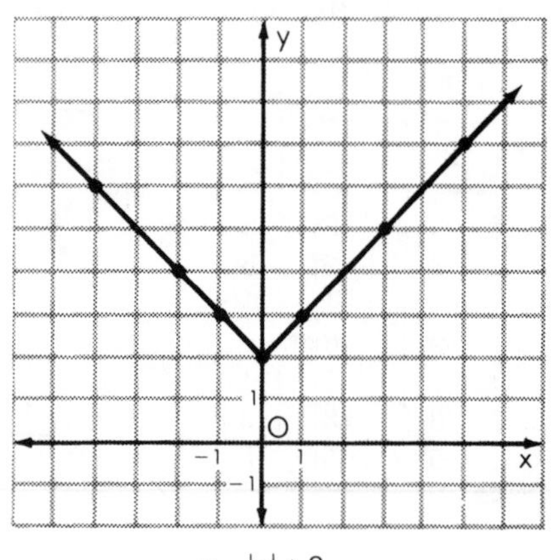

y=|x|+2

2. Draw the graph of $|x| + |y| = 3$.

Solution

By the definition of absolute value, $|x| = |-x|$ and $|y| = |-y|$.

Since (1, 2) is a solution, (−1, 2), (1, −2) and (−1, −2) are solutions.

Since (2, 1) is a solution, (−2, 1), (2, −1) and (−2, −1) are solutions.

Since (0, 3) is a solution, (0, −3) is a solution.

Since (3, 0) is a solution, (−3, 0) is a solution.

Plot these points that are solutions, and draw the line segments joining them.

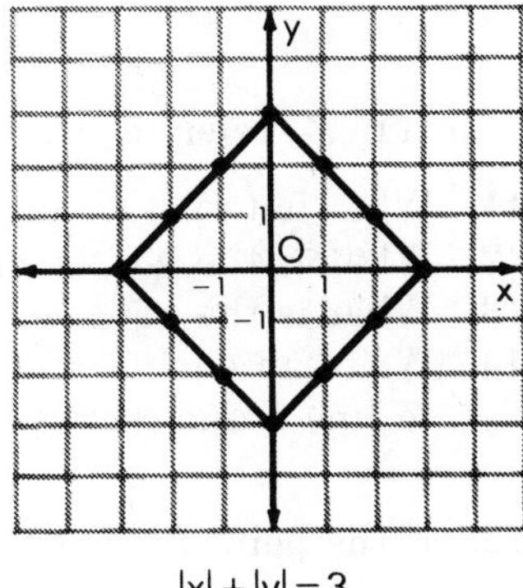

$|x| + |y| = 3$

EXERCISES

In 1–9, graph the equation.

1. $y = \|x\| - 1$	**2.** $y = \|x\| + 3$	**3.** $y = \|x - 1\|$
4. $y = \|x + 3\|$	**5.** $y = 2\|x\|$	**6.** $y = 2\|x\| + 1$
7. $\|x\| + \|y\| = 5$	**8.** $\|x\| + 2\|y\| = 7$	**9.** $\|y\| = \|x\|$

10. **a.** Draw the graph of $|x| + |y| = 4$.
b. Write the equations of two lines of symmetry for the graph drawn in part **a**.

17-14 REVIEW EXERCISES

1. What is the slope of the graph of $y = -2x + 5$?

2. Solve the equation for y in terms of x: $3x - 2y = 12$.

3. Write an equation of the line whose slope is −1 and whose y-intercept is 7.

4. What is the slope of the line that passes through the points (4, 5) and (6, 1)?

In 5–10, graph the equation or inequality.

5. $y = -x + 2$ **6.** $y = 3$ **7.** $y = \frac{2}{3}x$

8. $x + 2y = 8$ **9.** $y - x > 2$ **10.** $2x - y \geq 4$

In 11–14, refer to the coordinate graph.

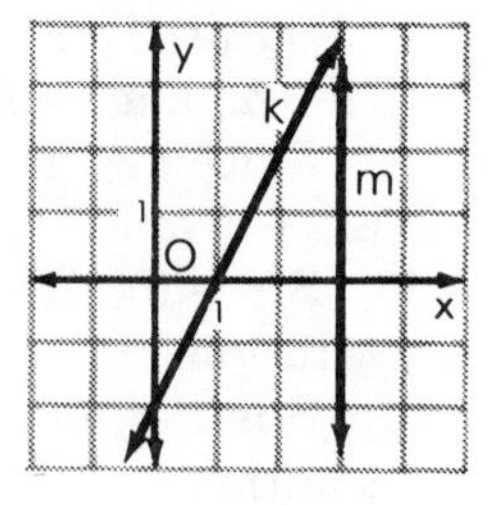

11. What is the slope of line k?
12. What is the y-intercept of line k?
13. What is the equation of line m?
14. Write an equation of the line that is parallel to line k and passes through the origin.

15. If the point $(d, 3)$ lies on the graph of $3x - y = 9$, find the value of d.

In 16–21, select the numeral preceding the correct answer.

16. Which point does *not* lie on the graph of $3x - y = 9$?
(1) $(1, -6)$ (2) $(2, 3)$ (3) $(3, 0)$ (4) $(0, -9)$

17. Which ordered pair is in the solution set of $y < 2x - 4$?
(1) $(0, -5)$ (2) $(2, 0)$ (3) $(3, 3)$ (4) $(0, 2)$

18. Which equation has a graph parallel to the graph of $y = 5x - 2$?
(1) $y = -5x$ (2) $y = 5x + 3$ (3) $y = -2x$ (4) $y = 2x - 5$

19. The graph of $2x + y = 8$ intersects the x-axis at
(1) $(0, 8)$ (2) $(8, 0)$ (3) $(0, 4)$ (4) $(4, 0)$

20. What is the slope of the graph of the equation $y = 4$?
(1) 1 (2) 0 (3) -4 (4) 4

21. In which ordered pair is the abscissa 3 more than the ordinate?
(1) $(1, 4)$ (2) $(1, 3)$ (3) $(3, 1)$ (4) $(4, 1)$

22. **a.** Plot the points $A(5, -2)$, $B(3, 3)$, $C(-3, 3)$, and $D(-5, -2)$.
b. Draw the polygon $ABCD$.
c. What kind of polygon is $ABCD$?
d. Find DA and BC.
e. Find the length of the altitude from C to DA.
f. Find the area of $ABCD$.
g. Write an equation for the line of symmetry of $ABCD$.

In 23–25, graph each equation.

23. $y = |x - 2|$ **24.** $y = |x| - 2$ **25.** $|x| + 2|y| = 6$

Chapter 18

Systems of Linear Open Sentences in Two Variables

18-1 GRAPHIC SOLUTION OF A SYSTEM OF LINEAR EQUATIONS IN TWO VARIABLES

Consistent Equations

Consider the problem of finding two numbers whose sum is 4 and whose difference is 2. Let us see whether we have enough information to find the numbers.

Let x = the larger number. Let y = the smaller number.
Since the sum of the numbers is 4, $x + y = 4$.
Since the difference of the numbers is 2, $x - y = 2$.

The two equations impose two conditions on the variables at the same time. The two equations are called a ***system of simultaneous equations***.

A ***solution*** of a system of simultaneous equations in two variables is an ordered pair of numbers that satisfies both equations. The set of all solutions of the system is called the ***solution set*** of the system.

The graph of a linear equation in two variables is a line. The graphs of $x + y = 4$ and $x - y = 2$ in a coordinate plane, using the same set of axes, are shown at the right. The coordinates of the point of intersection (3, 1) satisfy the equations of both lines.

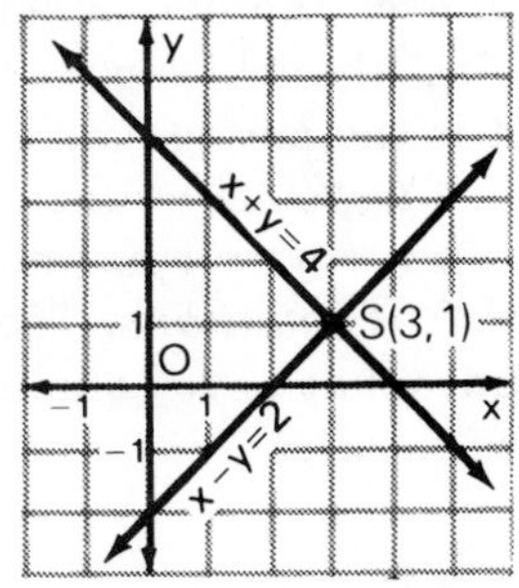

$x + y = 4$ [Let $x = 3$, $y = 1$.]
$3 + 1 \stackrel{?}{=} 4$
$4 = 4$ (True)

$x - y = 2$ [Let $x = 3$, $y = 1$.]
$3 - 1 \stackrel{?}{=} 2$
$2 = 2$ (True)

There is no other ordered pair that satisfies both equations because there is no point other than the point (3, 1) that lies on both graphs. Hence, the ordered pair (3, 1) is the solution of the system and {(3, 1)} is the solution set of the system.

Therefore, the larger number is 3 and the smaller number is 1.

When a pair of lines is graphed in the same coordinate plane on the same set of axes, one and only one of the following three possibilities can occur. The pair of lines will:

1. intersect in one point and have one ordered number pair in common; or
2. be parallel and have no ordered number pairs in common; or
3. coincide, that is, be the same line with an infinite number of ordered number pairs in common.

If a system of linear equations such as $x + y = 4$ and $x - y = 2$ has one common solution, it is called a ***system of consistent equations.***

Inconsistent Equations

Sometimes, when two linear equations are graphed in a coordinate plane using the same set of axes, the lines are parallel and fail to intersect, as in the case of $x + y = 2$ and $x + y = 4$. There is no common solution for the system of equations $x + y = 2$ and $x + y = 4$. It is obvious that there can be no ordered number pair (x, y) such that the sum of those numbers, $x + y$, is both 2 and 4. Since the solution set of the system has no members, it is the empty set.

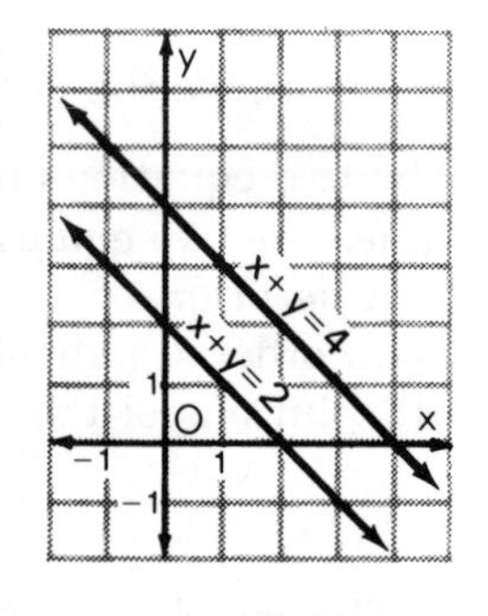

If a system of linear equations such as $x + y = 2$ and $x + y = 4$ has no common solution, it is called a ***system of inconsistent equations.*** The graphs of two inconsistent linear equations are lines that have equal slopes or lines that have no slopes. Such lines will be parallel.

Dependent Equations

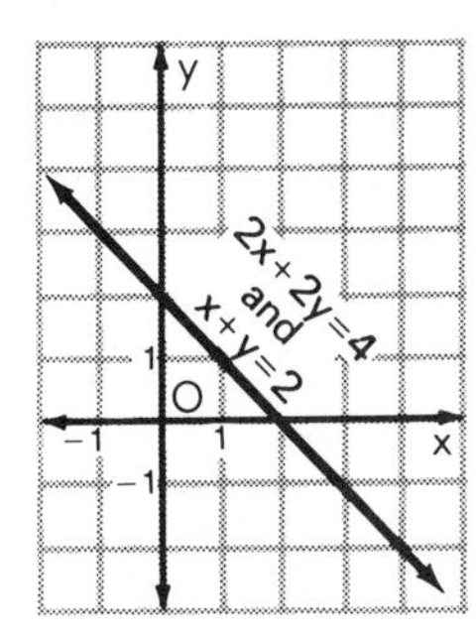

Sometimes, when two linear equations are graphed in a coordinate plane using the same set of axes, they turn out to be the same line; that is, they coincide. This happens in the case of the equations $x + y = 2$ and $2x + 2y = 4$. Every one of the infinite number of solutions of $x + y = 2$ is also a solution of $2x + 2y = 4$. Thus, we see that $2x + 2y = 4$ and $x + y = 2$ are equivalent equations with identical solutions. Note that when both members of the equation $2x + 2y = 4$ are divided by 2, the result is $x + y = 2$.

If a system of two linear equations, for example $x + y = 2$ and $2x + 2y = 4$, is such that every solution of one of the equations is also a solution of the other, it is called a ***system of dependent equations***. The graphs of two dependent linear equations are the same line.

PROCEDURE. To solve a pair of linear equations graphically:

1. Graph one equation in a coordinate plane.
2. Graph the other equation using the same set of coordinate axes.
3. Find the common solution, that is, the ordered number pair associated with the point of intersection of the two graphs. If the equations are not consistent, the graphs of the lines will not intersect.
4. Check the solution by verifying that the ordered pair satisfies both equations.

KEEP IN MIND

The solution set of a system of two linear equations is the intersection of the solution sets of the individual equations.

MODEL PROBLEM

Solve graphically and check: $2x + y = 8$
$y - x = 2$

Solution:

(1) Graph $2x + y = 8$, or $y = -2x + 8$.

x	−2x + 8	y
1	−2(1) + 8	6
3	−2(3) + 8	2
4	−2(4) + 8	0

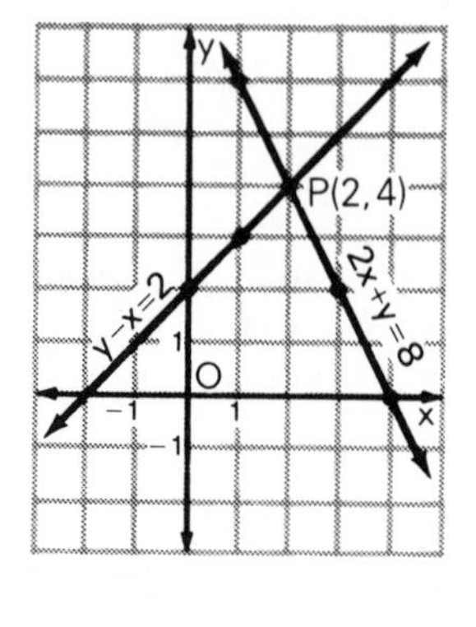

(2) Graph $y - x = 2$, or $y = x + 2$.

x	x + 2	y
0	0 + 2	2
1	1 + 2	3
2	2 + 2	4

(3) Read the coordinates of the point of intersection $P(2, 4)$.

(4) *Check:* $(x = 2, y = 4)$

$$2x + y = 8 \qquad\qquad y - x = 2$$
$$2(2) + 4 \stackrel{?}{=} 8 \qquad\qquad 4 - 2 \stackrel{?}{=} 2$$
$$8 = 8 \text{ (True)} \qquad\qquad 2 = 2 \text{ (True)}$$

Answer: The common solution is (2, 4). The solution set is {(2, 4)}.

EXERCISES

In 1–30, solve the systems of equations graphically. Check.

1. $y = 2x$
$y = 3x - 3$

2. $y = x + 4$
$y = 2x + 5$

3. $y = -2x + 3$
$y = \frac{1}{2}x + 3$

4. $x + y = 7$
$x - y = 1$

5. $x + y = 4$
$x - y = 0$

6. $x + y = -4$
$x - y = 6$

7. $x + y = 1$
$x + 3y = 9$

8. $y = 2x + 1$
$x + 2y = 7$

9. $x - y = 4$
$x + 2y = 10$

10. $x + 2y = 17$
$y = 2x + 1$

11. $y = 3x$
$2x + y = 10$

12. $x + 3y = 9$
$x = 3$

13. $y - x = -2$
$x - 2y = 4$

14. $3x + y = 6$
$y = 3$

15. $4x - y = 9$
$2x + y = 12$

16. $y = 2x + 4$
$x = y - 5$

17. $x + y = 3$
$2x - y = -9$

18. $y - 3x = 12$
$y = -3$

19. $2x - y = -1$
$x = y + 1$

20. $3x + y = -9$
$x + 3y = -11$

21. $x = 3$
$y = 4$

22. $y = \frac{1}{3}x - 3$
$2x - y = 8$

23. $3x + y = 13$
$x + 6y = -7$

24. $x = 0$
$y = -5$

25. $2x = y + 9$
$6x + 3y = 15$

26. $5x - 3y = 9$
$5y = 13 - x$

27. $x = 0$
$y = 0$

28. $x + y + 2 = 0$
$x = y - 8$

29. $y + 2x + 6 = 0$
$y = 2x$

30. $7x - 4y + 7 = 0$
$3x - 5y + 3 = 0$

In 31–36, graph both equations. Determine whether the system is consistent, inconsistent, or dependent.

31. $x + y = 1$
$x + y = 3$

32. $x + y = 5$
$2x + 2y = 10$

33. $y = 2x + 1$
$y = 3x + 3$

34. $2x - y = 1$
$2y = 4x - 2$

35. $y - 3x = 2$
$y = 3x - 2$

36. $x + 4y = 6$
$x = 2$

37. Are there any ordered number pairs that satisfy both equations $2x + y = 7$ and $2x = 5 - y$?

38. Are there any ordered number pairs that satisfy the equation $y - x = 4$ but which do not satisfy the equation $2y = 8 + 2x$?

In 39–42: **a.** Write a system of two first-degree equations involving the two variables x and y that represent the conditions stated in the problem. **b.** Solve the system graphically.

39. The sum of two numbers is 8. Their difference is 2. Find the numbers.

40. The sum of two numbers is 3. The larger number is 5 more than the smaller number. Find the numbers.

41. The perimeter of a rectangle is 12 meters. Its length is twice its width. Find the dimensions of the rectangle.

42. The perimeter of a rectangle is 14 centimeters. Its length is 3 centimeters more than its width. Find the length and the width.

18-2 ALGEBRAIC SOLUTION OF A SYSTEM OF SIMULTANEOUS LINEAR EQUATIONS BY USING ADDITION

In the previous section, graphs were used to find solutions of systems of simultaneous equations. Since most of the solutions were integers, the values of x and y were easily read from the graphs. This is not always possible, as shown below.

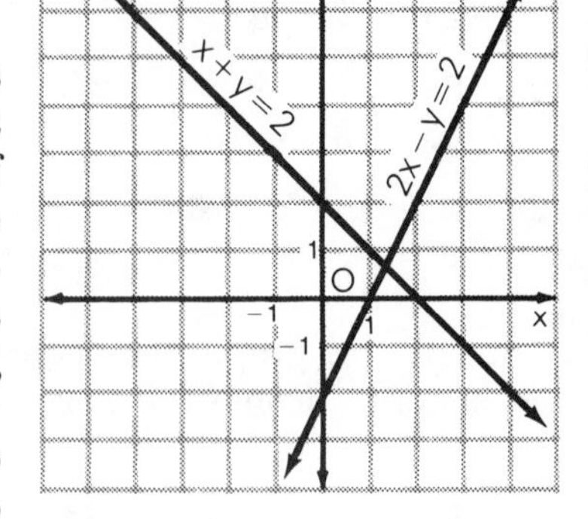

The graphs of the system of equations $2x - y = 2$ and $x + y = 2$ are shown at the right. The solution of this system of equations is not a pair of integers. We could approximate the solution and determine if our approximation is correct by checking. However, there are other, more direct methods of solution.

Algebraic methods can be used to solve a system of linear equations in two variables. Solutions by these methods often take less time and lead to more accurate results than the graphic method previously used.

Systems of equations that have the same solution set are called ***equivalent systems***. For example, the following two systems are equivalent systems since they have the same solution set, $\left\{\left(\frac{4}{3}, \frac{2}{3}\right)\right\}$.

System A	*System B*
$x + y = 2$	$x = \frac{4}{3}$
$2x - y = 2$	$y = \frac{2}{3}$

To solve a system of linear equations such as System A, whose solution set is not obvious, we transform it into an equivalent system such as System B, whose solution set is obvious. To do this, we make use of the properties of equality.

(1) The coefficients of y in the two equations are additive inverses. Therefore, adding the equations will result in an equation in one variable.

$$\begin{array}{r} 2x - y = 2 \\ \underline{x + y = 2} \\ 3x \quad = 4 \end{array}$$

(2) Solve the resulting equation for x.

$$x = \frac{4}{3}$$

(3) Replace x by its value in either of the given equations.

$$x + y = 2$$
$$\frac{4}{3} + y = 2$$

(4) Solve the resulting equation for y.

$$\underline{-\frac{4}{3} \qquad = -\frac{4}{3}}$$
$$y = \frac{2}{3}$$

(5) Check. Substitute $\frac{4}{3}$ for x and $\frac{2}{3}$ for y in each of the given equations, and show that these values make the given equations true.

$$\begin{array}{rl} 2x - y = 2 & \quad x + y = 2 \\ 2\left(\frac{4}{3}\right) - \frac{2}{3} \stackrel{?}{=} 2 & \quad \frac{4}{3} + \frac{2}{3} \stackrel{?}{=} 2 \\ \frac{8}{3} - \frac{2}{3} \stackrel{?}{=} 2 & \quad \frac{6}{3} \stackrel{?}{=} 2 \\ \frac{6}{3} \stackrel{?}{=} 2 & \quad 2 = 2 \\ 2 = 2 & \quad \text{(True)} \\ \text{(True)} & \end{array}$$

MODEL PROBLEMS

1. Solve the system of equations and check: $x + 3y = 13$
$x + y = 5$

How to Proceed | *Solution*

(1) The coefficients of the variable x are the same in both equations. Therefore, write an equation equivalent to equation [B] by multiplying both members of equation [B] by -1. Now, the coefficients of x are additive inverses. Add the two equations so the resulting equation involves one variable, y.

$$x + 3y = 13 \quad [A]$$
$$x + y = 5 \quad [B]$$

$$\begin{array}{r} x + 3y = 13 \\ \underline{-x - y = -5} \\ 2y = 8 \end{array}$$

(2) Solve the resulting equation for the variable, y.

$$y = 4$$

(3) Replace y by its value in any equation involving both variables.

$$x + y = 5 \quad [B]$$
$$x + 4 = 5$$

(4) Solve the resulting equation for the remaining variable, x.

$$x = 1$$

(5) *Check:* Substitute 1 for x and 4 for y in each of the given equations to verify that the resulting sentences are true.

$$\begin{aligned} x + 3y &= 13 \\ 1 + 3(4) &\stackrel{?}{=} 13 \\ 13 &= 13 \quad \text{(True)} \end{aligned} \qquad \begin{aligned} x + y &= 5 \\ 1 + 4 &\stackrel{?}{=} 5 \\ 5 &= 5 \quad \text{(True)} \end{aligned}$$

Answer: Since $x = 1$ and $y = 4$, the solution is (1, 4), or the solution set is {(1, 4)}.

2. Solve the system of equations and check:

$$5a + b = 13$$
$$4a - 3b = 18$$

How to Proceed / *Solution*

(1) Multiply both members of equation [A] by 3. This yields an equivalent equation [C] in which the coefficient of b is the additive inverse of the coefficient of b in equation [B].

$$\begin{aligned} 5a + b &= 13 \quad [A] \\ 4a - 3b &= 18 \quad [B] \\ 15a + 3b &= 39 \quad [C] \end{aligned}$$

(2) Add the corresponding members of equations [B] and [C] to eliminate the variable b.

$$19a = 57$$

(3) Solve the resulting equation for the variable a.

$$a = 3$$

(4) Replace a by its value in any equation involving both variables.

$$5a + b = 13 \quad [A]$$
$$5(3) + b = 13$$

(5) Solve the resulting equation for the remaining variable, b.

$$15 + b = 13$$
$$b = -2$$

(6) *Check:* Substitute 3 for a and -2 for b in each of the given equations to verify that the resulting sentences are true. This is left to the student.

Answer: Since $a = 3$ and $b = -2$, $(a, b) = (3, -2)$, or the solution set is {(3, −2)}.

3. Solve the system of equations and check: $7x = 5 - 2y$
$3y = 16 - 2x$

How to Proceed	*Solution*
(1) Transform each of the given equations [*A*] and [*B*] into equivalent equations [*C*] and [*D*] in which the terms containing the variables appear on one side and the constant appears on the other side.	$7x = 5 - 2y$ [*A*] $3y = 16 - 2x$ [*B*] $7x + 2y = 5$ [*C*] $2x + 3y = 16$ [*D*]
(2) To eliminate y, multiply both members of equation [*C*] by 3; multiply both members of equation [*D*] by -2. In the resulting equivalent equations [*E*] and [*F*], the coefficients of y are additive inverses.	$21x + 6y = 15$ [*E*] $-4x - 6y = -32$ [*F*]
(3) Add the members of equation [*F*] to the corresponding members of equation [*E*] to eliminate the variable y.	$17x = -17$
(4) Solve the resulting equation for the variable x.	$x = -1$
(5) Replace x by its value in any equation containing both variables.	$3y = 16 - 2x$ [*B*] $3y = 16 - 2(-1)$
(6) Solve the resulting equation for the remaining variable y.	$3y = 16 + 2$ $3y = 18$ $y = 6$

(7) *Check:* Substitute -1 for x and 6 for y in each of the given equation to verify that the resulting sentences are true. This is left to the student.

Answer: Since $x = -1$ and $y = 6$, the solution is $(-1, 6)$, or the solution set is $\{(-1, 6)\}$.

EXERCISES

In 1 and 2, state which of the given ordered pairs is the solution of the system of equations.

1. $x + y = 5$
   $x - y = -1$
   (2, 3) (2, −3) (3, −2) (3, 2)

2. $6x + 2y = 14$
   $3x + 2y = 8$
   (1, 2) (1, 4) (2, 1) (4, −2)

In 3–47, solve each system of equations by using addition to eliminate one of the variables. Check.

3. $x + y = 12$
   $x - y = 4$

4. $a + b = 13$
   $a - b = 5$

5. $r + s = -6$
   $r - s = -10$

6. $3x + y = 16$
   $2x + y = 11$

7. $c - 2d = 14$
   $c + 3d = 9$

8. $x + y = 10$
   $x - y = 0$

9. $a - 4b = -8$
   $a - 2b = 0$

10. $x + 2y = 8$
    $x - 2y = 4$

11. $8a + 5b = 9$
    $2a - 5b = -4$

12. $4x + 5y = 23$
    $4x - y = 5$

13. $-2m + 4n = 13$
    $6m + 4n = 9$

14. $3a - b = 3$
    $a + 3b = 11$

15. $3r + s = 6$
    $r + 3s = 10$

16. $4x - y = 10$
    $2x + 3y = 12$

17. $5m + 3n = 14$
    $2m + n = 6$

18. $2c - d = -1$
    $c + 3d = 17$

19. $2m + n = 12$
    $m + 2n = 9$

20. $r - 3s = -11$
    $3r + s = 17$

21. $a + 3b = 4$
    $2a - b = 1$

22. $3x + 4y = 26$
    $x - 3y = 0$

23. $5x + 8y = 1$
    $3x + 4y = -1$

24. $x - y = -1$
    $3x - 2y = 3$

25. $5a - 2b = 3$
    $2a - b = 0$

26. $5x - 2y = 20$
    $2x + 3y = 27$

27. $2x - y = 26$
    $3x - 2y = 42$

28. $2x + 3y = 6$
    $3x + 5y = 15$

29. $5r - 2s = 8$
    $3r - 7s = -1$

30. $3x + 7y = -2$
    $2x + 3y = -3$

31. $4x + 3y = -1$
    $5x + 4y = 1$

32. $4a - 6b = 15$
    $6a - 4b = 10$

33. $5y + x = -8$
    $x = 7$

34. $x + y = 4$
    $y = x$

35. $2x + y = 17$
    $5x = 25 + y$

36. $5r + 3s = 30$
    $2r = 12 - 3s$

37. $6r = s$
    $5r = 2s - 14$

38. $3a - 7 = 7b$
    $4a = 3b + 22$

39. $3x - 4y = 2$
$x = 2(7 - y)$

40. $3x + 5(y + 2) = 1$
$8y = -3x$

41. $\frac{1}{3}x + \frac{1}{4}y = 10$
$\frac{1}{3}x - \frac{1}{2}y = 4$

42. $\frac{1}{2}a + \frac{1}{3}b = 8$
$\frac{3}{2}a - \frac{4}{3}b = -4$

43. $c - 2d = 1$
$\frac{2}{3}c + 5d = 26$

44. $2a = 3b$
$\frac{2}{3}a - \frac{1}{2}b = 2$

45. $.04x + .06y = 26$
$x + y = 500$

46. $.03x + .05y = 17$
$x + y = 400$

47. $.03x = .06y + 9$
$x + y = 600$

18-3 ALGEBRAIC SOLUTION OF A SYSTEM OF SIMULTANEOUS LINEAR EQUATIONS BY USING SUBSTITUTION

Another algebraic method, called the ***substitution method***, can be used to eliminate one of the variables when solving a system of equations. When we use this method, we apply the substitution principle to transform one of the equations of the system into an equivalent equation that involves only one variable.

MODEL PROBLEMS

1. Solve the system of equations and check: $4x + 3y = 27$
$y = 2x - 1$

How to Proceed	*Solution*
(1) In equation [*B*], both y and $2x - 1$ name the same number. Therefore, eliminate y in equation [*A*] by replacing y with $2x - 1$.	$4x + 3y = 27$ [*A*] $y = 2x - 1$ [*B*] $4x + 3(2x - 1) = 27$
(2) Solve the resulting equation for x.	$4x + 6x - 3 = 27$ $10x - 3 = 27$ $10x = 30$ $x = 3$
(3) Replace x with its value in any equation involving both variables.	$y = 2x - 1$ [*B*] $y = 2(3) - 1$

(4) Solve the resulting equation for y.

$$y = 6 - 1$$
$$y = 5$$

(5) *Check:* Substitute 3 for x and 5 for y in each of the given equations to verify that the resulting sentences are true.

$$\begin{aligned} 4x + 3y &= 27 \\ 4(3) + 3(5) &\stackrel{?}{=} 27 \\ 12 + 15 &\stackrel{?}{=} 27 \\ 27 &= 27 \quad \text{(True)} \end{aligned} \qquad \begin{aligned} y &= 2x - 1 \\ 5 &\stackrel{?}{=} 2(3) - 1 \\ 5 &\stackrel{?}{=} 6 - 1 \\ 5 &= 5 \quad \text{(True)} \end{aligned}$$

Answer: Since $x = 3$ and $y = 5$, the solution is (3, 5), or the solution set is {(3, 5)}.

2. Solve the system of equations and check: $3x - 4y = 26$
$x + 2y = 2$

How to Proceed	*Solution*
(1) Transform one of the equations into an equivalent equation in which one of the variables is expressed in terms of the other. In equation [*B*], solve for x in terms of y, to obtain equation [*C*].	$3x - 4y = 26$ [*A*] $x + 2y = 2$ [*B*] $x = 2 - 2y$ [*C*]
(2) Eliminate x in equation [*A*] by replacing it with $2 - 2y$, the expression for x in equation [*C*].	$3(2 - 2y) - 4y = 26$
(3) Solve the resulting equation for y.	$6 - 6y - 4y = 26$ $6 - 10y = 26$ $-10y = 20$ $y = -2$
(4) Replace y by its value in any equation involving both variables.	$x = 2 - 2y$ [*C*] $x = 2 - 2(-2)$
(5) Solve the resulting equation for x.	$x = 2 + 4$ $x = 6$

(6) *Check:* Substitute 6 for x and -2 for y in each of the given equations to verify that the resulting sentences are true. This is left to the student.

Answer: Since $x = 6$ and $y = -2$, the solution is $(6, -2)$, or the solution set is $\{(6, -2)\}$.

EXERCISES

In 1–18, solve each system of equations by using substitution to eliminate one of the variables. Check.

1. $y = x$
$x + y = 14$

2. $x = y$
$5x - 4y = -2$

3. $y = 2x$
$x + y = 21$

4. $x = 4y$
$2x + 3y = 22$

5. $a = -2b$
$5a - 3b = 13$

6. $r = -3s$
$3r + 4s = -10$

7. $y = x + 1$
$x + y = 9$

8. $x = y - 2$
$x + y = 18$

9. $y = x + 3$
$3x + 2y = 26$

10. $y = 2x + 1$
$x + y = 7$

11. $a = 3b + 1$
$5b - 2a = 1$

12. $a + b = 11$
$3a - 2b = 8$

13. $3m - 2n = 11$
$m + 2n = 9$

14. $a - 2b = -2$
$2a - b = 5$

15. $7x - 3y = 23$
$x + 2y = 13$

16. $2x = 3y$
$4x - 3y = 12$

17. $4y = -3x$
$5x + 8y = 4$

18. $2x + 3y = 7$
$4x - 5y = 25$

In 19–33, solve each system of equations by using an algebraic method that seems convenient. Check.

19. $y = 3x$
$y - x = 18$

20. $s + r = 0$
$r - s = 6$

21. $3a - b = 13$
$2a + 3b = 16$

22. $m + 2n = 14$
$3n + m = 18$

23. $x = y$
$4x - 5y = -2$

24. $y = x - 2$
$3x - y = 16$

25. $-2c = d$
$6c + 5d = -12$

26. $3x + 8y = 16$
$5x + 10y = 25$

27. $y = 3x$
$\frac{1}{3}x + \frac{1}{2}y = 11$

28. $a - \frac{2}{3}b = 4$
$\frac{3}{5}a + b = 15$

29. $3(y - 6) = 2x$
$3x + 5y = 11$

30. $x + y = 300$
$.25x + .75y = 195$

31. $3d = 13 - 2c$
$\frac{3c + d}{2} = 8$

32. $3x = 4y$
$\frac{3x + 8}{5} = \frac{3y - 1}{2}$

33. $\frac{a}{3} + \frac{a + b}{6} = 3$
$\frac{b}{3} - \frac{a - b}{2} = 6$

18-4 SOLVING VERBAL PROBLEMS BY USING TWO VARIABLES

You have previously learned how to solve word problems using one variable. Frequently, a problem can be solved more easily by using two variables rather than one variable. This is done in the solutions of the problems that follow.

PROCEDURE. To solve word problems by using a system of two equations involving two variables:

1. Use two different variables to represent the different unknown quantities in the problem.
2. Translate two relationships in the problem into a system of two equations.
3. Solve the system of equations to determine the answer(s) to the problem.
4. Check the answer(s) in the original word problem.

Number Problems

The sum of two numbers is 10. Three times the larger decreased by twice the smaller is 15. Find the numbers.

How to Proceed	*Solution*
(1) Represent the two different unknown quantities by two different variables.	Let x = the larger number. Let y = the smaller number.
(2) Translate two given relationships in the problem into a system of two equations.	The sum of two numbers is 10. $x + y = 10$ [A] Three times the larger decreased by twice the smaller is 15. $3x - 2y = 15$ [B]
(3) Solve the system of equations. In equation [A], multiply both members by 2. Then, add the members of the resulting equation to the corresponding members of equation [B].	$x + y = 10$ [A] $3x - 2y = 15$ [B] $\underline{2x + 2y = 20}$ $5x = 35$

(4) Solve the resulting equation for x.	$x = 7$
(5) Replace x with its value in any equation involving both variables.	$x + y = 10$ [A] $7 + y = 10$
(6) Solve the resulting equation for y.	$y = 3$

Check: The sum of the larger number 7 and the smaller number 3 is 10. Three times the larger decreased by twice the smaller, $(3 \cdot 7) - (2 \cdot 3)$, equals $21 - 6$, or 15.

Answer: The larger number is 7; the smaller number is 3.

EXERCISES

In 1–7, solve the problem algebraically, using two variables.

1. The sum of two numbers is 36. Their difference is 24. Find the numbers.
2. The sum of two numbers is 77. The larger number is 3 more than the smaller number. Find the numbers.
3. The sum of two numbers is 104. The larger number is 1 less than twice the smaller number. Find the numbers.
4. The difference between two numbers is 34. The larger exceeds 3 times the smaller by 4. Find the numbers.
5. If 5 times the smaller of two numbers is subtracted from twice the larger, the result is 16. If the larger is increased by 3 times the smaller, the result is 63. Find the numbers.
6. One number is 15 larger than another. The sum of twice the larger and three times the smaller is 180. Find the numbers.
7. The sum of two numbers is 900. When 4% of the larger is added to 7% of the smaller, the sum is 48. Find the numbers.

Business Problems

The owner of a men's clothing store bought 6 belts and 8 hats for $140. A week later, at the same prices, he bought 9 belts and 6 hats for $132. Find the price of a belt and the price of a hat.

Solution: Let b = the price of a belt in dollars.
Let h = the price of a hat in dollars.

$$\underbrace{\textit{6 belts and 8 hats}}_{6b + 8h} \ \underbrace{\textit{cost}}_{=} \ \underbrace{\textit{\$140.}}_{140} \quad [A]$$

$$\underbrace{\textit{9 belts and 6 hats}}_{9b + 6h} \ \underbrace{\textit{cost}}_{=} \ \underbrace{\textit{\$132.}}_{132} \quad [B]$$

(1) In order to eliminate h, multiply both members of equation [B] by 4 and both members of equation [A] by -3. Then, add the equations and solve for b.

$$\begin{array}{rcr} 36b + 24h & = & 528 \\ -18b - 24h & = & -420 \\ \hline 18b & = & 108 \\ b & = & 6 \end{array}$$

(2) In equation [A], substitute 6 for b.

$$\begin{aligned} 36 + 8h &= 140 \\ 8h &= 104 \\ h &= 13 \end{aligned}$$

Check: 6 belts and 8 hats cost 6($6) + 8($13) = $36 + $104 = $140.
9 belts and 6 hats cost 9($6) + 6($13) = $54 + $78 = $132.

Answer: A belt costs $6; a hat costs $13.

EXERCISES

In 1–5, solve the problem algebraically, using two variables.

1. At a quick-lunch counter, 3 pretzels and 1 cup of soda cost $2.75. Two pretzels and 1 cup of soda cost $2.00. Find the cost of a pretzel and the cost of a cup of soda.

2. On one day, 4 gardeners and 4 helpers earned $360. On another day, working the same number of hours and at the same rate of pay, 5 gardeners and 6 helpers earned $480. How much does a gardener and how much does a helper earn each day?

3. A baseball manager bought 4 bats and 9 balls for $76.50. On another day, she bought 3 bats and 1 dozen balls at the same prices and paid $81.00. How much did she pay for each bat and each ball?

4. Mrs. Black bought 2 pounds of veal and 3 pounds of pork, for which she paid $20.00. Mr. Cook, paying the same prices, paid $11.25 for 1 pound of veal and 2 pounds of pork. Find the price of a pound of veal and the price of a pound of pork.
5. One day, Mrs. Rubero paid $18.70 for 4 kilograms of brown rice and 3 kilograms of white rice. On another day, Mrs. Rubero paid $13.30 for 3 kilograms of brown rice and 2 kilograms of white rice. If the prices were the same on each day, find the price per kilogram for each type of rice.

Geometry Problems

In 1–7, solve the problem algebraically, using two variables.

1. The perimeter of a rectangle is 50 cm. The length is 9 cm more than the width. Find the length and the width of the rectangle.
2. A rectangle has a perimeter of 38 ft. The length is 1 ft. less than 3 times the width. Find the dimensions of the rectangle.
3. Two angles are supplementary. The larger angle measures 120° more than the smaller. Find the degree measure of each angle.
4. Two angles are supplementary. The larger angle measures 15° less than twice the smaller. Find the degree measure of each angle.
5. Two angles are complementary. The measure of the larger angle is 30° more than the measure of the smaller angle. Find the degree measure of each angle.
6. The larger of two complementary angles measures 6° less than twice the smaller angle. Find the degree measure of each angle.
7. In an isosceles triangle, each base angle measures 30° more than the vertex angle. Find the degree measures of the three angles of the triangle.

Miscellaneous Problems

In 1–10, solve the problem algebraically, using two variables.

1. Tickets for a high school dance cost $1.00 each if purchased in advance of the dance, but $1.50 each if bought at the door. If 100 tickets were sold and $120 was collected, how many tickets were sold in advance and how many were sold at the door?
2. A dealer sold 200 tennis racquets. Some were sold at $18 each and the rest were sold at $33 each. The total receipts from these sales were $4,800. How many racquets did he sell at $18 each?

3. Mrs. Rinaldo changed a $100 bill in a bank. She received $20 bills and $10 bills. The number of $20 bills was 2 more than the number of $10 bills. How many bills of each kind did she receive?
4. Linda spent $3.60 for stamps to mail packages. Some were 30-cent stamps and the rest were 20-cent stamps. The number of 20-cent stamps was 2 less than the number of 30-cent stamps. How many stamps of each kind did Linda buy?
5. A dealer has some hard candy worth $2.00 a pound and some worth $3.00 a pound. He wishes to make a mixture of 80 pounds that he can sell for $2.20 a pound. How many pounds of each kind should he use?
6. At the Savemore Supermarket, 3 pounds of squash and 2 pounds of eggplant cost $2.85. The cost of 4 pounds of squash and 5 pounds of eggplant is $5.41. What is the cost of 1 pound of squash and what is the cost of 1 pound of eggplant?
7. One year, Roger Jackson and his wife Wilma together earned $47,000. If Roger earned $4,000 more than Wilma earned that year, how much did each earn?
8. Mrs. Moto invested $1,400, part at 5% and part at 8%. Her total annual income from both investments was $100. Find the amount she invested at each rate.
9. Mr. Stein invested a sum of money in bonds yielding 4% a year and another sum in bonds yielding 6% a year. In all, he invested $4,000. If his total annual income from the two investments was $188, how much did he invest at each rate?
10. Mr. May invested $21,000, part at 8% and the rest at 6%. If the annual incomes from both investments were equal, find the amount invested at each rate.

18-5 GRAPHING SOLUTION SETS OF SYSTEMS OF INEQUALITIES

In order to find the solution set of a system of inequalities, we must find the ordered pairs that satisfy the open sentences of the system. We do this by a graphic method that is similar to the method used in finding the solution set of a system of equations.

MODEL PROBLEMS

1. Graph the solution set of the system: $x > 2$

$$y < -2$$

Solution:

(1) Graph $x > 2$ by first graphing the plane divider $x = 2$. (In the figure, see the dashed line labeled l.) The half-plane to the right of this line is the graph of the solution set of $x > 2$.

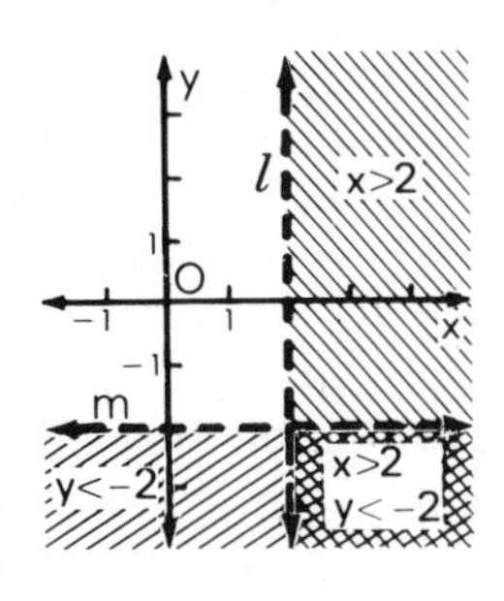

(2) Using the same set of axes, graph $y < -2$ by first graphing the plane divider $y = -2$. (In the figure, see the dashed line labeled m.) The half-plane below this line is the graph of the solution set of $y < -2$.

(3) The solution set of the system $x > 2$ and $y < -2$ consists of the intersection of the solution sets of $x > 2$ and $y < -2$.

Therefore, the crosshatched region, which is the intersection of the graphs made in steps (1) and (2), is the graph of the solution set of the system $x > 2$ and $y < -2$.

All points in this region, and no others, satisfy both sentences of the system. For example, the point $(4, -3)$, which lies in the region, satisfies both sentences of the system because its x-value satisfies one of the given inequalities, $4 > 2$, and its y-value satisfies the other inequality, $-3 < -2$.

2. Graph the solution set of $3 < x < 5$ in a coordinate plane.

Solution:

(1) The sentence $3 < x < 5$ means $3 < x$ and $x < 5$. This may be written $x > 3$ and $x < 5$.

Therefore, graph $x > 3$ by first graphing the plane divider $x = 3$. (In the figure, see the dashed line labeled l.)

The half-plane to the right of the line $x = 3$ is the graph of the solution set of $x > 3$.

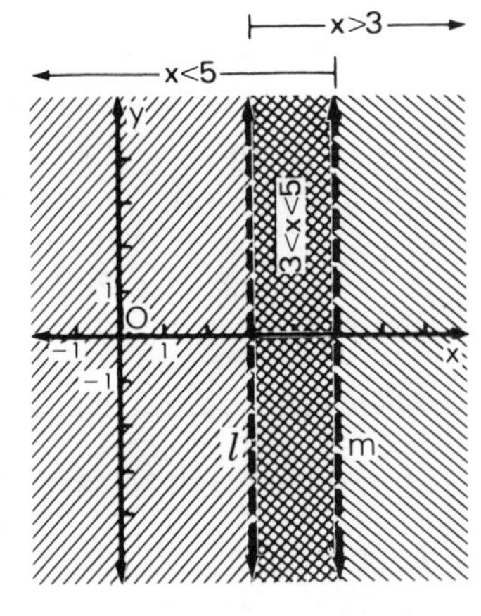

(2) Using the same set of axes, graph $x < 5$ by first graphing the plane divider $x = 5$. (In the figure, see the dashed line labeled m.) The half-plane to the left of the line $x = 5$ is the graph of the solution set of $x < 5$.

(3) The crosshatched region, which is the intersection of the graphs made in steps (1) and (2), is the graph of the solution set of $x > 3$ and $x < 5$, or $3 < x < 5$.

All points in this region, and no others, satisfy $3 < x < 5$. For example, the point (4, 3), which lies in the region and whose x-value is 4, satisfies $3 < x < 5$ because $3 < 4 < 5$ is a true statement.

3. Graph the following system of inequalities and label the solution set R.

$$x + y \geq 4$$
$$y \leq 2x - 3$$

Solution:

(1) Graph $x + y \geq 4$ by first graphing the plane divider $x + y = 4$. (In the figure, see the solid line labeled l.)

The line $x + y = 4$ and the half-plane above this line together form the graph of the solution set of $x + y \geq 4$.

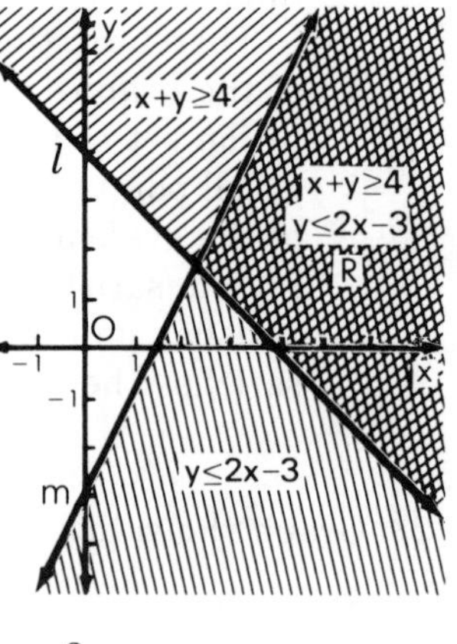

(2) Using the same set of axes, graph $y \leq 2x - 3$ by first graphing the plane divider $y = 2x - 3$. (In the figure, see the solid line labeled m.)

The line $y = 2x - 3$ and the half-plane below this line together form the graph of the solution set of $y \leq 2x - 3$.

(3) The crosshatched region labeled R, the intersection of the graphs made in steps (1) and (2), is the graph of the solution set of the system $x + y \geq 4$ and $y \leq 2x - 3$. Any point in the region R, such as (5, 2), will satisfy $x + y \geq 4$ because $5 + 2 \geq 4$, that is, $7 \geq 4$, is true, and will at the same time satisfy $y \leq 2x - 3$ because $2 \leq 2(5) - 3$, or $2 \leq 7$ is true.

EXERCISES

In 1–24, graph the system of inequalities and label the solution set S.

1. $x \geq 1$, $y > -2$

2. $x < 2$, $y \geq 3$

3. $x > 0$, $y > 0$

4. $x < 0$, $y > 0$

5. $y \geq x$, $x < 2$

6. $y \leq x$, $x \geq -1$

7. $y \geq 1$
$y < x - 1$

8. $y \leq 5$
$y < x + 3$

9. $y > x$
$y < 2x + 3$

10. $y \geq 2x$
$y > x + 3$

11. $y \leq 2x + 3$
$y \geq -x$

12. $y - x \geq 5$
$y - 2x \leq 7$

13. $y > x - 3$
$y > -x + 5$

14. $y \geq -2x + 4$
$y < x - 5$

15. $y < -x + 7$
$y \geq 2x + 1$

16. $y < x - 1$
$x + y \geq 2$

17. $x + y \leq 8$
$y > x - 4$

18. $y + 3x \geq 6$
$y < 2x - 4$

19. $x + y > 3$
$x - y < 6$

20. $x - y \leq -2$
$x + y \geq 2$

21. $2x + y \leq 6$
$x + y - 2 > 0$

22. $2x + 3y \geq 6$
$x + y - 4 \leq 0$

23. $y \geq x$
$x = 0$

24. $x + y \leq 3$
$y - 2x = 0$

In 25–34, graph the solution set in a coordinate plane.

25. $1 < x < 4$
26. $-5 \leq x \leq -1$
27. $-4 \leq x < 0$
28. $2 < y \leq 6$
29. $-2 \leq y \leq 3$
30. $0 < y \leq 4$
31. $(x > 2) \wedge (x \leq 7)$
32. $(x < 1) \vee (x \geq 4)$
33. $(y \geq -1) \wedge (y < 5)$
34. $(y < -2) \vee (y \geq 5)$

18-6 REVIEW EXERCISES

1. Solve the following system of equations for x:
$5x - 2y = 22$
$x + 2y = 2$

In 2–4, solve each system of equations *graphically* and check.

2. $x + y = 6$
$y = 2x - 6$

3. $y = -x$
$2x + y = 3$

4. $2y = x + 4$
$x - y + 4 = 0$

In 5–7, solve each system of equations by using *addition* to eliminate one of the variables. Check.

5. $2x + y = 10$
$x + y = 3$

6. $x + 4y = 1$
$5x - 6y = -8$

7. $3c + d = 0$
$c - 4d = 52$

In 8–10, solve each system of equations by using *substitution* to eliminate one of the variables. Check.

8. $x + 2y = 7$
$x = y - 8$

9. $3r + 2s = 20$
$r = -2s$

10. $x + y = 7$
$2x + 3y = 21$

In 11–16, solve each system of equations by using an algebraic method that seems convenient. Check.

11. $x + y = 0$
$3x + 2y = 5$

12. $5a + 3b = 17$
$4a - 5b = 21$

13. $t + u = 12$
$t = \frac{1}{3}u$

14. $3a = 4b$
$4a - 5b = 2$

15. $x + y = 1{,}000$
$.06x = .04y$

16. $10t + u = 24$
$t + u = \frac{1}{7}(10u + t)$

In 17–19, solve the problem algebraically, using two variables.

17. The sum of two numbers is 64. Their difference is 18. Find the numbers.

18. At a store, 3 notebooks and 2 pencils cost $2.80. At the same prices, 2 notebooks and 5 pencils cost $2.60. Find the cost of one notebook and one pencil.

19. Two angles are complementary. The larger angle measures 15° less than twice the smaller angle. Find the degree measure of each angle.

20. **a.** Solve the system of equations algebraically: $x - y = 3$
$x + 3y = 9$

b. On a set of coordinate axes, graph the system of equations given in part **a.**

In 21–23, graph the system of inequalities and label the solution set A.

21. $y > 2x - 3$
$y \leq 5 - x$

22. $y \leq \frac{1}{2}x$
$x \geq -4$

23. $2x + y < 4$
$x - y < -2$

24. Write the coordinates of one point in the solution set A for Exercise 23.

25. On the first day that tickets to the school play were sold, 100 tickets were sold for $230. Regular tickets cost $3 and student tickets cost $2. How many of each kind of ticket were sold?

Chapter 19

The Real Numbers

19-1 THE SET OF RATIONAL NUMBERS

The numbers with which you are familiar are zero, and the positive and negative integers and fractions. Examples of these numbers are 5, -3, $\frac{7}{4}$, and $\frac{-5}{4}$. Each of these numbers can be expressed in the form $y = \frac{a}{b}$ where a and b are integers and $b \neq 0$. Numbers that can be expressed in this form are called ***rational numbers***.

Remember that 5 may be expressed as $\frac{5}{1}$, -3 as $\frac{-3}{1}$, and 0 as $\frac{0}{1}$. In fact, every integer n is a rational number because $n = \frac{n}{1}$, which is a quotient of two integers.

Properties of the Set of Rational Numbers

The set of rational numbers has all the addition and multiplication properties of the set of integers. This set also has additional properties.

● **PROPERTY 1. The set of rational numbers is closed under division by nonzero rational numbers as well as under addition, multiplication, and subtraction.**

When we divide a rational number by a nonzero rational number, we get a unique rational number as the result. For example:

$$-3 \div 2 = \frac{-3}{2} = -\frac{3}{2}$$

$$5 \div -4 = \frac{5}{-4} = -\frac{5}{4}$$

$$-\frac{5}{2} \div -\frac{4}{3} = -\frac{5}{2} \cdot -\frac{3}{4} = \frac{15}{8}$$

In the set of rational numbers, division by a nonzero number is a binary operation.

● **PROPERTY 2. For every nonzero rational number, there is a unique corresponding number such that the product of these numbers is 1, the identity element of multiplication.**

For example, for the given number $\frac{2}{3}$, there is the unique corresponding number $\frac{3}{2}$ such that $\frac{2}{3} \cdot \frac{3}{2} = 1$. Recall that the number $\frac{3}{2}$ is called the *reciprocal*, or *multiplicative inverse*, of $\frac{2}{3}$.

The set of rational numbers shares the following two properties with the set of integers:

● **PROPERTY 3. The set of rational numbers can be associated with points on a number line.**

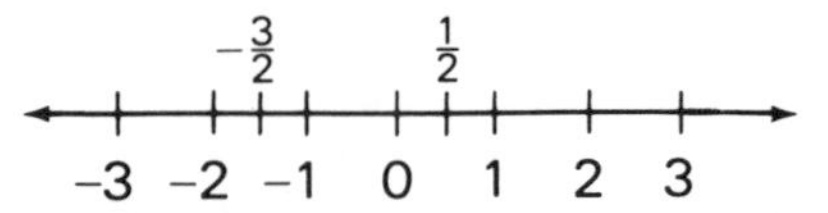

● **PROPERTY 4. The set of rational numbers is an ordered set. Given any two unequal rational numbers, we can tell which is the greater.**

Study the following model problems to see how different methods may be used to order rational numbers:

MODEL PROBLEMS

1. Which is the greater of the numbers $\frac{1}{2}$ and -1?

How to Proceed: Graph the numbers on a standard real number line. Then, determine which number is at the right. The number at the right is the greater number.

Solution:

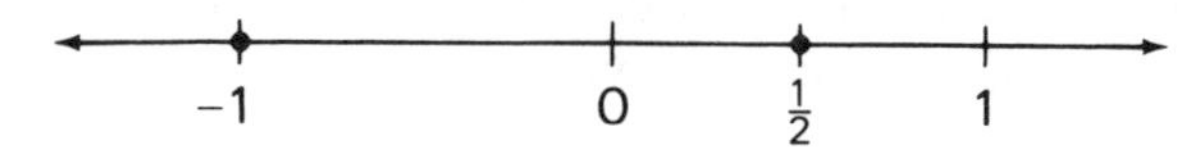

The number $\frac{1}{2}$ is to the right of the number -1.

Answer: $\frac{1}{2} > -1$

2. Which is the greater of the numbers $\frac{7}{9}$ and $\frac{8}{11}$?

How to Proceed: Express the numbers as equivalent fractions with a common denominator, and compare the numerators.

Solution: $\frac{7}{9} = \frac{7}{9} \cdot \frac{11}{11} = \frac{77}{99}$

$\frac{8}{11} = \frac{8}{11} \cdot \frac{9}{9} = \frac{72}{99}$

Since $\frac{77}{99} > \frac{72}{99}$, then $\frac{7}{9} > \frac{8}{11}$. $\frac{7}{9} > \frac{8}{11}$ *Ans.*

The next property of the set of rational numbers is not shared by the set of integers.

● **PROPERTY 5. The set of rational numbers is everywhere dense. That is, given any two unequal rational numbers, it is always possible to find a rational number between them.**

For example, some rational numbers between 1 and 2 are $1\frac{1}{2}$, $1\frac{1}{8}$, $1\frac{2}{3}$, and $1\frac{9}{10}$. In fact, there is an infinite number of rational numbers between two rational numbers.

One way to find a rational number between two rational numbers is to compute their average, the number midway between them. Thus, a rational number between $\frac{1}{4}$ and $\frac{3}{4}$ is: $\left(\frac{1}{4} + \frac{3}{4}\right) \div 2 = (1) \div 2 = \frac{1}{2}$.

Expressing a Rational Number as a Decimal

To express as a decimal a rational number named as a fraction, we simply perform the indicated division.

MODEL PROBLEM

Express as a decimal: **a.** $\frac{1}{2}$ **b.** $\frac{3}{4}$ **c.** $\frac{1}{16}$

Solution

a. $\frac{1}{2} = 2\overline{)1.000000}$ quotient $.500000$

b. $\frac{3}{4} = 4\overline{)3.000000}$ quotient $.750000$

c. $\frac{1}{16} = 16\overline{)1.000000}$ quotient $.062500$

In each of the examples $\frac{1}{2}$, $\frac{3}{4}$, and $\frac{1}{16}$, when we perform the division, we reach a point after which we continually obtain only zeros in the quotient. Decimals that result from such divisions, for example, .5, .75, and .0625, are called ***terminating decimals***.

Not all rational numbers can be expressed as terminating decimals, as you see in the following model problem.

MODEL PROBLEM

Express as a decimal: **a.** $\frac{1}{3}$ **b.** $\frac{2}{11}$ **c.** $\frac{1}{6}$ **d.** $\frac{3}{55}$

Solution

a. $\frac{1}{3} = 3\overline{)1.000000}$ quotient $.333333\ldots$

b. $\frac{2}{11} = 11\overline{)2.000000}$ quotient $.181818\ldots$

c. $\frac{1}{6} = 6\overline{)1.000000}$ quotient $.166666\ldots$

d. $\frac{3}{55} = 55\overline{)3.000000}$ quotient $.054545\ldots$

In each of the above examples, when we perform the division, we find, in the quotient, that the same group of digits is continually repeated in the same order. Decimals that keep repeating endlessly are called ***repeating decimals***, or ***periodic decimals***.

A repeating decimal may be written in an abbreviated form by placing a bar $\bar{}$ over the group of digits that is to be continually repeated. For example:

$$.333333\ldots = .\overline{3} \qquad .181818\ldots = .\overline{18} \qquad .166666\ldots = .1\overline{6}$$

The examples in the two preceding model problems illustrate the truth of the following statement:

● **Every rational number can be expressed as either a terminating decimal or a repeating decimal.**

Note that the equalities $.5 = .5\overline{0}$ and $.75 = .75\overline{0}$ illustrate the fact that every terminating decimal can be expressed as a repeating decimal which, after a point, repeats with all zeros. Therefore, we may say:

● **Every rational number can be expressed as a repeating decimal.**

Since every terminating decimal can be expressed as a repeating decimal, we will henceforth regard terminating decimals as repeating decimals.

Expressing a Decimal as a Rational Number

When you studied arithmetic, you learned how to express rational numbers named as terminating decimals as fractions.

MODEL PROBLEM

Express as a fraction: **a.** .3 **b.** .37 **c.** .139 **d.** .0777

Solution

a. $.3 = \frac{3}{10}$ **b.** $.37 = \frac{37}{100}$ **c.** $.139 = \frac{139}{1{,}000}$ **d.** $.0777 = \frac{777}{10{,}000}$

Study the following model problem to learn how to express repeating decimals as fractions.

MODEL PROBLEM

Find a fraction that names the same rational number as:
a. .6666 . . . **b.** .4141 . . .

Solution

a. Let $N = .6666\ldots$ [A]
Multiply both members of [A] by 10.
Then $10N = 6.6666\ldots$ [B]
Subtract [A] from [B].

$$\begin{array}{rl} 10N = 6.6666\ldots & [B] \\ N = \ .6666\ldots & [A] \\ \hline 9N = 6 & \end{array}$$

$$N = \frac{6}{9} = \frac{2}{3}$$

Answer: $.6666\ldots = \frac{2}{3}$

b. Let $N = .4141\ldots$ [A]
Multiply both members of [A] by 100.
Then $100N = 41.4141\ldots$ [B]
Subtract [A] from [B].

$$\begin{array}{rl} 100N = 41.4141\ldots & [B] \\ N = \ \ .4141\ldots & [A] \\ \hline 99N = 41 & \end{array}$$

$$N = \frac{41}{99}$$

Answer: $.4141\ldots = \frac{41}{99}$

The examples in the two preceding model problems illustrate the following statement:

- **Every repeating decimal is a rational number.**

Notice that this statement is the converse of:

- **Every rational number can be expressed as a repeating decimal.**

Since both the statement and its converse are true, we can make the following biconditional statement:

- **A number is a rational number if and only if it can be represented by a repeating decimal.**

EXERCISES

In 1–12, state which of the given numbers is the greater.

1. $\frac{5}{2}, \frac{7}{2}$ **2.** $\frac{-9}{3}, \frac{-11}{3}$ **3.** $\frac{5}{6}, -\frac{13}{6}$ **4.** $-\frac{1}{5}, -5$

5. $\frac{5}{2}, \frac{7}{4}$ **6.** $\frac{-10}{3}, \frac{-13}{6}$ **7.** $\frac{13}{6}, \frac{15}{10}$ **8.** $\frac{-5}{8}, \frac{-5}{12}$

9. $1.4, 1\frac{3}{5}$ **10.** $-3.4, -3\frac{1}{3}$ **11.** $.06, \frac{1}{6}$ **12.** $\frac{-15}{11}, \frac{-11}{15}$

In 13–22, find a rational number midway between the two given numbers.

13. $5, 6$ **14.** $-4, -3$ **15.** $-1, 0$ **16.** $\frac{1}{4}, \frac{1}{2}$

17. $\frac{1}{2}, \frac{7}{8}$ **18.** $\frac{-3}{4}, \frac{-2}{3}$ **19.** $-2.1, -2.2$

20. $2\frac{1}{2}, 2\frac{5}{8}$ **21.** $-1\frac{1}{3}, -1\frac{1}{4}$ **22.** $3.05, 3\frac{1}{10}$

In 23–29, write the rational number as a repeating decimal.

23. $\frac{5}{8}$ **24.** $\frac{9}{4}$ **25.** $-5\frac{1}{2}$ **26.** $\frac{13}{8}$

27. $-\frac{7}{12}$ **28.** $\frac{5}{3}$ **29.** $\frac{7}{9}$

In 30–35, find a fraction that names the same rational number as the decimal.

30. $.5$ **31.** $.555\ldots$ **32.** $-.\overline{2}$ **33.** $.125\overline{0}$ **34.** $.2525\ldots$ **35.** $.0\overline{7}$

19-2 THE SET OF IRRATIONAL NUMBERS

There are infinitely many decimals that are nonrepeating and nonterminating. An example of such a decimal is:

$$.03003000300003\ldots$$

Observe that, in this numeral, only the digits 0 and 3 appear. First, there is a 3 preceded by one 0, then a 3 preceded by two 0's, then a 3 preceded by three 0's, and so on. Since this numeral is not a repeating decimal, it is not a rational number.

A nonrepeating decimal is called an ***irrational number***. An irrational number cannot be expressed in the form $\frac{a}{b}$ where a and b are integers ($b \neq 0$).

When we use three dots, . . . , after a series of digits to represent a number, the three dots indicate that the number is nonterminating. In an *irrational number*, we are not always certain what the next digit will be when . . . is used. For example, the irrational number .3824 . . . may be .38240 . . . , or .38241 . . . , or .38242 . . . , and so on. In a *rational number*, we can identify a pattern. For example, the rational number .8333 . . . means .83333333 . . . , or simply $.8\overline{3}$.

Irrational numbers may be positive or negative. For example, .030030003 . . . represents a positive irrational number; −.030030003 . . . represents a negative irrational number.

The set of irrational numbers is not closed under addition. For example, the numbers a = .030030003 . . . and $-a$ = −.030030003 . . . are both irrational, but their sum, $a + (-a) = 0$, is rational.

Also, the set of irrational numbers is not closed under multiplication. It can be shown that the reciprocal of an irrational number is irrational, and the product of a number and its reciprocal is 1, which is rational.

There are infinitely many irrational numbers. Among them is the number π = 3.14159 . . . , a number that you have met in previous mathematics courses when you studied the geometry of the circle. Another irrational number is the number that represents the length of a diagonal of a square whose side is 1. As you will later learn, this irrational number is symbolized by $\sqrt{2}$.

It is interesting to note that the sum or difference of a rational number and an irrational number is an irrational number. For example, $10 + \pi$ is an irrational number and $10 - \pi$ is an irrational number.

EXERCISES

In 1–16, tell whether the number is rational or irrational.

1. .36 **2.** .363636 . . . **3.** .363363336 . . .

4. $.3\overline{6}$ **5.** $-.\overline{945}$ **6.** .989889888 . . .

7. $.36336333\overline{6}$ **8.** π **9.** $.8\overline{3}$

10. .6789101112 . . . **11.** .9856473512 . . . **12.** $\sqrt{2}$

13. 5.08 **14.** .125612561256 . . . **15.** .125125127 . . . **16.** $3 + \pi$

In 17–20, find a rational number between the given numbers.

17. .7777 . . . and .868686 . . .

18. .151551555 . . . and .161661666 . . .

19. 3.6464 . . . and $3.\overline{125}$

20. 2.343343334 . . . and 2.414114111 . . .

19-3 THE SET OF REAL NUMBERS

The set that consists of all rational numbers and all irrational numbers is called the ***set of real numbers.***

This definition can be stated in a more formal manner:

● **The union of the set of all rational numbers and the set of all irrational numbers is called the *set of real numbers.***

{real numbers} = {rational numbers} ∪ {irrational numbers}

You have seen that there is an infinite number of rational numbers. For each rational number, there is a corresponding point on the number line. However, not every point on the number line corresponds to a rational number.

There is also an infinite number of irrational numbers. For each irrational number, there is a corresponding point on the number line. All the points that are the graphs of the rational numbers and the irrational numbers, make up the number line, which we can now call the real number line. We say that the set of real numbers is complete. The ***completeness property of real numbers*** may be stated as follows:

● **Every point on the real number line corresponds to a real number, and every real number corresponds to a point on the real number line.**

Ordering Real Numbers

We can order real numbers in ways corresponding to ordering of rational numbers.

1. We can use a number line. The graph of the greater of two unequal numbers is always to the right of the graph of the smaller number on a standard real number line.
2. We can make use of decimals. Given any two unequal real numbers, we can determine which is the larger by first expressing each number as a decimal. Then, we compare the resulting decimals.

Properties of Real Numbers

The following properties are assumed for the set of real numbers under the operations of addition and multiplication. They are used in operations with real numbers.

In the following eleven statements, a, b, and c represent any numbers that are members of the set of real numbers.

Property	*In Symbols*
1. The set is closed under addition (or addition is a binary operation).	1. $a + b = c$ (c is a unique number in the set.)
2. Addition is commutative.	2. $a + b = b + a$
3. Addition is associative.	3. $(a + b) + c = a + (b + c)$
4. Zero is the additive identity.	4. $a + 0 = a$ and $0 + a = a$
5. Every number a has an additive inverse $-a$.	5. $a + (-a) = 0$
6. The set is closed under multiplication (or multiplication is a binary operation).	6. $ab = c$ (c is a unique number in the set.)
7. Multiplication is commutative.	7. $ab = ba$
8. Multiplication is associative.	8. $(ab)c = a(bc)$
9. The number 1 is the multiplicative identity.	9. $a \cdot 1 = a$ and $1 \cdot a = a$
10. Every nonzero number a has a unique multiplicative inverse $\frac{1}{a}$.	10. $a \cdot \frac{1}{a} = 1$ $(a \neq 0)$
11. Multiplication is distributive over addition.	11. $a(b + c) = ab + ac$

EXERCISES

In 1–6, determine which is the greater number.

1. 2 or 2.5 **2.** -5.7 or -5.9 **3.** .5353 or .535353

4. .7 or $.\overline{7}$ **5.** $-.53$ or $-.\overline{531}$ **6.** .2121 . . . or .212112111 . . .

In 7–16, state whether the statement is true or false.

7. Every real number is a rational number.
8. Every rational number is a real number.
9. Every irrational number is a real number.
10. Every real number is an irrational number.
11. Every rational number corresponds to a point on the real number line.

12. Every point on the real number line corresponds to a rational number.
13. Every irrational number corresponds to a point on the real number line.
14. Every point on the real number line corresponds to an irrational number.
15. There are some numbers that are both rational and irrational.
16. Every repeating decimal corresponds to a point on the real number line.

17. Which is an illustration of the commutative property of addition?
(1) $ab = ba$ (2) $a + 0 = a$
(3) $a + b = b + a$ (4) $(a + b) + c = a + (b + c)$
18. Which is an illustration of the associative property of multiplication?
(1) $ab = ba$ (2) $a(0) = 0$ (3) $a(1) = a$ (4) $(ab)c = a(bc)$
19. Which is an illustration of the distributive property of multiplication over addition?
(1) $a + b = b + a$ (2) $a(b + c) = ab + ac$
(3) $(a + b) + c = a + (b + c)$ (4) $a(b + c) = ab + c$
20. What is the additive inverse of the real number represented by n?
21. What is the multiplicative inverse of the nonzero real number represented by n?
22. What is the additive identity for the set of real numbers?
23. What is the multiplicative identity for the set of real numbers?

19-4 FINDING A RATIONAL ROOT OF A NUMBER

The area of a square whose side measures 3 units is equal to $3 \cdot 3$ or 9 square units. Notice that we used 3 as a factor twice, that is, we squared 3 to find the area.

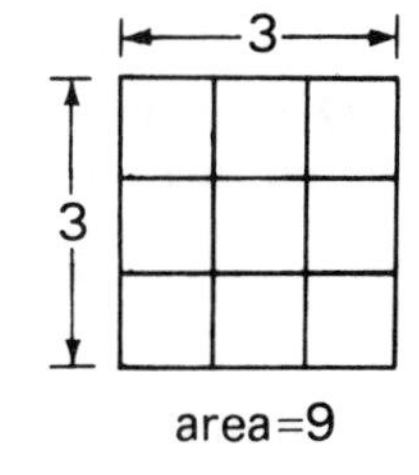

area=9

If we wish to find the length of a side of a square whose area is 16 square units, we have to find a number that, when squared (that is, used two times as a factor), would equal 16. Obviously, the number is 4 because $4 \cdot 4 = 16$. We call 4 a ***square root*** of 16.

area=16

To indicate a square root of a number, a ***radical sign***, $\sqrt{\ }$, is used. The symbol $\sqrt{25}$ is called a ***radical***, and 25, the number under the radical sign, is called the ***radicand***.

To find *a square root* of a number is to find one of its two equal factors. For example, a square root of 25, written $\sqrt{25}$, is 5 because $5 \cdot 5 = 25$, or $5^2 = 25$.

● **Finding a square root of a number is the inverse operation of squaring the number.**

In general, x is a square root of b if and only if $x^2 = b$.

Rational numbers such as 9, $\frac{4}{49}$, and 0 are called ***perfect squares*** because they are squares of rational numbers and, therefore, their square roots are rational numbers. For example, $\sqrt{9} = 3$, and 3 is a rational number; $\sqrt{\frac{4}{49}} = \frac{2}{7}$, and $\frac{2}{7}$ is a rational number; $\sqrt{0} = 0$, and 0 is a rational number.

Since a square root of 25 is a number whose square is 25, we can write $(\sqrt{25})^2 = 25$.

In general, for every real nonnegative number n: $(\sqrt{n})^2 = n$

Since $(+5)(+5) = 25$ and $(-5)(-5) = 25$, both $+5$ and -5 are square roots of 25. This example illustrates the following statement:

● **Every positive number has two square roots that have the same absolute value. One root is a positive number, and the other root is a negative number.**

The positive square root of a number is called the ***principal square root***. To indicate that the principal square root of a number is to be found, a radical sign is placed over the number. For example:

$$\sqrt{25} = 5 \qquad \sqrt{\frac{9}{16}} = \frac{3}{4} \qquad \sqrt{.49} = .7$$

To indicate that the negative square root of a number is to be found, we place a minus sign in front of the radical sign. For example:

$$-\sqrt{25} = -5 \qquad -\sqrt{\frac{9}{16}} = -\frac{3}{4} \qquad -\sqrt{.49} = -.7$$

To indicate that both square roots are to be found, we place a plus sign and a minus sign in front of the radical. For example:

$$\pm\sqrt{25} = \pm 5 \qquad \pm\sqrt{\frac{9}{16}} = \pm\frac{3}{4} \qquad \pm\sqrt{.49} = \pm .7$$

Observe that 0 is the only number whose square is 0. Thus, $\sqrt{0} = 0$.

Since the square of any real number is never negative, no negative number has a square root in the set of real numbers. For example, $\sqrt{-25}$ does not exist in the set of real numbers; there is no real number whose square is -25.

Finding a cube root of a number is to find one of its three equal factors. For example, 2 is a cube root of 8 because $2 \cdot 2 \cdot 2 = 8$, or $2^3 = 8$. A cube root of 8 is written $\sqrt[3]{8}$.

● **Finding a cube root of a number is the inverse operation of cubing a number. In general, x is a cube root of b if and only if $x^3 = b$.**

Likewise, for any positive integer n: If $x^n = b$, then x is an nth root of b, written $\sqrt[n]{b}$.

In the symbol $\sqrt[n]{b}$, n, the integer that indicates the root to be taken, is called the ***index*** of the radical. In $\sqrt[3]{8}$, the index is 3; in $\sqrt[4]{16}$, the index is 4. When no index appears, the index is understood to be 2. For example, $\sqrt{25}$ means a square root of 25.

We have said that $\sqrt{-25}$ does not exist in the set of real numbers. However, $\sqrt[3]{-8}$ does exist in the set of real numbers. Since $(-2)^3 = -8$, then $\sqrt[3]{-8} = -2$. These examples illustrate the following statement:

● **An even root of a negative number does not exist in the set of real numbers; an odd root of a negative number does exist.**

For example, $\sqrt{-4}$ and $\sqrt[4]{-16}$ do not exist in the set of real numbers. But $\sqrt[3]{-1} = -1$ since $(-1)^3 = -1$, and $\sqrt[5]{-32} = -2$ since $(-2)^5 = -32$.

MODEL PROBLEMS

1. Find the principal square root of 64.

Solution: Since $8 \cdot 8 = 64$, then $\sqrt{64} = 8$.

2. Find the value of $-\sqrt{.0016}$.

Solution: Since $.04 \cdot .04 = .0016$,
then $-\sqrt{.0016} = -.04$.

3. Find the value of $\sqrt[3]{27}$.

Solution: Since $3 \cdot 3 \cdot 3 = 27$, then $\sqrt[3]{27} = 3$.

4. Find the value of $(\sqrt{13})^2$.

Solution: Since $(\sqrt{n})^2 = n$, then $(\sqrt{13})^2 = 13$.

5. Solve for x: $x^2 = 36$.

Solution: If $x^2 = a$, then $x = \pm\sqrt{a}$ when a is a positive number.

$x^2 = 36$	*Check:* $x^2 = 36$	$x^2 = 36$
$x = \pm\sqrt{36}$	$(+6)^2 \stackrel{?}{=} 36$	$(-6)^2 \stackrel{?}{=} 36$
$x = \pm 6$	$36 = 36$ (True)	$36 = 36$ (True)

Answer: $x = +6$ or $x = -6$; the solution set is $\{+6, -6\}$.

EXERCISES

In 1–5, state the index and the radicand of the radical.

1. $\sqrt{36}$ **2.** $\sqrt[3]{125}$ **3.** $\sqrt[4]{81}$ **4.** $\sqrt[5]{32}$ **5.** $\sqrt[n]{1}$

In 6–15, find the principal square root of the number.

6. 81 **7.** 1 **8.** 121 **9.** 225 **10.** 900

11. $\frac{1}{9}$ **12.** $\frac{4}{25}$ **13.** .49 **14.** 1.44 **15.** .04

In 16–40, express the radical as integer(s), fraction(s), or decimal(s).

16. $\sqrt{16}$ **17.** $\sqrt{81}$ **18.** $\sqrt{121}$ **19.** $-\sqrt{64}$

20. $-\sqrt{144}$ **21.** $\sqrt{0}$ **22.** $\pm\sqrt{100}$ **23.** $\pm\sqrt{169}$

24. $\sqrt{400}$ **25.** $-\sqrt{625}$ **26.** $\sqrt{\frac{1}{4}}$ **27.** $-\sqrt{\frac{9}{16}}$

28. $\pm\sqrt{\frac{25}{81}}$ **29.** $\sqrt{\frac{49}{100}}$ **30.** $\pm\sqrt{\frac{144}{169}}$ **31.** $\sqrt{.64}$

32. $-\sqrt{1.44}$ **33.** $\pm\sqrt{.09}$ **34.** $-\sqrt{.01}$ **35.** $\pm\sqrt{.0004}$

36. $\sqrt[3]{1}$ **37.** $\sqrt[4]{81}$ **38.** $\sqrt[5]{32}$ **39.** $\sqrt[3]{-8}$

40. $-\sqrt[3]{-125}$

In 41–54, find the value of the expression.

41. $\sqrt{(8)^2}$ **42.** $\sqrt{\left(\frac{1}{2}\right)^2}$ **43.** $\sqrt{(.7)^2}$ **44.** $\sqrt{\left(\frac{9}{3}\right)^2}$

45. $\sqrt{\left(\frac{9}{5}\right)^2}$ **46.** $(\sqrt{4})^2$ **47.** $(\sqrt{36})^2$ **48.** $(\sqrt{11})^2$

49. $(\sqrt{39})^2$ **50.** $(\sqrt{97})(\sqrt{97})$

51. $\sqrt{36} + \sqrt{49}$ **52.** $\sqrt{100} - \sqrt{25}$

53. $(\sqrt{17})^2 + (\sqrt{7})(\sqrt{7})$ **54.** $\sqrt{(-9)^2} - (\sqrt{83})^2$

55. In each part, compare the given number with its square, using one of the symbols >, <, or =.
a. $\frac{1}{2}$ **b.** $\frac{3}{4}$ **c.** 1 **d.** $\frac{3}{2}$ **e.** 4 **f.** 100

56. Compare the positive real number n with its square n^2 when:
a. $n < 1$ **b.** $n = 1$ **c.** $n > 1$

57. In each part, compare the principal square root of the given number with the number.
a. $\frac{1}{9}$ **b.** $\frac{4}{25}$ **c.** 1 **d.** $\frac{49}{25}$ **e.** 4 **f.** 9

58. Compare $\sqrt{m}$, when m is a positive real number, with the number m, when:
a. $m < 1$ **b.** $m = 1$ **c.** $m > 1$

In 59–70, solve for the variable when the replacement set is the set of real numbers.

59. $x^2 = 4$ **60.** $y^2 = 100$ **61.** $z^2 = \frac{4}{81}$

62. $x^2 = .49$ **63.** $x^2 - 16 = 0$ **64.** $y^2 - 36 = 0$

65. $2x^2 = 50$ **66.** $3y^2 - 27 = 0$ **67.** $x^3 = 8$

68. $y^3 = 1$ **69.** $y^4 = 81$ **70.** $z^5 = 32$

In 71–74, **(a)** find the length of each side of a square that has the given area, and **(b)** find the perimeter of the square.

71. 36 sq. ft. **72.** 196 sq. yd. **73.** 121 cm^2 **74.** 225 m^2

75. Express in terms of x the perimeter of a square whose area is represented by x^2.

76. Write each of the integers from 101 to 110 as the sum of the smallest possible number of perfect squares.

19-5 SQUARE ROOTS THAT ARE IRRATIONAL NUMBERS

Suppose n is a nonnegative rational number that is not a perfect square, for example, 2. What kind of number is $\sqrt{n}$? What is the value of $\sqrt{2}$?

$(1)^2 = 1$	$(1.4)^2 = 1.96$	$(1.41)^2 = 1.9881$
$(\sqrt{2})^2 = 2$	$(\sqrt{2})^2 = 2$	$(\sqrt{2})^2 = 2$
$(2)^2 = 4$	$(1.5)^2 = 2.25$	$(1.42)^2 = 2.0164$

Note that:

$1 < 2 < 4$	$1.96 < 2 < 2.25$	$1.9881 < 2 < 2.0164$
$1 < \sqrt{2} < 2$	$1.4 < \sqrt{2} < 1.5$	$1.41 < \sqrt{2} < 1.42$

Regardless of how far we continue this work, we will never reach a point where the number $\sqrt{2}$ is expressed as a terminating or a repeating decimal. Therefore, we call $\sqrt{2}$ an ***irrational number***. We have been finding only approximations of $\sqrt{2}$; the value of $\sqrt{2}$ cannot be expressed as a rational number.

If a number cannot be expressed in the form $\frac{a}{b}$, where a and b are integers $(b \neq 0)$, it is an irrational number.

It can be proved that if n is a positive number that is not a perfect square, then $\sqrt{n}$ is an irrational number. It can also be shown that if n is a positive integer that is not the square of an integer, then $\sqrt{n}$ is irrational. Examples of irrational numbers are $\sqrt{2}$, $\sqrt{3}$, $\sqrt{5}$, $\sqrt{7}$, and $\sqrt{8}$.

Even though the value of an irrational number can be only approximated, every square root that is an irrational number can be associated with a point on the real number line.

In addition to irrational numbers of the form $\sqrt{n}$, where n is rational, there are infinitely many irrational numbers that cannot be expressed as a root of a rational number. π is such a number.

MODEL PROBLEMS

1. Between which consecutive integers is $\sqrt{42}$?

Solution: Since $6^2 = 36$ and $7^2 = 49$, then $\sqrt{42}$ is between 6 and 7.

Answer: $\sqrt{42}$ is between 6 and 7, or $6 < \sqrt{42} < 7$.

2. State whether $\sqrt{56}$ is a rational or an irrational number.

Solution: Since 56 is a positive integer that is not the square of an integer, $\sqrt{56}$ is an irrational number.

Answer: $\sqrt{56}$ is an irrational number.

EXERCISES

In 1–10, between which consecutive integers is each given number?

1. $\sqrt{5}$ **2.** $\sqrt{13}$ **3.** $\sqrt{40}$ **4.** $-\sqrt{2}$ **5.** $-\sqrt{14}$

6. $\sqrt{52}$ **7.** $\sqrt{73}$ **8.** $-\sqrt{125}$ **9.** $\sqrt{143}$ **10.** $-\sqrt{150}$

In 11–16, order the given numbers, starting with the smallest.

11. $2, \sqrt{3}, -1$ **12.** $4, \sqrt{17}, 3$ **13.** $-\sqrt{15}, -3, -4$

14. $0, \sqrt{7}, -\sqrt{7}$ **15.** $5, \sqrt{21}, \sqrt{30}$ **16.** $-\sqrt{11}, -\sqrt{23}, -\sqrt{19}$

In 17–26, state whether the number is rational or irrational.

17. $\sqrt{25}$ **18.** $\sqrt{40}$ **19.** $-\sqrt{36}$ **20.** $-\sqrt{54}$ **21.** $-\sqrt{150}$

22. $\sqrt{400}$ **23.** $\sqrt{\frac{1}{2}}$ **24.** $-\sqrt{\frac{4}{9}}$ **25.** $\sqrt{.36}$ **26.** $\sqrt{.1}$

19-6 USING A TABLE TO FIND SQUARES AND SQUARE ROOTS

When computing the square or the square root of a number, time can be saved by using a table of squares and square roots such as the one that appears on page 706.

MODEL PROBLEMS

1. Find the square of 58.

Solution: In the table on page 706, in the column headed "No.," find 58. Look to the right of 58 in the column headed "Square" and find 3,364.

Answer: $(58)^2 = 3{,}364$

2. Approximate $\sqrt{48}$: **a.** to the *nearest tenth*
b. to the *nearest hundredth*

Solution: In the table on page 706, in the column headed "No.," find 48. Look to the right of 48 in the column headed "Square Root" and find 6.928. Then, round off the decimal.

Answer: **a.** $\sqrt{48} \approx 6.9$ to the nearest tenth
b. $\sqrt{48} \approx 6.93$ to the nearest hundredth

Note: The symbol $\approx$ means *is approximately equal to.*

EXERCISES

In 1–6, use the table on page 706 to find the square of the number.

1. 27 **2.** 68 **3.** 94 **4.** 119 **5.** 132 **6.** 147

In 7–21, use the table on page 706 to approximate the expression: **a.** to the *nearest tenth* **b.** to the *nearest hundredth*

7. $\sqrt{13}$ **8.** $\sqrt{53}$ **9.** $-\sqrt{63}$ **10.** $\sqrt{135}$
11. $-\sqrt{87}$ **12.** $\sqrt{5}$ **13.** $\sqrt{91}$ **14.** $\sqrt{85}$
15. $-\sqrt{111}$ **16.** $\sqrt{141}$ **17.** $2 + \sqrt{3}$ **18.** $9 - \sqrt{17}$
19. $\sqrt{55} + 7$ **20.** $\sqrt{120} - 4$ **21.** $-2 - \sqrt{13}$

In 22–25, find c if $c = \sqrt{a^2 + b^2}$ and a and b have the given values. Use the table on page 706.

22. $a = 6, b = 8$ **23.** $a = 5, b = 12$
24. $a = 15, b = 20$ **25.** $a = 15, b = 36$

In 26–29, compute the perimeter of the figure. Use the table on page 706 to find approximations to two decimal places. Round the answer to the *nearest tenth.*

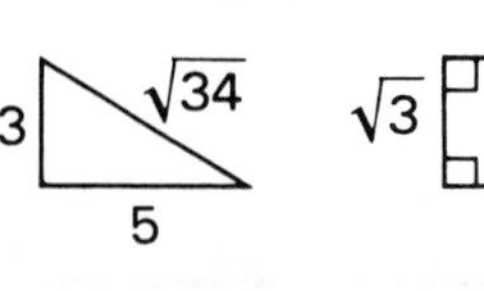

Ex. 26

√3
√7
Ex. 27

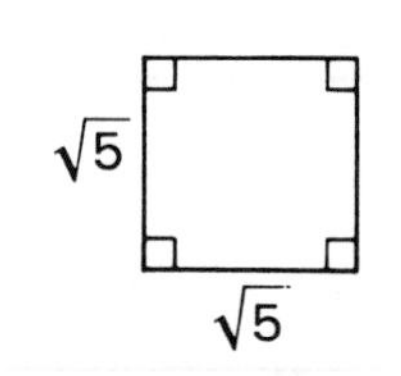

Ex. 28

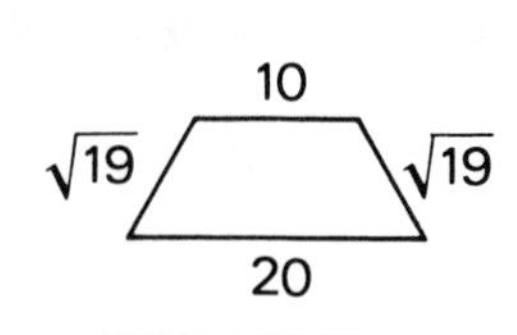

Ex. 29

19-7 USING DIVISION TO FIND APPROXIMATE SQUARE ROOTS

The following principles enable us to approximate square roots.

● PRINCIPLE 1. **When a divisor of a number and the quotient are equal, the square root of the number is either the divisor or the quotient.**

Since $144 \div 12 = 12$, $12 \cdot 12 = 144$ and $\sqrt{144} = 12$, which is either the divisor or the quotient.

● PRINCIPLE 2. **When a divisor of a number and the quotient are unequal, the square root of the number lies between the divisor and the quotient.**

For example, $144 \div 9 = 16$. We see that $\sqrt{144}$, which equals 12, lies between the divisor 9 and the quotient 16.

Also, $144 \div 18 = 8$. We see that $\sqrt{144}$, which is 12, lies between the divisor 18 and the quotient 8.

The square root of a number may be approximated to any number of decimal places by applying the two preceding principles and using estimates, divisions, and averages.

MODEL PROBLEM

Approximate $\sqrt{14}$ to the nearest:
a. integer **b.** tenth **c.** hundredth

How to Proceed	*Solution*
(1) Approximate the square root of the number 14 by estimation.	Since $3^2 = 9$ and $4^2 = 16$, then $\sqrt{14}$ lies between 3 and 4, closer to 4. Estimate $\sqrt{14} \approx 3.8$.
(2) Divide the number 14 by the estimate 3.8, finding the quotient to one more decimal place than there is in the divisor.	$$\begin{array}{r} 3.68 \\ 3.8\overline{)\,14.0\,00} \\ \underline{11\,4} \\ 2\,6\,0 \\ \underline{2\,2\,8} \\ 3\,20 \\ \underline{3\,04} \\ 16 \end{array}$$

(3) Since the divisor and quotient are not equal, find a number between the divisor and the quotient by finding the average of the divisor 3.8, and the quotient 3.68.

$$\frac{3.8 + 3.68}{2} = \frac{7.48}{2} = 3.74$$

(4) Divide the number 14 by the average 3.74. Find the quotient to one more decimal place than there is in the divisor.

$3.74\overline{)14.} \approx 3.743$
(The division is left to the student.)

(5) Find the average of the divisor 3.74 and the quotient 3.743.

$$\frac{3.74 + 3.743}{2} = \frac{7.483}{2} = 3.7415$$

Note: Repeat steps (4) and (5) as often as is necessary to obtain as close an approximation as is desired.

Since $\sqrt{14} \approx 3.7415$, we obtain the following approximations when we round off:

Answer: **a.** $\sqrt{14} \approx 4$ (nearest integer)
b. $\sqrt{14} \approx 3.7$ (nearest tenth)
c. $\sqrt{14} \approx 3.74$ (nearest hundredth)

EXERCISES

In 1–15, approximate the expression to the nearest:
a. integer **b.** tenth **c.** hundredth

1. $\sqrt{2}$ **2.** $\sqrt{3}$ **3.** $\sqrt{21}$ **4.** $\sqrt{39}$ **5.** $\sqrt{80}$

6. $\sqrt{90}$ **7.** $\sqrt{108}$ **8.** $\sqrt{23.5}$ **9.** $\sqrt{88.2}$ **10.** $-\sqrt{115.2}$

11. $\sqrt{28.56}$ **12.** $\sqrt{67.24}$ **13.** $\sqrt{4389}$ **14.** $\sqrt{123.7}$ **15.** $\sqrt{134.53}$

In 16–20, find to the *nearest tenth* of a centimeter the length of a side of a square whose area is the given measure.

16. 8 cm^2 **17.** 29 cm^2 **18.** 96 cm^2 **19.** 140 cm^2 **20.** 200 cm^2

19-8 FINDING THE PRINCIPAL SQUARE ROOT OF A MONOMIAL

Since $(6a^2)(6a^2) = 36a^4$, then $\sqrt{36a^4} = 6a^2$. Observe that $6a^2$ is the product of two factors. The first factor 6 is $\sqrt{36}$ and the second factor a^2 is $\sqrt{a^4}$. Therefore, $\sqrt{36a^4} = \sqrt{36} \cdot \sqrt{a^4} = 6a^2$.

PROCEDURE. To find the square root of a monomial that has more than one factor, write the indicated product of the square roots of its factors.

Note: In our work, we limit the domain of the variables that appear under a radical sign to nonnegative numbers only.

MODEL PROBLEMS

In 1–3, find the principal square root. Assume y and m represent positive numbers.

1. $\sqrt{25y^2}$ **2.** $\sqrt{16m^6}$ **3.** $\sqrt{.81y^8}$

Solution:

1. $\sqrt{25y^2} = (\sqrt{25})(\sqrt{y^2}) = 5y$

2. $\sqrt{16m^6} = (\sqrt{16})(\sqrt{m^6}) = 4m^3$

3. $\sqrt{.81y^8} = (\sqrt{.81})(\sqrt{y^8}) = .9y^4$

EXERCISES

In 1–18, find the indicated root. Assume that all variables represent positive numbers.

1. $\sqrt{4a^2}$ **2.** $\sqrt{16d^2}$ **3.** $\sqrt{49z^2}$ **4.** $\sqrt{\frac{16}{25}r^2}$

5. $\sqrt{.81w^2}$ **6.** $\sqrt{9c^2}$ **7.** $\sqrt{36y^4}$ **8.** $\sqrt{c^2d^2}$

9. $\sqrt{x^4y^2}$ **10.** $\sqrt{r^8s^6}$ **11.** $\sqrt{4x^2y^2}$ **12.** $\sqrt{36a^6b^4}$

13. $\sqrt{144a^4b^2}$ **14.** $\sqrt{169x^4y^2}$ **15.** $\sqrt{.36m^2}$ **16.** $\sqrt{.49a^2b^2}$

17. $\sqrt{.04y^2}$ **18.** $\sqrt{.01x^4y^2}$

In 19–22: **a.** Represent each side of the square whose area is given. **b.** Represent the perimeter of that square.

19. $49c^2$ **20.** $64x^2$ **21.** $100x^2y^2$ **22.** $144a^2b^2$

19-9 SIMPLIFYING A SQUARE-ROOT RADICAL

Since $\sqrt{4 \cdot 9} = \sqrt{36} = 6$

and $\sqrt{4} \cdot \sqrt{9} = 2 \cdot 3 = 6$, then $\sqrt{4 \cdot 9} = \sqrt{4} \cdot \sqrt{9}$.

Since $\sqrt{16 \cdot 25} = \sqrt{400} = 20$

and $\sqrt{16} \cdot \sqrt{25} = 4 \cdot 5 = 20$, then $\sqrt{16 \cdot 25} = \sqrt{16} \cdot \sqrt{25}$.

These examples illustrate the following property of square-root radicals:

● **The square root of a product of nonnegative numbers is equal to the product of the square roots of the numbers.**

In general, if a and b are nonnegative numbers:

$$\sqrt{a \cdot b} = \sqrt{a} \cdot \sqrt{b} \quad \textit{and} \quad \sqrt{a} \cdot \sqrt{b} = \sqrt{a \cdot b}$$

This rule can be used to transform a square-root radical into an equivalent radical. For example:

$$\sqrt{50} = \sqrt{25 \cdot 2} = \sqrt{25} \cdot \sqrt{2} = 5\sqrt{2}$$

Notice that we expressed 50 as the product of 25 (the greatest perfect-square factor of 50) and 2. Then, we expressed $\sqrt{25}$ as 5.

When we expressed $\sqrt{50}$ as $5\sqrt{2}$, we simplified $\sqrt{50}$.

A square-root radical is said to be in simplest form when:

1. the radicand has no perfect-square factor other than 1, and
2. there is no fraction under the radical sign.

PROCEDURE. To simplify the square root of a product:

1. Find, if possible, two factors of the radicand, one of which is the largest perfect-square factor of the radicand.
2. Express the square root of the product as the product of the square roots of the factors.
3. Find the square root of the factor that is a perfect square.

If we wish to find $\sqrt{200}$, correct to the nearest hundredth, by the use of the table on page 706, we can first simplify the radical as follows:

$$\sqrt{200} = \sqrt{100 \cdot 2} = \sqrt{100} \cdot \sqrt{2} = 10\sqrt{2} \approx 10(1.414) \approx 14.14$$

Note that 100 is the largest perfect-square factor of 200.

MODEL PROBLEM

In each part, simplify the expression. Assume that $y > 0$.

a. $\sqrt{18}$ **b.** $4\sqrt{50}$ **c.** $\frac{1}{2}\sqrt{48}$ **d.** $\sqrt{4y^3}$

Solution:

a. $\sqrt{18} = \sqrt{9 \cdot 2} = \sqrt{9} \cdot \sqrt{2} = 3\sqrt{2}$

b. $4\sqrt{50} = 4\sqrt{25 \cdot 2} = 4\sqrt{25} \cdot \sqrt{2} = 4 \cdot 5\sqrt{2} = 20\sqrt{2}$

c. $\frac{1}{2}\sqrt{48} = \frac{1}{2}\sqrt{16 \cdot 3} = \frac{1}{2}\sqrt{16} \cdot \sqrt{3} = \frac{1}{2} \cdot 4\sqrt{3} = 2\sqrt{3}$

d. $\sqrt{4y^3} = \sqrt{4y^2 \cdot y} = \sqrt{4y^2} \cdot \sqrt{y} = 2y\sqrt{y}$

EXERCISES

In 1–32, simplify the expression. Assume that all variables represent positive numbers.

1. $\sqrt{8}$ **2.** $\sqrt{12}$ **3.** $\sqrt{20}$ **4.** $\sqrt{28}$

5. $\sqrt{40}$ **6.** $\sqrt{27}$ **7.** $\sqrt{54}$ **8.** $\sqrt{63}$

9. $\sqrt{90}$ **10.** $\sqrt{98}$ **11.** $\sqrt{99}$ **12.** $\sqrt{108}$

13. $\sqrt{162}$ **14.** $\sqrt{175}$ **15.** $\sqrt{300}$ **16.** $3\sqrt{8}$

17. $4\sqrt{12}$ **18.** $2\sqrt{20}$ **19.** $4\sqrt{90}$ **20.** $2\sqrt{45}$

21. $3\sqrt{200}$ **22.** $\frac{1}{2}\sqrt{72}$ **23.** $\frac{1}{4}\sqrt{48}$ **24.** $\frac{3}{4}\sqrt{96}$

25. $\frac{2}{3}\sqrt{63}$ **26.** $5\sqrt{24}$ **27.** $2\sqrt{80}$ **28.** $7\sqrt{45}$

29. $8\sqrt{9x}$ **30.** $\sqrt{3x^3}$ **31.** $\sqrt{49x^5}$ **32.** $\sqrt{36r^2s}$

33. The expression $\sqrt{48}$ is equivalent to
(1) $2\sqrt{3}$ (2) $4\sqrt{12}$ (3) $4\sqrt{3}$ (4) $16\sqrt{3}$

34. The expression $4\sqrt{2}$ is equivalent to
(1) $\sqrt{8}$ (2) $\sqrt{42}$ (3) $\sqrt{32}$ (4) $\sqrt{64}$

35. The expression $3\sqrt{18}$ is equivalent to
(1) $\sqrt{54}$ (2) $3\sqrt{2}$ (3) $9\sqrt{2}$ (4) $3\sqrt{6}$

36. The expression $3\sqrt{3}$ is equivalent to
(1) $\sqrt{9}$ (2) $\sqrt{6}$ (3) $\sqrt{12}$ (4) $\sqrt{27}$

In 37–40, use the table on page 706 to find the approximate value of the expression, correct to the nearest tenth.

37. $\sqrt{300}$ **38.** $\sqrt{180}$ **39.** $2\sqrt{288}$ **40.** $\frac{1}{3}\sqrt{252}$

41. **a.** Does $\sqrt{9 + 16} = \sqrt{9} + \sqrt{16}$? Why?
b. Is finding a square root always distributive over addition?

42. **a.** Does $\sqrt{25 - 9} = \sqrt{25} - \sqrt{9}$? Why?
b. Is finding a square root always distributive over subtraction?

19-10 ADDING AND SUBTRACTING RADICALS

Adding and Subtracting Like Radicals

Like radicals are radicals that have the same index and the same radicand. For example, $7\sqrt{3}$ and $5\sqrt{3}$ are like radicals, as are $4\sqrt[3]{7}$ and $9\sqrt[3]{7}$. However, $3\sqrt{5}$ and $5\sqrt{2}$ are unlike radicals because the radicands are different. Also, $\sqrt[3]{2}$ and $\sqrt{2}$ are unlike radicals because the index in one radical is 3 and the index in the second radical is 2.

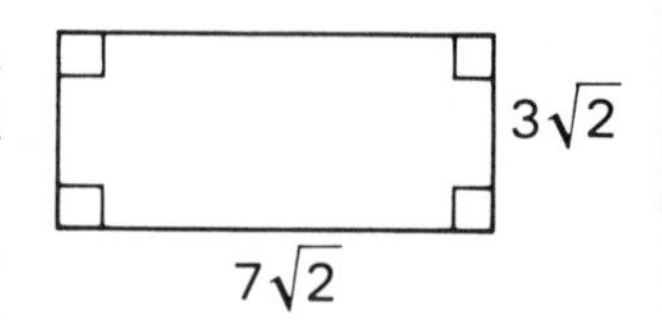

To find the sum of the length and the width of the rectangle pictured at the right, we add $7\sqrt{2}$ and $3\sqrt{2}$. The sum is $7\sqrt{2} + 3\sqrt{2}$.

We can express this sum of two like radicals as a single term by using the distributive property as follows:

$$7\sqrt{2} + 3\sqrt{2} = (7 + 3)\sqrt{2} = 10\sqrt{2}$$

Similarly, $5\sqrt{3} - \sqrt{3} = 5\sqrt{3} - 1\sqrt{3} = (5 - 1)\sqrt{3} = 4\sqrt{3}$.

PROCEDURE. To add or subtract like radicals:

1. Add or subtract the coefficients of the radicals.
2. Multiply the sum or difference obtained by the common radical.

Adding and Subtracting Unlike Radicals

The sum of the unlike radicals $\sqrt{5}$ and $\sqrt{2}$ is $\sqrt{5} + \sqrt{2}$, which cannot be expressed as a single term. Similarly, the difference of $\sqrt{5}$ and $\sqrt{2}$ is $\sqrt{5} - \sqrt{2}$.

However, when it is possible to transform unlike radicals into equivalent radicals that are like radicals, the resulting like radicals can be added or subtracted. For example, it would appear that the sum of $2\sqrt{3}$ and $\sqrt{27}$ can be indicated only as $2\sqrt{3} + \sqrt{27}$ since $2\sqrt{3}$ and $\sqrt{27}$ are unlike radicals. However, since $\sqrt{27} = \sqrt{9 \cdot 3} = \sqrt{9} \cdot \sqrt{3} = 3\sqrt{3}$, we can express $2\sqrt{3} + \sqrt{27}$ as like radicals, and then add:

$$2\sqrt{3} + \sqrt{27} = 2\sqrt{3} + 3\sqrt{3} = (2 + 3)\sqrt{3} = 5\sqrt{3}$$

PROCEDURE. To combine unlike radicals:

1. Simplify each radical.
2. Combine like radicals by using the distributive property.
3. Indicate the sum or difference of the unlike radicals.

MODEL PROBLEMS

1. Combine: $8\sqrt{5} + \sqrt{5} - 2\sqrt{5}$

Solution: $8\sqrt{5} + \sqrt{5} - 2\sqrt{5} = (8 + 1 - 2)\sqrt{5} = 7\sqrt{5}$

2. Combine: $5\sqrt{3} + 4\sqrt{12}$

Solution:
$$\begin{aligned} 5\sqrt{3} + 4\sqrt{12} &= 5\sqrt{3} + 4\sqrt{4 \cdot 3} \\ &= 5\sqrt{3} + 4\sqrt{4} \cdot \sqrt{3} \\ &= 5\sqrt{3} + 4 \cdot 2 \cdot \sqrt{3} \\ &= 5\sqrt{3} + 8\sqrt{3} \\ &= (5 + 8)\sqrt{3} \\ &= 13\sqrt{3} \quad \textit{Ans.} \end{aligned}$$

EXERCISES

In 1–28, combine the radicals. Assume that all variables represent positive numbers.

1. $8\sqrt{2} + 7\sqrt{2}$
2. $8\sqrt{5} + \sqrt{5}$
3. $5\sqrt{3} + 2\sqrt{3} + 8\sqrt{3}$
4. $14\sqrt{6} - 2\sqrt{6}$
5. $7\sqrt{2} - \sqrt{2}$
6. $4\sqrt{3} + 2\sqrt{3} - 6\sqrt{3}$
7. $5\sqrt{3} + \sqrt{3} - 2\sqrt{3}$
8. $4\sqrt{7} - \sqrt{7} - 5\sqrt{7}$
9. $3\sqrt{5} + 6\sqrt{2} - 3\sqrt{2} + \sqrt{5}$
10. $9\sqrt{x} + 3\sqrt{x}$
11. $15\sqrt{y} - 7\sqrt{y}$
12. $\sqrt{2} + \sqrt{50}$
13. $\sqrt{27} + \sqrt{75}$
14. $\sqrt{80} - \sqrt{5}$
15. $\sqrt{72} - \sqrt{50}$
16. $\sqrt{12} - \sqrt{48} + \sqrt{3}$
17. $3\sqrt{32} - 6\sqrt{8}$
18. $5\sqrt{27} - \sqrt{108} + 2\sqrt{75}$
19. $3\sqrt{8} - \sqrt{2}$
20. $5\sqrt{8} - 3\sqrt{18} + \sqrt{3}$
21. $3\sqrt{50} - 5\sqrt{18}$
22. $\sqrt{98} - 4\sqrt{8} + 3\sqrt{128}$
23. $\frac{2}{3}\sqrt{18} - \sqrt{72}$
24. $\sqrt{7a} + \sqrt{28a}$
25. $\sqrt{81x} + \sqrt{25x}$
26. $\sqrt{100b} - \sqrt{64b} + \sqrt{9b}$
27. $3\sqrt{3x} - \sqrt{12x}$
28. $\sqrt{3a^2} + \sqrt{12a^2}$

29. Express in simplest radical form the sum of:
a. $3\sqrt{2}$ and $\sqrt{98}$ **b.** $5\sqrt{3}$ and $\sqrt{27}$ **c.** $2\sqrt{12}$ and $4\sqrt{75}$

30. Express in simplest radical form the difference between:
a. $4\sqrt{2}$ and $\sqrt{18}$ **b.** $\sqrt{75}$ and $\sqrt{12}$ **c.** $4\sqrt{48}$ and $8\sqrt{12}$

31. The difference $5\sqrt{2} - \sqrt{32}$ is equivalent to
(1) $\sqrt{2}$ (2) $9\sqrt{2}$ (3) $-11\sqrt{2}$ (4) $18\sqrt{2}$

32. The sum $3\sqrt{8} + 6\sqrt{2}$ is equivalent to
(1) $9\sqrt{10}$ (2) $\sqrt{72}$ (3) $18\sqrt{10}$ (4) $12\sqrt{2}$

33. The sum of $\sqrt{12}$ and $\sqrt{27}$ is equivalent to
(1) $\sqrt{39}$ (2) $5\sqrt{6}$ (3) $13\sqrt{3}$ (4) $5\sqrt{3}$

In 34 and 35:
a. Express the perimeter of the figure in simplest radical form.
b. Using the table on page 706, approximate the expression obtained in part **a** correct to the nearest tenth.

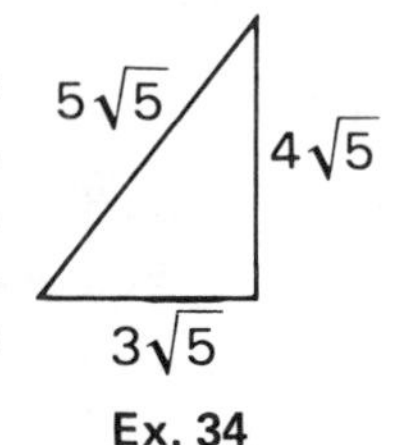

Ex. 34

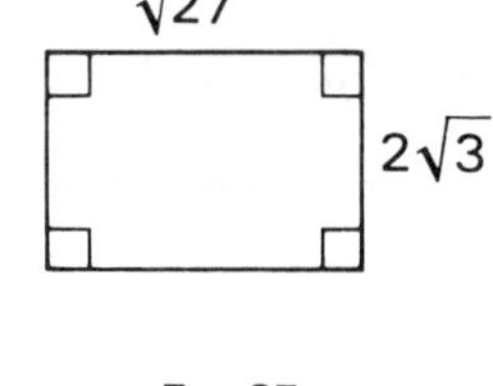

Ex. 35

19-11 MULTIPLYING SQUARE-ROOT RADICALS

To find the area of the rectangle pictured at the right, we multiply $5\sqrt{3}$ by $4\sqrt{2}$.

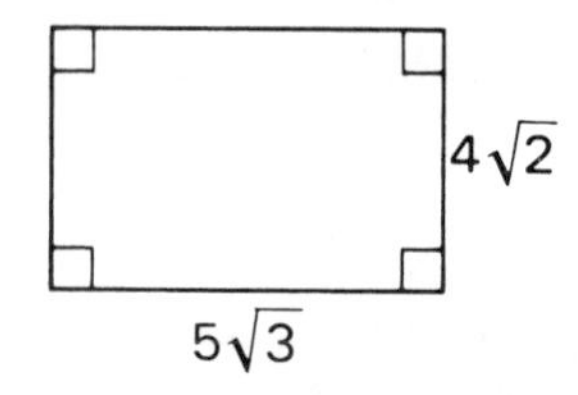

You have learned that $\sqrt{a} \cdot \sqrt{b} = \sqrt{ab}$ when a and b are nonnegative numbers. For example, $\sqrt{3} \cdot \sqrt{7} = \sqrt{3 \cdot 7} = \sqrt{21}$.

To multiply $4\sqrt{2}$ by $5\sqrt{3}$, we use the commutative and associative laws of multiplication as follows:

$$(4\sqrt{2})(5\sqrt{3}) = (4)(5)(\sqrt{2})(\sqrt{3}) = (4 \cdot 5)(\sqrt{2 \cdot 3}) = 20\sqrt{6}$$

In general, if a and b are nonnegative numbers:

$$x\sqrt{a} \cdot y\sqrt{b} = xy\sqrt{ab}$$

PROCEDURE. To multiply two monomial square roots:

1. Multiply the coefficients to find the coefficient of the product.
2. Multiply the radicands to find the radicand of the product.
3. If possible, simplify the result.

MODEL PROBLEMS

1. Multiply: $3\sqrt{6} \cdot 5\sqrt{2}$

Solution: $3\sqrt{6} \cdot 5\sqrt{2} = 3 \cdot 5\sqrt{6 \cdot 2} = 15\sqrt{12} = 15\sqrt{4} \cdot \sqrt{3}$
$= 15 \cdot 2 \cdot \sqrt{3}$
$= 30\sqrt{3}$ *Ans.*

2. Find the value of $(2\sqrt{3})^2$.

Solution: $(2\sqrt{3})^2 = 2\sqrt{3} \cdot 2\sqrt{3}$
$= 2 \cdot 2\sqrt{3 \cdot 3} = 4\sqrt{9} = 4 \cdot 3 = 12$ *Ans.*

3. Find the indicated product: $\sqrt{3x} \cdot \sqrt{6x}$ when $x > 0$.

Solution: $\sqrt{3x} \cdot \sqrt{6x} = \sqrt{3x \cdot 6x}$
$= \sqrt{18x^2}$
$= \sqrt{9x^2 \cdot 2} = \sqrt{9x^2} \cdot \sqrt{2} = 3x\sqrt{2}$ *Ans.*

EXERCISES

In 1–24: Multiply, or raise to the power, as indicated. Then, simplify the result. Assume that all variables represent positive numbers.

1. $\sqrt{3} \cdot \sqrt{3}$
2. $\sqrt{7} \cdot \sqrt{7}$
3. $\sqrt{a} \cdot \sqrt{a}$
4. $\sqrt{2x} \cdot \sqrt{2x}$
5. $\sqrt{12} \cdot \sqrt{3}$
6. $2\sqrt{18} \cdot 3\sqrt{8}$
7. $\sqrt{14} \cdot \sqrt{2}$
8. $\sqrt{60} \cdot \sqrt{5}$
9. $3\sqrt{6} \cdot \sqrt{3}$
10. $5\sqrt{8} \cdot 7\sqrt{3}$
11. $\frac{2}{3}\sqrt{24} \cdot 9\sqrt{3}$
12. $5\sqrt{6} \cdot \frac{2}{3}\sqrt{15}$
13. $(-4\sqrt{a})(3\sqrt{a})$
14. $(-\frac{1}{2}\sqrt{y})(-6\sqrt{y})$
15. $(\sqrt{2})^2$
16. $(\sqrt{y})^2$
17. $(\sqrt{t})^2$
18. $(3\sqrt{6})^2$
19. $\sqrt{25x} \cdot \sqrt{4x}$
20. $\sqrt{27a} \cdot \sqrt{3a}$
21. $\sqrt{15x} \cdot \sqrt{3x}$
22. $\sqrt{9a} \cdot \sqrt{ab}$
23. $(\sqrt{5x})^2$
24. $(2\sqrt{t})^2$

In 25–30: **a.** Perform the indicated operation. **b.** State whether the product is an irrational number or a rational number.

25. $(5\sqrt{12})(4\sqrt{3})$
26. $(3\sqrt{2})(2\sqrt{32})$
27. $(4\sqrt{6})(9\sqrt{3})$
28. $(8\sqrt{5})\left(\frac{1}{2}\sqrt{10}\right)$
29. $\left(\frac{2}{3}\sqrt{5}\right)^2$
30. $\left(\frac{1}{6}\sqrt{8}\right)\left(\frac{1}{2}\sqrt{18}\right)$

In 31–34, find the area of the square in which the length of each side is the given number.

31. $\sqrt{2}$
32. $2\sqrt{3}$
33. $6\sqrt{2}$
34. $5\sqrt{3}$

In 35 and 36:

a. Express the area of the figure in simplest radical form.

b. If the answer to part **a** contains a radical, use the table on page 706 to approximate the result correct to the nearest tenth.

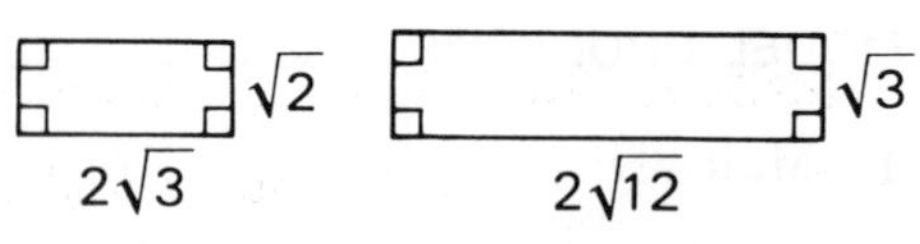

Ex. 35 Ex. 36

19-12 DIVIDING SQUARE-ROOT RADICALS

$$\text{Since } \sqrt{\frac{4}{9}} = \frac{2}{3} \text{ and } \frac{\sqrt{4}}{\sqrt{9}} = \frac{2}{3}, \text{ then } \sqrt{\frac{4}{9}} = \frac{\sqrt{4}}{\sqrt{9}}.$$

$$\text{Since } \sqrt{\frac{16}{25}} = \frac{4}{5} \text{ and } \frac{\sqrt{16}}{\sqrt{25}} = \frac{4}{5}, \text{ then } \sqrt{\frac{16}{25}} = \frac{\sqrt{16}}{\sqrt{25}}.$$

These examples illustrate the following property of square-root radicals:

- **The square root of a fraction that is the quotient of a nonnegative number and a positive number is equal to the square root of its numerator divided by the square root of its denominator.**

In general, if a is nonnegative and b is positive:

$$\sqrt{\frac{a}{b}} = \frac{\sqrt{a}}{\sqrt{b}} \quad \textit{and} \quad \frac{\sqrt{a}}{\sqrt{b}} = \sqrt{\frac{a}{b}}$$

See how we use this principle when we divide $\sqrt{72}$ by $\sqrt{8}$:

$$\frac{\sqrt{72}}{\sqrt{8}} = \sqrt{\frac{72}{8}} = \sqrt{9} = 3$$

To divide $6\sqrt{10}$ by $3\sqrt{2}$, we use the property of fractions, $\frac{ac}{bd} = \frac{a}{b} \cdot \frac{c}{d}$. See how the division is performed:

$$\frac{6\sqrt{10}}{3\sqrt{2}} = \frac{6}{3} \cdot \frac{\sqrt{10}}{\sqrt{2}} = \frac{6}{3} \cdot \sqrt{\frac{10}{2}} = 2\sqrt{5}$$

In general, if a is nonnegative, b is positive, and $y \neq 0$:

$$\frac{x\sqrt{a}}{y\sqrt{b}} = \frac{x}{y}\sqrt{\frac{a}{b}}$$

PROCEDURE: To divide two monomial square roots:

1. Divide the coefficients to find the coefficient of the quotient.
2. Divide the radicands to find the radicand of the quotient.
3. If possible, simplify the result.

MODEL PROBLEM

Divide: $8\sqrt{48} \div 4\sqrt{2}$

Solution: $8\sqrt{48} \div 4\sqrt{2} = \frac{8}{4}\sqrt{\frac{48}{2}} = 2\sqrt{24} = 2\sqrt{4 \cdot 6} = 2 \cdot \sqrt{4} \cdot \sqrt{6}$

$= 2 \cdot 2 \cdot \sqrt{6} = 4\sqrt{6}$ *Ans.*

EXERCISES

In 1–16, divide. Then simplify the quotient.

1. $\sqrt{72} \div \sqrt{2}$ **2.** $\sqrt{75} \div \sqrt{3}$ **3.** $\sqrt{70} \div \sqrt{10}$ **4.** $\sqrt{14} \div \sqrt{2}$

5. $8\sqrt{48} \div 2\sqrt{3}$ **6.** $\sqrt{24} \div \sqrt{2}$ **7.** $\sqrt{150} \div \sqrt{3}$

8. $21\sqrt{40} \div \sqrt{5}$ **9.** $9\sqrt{6} \div 3\sqrt{6}$ **10.** $7\sqrt{3} \div 3\sqrt{3}$

11. $2\sqrt{2} \div 8\sqrt{2}$ **12.** $\sqrt{9y} \div \sqrt{y}$ **13.** $\frac{12\sqrt{20}}{3\sqrt{5}}$

14. $\frac{20\sqrt{50}}{4\sqrt{2}}$ **15.** $\frac{25\sqrt{24}}{5\sqrt{2}}$ **16.** $\frac{3\sqrt{54}}{6\sqrt{3}}$

In 17–22, state whether the quotient is a rational number or an irrational number.

17. $\frac{\sqrt{7}}{5}$ **18.** $\frac{\sqrt{50}}{\sqrt{2}}$ **19.** $\frac{\sqrt{18}}{\sqrt{3}}$ **20.** $\frac{\sqrt{49}}{\sqrt{7}}$ **21.** $\frac{\sqrt{9}}{\sqrt{16}}$ **22.** $\frac{\sqrt{18}}{\sqrt{25}}$

In 23–28, simplify the given expression.

23. $\sqrt{\frac{36}{49}}$ **24.** $\sqrt{\frac{3}{4}}$ **25.** $4\sqrt{\frac{5}{16}}$

26. $\sqrt{\frac{8}{49}}$ **27.** $10\sqrt{\frac{8}{25}}$ **28.** $\frac{2}{3}\sqrt{\frac{144}{64}}$

29. Each of five real numbers is written on a separate piece of paper. The numbers are:

$$\sqrt{\frac{1}{4}} \qquad \sqrt{3} \qquad \sqrt{.4} \qquad \sqrt{.64} \qquad \sqrt{\frac{2}{9}}$$

These five pieces of paper are placed into a bag and one is chosen at random. Find the probability that the paper chosen contains:

a. a rational number
b. an irrational number
c. a number equal to $\frac{1}{2}$
d. a number equal to .2
e. a number equal to .8 or to $\frac{1}{2}$
f. a real number

19-13 THE GEOMETRY OF THE CIRCLE

A ***circle*** is a set of points in a plane such that each point is the same distance from a fixed point called the ***center***. When the center is named by the capital letter O, the circle is called circle O.

A ***radius*** of a circle (plural, ***radii***) is a line segment from the center of the circle to any point of the circle. A radius of circle O in the diagram is $\overline{OA}$.

A ***diameter*** of a circle is a line segment that contains the center of the circle and whose endpoints are points of the circle. A diameter of circle O in the diagram is $\overline{CB}$.

A ***chord*** is a line segment whose endpoints are points of the circle. A chord of circle O in the diagram is $\overline{CD}$.

The words radius and diameter mean both the line segments and the lengths of these line segments. Thus, it is correct to say either that the length of a radius is 5 or that the radius is 5.

The distance around a circle, or the perimeter of the circle, is called its ***circumference***. If you measure different circles, no matter how large or small the circle, you will discover that the circumference of a circle will always be slightly more than 3 times the diameter of the circle. Mathematicians have shown that for every circle, the ratio of the circumference C to the diameter d is always the same value. This constant value is represented by the Greek letter π (read pi). Therefore:

$$\frac{\textit{circumference}}{\textit{diameter}} = \textit{a constant} \qquad \textit{or} \qquad \frac{C}{d} = \pi$$

The Value of π

The number π is an irrational number. Therefore, π cannot be written exactly as a terminating or repeating decimal. Computers have calculated the value of π to millions of decimal places. Rounded to only ten decimal places, π is 3.1415926536. Approximate values of π such as 3.14 and $\frac{22}{7}$ are commonly used. When you use these values in place of π in your computations, your answers will be approximations. The exact value of π is between 3.14 and $\frac{22}{7}$.

$$3.14 < \pi < \frac{22}{7}$$

Finding the Circumference of a Circle

Since $\frac{C}{d} = \pi$, when we multiply both members of the equation by d, we have $\frac{C}{d} \cdot d = \pi \cdot d$. Hence, we obtain the formula:

$$\mathbf{C = \pi d}$$

Since d, the length of a diameter of a circle, is equal to twice r, the length of a radius of the circle, that is $d = 2r$, we have $C = \pi d$ or $C = \pi \cdot 2r$. Hence, we obtain the formula:

$$\mathbf{C = 2\pi r}$$

The circumference of a circle can be found by:

1. multiplying the length of its diameter by the constant value π, or
2. multiplying twice the length of its radius by the constant value π.

For example, if the diameter of a circle is 7 feet, its circumference is $\pi \cdot 7$, or 7π feet.

Here, the circumference has been given in terms of π. Note that 7π is the *exact* numerical solution. Also, note that circumference is a linear measure, here given in feet.

If we had used an approximate value for π, such as $\pi = 3.14$, then the circumference could be found to be approximately (7)(3.14), or 21.98 feet. By using $\pi = \frac{22}{7}$, the circumference could be found to be approximately $(7)\left(\frac{22}{7}\right)$, or 22 feet. These rational approximations produce approximate answers.

MODEL PROBLEMS

1. In the diagram, two circles are shown with center O. Points A, B, and G are on the smaller circle; points C, D, E, and F are on the larger circle. $\overline{DF}$ is a line segment containing points B, O, and G; $\overline{OC}$ is a line segment containing point A; $\overline{EC}$ is a line segment containing point B.

   Name the points or segments that describe each condition.

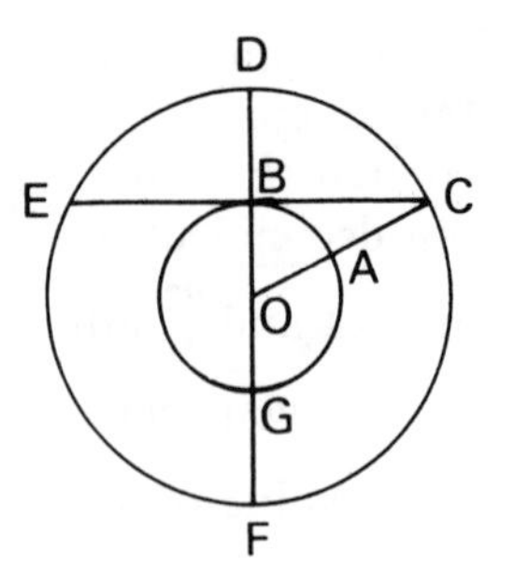

Answers

a. The radii of the smaller circle — a. $\overline{OA}$; $\overline{OB}$; $\overline{OG}$

b. The radii of the larger circle — b. $\overline{OC}$; $\overline{OD}$; $\overline{OF}$

c. The chords shown for the larger circle — c. $\overline{EC}$; $\overline{DF}$

d. The diameter of the larger circle — d. $\overline{DF}$

2. Find the circumference of a circle whose radius is 14 mm:

   a. expressed in terms of π **b.** using $\pi = \frac{22}{7}$ **c.** using $\pi = 3.14$

Solution

Method 1

a. Substitute radius $r = 14$ in the formula $C = 2\pi r$:

$$C = 2\pi \cdot 14$$
$$C = 28\pi$$

Method 2

If the radius $r = 14$, then the diameter $d = 28$. Substitute $d = 28$ in the formula $C = \pi d$:

$$C = \pi \cdot 28$$
$$C = 28\pi$$

Answer: $C = 28\pi$ mm

b. Using the approximation $\pi = \frac{22}{7}$, substitute and multiply:

$$C = 28\pi = 28 \cdot \frac{22}{7} = \frac{\overset{4}{\cancel{28}}}{1} \cdot \frac{22}{\underset{1}{\cancel{7}}} = 88 \text{ mm} \quad \textit{Ans.}$$

c. Using the approximation $\pi = 3.14$, substitute and multiply:

$$C = 28\pi = 28(3.14) = 87.92 \text{ mm} \quad \textit{Ans.}$$

3. If the circumference of a circle is 80π in., find the length of a radius of this circle.

Solution

(1) Write the formula for the circumference that involves the radius: $C = 2\pi r$

(2) Substitute 80π for the circumference C: $80\pi = 2\pi r$

(3) Solve for radius r by dividing both members of the equation by 2π:

$$\frac{\overset{40}{\cancel{80\pi}}}{\underset{1}{\cancel{2\pi}}} = \frac{\overset{1}{\cancel{2\pi}} r}{\underset{1}{\cancel{2\pi}}}$$

$$40 = r$$

Answer: $r = 40$ in.

4. If the radius of a circle is tripled, what change takes place in the circumference?

Solution

In the original circle, $C = 2\pi r$.
In the new circle, $C' = 2\pi r'$. Let $r' = 3r$ to show that the radius is triple that of the original circle. Then, $C' = 2\pi(3r) = 6\pi r$.

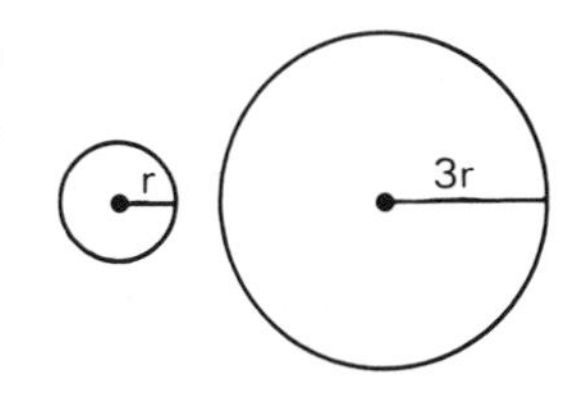

Compare the circumferences: $\frac{C'}{C} = \frac{6\pi r}{2\pi r}$

Simplifying, $\frac{6\cancel{\pi r}}{2\cancel{\pi r}} = \frac{6}{2} = \frac{3}{1}$.

Answer: The new circumference is 3 times the original circumference, or the circumference is tripled.

EXERCISES

1. Find the diameter of a circle whose radius measures:

a. 2 inches **b.** 13 feet **c.** $2\frac{1}{8}$ inches
d. 3.5 meters **e.** 2.75 centimeters

2. Find the radius of a circle whose diameter measures:

a. 4 feet **b.** 7 inches **c.** $3\frac{1}{2}$ inches
d. 1.4 meters **e.** 3.9 centimeters

3. What is the ratio of the radius of a circle to its diameter?

4. What is the ratio of the diameter of a circle to its radius?

5. Find the circumference of a circle, expressed in terms of π, when its diameter is:

a. 10 **b.** 16 **c.** 4.7 **d.** $\frac{1}{3}$ **e.** .08 **f.** $1\frac{3}{8}$

6. Find the circumference of a circle, expressed in terms of π, when its radius is:

a. 4 **b.** 9 **c.** 3.8 **d.** $\frac{1}{5}$ **e.** .06 **f.** $1\frac{3}{4}$

7. Find the circumference of a circle, expressed in terms of π, for the given radius r or diameter d.

a. $r = 5$ cm **b.** $d = 9$ ft. **c.** $r = 2.4$ mm **d.** $d = \frac{1}{3}$ in.

8. Using $\pi = 3.14$, find the approximate length of the circumference of a circle when:

a. $r = 8$ cm **b.** $d = 8$ in. **c.** $r = 2.5$ mm **d.** $d = 3\frac{1}{2}$ in.

9. Using $\pi = \frac{22}{7}$, find the approximate length of the circumference of a circle when:

a. $r = 14$ cm **b.** $d = 7$ ft. **c.** $r = 3.5$ mm **d.** $d = 10\frac{1}{2}$ in.

10. How many inches does Joan's bicycle go in one turn of the wheels if the diameter of each wheel is 28 inches? Express your answer to the nearest inch.

11. A circular flower bed has a diameter of 4.2 meters. How many meters of fencing will be needed to enclose this garden? Express your answer to the nearest tenth of a meter.

In 12–16, for the circle whose circumference is given, **(a)** find the length of the diameter of the circle and **(b)** find the length of the radius of the circle.

12. $C = 30\pi$ **13.** $C = 25\pi$ **14.** $C = \pi$

15. $C = 4.2\pi$ **16.** $C = \frac{1}{3}\pi$

17. The distance around a circular track is 440 yards. Using $\pi = \frac{22}{7}$, find the diameter of the circular track.

In 18–22, select the best answer from the four choices given.

18. If the diameter of a circle is 7, the circumference of the circle is:
(1) exactly 21.98 (2) exactly 22
(3) between 21 and 22 (4) exactly 44

19. If the radius of a circle is 14, the circumference of the circle is:
(1) between 43 and 44 (2) exactly 43.96
(3) between 87 and 88 (4) exactly 88

20. If the circumference of a circle is exactly 10, then its diameter is:
(1) $\frac{10}{\pi}$ (2) $\frac{5}{\pi}$ (3) 10π (4) 5π

21. If the circumference of a circle is exactly 12, then its radius is:
(1) $\frac{12}{\pi}$ (2) $\frac{6}{\pi}$ (3) 12π (4) 6π

22. If the diameter of a circle is tripled, then its circumference is multiplied by:
(1) π (2) 9 (3) 3 (4) 6

23. If the radius of a circle is doubled, what change takes place in its diameter?

24. If the radius of a circle is multiplied by 4, by what number is its circumference multiplied?

25. If the radius of a circle is divided by 2, what change takes place in its circumference?

26. Find the number of inches in the circumference of a circle in which the length of the longest chord that can be drawn is 5 inches. Answer may be expressed in terms of π.

27. Solve for r in terms of C and π: $C = 2\pi r$

28. Solve for d in terms of C and π: $C = \pi d$

29. Solve for π in terms of C and d: $C = \pi d$

19-14 AREA AND VOLUME RELATED TO THE CIRCLE

Area of a Circle

The area of a circle means the area of the region enclosed by the circle. Mathematicians have discovered that the area A of a circle is equal to π multiplied by the square of the radius r. The formula for the area of a circle is:

$$A = \pi r^2$$

If the radius of a circle is 3 feet, then the area is:

$$\begin{aligned} A &= \pi(3)^2 \\ &= \pi(9) \\ &= 9\pi \text{ square feet} \end{aligned}$$

The irrational number 9π is the exact number of square feet in the area of the circle. Since the rational number 3.14 is close to the value of π, a rational approximation of the area 9π is 9(3.14), or 28.26 square feet. Since the rational number $\frac{22}{7}$ is also close to the value of π, another rational approximation of the area 9π is $9\left(\frac{22}{7}\right)$, or $28\frac{2}{7}$ square feet.

The Right Circular Cylinder

The upper and lower bases of a ***right circular cylinder*** are parallel, congruent circles. The perpendicular distance between the two bases is the height of the right circular cylinder.

The volume, V, of a right circular cylinder is equal to the area of the base, B, multiplied by the height, h. This rule gives the following formula:

$$V = Bh$$

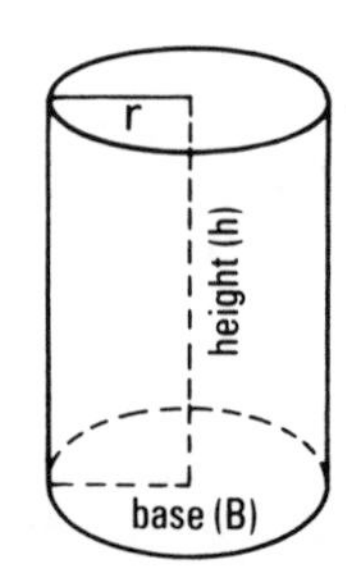

Right Circular Cylinder

Since the base is a circle whose area is πr^2, another formula for the volume of a right circular cylinder is:

$$V = \pi r^2 h$$

The Cone

In the ***cone*** and the *right circular cylinder* pictured at the right, the bases are congruent circles and the heights are also congruent. The volume, or capacity, of the cone is $\frac{1}{3}$ of the volume of the right circular cylinder.

Since the formula for the volume of a right circular cylinder is $V = Bh$, or $V = \pi r^2 h$, the formula for the volume of the cone is:

$$V = \frac{1}{3}Bh \quad or \quad V = \frac{1}{3}\pi r^2 h$$

The Sphere

A circle has been described as a set of points on a *plane* that are equally distant from one fixed point called the center.

A ***sphere*** is the set of all points in *space* that are equally distant from one fixed point called the center. A line segment, such as $\overline{OC}$, that joins the center O to any point on the sphere, is called a ***radius*** of the sphere. A line segment, such as $\overline{AB}$, that joins two points of the sphere and passes through its center, is called a ***diameter*** of the sphere.

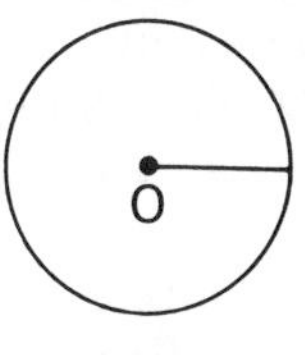

Circle

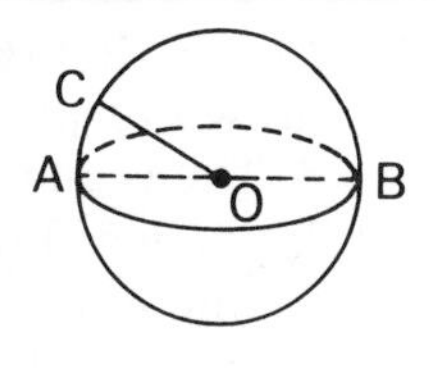

Sphere

The volume of a sphere is found by multiplying $\frac{4}{3}\pi$ by the cube of the radius:

$$V = \frac{4}{3}\pi r^3$$

MODEL PROBLEMS

1. Find the area of a circle whose radius is 14:

a. expressed in terms of π **b.** using $\pi = 3.14$ **c.** using $\pi = \frac{22}{7}$

Solution

a. $A = \pi r^2$
$A = \pi \cdot (14)^2$
$A = \pi \cdot 14 \cdot 14$
$A = \pi \cdot 196$
$A = 196\pi$ *Ans.*

b. $A = \pi r^2$
$A = (3.14) \cdot (14)^2$
$A = (3.14) \cdot (196)$
$A = 615.44$ *Ans.*

c. $A = \pi r^2$
$A = \frac{22}{7} \cdot (14)^2$
$A = \frac{22}{\cancel{7}_1} \cdot \cancel{14}^{2} \cdot 14$
$A = 616$ *Ans.*

2. If the radius of a circle is tripled, what change takes place in the area of the circle?

Solution: In the original circle, area $A = \pi r^2$.

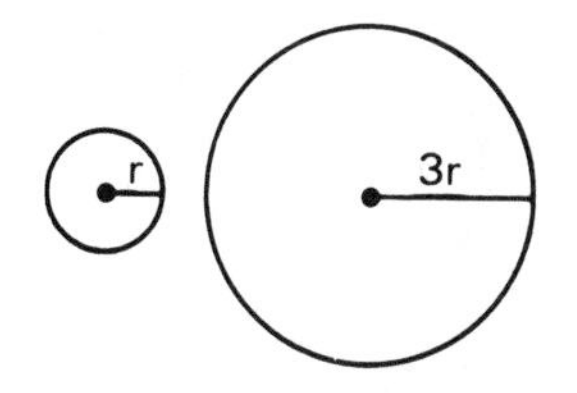

In the new circle, let the radius $r' = 3r$ to show that the radius is tripled. Then:

$$A' = \pi(3r)^2 = \pi \cdot (3r)(3r)$$
$$= \pi \cdot (9r^2) = 9\pi r^2$$

Compare the areas: $\frac{A'}{A} = \frac{9\pi r^2}{\pi r^2} = \frac{9}{1}$.

Answer: The new area is 9 times the original area, or the area is multiplied by 9.

3. In the figure at the right, square $ABCD$ is drawn about circle O. The radius of circle O is 5. The region between the square and the circle is shaded. Find:

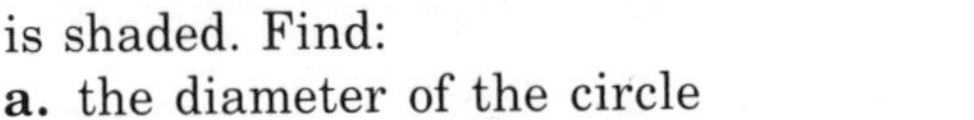

a. the diameter of the circle
b. the length of a side of the square
c. the area of the square
d. the area of the circle
e. the area of the shaded region.

Solution

a. The length of the diameter d is twice the length of the radius 5. Hence, $d = 2(5) = 10$.

b. The length s of the side of the square is equal to the length of the diameter of the inscribed circle. Hence, $s = 10$.

c. The area of a square is found by the formula $A = s^2$. Thus, $A = 10^2 = 100$.

d. The area of a circle is found by the formula $A = \pi r^2$. Here, $A = \pi r^2 = \pi \cdot (5)^2 = \pi \cdot 25 = 25\pi$.

e. The area of the shaded region is the difference between the area of the square and the area of the circle, namely, $100 - 25\pi$. Notice that the smaller area must be subtracted from the larger area. *Answer:* $100 - 25\pi$

4. A cylindrical can has a radius of $3\frac{1}{2}$ inches and a height 8 inches. Express the volume:

a. in terms of π

b. as a rational approximation, using $\pi = \frac{22}{7}$

Solution:		*Answers*

a. $V = \pi r^2 h = \pi \cdot \frac{7}{\cancel{2}_{}} \cdot \frac{7}{\cancel{2}} \cdot \frac{\overset{2}{\cancel{\overset{4}{\cancel{8}}}}}{1} = 98\pi$ — 98π cu. in.

b. $V = \pi r^2 h = \frac{22}{\underset{1}{\cancel{7}}} \cdot \frac{\overset{1}{\cancel{7}}}{\underset{1}{\cancel{2}}} \cdot \frac{7}{\underset{1}{\cancel{2}}} \cdot \frac{\overset{2}{\cancel{\overset{4}{\cancel{8}}}}}{1} = 308$ — 308 cu. in.

EXERCISES

1. Find the area of a circle, expressed in terms of π, when its radius is:

   a. 4 **b.** 9 **c.** .8 **d.** .3 **e.** $\frac{1}{3}$ **f.** $1\frac{1}{2}$

2. Find the area of a circle, expressed in terms of π, when its diameter is:

   a. 12 **b.** 2 **c.** 2.4 **d.** .4 **e.** $\frac{2}{5}$ **f.** $2\frac{1}{2}$

3. Find the area of a circle, expressed in terms of π, for the given radius r or diameter d:

   a. $r = 8$ in. **b.** $d = 18$ cm **c.** $r = \frac{1}{4}$ in.

   d. $d = 0.6$ mm **e.** $r = 1.5$ m **f.** $d = 1\frac{1}{3}$ ft.

4. Using $\pi = 3.14$, find the approximate area of a circle when:

   a. $r = 10$ cm **b.** $d = 20$ ft. **c.** $r = .4$ mm

   d. $d = 3$ in. **e.** $r = .1$ m **f.** $d = 1\frac{1}{2}$ cm

5. Using $\pi = \frac{22}{7}$, find the approximate area of a circle when:

   a. $r = 7$ ft. **b.** $d = 140$ mm **c.** $r = 21$ cm

   d. $d = 7$ in. **e.** $r = \frac{1}{2}$ in. **f.** $d = 1.5$ mm

In 6–9, find the volume of a right circular cylinder that has the given dimensions: **a.** in terms of π **b.** using $\pi = \frac{22}{7}$

6. $r = 21$ in., $h = 10$ in.
7. $r = 28$ ft., $h = \frac{3}{4}$ ft.
8. $r = 10$ cm, $h = 4.2$ cm
9. $r = 1.4$ m, $h = 6$ m

In 10–13, find the volume of a cone that has the given dimensions: **a.** in terms of π **b.** using $\pi = 3.14$

10. $r = 10$ in., $h = 12$ in.
11. $r = 6$ ft., $h = \frac{1}{3}$ ft.
12. $r = 1$ mm, $h = 3$ mm
13. $r = .2$ cm, $h = 2.4$ cm

14. Find the volume of a sphere with radius 10:

    a. in terms of π **b.** using $\pi = 3.14$

15. Find the volume of a sphere with radius $3\frac{1}{2}$:

    a. in terms of π **b.** using $\pi = \frac{22}{7}$

16. Find the length of the radius of a circle when its area is:

a. 16π **b.** 81π **c.** 225π **d.** $\frac{1}{64}\pi$ **e.** 1.96π

17. If the area of a circle is 100π, find its:
a. radius **b.** diameter **c.** circumference

18. If the circumference of a circle is 16π, find its:
a. diameter **b.** radius **c.** area

19. The radius of a circular flower bed is 3.5 meters. **a.** Express its area in terms of π. **b.** Use $\pi = \frac{22}{7}$ to find its approximate area.

20. A circular mirror has a diameter of 18 inches. Use $\pi = 3.14$ to find its approximate area.

21. A tank in the form of a right circular cylinder is used for storing water. It has a diameter of 12 feet and a height of 14 feet. How many gallons of water will it hold? Use $\pi = \frac{22}{7}$. [1 cubic foot contains 7.5 gallons.]

22. If the diameter of a circle is tripled, state the change that takes place in: **a.** its area **b.** its circumference

In 23–31, select the best answer from the four choices given.

23. If the radius of a circle is 7, then the area of the circle is:
(1) exactly 44 (2) exactly 153.86
(3) between 153.86 and 154 (4) exactly 154

24. If the diameter of a circle is 20, then the area of the circle is:
(1) exactly 314 (2) exactly $314\frac{2}{7}$
(3) between 314 and 315 (4) between 1,256 and 1,257

25. If the radius of a circle is tripled, then its area is multiplied by:
(1) 27 (2) 9 (3) 3 (4) 6

26. If the radius of a circle is doubled, then its area is:
(1) doubled (2) multiplied by 2
(3) squared (4) multiplied by 4

27. The formula for the volume of a sphere is $\frac{4}{3}\pi r^3$. If its radius is doubled, the volume is:
(1) doubled (2) squared (3) cubed (4) multiplied by 8

28. The formula for the volume of a cone is $\frac{1}{3}\pi r^2 h$. If its radius is doubled and the height is unchanged, the volume is:
(1) doubled (2) squared
(3) multiplied by 4 (4) increased by 4

29. The radius of a cone is tripled while its height is held constant. Then its volume is:
(1) tripled (2) cubed
(3) multiplied by 6 (4) multiplied by 9

30. The radius and height of a right circular cylinder are each doubled. Its volume is then multiplied by:
(1) 8 (2) 2 (3) 6 (4) 4

31. If the height of a right circular cylinder is divided by 4 while its radius is doubled, the new cylinder will have a volume that is:
(1) half the original volume (2) twice the original volume
(3) the same as the original (4) 4 times the original volume

32. In the figure at the right, a circle is drawn in a square whose side has a length of 12.

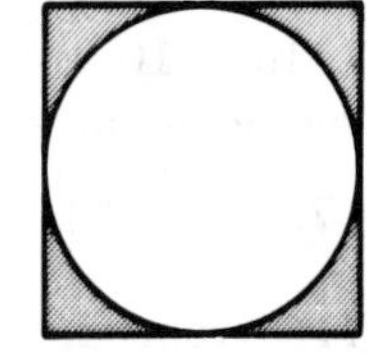

a. Find the diameter of the circle.
b. Find the radius of the circle.
c. Find the area of the square.
d. Find the area of the circle in terms of π.
e. Using the results of **c** and **d**, express the area of the shaded portion of the diagram in terms of π.

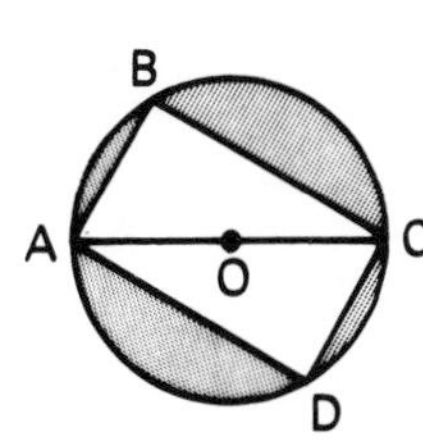

33. In the diagram at the left, rectangle $ABCD$ is drawn in circle O. $AB = 6$, $BC = 8$, and $AC = 10$. The diagonal of the rectangle, $\overline{AC}$, is also the diameter of the circle. [Wherever possible, answers may be left in terms of π.]
a. Find OC.
b. Find the circumference of circle O.
c. Find the perimeter of the rectangle.
d. Find the area of circle O.
e. Find the area of the rectangle.
f. Find the area of the shaded region.

34. An 8-inch pizza has a diameter of 8 inches; a 12-inch pizza has a diameter of 12 inches. Two boys purchased an 8-inch pizza for $3.20 and shared the pie equally. Four girls purchased a 12-inch pizza for $5.40 and shared the pie equally. Assume that both pies are of the same thickness.
a. Who got a larger piece of pizza, a boy or a girl? Explain why. [*Hint:* Relate the problem to the area of a circle.]
b. How much did each boy pay for his share of the pie?
c. How much did each girl pay for her share of the pie?
d. Which size pie is the better buy? Explain why.

19-15 REVIEW EXERCISES

1. Tell whether the number is rational or irrational.

a. $.12$ **b.** $.\overline{12}$ **c.** $\sqrt{12}$ **d.** $\sqrt{(12)^2}$ **e.** $\sqrt{169}$

f. 2π **g.** $\frac{1}{2}\sqrt{8}$ **h.** $0.3691215182124\ldots$

2. For the real number represented by r, where $r \neq 0$, find:
a. the additive inverse of r **b.** the multiplicative inverse of r

In 3–6, write the given number without a radical.

3. $\sqrt{\frac{9}{25}}$ **4.** $-\sqrt{49}$ **5.** $\sqrt[3]{-27}$ **6.** $\pm\sqrt{1.21}$

In 7–10, solve for the variable, using the set of real numbers as the replacement set.

7. $y^2 = 81$ **8.** $m^2 = .09$ **9.** $3x^2 = 300$ **10.** $k^2 - 144 = 0$

11. Approximate $\sqrt{42}$ to the nearest tenth.

In 12–15, find the indicated root. Assume that all variables represent positive numbers.

12. $\sqrt{400x^2}$ **13.** $\sqrt{4y^4}$ **14.** $\sqrt{c^{10}d^2}$ **15.** $\sqrt{.01m^{16}}$

In 16–19, simplify the expression. Assume $b > 0$.

16. $\sqrt{180}$ **17.** $3\sqrt{18}$ **18.** $\frac{1}{2}\sqrt{28}$ **19.** $\sqrt{48b}$

In 20–23, combine the radicals.

20. $\sqrt{18} + \sqrt{8} - \sqrt{32}$ **21.** $3\sqrt{20} - 2\sqrt{45}$

22. $2\sqrt{50} - \sqrt{98} + \frac{1}{2}\sqrt{72}$ **23.** $\sqrt{75} - 3\sqrt{12}$

In 24–29, multiply or divide, and then simplify.

24. $8\sqrt{2} \cdot 2\sqrt{2}$ **25.** $(3\sqrt{5})^2$ **26.** $2\sqrt{7} \cdot \sqrt{70}$

27. $\sqrt{98} \div \sqrt{2}$ **28.** $\frac{16\sqrt{21}}{2\sqrt{7}}$ **29.** $\frac{5\sqrt{162}}{9\sqrt{50}}$

In 30 and 31, select the numeral preceding the correct choice.

30. The expression $\sqrt{108} - \sqrt{3}$ is equivalent to
(1) $\sqrt{105}$ (2) $35\sqrt{3}$ (3) $5\sqrt{3}$ (4) 6

31. The sum of $9\sqrt{2}$ and $\sqrt{32}$ is
(1) $9\sqrt{34}$ (2) $13\sqrt{2}$ (3) $10\sqrt{34}$ (4) 15

32. In the figure, F, G, and H are points on circle O. Name all segments that are:

a. radii

b. diameters

c. chords

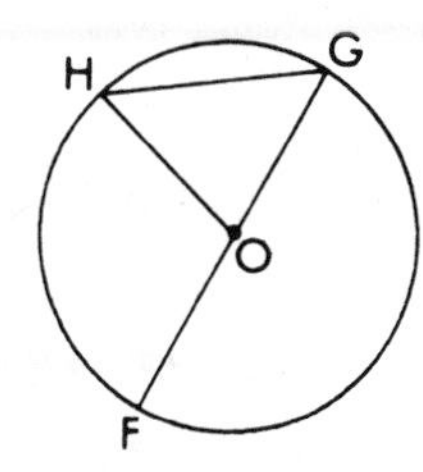

33. If the circumference of a circle is 28π, find the radius of the circle.

34. The diameter of a half-dollar is 30 mm. Find the area in square millimeters of one face of the coin. [Express answer in terms of π.]

35. Using $\pi = 3.14$, find the approximate area of a circle whose radius measures 3 inches.

36. A circle is drawn in a square whose side has a length of 40. Expressing answers in terms of π, find:

a. the circumference of the circle

b. the area of the circle

c. the area of the shaded region

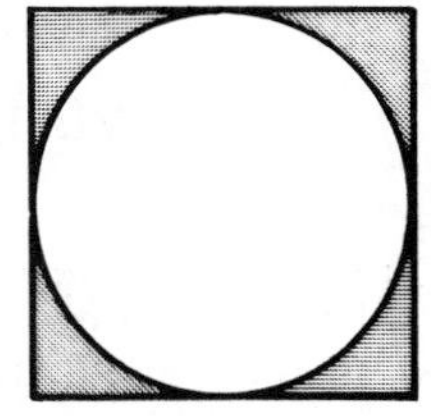

37. The formula for the volume of a right circular cylinder is $V = \pi r^2 h$.

a. Find the height h if $r = 5$ and $V = 100\pi$.

b. Express h in terms of V, π, and r.

c. What change occurs in the volume V if the radius r is doubled and the height h remains constant?

In 38 and 39, select the numeral preceding the correct answer.

38. What is the radius of a circle whose area is 144π?
(1) 6 (2) 12 (3) 24 (4) 72

39. In a circle, if one radius measures $x + 3$ and another radius measures $3x - 5$, what is the length of a diameter of the circle?
(1) 14 (2) 8 (3) 7 (4) 4

40. A dog is fastened by a 6-foot chain to an outside corner of a shed that measures 8 feet by 12 feet. Find the area of the space in which the dog is confined.

Chapter 20

Quadratic Equations

20-1 THE STANDARD FORM OF A QUADRATIC EQUATION

The equation $x^2 - 3x - 10 = 0$ is a *polynomial equation* in one variable. This equation is of degree two, or second degree, because the greatest exponent of the variable x is 2. The equation is in ***standard form*** because all terms of the equation are collected, in descending order of exponents, in one member, and the other member is 0.

An equation of the second degree, such as $x^2 - 3x - 10 = 0$, is called a ***quadratic equation***.

In general, the standard form of a quadratic equation in one variable is:

$$ax^2 + bx + c = 0$$

where a, b, and c are real numbers and $a \neq 0$.

MODEL PROBLEM

Transform the equation $x(x - 4) = 5$ into an equivalent quadratic equation in standard form.

Solution

$$x(x - 4) = 5$$
$$x^2 - 4x = 5$$
$$x^2 - 4x - 5 = 0 \quad Ans.$$

EXERCISES

In 1–9, transform the equation into an equivalent equation in standard form.

1. $x^2 + 9x = 10$ **2.** $2x^2 + 7x = 3x$ **3.** $x^2 = 3x - 8$

4. $4x + 3 = x^2$ **5.** $3x^2 = 27x$ **6.** $x(x - 3) = 10$

7. $x^2 = 5(x + 4)$ **8.** $\frac{x^2}{2} + 3 = \frac{x}{4}$ **9.** $x^2 = 6 - \frac{x}{2}$

20-2 USING FACTORING TO SOLVE A QUADRATIC EQUATION

Consider the following products:

$$5 \times 0 = 0 \qquad (-2) \times 0 = 0 \qquad \frac{1}{2} \times 0 = 0$$

$$0 \times 7 = 0 \qquad 0 \times (-3) = 0 \qquad 0 \times 0 = 0$$

These examples illustrate that whenever the product is 0, at least one of the factors is 0.

In general, **if a and b are real numbers, then:**

$$\mathbf{ab = 0 \text{ if and only if } a = 0 \text{ or } b = 0}$$

This principle is used to solve quadratic equations. For example, in order to solve the quadratic equation $x^2 - 3x + 2 = 0$, it can be written as $(x - 2)(x - 1) = 0$. The factors $(x - 2)$ and $(x - 1)$ represent real numbers whose product is 0. The equation will be true if either one of the factors is 0, that is, if $(x - 2)$ is 0 or if $(x - 1)$ is 0.

$$(x - 2)(x - 1) = 0$$

$$\begin{array}{rclcrcl} x - 2 &=& 0 & \text{or} & x - 1 &=& 0 \\ +2 &=& +2 & & +1 &=& +1 \\ \hline x &=& 2 & \text{or} & x &=& 1 \end{array}$$

A check will show that both 2 and 1 are values of x for which the equation is true.

Check for $x = 2$

$$\begin{aligned} x^2 - 3x + 2 &= 0 \\ (2)^2 - 3(2) + 2 &\stackrel{?}{=} 0 \\ 4 - 6 + 2 &\stackrel{?}{=} 0 \\ 0 &= 0 \quad \text{(True)} \end{aligned}$$

Check for $x = 1$

$$\begin{aligned} x^2 - 3x + 2 &= 0 \\ (1)^2 - 3(1) + 2 &\stackrel{?}{=} 0 \\ 1 - 3 + 2 &\stackrel{?}{=} 0 \\ 0 &= 0 \quad \text{(True)} \end{aligned}$$

Since both 2 and 1 satisfy the equation $x^2 - 3x + 2 = 0$, the solution set of this equation is {2, 1}. The roots of the equation are 2 and 1.

Every quadratic equation has two roots, but the two roots of a quadratic equation are not always different numbers. Sometimes, the roots are the same number, as in the case of $x^2 - 2x + 1 = 0$, which can be written $(x - 1)(x - 1) = 0$. Both roots are 1. Such a root is called a ***double root*** and is written only once in the solution set.

PROCEDURE. To solve a quadratic equation by using factoring:

1. If necessary, transform the equation into standard form.
2. Factor the left member of the equation.
3. Set each factor containing the variable equal to zero.
4. Solve each of the resulting equations.
5. Check by substituting each value of the variable in the original equation.

MODEL PROBLEMS

1. Solve and check: $x^2 - 7x = -10$

How to Proceed	*Solution*
	$x^2 - 7x = -10$
(1) Transform into standard form.	$x^2 - 7x + 10 = 0$
(2) Factor the left member.	$(x - 2)(x - 5) = 0$
(3) Let each factor = 0.	$x - 2 = 0 \mid x - 5 = 0$
(4) Solve each equation.	$x = 2 \mid x = 5$
(5) Check both values in the original equation.	

Check for x = 2

$$x^2 - 7x = -10$$
$$(2)^2 - 7(2) \stackrel{?}{=} -10$$
$$4 - 14 \stackrel{?}{=} -10$$
$$-10 = -10 \quad \text{(True)}$$

Check for x = 5

$$x^2 - 7x = -10$$
$$(5)^2 - 7(5) \stackrel{?}{=} -10$$
$$25 - 35 \stackrel{?}{=} -10$$
$$-10 = -10 \quad \text{(True)}$$

Answer: $x = 2$ or $x = 5$; the solution set is {2, 5}.

2. List the members of the set $\{x \mid 2x^2 = 3x\}$.

How to Proceed	*Solution*
	$2x^2 = 3x$
(1) Transform into standard form.	$2x^2 - 3x = 0$
(2) Factor the left member.	$x(2x - 3) = 0$
(3) Let each factor = 0.	$x = 0 \mid 2x - 3 = 0$
(4) Solve each equation.	$2x = 3$
	$x = \frac{3}{2}$

The check is left to the student.

Answer: $\left\{0, \frac{3}{2}\right\}$

Caution: Never transform an equation by dividing both members of the equation by an expression involving the variable. If you had divided both members of the equation $2x^2 = 3x$ by x, you would have obtained the equation $2x = 3$ whose solution is $x = \frac{3}{2}$. You would have lost the solution $x = 0$.

3. Solve and check: $x(x + 2) = 25 + 2x$

How to Proceed	*Solution*
	$x(x + 2) = 25 + 2x$
(1) Use the distributive property.	$x^2 + 2x = 25 + 2x$
(2) Transform into standard form.	$x^2 - 25 = 0$
(3) Factor the left member.	$(x - 5)(x + 5) = 0$
(4) Let each factor = 0.	$x - 5 = 0 \mid x + 5 = 0$
(5) Solve each equation.	$x = 5 \mid x = -5$

The check is left to the student.

Answer: $x = 5$ or $x = -5$; the solution set is $\{5, -5\}$.

4. The areas of two similar triangles are in the ratio of 4:1. The length of a side of the smaller triangle is 5 inches. Find the length of the corresponding side in the larger triangle.

Solution: Recall that the ratio of the areas of two similar polygons is the square of the ratio of the measures of the sides.

$$\frac{A}{A'} = \frac{(s)^2}{(s')^2}$$

$$\frac{4}{1} = \frac{s^2}{25}$$

$$s^2 = 100$$

$$s^2 - 100 = 0$$

$$(s + 10)(s - 10) = 0$$

$$s + 10 = 0 \quad | \quad s - 10 = 0$$

$$s = -10 \quad | \quad s = 10$$

[Reject the negative value because the length of the side cannot be a negative number.]

Check

$$\frac{4}{1} \stackrel{?}{=} \frac{10^2}{25}$$

$$\frac{4}{1} \stackrel{?}{=} \frac{100}{25}$$

$$\frac{4}{1} = \frac{4}{1} \text{ (True)}$$

Answer: The length of the corresponding side is 10 inches.

KEEP IN MIND

To solve a quadratic equation by using factoring, one member of the equation must be zero.

EXERCISES

In 1–48, solve and check the equation.

1. $x^2 - 3x + 2 = 0$
2. $z^2 - 5z + 4 = 0$
3. $x^2 - 8x + 16 = 0$
4. $r^2 - 12r + 35 = 0$
5. $c^2 + 6c + 5 = 0$
6. $m^2 + 10m + 9 = 0$
7. $x^2 + 2x + 1 = 0$
8. $y^2 + 11y + 24 = 0$
9. $x^2 - 4x - 5 = 0$
10. $x^2 - 5x - 6 = 0$
11. $x^2 + x - 6 = 0$
12. $x^2 + 2x - 15 = 0$
13. $r^2 - r - 72 = 0$
14. $x^2 - x - 12 = 0$
15. $x^2 - 49 = 0$
16. $z^2 - 4 = 0$
17. $m^2 - 64 = 0$
18. $3x^2 - 12 = 0$
19. $d^2 - 2d = 0$
20. $s^2 - s = 0$

21. $x^2 + 3x = 0$
22. $z^2 + 8z = 0$
23. $2x^2 - 5x + 2 = 0$
24. $3x^2 - 10x + 3 = 0$
25. $3x^2 - 8x + 4 = 0$
26. $5x^2 + 11x + 2 = 0$
27. $x^2 - x = 6$
28. $y^2 - 3y = 28$
29. $c^2 - 8c = -15$
30. $2m^2 + 7m = -6$
31. $r^2 - 4 = 0$
32. $x^2 = 121$
33. $y^2 = 6y$
34. $s^2 = -4s$
35. $y^2 = 8y + 20$
36. $2x^2 - x = 15$
37. $x^2 = 9x - 20$
38. $30 + x = x^2$
39. $x^2 + 3x - 4 = 50$
40. $2x^2 + 7 = 5 - 5x$
41. $\frac{1}{3}x^2 + \frac{4}{3}x + 1 = 0$
42. $\frac{1}{2}x^2 - \frac{7}{6}x = 1$
43. $x(x - 2) = 35$
44. $y(y - 3) = 4$
45. $x(x + 3) = 40$
46. $\frac{x + 2}{2} = \frac{12}{x}$
47. $\frac{y + 3}{3} = \frac{6}{y}$
48. $\frac{x}{3} = \frac{12}{x}$
49. The areas of two similar polygons are in the ratio 36:1. The length of a side of the smaller polygon is 2 cm. Find the length of the corresponding side of the larger polygon.
50. Two similar triangles have areas of 16 and 36. The length of a side of the smaller triangle is 8. Find the length of the corresponding side of the larger triangle.

20-3 SOLVING INCOMPLETE QUADRATIC EQUATIONS

A quadratic equation in which the first-degree term is missing is called an ***incomplete quadratic equation***, or a ***pure quadratic equation***. For example, $x^2 - 25 = 0$ (in general, $\boldsymbol{ax^2 + c = 0}$ when $a \neq 0$) is an incomplete quadratic equation.

One method of solving $x^2 - 25 = 0$ is to factor the left member, proceeding as in Section 20-2 (see Model Problem 3 on page 683).

Another method that we may use to solve $x^2 - 25 = 0$ makes use of the following principle:

● **Every positive real number has two real square roots, one of which is the opposite of the other.**

Solution

$$x^2 - 25 = 0$$
$$x^2 = 25$$
$$x = \sqrt{25} \text{ or } x = -\sqrt{25}$$
$$x = 5 \text{ or } x = -5$$

Check

$$x^2 - 25 = 0$$
$$\text{If } x = 5,\ (5)^2 - 25 \stackrel{?}{=} 0$$
$$25 - 25 \stackrel{?}{=} 0$$
$$0 = 0 \quad \text{(True)}$$

$$x^2 - 25 = 0$$
$$\text{If } x = -5,\ (-5)^2 - 25 \stackrel{?}{=} 0$$
$$25 - 25 \stackrel{?}{=} 0$$
$$0 = 0 \quad \text{(True)}$$

Answer: $x = 5$ or $x = -5$, which may be written $x = \pm 5$.
The solution set is $\{5, -5\}$.

PROCEDURE. To solve an incomplete quadratic equation:

1. Transform the equation into the form $x^2 = n$, where n is a non-negative real number.
2. Let $x = \sqrt{n}$ and let $x = -\sqrt{n}$.
3. Check the resulting values for x in the original equation.

MODEL PROBLEMS

1. Find the solution set:
$7y^2 = 3y^2 + 36$

Solution

$$7y^2 = 3y^2 + 36$$
$$7y^2 - 3y^2 = 36$$
$$4y^2 = 36$$
$$y^2 = 9$$
$$y = \sqrt{9} = 3 \text{ or}$$
$$y = -\sqrt{9} = -3$$

Answer: The solution set is $\{3, -3\}$.
(The roots are rational numbers.)

2. Solve: $4x^2 - 14 = 2x^2$

Solution

$$4x^2 - 14 = 2x^2$$
$$4x^2 - 2x^2 - 14 = 0$$
$$2x^2 - 14 = 0$$
$$2x^2 = 14$$
$$x^2 = 7$$
$$x = \sqrt{7} \text{ or}$$
$$x = -\sqrt{7}$$

Answer: $x = \pm\sqrt{7}$
(The roots are irrational numbers.)

The checks in Model Problems 1 and 2 are left to the student.

EXERCISES

In 1–15, solve and check the equation.

1. $x^2 = 4$
2. $a^2 - 36 = 0$
3. $\frac{1}{2}x^2 = 50$
4. $5y^2 = 45$
5. $3k^2 = 147$
6. $2x^2 - 8 = 0$
7. $r^2 - 11 = 70$
8. $4x^2 + 5 = 21$
9. $2x^2 - 11 = 39$
10. $2x^2 + 3x^2 = 45$
11. $6x^2 - 4x^2 = 98$
12. $4y^2 - 13 = y^2 + 14$
13. $\frac{y^2}{3} = 12$
14. $\frac{x}{9} = \frac{4}{x}$
15. $\frac{4x}{25} = \frac{4}{x}$

In 16–24, solve for x in simplest radical form.

16. $x^2 = 10$
17. $x^2 = 27$
18. $3x^2 = 6$
19. $2x^2 - 16 = 0$
20. $x^2 + 25 = 100$
21. $x^2 - 4 = 4$
22. $8x^2 - 6x^2 = 54$
23. $3x^2 - 28 = 2x^2 + 33$
24. $\frac{2x}{9} = \frac{6}{x}$

In 25–27, find the positive value of x, correct to the *nearest tenth*. Use the table on page 706.

25. $x^2 = 24$
26. $4x^2 - 160 = 0$
27. $7x^2 = x^2 + 198$

In 28–33, solve for x in terms of the other variable(s).

28. $x^2 = b^2$
29. $x^2 = 25a^2$
30. $9x^2 = r^2$
31. $4x^2 - a^2 = 0$
32. $x^2 + a^2 = c^2$
33. $x^2 + b^2 = c^2$

In 34–39, solve for the indicated variable in terms of the other variable(s).

34. Solve for s: $s^2 = A$
35. Solve for r: $A = \pi r^2$
36. Solve for r: $S = 4\pi r^2$
37. Solve for r: $V = \pi r^2 h$
38. Solve for t: $s = \frac{1}{2}gt^2$
39. Solve for v: $F = \frac{mv^2}{gr}$

40. How many feet of fencing would be needed to enclose a square garden that has an area of 36 square feet?

41. The length of a rectangular flower bed is 3 times the width. The area of the bed is 108 square meters. What are the dimensions of the bed?
42. The height of a triangular metal plate is 6 times the measure of the base. The area of the plate is 120 square inches. In simplest radical form, what is the measure of the base?

20-4 THE QUADRATIC FORMULA

Not every quadratic equation can be solved by factoring. The left member of an equation such as $x^2 - 6x - 3 = 0$ cannot be factored over the set of integers and, therefore, cannot be solved by that method.

In trying to understand a method that can be used to solve such an equation, begin with the idea that you saw in Section 20-3, which is that an incomplete quadratic equation can be solved by taking the square root of each member of the equation. Note that the constant member can be a perfect square as in Example 1 below, but the constant member need not be a perfect square, as in Example 2.

Example **1.**
$$x^2 = 16$$
$$x = \pm\sqrt{16}$$
$$x = \pm 4$$

Example **2.**
$$x^2 = 12$$
$$x = \pm\sqrt{12}$$
$$x = \pm 2\sqrt{3}$$

Now, let us extend the idea of taking the square root of each member of an equation to the case when one member is a trinomial. The trinomial could already be a perfect square. For example, the trinomial $x^2 + 2x + 1$ is a perfect square: $(x + 1)^2$. When such a perfect square trinomial is one member of an equation, we can solve that equation by taking the square root of each member, as follows:

Solve:
$$x^2 + 2x + 1 = 4$$
$$(x + 1)^2 = 4$$
$$x + 1 = \pm\sqrt{4}$$
$$x + 1 = \pm 2$$
$$x + 1 = 2 \quad \text{or} \quad x + 1 = -2$$
$$x = 1 \quad \text{or} \quad x = -3$$

Answer: $\{-3, 1\}$

Note that in the preceding example, the right-hand member, 4, is also a perfect square. This condition is not necessary for the procedure. It is only necessary that the trinomial be a perfect square.

Suppose, now, that the trinomial is not given as a perfect square. Consider the first equation mentioned, $x^2 - 6x - 3 = 0$. We can create our own perfect square trinomial.

(1) Move the given constant out of the way so that we can supply the number necessary to complete the square.

$$x^2 - 6x - 3 = 0$$
$$x^2 - 6x = 3$$

(2) If 9 were the third term of the trinomial, the trinomial would be a perfect square.

$$x^2 - 6x + 9$$
$$(x - 3)^2$$

(3) We can add 9 to the left member of the equation if we also add 9 to the right member.

$$x^2 - 6x + 9 = 3 + 9$$

(4) Now, we have our perfect square trinomial and can proceed to take the square root of each member, thus leading to the solution.

$$x^2 - 6x + 9 = 12$$
$$(x - 3)^2 = 12$$
$$x - 3 = \pm\sqrt{12}$$
$$x - 3 = \pm 2\sqrt{3}$$
$$x - 3 = 2\sqrt{3} \quad \text{or} \quad x - 3 = -2\sqrt{3}$$
$$x = 3 + 2\sqrt{3} \quad \text{or} \quad x = 3 - 2\sqrt{3}$$

Answer: $\{3 + 2\sqrt{3}, 3 - 2\sqrt{3}\}$

This procedure can be applied to the general quadratic equation to obtain a formula for the solution of *any* quadratic equation.

● **Given any quadratic equation of the form $ax^2 + bx + c = 0$, where a, b, and c are real numbers and $a \neq 0$, we can find the roots of the equation by the *quadratic formula:***

$$x = \frac{-b \pm \sqrt{b^2 - 4ac}}{2a}$$

MODEL PROBLEMS

1. Solve by the quadratic formula: $2x^2 + x = 6$

How to Proceed	*Solution*
1. Write the equation in standard form.	$2x^2 + x - 6 = 0$
2. Compare the equation to $ax^2 + bx + c = 0$ to find the values of a, b, and c.	$a = 2, b = 1, c = -6$

3. Write the quadratic formula. $$x = \frac{-b \pm \sqrt{b^2 - 4ac}}{2a}$$

4. Substitute. $$x = \frac{-(1) \pm \sqrt{(1)^2 - 4(2)(-6)}}{2(2)}$$

5. Simplify. $$x = \frac{-1 \pm \sqrt{1 + 48}}{4}$$

$$x = \frac{-1 \pm \sqrt{49}}{4} = \frac{-1 \pm 7}{4}$$

6. Determine the two roots. (The check is left to the student.)

$$x = \frac{-1 + 7}{4} \quad \Big| \quad x = \frac{-1 - 7}{4}$$

$$x = \frac{6}{4} = \frac{3}{2} \quad \Big| \quad x = \frac{-8}{4} = -2$$

Answer: $x = -2$ or $x = \frac{3}{2}$

2. Find the irrational roots of $x^2 + 2x - 1 = 0$.

Solution

The equation $x^2 + 2x - 1 = 0$ is in standard form.
$a = 1, b = 2, c = -1$

$$x = \frac{-b \pm \sqrt{b^2 - 4ac}}{2a}$$

$$x = \frac{-(2) \pm \sqrt{(2)^2 - 4(1)(-1)}}{2(1)}$$

$$x = \frac{-2 \pm \sqrt{4 + 4}}{2}$$

$$x = \frac{-2 \pm \sqrt{8}}{2}$$

$$x = \frac{-2 \pm 2\sqrt{2}}{2}$$

$x = -1 \pm \sqrt{2}$ *Ans.*

EXERCISES

In 1–18, solve the given equation by using the quadratic formula.

1. $x^2 - 7x + 6 = 0$
2. $x^2 + 4x - 5 = 0$
3. $x^2 + 3x + 2 = 0$
4. $2x^2 + x - 1 = 0$
5. $3x^2 + 5x - 2 = 0$
6. $3x^2 + 5x + 2 = 0$
7. $x^2 + 6x + 9 = 0$
8. $4x^2 - 4x + 1 = 0$
9. $x^2 + 10x = -25$
10. $x^2 + x = 12$
11. $x^2 + 2x = 24$
12. $x^2 = x + 2$
13. $x^2 + 8 = 6x$
14. $2x^2 - 10 = x$
15. $x^2 - 9 = 0$
16. $5x^2 = 20$
17. $x^2 - 3x + 1 = 1$
18. $x^2 = 5x$

In 19–27, solve the given equation by using the quadratic formula. Express the answers in simplest radical form.

19. $x^2 - 2x - 2 = 0$
20. $x^2 - 10x + 4 = 0$
21. $x^2 + 2x - 4 = 0$
22. $x^2 - 2 = 4x$
23. $2x^2 - 8x + 7 = 0$
24. $4x^2 = 2x + 1$
25. $9x^2 + 1 = 12x$
26. $x^2 = 20$
27. $2x^2 = 5$

20-5 USING THE THEOREM OF PYTHAGORAS

You are now ready to study and apply a most useful relationship that exists among the sides of a right triangle.

The figure at the right represents a *right triangle*. Recall that such a triangle contains one and only one right angle. In right triangle ABC, side $\overline{AB}$, which is opposite the right angle, is called the *hypotenuse*. The hypotenuse is the longest side of the triangle. The other two sides of the triangle, $\overline{BC}$ and $\overline{AC}$, form the right angle. They are called the *legs* of the right triangle.

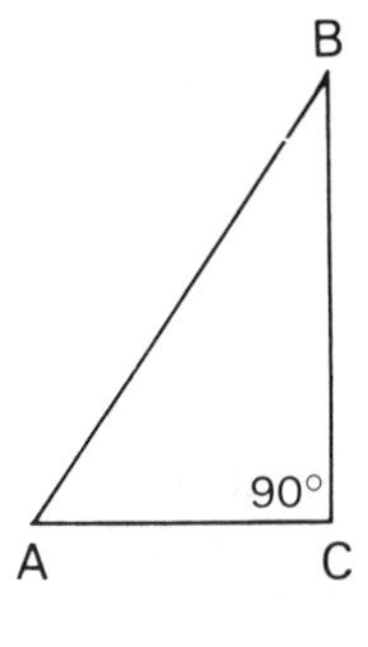

More than 2,000 years ago, the Greek mathematician Pythagoras demonstrated the following property of the right triangle, which is called the ***Pythagorean Theorem:***

● **In a right triangle, the square of the length of the hypotenuse is equal to the sum of the squares of the lengths of the other two sides.**

If we represent the length of the hypotenuse of right triangle ABC by c and the lengths of the other two sides by a and b, the Theorem of Pythagoras may be written as the following formula:

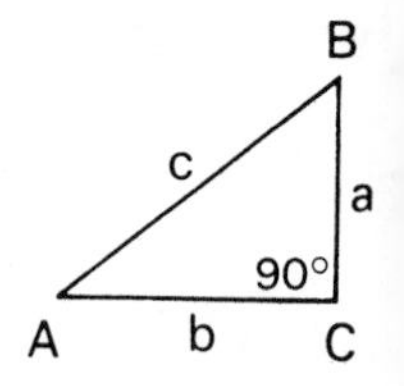

$$c^2 = a^2 + b^2$$

Geometrically, this relation means that if squares are built on the hypotenuse and on each leg of a right triangle, then the area of the square on the hypotenuse is equal to the sum of the areas of the squares on the legs. For example:

Triangle ABC represents a right triangle in which the length of side $\overline{AC}$ = 4 cm, the length of side $\overline{BC}$ = 3 cm, and the length of the hypotenuse $\overline{AB}$ = 5 cm.

The area of the square on side $\overline{BC}$ = 9 cm^2.

The area of the square on side $\overline{AC}$ = 16 cm^2.

The area of the square on hypotenuse $\overline{AB}$ = 25 cm^2.

$$25 \text{ cm}^2 = 9 \text{ cm}^2 + 16 \text{ cm}^2$$

$$5^2 = 3^2 + 4^2$$

Similar triangles can be used to demonstrate the truth of the Pythagorean Theorem. Recall that two triangles are similar if two angles of one are congruent to two angles of the other. Recall also that the measures of the corresponding sides of similar triangles are in proportion.

$\triangle ABC$ is a right triangle with $\angle ACB$ the right angle.

Altitude $\overline{CD}$ has been drawn to side $\overline{AB}$, thus producing right angles ADC and BDC.

Let $BC = a$, $AC = b$, $AB = c$, $AD = x$, and $DB = c - x$.

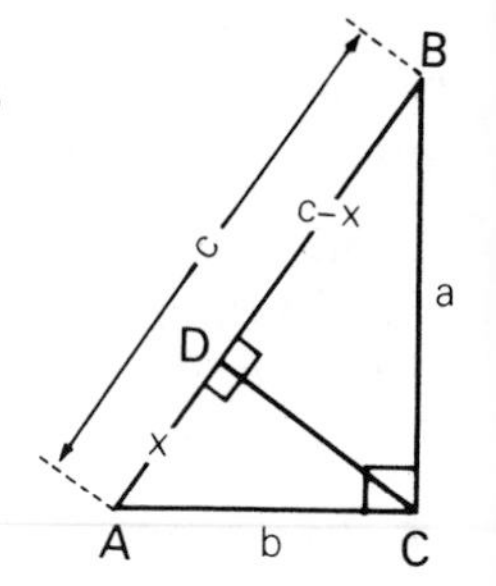

Since $\angle A \cong \angle A$
and $\angle ADC \cong \angle ACB$,
$\triangle ACD \sim \triangle ABC$

$$\frac{AC}{AB} = \frac{AD}{AC}$$

$$\frac{b}{c} = \frac{x}{b}$$

$$b^2 = cx$$

Since $\angle B \cong \angle B$
and $\angle BDC \cong \angle ACB$,
$\triangle CBD \sim \triangle ABC$

$$\frac{CB}{AB} = \frac{BD}{BC}$$

$$\frac{a}{c} = \frac{c - x}{a}$$

$$a^2 = c^2 - cx$$

Now, add these two equations

$$a^2 + b^2 = c^2 - cx + cx$$
$$a^2 + b^2 = c^2$$

Three logical statements can be made for any right triangle where c represents the length of the hypotenuse, and a and b represent the lengths of the other two sides. All of the following statements are true.

1. The *conditional* form of the Pythagorean Theorem:
   If a triangle is a right triangle, then the square of the length of the hypotenuse is equal to the sum of the squares of the lengths of the other two sides. (If a triangle is a right triangle, then $c^2 = a^2 + b^2$.)
2. The *converse* of the Pythagorean Theorem:
   If the square of the length of the longest side of a triangle is equal to the sum of the squares of the lengths of the other two sides, then the triangle is a right triangle. (If $c^2 = a^2 + b^2$ in a triangle, then the triangle is a right triangle.)
3. The *biconditional* form:
   A triangle is a right triangle if and only if $c^2 = a^2 + b^2$.

MODEL PROBLEMS

1. A ladder is placed 5 feet from the foot of a wall. The top of the ladder reaches a point 12 feet above the ground. Find the length of the ladder.

Solution: Let the length of the hypotenuse $= c$, the length of side $a = 5$, the length of side $b = 12$.

$c^2 = a^2 + b^2$ [Theorem of Pythagoras]
$c^2 = 5^2 + 12^2$
$c^2 = 25 + 144$
$c^2 = 169$
$c = \sqrt{169} = 13$ or
$c = -\sqrt{169} = -13$ [Reject the negative value.]

Answer: The length of the ladder is 13 feet.

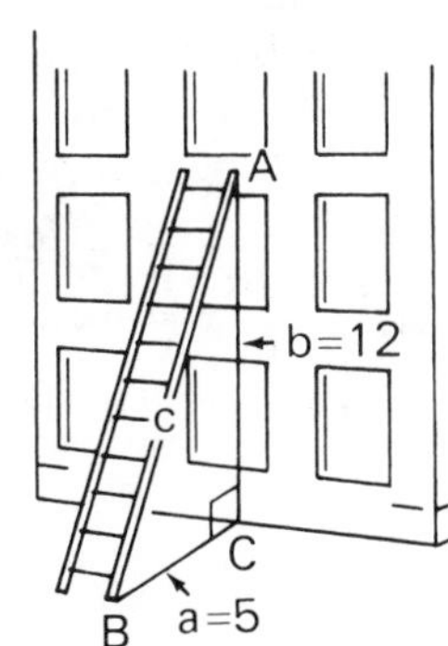

2. The hypotenuse of a right triangle is 20 centimeters long and one leg is 16 centimeters long. **a.** Find the length of the other leg. **b.** Find the area of the triangle.

Solution

a. Let the length of the unknown leg = a, the length of the hypotenuse $c = 20$, the length of side $b = 16$.

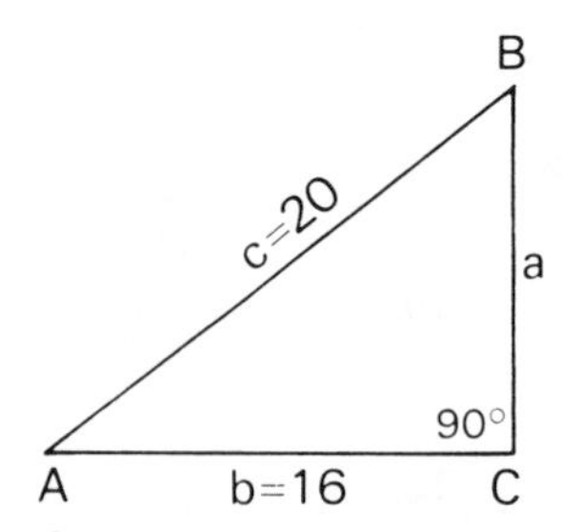

$$\begin{aligned} c^2 &= a^2 + b^2 \quad \text{[Theorem of Pythagoras]} \\ 20^2 &= a^2 + 16^2 \\ 400 &= a^2 + 256 \\ 144 &= a^2 \\ 12 &= a \text{ or} \\ -12 &= a \quad \text{[Reject the negative value.]} \end{aligned}$$

Answer: The length of the other leg is 12 feet.

b. Area of $\triangle ABC = \frac{1}{2}bh$

$$= \frac{1}{2}(16)(12)$$

$$= 96$$

Answer: Area of $\triangle ABC$ = 96 cm^2

3. Express in simplest radical form the length of a side of a square whose diagonal is 10.

Solution: In square $QRST$, angle QRS is a right angle. Therefore, triangle QRS is a right triangle. Let x = the length of each side of the square.

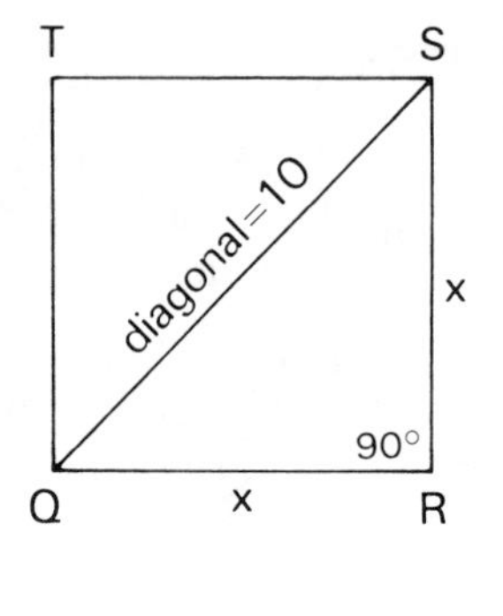

$$\begin{aligned} a^2 + b^2 &= c^2 \\ x^2 + x^2 &= 10^2 \\ 2x^2 &= 100 \\ x^2 &= 50 \\ x &= \pm\sqrt{50} \\ x &= \pm\sqrt{25 \cdot 2} \quad \text{[Reject the negative value.]} \\ x &= 5\sqrt{2} \end{aligned}$$

Answer: The length of the side of the square is $5\sqrt{2}$.

4. Is a triangle whose sides measure 8 centimeters, 7 centimeters, and 4 centimeters a right triangle?

Solution: If the triangle is a right triangle, the longest side, whose measure is 8, must be the hypotenuse.

$$c^2 = a^2 + b^2$$
$$8^2 \stackrel{?}{=} 7^2 + 4^2$$
$$64 \stackrel{?}{=} 49 + 16$$
$$64 \neq 65$$

Answer: The triangle is not a right triangle.

EXERCISES

In 1–9, let c represent the length of the hypotenuse in a right triangle, and let a and b represent the lengths of the other two sides. Find the length of the side of the right triangle whose measure is not given.

1. $a = 3, b = 4$ **2.** $a = 8, b = 15$ **3.** $c = 10, a = 6$

4. $c = 13, a = 12$ **5.** $c = 17, b = 15$ **6.** $c = 25, b = 20$

7. $a = \sqrt{2}, b = \sqrt{2}$ **8.** $a = 4, b = 4\sqrt{3}$ **9.** $a = 5\sqrt{3}, c = 10$

In 10–15: **a.** Express in simplest radical form the length of the third side of the right triangle the length of whose hypotenuse is represented by c and the lengths of whose other sides are represented by a and b. **b.** Approximate the length of this third side correct to the *nearest tenth.*

10. $a = 2, b = 3$ **11.** $a = 3, b = 3$ **12.** $a = 4, c = 8$

13. $a = 7, b = 1$ **14.** $b = \sqrt{3}, c = \sqrt{15}$ **15.** $a = 4\sqrt{2}, c = 8$

In 16–19, find x and express irrational results in simplest radical form.

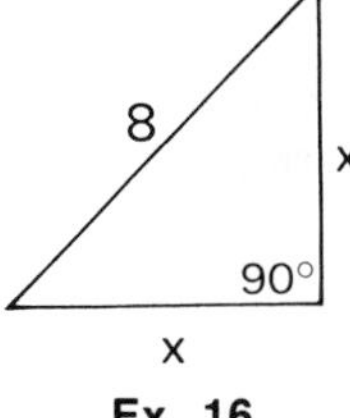

Ex. 16

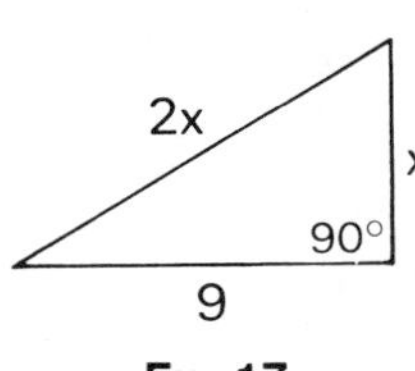

Ex. 17

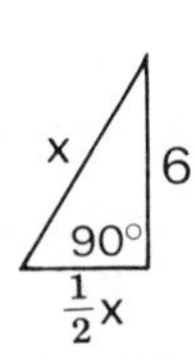

Ex. 18

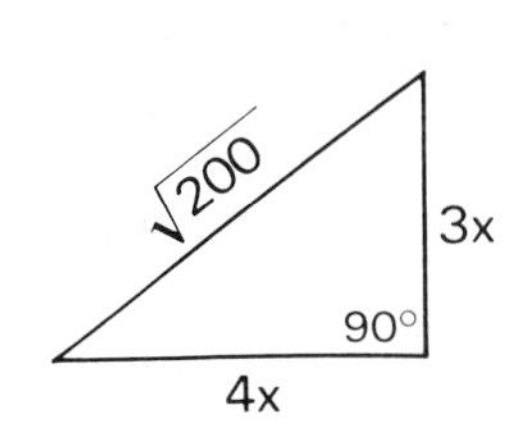

Ex. 19

20. A ladder 39 feet long leans against a building and reaches the ledge of a window. If the foot of the ladder is 15 feet from the foot of the building, how high is the window ledge above the ground?

21. What must be the length of a ladder that Mr. Rizzo can use if he wishes to place the bottom of the ladder 5 feet from a wall and he wishes to reach a window that is 15 feet above the ground? Give the answer, correct to the *nearest tenth* of a foot.

22. Miss Murray traveled 24 kilometers north and then 10 kilometers east. How far was she from her starting point?

23. One day, Ronnie walked from this home at A to his school at C by walking along $\overline{AB}$ and $\overline{BC}$, the sides of a rectangular open field that was muddy. When he returned home, the field was dry and Ronnie decided to take a shortcut by walking diagonally across the field along $\overline{CA}$. How much shorter was the trip home than the trip to school?

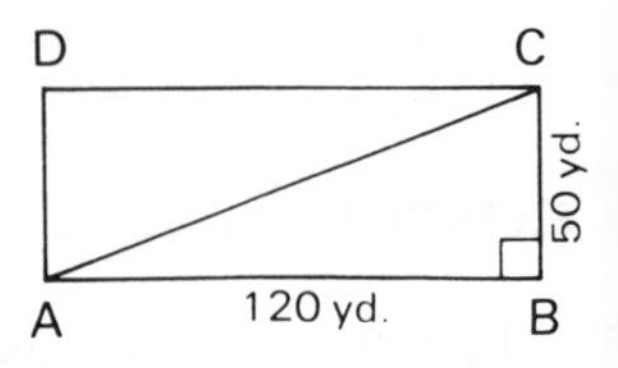

In 24–27, find the length of a diagonal of a rectangle whose sides are the given measurements.

24. 8 inches and 15 inches
25. 15 centimeters and 20 centimeters
26. 10 feet and 24 feet
27. 30 meters and 40 meters

28. The diagonal of a rectangle measures 13 centimeters. One side is 12 centimeters long. **a.** Find the length of the other side. **b.** Find the area of the rectangle.

29. Find the area of a rectangle in which the diagonal measures 26 and one side measures 10.

30. Approximate, to the *nearest inch*, the measure of the base of a rectangle whose diagonal measures 25 inches and whose altitude measures 18 inches.

In 31–35, approximate, to the *nearest tenth of a meter*, the length of a diagonal of a square whose side has the given measurement.

31. 2 meters **32.** 4 meters **33.** 5 meters
34. 6 meters **35.** 7 meters

36. A baseball diamond has the shape of a square 90 feet on each side. Approximate, to the *nearest tenth of a foot*, the distance from home plate to second base.

37. In the figure, ABC is an isosceles triangle in which sides $\overline{AC}$ and $\overline{CB}$ are of equal length, $AC = CB$. $\overline{CD}$, which is drawn so that angle CDB is a right angle, is called the altitude drawn to the base of the triangle. When altitude $\overline{CD}$ is drawn, it divides the base $\overline{AB}$ into two segments of equal measure, $AD = DB$. Each of the sides $\overline{AC}$ and $\overline{CB}$ measures 26 centimeters and the base $\overline{AB}$ measures 20 centimeters.

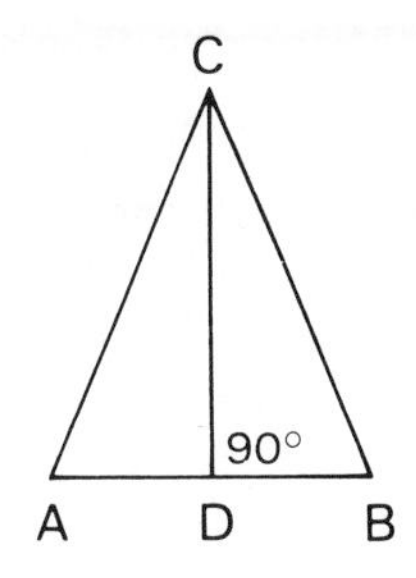

a. Find the length of the altitude drawn to the base.
b. Find the area of triangle ABC.

38. In an isosceles triangle, each of the equal sides measures 25 in. and the altitude drawn to the base measures 15 in.
a. Find the length of the base. **b.** Find the area of the triangle.

In 39–43, find the length of the altitude of equilateral triangle ABC when each of its sides has the given measure. (Express the answer in simplest radical form.)

39. 4 cm **40.** 6 cm **41.** 8 cm **42.** 10 cm **43.** 5 cm

44. **a.** On graph paper, draw the triangle whose vertices are $A(0, 5)$, $B(4, 5)$, $C(4, 9)$.
b. Find the length of each side of $\triangle ABC$.
c. Find the area of $\triangle ABC$.

45. The length of a side of a square is represented by s and the length of a diagonal by d. Show that $d = s\sqrt{2}$.

In 46–49, tell whether or not the measurements can be the lengths of the sides of a right triangle.

46. 6 yards, 10 yards, 8 yards **47.** 7 m, 4 m, 5 m

48. 12 cm, 16 cm, 20 cm **49.** 10 feet, 15 feet, 20 feet

20-6 USING QUADRATIC EQUATIONS TO SOLVE PROBLEMS

The use of an algebraic strategy to solve some problems requires the solution of a quadratic equation. The most convenient method of solving the quadratic equation should be used.

Number Problems

1. The square of a number decreased by 4 times the number equals 21. Find the number.

Solution: Let x = the number.

The square of the number decreased by 4 times the number equals 21.

$$x^2 - 4x = 21$$
$$x^2 - 4x - 21 = 0$$
$$(x - 7)(x + 3) = 0$$
$$x - 7 = 0 \quad \Big| \quad x + 3 = 0$$
$$x = 7 \quad \Big| \quad x = -3$$

Check for the number 7:

$$(7)^2 - 4(7) \stackrel{?}{=} 21$$
$$49 - 28 \stackrel{?}{=} 21$$
$$21 = 21 \quad \text{(True)}$$

Check for the number -3:

$$(-3)^2 - 4(-3) \stackrel{?}{=} 21$$
$$9 + 12 \stackrel{?}{=} 21$$
$$21 = 21 \quad \text{(True)}$$

Answer: The number is 7 or -3.

2. The product of two positive consecutive even integers is 80. Find the integers.

Solution:

Let x = the first positive even integer.
Then, $x + 2$ = the next consecutive positive even integer.

The product of two positive consecutive even integers is 80.

$$x(x + 2) = 80$$
$$x^2 + 2x = 80$$
$$x^2 + 2x - 80 = 0$$
$$(x - 8)(x + 10) = 0$$

$$x - 8 = 0 \quad \Big| \quad x + 10 = 0$$
$$x = 8 \quad \Big| \quad x = -10$$
$$x + 2 = 10 \quad \Big|$$

[Reject. The required integer is positive.]

Check: The product of the two positive consecutive even integers 8 and 10 is (8)(10), or 80.

Answer: The integers are 8 and 10.

EXERCISES

1. The square of a number increased by 3 times the number is 28. Find the number.
2. When the square of a certain number is diminished by 9 times the number, the result is 36. Find the number.

3. A certain number added to its square is 30. Find the number.
4. The square of a number exceeds the number by 72. Find the number.
5. Find a positive number that is 20 less than its square.
6. If the square of a positive number is added to 5 times the number, the result is 36. Find the number.
7. The square of a number decreased by 15 is equal to twice the number. Find the number.
8. Find two positive numbers whose ratio is 2:3 and whose product is 600.
9. The larger of two positive numbers is 5 more than the smaller. The product of the numbers is 66. Find the numbers.
10. One number is 5 more than another. Their product is 14. Find the numbers.
11. One number is 6 less than another. The sum of the squares of these numbers is 20. Find the numbers.
12. One number is 7 more than another. The sum of the squares of these numbers is 29. Find the numbers.
13. The product of two consecutive integers is 56. Find the integers.
14. The product of two consecutive odd integers is 99. Find the integers.
15. Find two consecutive positive integers such that the square of the smaller increased by 4 times the larger is 64.
16. Find two consecutive positive integers such that the square of the first decreased by 17 equals 4 times the second.
17. Find three consecutive positive integers such that the product of the second integer and the third integer is 20.
18. Find three consecutive odd integers such that the square of the first increased by the product of the other two is 224.
19. Find two consecutive integers such that the sum of their squares is 61.
20. Find the smallest of three consecutive positive integers such that the product of the two smaller integers is 38 more than twice the largest.
21. Nine times a certain number is 5 less than twice the square of the number. Find the number.
22. The sum of two numbers is 10. The sum of their squares is 52. Find the numbers.
23. The sum of a number and its reciprocal is $\frac{5}{2}$. Find the number.
24. If a positive number is decreased by its reciprocal, the result is $\frac{8}{3}$. Find the number.
25. The sum of a number and the square of its additive inverse is 42. Find the number.

Geometric Problems

The base of a parallelogram measures 7 centimeters more than its altitude. If the area of the parallelogram is 30 square centimeters, find the measure of its base and the measure of its altitude.

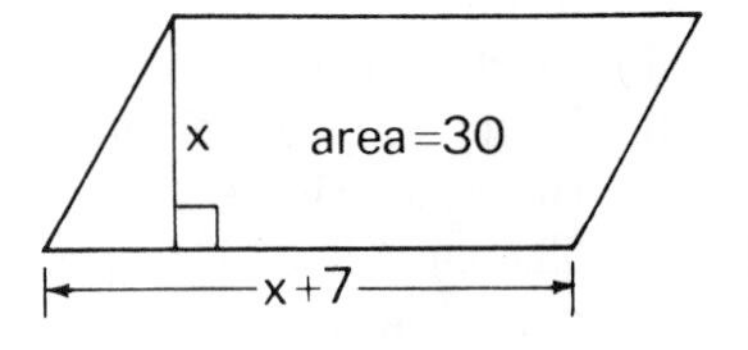

Solution

Let x = the number of centimeters in the altitude.
Then, $x + 7$ = the number of centimeters in the base.

The area of the parallelogram is 30.

$$\begin{aligned} \text{base} \cdot \text{altitude} &= \text{Area} \\ (x + 7)x &= 30 \\ x^2 + 7x &= 30 \\ x^2 + 7x - 30 &= 0 \\ (x - 3)(x + 10) &= 0 \end{aligned}$$

$x - 3 = 0$	$x + 10 = 0$
$x = 3$	$x = -10$ [Reject. The altitude cannot be negative.]
$x + 7 = 10$	

Check: When the base of a parallelogram measures 10 and its altitude measures 3, the area is $10 \cdot 3$, or 30. Also, the measure of the base, 10, is 7 more than the measure of the altitude, 3.

Answer: The altitude measures 3 centimeters;
the base measures 10 centimeters.

EXERCISES

1. The length of a rectangle is 2 times its width. The area of the rectangle is 72 square centimeters. Find the dimensions of the rectangle.
2. The ratio of the measures of the base and altitude of a parallelogram is 3:4. The area of the parallelogram is 1,200 square centimeters. Find the measure of the base and of the altitude of the parallelogram.
3. The length of a rectangular garden is 4 meters more than its width. The area of the garden is 60 square meters. Find the dimensions of the garden.
4. The altitude of a parallelogram measures 11 cm less than its base. The area of the parallelgram is 80 cm^2. Find the measure of its base and the measure of its altitude.

5. The perimeter of a rectangle is 20 inches and its area is 16 square inches. Find the dimensions of the rectangle.
6. If the measure of one side of a square is increased by 2 centimeters and the measure of an adjacent side is decreased by 2 centimeters, the area of the resulting rectangle is 32 square centimeters. Find the measure of one side of the square.
7. The length of a rectangle is 3 times its width. If the width is diminished by 1 meter and the length is increased by 3 meters, the area of the rectangle that is formed is 72 square meters. Find the dimensions of the original rectangle.
8. A rectangle is 6 feet long and 4 feet wide. If each dimension is increased by the same number of feet, a new rectangle is formed whose area is 39 square feet more than the area of the original rectangle. By how many feet was each dimension increased?
9. Joan's rectangular garden is 6 meters long and 4 meters wide. She wishes to double the area of her garden by increasing its length and width by the same amount. Find the number of meters by which each dimension must be increased.
10. The length of the base of a parallelogram is twice the length of its altitude. The area of the parallelogram is 50 square centimeters. Find the length of its base and altitude.
11. The altitude of a triangle measures 5 centimeters less than its base. The area of the triangle is 42 square centimeters. Find the lengths of its base and altitude.
12. A baseball diamond has the shape of a square 90 feet on each side. The pitcher's mound is 60.5 feet from home plate on the segment joining home plate and second base. Find the distance from the pitcher's mound to second base to the nearest tenth of a foot.
13. One leg of a right triangle is 1 cm longer than the other leg. The hypotenuse measures 5 cm. Find the measure of each leg of the triangle.
14. The measure of one leg of a right triangle exceeds the measure of the other leg by 7 meters. The hypotenuse of the triangle is 13 meters long. Find the measurements of the legs of the triangle.
15. The hypotenuse of a right triangle is 2 centimeters longer than one leg and 4 centimeters longer than the other leg. Find the length of each side of the triangle.
16. The length of the hypotenuse of a right triangle is 25 centimeters. One of the legs is 5 centimeters longer than the other.
    a. Find the length of each leg.
    b. Find the area of the triangle.
17. The ratio of the lengths of the two legs of a right triangle is 3:4. **(a)** Find the length of each leg when the length of the hypotenuse is the given measure. **(b)** Find the area of the triangle.
    a. 10 cm **b.** 20 cm **c.** 25 cm **d.** 100 cm

18. The perimeter of a right triangle is 30 centimeters. If the hypotenuse measures 13 centimeters:
a. Find the length of each leg.
b. Find the area of the triangle.

19. Denise and Dawn start from the same point and travel at the rates of 30 mph and 40 mph along straight roads that are at right angles to each other. In how many hours will they be 100 miles apart?

20-7 GRAPHING $y = ax^2 + bx + c$

The graph of every first-degree equation in two variables is a straight line. For example, the graph of $x + y = 6$ is a straight line. Two or three points are sufficient to draw this graph. The graph of a quadratic equation of the form $y = ax^2 + bx + c$, however, is not a straight line and a larger number of points must be used to draw this graph.

MODEL PROBLEMS

1. Graph the quadratic equation $y = 2x^2$ using integral values for x from $x = -3$ to $x = 3$ inclusive, that is, $-3 \leq x \leq 3$.

Solution

(1) Develop the following table of values:

x	$2x^2$	y
−3	$2(-3)^2$	18
−2	$2(-2)^2$	8
−1	$2(-1)^2$	2
0	$2(0)^2$	0
1	$2(1)^2$	2
2	$2(2)^2$	8
3	$2(3)^2$	18

(2) Plot the point associated with each ordered number pair (x, y): $(-3, 18)$, $(-2, 8)$, $(-1, 2)$, and so on.

(3) Draw a smooth curve through the points. Notice that the graph of $y = 2x^2$ is a curve; it is not a straight line. This curve is called a ***parabola***.

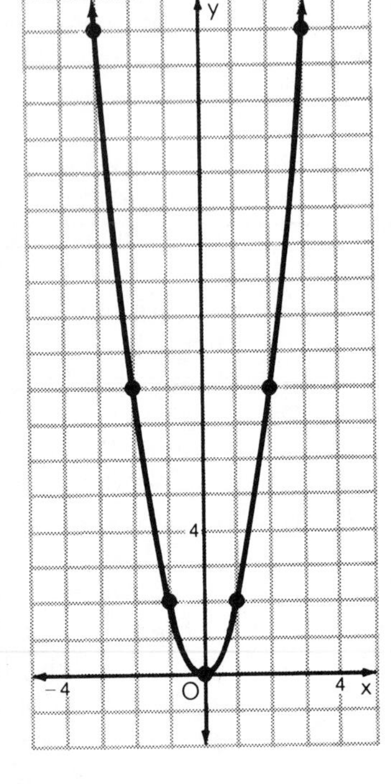

● **The graph of a quadratic equation of the form $y = ax^2 + bx + c$ (where a, b, and c are real numbers and $a \neq 0$) is a parabola.**

2. Graph the quadratic equation $y = x^2 - 2x - 8$ using integral values of x from $x = -3$ to $x = 5$ inclusive, that is, $-3 \leq x \leq 5$.

Solution

(1) Develop the following table of values:

x	$x^2 - 2x - 8$	y
−3	9 + 6 − 8	7
−2	4 + 4 − 8	0
−1	1 + 2 − 8	−5
0	0 − 0 − 8	−8
1	1 − 2 − 8	−9
2	4 − 4 − 8	−8
3	9 − 6 − 8	−5
4	16 − 8 − 8	0
5	25 − 10 − 8	7

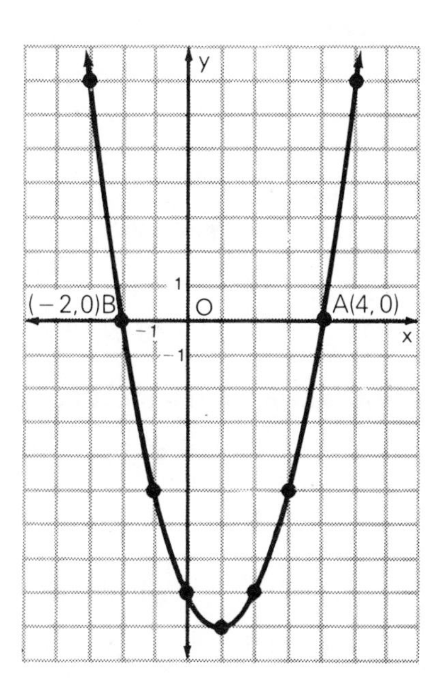

(2) Plot the point associated with each ordered number pair (x, y): (5, 7), (4, 0), etc. Draw a smooth curve through the points. The curve is a parabola.

EXERCISES

In 1–20, graph the quadratic equation. Use the integral values for x indicated in parentheses to prepare the necessary table of values.

1. $y = x^2 \quad (-3 \leq x \leq 3)$

2. $y = 3x^2 \quad (-2 \leq x \leq 2)$

3. $4x^2 = y \quad (-2 \leq x \leq 2)$

4. $5x^2 = y \quad (-2 \leq x \leq 2)$

5. $y = -x^2 \quad (-3 \leq x \leq 3)$

6. $y = -2x^2 \quad (-2 \leq x \leq 2)$

7. $y = \frac{1}{2}x^2 \quad (-4 \leq x \leq 4)$

8. $-\frac{1}{2}x^2 = y \quad (-2 \leq x \leq 2)$

9. $y = x^2 + 1 \quad (-3 \leq x \leq 3)$
10. $x^2 - 1 = y \quad (-3 \leq x \leq 3)$
11. $y = x^2 - 4 \quad (-3 \leq x \leq 3)$
12. $-x^2 + 4 = y \quad (-3 \leq x \leq 3)$
13. $y = x^2 - 2x \quad (-1 \leq x \leq 3)$
14. $x^2 + 2x = y \quad (-3 \leq x \leq 1)$
15. $y = -x^2 + 2x \quad (-1 \leq x \leq 3)$
16. $y = x^2 - 6x + 8 \quad (0 \leq x \leq 6)$
17. $y = x^2 - 4x + 3 \quad (-1 \leq x \leq 5)$
18. $x^2 - 2x - 3 = y \quad (-2 \leq x \leq 4)$
19. $x^2 - 2x + 1 = y \quad (-2 \leq x \leq 4)$
20. $y = x^2 - 3x + 2 \quad (-1 \leq x \leq 4)$

20-8 REVIEW EXERCISES

In 1–6, solve the equation and check all roots.

1. $x^2 - 11x + 18 = 0$
2. $y^2 - 36 = 0$
3. $x^2 - 2x = 15$
4. $2k^2 + 5k + 3 = 3$
5. $w(w + 7) = 18$
6. $\frac{t - 1}{3} = \frac{4}{t}$
7. What is the positive root of $m^2 = 16 - 6m$?
8. Solve for x in simplest radical form: $2x^2 - 36 = 0$

In 9–11, a and b represent the lengths of the legs of a right triangle, and c represents the length of the hypotenuse. Find the missing length.

9. $a = 5$, $b = 12$
10. $c = 41$, $b = 40$
11. $a = 2\sqrt{5}$, $b = 4$
12. The legs of a right triangle measure 2 and 6. Find, in radical form, the length of the hypotenuse of the triangle.
13. The diagonal path in a rectangular garden measures 34 meters. If the width of the garden is 16 meters, find the number of meters in the length of the garden.
14. In rectangle $QUAD$, P is a point on $\overline{QD}$, $PD = 3$, $PA = 5$, and QP equals the sum of PD and PA.
   a. Find QP. b. Find AD. c. Find AU.
   d. Find the area of $\triangle PAD$.
   e. Find the area of trapezoid $QUAP$.

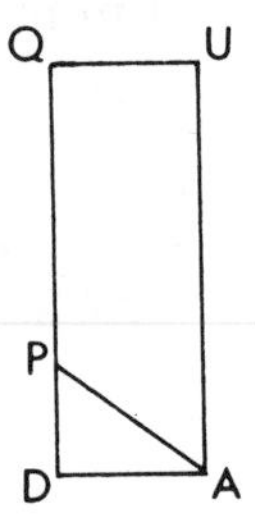

15. The square of a positive number is 6 more than 5 times the number. Find the number.

16. A rectangle has an area of 70. If the sides of the rectangle are represented by x and $3x - 1$, find the lengths of these sides.

17. The sum of the squares of two consecutive positive integers is 85. Find the integers.

18. Graph the quadratic equation $y = x^2 + 2x - 3$, using integral values of x from $x = -4$ to $x = 2$, inclusive.

19. Two square pieces are cut from a rectangular piece of carpet as shown in the diagram. The area of the original piece is 136 square feet and the width of the small rectangle that is left is 1 foot. Find the dimensions of the original piece of carpet.

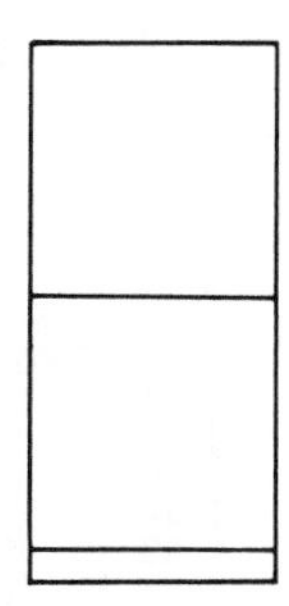

20. Find the irrational roots of $x^2 - 2x - 2 = 0$, using the quadratic formula. Express the solutions in simplest radical form.

Squares and Square Roots

No.	Square	Square Root	No.	Square	Square Root	No.	Square	Square Root
1	1	1.000	51	2,601	7.141	101	10,201	10.050
2	4	1.414	52	2,704	7.211	102	10,404	10.100
3	9	1.732	53	2,809	7.280	103	10,609	10.149
4	16	2.000	54	2,916	7.348	104	10,816	10.198
5	25	2.236	55	3,025	7.416	105	11,025	10.247
6	36	2.449	56	3,136	7.483	106	11,236	10.296
7	49	2.646	57	3,249	7.550	107	11,449	10.344
8	64	2.828	58	3,364	7.616	108	11,664	10.392
9	81	3.000	59	3,481	7.681	109	11,881	10.440
10	100	3.162	60	3,600	7.746	110	12,100	10.488
11	121	3.317	61	3,721	7.810	111	12,321	10.536
12	144	3.464	62	3,844	7.874	112	12,544	10.583
13	169	3.606	63	3,969	7.937	113	12,769	10.630
14	196	3.742	64	4,096	8.000	114	12,996	10.677
15	225	3.873	65	4,225	8.062	115	13,225	10.724
16	256	4.000	66	4,356	8.124	116	13,456	10.770
17	289	4.123	67	4,489	8.185	117	13,689	10.817
18	324	4.243	68	4,624	8.246	118	13,924	10.863
19	361	4.359	69	4,761	8.307	119	14,161	10.909
20	400	4.472	70	4,900	8.367	120	14,400	10.954
21	441	4.583	71	5,041	8.426	121	14,641	11.000
22	484	4.690	72	5,184	8.485	122	14,884	11.045
23	529	4.796	73	5,329	8.544	123	15,129	11.091
24	576	4.899	74	5,476	8.602	124	15,376	11.136
25	625	5.000	75	5,625	8.660	125	15,625	11.180
26	676	5.099	76	5,776	8.718	126	15,876	11.225
27	729	5.196	77	5,929	8.775	127	16,129	11.269
28	784	5.292	78	6,084	8.832	128	16,384	11.314
29	841	5.385	79	6,241	8.888	129	16,641	11.358
30	900	5.477	80	6,400	8.944	130	16,900	11.402
31	961	5.568	81	6,561	9.000	131	17,161	11.446
32	1,024	5.657	82	6,724	9.055	132	17,424	11.489
33	1,089	5.745	83	6,889	9.110	133	17,689	11.533
34	1,156	5.831	84	7,056	9.165	134	17,956	11.576
35	1,225	5.916	85	7,225	9.220	135	18,225	11.619
36	1,296	6.000	86	7,396	9.274	136	18,496	11.662
37	1,369	6.083	87	7,569	9.327	137	18,769	11.705
38	1,444	6.164	88	7,744	9.381	138	19,044	11.747
39	1,521	6.245	89	7,921	9.434	139	19,321	11.790
40	1,600	6.325	90	8,100	9.487	140	19,600	11.832
41	1,681	6.403	91	8,281	9.539	141	19,881	11.874
42	1,764	6.481	92	8,464	9.592	142	20,164	11.916
43	1,849	6.557	93	8,649	9.644	143	20,449	11.958
44	1,936	6.633	94	8,836	9.695	144	20,736	12.000
45	2,025	6.708	95	9,025	9.747	145	21,025	12.042
46	2,116	6.782	96	9,216	9.798	146	21,316	12.083
47	2,209	6.856	97	9,409	9.849	147	21,609	12.124
48	2,304	6.928	98	9,604	9.899	148	21,904	12.166
49	2,401	7.000	99	9,801	9.950	149	22,201	12.207
50	2,500	7.071	100	10,000	10.000	150	22,500	12.247

Index

B

C

D

E

F

G

H

I

L

M

N

O

P

Q

R

S

T

U

V

X

Y

Z